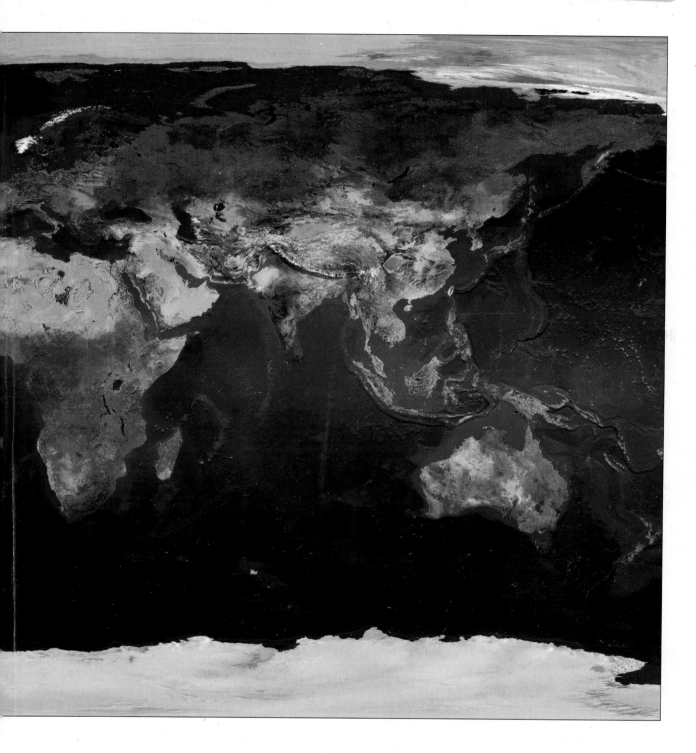

"The Living Earth," Copyright © 1995, Earth Imaging, Technical Director Erik Bruhwiler.
Earth Imaging, 1526 14th Street, #106, Santa Monica, CA 90404.
Used by Permission.

# Geosystems

## An Introduction
## to Physical Geography

# Geosystems

## An Introduction to Physical Geography

### THIRD EDITION

## Robert W. Christopherson

American River College

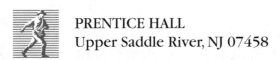
PRENTICE HALL
Upper Saddle River, NJ 07458

Library of Congress Cataloging-in-Publication Data

Christopherson, Robert W.
    Geosystems : an introduction to physical geography /
  Robert W. Christopherson. — 3rd ed.
        p.   cm.
    Includes bibliographical references and index.
    ISBN 0-13-505314-5
    1. Physical geography.   I. Title.
  GB54.5.C48   1997
  910.02—dc20                                          96-16334
                                                          CIP

> *To all the students and teachers of Earth, our home
> planet, and its sustainable future.
> And to our Dad, who gave us the world.*

**Acquisitions Editor:** *Daniel Kaveney*
**Editor-in-Chief:** *Paul F. Corey*
**Editorial Director:** *Tim Bozik*
**Development Editor:** *Fred Schroyer*
**Assistant Vice President of Production and Manufacturing:** *David W. Riccardi*
**Executive Managing Editor:** *Kathleen Schiaparelli*
**Assistant Managing Editor, Production:** *Shari Toron*
**Production Editor:** *Alison Aquino*
**Production Assistant:** *Bobbé Christopherson*
**Senior Marketing Manager:** *Leslie Cavaliere*
**Manufacturing Manager:** *Trudy Pisciotti*
**Editor-in-Chief of Development:** *Ray Mullaney*
**Creative Director:** *Paula Maylahn*
**Art Director:** *Joseph Sengotta*
**Cover Designer:** *Joseph Sengotta*
**Photo Editors:** *Lorinda Morris-Nantz and Melinda Reo*
**Photo Researcher:** *Tobi Zausner*
**Copy Editor:** *Margo Quinto*
**Art Studios:** *Precision Graphics, Maryland CartoGraphics, Inc., Tasa Graphic Arts Inc., Earth Imaging, Rolin Graphics and Biographics*
**Text Composition:** *Molly Pike Riccardi/Lido Graphics*
**Front Cover Photo:** *The dramatic ramparts of Cuernos (horns) del Paine (pronunced "pinay"), Torres del Paine National Park, Chile. The highest peak is 3050 m (10,000 ft); near the southern tip of the Andes Mountains in Chilean Patagonia. Photo by Alan Kearney/Tony Stone Images.*
**Back Cover Photo:** *Full Earth photo by Apollo 17 astronauts, December 1972, from NASA.*
**Frontispiece:** *Clinton Valley, Milford Track, near Fjordland National Park, South Island, New Zealand (45° S 168° E). The mark of the glaciers on this landscape is evident in the U-shaped valley and the small tarn (lake). This scene is inland from the cover photo on our second edition. Photo by Mark A. Johnson/Stock Market.*

©1997 by Prentice-Hall, Inc.
Simon & Schuster/A Viacom Company
Upper Saddle River, New Jersey 07458

Earlier editions ©1994, 1992, by Macmillan Publishing Company, a division of Macmillian.

Printed in the United States of America
10  9  8  7  6  5  4  3  2  1

ISBN 0-13-505314-5

Prentice-Hall International (UK) Limited, *London*
Prentice-Hall of Australia Pty. Limited, *Sydney*
Prentice-Hall Canada Inc., *Toronto*
Prentice-Hall Hispanoamericana, S.A., *Mexico*
Prentice-Hall of India Private Limited, *New Delhi*
Prentice-Hall of Japan, Inc., *Tokyo*
Simon & Schuster Asia Pte. Ltd., *Singapore*
Editora Prentice-Hall do Brasil, Ltda., *Rio de Janeiro*

# Brief Contents

# Contents

## Part 2

### The Water, Weather, and Climate Systems 172

# *Part 3*
The Earth-Atmosphere Interface    316

## *11*   The Dynamic Planet    319

## *12*   Tectonics, Earthquakes, and Volcanism    351

## *13*   Weathering, Karst Landscapes, and Mass Movement    391

## *Part 4*
Soils, Ecosystems, and Biomes    546

## *18*  The Geography of Soils    549

## *19*  Ecosystem Essentials    581

## *20*  Terrestrial Biomes    617

## *21* Earth, Humans, and the New Millennium   645

# Preface

Earth is a place of great physical and cultural diversity, yet people generally know little of it. An informed citizenry requires meaningful education about our life-sustaining environment, and that is the purpose of this book.

The third edition of *Geosystems* builds on the qualities that made the first two editions such a success throughout the United States and Canada. Students and teachers alike have expressed appreciation for the systems organization, content breadth, readability, scientific accuracy, clarity of the glossary, functional beauty of the photographs, art, and cartography, and our up-to-date coverage. The third edition improves on these established qualities, respectfully presenting physical geography as the important Earth science that it is.

We now live on a planet served by the Internet and World Wide Web, a resource that weaves threads of information from around the globe into a vast fabric. The fact that we now have Internet access into almost all the compartments aboard Spaceship Earth is clearly evident in *Geosystems*. Such ready availability of worldwide information allowed your author to illustrate content with a fascinating array of up-to-date examples and to further verify accuracy.

For instance, to better acquaint you with what it is like to live and work in the harsh climate at the South Pole, Chapter 10 presents an account from people stationed there, as carried on the Internet (http://205.174.118.254/nspt/home.htm). This third edition reflects the richness of the Internet as a resource and useful supplement to the scientific literature. And to reinforce our connection—student-teacher-author—we are linked through our own *Geosystems Home Page* at http://www.prenhall.com/geosystm).

## *Geosystems* Communicates the Science of Physical Geography

The goal of physical geography is to explain the spatial dimension of Earth's dynamic systems—its energy, air, water, weather, climate, tectonics, landforms, rocks, soils, plants, ecosystems, and biomes. Understanding human-Earth relations is part of the challenge of physical geography—to create a holistic (or complete) view of the planet and its inhabitants. A goal of *Geosystems* is to serve this need by demonstrating that physical geography is a critical Earth systems science.

*Geosystems* is well suited to your study of physical geography, whether you are a geography major or are taking this class as a science elective. Clear graphics and explanations are designed to help you understand essential concepts of how Earth works. This attention to your needs as a student makes *Geosystems* an accessible science text.

In Chapter 2, two brief essays introduce you to scientific thinking and processes—a Focus Study explains the scientific method and a News Report describes "The Nature of Order is Chaos." Throughout the text, the latest scientific discoveries and techniques are employed as we explore physical geography. Take a moment to sample the pages of the text and you will find an excitement in the presentation of our remarkable planet.

*Geosystems* analyzes the worldwide impact of an event, synthesizing many physical factors into a complete picture. A good example is the 1991 eruption of Mount Pinatubo in the Philippines. The global implications of this major event (one of the largest eruptions this century) are woven throughout the book (Chapters 1, 3, 4, 5, 10, 12, and 21), not just in the section on volcanoes. Our update on global climate change and its related potential effects is part of the fabric in six chapters. Information from current research includes the 1990, 1992, and 1995 reports from the Intergovernmental Panel on Climate Change (IPCC). These content threads tie together the many interesting and diverse topics crucial to a thorough understanding of physical geography.

## Systems Organization Makes *Geosystems* Flow

Each section of this book is organized around the flow of energy, materials, and information. *Geosystems* presents subjects in the same sequence in which they occur in nature. In this way you and your teacher logically progress through topics which unfold according to the flow of individual systems, or in accord with time and the flow of events. For flexibility, *Geosystems* is structured in four parts, each containing chapters that link content in logical groupings.

Chapter 1 presents the essentials of physical geography as a foundation, including an overview of geography, systems analysis, latitude, longitude, time, the science of mapmaking (cartography), remote sensing, and geographic information systems (GIS). With these essentials learned, each of the four parts can then be covered in their present order or in any sequence preferred by the teacher.

*Part One*, Chapters 2 through 6, shows the systems organization of the text, beginning with solar energy flowing across space to Earth's atmosphere, varying in intensity by latitude and season as the solar angle and daylength change. Energy is traced through the atmosphere to Earth's surface, where patterns of temperature and air pressure are generated. Part 1 concludes with general and local atmospheric and oceanic circulations. Human-Earth themes are woven through the presentation in these chapters: the new UV Index, 1995 Nobel Chemistry Prize for ozone research, solar and wind energy resources, air pollution, acid deposition, and heat-index deaths in Chicago in 1995. Ocean currents take on new meaning with a map that shows the journey of a message in a bottle and the odyssey of toys spilled from a cargo ship.

*Part Two* presents water, weather, violent storms, water resources, climate, and climate change, in a flowing sequence from Chapters 7 to 10. The section on water balance explains water resources from varied perspectives, from a house plant or front lawn to vast regional water projects.

In *Part Three*, physical geography is linked to other Earth sciences, an influence seen in Chapters 11 through 17 where we discuss the physical planet and related processes that affect the crust. Earth's surface is a place of enormous struggle between tectonic processes that build, warp, and fault the landscape and those that wear it down through the action of rivers, wind, waves, and ice. Dramatic recent earthquakes in Northridge, California, and Kobe, Japan, are highlighted in Chapter 12.

Finally, *Part Four* integrates the content of the first three parts in biogeography: including soils, ecosystems, and Earth's major terrestrial biomes (Chapters 18 through 20). Critical topics are covered: soil fertility and soil loss, biodiversity, fire ecology, the Great Lakes ecosystem, and desertification, among others.

The text culminates with Chapter 21, "Earth, Humans, and the New Millennium," a unique capstone chapter that summarizes physical geography as an important discipline to help us understand Earth's present status and possible future. This chapter is sure to stimulate further thought and discussion, dealing as it does with the most profound issue of our time, *Earth's stewardship.* An important new study is summarized that examines the costs and benefits of the Clean Air Act. The study reports a 20 to 1 net benefit ratio from regulations directed at cleaning the air. The five agreements reached at the historic 1992 Earth Summit—the United Nations Conference on Environment and Development, held in Rio de Janeiro—are detailed in a focus study.

## *Geosystems* is a Text That Teaches

*Geosystems* is written to assist you in the learning process. Three heading levels are used throughout the text and precise topic sentences initiate each paragraph to help you outline and review material. **Boldface** words are defined where they first appear in the text.

They also are collected in the Glossary and are identified alphabetically and by chapter number. *Italics* are used to emphasize other words and concepts. Also, every figure now has a title that summarizes the caption.

An important new feature in this third edition is the 5 or 6 *Key Learning Concepts* that open each chapter, stating what you should be able to do upon completing the chapter. These objectives are keyed to the main headings in the chapter. At each chapter's end is a unique *Summary and Review* section that corresponds to the *Key Learning Concepts*. Grouped under each learning concept is a narrative review that redefines the boldfaced terms, a key terms list, and specific review questions for that concept. You can conveniently review each concept, test your understanding with review questions, and check key terms in the glossary, then return to the chapter and the next learning concept. In this way, the chapter content is woven together with specific *concept threads*.

Critical thinking is activated by the book's structure and presentation. The key learning concepts help you determine what you want to learn, the text helps you develop more questions and answers, and the summary and review helps you assess what you have learned and what more you might want to know about the subject. The text and our home page give you many facets of a subject with which to develop your own point of view. The *Geosystems* home page adds important dimension to interactivity and critical thinking.

## Features of this Fully Revised Third Edition

Every chapter of the third edition is thoroughly revised, updated, and expanded, with new or improved maps, photographs, art, and tables. There is expanded coverage of global change, climate, the Great Lakes, weather models, and Canada. (Instructors: the *Instructor's Resource Manual* details these changes by chapter.) The many new figures in this edition are fully integrated within the text discussions. Here are a few of the features of the third edition:

- Our widely praised cartography program is updated to reflect the rapid pace of change in political boundaries and physical systems. Of all the maps in the text, 40% are new to this edition. Locator maps accompany most remote sensing images and photographs. Also, numerous topographic maps, many new to this edition, are used to illustrate key features of the landscape.

- Photographs from every continent are included in our 375 photos; 60% are new to this edition. Many photos are integrated with art so you can compare the concept shown in the art with a representative scene from the real world. For example: the four seasons on land and in orbit (Figure 2-21 and 2-22), limestone landscapes and caverns (Figure 13-18), Earth's major deserts (Figure 15-14), glacial features (Figure 17-9 and 17-13), and the Great Lakes shorelines (Focus Study 19-1).

- New, expanded coverage of the Great Lakes: lake-effect snow (Chapter 8), formation of the Great Lakes (Chapter 17), and a new focus study on the lakes and their ecosystems (Chapter 20).

- Continued coverage of Canadian physical geography includes text, figures, and maps of periglacial landscapes (a new focus study in Chapter 17) and Canadian soils (a new Appendix C). Canadian data on a variety of subjects are portrayed on 25 different maps in combination with the United States—physical geography does not stop at the United States–Canadian border!

 Twenty-one "Focus Study" essays, some completely revised and several new to this edition, provide additional explanation of key topics as diverse as the stratospheric ozone predicament, wind power, forecasting the near-record 1995 hurricane season, the El Niño phenomenon, an environmental approach to shoreline planning, the Mount St. Helens eruption, the status of the Colorado River, and the loss of biodiversity.

 Seventy "News Reports" relate topics of special interest. For example: careers in GIS, jet streams and airline flight times, how one culture harvests fog, the new UV Index, using Earthshine to study global energy budgets, the disappearing Nile Delta, humans eat clay, how sea turtles read Earth's magnetic field, and drilling for oil in the rain forest.

The *Geosystems Home Page* provides on-line resources for each chapter on the World Wide Web. You will find review exercises, specific updates for items in the chapter, suggested readings, and links to interesting related pathways on the Internet. A click on the Table of Contents link and selection of a chapter launches you into this new dimension of physical geography.

- The text and all figures show both metric and English measurement. This is done to help us through this transition period as we change from English units to metric. A complete set of measurement conversions is presented inside the back cover.

## The *Geosystems* Learning/Teaching Package

The third edition provides a *complete* physical geography program for you and your teacher.

### *For You the Student:*

- ***Student Workbook*** (ISBN: 0-13-505405-2), by Robert Christopherson. The workbook includes additional learning objectives, a complete chapter outline, critical thinking exercises, problems and short essay work using actual figures from the text, and a self-test with answer key in the back.

 ***Internet Support*** by Duane Griffin and James E. Burt, University of Wisconsin-Madison, and Robert Christopherson. Our ***Geosystems Home Page*** (http://www.prenhall.com/geosystm) gives you on-line review exercises, opportunities to delve deeper into subjects out on the Net, follow-up answers to specific items in the text (for example, the forecast outcome of the 1996 and 1997 Atlantic hurricane seasons), and links to a wealth of interesting sites that relate to each chapter.

 ***Life on the Internet: Geosciences*** by Andrew T. Stull and Duane Griffin is a student's guide to the Internet and World Wide Web specific to geography. It is available *free* as a shrink-wrap with the text.

***For Your Teacher:*** *Geosystems* is designed to give you flexibility in presenting your course. The text is comprehensive in that it is true to each subject area from which it draws content. This diversity is a strength of physical geography, yet it makes it difficult to cover an entire book in a school term. You should feel free to customize use of the text based on your specialty or emphasis. The systems organization within each chapter, four-part structure of chapters, focus study and news report features, all will assist you in sampling some chapters while covering others to greater depth. The following materials are available to assist you—have a great class!

- ***Instructor's Resource Manual*** (ISBN: 0-13-505348-X), by Robert Christopherson and Cecilia Huddleson of Foothill College. The *Instructors Resource Manual*, intended as a resource for both new and experienced teachers, includes a variety of lecture outlines, additional source materials, teaching tips, advice on how to integrate visual supplements, and various other ideas for the classroom.

- *Geosystems Test Bank* (ISBN: 0-13-505330-7), by Robert Christopherson and Marcus Gillespie of Northwest Missouri State University. This collaboration has produced the most extensive test item file available in physical geography. (Mac Test Manager 013-505389-7 and IBM/DOS Test Manager 013-505371-4 are available).

- *Overhead Transparencies* (ISBN: 0-13-505355-2) include 105 illustrations from the text, all enlarged for excellent classroom visibility. And, *Slide Set* (ISBN: 0-13-505363-3) includes illustrations and some photographs from the text.

- *Applied Physical Geography-Geosystems in the Laboratory*, 2nd edition (ISBN: 0-13-505405-2), by Robert Christopherson and Gail Hobbs of Pierce College. This new edition is greatly expanded and improved over the original. Twenty-two lab exercises, divided into logical steps, allow flexibility in presentation of labs. A solutions manual is available.

- *Prentice Hall GEODISC* © (ISBN: 0-13-304163-8) is available to adopters to supplement the text with over 900 pictures and illustrations, 12 minutes of animations, and 50 minutes of motion video segments on an interactive laser disk with bar codes.

**ABCNEWS** *Prentice Hall–ABC News Video Library* is a collection of broadcast segments that highlight many related current events.

 *Prentice Hall–New York Times* **Themes of the Times** supplements, *Geography* and *The Changing Earth*, reprint significant recent articles on related topics.

## Acknowledgments

As in past editions, I must first recognize my family, for they have endured my absence while I worked on *Geosystems*, yet they never wavered in their loving support—my Mom and Dad, my sister Lynne, my brothers Randy and Marty, and our children Keri, Matt, Reneé, and Steve. And now the next generation-Chavon, Bryce, and our newest, Payton. When I look into our grandchildren's faces it tells me why we need to work toward a sustainable future; one for the children.

My thanks go to the many authors and scientists who published research, articles and books, and shared with me over the Internet, e-mail, FAX, and phone. And I give special gratitude to all the students over these past 27 years at American River College for defining the importance of Earth's future, for their questions, and their enthusiasm. *To all students and teachers this text is dedicated.*

I owe many thanks to my colleagues who variously served as reviewers on one or more editions, who participated in our focus groups, or who offered helpful suggestions at our national and regional geography meetings. I am grateful to all of them for their generosity of ideas and sacrifice of time:

Ted J. Alsop, Utah State University
Ward Barrett, University of Minnesota
David R. Butler, University of North Carolina
Ian A. Campbell, University of Alberta–Edmonton
Richard A. Crooker, Kutztown University
Armando M. da Silva, Towson State University
Dirk H. de Boer, University of Saskatchewan
Mario P. Delisio, Boise State University
Joseph R. Desloges, University of Toronto
Lee R. Dexter, Northern Arizona University
Don W. Duckson, Jr., Frostburg State University
Christopher H. Exline, University of Nevada–Reno
Michael M. Folsom, Eastern Washington University
Mark Francek, Central Michigan University
Glen Fredlund, University of Wisconsin–Milwaukee
David E. Greenland, University of Oregon
John W. Hall, Louisiana State University–Shreveport
Vern Harnapp, University of Akron
Gail Hobbs, L.A. Pierce College
David A. Howarth, University of Louisville
Patricia G. Humbertson, Youngstown State University
David W. Icenogle, Auburn University
Philip L. Jackson, Oregon State University
J. Peter Johnson, Jr., Carleton University
Guy King, California State University–Chico
Ronald G. Knapp, SUNY–The College at New Paltz
Peter W. Knightes, Central Texas College
Thomas Krabacher, California State University–Sacramento
Richard Kurzhals, Grand Rapids Junior College
Steve Ladochy, California State University–Los Angeles
Robert D. Larson, Southwest Texas State University
Joyce Lundberg, Carleton University
W. Andrew Marcus, Montana State University
Elliot G. McIntire, California State University–Northridge
Norman Meek, California State University–San Bernardino
Lawrence C. Nkemdirim, University of Calgary
John E. Oliver , Indiana State University
Bradley M. Opdyke, Michigan State University
James Penn, Southeastern Louisiana University
Greg Pope, Montclair State University
Robin J. Rapai, University of North Dakota
Gary Rees, Santa Barbara
Philip D. Renner, American River College
William C. Rense, Shippensburg University
Dar Roberts, University of California–Santa Barbara
Wolf Roder, University of Cincinnati
Bill Russell, L.A. Pierce College
Dorothy Sack, Ohio University
Glenn R. Sebastian, University of South Alabama

Daniel A. Selwa, U.S.C. Coastal Carolina College
Thomas W. Small, Frostburg State University
Daniel J. Smith, University of Victoria
Stephen J. Stadler, Oklahoma State University
Susanna T.Y. Tong, University of Cincinnati
David Weide, University of Nevada–Las Vegas
Brenton M. Yarnal, Pennsylvania State University

I extend my continuing gratitude to the editorial, production, and sales staff of Prentice Hall. Thanks to Editor-in-Chief Paul Corey for his friendship and continuity of leadership from before the first edition. My appreciation to Dan Kaveney, an energetic geography editor and new friend, for playing such a positive role during this project; always encouraging and supportive and always willing to let me participate in the entire publishing process.

My thanks to Production Editor Alison Aquino and Assistant Managing Editor Shari Toron for coordinating this complex project, for understanding our 72-page graphics log, and for bringing together diverse elements that produced this book. My appreciation for thoroughness and attention to detail goes to copy editor Margo Quinto. Also, thanks go to Tobi Zausner for her photo research and friendship. I much appreciate Joe Sengotta's powerful cover and beautiful text design and for letting us in on many decisions. On the marketing side there is great comfort in knowing that such a tireless worker as Leslie Cavaliere is in there pitching as our Marketing Manager! My thanks to Editor-in-Chief, Engineering, Science, and Mathematics Development Department Ray Mullaney, for coordinating schedules like a symphony conductor. A special appreciation to our Developmental Editor Fred Schroyer for his love of words, skill with ideas, humor in the heat of battle, and sense of the student's point of view.

Most importantly, I am blessed with the continuing partnership of my special collaborator, Bobbé Christopherson. My wife is the production assistant for all *Geosystems* projects, which truly would not be possible without her effort. I express extraordinary gratefulness, although words fail me here, for her endless hours researching and cataloging photographs, preparing and controlling our figure logs and photo request logs, copy editing, proofing art, obtaining permissions, and reading through page proofs. Her natural sense of the Home Planet, understanding of *Gaia*, and love for the smallest of living things is a strong force in meeting the challenges of our text projects.

Physical geography teaches us a holistic view of the intricate supporting web that is Earth's environment and our place in it. Dramatic changes that demand our understanding are occurring in many human-Earth relations as we approach the new millennium. All things considered, this is an important time for you to be enrolled in a physical geography course! The best to you in your studies—and carpe diem!

*Robert W. Christopherson*
Folsom, California (bobobbe@aol.com)

# 1

# Essentials of Geography

**The Science of Geography**

**Earth Systems Concepts**

**A Spherical Planet**

**Location and Time on Earth**

**Maps, Scales, and Projections**

**Remote Sensing and GIS**

**Summary and Review**

## Key Learning Concepts

After reading the chapter, you should be able to:

- *Define* geography and physical geography in particular.
- *Describe* systems analysis, open and closed systems, feedback information, and system operations and *relate* those concepts to Earth systems.
- *Explain* Earth's reference grid: latitude, longitude, and latitudinal geographic zones and time.
- *Define* cartography and mapping basics: map scale and map projections.
- *Describe* remote sensing and *explain* geographic information system (GIS) methodology as a tool used in geographic analysis.

*Early geographers and surveyors gathering data for maps.* [From a German engraving, 1954. The Granger Collection, New York.]

# Welcome to physical geography!

Physical geography is an exciting subject. It deals with our environment and the powerful forces and events that influence our lives. Consider the following:

- The costliest natural disasters in history—Hurricane Andrew in Florida and Louisiana (1992), the U.S. Midwest floods of 1993, the Northridge (Reseda) earthquake in Los Angeles (1994), and the 1995 Atlantic hurricane season—collectively exceeded $85 billion in damage.

- Devastating earthquakes struck Kobe, Japan (5000 dead, $100 billion in damage), the Sakhalin Islands, Cyprus, Turkey, Colombia, Mexico City, and eastern Russia, all in the past several years.

- From 1991 to 1996, record numbers of tornadoes struck the United States, which is the most tornado-prone country on Earth.

- Worldwide, lake and ocean levels are changing. Since 1980, Lake Eyre in central Australia has filled to unusual depths, Lake Chad in western Africa has receded, and Utah's Great Salt Lake has risen to record levels. Water levels in California's Mono Lake and the Aral Sea in Uzbekistan dropped as inflowing rivers were diverted for irrigation and domestic water supplies. Around the globe, sea level is rising at a rate faster than at any time during the past thousand years.

- Rocks discovered in northwestern Canada have been measured to be 3.96 billion years old, predating all others found so far on our 4.6 billion-year-old planet.

- The Appalachian Mountains of eastern North America and the Atlas Mountains in North Africa, which together formed a continuous chain millions of years ago, are still drifting apart as vast plates of continental crust continue to slowly move.

- Record floods and droughts plague many regions. Severely flooded were Norway, China, Bangladesh, Ghana, the Netherlands, southern Florida, and the Mississippi-Missouri River systems in 1993 and again in 1995. Intense drought afflicted Spain, Mexico, southern Europe, Cambodia, the northeastern United States, and parts of China.

- Volcanoes erupted in Chile, Nicaragua, Indonesia, Zaire, Japan, and Alaska. In 1991, Mount Pinatubo exploded in the Philippine Islands, spewing dust and gases into the upper atmosphere, causing colorful sunsets worldwide, lowering global temperatures slightly, and spreading sulfuric acid mist.

- Earth's diversity continues to fascinate us: lush rain forests straddle the equator, stark deserts bake in the subtropics, cool moist climates dominate northwestern coastlines, and perpetual cold hugs the polar regions.

- Record high atmospheric and ocean temperatures and killer heat waves dominated the last two decades in various parts of the world. In the United States, nearly 1000 people died from the heat, more than 700 in Chicago alone during the 1995 summer season.

- Most natural ecosystems now bear the imprint of civilization: Only a dozen white rhinos remain in the northern portion of their African habitat; record numbers of species face extinction in the tropics, agricultural soils lose productivity; ocean fishery yields continue to fall; and forests decline because of logging and acid rain. Persistently high air temperatures affect habitats and ecosystems across the globe. Scientists and governments scramble to understand the impact of these global changes.

Why do all of these conditions and events occur? Where are all the places just mentioned? Why does the environment differ from equator to midlatitudes, between deserts and polar regions? How does solar energy influence the distribution of trees, soils, climates, and lifestyles? How does it produce the patterns of wind, weather, and ocean currents? How do natural systems affect human populations, and, in turn, what impact are humans having on natural systems? In this book, we explore those questions, and others, through geography's unique perspective. Once again, welcome to physical geography!

We live in an extraordinary era of **Earth systems science**. This science contributes to our emerging view of Earth as a complete entity—an interacting set of physical, chemical, and biological systems that produce the processes and conditions of the whole Earth. Earth systems scientists are monitoring and analyzing changes that are occurring in natural systems and speculating on how these changes might affect life on Earth. As you will see, physical geography is key to studying entire Earth systems because of its integrative approach.

A knowledge of physical geography helps you answer the *where*, the *why*, and the *how* questions about Earth's physical systems and their interaction with living things. Thus, our study of geosystems—Earth systems—begins with a look at the science of physical geography and the geographic tools we use (Figure 1-1).

Physical geographers analyze systems to study the environment. Therefore, we introduce Earth's interrelated systems, natural and human. We then consider location, a key theme of geographic inquiry—the latitude, longitude, and time coordinates that delimit Earth's surface. The study of longitude and a universal time system provide us with interesting insights into geography. Next, we examine maps as critical tools that geographers create to portray physical and cultural information. This chapter concludes with an overview of some advanced technology that is adding new and exciting dimensions to geography: remote sensing from space and computer-based geographic information systems (GIS).

### Location

Absolute and relative location on Earth. Location answers the question *Where?*—the specific planetary address of a location. This road sign is posted along the I–5 freeway in Oregon.

### Region

Portions of Earth's surface that have uniform characteristics; how they form and change; how they are related to other regions. The great grain-producing areas on the midlatitude grasslands of North America form a vast agricultural region.

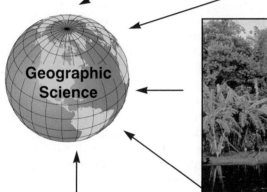

### Human-Earth Relationships

Humans and the environment: resource exploitation, hazard perception, and environmental modification. The oldest theme of geographic inquiry. The village on the Amazon River is constructed form materials derived form the surrounding area.

**Geographic Science**

### Place

Tangible and intangible living and nonliving characteristics that make each place unique. No two places on Earth are exactly alike. The power and beauty of Victoria Falls on the Zambesi River, along the Zambia-Zimbabwe border, make a unique place.

### Movement

Communication, movement, circulation, and diffusion across Earth's surface. Global interdependence links all regions and places—both physical and human systems. Winds and ocean currents form circulations of energy and water. The oceans provide transportation connections.

**FIGURE 1-1**

**Five themes of geographic science.**

Definitions of five fundamental themes in geographic science with examples of each—location, region, human-Earth relationships, movement, and place. [Regions photo by G. R. Roberts/Photo Researchers, Inc.; Place photo by George Holton; Human-Earth relationship photo by Gael Summer-Hebden; Movement and Location photos by author.]

# The Science of Geography

**Geography** (from *geo*, "Earth," and *graphein*, "to write") is the science that studies the relationships among geographic areas, natural systems, society, cultural activities, and the interdependence of all these *over space*. The term *spatial* refers to the nature and character of physical space and to measurements, relations, and the distribution of things. For example, think of your own route to the classroom or library today and how you used your knowledge of street patterns, traffic trouble spots, parking spaces, or bike rack locations to minimize walking distance. These are spatial considerations. Human beings are spatial actors, both affecting and being affected by Earth. We profoundly influence vast areas because of our mobility. Earth's systems influence our activities in a most obvious way—they give us life—and in less obvious ways—they may affect our lifestyle choices.

New "National Geography Standards" for learning have been prepared by the Association of American Geographers (AAG) and the National Council for Geographic Education (NCGE) in response to *Goals 2000*, the Educate America Act of 1994. Geographic science is divided into six essential elements, five critical skills, and eighteen specific geographic standards (see *Geography for Life*, prepared by the Geography Education Project for AAG, NCGE, American Geographical Society, and National Geographic Society, 1994).

We can simplify the supporting framework of geographic science using five key themes: **location**, **place**, **movement**, **region**, and **human-Earth relationships**, illustrated and defined in Figure 1-1. *Geosystems* draws on each theme.

Geography is governed by a *method* rather than by a specific body of knowledge, and this method is **spatial analysis**. Using this method, geography synthesizes knowledge from many fields, integrating information to form a coherent picture of Earth. Geographers view phenomena as occurring in spaces and areas that have distinctive characteristics. The language of geography reflects this view, using the words *space, territory, zone, pattern, distribution, place, location, region, sphere, province*, and *distance*. Geographers analyze the differences and similarities among places and locations.

**Process**, a set of actions or mechanisms that operate in some special order, is central to geographic analysis. As an example, numerous processes are involved in Earth's vast water-atmosphere-weather system or in continental crust movements and earthquake occurrences. Spatial analysis examines how Earth's processes interact over space or area.

**Physical geography** thus centers on spatial analysis of all the physical elements and processes that make up the environment: energy, air, water, weather, climate, landforms, soils, animals, plants, and Earth itself. We add to this the oldest theme in the geographic tradition, that of human activity and impact, the result of which is the shared human-Earth relationship.

## The Geographic Continuum

Geography is eclectic, integrating a wide range of subject matter from diverse fields; virtually any subject can be examined geographically. Figure 1-2 shows a continuous distribution—a continuum—along which the content of geography is arranged. Disciplines in the physical and life sciences are at one end, and those in the human and cultural sciences are at the other. As the figure shows, various specialties within geography draw from these disciplines.

The continuum in Figure 1-2 reflects a basic duality within geography—*physical geography* versus *human and cultural geography*. This duality is paralleled in society by the tendency of we who live in developed countries to distance ourselves from our supportive life-sustaining environment and to think of ourselves as exempt from the physical functions of Earth. In contrast, many people in developing countries live closer to nature and are acutely aware of its importance in their daily lives. Regardless of our philosophies toward Earth, we all depend on Earth's systems to provide oxygen, water, nutrients, energy, and materials to sustain life. Our modern world requires that we shift our study of geographic processes, and perhaps our philosophies, toward the center of the continuum to attain a more *holistic*, or complete, perspective—such is the thrust of Earth systems science.

Modern environmental problems call for geographers to examine relations among Earth, society, and cultures throughout space and time. A sample of critical environmental concerns that are integral to physical geography includes:

1. *Global ozone depletion* in the upper atmosphere that allows increasing amounts of ultraviolet radiation to reach Earth's surface.
2. *Possible global warming* through human-caused increases of carbon dioxide in the atmosphere.
3. *Natural hazards* that threaten society, such as hurricanes, earthquakes, the 1991 eruption of Mount Pinatubo, tsunami, landslides, droughts, and floods.
4. *Human-induced hazards* that threaten the sustainability of Earth's life-support systems, such as the disposal of hazardous waste, radiation added to the environment, settlement in hazardous places, the worsening problems of air, water, and ocean pollution, and the emerging patterns of global environmental change.
5. *Deliberate destruction of Earth's forests* beyond levels that are sustainable.
6. *Increasing losses of plant and animal diversity* as habitats change or disappear.
7. *Human-made disasters*, such as the 1986 Chornobyl' nuclear power station explosion, the thousands of oil tanker spills in oceans, seas, and rivers each year, the

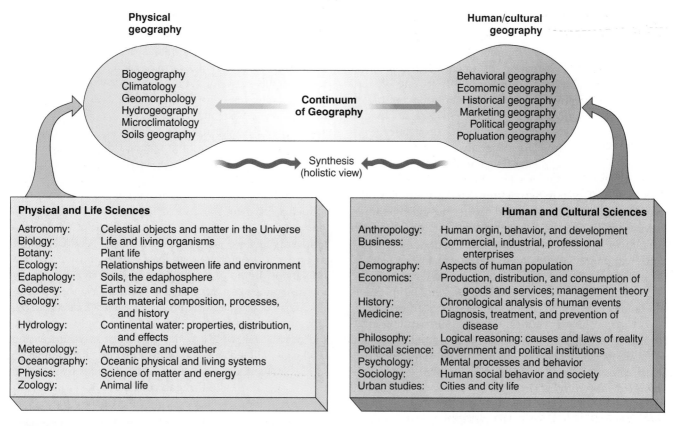

**FIGURE 1-2**
**The content of geography.**
Distribution of geographic content along a continuum (a continuous distribution). Geography derives subject matter from many different sciences. Do you find any subjects you have taken? The focus of this book is physical geography, but we also integrate some of the human and cultural component. Synthesis of Earth topics and human topics is suggested by movement toward the middle of the continuum—a "holistic" view.

environmental impact of the 1991 Persian Gulf War, and patterns of widespread groundwater contamination.

8. *Natural resources* as environmental assets on national economic balance sheets.

Geography is in a unique position to synthesize the environmental, spatial, and human aspects of all these concerns.

Global climate change, biodiversity, sustainable forestry, and an "Earth Charter"—themes at the heart of geographic studies—were the focus of the United Nations Conference on Environment and Development, held at Rio de Janeiro in 1992. This Earth summit brought together tens of thousands of delegates from more than 160 countries and 100 nongovernmental organizations (see Focus Study 21-1).

As we approach the new millennium (the year 2001) the spatial patterns of change we face are unique, for we are taxing Earth's systems in new ways. Some past civilizations adapted to crises, whereas others failed. Perhaps this ability to adapt is the key. If so, understanding our relationship to Earth's physical geography is of great importance to human survival, for the innumerable physical processes in the environment constitute *our* life-support systems.

# Earth Systems Concepts

The word *system* pervades our lives daily: "Check the car's cooling system"; "How does the grading system work?"; "There is a weather system approaching." Systems of many kinds surround us. Not surprisingly, *systems analysis* has moved to the forefront as a method for understanding operational behavior in many disciplines. The technique began with studies of energy and temperature (thermodynamics) in systems in the nineteenth century and was further developed by engineering disciplines during World War II. Today, geographers use systems methodology as an analytical tool. In this book, content is organized along logical flow paths consistent with systems thinking.

## *Systems Theory*

Simply stated, a **system** is any ordered, interrelated set of things and their attributes, linked by flows of energy and matter, as distinct from the surrounding environment outside the system. The elements within a system may be arranged in a series or interwoven with one another. A system comprises any number of subsystems. Within Earth's systems, both matter and energy are stored and retrieved, and energy is transformed from one type to another. (Remember; *matter* is an entity that assumes a physical shape and occupies space; *energy* is a capacity to change the motion of, or to do work on, matter.)

Systems in nature are generally not self-contained: Inputs of energy and matter flow into the system, and outputs of energy and matter flow from the system. Such a system is called an **open system.** Within a system, the parts function in an interrelated manner, acting together in a way that gives each system its character. Earth is an open system *in terms of energy,* for solar energy enters freely and heat energy leaves freely and goes back into space. Most natural systems are open in terms of energy. Figure 1-3 schematically illustrates an open system and presents the inputs and outputs of an automobile as an example.

A system that is shut off from the surrounding environment so that it is self-contained is a **closed system.** Although such closed systems are rarely found in nature, Earth is essentially a closed system *in terms of physical matter and resources.* This "closed" status applies despite the quantity of cosmic dust that enters the atmosphere from space and volumes of hydrogen gas that exit to space from the atmosphere.

*Example.* Figure 1-4 illustrates a simple open-flow system, using plant photosynthesis and respiration as an example. In photosynthesis (Figure 1-4a), plants use an energy input (certain wavelengths of sunlight) and material inputs of water, nutrients, and carbon dioxide. The photosynthetic process converts these inputs to stored chemical energy in

the form of plant sugars (carbohydrates). The process also releases an output: the oxygen we breathe.

Reversing the process, plants derive energy for their operations from *respiration.* In respiration, the plant consumes inputs of chemical energy (carbohydrates) and oxygen and releases outputs of carbon dioxide, water, and heat into the environment (Figure 1-4b). Thus, a plant acts as an open system, in which both energy and materials freely flow into and out of the plant. (Photosynthesis and respiration processes are discussed further in Chapter 19.)

*Feedback.* As a system operates, it generates outputs that influence its own operations. These outputs function as "information" that is returned to various points in the system via pathways called **feedback loops.** Feedback information can control (or at least guide) further system operations. In the photosynthetic system (Figure 1-4), any increase or decrease in daylength (sunlight availability), carbon dioxide, or water will produce feedback that elicits specific responses in plant operations. For example, decreasing the water input will slow the growth process; increasing daylength will increase the growth process within limits.

If the feedback *information* discourages response in the system, it is called **negative feedback.** *Further production in the system decreases the growth of the system.* Such negative feedback causes self-regulation in a natural system, stabilizing and maintaining the system. Think of a weight-loss diet—the human body is an open system. When you stand on the scales, the good or bad news reported about your weight acts as feedback to you. This, in turn, provides you with guidance (negative feedback) as to how inputs (food calories) need to be adjusted.

If feedback *information* encourages increased response in the system, it is called **positive feedback.**

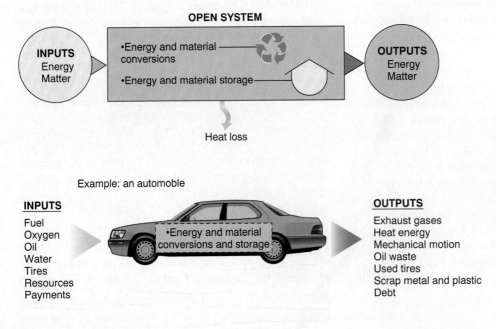

FIGURE 1-3
An open system.

In an open system, inputs of energy and matter undergo conversions and are stored as the system operates. Outputs include energy and matter and waste (heat energy) that flow from the system. See how the various inputs and outputs are related to the operation of a car: Matter and energy are converted, stored, and produced in an automobile—an open system.

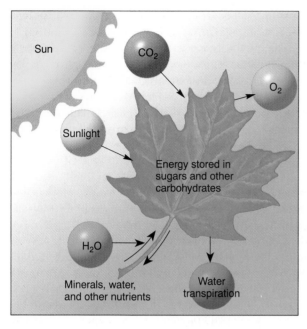

(a) Plant photosynthesis

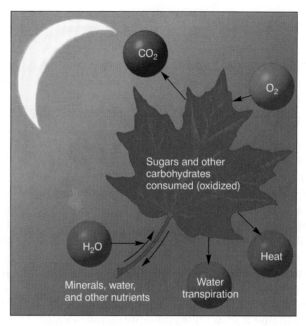

(b) Plant respiration

**FIGURE 1-4**

**A leaf is a natural open system.**

A plant leaf provides an example of a natural open system. (a) In the process of photosynthesis, plants consume light, carbon dioxide ($CO_2$), water ($H_2O$), and produce outputs of oxygen ($O_2$) and carbohydrates (sugars) as stored chemical energy. (b) Plant respiration, illustrated here at night, approximately reverses this process and produces outputs of carbon dioxide ($CO_2$) and water ($H_2O$), using oxygen and consuming (oxidizing) carbohydrates to produce energy for cell operations.

*Further production in the system stimulates the growth of the system.* In finance, a compound interest–bearing account provides an example: The larger the account becomes, the more interest it earns, thus, the larger the account becomes, and so on. Unchecked positive feedback in a system can create a runaway ("snowballing") condition. In natural systems, such unchecked growth will reach a critical limit, leading to instability, disruption, or death of organisms. An example is an exploding population of bacteria where a sewage pipe drains into a stream. The bacteria flourish as more sewage (nutrient) enters the system. The population continues to grow until the production of waste by the bacteria and declining nutrient levels make the environment toxic and unlivable; the subsequent die-off dampens further runaway growth.

One of the numerous wildfires that occur in California each summer gives us an example of positive feedback. Control of the input of flammable fuel and oxygen is the key to extinguishing a fire. The fire itself dries wet shrubs and green wood around the fire, thus providing more fuel for combustion. The greater the fire grows, the greater becomes the availability of fuel, and thus more fire is possible. Knowing this process allows us to devise strategies for controlling fuel availability, regulating excessive landscaping in vulnerable urban areas, and practicing "control burns."

Can you describe possible negative and positive feedbacks in the automobile illustration in Figure 1-3? For instance, how would the inputs and outputs be affected if tailpipe exhaust emission standards are weakened or strengthened? If the automobile is operated at high elevation? Or if the price of fuel increases or decreases?

***Equilibrium.*** Most systems maintain structure and character over time. An energy and material system that remains balanced over time, where conditions are constant or recur, is in a *steady-state condition.* When the rates of inputs and outputs in the system are equal and the amounts of energy and matter in storage within the system are constant (or more realistically, as they fluctuate around a stable average), the system is in **steady-state equilibrium**.

However, a steady-state system fluctuating around an average value may demonstrate a changing trend over time, a condition described as **dynamic equilibrium**. These changing trends of either increasing or decreasing system operations may appear gradual over time. Examples include long-term climatic changes and the present pattern of increasing temperatures in the atmosphere and ocean. The present rate of species extinction exhibits a downward trend in numbers of living species and change in the "balance of nature." Figure 1-5 illustrates these two states.

Note that given the nature of systems to maintain their operations, they tend to resist abrupt change. However, a system may reach a *threshold* at which it can no longer maintain its character, so it lurches to a new operational level. This abrupt change places the system in a *metastable equilibrium.* An example of such a condition is a landscape, such as a hillside, that adjusts after a sudden landslide. A new equilibrium is eventually achieved among slope, materials, and energy over time. This "threshold" concept raises concern in the scientific community, especially if some natural systems reach such threshold limits. A sudden change to a new equilibrium arrangement may not be as desirable or supportive as present environmental conditions.

A particularly interesting example of such system fluctuations is the interaction between volcanic eruptions and the atmosphere. Mount Pinatubo in the Philippines erupted violently in 1991, injecting 15 to 20 million tons of ash and sulfuric acid mist into the upper atmosphere (Figure 1-6). These materials increased the reflection of sunlight from the atmosphere worldwide and thus reduced the amount of solar energy that passes through the atmosphere, temporarily lowering global temperatures an average of 0.5 C° (0.9 F°).

As you progress through this book, you will see the story of Mount Pinatubo and its implications unfold in seven chapters: Chapter 4 (effects on energy budgets in the atmosphere), Chapter 6 (the eruption, plus satellite images of the spread of debris), Chapter 10 (temporary effect on atmospheric temperature), Chapters 11 and 12 (volcanic processes), Chapter 17 (past climatic effects of volcanoes), and Chapter 21 (a comparison of the eruption effects with the "nuclear winter" hypothesis).

***Models of Systems.*** A **model** is a simplified, idealized representation of part of the real world. Models are designed with varying degrees of abstraction. Physical geographers construct simple system models to demonstrate complex associations in the environment. A good example is the *hydrologic system,* which models Earth's entire water system, its related energy flows, and atmosphere, surface, and subsurface environments through which water flows. (If water alone is considered and not the complete system, the term *cycle* is sometimes applied, as in the "hydrologic cycle.") The simplicity of a model makes a system easier to comprehend and to simulate in experiments (for example, the leaf model in Figure 1-4).

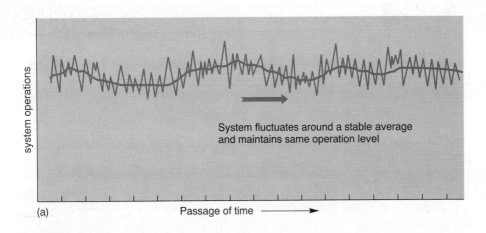

System fluctuates around a stable average and maintains same operation level

(a)    Passage of time ⟶

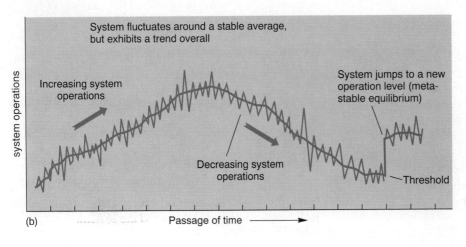

System fluctuates around a stable average, but exhibits a trend overall

Increasing system operations

System jumps to a new operation level (metastable equilibrium)

Decreasing system operations

Threshold

(b)    Passage of time ⟶

**FIGURE 1-5**

**System equilibria: steady-state and dynamic.**

Some systems exhibit a stead-state equilibrium over time; system operations fluctuate around a stable average (a). Other systems are in a condition of dynamic equilibrium, with an increasing or decreasing operational trend (b). Rather than changing gradually, some systems may reach a threshold at which system operations jump (change abruptly) to a new set of relations.

**FIGURE 1-6**
**The eruption of Mount Pinatubo.**
The 1991 Mount Pinatubo eruption, one of the largest volcanic eruptions in this century, widely affected the Earth-atmosphere system. Geographers and other scientists use the latest technology to study how such eruptions affect the atmosphere's dynamic equilibrium. [Photo by Tim Alipano, Reuters/Bettmann.]

Adjusting the variables in a model produces differing conditions and allows predictions of possible system operations to be made. A general circulation model of the atmosphere, operated by the Goddard Institute, accurately predicted the effect of Mount Pinatubo's ash on the atmosphere, the lowering of global temperatures, and subsequent recovery of temperatures to record-high levels. However, predictions are only as good as the assumptions and accuracy built into the model. Imposing a model too rigidly on a natural setting can lead to misinterpretation, so it is best to view a model for what it is—a simplification to help us understand a complex process.

We discuss many system models in this text, including the hydrologic system, water balance, surface energy budgets, earthquakes and faulting as outputs of Earth systems, glacier mass budgets, soil profiles, and various ecosystems. Computer-based models are in use to study most natural systems—from the upper atmosphere, to climate change, to Earth's interior.

## Earth as a System

Because it receives energy from an outside source—the Sun—and radiates energy back into space, Earth operates as an open system. In terms of matter, however, Earth is essentially a closed system, discounting the exceptions mentioned earlier.

***Earth's Energy Equilibrium: An Open System.*** Most Earth systems are dynamic because of the tremendous infusion of radiant energy from thermonuclear reactions deep within the Sun. This energy penetrates the outermost edge of Earth's atmosphere and cascades through the terrestrial systems, being transformed along the way into various forms of energy, such as kinetic energy (of motion), potential energy (of position), or other expressions as chemical or mechanical energy. Eventually, Earth radiates this energy back to the cold vacuum of space in an amount essentially equal to that which entered the system.

Researchers are examining the dynamics of Earth's energy equilibrium in more detail than ever before to distinguish natural changes from those forced by human activities. Tremendous breakthroughs in understanding and prediction are occurring as general circulation models (GCMs) of Earth's energy-atmosphere-water system become even more accurate. By the late 1990s, at least four polar-orbiting satellites—part of an Earth Observation System, or EOS—will introduce a new era in the monitoring of Earth's open energy system.

***Earth's Physical Matter: A (Nearly) Closed System.*** In terms of physical matter—air, water, and material resources—Earth is nearly a closed system. The only exceptions are the slow escape of lightweight gases (such as hydrogen) from the atmosphere into space and the input of frequent but tiny meteors and cosmic and meteoric dust. Since the initial formation of our planet, no significant quantities of matter have entered the system. Just as important, no significant quantities have left the system, either. This is it! Earth's physical materials are finite. No matter how numerous and daring the technological reorganizations of matter become, our physical base is, for all practical purposes, fixed and finite.

Society loses track of its resources in what is essentially a once-through resource stream from virgin materials to the landfill (Figure 1-7). As people come to understand this fact of finite resources, a serious effort to recycle and to make efficient use of energy and materials can begin. The fact that Earth is a closed material system makes such recycling efforts inevitable.

## Earth's Four "Spheres"

Earth's surface is a vast area of 500 million square kilometers (193 million square miles) where four immense open systems interact. Figure 1-8 shows three **abiotic** (nonliving) systems overlapping to form the realm of the **biotic** (living) system. Each system loosely occupies a "shell" around Earth, so each is called a "sphere." The abiotic spheres are the **atmosphere**, **hydrosphere**, and **lithosphere**. The biotic sphere is called the **biosphere**. Because these four systems are not independent units in nature, their boundaries must be understood as transition zones rather than sharp delimitations.

***Atmosphere.*** The atmosphere is a thin, gaseous veil surrounding Earth, held to the planet by the force of gravity. Formed by gases arising from within Earth's crust and interior, and the exhalations of all life over time, the lower atmosphere is unique in the Solar System. It is a combination of nitrogen, oxygen, argon, carbon dioxide, water vapor, and trace gases.

***Hydrosphere.*** Earth's waters exist in the atmosphere, on the surface, and in the crust near the surface. Collectively, these three areas are home to the hydrosphere. Water of

**FIGURE 1-7**
**An urban landfill.**
A typical urban landfill is essentially an "urban ore mine" of displaced or set-aside resources. Recycling costs for some materials may be prohibitive, but about half the contents of an average landfill are extractable economically—glass, newspaper, aluminum and some other metals, and many plastics. [Photo by author.]

the hydrosphere exists in all three states: liquid, solid (the frozen cryosphere), and gaseous (water vapor). Water occurs in two general chemical conditions, fresh and saline (salty). It exhibits important heat-storage properties. And water is an extraordinary solvent. Among the planets in the Solar System, only Earth possesses water in any quantity, adding to Earth's uniqueness among the planets.

***Lithosphere.*** Earth's crust and a portion of the upper mantle directly below the crust form the lithosphere. The crust is quite brittle compared with the layers deep beneath the surface, which are more plastic and move slowly in response to an uneven distribution of heat and pressure. An important component of the lithosphere is soil, which generally covers Earth's land surfaces; the soil layer sometimes is referred to as the *edaphosphere*. (In a broad sense, the term *lithosphere* sometimes refers to the entire solid planet.)

***Biosphere.*** The intricate, interconnected web that links all organisms with their physical environment is the biosphere. Sometimes called the **ecosphere**, the biosphere is the area in which physical and chemical factors form the context of life. The biosphere exists in the overlap among the abiotic spheres, extending from the sea floor to about 8 km (5 mi) into the atmosphere. Life is sustainable within these natural limits. In turn, life processes have powerfully shaped the other three spheres through various interactive processes. The biosphere evolves, reorganizes itself at times, faces some extinctions, gains new vitality, and manages to flourish overall. Earth's biosphere is the only one known in the Solar System; thus, life as we know it is unique to Earth.

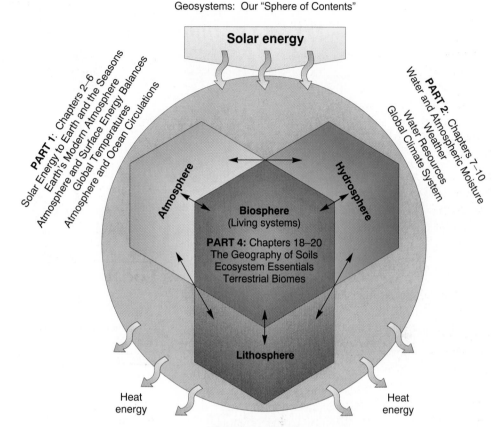

**FIGURE 1-8**

**Earth's four spheres.**

Each sphere is a model of vast Earth systems. This general model further provides the organizational framework for the four-part structure of *Geosystems*.

Part 1–Atmosphere: The Energy-Atmosphere System.

Part 2–Hydrosphere: The Water, Weather, and Climate System.

Part 3–Lithosphere: The Earth-Atmosphere Interface.

Part 4–Biosphere: Soils, Ecosystems, and Biomes.

Geosystems: Our "Sphere of Contents"

**Solar energy**

PART 1: Chapters 2–6
Solar Energy to Earth and the Seasons
Earth's Modern Atmosphere
Atmosphere and Surface Energy Balances
Global Temperatures
Atmosphere and Ocean Circulations

PART 2: Chapters 7–10
Water and Atmospheric Moisture
Weather
Water Resources
Global Climate System

**Atmosphere**

**Hydrosphere**

**Biosphere**
(Living systems)

**PART 4:** Chapters 18–20
The Geography of Soils
Ecosystem Essentials
Terrestrial Biomes

**Lithosphere**

Heat energy

Heat energy

**PART 3:** Chapters 11–17
The Dynamic Planet
Tectonics, Earthquakes, and Volcanism
Weathering, Karst Landscapes, and Mass Movement
River Systems and Landforms
Eolian Processes and Arid Landscapes
The Oceans, Coastal Processes and Landforms
Glacial and Periglacial Processes and Landforms

# A Spherical Planet

We have all heard that some people used to believe that Earth is flat. Yet Earth's sphericity is not as modern a concept as many think. For instance, more than two millennia ago, the Greek mathematician and philosopher Pythagoras (ca. 580–500 B.C.) determined through observation that Earth is spherical. We do not know what observations led Pythagoras to this conclusion, but we can guess. He might have noticed ships sailing beyond the horizon and apparently sinking below the water's surface, only to arrive back at port with dry decks. Perhaps he noticed Earth's curved shadow cast on the lunar surface during an eclipse of the Moon. He might have deduced that the Sun and Moon are not just the flat disks they appear to be in the sky but are spherical and that Earth must be a sphere as well.

Earth's sphericity was generally accepted by the educated populace as early as the first century A.D.

Christopher Columbus, for example, knew he was sailing around a sphere in 1492; that is why he thought he had reached the East Indies.

Until 1687, the spherical-perfection model was a basic assumption of **geodesy**, the science that determines Earth's shape and size by surveys and mathematical calculations. But in that year, Sir Isaac Newton postulated that the round Earth, along with the other planets, could not be perfectly spherical. Newton reasoned that the more-rapid rotational speed at the equator—the equator being farthest from the central axis of the planet and therefore moving faster—would produce an equatorial bulge in response to a greater centrifugal force, which, in effect, pulls Earth's surface outward. He was convinced that Earth is slightly misshapen into what he termed an *oblate spheroid, or more correctly an* oblate ellipsoid (*oblate* means "flattened"), with the oblateness occurring at the poles.

Today, Earth's equatorial bulge and its polar oblateness are universally accepted and confirmed by satellite observations. Our modern era of Earth measurement is

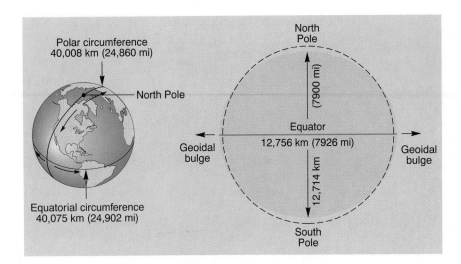

**FIGURE 1-9**
**Earth's dimensions.**
Earth's diameter and circumference—equatorial and polar—are shown. The dashed line is a perfect circle for reference to Earth's geoid.

one of tremendous precision and is called the "geoidal epoch" because Earth is considered a **geoid**, meaning literally that "the shape of Earth is Earth-shaped." Imagine Earth's geoid as a sea-level surface that is extended uniformly worldwide, beneath the continents. Both heights on land and depths in the oceans are measured from this hypothetical surface. Think of the geoid surface as a balance between the gravitational attraction of Earth's mass and the centrifugal pull caused by Earth's rotation.

Figure 1-9 gives Earth's polar and equatorial circumferences and diameters. Earth's polar circumference was first measured over 2200 years ago by the Greek geographer, astronomer, and librarian Eratosthenes (ca. 276–195 B.C.). The ingenious reasoning by which he arrived at his calculation is presented in Focus Study 1-1.

# Location and Time on Earth

Determining specifically where something is located on Earth's surface requires a coordinated grid system, one that is agreed to by all peoples. The terms *latitude* and *longitude* were in use on maps as early as the first century A.D., with the concepts themselves dating back to Eratosthenes and others.

The geographer, astronomer, and mathematician Ptolemy (ca. A.D. 90–168) contributed greatly to modern maps, and many of his terms and configurations are still used today. On his map of the known world (Figure 1-10), Ptolemy described 8000 places according to their north-south and east-west coordinates. He also placed north at the top and east at the right, orienting maps in a manner familiar to us.

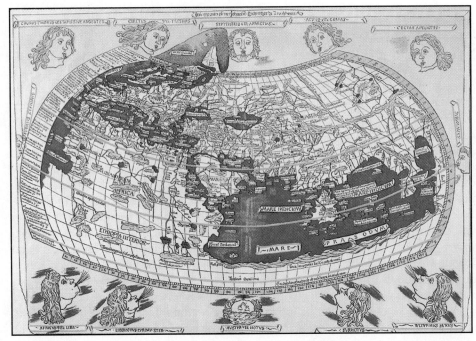

**FIGURE 1-10**
**Ptolemy's world map.**
Ptolemy published an early version of this map of the known world—Europe, Asia, and North Africa—in A.D. 140. Shown is a Renaissance interpretation of Ptolemy's original map. [The Granger Collection, New York.]

## Focus Study 1-1

## Measuring Earth in 247 B.C.

Eratosthenes, foremost among early geographers, served as the librarian of Alexandria in Egypt during the third century B.C. He was in a position of scientific leadership, for Alexandria's library was the finest in the ancient world. Among his achievements was calculation of Earth's circumference to a high level of accuracy, quite a feat for 247 B.C.

Travelers told Eratosthenes that on June 21 they had seen the Sun's rays shine directly to the bottom of a well at Syene, the location of present-day Aswan, Egypt (Figure 1). This meant that the Sun had to be directly overhead. North of Syene in Alexandria, Eratosthenes knew from his own observations that the Sun's rays never were directly overhead at Alexandria, even at noon on June 21, the longest day of the year and the day on which the Sun is at its northernmost position in the sky. Unlike objects in Syene, objects in Alexandria always cast a daytime shadow. Using the considerable geometric knowledge of the era, Eratosthenes conducted an experiment.

In Alexandria at noon on June 21, he carefully measured the angle of a shadow cast by an obelisk, a perpendicular column used for telling time by the Sun. Knowing the height of the obelisk and measuring the length of the shadow from its base, he solved the triangle for the angle of the Sun's rays, which he determined to be 7.2°. However, at Syene on the same day, the angle of the Sun's rays was 0° from a perpendicular—that is, the Sun was directly overhead.

Geometric principles told Eratosthenes that the distance on the ground between Alexandria and Syene formed an arc of Earth's circumference equal to the angle of the Sun's rays at Alexandria. Since 7.2° is roughly 1/50 of the 360° in

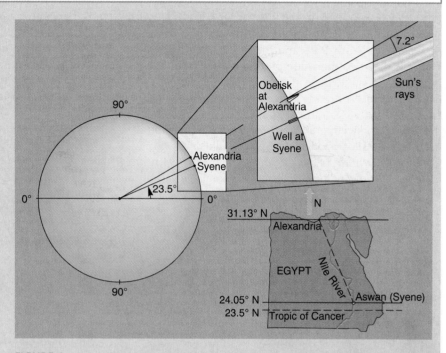

**FIGURE 1**
**Eratosthenes' calculation.**
Eratosthenes' determination of Earth's circumference. This remarkably accurate measurement was based on his precise work in geometry and approximation of the speed of camel caravans.

Earth's total circumference, the distance between Alexandria and Syene must represent approximately 1/50 of Earth's total circumference. Next, he needed to know the distance between the two cities.

Camel caravans took 50 days to trek from Syene to Alexandria, covering about 100 stadia a day—5000 stadia one way. Eratosthenes determined the surface distance between the two cities as 5000 stadia. He then multiplied 5000 stadia by 50 to determine that Earth's circumference is about 250,000 stadia. A stadium, a Greek unit of measure, equals approximately 185 m (607 ft), so Eratosthenes'

calculations convert to roughly 46,250 km (28,738 mi), which is remarkably close to the correct value of 40,075 km (24,902 mi) for Earth's equatorial circumference. (Several values can be used for the distance of a stadium—the 185 m used here represents an average value.)

Eratosthenes' work teaches the value of observing carefully and integrating all observations with previous learning. Measuring Earth's circumference required application of his knowledge of Earth-Sun relationships, geometry, and geography to his keen observations to prove his perception about Earth's size.

Ptolemy divided the circle into 360 degrees (360°), with each degree comprising 60 *minutes* (60′) and each minute including 60 *seconds,* (60″) in a manner adapted from the Babylonians. He located places using these degrees, minutes, and seconds. However, the precise length of a degree of

latitude and a degree of longitude on Earth's surface remained unresolved for the next 17 centuries. Ptolemy's values for longitude were in error because he accepted an estimate of Earth's circumference that was too short. Let us now examine the grid coordinate elements, latitude and longitude.

## Latitude

**Latitude** *is an angular distance north or south of the equator*, measured from the center of Earth (Figure 1-11a). On a map or globe, the lines designating these angles of latitude run east and west, parallel to the equator (Figure 1-11b). Because Earth's equator divides the distance between the North Pole and the South Pole exactly in half, it is assigned the value of 0° latitude. Thus, latitude increases in value from the equator northward to the North Pole, at 90° north latitude, and southward to the South Pole, at 90° south latitude.

A line connecting all points along the same latitudinal angle is called a **parallel**. Thus, as illustrated in Figure 1-11b, *latitude* is the name of the angle (49° north latitude), *parallel* names the line (49th parallel), and both

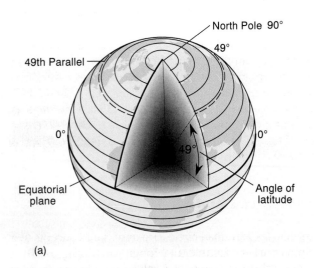

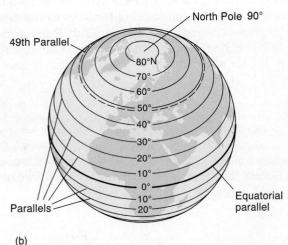

(b)

**FIGURE 1-11**
**Parallels of latitude.**
(a) Latitude is measured in degrees north or south of the equator, which is 0°. Each pole is at 90°. (b) These angles of latitude determine parallels along Earth's surface.

indicate distance north of the equator. In the figure, an angle of 49° north latitude is measured, and, by connecting all points at this latitude, the 49th parallel is designated. The 49th parallel is a significant one in the Western Hemisphere, for it forms the boundary between Canada and the United States from Minnesota to the Pacific.

Latitude is readily determined by reference to *fixed celestial objects* such as the Sun or the stars, a method dating to ancient times. During daylight hours, the angle of the Sun above the horizon indicates the observer's latitude, after adjustment is made for the seasonal tilt of Earth and for the time of day. Because Polaris (the North Star) is almost directly overhead at the North Pole, persons anywhere in the Northern Hemisphere can determine their latitude at night simply by sighting Polaris and measuring its angle above the local horizon (Figure 1-12). The angle of elevation of Polaris above the horizon equals the latitude of the observation point.

In the Southern Hemisphere, Polaris cannot be seen, because it is below the horizon. Instead, latitude measurement south of the equator is accomplished by sighting on a constellation that points to a celestial location above the South Pole. This indicator constellation is the Southern Cross (Crux Australis).

***Latitudinal Geographic Zones.*** Natural environments differ dramatically from the equator to the poles, in both their processes and their appearance. These differences result from the amount of solar energy received, which varies by latitude and season of the year. As a convenience, geographers identify *latitudinal geographic zones* as regions with fairly consistent qualities. Figure 1-13 portrays these zones, their locations, and their names: *equatorial* and *tropical*, *subtropical*, *midlatitude*, *subarctic* or *subantarctic*, and *arctic* or *antarctic*. These generalized latitudinal zones are useful for reference, but they are not rigid delineations. "Lower latitudes" are those nearer the equator, whereas "higher latitudes" are to those nearer the poles.

The *Tropic of Cancer* (23.5° north parallel) and the *Tropic of Capricorn* (23.5° south parallel) are the most extreme northern and southern parallels that experience perpendicular (directly overhead) rays of the Sun at local noon, marking the first day of summer in each hemisphere. (The tropics are discussed further in Chapter 2.) The *Arctic Circle* (66.5° north parallel) and the *Antarctic Circle* (66.5° south parallel) are the parallels farthest from the poles that experience 24 uninterrupted hours of night, during local winter, or of day, during local summer.

## Longitude

**Longitude** *is an angular distance east or west of a point on Earth's surface*, measured from the center of Earth (Figure 1-14a). On a map or globe, the lines designating these

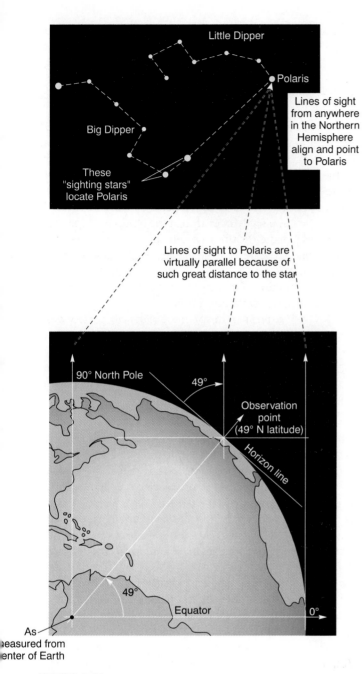

**FIGURE 1-12**

**Determining latitude by using Polaris (the North Star).**
To locate Polaris from anywhere in the Northern Hemisphere you can use the "sighting stars" in the Big Dipper constellation (above). Next, measure the angular distance between Polaris and the local horizon (below). This angular distance of Polaris above the horizon is the same as your latitude. In this case Polaris appears 49° above the horizon, so you are standing at 49° north latitude. (Note that Polaris is at such a great distance from Earth that lines of sight from anywhere in the Northern Hemisphere can be considered parallel.)

the angle, *meridian* names the line, and both indicate distance east or west of an arbitrary *prime meridian* (Figure 1-14b). Earth's prime meridian was not generally agreed to by most nations until 1884, when one was finally selected: 0° passes through the old Royal Observatory at Greenwich, England.

Because meridians of longitude converge toward the poles, the actual distance on the ground spanned by a degree of longitude is greatest at the equator (where meridians separate to their widest distance apart) and diminishes to zero at the poles (where meridians converge). The distance covered by a degree of latitude, however, varies only slightly, owing to deviations in Earth's shape. Table 1-1 compares latitude and longitude degree length. It also shows the similarity of ground distance for a degree of latitude and a degree of longitude at the equator. Note the consistent distance represented by a degree of latitude from equator to poles, yet the decreasing value covered by a degree of longitude as meridians converge toward each pole.

We have noted that latitude is easily determined by sighting the Sun, the North Star, or using the Southern Cross as a pointer, but a method of accurately determining longitude, especially at sea, remained a major difficulty in navigation until the mid-1700s. The key to measuring the longitude of a place lies in the determination of time. The relation between time and longitude is the topic of Focus Study 1-2.

## Great Circles and Small Circles

Great circles and small circles are important concepts that help summarize latitude and longitude (Figure 1-15). A **great circle** is any circle of Earth's circumference whose center coincides with the center of Earth. An infinite number of great circles can be drawn on Earth. Every meridian is one-half of a great circle that passes through the poles. On flat maps, airline and shipping routes appear to arch their way across oceans and landmasses. These are *great circle routes*, the shortest distance between two points on Earth.

Only one parallel is a great circle—the equatorial parallel. All other parallels diminish in length toward the poles and, along with any other non-great circles that one might draw, constitute **small circles**—circles whose centers do not coincide with Earth's center.

## Prime Meridian and Standard Time

Today we take for granted standard time zones and an agreed-upon prime meridian. If it is 3:00 P.M. in Oklahoma City, it is 4:00 P.M. in Baltimore, 2:00 P.M. in Salt Lake City, 1:00 P.M. in Seattle and Los Angeles, noon in

angles of longitude run north and south (Figure 1-14b). They run at right angles (90°) to all parallels, including the equator. A line connecting all points along the same longitude is a **meridian**. Thus, *longitude* is the name of

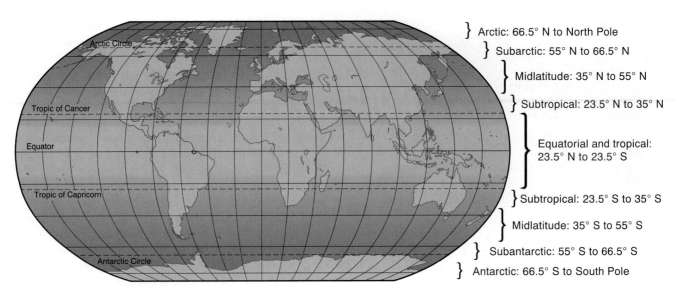

**FIGURE 1-13**
**Latitudinal geographic zones.**
Geographic zones are generalizations that characterize various regions by latitude. Think of these as transitional into one another over broad areas.

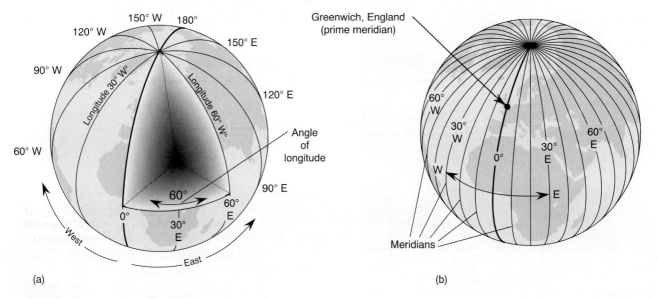

(a)                                                                    (b)

**FIGURE 1-14**
**Meridians of longitude.**
(a) Longitude is measured in degrees east or west of a 0° starting line, the prime meridian. (b) Angles of longitude measured from this prime meridian determine other meridians. The prime meridian is drawn from the North Pole through the Royal Observatory in Greenwich, England, to the South Pole. North America is west of Greenwich, therefore it is in the Western Hemisphere.

Anchorage, and 11:00 A.M. in Honolulu. It is 9:00 P.M. in London and midnight in Ar Riyāḍ, Saudi Arabia. (The designation A.M. is for *ante meridiem*, "before noon," whereas P.M. is for *post meridiem*, "after noon.") Coordination of international trade, airline schedules, business and agricultural activities, and daily living depends on this worldwide time system.

Such a standard time system is based on a **prime meridian**, which, as noted earlier, was not agreed upon until 1884. Before that year, most nations used their own national prime meridians for land maps, plus different meridians for marine navigation charts, as shown in Table 1-2. The importance of Great Britain as a world power at the time is evident in that the Greenwich Observatory near

**TABLE 1-1**

| Physical Distances Represented by Degrees of Latitude and Longitude | | |
|---|---|---|
| **Latitudinal Location** | **Latitude Degree Length km (mi)** | **Longitude Degree Length km (mi)** |
| 90° (poles) | 111.70 (69.41) | 0 (0) |
| 60° | 111.42 (69.23) | 55.80 (34.67) |
| 50° | 111.23 (69.12) | 71.70 (44.55) |
| 40° | 111.04 (69.00) | 85.40 (53.07) |
| 30° | 110.86 (68.89) | 96.49 (59.96) |
| 0° (equator) | 110.58 (68.71) | 111.32 (69.17) |

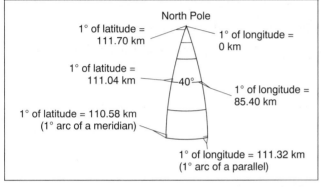

London was used on many nations' maritime charts as the prime meridian. The lack of a single prime meridian and standard time zones within and among countries created great confusion in global mapping and telling clock time.

Earth revolves 360° every 24 hours, or 15° per hour (360° ÷ 24 = 15°). Thus, a time zone of 1 hour is established for each 15° increment of longitude. Setting time was not a great problem in small European countries, most of which are less than 15° wide. But in North America, which spans more than 90° of longitude (the equivalent of six 15° time zones), the problem was serious. In 1870, railroad travelers going from Maine to San Francisco made 22 adjustments to their watches to stay consistent with local time!

In Canada, Sir Sanford Fleming led the fight for standard time and for an international agreement on a prime meridian. His struggle led the United States and Canada to adopt a standard time in 1883. Today, only three adjustments are needed—from Eastern Standard Time to Central, Mountain, and Pacific—in the continental United States, and four across Canada.

In 1884 the International Meridian Conference was held in Washington, D.C., attended by 27 nations. After lengthy debate, most participating nations chose the Royal

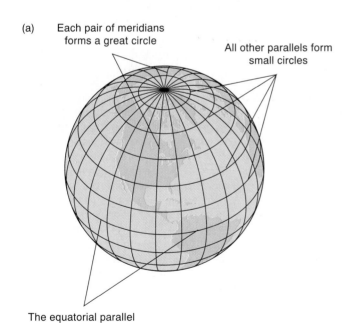

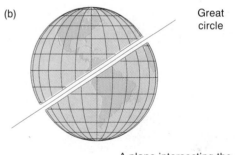

A plane intersecting the globe along a great circle divides the globe into equal halves and passes through its center

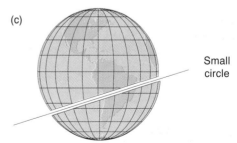

A plane that intersects the globe along a small circle splits the globe into unequal sections—this plane does not pass through the center of the globe

**FIGURE 1-15**

**Great circles and small circles.**

(a) Examples of great circles and small circles on Earth. (b) Any plane that divides Earth into equal halves will intersect the globe along a great circle; this great circle is a full circumference of the globe and is the shortest distance between any two surface points. (c) Any plane that splits the globe into unequal portions will intersect the globe along a small circle.

## Focus Study 1-2

### The Timely Search for Longitude

Unlike latitude, longitude cannot be determined readily from fixed celestial bodies. The problem is Earth's rotation, which constantly changes the apparent position of the Sun and stars. Determining longitude is particularly critical at sea, where no landmarks are visible and, until late in the eighteenth century, there was no device to measure time. Amerigo Vespucci, the explorer for whom America is named, made this note in his diary, as quoted by Daniel J. Boorstin in *The Discoverers*: "5 September 1499—As to longitude, I declare that I found so much difficulty in determining it that I was put to great pains to ascertain the east-west distance I had covered."*

In his historical novel *Shogun*, author James Clavell expressed a similar perspective through his pilot, Blackthorn: "Find how to fix longitude and you're the richest man in the world....The Queen, God bless her, 'll give you ten thousand pound and dukedom for answer to the riddle....Out of sight of land you're always lost, lad."†

In the early 1600s, Galileo explained that longitude could be measured by using two clocks. Any point on Earth takes 24 hours to travel around the full 360° of one rotation (one day). If you divide 360° by 24 hours, you find that any point on Earth travels through 15° of longitude every hour. Thus, if there were a way to measure time accurately at sea, a comparison

*D. J. Boorstin, *The Discoverers* (New York: Random House, 1983, p. 247).
†From *Shogun* by J. Clavell, p. 10. Copyright © 1975 by James Clavell, Delacorte Press, a division of Bantam, Doubleday, Dell Publishing Group, Inc.

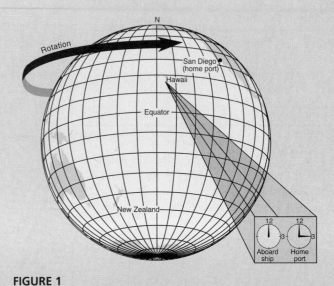

**FIGURE 1**

**Clock time determines longitude.**

Using two clocks on a ship to determine longitude. The 3-hour time difference, at 15° per hour, indicates that the ship is 45° west of port.

of two clocks could give a value for longitude. One clock would indicate the time back at home port (Figure 1). The other clock would be reset at local noon each day, as determined by the highest Sun position in the sky (solar zenith). The time difference then would indicate the longitudal difference traveled: 1 hour for each 15° of longitude.

For example, if the shipboard clock reads local noon and the clock set for home port reads 3:00 P.M., ship time is 3 hours earlier than home time. Therefore, calculating 3 hours at 15° per hour puts the ship at 45° west longitude from home port (Figure 1). The principle was sound; all that was needed were accurate clocks. Unfortunately, the pendulum clock invented by Christian Huygens in 1656 did not work on the rolling deck of a ship at sea!

In 1707, the British lost four ships and 2000 men in a sea tragedy that was blamed specifically on the longitude problem. In response, Parliament passed an act in 1714—"Publik Reward...to Discover the Longitude at Sea"—and authorized a prize of £20,000 sterling (equal to over $1 million today) to the first successful inventor of an accurate seafaring clock. The Board of Longitude was established to judge any devices submitted.

John Harrison, a self-taught country clockmaker, began work on the problem in 1728 and finally produced his marine chronometer, known as Number 4, in 1760. The clock was tested during a 9-week trip to Jamaica in November 1761. When taken ashore for testing against Jamaica's land-based longitude, Harrison's Number 4 was only 5 seconds slow,

Observatory at Greenwich as the place for the prime meridian of 0° longitude. The Observatory was highly respected, and more than 70% of the world's merchant ships already used the London prime meridian. Thus, a world standard was established—**Greenwich Mean Time (GMT)**—and a consistent Universal Time was established for the first time in history. Each time zone basically spans 15°, covering 7.5°

on either side of a central meridian and representing 1 hour of time. As you can see from Figure 1-16, national boundaries and political considerations distort time zone boundaries here and there. For example, China spans four time zones, but its government decided to keep the entire country operating at the same time. Thus in some parts of China clocks are several hours off from what the Sun is doing.

an error that translates to only 1.25′ or 2.3 km (1.4 mi), well within Parliament's standard. After many delays, Harrison finally received most of the prize money in his last years of life.

From that time it was possible to determine longitude accurately on land and sea, as long as everyone agreed upon a meridian to use as a reference for time comparisons. Figure 2 shows the Royal Observatory in Greenwich, England, established by international treaty in 1884 as Earth's prime meridian.

In this modern era of atomic clocks and satellites in mathematically precise orbits, we have far greater accuracy available for the determination of longitude on Earth's surface. Measurements of Earth's surface from satellites are accurate within millimeters, and surface laser surveys are precise within micrometers.

**FIGURE 2**
**The prime meridian at the Royal Observatory.**
Courtyard of the old Royal Observatory, Greenwich, which is still used as Earth's prime meridian—0° longitude. [© The National Maritime Museum, Greenwich, England.]

**TABLE 1-2**

| Sample of Prime Meridians Used in the 1800s | | |
|---|---|---|
| | **Prime Meridians on** | |
| **Country** | **Land Maps** | **Marine Charts** |
| Austria | Ferro* | Greenwich |
| Belgium | Brussels | Greenwich |
| Brazil | Rio de Janeiro | Rio de Janeiro |
| France | Paris | Ferro |
| Italy | Rome | Greenwich |
| Portugal | Lisbon | Lisbon |
| Spain | Madrid | Cádiz |
| United States | Washington | Greenwich |

*Ferro (Spanish: Hierro) Island is the westernmost of the Canary Islands in the eastern Atlantic Ocean. Ferro was the most westerly place known to early geographers. Ptolemy chose it as an initial prime meridian, ca A.D. 150. France used the Ferro prime meridian until 1911.

The Royal Observatory at Greenwich is pictured in Focus Study 1-2. Note the strip in the courtyard dividing the Eastern and Western Hemispheres and the open roof that was used for sighting on the Sun.

*International Date Line.* An important corollary of the prime meridian is the 180° meridian on the opposite side of the planet. This meridian is called the **International Date Line** and marks the place where each day officially begins (at 12:01 A.M.). From this "line" the new day sweeps westward. This *westward* movement of time is created by the planet's turning *eastward* on its axis. At the International Date Line, the west side of the line is always one day ahead of the east side. No matter what time of day it is when the line is crossed, the calendar changes a day (Figure 1-17). Note in the illustration how the I. D. L. deviates from the 180° meridian; this deviation is due to local administrative and political preferences.

Locating the date line in the sparsely populated Pacific Ocean minimizes most local confusion. However, the consternation of early explorers before the date-line concept was understood is interesting. For example, Magellan's crew returned from the first circumnavigation of Earth in 1522, confident from their ship's log that the day of their arrival was a Wednesday, September 7. They were shocked when informed by insistent local residents that it was actually a Thursday, September 8! Of course, they had no idea that they must advance a day when sailing around the world.

# News Report 1

## GPS: A Personal Locator

The Global Positioning System (GPS) comprises 25 orbiting satellites that transmit navigational signals for Earth-bound use. Originally devised in the 1970s by the Department of Defense for military purposes, the present system is commercially available worldwide.

A hand-held receiver about the size of a pocket radio receives signals from three satellites at the same time, calculates latitude and longitude within 40 to 100 m accuracy (130 to 330 ft), and displays the results. Some units also indicate elevation. (For military applications, precision down to millimeters is possible.)

GPS is useful for diverse applications, such as ocean navigation, land surveying, managing the movement of fleets of trucks, mining and resource mapping, and environmental planning. As of 1995, some commercial airlines began using GPS to improve accuracy of routes flown and thus increase fuel efficiency. GPS also is useful to the backpacker and sportsperson. Farmers are using GPS to determine crop yields on specific parts of their farms. A detailed plot map is made to guide them as to where more fertilizer or other work is needed. (This is the science of *variable-rate technology*, made possible by GPS.)

The importance of GPS to geography is obvious because this precise technology reduces the need to maintain ground control points for location, mapping, and spatial analysis. Instead, geographers working in the field can determine their position accurately as they work. Boundaries and data points in a study area can be easily determined and entered into a data base, and the need for traditional surveys is reduced. For this and myriad other applications, GPS system sales are expected to reach $10 billion by the year 2000.

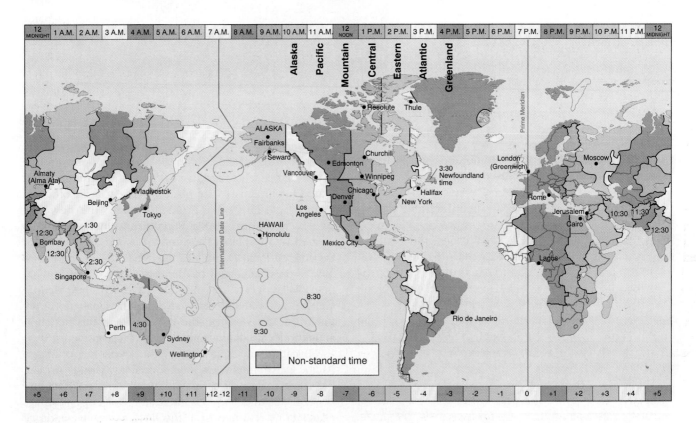

## FIGURE 1-16
**Modern international standard time zones.**

Numbers along the bottom of the map indicate how many hours each zone is earlier (minus sign) or later (plus sign) than the Coordinated Universal Time (UTC) standard at the prime meridian. The United States has five time zones; Canada is divided into six. Try to determine the present time in Moscow, London, Halifax, Chicago, Winnipeg, Denver, Los Angeles, Fairbanks, Honolulu, Tokyo, and Singapore. After the breakup of the Soviet Union in the early 1990s, the states of Eastern Europe changed their time zone designation by 1 hour to +2 hours.

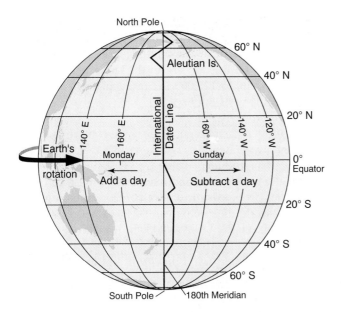

**FIGURE 1-17**
**International Date Line.**
International Date Line location, approximately along the 180th meridian (see also Figure 1-16). Note that it is officially one day later west of the I. D. L.

***Coordinated Universal Time.*** For decades, Greenwich Mean Time from the Royal Observatory's astronomical clocks was the world's Universal Time standard for accuracy. GMT was broadcast using radio time signals as early as 1910. The French government took the initiative in 1912 and called a gathering of nations to better coordinate the various radio time signals that were then originating in many countries. From this conference, time signals became more unified, GMT was made standard, and a new organization was established to be the custodian of the most "exact" time—the Bureau International de l'Heure (BIH) outside Paris. Progress in accurately measuring time progressed rapidly with the invention of a quartz clock in 1939 and atomic clocks in the early 1950s.

GMT universal time was updated in 1972 when the time signal system of **Coordinated Universal Time (UTC)** became legal time worldwide. Today, UTC is the reference for official time in all countries. Although the prime meridian still runs through Greenwich, UTC is based on average time calculations collected by the BIH near Paris and broadcast worldwide.

Today, UTC and the official length of the second are measured by the very regular vibrations (natural frequency) of cesium atoms in six *primary standard clocks*. The newest is operated by the Time and Frequency Services of the National Institute for Standards and Technology (this new clock, named *NIST-7*, became operational in 1994). Three clocks are operated in Ottawa, Ontario, by the Time Standards group of the Institute for Measurement Standards, National Research Council. The other

two are in Germany, operated by the Physikalisch-Technische Bundesanstalt (PTB).

Added to these primary clocks, 200 *normal standard clocks* of commercial accuracy operate in 70 nations and report to the BIH in Paris, where computers bring together all measurements for "keeping the time." This level of precision has permitted scientists to add a "leap second" each year since 1972 to maintain UTC accuracy in coordination with Earth's slightly irregular rate of rotation.

***Daylight Saving Time.*** In many countries, time is set ahead 1 hour in the spring and set back 1 hour in the fall—a practice known as **daylight saving time**. The idea to extend daylight for early evening activities (at the expense of daylight in the morning) was first proposed by Benjamin Franklin. It was not adopted until World War I, when Great Britain, Australia, Germany, Canada, and the United States used the practice to save energy (one less hour of artificial lighting needed).

During World War II, clocks in the United States were kept advanced an hour year-round throughout 1942–1945 for energy savings. For the same reason, a year-round daylight saving time was in force from January 1974 to October 1975.

In 1986, the United States and Canada increased the extent of daylight saving time. Time now "springs forward" 1 hour on the first Sunday in April and "falls back" an hour on the last Sunday in October, except in a few places that do not use it (Hawaii, Arizona, portions of Indiana, and Saskatchewan). In Europe, the last Sundays in March and September generally are used to begin and end what they call "summer time."

# Maps, Scales, and Projections

The earliest known graphic map presentations date to 2300 B.C., when the Babylonians used clay tablets to record information about the region of the Tigris and Euphrates Rivers (the area of modern-day Iraq). Much later, when sailing vessels dominated the seas, a pilot's *rutter*—a descriptive diary of locations, places, coastlines, and collected maps—became a critical reference. *Portolan* charts, describing harbor locations, coastal features, and showing compass directions, also were prepared from these many navigational experiences.

Today, the making of maps and charts is a specialized science as well as an art, blending aspects of geography, engineering, mathematics, graphics, computer science, and artistic specialties. It is similar in ways to architecture, in which aesthetics and utility are combined to produce an end product.

A *map* is a generalized view of an area, usually some portion of Earth's surface, as seen from above and greatly reduced in size. The part of geography that embodies mapmaking is called **cartography**. Maps are critical tools with which geographers depict spatial information and analyze spatial relationships. We all use maps at some time to visualize our location and our relationship to other places, or maybe to plan a trip, or to coordinate commercial and economic activities. Have you found yourself looking at a map, planning real and imagined adventures to far-distant places? Maps are wonderful tools! Learning a few basics about maps is essential to our study of physical geography.

## The Scale of Maps

Architects, toy designers, and mapmakers have something in common: they all create *scale models*. They reduce real things and places to the more convenient scale of a drawing, a model car, a train, or plane, a diagram, or a map. An architect renders a blueprint of a building to guide the contractors, selecting a scale so that one centimeter (or inch) on the drawing represents so many meters (or feet) on the proposed building. Often, the drawing is 1/50 to 1/100 of real size.

The cartographer does the same thing in preparing a map. The ratio of the image on a map to the real world is called **scale**; it relates a unit on the map to a similar unit on the ground. A 1:1 scale means that a centimeter on the map represents a centimeter on the ground (although this certainly is an impractical map scale, for the map is as large as the area mapped!). A more appropriate scale for a local map is 1:24,000, in which 1 unit on the map represents 24,000 identical units on the ground.

Map scales are presented in several ways: as a written scale, a representative fraction, or a graphic scale (Figure 1-18). A *written scale* simply states the ratio—for example, "one centimeter to one kilometer" or "one inch to one mile." A *representative fraction* (RF, or fractional

scale) is expressed with either a colon or a slash, as in 1:125,000 or 1/125,000. No actual units of measurement are mentioned because any unit is applicable as long as both parts of the fraction are in the same unit: 1 cm to 125,000 cm, 1 in. to 125,000 in., or even 1 arm length to 125,000 arm lengths, and so on.

A *graphic scale*, or bar scale, is a bar graph with units to allow measurement of distances on the map. An important advantage of a graphic scale is that, if the map is enlarged or reduced, the graphic scale enlarges or reduces along with the map. In contrast, written and fractional scales become incorrect with enlargement or reduction: You can shrink a map from 1:24,000 to 1:63,360, but the scale will still say "1 in. to 2000 ft," instead of the new correct scale of 1 in. to 5280 ft (1 mi).

Scales are called *small*, *medium*, and *large*, depending on the ratio described. Thus, in relative terms, a scale of 1:24,000 is a *large scale*, whereas a scale of 1:50,000,000 is a *small scale*. The greater the denominator in a fractional scale (or the number on the right in a ratio expression), the smaller the scale and the more abstract the map must be in relation to what is being mapped. Examples of selected representative fractions and written scales are listed in Table 1-3 for small-, medium-, and large-scale maps. In Chapter 12, a figure presents a small-scale map of a portion of Pennsylvania at a 1:3,500,000 scale. Enlarged from this in the figure is a *Landsat* image and topographic map at a medium scale of 1:250,000. Note the increased detail at the larger scale.

Let's say you find a world globe in the library that is 61 cm (24 in.) in diameter. We know that Earth has an equatorial diameter of 12,756 km (7926 mi), so the scale of the globe is the ratio of 61 cm to 12,756 km. We divide Earth's actual diameter by the globe's diameter (12,756 km ÷ 61 cm) and determine that 1 cm of the globe's diameter equals about 20,900,000 cm of Earth's diameter. Thus, the representative fraction for the globe is expressed in centimeters as 1:20,900,000. This representative fraction can now be expressed in *any* unit of measure, metric or English, as long as both numbers are in the same units. This 61-cm globe is a small-scale representation of Earth with little local detail.

If there is a globe available in your library or classroom, check to see the scale at which it was drawn. The scale will be printed somewhere on the globe. See if you can find examples of written, representative, and graphic scales on wall maps, highway maps, and in atlases.

## Map Projections

A globe is not always a helpful representation of Earth. When you go on a trip you need more-detailed information than a globe can provide, and large globes don't fit

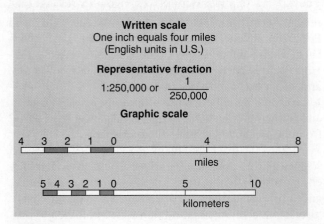

**FIGURE 1-18**
**Map scale.**
Three common expressions of map scale.

**TABLE 1-3**

| Sample Representative Fractions and Written Scales for Small-, Medium-, and Large-Scale Maps | | | |
|---|---|---|---|
| *System* | *Scale Size* | *Representative Fraction* | *Written Scale* |
| English | Small | 1:3,168,000 | 1 in. = 50 mi |
| | | 1:2,500,000 | 1 in. = 40 mi |
| | | 1:1,000,000 | 1 in. = 16 mi |
| | | 1:500,000 | 1 in. = 8 mi |
| | | 1:250,000 | 1 in. = 4 mi |
| | Medium | 1:125,000 | 1 in. = 2 mi |
| | | 1:63,360 (or 1:62,500) | 1 in. = 1 mi |
| | | 1:31,680 | 1 in. = 0.5 mi |
| | | 1:30,000 | 1 in. = 2500 ft |
| | Large | 1:24,000 | 1 in. = 2000 ft |

| *System* | *Representative Fraction* | *Written Scale* |
|---|---|---|
| Metric | 1:1,100,000 | 1 cm = 10.0 km |
| | 1:50,000 | 1 cm =  0.50 km |
| | 1:25,000 | 1 cm =  0.25 km |
| | 1:20,000 | 1 cm =  0.20 km |

well in your car. Consequently, to provide localized detail, cartographers prepare large-scale flat maps, which are two-dimensional representations (scale models) of our three-dimensional Earth. Unfortunately, this conversion from three dimensions to two causes distortion.

A globe is the only true representation of *distance, direction, area, shape,* and *proximity.* A flat version distorts those properties. Therefore, in preparing a flat map, the cartographer must decide which characteristic to preserve, which to distort, and how much distortion is acceptable. To understand this problem, consider these important properties of a *globe*:

- Parallels always are parallel to each other, always are evenly spaced along meridians, and always decrease in length toward the poles.

- Meridians converge at both poles and are evenly spaced along any individual parallel.

- The distance between meridians decreases toward poles, with the spacing between meridians at the 60th parallel equal to one-half the equatorial spacing.

- Parallels and meridians always cross each other at right angles.

*The problem is that all these qualities cannot be reproduced on a flat surface.* Simply taking a globe apart and laying it flat on a table illustrates the problem faced by cartographers (Figure 1-19). You can see the empty spaces that open up between the sections, or *gores,* of the globe. This reduction of the spherical Earth to a flat surface is called a **map projection**. Thus, no flat map projection of Earth can ever have all the features of a globe. Flat maps always possess some degree of distortion—much less for large-scale maps representing a few kilometers; much more for small-scale maps covering individual countries, continents, or the entire world.

***Properties of Projections.*** There are many projections, four of which are shown in Figure 1-20. *The best projection to use is always determined by its intended use.* The major decisions in selecting a map projection involve the properties of **equal area** (equivalence) and **true shape** (conformality). Two other properties of map projections that can also be considered are true direction (azimuth) and true distance (equidistance). If a cartographer selects equal area as the desired trait—for example, for a map showing the distribution of world climates—then true shape must be sacrificed by *stretching* and *shearing,* which allows parallels and meridians to cross at other than right angles. On an equal-area map, a coin covers the same amount of surface area no matter where you place it on the map.

If, on the other hand, a cartographer selects the property of true shape, as for a map used for navigational purposes, then equal area must be sacrificed, and the scale will actually change from one region of the map to another.

***The Nature and Classes of Projections.*** Despite the fact that modern cartographic technology uses mathematical constructions and computer-assisted graphics, the word *projection* is still used. The term comes from times

**FIGURE 1-19**

**From globe to flat map.**
Conversion of the globe to a flat map projection
requires decisions about which properties to
preserve and the amount of distortion that is
acceptable. [NASA photo.]

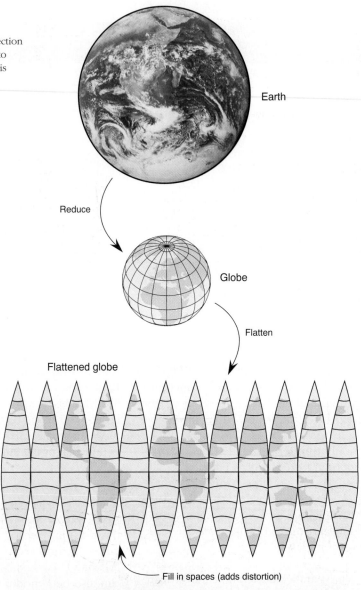

Earth

Reduce

Globe

Flatten

Flattened globe

Fill in spaces (adds distortion)

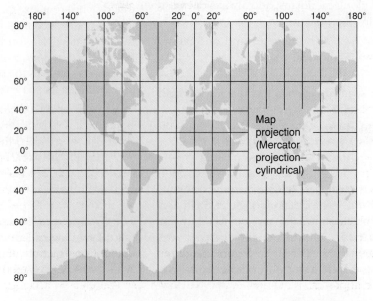

Map
projection
(Mercator
projection–
cylindrical)

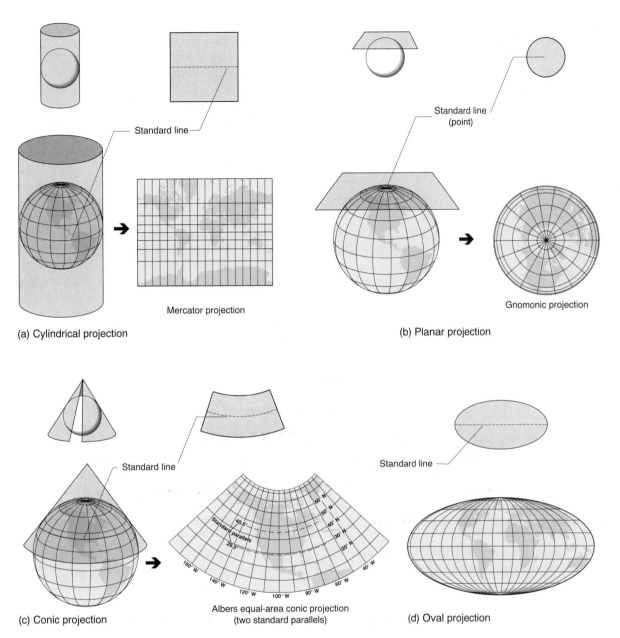

**FIGURE 1-20**
**Classes of map projections.**
Four general classes and perspectives of map projections.

past, when geographers actually projected the shadow of a wire-skeleton globe onto a geometric surface. The wires represented parallels, meridians, and outlines of the continents. A light source then cast a shadow pattern of latitude and longitude lines from the globe onto various geometric surfaces, such as a cylinder, plane, or, cone.

Figure 1-20 illustrates the derivation of the general classes of map projections and the perspectives from which they are generated. The classes shown include the *cylindrical, planar* (or azimuthal), and *conic.* Another class of projections, which cannot be derived from this physical-perspective approach, is the nonperspective *oval*

shape. Still other projections are derived from purely mathematical calculations.

With projections, the contact line or contact point between the wire globe and the projection surface—called a *standard line* or *standard point*—is *the only place where all globe properties are preserved.* Thus, a *standard parallel* or *standard meridian* is a standard line true to scale along its entire length without any distortion. Areas away from this critical tangent line or point become increasingly distorted. Consequently, this area of optimum spatial properties should be centered on the region of greatest interest so that greatest accuracy is preserved there.

The commonly used Mercator projection (from Gerardus Mercator, A.D. 1569) is a cylindrical projection (Figure 1-20a). The Mercator is a true-shape projection, with meridians appearing as equally spaced straight lines and parallels appearing as straight lines that are spaced closer together near the equator. The poles are infinitely stretched, with the 84th north parallel and 84th south parallel fixed at the same length as that of the equator. Note in Figure 1-20a that the Mercator projection is cut around the 80th parallel in each hemisphere because of the severe distortion at higher latitudes.

Unfortunately, Mercator classroom maps present false notions of the size (area) of midlatitude and poleward landmasses. A dramatic example on the cylindrical projection in Figure 1-20a is Greenland, which looks bigger than all of South America. In reality, Greenland is only one-eighth the size of South America and is actually 20% smaller than Argentina alone! The influence of such distortion in the Cold War mind game is impossible to determine, but generations of North Americans grew up looking at an enormously enlarged (now former) Soviet Union.

The advantage of the Mercator projection is that, on it, the lines of constant direction, called **rhumb lines**, are straight and thus facilitate plotting directions between two points (see Figure 1-21). Thus, the Mercator projection is useful in navigation and has been the standard for nautical charts prepared by the National Ocean Service (formerly U.S. Coast and Geodetic Survey) since 1910.

The gnomonic, or planar projection in Figure 1-20b is generated by projecting a light source at the center of a globe onto a plane tangent to (touching) the globe's surface. The resulting severe distortion prevents showing a full hemisphere on one projection. However, a valuable feature is derived: All great circle routes, which are the shortest distance between two points on Earth's surface, are projected as straight lines (Figure 1-21a). The great circle routes plotted on a gnomonic projection then can be transferred to a true-direction projection, such as the Mercator, for determination of precise compass headings (Figure 1-21b).

***Maps Used in This Text.*** Several projections are used in this text: Goode's homolosine, Robinson, Miller cylindrical, and polyconic. Each was chosen to best present specific types of data. **Goode's homolosine projection** is an interrupted world map designed in 1923 by Dr. J. Paul Goode of the University of Chicago. It was first used in Rand McNally *Goode's Atlas* in 1925. Goode's homolosine equal-area projection (Figure 1-22) is a combination of two oval projections (*homolo*graphic and *sin*usoidal projections).

Two equal-area projections are cut and pasted together to improve the rendering of landmass shapes. A *sinusoidal projection* is used between 40° N and 40° S latitudes. Its central meridian is a straight line; all other meridians are drawn as sinusoidal curves (based on sine-wave curves) and parallels are evenly spaced. A *Mollweide projection*, also called a *homolographic projection*, is used from 40° N to the North Pole and from 40° S to the South Pole. Its central meridian is a straight line; all other meridians are drawn as elliptical arcs and parallels are unequally spaced—farther apart at the equator, closer together poleward. This technique of combining two projections preserves areal size relationships, making the projection excellent for mapping spatial distributions when interruptions of oceans or continents do not pose a problem. We use Goode's homolosine projection in this book for the world climate map in Chapter 10, the soils map in Chapter 18, and the terrestrial ecosystems map in Chapter 20.

Another projection we often use is the **Robinson projection**, designed by Arthur Robinson in 1963 (Figure 1-23). This projection is neither equal-area nor true-shape,

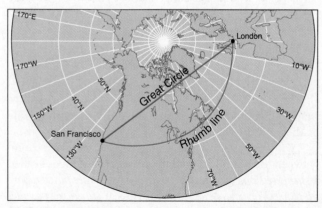

(a) Gnomonic Projection

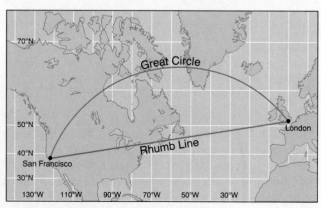

(b) Mercator Projection (conformal, true shape)

**FIGURE 1-21**

**Determining great circle routes.**

A gnomonic projection (a) is used to determine the shortest distance—great circle route—between San Francisco and London, because on this projection the arc of a great circle is a straight line. This great circle route is then plotted on a Mercator projection (b), which has true compass direction. Note that lines of constant direction (or bearing) on a Mercator projection—rhumb lines—are not the most efficient route in terms of distance.

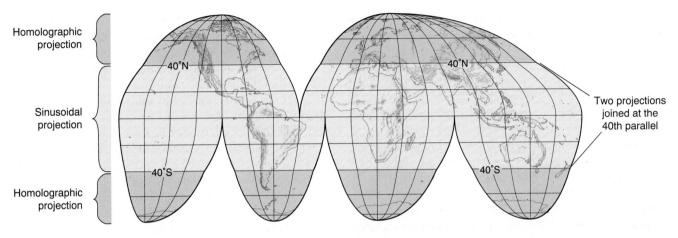

**FIGURE 1-22**
**Goode's homolosine projection.**
An equal-area map. [Copyright by the University of Chicago. Used by permission of the University of Chicago Press.]

but is a compromise between the two. The North and South Poles appear as lines slightly more than half the length of the equator; thus higher latitudes are exaggerated less than on other oval and cylindrical projections. Examples in this text are the latitudinal geographic zones map (Figure 1-13), the world temperature range map in (Figure 5-14), the maps of lithospheric plates of crust, and volcanoes and earthquakes in Chapter 11. The National Geographic Society adopted the Robinson projection for their primary world map in 1988.

Another compromise map, the **Miller cylindrical projection**, is also used in this text (Figure 1-24). Examples of this projection include the world time zone map (Figure 1-16), global temperature maps (Figures 5-11 and 5-13), and two global pressure maps in Figure 6-13. This projection is neither true shape nor true area but is a compromise that avoids the severe scale distortion of the Mercator. The Miller projection frequently appears in world atlases. This projection was first presented in 1942 by Osborn Miller of the American Geographical Society. Miller stated that his goal was "to find a system of spacing the parallels of latitude such that an acceptable balance is reached between shape and area distortion. . . . [that] reduces areal distortion as far as possible."*

For more information on maps and standard map symbols turn to Appendix B, Mapping, Quadrangles, and Topographic Maps. Topographic maps are essential tools of landscape analysis. They are used by geographers, other scientists, travelers, and anyone visiting the outdoors. USGS (U.S. Geological Survey) topographic maps appear in several chapters of this text because they are so useful in depicting the tremendously varied features of the physical landscape.

*O. M. Miller, "Notes on cylindrical world map projections," *Geographical Review* 32 (1942): 424.

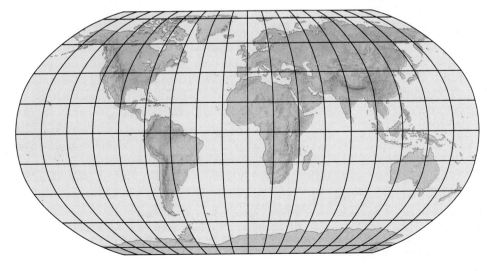

**FIGURE 1-23**
**Robinson projection.**
A compromise between equal area and true shape. [Developed by Arthur H. Robinson, 1963.]

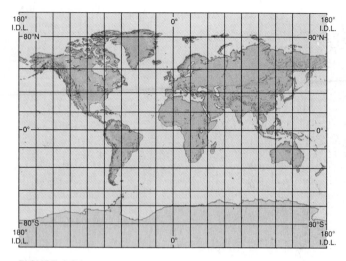

**FIGURE 1-24**
**Miller cylindrical projection.**
A compromise map projection between equal area and true shape.
[Developed by Osborn M. Miller, American Geographical Society, 1942.]

# Remote Sensing and GIS

Geographers now are probing, analyzing, and mapping our home planet through remote sensing and geographic information systems. These technologies are enhancing our understanding of Earth. Geographers use remote-sensing data to study humid and arid lands, vegetation, snow and ice, the seasonal variation of atmospheric and oceanic circulation, and the human activities that are producing global change.

## *Remote Sensing*

In this era of observations from orbit outside the atmosphere and from aircraft within it, scientists are obtaining a wide array of *remotely sensed* data (Figure 1-25). Remote sensing is nothing new to humans; we do it with our eyes all the time. When we scan the environment with our eyes, we are sensing the shape, size, and color of objects from a distance, registering energy from the visible-wavelength portion of the electromagnetic spectrum. Similarly, when a camera views the wavelengths for which its film or sensor is designed (visible light or infrared), it remotely senses energy that is reflected or emitted from a scene.

Our eyes and cameras are familiar means of obtaining **remote-sensing** information about a distant subject without having physical contact. Aerial photographs have been used for years to improve the accuracy of surface maps faster and more cheaply than can be done by on-site surveys. Deriving accurate measurements from photographs is the realm of *photogrammetry*, an important application of remote sensing.

Remote sensors on satellites and other craft sense a broader range of wavelengths than can our eyes. They can be designed to "see" wavelengths shorter than visible light (ultraviolet) and wavelengths longer than visible light (infrared and microwave radar).

Satellites do not take conventional-film photographs. Rather, they record *images* that are transmitted to Earth-based receivers in a manner similar to television broadcasts. A scene is scanned and broken down into *pixels* (*pic*ture *el*ements) each identified by coordinates known as *lines* (horizontal rows) and *samples* (vertical columns). For example, a grid of 6000 lines and 7000 samples forms 42,000,000 pixels, providing great detail and the ability to resolve ground features less than 100 m (330 ft) across, depending on the lenses used and the sensor's altitude. The large amount of data needed to produce a single image requires computer processing and data storage.

Further, digital data are processed in many ways to enhance their utility: simulated natural color, "false" color to highlight a particular feature, enhanced contrast, signal filtering, and different levels of sampling and resolution. Various wavelengths are recorded, and numerical values are assigned (see Figure 1-25). Many of the remotely sensed images used in this text were initially recorded in digital form for later processing, enhancement, and generation. Two types of remote-sensing systems are used: active and passive.

***Active Remote Sensing.*** Active systems direct a beam of energy at a surface and analyze the energy that is reflected back. An example is *radar* (*ra*dio *d*etection *a*nd *r*anging). A radar transmitter emits short bursts of energy that have relatively long wavelengths (1 to 10 m) toward the subject terrain, penetrating clouds and darkness. Energy that is reflected back, known as *backscatter*, is received by a radar receiver and analyzed. An example is the computer image of wind and sea-surface patterns over the Pacific in Figure 6-4, developed from 150,000 radar-derived measurements made on a single day by the *Seasat* satellite.

In addition, NASA sent imaging radar systems into orbit on three Space Shuttles in 1981, 1984, and 1994. The subjects of study included oceanography, landforms and geology, and biogeography. Shuttle missions in 1994 by *Endeavour* and *Atlantis* marked dramatic contributions to Earth observations using radar and other sensors to study stratospheric ozone, weather, volcanic activity, earthquakes, and water resources, among many subjects.

One Space Shuttle mission in September 1994 was appropriately loaded with radar sensors to study volcanoes. Only 8 hours after launch, the Kliuchevskoi Volcano on the Kamchatka Peninsula of Russia erupted unexpectedly. The shuttle radar was able to see through ash and smoke and expose lava flows and the volcanic eruption in dramatic images that were broadcast live

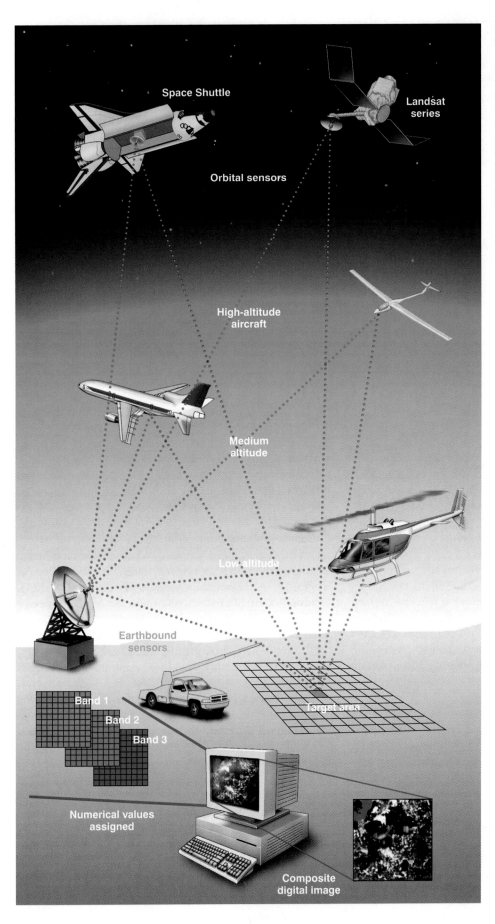

**FIGURE 1-25**
**Remote-sensing technologies.**
Remote-sensing technology is used to measure and monitor Earth's systems from orbiting spacecraft, aircraft in the atmosphere, and ground-based sensors. Computers process the collected data to produce digital images for analysis. Many of the physical systems discussed in this text are studied using this technology. The Space Shuttle is shown in its inverted orbital flight mode. (Illustration is not to scale.)

**FIGURE 1-26**
**A volcanic eruption seen from orbit.**
Image and photograph of the eruption of the Kliuchevskoi Volcano on the Kamchatka Peninsula of
Siberia, Russia, as captured by the Space Shuttle *Endeavour*, September 1994. [JPL photo/NASA.]

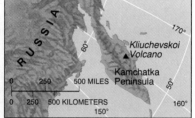

(Figure 1-26). Previously this volcano had erupted in 1737
and 1945. Volcanic processes are discussed in Chapter 12.

*Side-looking airborne radar (SLAR)* is another active
remote sensing system. Its radar energy produces high-res-
olution images of the surfaces it scans. Chapter 8 presents
an analysis of a hurricane using a SLAR image (Figure 8-33).

The European Space Agency (ESA) now operates two
Earth resource satellites (*ERS 1* and *2*). They work in tan-
dem, producing a spectacular 10-cm (3.9-in.) resolution,
imaging the same area at different times. The images produce a digital three-dimensional data set. These satellites
utilize a radar altimeter to provide ocean and ice-sheet
elevation. Also, a scanning radiometer records vegetation
cover, and a new instrument is measuring and mapping
stratospheric ozone depletion—both active and passive
systems are on board.

***Passive Remote Sensing.*** Passive remote-sensing systems record energy radiated from a surface, particularly
light and infrared. Our own eyes are passive remote sensors, as was the camera that took the picture of Earth on
the back cover of this book.

Passive remote sensors on five *Landsat* satellites,
launched by the United States between 1972 and 1984,
provided a variety of data, as shown in images of the
Appalachian Mountains in Chapter 12, river deltas in
Chapter 14, and the Malaspina and Hubbard glaciers in
Alaska in Chapter 17. All *Landsats* carried a *multispectral*

*scanner* (*MSS*) device that viewed Earth in spectral bands
of visible and infrared wavelengths. The two *Landsats*
that remain operational (4 and 5) are equipped with the
*thematic mapper* (*TM*) as a principal sensor. The TM provides high resolution from spectral bands of visible and
infrared wavelengths.

The National Oceanic and Atmospheric Administration (NOAA) polar-orbiting satellites carry the *advanced
very high resolution radiometer* (*AVHRR*) sensors. *NOAA
10* and *NOAA 11* are currently operating. AVHRR is sensitive in visible and infrared wavelengths. The incredible
images of Hurricane Andrew (Figure 8-34), among others
in this text, were produced by an AVHRR system. AVHRR
is primarily used for sensing day or night clouds, snow,
and ice; for monitoring forest fires, clouds, and surface
temperature; and for determining natural and planted vegetation patterns. In Chapter 19, an AVHRR image portrays
clear-cutting of trees in the Pacific Northwest and production of biomass. These examples of resource analysis were
impossible to perform at such a scale just a few years ago.

The highest-resolution commercial system, the French
satellite called *SPOT* (Systeme Probatoire d'Observation
de la Terre), can resolve objects on Earth down to 10 to
20 m (33 to 66 ft), depending on which of its sensors is
used. The dramatic image in Chapter 15 of sand dunes in
the Ar Rub'al Khālī Erg, Saudi Arabia (Figure 15-7), and in
Chapter 17 of floating ice in the Weddell Sea, Antarctica
(Figure 17-26), are examples of *SPOT* images. Diverse

applications of such remote sensing include agricultural and crop monitoring, terrain analysis, documentation of change in an ecosystem over time, river studies to reduce flood hazards, analysis of logging practices, identification of development and construction impacts, and surveillance.

In addition, with the end of the Cold War, previously unavailable intelligence ("spy") satellite images are now becoming available. An analysis of Earth's surface is a potentially valuable use of this data. These images are being released by several countries, including the United States.

The Geostationary Operational Environmental Satellite, known as *GOES-8*, became operational in late 1994 and provides the daily infrared and visible images of weather in the Western Hemisphere that you see on television. *GOES-9*, a new generation of high-tech satellites, is positioned to monitor central and eastern North America and the western Atlantic. *GOES-9* was in position for coverage of the record 1995 Atlantic hurricane season. Geostationary satellites stay in semipermanent positions because they keep pace with Earth's rotational speed at their altitude of 35,400 km (22,000 mi).

Three *GOES* images are used with the weather maps that appear in Chapter 8 (Figure 8-24). Other specific weather satellites include the *GMS* weather satellite, run by the Japan Weather Association and covering the Far East, and *METEOSAT* for Europe and Africa, operated by the European Space Agency. *METEOSAT 3* monitors conditions in the Atlantic.

## Geographic Information System (GIS)

Remote sensing is an important tool for acquiring large volumes of spatial data. The next step is storing, processing, and retrieving those data in useful ways. The value of remote sensing rests on the ability to provide data to powerful information-handling systems. Computers have allowed integration of geographic information from direct surveys (on-the-ground mapping) and remote sensing in complex ways never before possible. A **geographic information system (GIS)** is a computer-based, data-processing tool for gathering, manipulating, and analyzing geographic information. An example is shown in Figure 1-27.

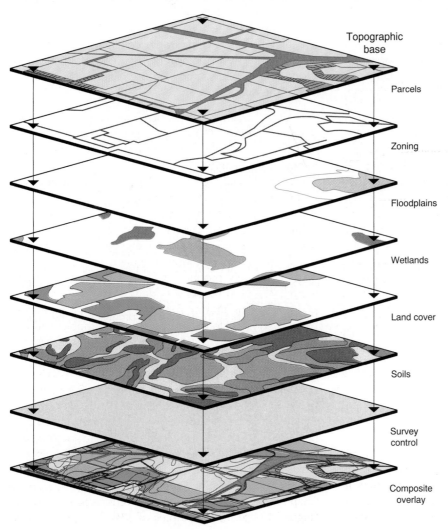

Topographic base

Parcels

Zoning

Floodplains

Wetlands

Land cover

Soils

Survey control

Composite overlay

**FIGURE 1-27**

**A geographic information system GIS model.**

Layered spatial data in a geographic information system (GIS) format. [After USGS.]

# News Report 2

## Careers in GIS

Geographic information system (GIS) methodology offers great career opportunities in industry, government, business, marketing, teaching, sales, military, and other fields. Right now, geographers trained in GIS are analyzing ozone depletion, deforestation, soil erosion, and acid deposition. They are mapping ecosystems and monitoring the declining diversity of plant and animal species. Geographers are planning, designing, and surveying urban developments, following the trends of global warming, studying the impact of human population, and analyzing air and water pollution.

GIS degree programs are available at many colleges and universities. Three schools have National Centers for Geographic Information and Analysis (NCGIA): Department of Geography, University of California Santa Barbara, Santa Barbara, California 93106; Department of Surveying and Engineering, University of Maine, Orono, Maine 04669; and State University of New York–Buffalo, Buffalo, New York 14260. Contact any of them for the brochure, *Geographic Information and Your Future— Careers for a Fragile Planet.*

The beginning component for any GIS is a coordinate system such as latitude-longitude, which establishes reference points against which to position data. The coordinate system is digitized, along with all areas, points, and lines. Remotely sensed imagery and data are then added on the coordinate system.

A GIS is capable of analyzing patterns and relationships within a single data plane, such as the floodplain or soil layer in the Figure 1-27. The GIS also can generate an *overlay analysis* where two or more data planes interact. Various assumptions, comparisons, and policies can be tested. When the layers are combined, the resulting synthesis—*a composite overlay*—is a valuable product, ready for use in analyzing complex problems. A study may follow specific points or areas through the complex of overlay planes. The utility of a GIS compared with that of a fixed map is the ability of the system to manipulate the variables in a study.

Before the advent of computers, an environmental-impact analysis required someone to gather data and painstakingly hand-produce overlays of information to determine positive and negative impacts of a project or event. Today, this layered information can be handled by a computer-driven GIS, which assesses the complex interconnections among different components. In this way, subtle changes in one element of a landscape may be identified as having a powerful impact elsewhere.

GIS applications are useful for the analysis of environmental events, both natural and human-caused. For example, the USGS completed a GIS to help analyze the spatial impact of the 1989 *Exxon Valdez* oil spill in Alaska. Scientists at NASA's Goddard Space Flight Center recently completed a 3-year comprehensive GIS of Brazil in an effort to better understand land-use patterns—specifically, loss of the rain forest.

GIS is particularly helpful in analyzing natural hazards and society. An example is the European Earthquake catalogue that records over 20,000 earthquakes, dating back to 500 B.C. Having this data base installed in a GIS permits detailed spatial analysis of these events along with country boundaries, human settlements, rivers, lakes, and seas, nuclear power plant locations, hazardous storage sites, and other economic considerations.

One of the most extensive and longest-operating systems is the Canada Geographic Information System (CGIS). Environmental data about natural features, resources, and land use were taken from maps, aerial photographs, and orbital sources, reduced to map segments, and entered into the CGIS. The development of this system has progressed hand in hand with the ongoing Canada land-inventory project.

# Summary and Review — Essentials of Geography

*Here is a review of the Key Learning Concepts for this chapter, in handy summary form. Each concept review concludes with a list of the key terms from the chapter and review questions. Such summary and review sections follow each chapter in the book.*

3. Suggest a representative example for each of the five geographic themes and use that theme in a sentence.

4. Have you made decisions today that involve geographic concepts discussed within the five themes presented? Explain briefly.

---

✔ **Define geography and physical geography in particular.**

**Geography** is a science of method, a special way of analyzing phenomena. Geography integrates a wide range of subject matter. Geography brings together disciplines from the physical and life sciences with the cultural and human sciences to attain a holistic view of Earth—an essential aspect of the emerging **Earth systems sciences**. Geographic education recognizes five major themes: **location**, **place**, **movement**, **region**, and **human-Earth relationships** (including environmental concerns). Geography's method is **spatial analysis**, used to study the interdependence among geographic areas, natural systems, society, and cultural activities over space. **Process**—that is, analyzing a set of actions or mechanisms that operate in some special order—is central to geographic synthesis.

**Physical geography** applies spatial analysis to all the physical elements and processes that make up the environment: energy, air, water, weather, climate, landforms, soils, animals, plants, and Earth itself. Understanding the complex relations among these elements is important to human survival because Earth's physical systems and human society are so intertwined.

> geography (p. 4)
> Earth systems science (p. 2)
> location (p. 4)
> place (p. 4)
> movement (p. 4)
> region (p. 4)
> human-Earth relationships (p. 4)
> spatial analysis (p. 4)
> process (p. 4)
> physical geography (p. 4)

1. What is unique about the science of geography? On the basis of information in this chapter, define physical geography and review the geographic approach.

2. Assess your geographic literacy by examining atlases and maps. What types of maps have you used—political? physical? topographic? Do you know what projections they employed? Do you know the names and locations of the four oceans, seven continents, and individual countries? Can you identify the new countries that have emerged since 1990?

---

✔ **Describe systems analysis, open and closed systems, feedback information, and system operations and *relate* those concepts to Earth systems.**

A **system** is any ordered, related set of things and their attributes, as distinct from their surrounding environment. Systems analysis is an important organizational and analytical tool used by geographers. Earth is an **open system** in terms of energy, receiving energy from the Sun, but essentially a **closed system** in terms of matter and physical resources.

As a system operates, "information" is returned to various points in the system via pathways called **feedback loops**. If the feedback information discourages response in the system, it is called **negative feedback**. (Further production in the system decreases the growth of the system.) If feedback information *encourages* response in the system, it is called **positive feedback**. (Further production in the system stimulates the growth of the system.) When the rates of inputs and outputs in the system are equal and the amounts of energy and matter in storage within the system are constant (or as they fluctuate around a stable average), the system is in **steady-state equilibrium**. A system that demonstrates a steady increase or decrease in system operations—a trend over time—is in **dynamic equilibrium**. Geographers often construct simplified **models** of natural systems to better understand them.

Four immense open systems powerfully interact at Earth's surface: three nonliving **abiotic** systems (**atmosphere**, **hydrosphere**, and **lithosphere**) and a living biotic system (**biosphere**, or **ecosphere**).

> system (p. 5)
> open system (p. 6)
> closed system (p. 6)
> feedback loops (p. 6)
> negative feedback (p. 6)
> positive feedback (p. 6)
> steady-state equilibrium (p. 7)
> dynamic equilibrium (p. 7)
> model (p. 8)
> abiotic (p. 10)
> atmosphere (p. 10)
> hydrosphere (p. 10)
> lithosphere (p. 10)

biotic (p. 10)
biosphere (p. 10)
ecosphere (p. 10)

5. Define systems theory as an organizational strategy. What are open systems, closed systems, and negative feedback? When is a system in a steady-state equilibrium condition? What type of system (open or closed) is a human body? A lake? A wheat plant?

6. Describe Earth as a system in terms of both energy and matter.

7. What are the three abiotic spheres (nonliving) that make up Earth's environment? Relate these to the biotic (living) sphere: the biosphere.

---

✔ *Explain* **Earth's reference grid: latitude, longitude, and latitudinal geographic zones and time.**

The science that studies Earth's shape and size is **geodesy**. Earth bulges slightly through the equator and is oblate (flattened) at the poles, producing a misshapen spheroid called a **geoid**. Absolute location on Earth is described with a specific reference grid of **parallels** of **latitude** (measuring distances north and south of the equator) and **meridians** of **longitude** (measuring distances east and west of a prime meridian). A **great circle** is any circle of Earth's circumference whose center coincides with the center of Earth. Great circle routes are the shortest distance between two points on Earth. **Small circles** are those whose centers do not coincide with Earth's center.

A historic breakthrough in navigation and timekeeping occurred with the establishment of an international **prime meridian** (0° through Greenwich, England) and the invention of precise chronometers that enabled accurate measurement of longitude. This prime meridian provided the basis for **Greenwich Mean Time (GMT)**, the world's first universal time system. A corollary of the prime meridian is the 180° meridian, the **International Date Line**, which marks the place where each day officially begins. Today, **Coordinated Universal Time (UTC)** is the worldwide standard and the basis for international time zones. **Daylight saving time** is a seasonal change of clocks by 1 hour in summer months.

geodesy (p. 11)
geoid (p. 12)
parallel (p. 14)
latitude (p. 14)
longitude (p. 14)
meridian (p. 15)
great circle (p. 15)
small circles (p. 15)
prime meridian (p. 16)
Greenwich Mean Time (GMT) (p. 18)
International Date Line (p. 19)
Coordinated Universal Time (UTC) (p. 21)
daylight saving time (p. 21)

8. Draw a simple sketch describing Earth's shape and size.

9. What are the latitude and longitude coordinates (in degrees, minutes, and seconds) of your present location? Where can you find this information?

10. Define latitude and parallel and define longitude and meridian using a simple sketch with labels.

11. Identify the various latitudinal geographic zones that roughly subdivide Earth's surface. In which zone do you live?

12. What does timekeeping have to do with longitude? Explain this relationship. How is Coordinated Universal Time (UTC) determined on Earth?

13. What and where is the prime meridian? How was the location originally selected? Describe the meridian that is opposite the prime meridian on Earth's surface.

14. Define a great circle, great circle routes, and a small circle. In terms of these concepts, describe the equator, other parallels, and meridians.

---

✔ *Define* **cartography and mapping basics: map scale and map projections.**

The science and art of mapmaking is called **cartography**. Maps are used by geographers for the spatial portrayal of Earth's physical systems. **Scale** is the ratio of the image on a map to the real world; it relates a unit on the map to an identical unit on the ground. Cartographers create **map projections** for specific purposes, selecting the best compromise of projection for each application. Compromise is always necessary because Earth's round, three-dimensional surface cannot be exactly duplicated on a flat, two-dimensional map. **Equal area** (equivalence), **true shape** (conformality), true direction, and true distance are all considerations in selecting a projection. **Rhumb lines** are lines of constant direction and appear as straight lines on the Mercator projection.

Several map projections are used in this text: **Goode's homolosine projection** (an equal-area projection created by combining two other projections), **Robinson projection** (a compromise projection), and **Miller cylindrical projection** (also a compromise projection).

cartography (p. 22)
map projections (p. 23)
scale (p. 22)
equal area (p. 23)
true shape (p. 23)
rhumb lines (p. 26)
Goode's homolosine projection (p. 26)
Robinson projection (p. 26)
Miller cylindrical projection (p. 27)

15. Define cartography. Explain why it is an integrative discipline.

16. What is map scale? In what three ways is it expressed on a map?

17. State whether each of the following ratios is a large scale, medium scale, or small scale: 1:3,168,000, 1:24,000, 1:250,000.

18. Describe the differences between the characteristics of a globe and those that result when a flat map is prepared.

19. What type of map projection is used in Figure 1-13? In Figure 1-16?

---

✔ *Describe* **remote sensing and** *explain* **geographic information system (GIS) methodology as a tool used in geographic analysis.**

The operation of Earth's systems is being disclosed through orbital and aerial **remote sensing**. Satellites do not take photographs but record images that are transmitted to Earth-based receivers. Satellite images are recorded in digital form for later processing, enhancement, and generation.

The mountain of data being collected has led to the development of **geographic information system (GIS)** technology. Computers process geographic information from direct surveys and remote sensing in complex ways never before possible. GIS methodology is an important step in better understanding Earth's systems and is a vital career opportunity for geographers.

The science of physical geography is in a unique position to synthesize the spatial, environmental, and human aspects of our increasingly complex relationship with our home planet—Earth.

remote sensing (p. 28)
geographic information system (GIS) (p. 31)

20. What is remote sensing? What are you viewing when you observe a weather satellite image on TV or in the newspaper? Explain.

21. If you were in charge of planning for development of a large tract of land, how would GIS methodologies assist you? How might planing and zoning be affected if a portion of the tract in the GIS is a floodplain or prime agricultural land?

---

 **NetWork**

The *Geosystems Home Page* provides on-line resources for this chapter on the World Wide Web. You will find review exercises, specific updates for items in the chapter, suggested readings, and links to interesting related pathways on the Internet (click on the Table of Contents link and select this chapter). *Geosystems* is at: **http://www.prenhall.com/geosystm**

# *Part 1*

# The Energy-Atmosphere System

*Sunset, Denali National Park.* [Photo by Kennan Ward/Stockmarket.]

O ur planet and our lives are powered by radiant energy from the star that is closest to Earth—the Sun. For more than 4.6 billion years, solar energy has traveled across interplanetary space to Earth, where a small portion of the solar output is intercepted. Because of Earth's curvature, the energy at the top of the atmosphere is unevenly distributed, creating imbalances from the equator to each pole—the equatorial region experiences energy surpluses; the polar regions experience energy deficits. This unevenness of energy receipt empowers circulations in the atmosphere and across Earth's surface. The slow pulse of seasonal change varies the distribution of energy during the year.

Earth's atmosphere acts as an efficient filter, absorbing most harmful radiation, charged particles, and space debris so that they do not reach Earth's surface. Surface energy balances are established with the daily receipt of solar energy, giving rise to global patterns of temperature, winds, and ocean currents. Each of us depends on many systems that are set into motion by energy from the Sun. These systems are the subject of Part 1.

# 2

# Solar Energy to Earth and the Seasons

**The Solar System, Sun, and Earth**

**Solar Energy: From Sun to Earth**

**Energy at the Top of the Atmosphere**

**The Seasons**

**Summary and Review**

## Key Learning Concepts

After reading the chapter, you should be able to:

- *Distinguish* among galaxies, stars, and planets and locate Earth.
- *Overview* the origin, formation, and development of Earth and the atmosphere. Specifically, *construct* Earth's annual orbit about the Sun.
- *Describe* the Sun's operation and *explain* the characteristics of the solar wind and the electromagnetic spectrum of radiant energy.
- *Portray* the intercepted solar energy and its uneven distribution at the top of the atmosphere.
- *Define* solar altitude, solar declination, and daylength and *describe* the annual variability of each—Earth's seasonality.

*Aurora borealis over Prince Edward Island, Canada.*
[Photo by Lionel F. Stevenson/Photo Researchers, Inc.]

The Universe is populated with millions of galaxies. One of these is our own Milky Way Galaxy, consisting of billions of stars. Among these stars is an average yellow star we call the Sun. Our sun radiates energy in all directions and upon its family of orbiting planets. Of special interest to us is the solar energy that falls on the third planet out from the Sun.

Incoming solar energy that strikes Earth's atmosphere sets into motion the winds, weather systems, and ocean currents that daily influence our lives. This solar energy input to the atmosphere, plus Earth's tilt and rotation, produce daily, seasonal, and annual patterns of changing daylength and Sun angle. The Sun is the ultimate energy source for most life processes in our biosphere.

## The Solar System, Sun, and Earth

Our Solar System is located on a remote, trailing edge of the **Milky Way Galaxy**, a flattened, disk-shaped mass estimated to contain up to 400 billion stars (Figure 2-1). Our Solar System is embedded more than halfway out from the center, in one of the Milky Way's spiral arms, called the Orion Arm. From our Earth-bound perspective in the Milky Way, the Galaxy appears to stretch across the night sky like a narrow band of hazy light. On a clear night, the unaided eye can see only a few thousand of these billions of stars.

### Solar System Formation and Structure

According to prevailing theory, our Solar System condensed from a large, slowly rotating and collapsing cloud of dust and gas called a *nebula*. **Gravity**, the mutual attracting force exerted by the mass of an object upon all other objects, was the key force in this condensing solar nebula. As the nebular cloud organized and flattened into a disk shape, the early proto-Sun grew in mass at the center, drawing more matter to it. Small accretion (accumulation) eddies swirled at varying distances from the center of the solar nebula; these were the *protoplanets*.

The early protoplanets, called *planetesimals*, were orbiting at approximately the same distances from the Sun as the planets are today. The beginnings of the Sun and its Solar System are estimated to have occurred more than 4.6 billion years ago. The Sun finally achieved its present

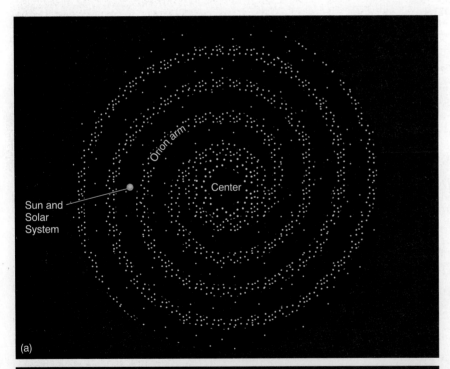

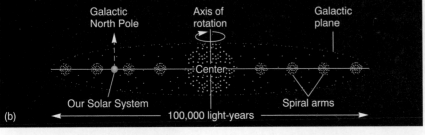

**FIGURE 2-1**

**Location of the Solar System in the Galaxy.**

The Milky Way Galaxy viewed from above (a) and cross-sectional side view (b). Our Solar System is some 30,000 light-years from the center of the Galaxy.

brightness only after a long period of development. The explanation of how suns condense from nebular clouds with planetesimals forming in orbits around their central masses is called the **planetesimal hypothesis**, or dust-cloud hypothesis. Astronomers are observing this formation process under way in other parts of the Galaxy, especially using the *Hubble Space Telescope* in orbit.

The development of such hypotheses and theories is an exercise of the **scientific method**, a methodology important to physical geography research. Focus Study 2-1 explains this essential process of science.

***Dimensions and Distances.*** The **speed of light** is 300,000 kmps (kilometers per second), or 186,000 mps (miles per second), which is about 9.5 trillion kilometers, or nearly 6 trillion miles, per year. (In more precise numbers, light speed is 299,792 kmps, or 186,282 mps.) This tremendous distance that light travels in a year is known as a *light-year*, and it is used as a unit of measurement for the vast Universe. For spatial comparison, our Moon is an average distance of 384,400 km (238,866 mi) from Earth, or about 1.28 seconds in terms of light speed. Our entire Solar System is approximately 11 hours in diameter,

## Focus Study 2-1

## The Scientific Method

The term *scientific method* may have an aura of complexity that it should not. The scientific method is simply the application of common sense in an organized and objective manner. A scientist observes, makes a general statement to summarize the observations, formulates a hypothesis, conducts experiments to test the hypothesis, and develops a theory and governing scientific laws. Sir Isaac Newton (1642–1727) developed this method of discovering the patterns of nature, although the term *scientific method* was applied later. Follow the illustration in Figure 1 as you read about the scientific method.

The scientific method begins with our perception of the real world and a determination of what we know, what we want to know, and the many unanswered questions that exist. Scientists who study the physical environment turn to nature for clues that they can observe and measure. They discern what data are needed and begin to collect those data. Then, these observations and data are analyzed to identify coherent patterns that may be present. This search for patterns requires *inductive reasoning*, or the process of drawing generalizations from specific facts. This step is important in modern Earth systems sciences, in which the goal is to understand *a whole functioning Earth*, rather than isolated, small compartments of specialization. Such understanding allows the scientist to construct models that simulate general operations of Earth systems.

If patterns are discovered, the researcher may formulate a hypothe-sis—a formal generalization of a principle. Examples include the planetesimal hypothesis, nuclear-winter hypothesis, and moisture-benefits-from-hurricanes hypothesis.

Further observations are related to the general principles established by the hypothesis. Further data gathered may support or refute the hypothesis, or predictions made according to it may prove accurate or inaccurate. All these findings provide feedback to adjust data collection and model building and to refine the hypothesis statement. Verification of the hypothesis after exhaustive testing may lead to its elevation to the status of a *theory*.

A theory is constructed on the basis of several hypotheses that have been extensively tested. Theories represent truly broad general principles—unifying concepts that tie together the laws that govern nature (e.g., the theory of relativity, theory of evolution, atomic theory, Big Bang theory, stratospheric ozone depletion theory, or plate tectonics theory). A theory is a powerful device with which to understand both the order and chaos in nature. Using a theory allows predictions to be made about things not yet known, the effects of which can be tested and verified or disproved through tangible evidence. The value of a theory is the continued observation, testing, understanding, and pursuit of knowledge that the theory stimulates. A general theory reinforces our perception of the real world, acting as positive feedback.

Important to consider is that pure science does not make value judgments. Instead, pure science provides people and their institutions with objective information on which to base their own value judgments. Social and political judgments about the applications of science are increasingly critical as Earth's natural systems respond to the impact of modern civilization.

The growing awareness that human activity is producing global change places increasing pressure on scientists to participate in decision making. Numerous editorials in scientific journals have called for such involvement. An example, discussed in Chapter 3, is provided by F. Sherwood Rowland and Mario Molina, who first proposed a hypothesis that certain human-made chemicals caused damaging reactions in the stratosphere. They hypothesized in 1974 that protective ozone ($O_3$) would be increasingly depleted by chlorine-containing products such as chlorofluorocarbons (CFCs), commonly used as aerosol propellants, refrigerants, foaming agents, and cleaning solvents. Subsequently, surface, atmosphere, and satellite orbital measurements confirmed the reactions and provided data to map the losses that were occurring. International treaties and agreements to ban the chemical culprits ensued. In 1995, the Royal Swedish Academy of Sciences awarded these two scientists, along with Paul Crutzen, who also contributed to these discoveries, the Nobel Prize for chemistry for their pioneering work. Such successful applied science is strengthening resolve and treaties to ban the problem chemicals and to introduce benign substitutes.

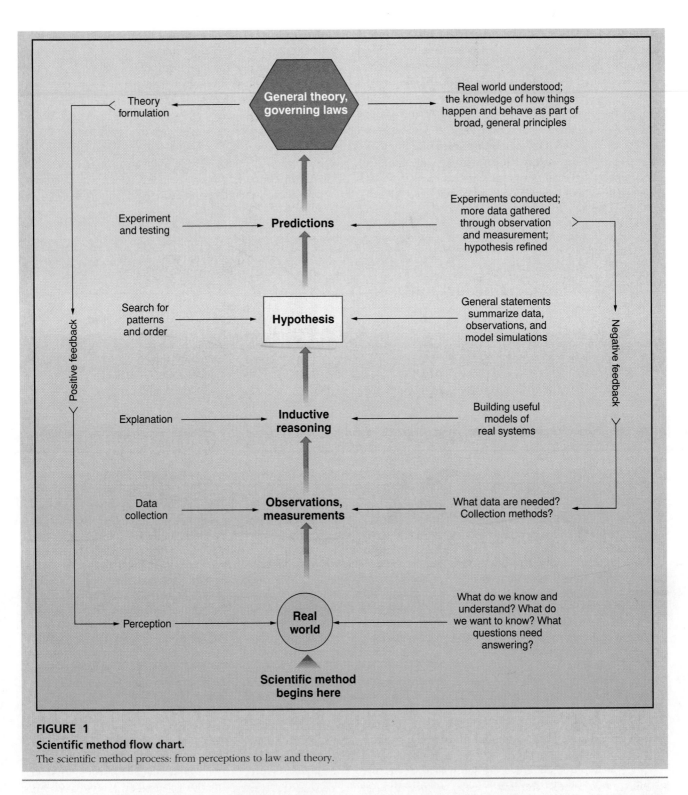

**FIGURE 1**
**Scientific method flow chart.**
The scientific method process: from perceptions to law and theory.

measured by light speed. In contrast, the Milky Way is about 100,000 light-years from side to side (Figure 2-1b). The known Universe that is observable from Earth stretches approximately 12 billion light-years in all directions.

***Earth's Orbit.*** Earth's orbit around the Sun is presently elliptical—a closed, oval path (Figure 2-2). Earth's average distance from the Sun is approximately 150 million

kilometers (93 million miles), which means that light reaches Earth from the Sun in an average of 8 minutes and 20 seconds. Earth is at **perihelion** (its closest position to the Sun) during the Northern Hemisphere winter (January 3 at 147,255,000 km, or 91,500,000 mi). It is at **aphelion** (its farthest position from the Sun) during the Northern Hemisphere summer (July 4 at 152,083,000 km, or 94,500,000 mi). This seasonal difference in distance from

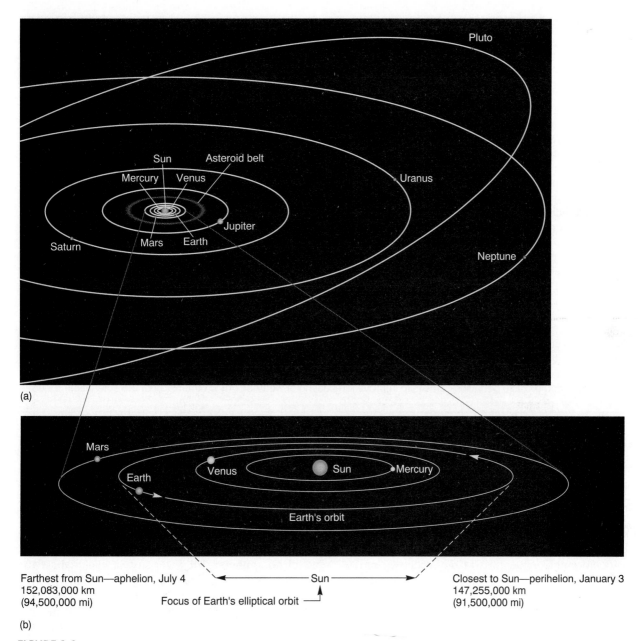

**FIGURE 2-2**

**Earth's orbit and the Solar System.**

(a) The Solar System of nine planets and asteroids. All of the planets except Pluto have orbits closely aligned to the plane of the ecliptic. (b) The four inner terrestrial planets and the structure of Earth's elliptical orbit, illustrating perihelion (closest) and aphelion (farthest) positions during the year.

the Sun causes a slight variation in the solar energy intercepted by Earth.

A plane including all points of Earth's orbit is termed the **plane of the ecliptic**. Earth's axis, or "tilt," remains fixed relative to this plane as Earth revolves around the Sun. The plane of the ecliptic is important to our discussion of Earth's seasons.

The structure of Earth's orbit is not a constant but instead exhibits changes over long periods. As shown in Chapter 17, Figure 17-23, Earth's distance from the Sun varies more than 17.7 million kilometers (11 million miles) during a 100,000-year cycle, placing it closer or farther at

different periods in the cycle. This variation is thought to be one of several factors that create Earth's cyclical pattern of glaciations (colder) and interglacial (warmer) periods. Figure 2-3 compares Earth's size and other measurements with those of the other planets in the Solar System.

## Earth's Development

The development of Earth is closely related to the growth of the Sun, and the long-term future of Earth likewise is tied closely to the Sun's life cycle. The evolution of Earth's atmosphere and surface, the formation of free oxygen gas, and the

## News Report 1

## The Nature of Order is Chaos

In 1960, Edward Lorenz, an MIT scientist, rocked the scientific world with the statement that a butterfly flapping its wings in Brazil might produce a tornado in Texas. He used this strange example to suggest that the interaction of orderly and deterministic systems may produce chaotic and unpredictable results. For example, ice has a rigid internal structure, forced by bonding between water molecules. This structure dictates that all ice crystals are six-sided, yet no two ice crystals, despite this similarity, are identical. Beneath this chaos of design exists an order dictated by physical principles (see Figure 7-6).

A major shift in our real-world view occurred with the advent of *chaos theory*, a revolution in science that considers the nonlinear, unpredictable behavior of operational systems. This theory suggests that the scientific method must consider the coexistence of disorder and order, randomness and pattern, and symmetry and chaos in natural systems.

Consider the weather. Mathematical models and numerical equations describe the behavior of water vapor, temperature, and pressure patterns. Yet weather systems are sensitive to very small fluctuations in any of those ingredients. Therefore, it is difficult to exactly predict how a weather system will develop, what track it will follow, or how severe it might be (see Chapter 8). Two similar chaotic weather systems will produce a similar result, although it is not possible to say exactly what the output will be. This understanding is helping scientists to improve forecasting of weather phenomena. In a dynamic weather system, chaos is the rule, just as it is in chemical and biological systems.

Chaos theory is useful in studying all of Earth's physical environments. For example, a river flows in branched channels over a floodplain in a pattern to reduce expended energy. The channels constantly shift in a deterministic randomness that is irregular and difficult to predict. See the Space Shuttle photograph of the many mouths of the Ganges River in Figure 14-24 for an example of such *fractal* branching—irregular, curving channels that may or may not repeat their pattern.

It may seem strange that the nature of order is the result of chaos! But apparently, this is the rule. Chaos theory is a new dimension of the scientific method.

development of the biosphere all represent complex interactions that were in operation from the beginning of Earth's environment. Processes set into motion as the solar nebula condensed and the planetesimals congealed to form our planet proceeded at rates imperceptible in human terms.

***Earth's Past Atmospheres.*** A principal component of Earth's history is the evolution of its modern atmosphere. This evolution occurred in four broad stages with long transitions, one into the next. The constituents of Earth's **primordial atmosphere** were derived from the original solar nebula. This atmosphere and the second stage—called the **evolutionary atmosphere**—are thought to have persisted for relatively short periods. The third and fourth stages—the **living atmosphere** and the **modern atmosphere**—have existed over much greater time spans. Table 2-1 summarizes present scientific views regarding the duration, composition, and dominant features of all four atmospheres. The times and durations listed in the table are estimates that represent gradual periods of transition.

A long process of chemical evolution beneath the shield of newly collected surface waters was a necessary forerunner to biological evolution. Still pools of water provided protection from the high radiation levels that existed in the environment at that time (both solar and terrestrial in origin). What a fascinating ecology was involved in the development of the atmosphere and hydrosphere, with organisms modifying their environment and in turn being modified by it!

Evidence suggests that primitive cells, called purple-sulfur bacteria, were functioning 3.6 billion years ago, producing organic materials from inorganic elements in the nonoxygen environment of the time. That process is known as *chemosynthesis*, in which an organism uses chemical energy to synthesize organic compounds. Similar chemosynthetic activity is observed today in certain areas of the cold, dark environment on the ocean floor. Vents of hot, mineral-rich water sustain bacterial activity that provides the basis of simple food chains through chemosynthesis in the sunless ocean depths.

Approximately 3.3 billion years ago, the first organism to perform photosynthesis evolved in Earth's shallow waters. *Photosynthesis* released oxygen into the air, marking the beginning of the third distinct atmosphere, the living atmosphere. Cyanobacteria (blue-green algae) began this production of oxygen, but oxygen did not reach levels comparable to today's level until about 500 million years ago. The modern atmosphere is the setting for Chapters 3 through 10.

## Solar Energy: From Sun to Earth

Our Sun is unique to us, but a commonplace star in our Galaxy. It is only average in temperature, size, and color when compared with other stars, yet it is the ultimate energy source for most life processes in our biosphere.

The Sun captured about 99.9% of the matter from the original nebula. The remaining 0.1% of the matter formed

**FIGURE 2-3**

**Comparison of the nine planets in our Solar System.**

The sizes of the planets and of the Sun are approximately to scale. Note that the four largest planets have ring systems. [Planet illustration after Chaisson and McMillan, *Astronomy-A Beginner's Guide to the Universe*, Figure 5-7, Prentice Hall, Inc., © 1995.]

| Planet | Average Distance from Sun (millions of km (mi)) | | Diameter (km (mi)) | | Tilt of Axis (degrees) | Number of Revolutions (in Earth years) | Rotation (in hours or Earth days) | Number of Natural Satellites |
|---|---|---|---|---|---|---|---|---|
| Mercury | 58 | (36) | 4,878 | (3,032) | 0 (?) | 0.24 | 59 d | 0 |
| Venus | 108 | (67) | 12,100 | (7,520) | 3 | 0.62 | 243 d | 0 |
| Earth | 150 | (93) | 12,756 | (7,926) | 23.5 | 1.00 | 1 d | 1 |
| Mars | 228 | (142) | 6,796 | (4,225) | 25 | 1.88 | 1 d | 2 |
| Jupiter | 778 | (483) | 142,800 | (88,730) | 3.1 | 11.86 | 9.8 h | 16+ |
| Saturn | 1472 | (913) | 120,660 | (74,975) | 26.7 | 29.46 | 10.6 h | 20+ |
| Uranus | 2870 | (1783) | 51,400 | (31,940) | 82 | 84.01 | 17 h | 15 |
| Neptune | 4486 | (2787) | 50,950 | (31,660) | 28.8 | 164.97 | ≅ 14 h | 8 |
| Pluto | 5900 | (3666) | 3,500 | (2,170) | unknown | 248.40 | 6.4 d | 1 |

all the planets, their satellites, asteroids, comets, and debris. Consequently, the dominant object in our region of space is the Sun. In the entire Solar System, it is the only object having the enormous mass needed to create the internal temperature and pressure to sustain a nuclear reaction and produce radiant energy.

## Solar Fusion

The solar mass produces tremendous pressure and high temperatures deep in its dense interior. Under these conditions, the Sun's abundant hydrogen atoms, the lightest of all the natural elements, are forced together, and pairs of

**TABLE 2-1**

| Summary of Earth's Past Atmospheres | | | |
| --- | --- | --- | --- |
| *Name* | *Approximate Duration (billions of years ago)* | *Composition (slow transitionary phases)* | *Probable Dominant Features* |
| Primordial atmosphere | 4.6–4.0 | Water ($H_2O$), hydrogen cyanide (HCN), ammonia ($NH_3$), methane ($CH_4$), sulfur, iodine, bromine, chlorine | Character derived from the nebula. Lighter gases of hydrogen and helium escaping to space. An unstable, hot surface with no liquid water collecting |
| Evolutionary atmosphere | 4.0–3.3 | At 4.0 billion years ago: $H_2O$, carbon dioxide ($CO_2$), nitrogen ($N_2$), sulfurous fumes, hydrocarbons, little or no free oxygen ($O_2$) | Terrigenic origins (outgassing from Earth). Surface water accumulation. Earth thought to have been shrouded in clouds. Anaerobic (nonoxygen) environment. Chemosynthetic bacteria (at 3.6 billion years ago) |
| Living atmosphere | 3.3–0.6 | At 3.0 billion years ago: $CO_2$, $H_2O$ vapor, $N_2$, <1% $O_2$ | Continued outgassing. First photosynthesis in cyanobacteria (at 3.3 billion years ago). Heavy global rains, ocean basins filling. Slow evolution of gaseous constituents toward the modern atmosphere; increasing oxygen levels |
| Modern atmosphere | 0.6–present | Today: 78% $N_2$, 21% $O_2$, 0.9% argon, 0.036% $CO_2$, trace gases | Gradual development to the modern atmosphere. Abundance of life. Moderately fluctuating climate. Beginning of anthropogenic atmosphere (human impact) |

hydrogen nuclei are joined in a process called **fusion.** In the fusion reaction, hydrogen nuclei form helium, the second-lightest element in nature, and enormous quantities of energy are liberated. During each second of operation, the Sun consumes 657 million tons of hydrogen, converting it into 652.5 million tons of helium. The difference of 4.5 million tons is the quantity that is converted directly to energy—literally, disappearing solar mass becomes energy.

A sunny day can seem so peaceful, certainly belying the violence proceeding on the Sun. Before the lunar voyages of the late 1960s, there was serious scientific concern that humans traveling above the protective layers of the atmosphere would be killed by solar radiation. Of course, this fear proved to be unfounded. The Sun's principal outputs consist of the solar wind and radiant energy in portions of the electromagnetic spectrum. Let us trace each of these emissions across space to Earth.

## Solar Wind

The Sun constantly emits clouds of electrically charged particles (principally hydrogen nuclei and free electrons) that surge outward in all directions from the Sun's surface. This stream of energetic material travels much more slowly than light—only about 50 million kilometers (31 million miles) a day—taking approximately 3 days to reach Earth. The term **solar wind** was first applied to this phenomenon in 1958. Solar wind extends from the Sun to a distance beyond Pluto's orbit. The *Voyager* and *Pioneer* spacecraft launched in the 1970s are now far beyond our Solar System and have yet to escape the solar wind.

***Sunspots and Solar Activity.*** The Sun's most conspicuous features are large **sunspots**, caused by magnetic storms on the Sun. Individual sunspots may range in diameter from

10,000 to 50,000 km (6200 to 31,000 mi), with some growing as large as 160,000 km (100,000 mi), more than 12 times Earth's diameter (Figure 2–4). Although sunspots have been described and recorded in some detail for almost 400 years, a full explanation of their occurrence is still evolving. These surface disturbances produce flares and prominences. In addition, outbursts of charged material, referred to as *coronal mass ejections*, contribute to the flow of material to space as the solar wind.

***Sunspot Cycles.*** The solar wind grows stronger during periods of increased sunspot activity and weaker during times of less activity. A regular cycle exists for sunspot occurrences, averaging 11 years from maximum to maximum; however, the cycle may vary from 7 to 17 years. In recent cycles, a solar minimum occurred in 1976 and a solar maximum took place during 1979, with over 100 sunspots visible. Another minimum was reached in 1986, and an extremely active solar maximum followed in 1990, with over 200 sunspots visible at some time during the year. In fact, the 1990–1991 maximum was the most intense ever observed. This happened 11 years after the previous maximum. The next sunspot minimum, in 1997, will further maintain the average.

## Earth's Magnetosphere

The charged particles of the solar wind first interact with Earth's magnetic field as they approach Earth. The **magnetosphere** is a magnetic field surrounding Earth, generated by dynamo-like motions within our planet. The magnetosphere deflects the solar wind toward both of Earth's poles so that only a small portion of it enters the atmosphere (Figure 2-5). When the magnetosphere was first discovered in 1958, scientists thought that it was doughnut-shaped and symmetrical. But sensitive instruments aboard space probes found it to be teardrop-shaped.

Figure 2-5 shows that the extreme northern and southern polar regions of the upper atmosphere are the points of entry for the solar wind stream, far above Earth's protected surface. Because the solar wind does not reach Earth's surface, research on this phenomenon must be conducted in space. In 1969, the *Apollo XI* astronauts exposed a piece of foil on the lunar surface as a solar wind experiment (Figure 2-6). When examined back on Earth, the exposed foil exhibited particle impacts that confirmed the presence and character of the solar wind.

## Solar Wind Effects

The interaction of the solar wind and the upper layers of Earth's atmosphere produces some remarkable phenomena: the auroras that occur toward both poles, disruption of certain radio broadcasts and some satellite transmissions, overloads on Earth-based electrical systems, and possible effects on weather patterns. Our understanding of the solar wind is increasing dramatically as data are collected by a variety of satellites: the new *Ulysses* and *Wind*, the *Dynamics Explorer*, and the *Voyager-2* and *Pioneers-10* and *11* launched during the 1970s.

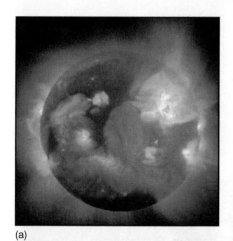

(a)

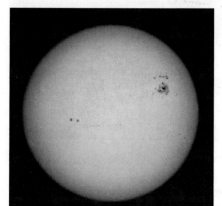

(b)

**Sunspot group (enlargement)**

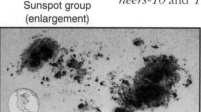

(c) **Earth (to scale)**

**FIGURE 2-4**
**Images of the Sun.**
The Sun, imaged from X-ray wavelengths (a) and visible light wavelengths (b) on the same day in 1992 by the *Yohkoh* satellite. Sunspots appear as visible dark patches and as areas of intense X-ray activity. Earth is shown for scale next to a close-up of a sunspot group (c). *Yohkoh* is a joint effort of the National Astronomical Observatory of Japan, the University of Tokyo, Lockheed Palo Alto Research Laboratory, and NASA. [Images (a) and (b) courtesy of Dr. Keith T. Strong of Lockheed and Dr. Yutaka Uchida of the Yohkoh Science Committee and the University of Tokyo. (c) Courtesy of the observatories of the Carnegie Institute of Washington.]

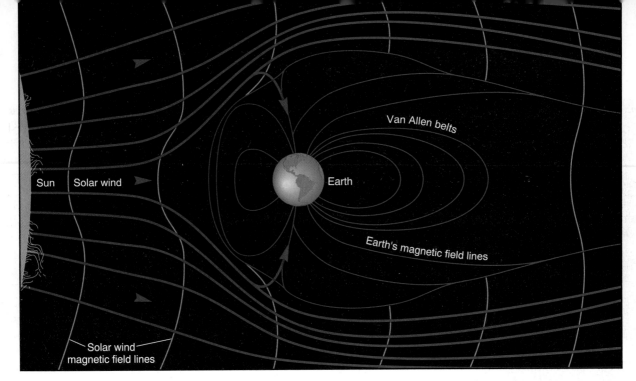

**FIGURE 2-5**
**Earth's magnetosphere and solar wind interaction.**
Streams of charged particles (purple arrows) are swept toward the poles by Earth's magnetic field, entering the upper atmosphere above the North and South Poles.

***Auroras.*** The solar wind creates auroras 80–500 km (50–300 mi) above Earth's surface. The incoming solar wind electrically charges (ionizes) certain atoms and molecules in this region of the atmosphere, and the ions reradiate the energy as light of varying colors, depending on which atoms and molecules are stimulated. These lighting effects are the **auroras**: *aurora borealis* (northern lights) and *aurora australis* (southern lights).

The auroras generally are visible poleward of 65° latitude (north and south) when the solar wind is active. In Figure 2-7a, the auroral halo crowning Earth is a dramatic example, seen by a polar-orbiting satellite. Ground-based observations are equally dramatic, as folded sheets of green, yellow, blue, and red light undulate across the skies of high latitudes (Figure 2-7b). Occasionally, Space Shuttle astronauts are treated to a view of the auroras from orbit (Figure 2-7c). At times of intense solar activity, these glowing colored lights in the night sky can be seen far into the middle latitudes. In March 1989, after an enormous solar outbreak, the aurora borealis was visible as far south as Key West, Florida, and Kingston, Jamaica, below 20° N latitude.

***Weather Effects.*** Another effect of the solar wind in the atmosphere is its possible influence on weather and climate cycles. Why do wetter periods in some midlatitude areas tend to coincide with every other solar maximum? Why do droughts often occur near the time of every other solar minimum? For example, sunspot cycles during the 250 years from 1740 to 1994 coincided with periods of wetness and drought, as estimated by an analysis of tree growth rings for that period throughout the western United States and elsewhere. These variations in weather tend to lag 2 or 3 years behind the solar maximum or minimum.

The correlation is interesting and is still undergoing research to confirm the connection. Speculation as to

**FIGURE 2-6**
**Astronaut and solar wind experiment.**
Without a protective atmosphere, the lunar surface allows direct charged particles of the solar wind and all of the Sun's electromagnetic radiation to reach the surface. The unrolled sheet of foil is a solar wind experiment being deployed by an *Apollo XI* astronaut. [NASA photo.]

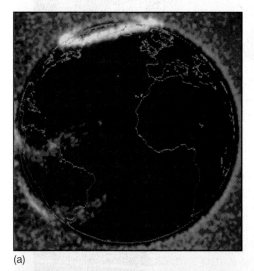

(a)

(b)

(c)

**FIGURE 2-7**

**Auroras from space, the surface, and orbital perspectives.**

(a) Satellite image of auroral halo over Earth's North Pole. (b) Surface view of the aurora borealis in the night sky over Alaska. (c) Space Shuttle astronauts look down on the auroras from their orbital perspective. [(a) *Dynamics Explorer 1* satellite image courtesy of L. A. Frank and J. D. Craven, The University of Iowa; (b) aurora photo by Johnny Johnson/Tony Stone Images, Inc. (c) NASA photo.]

the weather connection is centered on heating and wind changes in the upper atmosphere that may be attributable to a conversion of energy from the solar wind. Other explanations also are proposed. Researchers are investigating these apparent relations between solar wind and Earth's weather.

Regardless of the *cause* for the cyclical patterns of drought and wetness that do occur, a remarkable failure in current planning worldwide is the lack of *attention* given to them. Preparing for such patterns could reduce property loss and casualties. Cyclical drought could be offset through widespread water conservation and more efficient water use. Wet spells might require strengthening of levees along river channels, floodplain zoning against development, and better reservoir management to reduce flooding. As knowledge of the solar wind–weather relation improves, the ability to forecast probable effects on Earth systems will also improve.

## Electromagnetic Spectrum of Radiant Energy

The key essential solar input to life is electromagnetic energy of various wavelengths. Solar radiation occupies a portion of the **electromagnetic spectrum** of radiant energy. This radiant energy travels at the speed of light to Earth. The total spectrum of this radiant energy is made up of different wavelengths. Figure 2-8 shows that a **wavelength** is the distance between corresponding points on any two successive waves. The number of waves passing a fixed point in one second is the *frequency*.

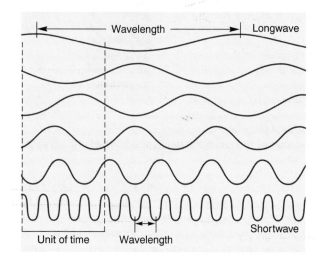

**FIGURE 2-8**

**Wavelength and frequency.**

Wavelength and frequency are two ways of describing the same phenomenon—electromagnetic wave motion. More short wavelengths pass a given point during a unit of time, so they are higher in frequency, whereas fewer long wavelengths pass a point in a unit of time, so they are lower in frequency.

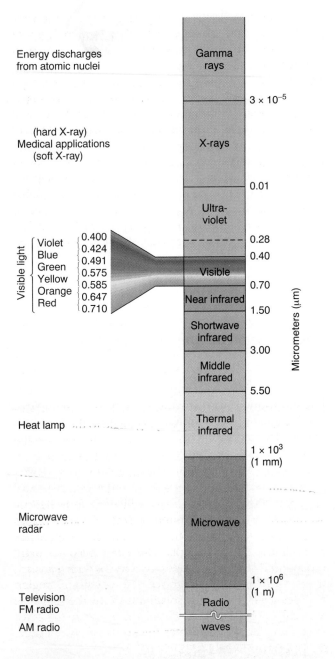

**FIGURE 2-9**
**A portion of the electromagnetic spectrum of radiant energy.**

and Earth. Figure 2-10 shows that the hot Sun radiates shorter-wavelength energy, concentrated around 0.4–0.5 μm (micrometer).

The Sun's surface temperature is about 6000°C (11,000°F), and its emission curve shown in the figure is similar to that predicted for an idealized 6000°C surface, or *black body radiator.* An ideal black body emits as much radiant energy as it absorbs—(the hotter the black body, the more radiation it emits at all wavelengths, with shorter wavelengths dominant at higher temperatures. The Sun emits a much greater amount of energy per unit area of its surface than does a similar area of Earth's environment.

Earth is a cooler radiating body, so longer wavelengths are emitted. In comparison to a hot body, lower temperatures at Earth's surface produce radiation mostly in the infrared portion of the spectrum. Figure 2-10 shows that the radiation emitted by Earth occurs in longer wavelengths, centered around 10 μm (micrometers) and entirely within the infrared portion of the spectrum.

To summarize, the solar spectrum is *shortwave radiation* that peaks in the short visible wavelengths, whereas Earth's radiated energy is *longwave radiation* concentrated in infrared wavelengths (Figure 2-11). In Chapter 4, we will see that Earth, clouds, sky, ground, and all things that are terrestrial are cool body radiators in contrast to the Sun.

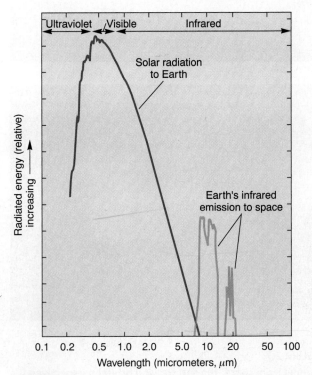

**FIGURE 2-10**
**Solar and terrestrial energy distribution by wavelength.**
The solar output peaks in shorter wavelenghts of visible light in relation to its high surface temperature, whereas Earth's emissions are concentrated in the infrared portion of the spectrum.

The Sun emits radiant energy composed of 8% ultraviolet, X-ray, and gamma ray wavelengths; 47% visible light wavelengths; and 45% infrared wavelengths. A portion of the electromagnetic spectrum is illustrated in Figure 2-9, with wavelengths increasing from the top of the illustration to the bottom. Note the wavelengths at which various phenomena and human applications of energy occur.

An important physical law states that all objects radiate energy in wavelengths related to their individual surface temperatures: the hotter the object, the shorter the wavelengths emitted. This law holds true for the Sun

# Energy at the Top of the Atmosphere

The region at the top of the atmosphere, approximately 500 km (300 mi) above Earth's surface, is termed the **thermopause**. It is the outer boundary of Earth's energy system and provides a useful point at which to assess the arriving solar radiation before it is diminished by scattering and absorption in passage through the atmosphere.

## Intercepted Energy

Earth's distance from the Sun results in its interception of only one two-billionth of the Sun's total energy output. Nevertheless, this tiny fraction of the Sun's overall output is an enormous amount of energy input to Earth's systems. Solar radiation that reaches a horizontal plane at Earth is called **insolation**, a term specifically applied to radiation arriving at Earth's surface and atmosphere. Insolation at the top of the atmosphere is expressed as the *solar constant*.

**Solar Constant.** Knowing the amount of insolation intercepted by Earth is important to climatologists and other scientists. The **solar constant** is the average insolation received at the thermopause when Earth is at its average distance from the Sun. That value is 1372 W/m² (watts per square meter).* As we follow insolation through the atmosphere to Earth's surface (Chapters 3 and 4), we will see that the value of the solar constant is reduced by half or more through reflection, scattering, and absorption of shortwave radiation.

---

*A *watt* is equal to 1 joule (a unit of energy) per second and is the standard unit of power in the SI-metric system. (See the inside back cover of this text for more information on measurement conversions.) In nonmetric calorie heat units, the solar constant is expressed as approximately 2 calories per square centimeter per minute, or 2 *langleys* per minute (a langley is 1 cal/cm²). A *calorie* is the amount of energy required to raise the temperature of 1 g of water (at 15°C) 1 degree Celsius and is equal to 4.184 joules.

The constancy of the solar constant over time is important, for small variations of even 0.5% or 1.0% could prove dramatic for Earth's energy system. Paleontologists, who deal with the life of past geologic periods, estimate that solar energy levels have varied slightly, perhaps less than 10%, over the past several billion years. The Solar Maximum Mission (1980–1989), dubbed Solar Max, was launched by NASA to measure total solar output. Solar Max found average variations of ±0.04% in the solar constant, with the largest changes ranging up to 0.3%, essentially in correlation with natural sunspot activity. Scientists have dismissed solar variability as a factor in the changing global temperature trends that are occurring during this century.

Since 1978, the Earth-atmosphere energy budget has been monitored by the Earth Radiation Budget package on board the *Nimbus-7* satellite. This satellite orbits at an altitude of 955 km (595 mi). All instruments on board are still operating, including the Earth Radiation Budget (ERB) package. ERB measurements determined that solar irradiance, or the luminous brightness of solar radiation, is directly correlated to the sunspot cycle discussed earlier. During the solar maximums of 1979 and 1991, the solar constant exceeded 1374 W/m²; the 1986 minimum produced a constant of 1371 W/m².

***Uneven Distribution of Insolation.*** Earth's curved surface presents a continually varying angle to the incoming parallel rays of insolation (Figure 2-12). Differences in the angle of solar rays at each latitude result in an uneven distribution of insolation and heating. The place receiving maximum insolation is the point where insolation rays are perpendicular to the surface (radiating from directly overhead), called the **subsolar point**. All other places receive insolation at less than 90° and thus experience more-diffuse energy receipts. This effect becomes more pronounced at higher latitudes. As a result, during a year's time, the thermopause above the equatorial region receives 2.5 times more insolation than the thermopause above the poles. Lower-angle solar rays toward the poles must pass through a greater thickness of atmosphere,

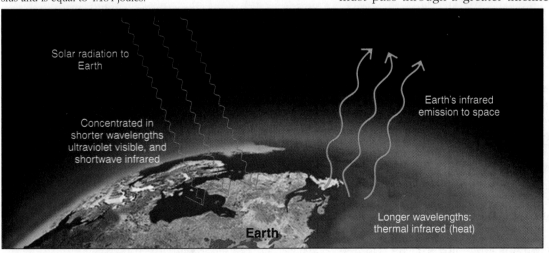

**Solar radiation to Earth**

**Concentrated in shorter wavelengths ultraviolet visible, and shortwave infrared**

**Earth's infrared emission to space**

**Longer wavelengths: thermal infrared (heat)**

**Earth**

**FIGURE 2-11**
**Earth's energy budget simplified.** Shorter wavelengths arrive at Earth from the Sun. Longer wavelengths of infrared radiate to space from Earth.

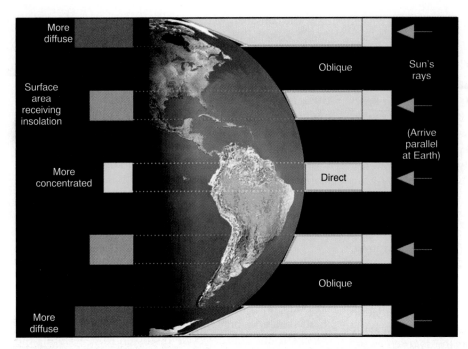

**FIGURE 2-12**
**Insolation receipts and Earth's curved surface.**
Solar insolation angles determine the concentration of energy receipts by latitude. Lower latitudes receive more-concentrated energy from a more-direct solar beam. Higher latitudes receive slanting (oblique) rays and more-diffuse energy. Note the area covered by identical columns of solar energy arriving at Earth's surface at different latitudes.

resulting in further losses of energy due to scattering, absorption, and reflection.

Figure 2-13 illustrates the daily variation in energy at the top of the atmosphere for various latitudes in watts per square meter ($W/m^2$) per day. The chart shows a decrease in insolation from the equatorial regions northward and southward toward the poles. However, in June, the North Pole receives more than 500 $W/m^2$ per day, which is more than is ever received at 40° N latitude or at the equator. Such high values result from the duration of exposure: 24 hours a day, compared with only 15 hours of daylight at 40° N latitude and 12 hours at the equator. However, the summertime Sun at noon is low in the sky at the poles, so a daylength twice that of the equator yields only about 100 $W/m^2$ difference.

In December, the pattern reverses. But note that the top of the atmosphere at the South Pole receives even more insolation than the North Pole does in June (over 550 $W/m^2$). This is a function of Earth's closer location to the Sun at perihelion (January 3 on Figure 2-2).

Along the equator, two maximum periods of approximately 430 $W/m^2$ occur at the spring and fall equinoxes, when the subsolar point is at the equator. Find your latitude on the graph and follow across the months to determine the seasonal variation of insolation where you live.

***Global Net Radiation.*** The Earth Radiation Budget (ERB) instrument aboard the *Nimbus-7* satellite measures shortwave and longwave flows of energy at the top of the atmosphere. ERB sensors collected the data used to develop the map in Figure 2-14. This map shows *net radiation*, or the balance between incoming shortwave and outgoing longwave radiation.

First, note the latitudinal energy imbalance in net radiation—positive values in lower latitudes and negative values toward the poles. In middle and high latitudes, approximately poleward of 36° north and south latitudes, net radiation is negative. The reason is that Earth's climate system loses more energy to space than it gains from the Sun, as measured at the top of the atmosphere for these higher latitudes. In the lower atmosphere, these polar energy deficits are offset by flows of energy from tropical energy surpluses (as we will see in Chapters 4 and 6). The largest net radiation values are above the tropical oceans along a narrow equatorial zone, averaging 80 $W/m^2$. Net radiation minimums are lowest over Antarctica.

Of interest is the –20 $W/m^2$ area over the Sahara region of North Africa. Here, typically clear skies—which permit great longwave radiation losses from Earth's surface—and light-colored reflective surfaces work together to reduce net radiation values at the thermopause. In other regions, clouds and atmospheric pollution in the lower atmosphere also affect net radiation patterns at the top of the atmosphere by reflecting more shortwave energy to space.

This overall imbalance of insolation and heating from equator to the poles is the cause of major circulations within the lower atmosphere and in the ocean. These circulations include global winds, ocean currents, and weather systems—subjects to follow in Chapters 6 and 8. The atmosphere and ocean form a giant heat engine, driven by differences in energy from place to place. As you go about your daily activities, let these dynamic natural systems remind you of the constant flow of solar energy through the environment.

Having examined the flow of solar energy to Earth, let us now look at seasonal changes as Earth orbits the Sun each year.

# The Seasons

Earth's periodic rhythms of warmth and cold, rain and drought, dawn and daylight, twilight and night have fascinated humans for centuries. In fact, many ancient societies demonstrated a greater awareness of seasonal change than modern peoples and formally commemorated natural energy rhythms with festivals and monuments. Figure 2-15 shows such a monument at Stonehenge, in England. Here, rocks weighing 25 metric tons (28 tons) were hauled about 500 km (300 mi) and placed in patterns that evidently mark seasonal changes.

Specifically, the large stones mark sunrise on the day of the summer solstice, on or about June 21. Other seasonal events and eclipses of the Sun and the Moon are apparently predicted by this 3500-year-old "calendar" monument. Such ancient seasonal monuments and markings occur worldwide, including thousands of sites in North America.

## *Seasonality*

Seasonality refers to both the seasonal variation of the Sun's position above the horizon and changing daylengths during the year. Seasonal variations are a response to

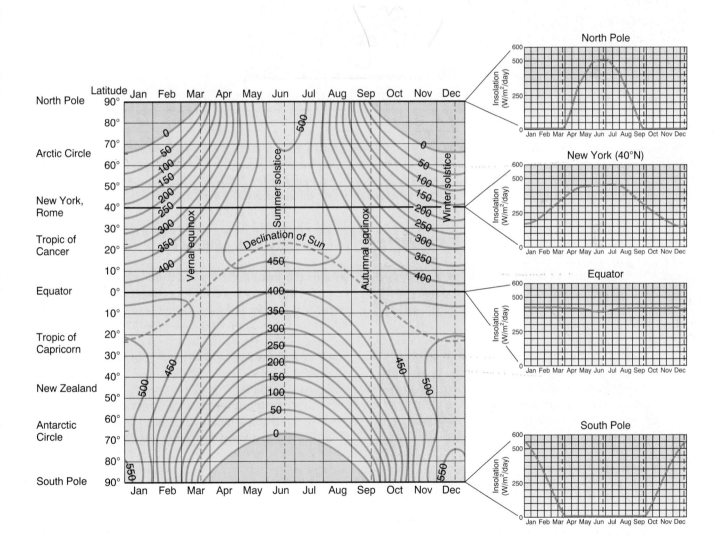

**FIGURE 2-13**

**Daily insolation received at the top of the atmosphere.**

The total daily insolation received at the top of the atmosphere is charted in watts per square meter per day by latitude and month (1 W/m²/day = 2.064 cal/cm²/day). A profile of annual energy receipts is graphed for the equator, 40° north latitude, and the North and South Poles. [Reproduced by permission of the Smithsonian Institution Press from *Smithsonian Miscellaneous Collections: Smithsonian Meteorological Tables*, vol. 114, 6th Edition. Robert List, ed. Smithsonian Institution, Washington, DC, 1984, p. 419, Table 134.]

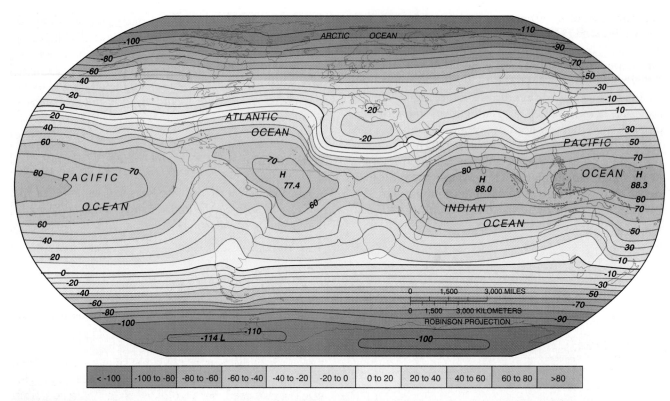

**FIGURE 2-14**
**Daily net radiation patterns.**
Averaged daily net radiation flows for a 9-year period (1979 to 1987), measured at the top of the atmosphere by the Earth Radiation Budget (ERB) instrument aboard the *Nimbus-7* satellite. Units are W/m². [Data for map courtesy of Dr. H. Lee Kyle, Goddard Space Flight Center, NASA.]

**FIGURE 2-15**
**Ancient calendars in stone.**
Stonehenge, Salisbury Plain in Wiltshire, England. [Photo by Robert Llewellyn.]

changes in the Sun's **altitude**, or the angle between the horizon and the Sun. At sunrise or sunset, the Sun is at the horizon, so its altitude is 0°. During the day, if the Sun reaches halfway between the horizon and directly overhead, it is at 45° altitude. If the Sun reaches the point directly overhead, it is at 90° altitude. The Sun is directly overhead (90° altitude, or its *zenith*) only at the *subsolar point*, where insolation is at maximum. At all other surface points, the Sun is at a lower angle, producing more-diffuse insolation.

The Sun's **declination** is the latitude of the subsolar point. Declination annually migrates through 47° of latitude, moving between the *Tropic of Cancer* at 23.5° N and the *Tropic of Capricorn* at 23.5° S latitude. The subsolar point does not reach the continental United States or Canada. Other than Hawaii, which is between 19° N and 22° N, and therefore south of the Tropic of Cancer, all other states and provinces are too far north.

Seasonality means a changing **daylength**, or duration of exposure. Daylength varies during the year, depending on latitude. The equator always receives equal hours of day and night: If you live in Ecuador, Kenya, or Singapore, every day and night is 12 hours long, year-round. People living along 40° N latitude (Philadelphia, Denver, Madrid, Beijing), or 40° S latitude (Buenos Aires,

Capetown, Melbourne), experience about 6 hours' difference in daylight between winter and summer. Those at 50° N or S latitude (Winnipeg, Paris, Falkland Islands) experience almost 8 hours of annual daylength variation.

At the North and South Poles, the range of daylength is extreme and extends from a 6-month period of no insolation (ranging from twilight to darkness to dawn) to a 6-month period of continuous 24-hour insolation (daylight). This is evident in Figure 2-18, if you note the illumination of the North Pole in June and South Pole in December. Can you determine the month during which the Earth photo on the back cover of this book was taken?

## *Reasons for Seasons*

Seasons result from variations in the Sun's altitude above the horizon, declination latitude, and daylength. These in turn are created by several physical factors that operate in concert: Earth's *revolution* in orbit around the Sun, its daily *rotation* on its axis, its *tilted axis*, the unchanging *orientation of its axis*, and its *sphericity* (Table 2-2). Of course, the essential ingredient is having a single source of radiant energy—the Sun. We now look at each of these factors individually. As we do, please note the distinction between *revolution*—Earth's travel around the Sun—and *rotation*—Earth's spinning on its axis (Figure 2-16).

**Revolution.** Earth's orbital **revolution** about the Sun is shown in Figure 2-2. Earth's speed in orbit averages 107,280 kmph (66,660 mph). This speed, together with Earth's distance from the Sun, determines the time required for one revolution around the Sun, and therefore

### TABLE 2-2

| Five Reasons for Seasons | |
| --- | --- |
| *Factor* | *Description* |
| Revolution | Orbit around the Sun; requires 365.24 days to complete at 107,280 kmph (66,660 mph) |
| Rotation | Earth turning on its axis; takes approximately 24 hours to complete at 1675 kmph (1041 mph) at the equator |
| Tilt | Axis is aligned at a 23.5° angle from a perpendicular to the plane of the ecliptic (the plane of Earth's orbit) |
| Axial parallelism | Remains in a fixed alignment, with Polaris directly overhead at the North Pole throughout the year |
| Sphericity | Appears as an oblate spheroid to the Sun's parallel rays; the geoid |

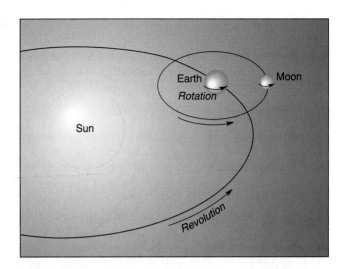

### FIGURE 2-16
**Earth's revolution and rotation.**
Earth's revolution about the Sun and rotation on its axis. The Moon's rotation on its axis and revolution about Earth, as viewed from above Earth's North Pole.

the length of the year and duration of the seasons. Earth completes its annual revolution in 365.2422 days. This number is based on a *tropical year*, measured from equinox to equinox, or the elapsed time between two crossings of the equator by the Sun.

The Earth-to-Sun distance might seem a seasonal factor, but it is not, even though it varies about 3% (4.8 million kilometers or 3 million miles) during the year. Remember that the distance averages 150 million kilometers (93 million miles).

**Rotation.** Earth's **rotation**, or turning on its axis, is a complex motion that averages 24 hours in duration. Rotation determines daylength, creates the apparent deflection of winds and ocean currents, and produces the twice-daily rise and fall of the ocean tides in relation to the gravitational pull of the Sun and the Moon.

Earth rotates about its **axis**, an imaginary line extending through the planet from the geographic North Pole to the South Pole. When viewed from above the North Pole, Earth rotates counterclockwise around this axis. Viewed from above the equator, Earth rotates west to east, or eastward. This eastward rotation creates the Sun's *apparent* daily journey from sunrise in the east to sunset in the west. Of course, the Sun actually remains in a fixed position in the center of the Solar System (Figure 2-16). (Note in the figure that the Moon also revolves around Earth and rotates counterclockwise on its axis.)

Although every point on Earth takes the same 24 hours to complete one rotation, the linear velocity of rotation at any point on Earth's surface varies dramatically with latitude. The equator is 40,075 km (24,902 mi) long; therefore, rotational velocity at the equator must be approximately 1675 kmph (1041 mph) to cover that distance in one day.

At 60° latitude, a parallel is only half the length of the equator, or 20,038 km (12,451 mi) long, so the rotational velocity there is 838 kmph (521 mph). At the poles, the velocity is 0. (This variation in rotational velocity establishes the effect of the Coriolis force, discussed in Chapter 6.) Table 2-3 lists the speed of rotation for several selected latitudes.

Earth's rotation produces the continually changing daily pattern of day and night. Half of Earth is in sunlight at any moment, and the traveling boundary that divides daylight and darkness is called the **circle of illumination** (as illustrated in Figure 2-18). Because this day-night dividing circle of illumination is a great circle that intersects the equator, which is another great circle, daylength at the equator is always evenly divided—12 hours of day and 12 hours of night. (Any two great circles on a sphere bisect one another.) Except for 2 days a year, on the equinoxes, all other latitudes experience uneven daylength throughout the seasons.

A true day varies slightly from 24 hours, but by international agreement a day is defined as exactly 24 hours, or 86,400 seconds. This average, called *mean solar time*, eliminates predictable variations in rotation and revolution that cause the solar day to change slightly in length throughout the year. The complexity of Earth's rotation is now exactly measured by systems in satellites that are in precise mathematical orbits: GPS (global positioning system; see News Report 1, Chapter 1), SLR (satellite-laser ranging), and VLBI (very-long baseline interferometry). All contribute to our knowledge of Earth's rotation.

*Tilt of Earth's Axis.* To understand Earth's **axial tilt**, imagine a plane (a flat surface) that intersects Earth's elliptical orbit about the Sun, with half of the Sun and Earth

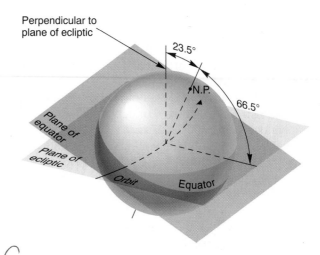

**FIGURE 2-17**
**The plane of Earth's orbit—the ecliptic—and Earth's axial tilt.**
Note that the plane of the equator is inclined at 23.5° to the plane of the ecliptic.

above the plane and half below. (It may help to envision two spheres floating half submerged in water, where the water's surface forms a plane.) This flat surface is termed the *plane of the ecliptic*. Now, imagine a perpendicular (at a 90° angle) line passing through the plane. Earth's axis and equatorial plane are tilted 23.5° from this perpendicular to the plane of the ecliptic. Another way of looking at it is that Earth's axis forms a 66.5° angle from the plane itself (Figure 2-17).

Hypothetically, if Earth were tilted on its side, with its axis parallel to the plane of the ecliptic, we would experience a maximum variation in seasons worldwide. On the other hand, if Earth's axis were perpendicular to the plane of its orbit—that is, with no tilt—we would experience no seasonal changes, just a perpetual spring or fall season, and all latitudes would experience 12-hour days and nights.

*Axial Parallelism.* Throughout our annual journey around the Sun, Earth's axis *maintains the same alignment* relative to the plane of the ecliptic and to Polaris and other stars. You can see this consistent alignment in Figure 2-18. If we compared the axis in different months, it would always appear parallel to itself, a condition known as **axial parallelism**.

*Other Factors.* Earth's *sphericity*, discussed in Chapter 1, is a part of seasonality, for it produces the uneven receipt of insolation from pole to pole. All five reasons for seasons are summarized in Table 2-2: revolution, rotation, tilt, axial parallelism, and sphericity. Now, considering all these factors operating together, let us examine the annual march of the seasons.

## TABLE 2-3

| Speed of Rotation at Selected Latitudes | | |
|---|---|---|
| **Latitude** | **Speed kmph (mph)** | **Cities at Approximate Latitudes** |
| 0° | 1675 (1041) | Quito, Ecuador; Pontianak, Indonesia |
| 30° | 1452 (902) | Pôrto Alegre, Brazil; New Orleans |
| 40° | 1284 (798) | Valdivia, Chile; Columbus, Ohio; Bejing |
| 50° | 1078 (670) | Chibougamau, Québec; Kyyîv (Kiev), Ukraine |
| 60° | 838 (521) | Seward, Alaska; Oslo, Norway; Saint Petersburg, Russia |
| 90° | 0 (0) | North Pole |

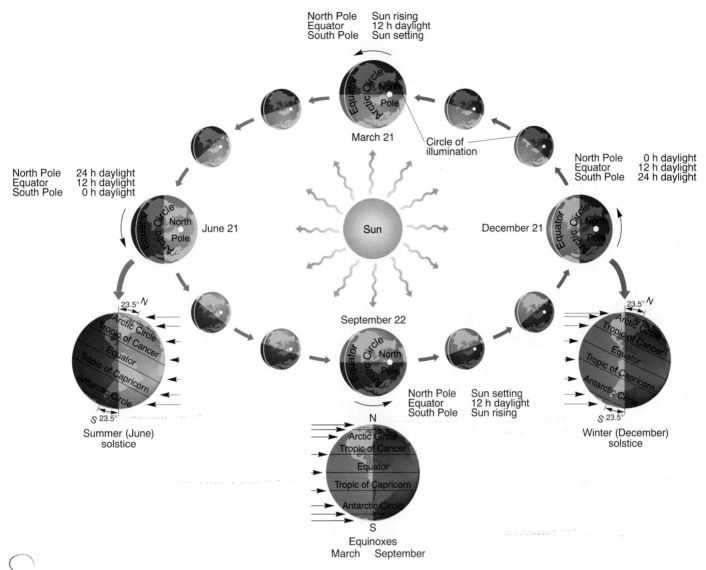

**FIGURE 2-18**

**Annual march of the seasons.**

Annual march of the seasons as Earth revolves about the Sun. Shading indicates the changing position of the circle of illumination. Note the hours of daylight for the equator and the poles.

## *Annual March of the Seasons*

During the annual march of the seasons on Earth, daylength is the most obvious way of sensing changes in season at latitudes away from the equator. Daylength is the interval between **sunrise**, the moment when the disk of the Sun first appears above the horizon in the east, and **sunset**, that moment when it totally disappears below the horizon in the west. Table 2-4 lists the times of sunrise and sunset and the daylength for various latitudes and seasons in the Northern Hemisphere.

The extremes of daylength occur in December and June. The times around December 21 and June 21 are termed *solstices*. Strictly speaking, however, the solstices are specific points in time at which the Sun's declination

is its position farthest north (Tropic of Cancer at 23.5° N) or south (Tropic of Capricorn at 23.5° S). "Tropic" is from *tropicus*, meaning a turn or change, so a tropic latitude is where the sun's declination appears to stand still briefly (Sun stance, or *sol stice*); then it "turns" and heads toward the other tropic.

Table 2-5 presents the key seasonal anniversary dates during which the specific time of the equinoxes or solstices occur, their names, and the subsolar point location (declination). During the year, places on Earth outside of the equatorial region experience a continuous but gradual shift in daylength, a few minutes each day, and the Sun's altitude increases or decreases a portion of a degree daily for any particular location. You may have noticed that these daily variations become more pronounced in

**TABLE 2-4**

| Daylength Times (Sunrise and Sunset) at Selected Latitudes (Northern Hemisphere) | | | | | | | | | | | | |
|---|---|---|---|---|---|---|---|---|---|---|---|---|
| Latitude | Winter Solstice (December Solstice) December 21–22 | | | Vernal Equinox (March Equinox) March 20–21 | | | Summer Solstice (June Solstice) June 20–21 | | | Autumnal Equinox (September Equinox) September 22–23 | | |
| | A.M. | P.M. | Daylength | A.M. | P.M. | Daylength | A.M. | P.M. | Daylength | A.M. | P.M | Daylength |
| 0° | 6:00 | 6:00 | 12:00 | 6:00 | 6:00 | 12:00 | 6:00 | 6:00 | 12:00 | 6:00 | 6:00 | 12:00 |
| 30° | 6:58 | 5:02 | 10:04 | 6:00 | 6:00 | 12:00 | 5:02 | 6:58 | 13:56 | 6:00 | 6:00 | 12:00 |
| 40° | 7:26 | 4:34 | 9:00 | 6:00 | 6:00 | 12:00 | 4:34 | 7:26 | 15:00 | 6:00 | 6:00 | 12:00 |
| 50° | 8:05 | 3:55 | 8:00 | 6:00 | 6:00 | 12:00 | 3:55 | 8:05 | 16:00 | 6:00 | 6:00 | 12:00 |
| 60° | 9:15 | 2:45 | 5:30 | 6:00 | 6:00 | 12:00 | 2:45 | 9:15 | 18:30 | 6:00 | 6:00 | 12:00 |
| 90° | No sunlight | | | Rising Sun | | | Continuous sunlight | | | Setting Sun | | |

*Note*: All times are standard and do not consider the local option of daylight saving time.

**TABLE 2-5**

| Annual March of the Seasons | | |
|---|---|---|
| Approximate Date | Northern Hemisphere Name | Location of the Subsolar Point |
| December 21–22 | Winter solstice (December solstice) | 23.5° S latitude (Tropic of Capricorn) |
| March 20–21 | Vernal equinox (March equinox) | 0° (equator) |
| June 20–21 | Summer solstice (June solstice) | 23.5° N latitude (Tropic of Cancer) |
| September 22–23 | Autumnal equinox (September equinox) | 0° (equator) |

spring and autumn, when the declination is changing at a faster rate.

Figure 2-18 demonstrates the annual march of the seasons and illustrates Earth's relationship to the Sun during the year. Let us begin with December. On December 21 or 22, at the moment of the **winter solstice** ("winter Sun stance"), or **December solstice**, the circle of illumination excludes the North Pole region from sunlight but includes the South Pole region. The subsolar point is at 23.5° S latitude, the parallel called the **Tropic of Capricorn**. The Northern Hemisphere is tilted away from these more direct rays of sunlight, thereby creating a lower angle for the incoming solar rays and thus a more diffuse pattern of insolation, thus causing our northern winter.

From 66.5° N latitude to 90° N (the North Pole), the Sun remains below the horizon the entire day. This latitude (66.5° N) marks the *Arctic Circle*, the southernmost parallel (in the Northern Hemisphere) that experiences a 24-hour period of darkness. During the following 3 months, daylength and solar angles gradually increase in the Northern Hemisphere as Earth completes one-fourth of its orbit.

The moment of the **vernal equinox**, or **March equinox**, occurs during March 20 or 21. At that time, the circle of illumination passes through both poles so that all

locations on Earth experience a 12-hour day and a 12-hour night. People living around 40° N latitude (New York, Denver) have gained 3 hours of daylight since the December solstice. At the North Pole, the Sun peeks above the horizon for the first time since the previous September; at the South Pole the Sun is setting. Figure 2-13 illustrates these moments of beginning and ending insolation at the poles.

From March, the seasons move on to June 20 or 21 and the moment of the **summer solstice**, or **June solstice**. The subsolar point now has shifted from the equator to 23.5° N latitude, the **Tropic of Cancer**. Because the circle of illumination now includes the North Polar region, everything north of the Arctic Circle receives 24 hours of daylight—the "midnight Sun." Figure 2-19 is a multiple-image photo of the midnight Sun as seen north of the Arctic Circle. In contrast, the region from the Antarctic Circle to the South Pole (66.5°–90° S latitude) is in darkness the entire day.

Note that the orientation of Earth's axis has remained fixed relative to the heavens and parallel to its position 6 months earlier (axial parallelism). Consequently, the Northern Hemisphere in June is tilted toward the Sun and receives higher Sun angles and longer days—and therefore more insolation—than the Southern Hemisphere. The result is summertime for northern latitudes. People living at 40° N latitude now receive more than 15 hours of sunlight, 6 hours longer than in December.

September 22 or 23 is the moment in time of the **autumnal equinox**, or **September equinox**, when Earth's orientation is such that the circle of illumination again passes through both poles so that all parts of the globe experience a 12-hour day and a 12-hour night. The subsolar point has returned to the equator, with days growing shorter to the north and longer to the south. Researchers stationed at the South Pole see the disk of the Sun just rising, ending their 6 months of night. (For a dramatic account of these seasonal changes, see the diary from the South Pole station in News Report 3, Chapter 10.) In the Northern Hemisphere, autumn arrives, a time of many colorful changes in the landscape, whereas in the Southern Hemisphere it is spring.

***Seasonal Observations.*** In the Northern Hemisphere mid-latitudes, the position of sunrise on the horizon migrates from day to day, from the southeast in December to the northeast in June. Over the same period, the point of sunset migrates from the southwest to the northwest. The Sun's altitude at local noon at 40° N latitude increases from a 26° angle above the horizon at the winter (December) solstice to a 73° angle above the horizon at the summer (June) solstice—a range of 47° (Figure 2-20).

Seasonal change is quite noticeable across the landscape away from the equator as shown by the four photos in Figure 2-21. Figure 2-22 presents two composite images of seasonal change in vegetation cover in winter and late summer as recorded by the AVHRR sensors aboard polar-orbiting satellites. Think back over the past year. What seasonal changes have you observed in vegetation, temperatures, and weather? Remote sensing is now revealing seasonal change in dramatic fashion and providing data for detailed study.

***Dawn and Twilight.*** *Dawn* is the period of diffused light that occurs before sunrise. The corresponding evening period after sunset is *twilight.* During both periods, light is scattered by molecules of atmospheric gases and reflected by dust and moisture so that the atmosphere is illuminated. This effect may be enhanced by the presence of pollution aerosols and suspended particles from volcanic eruptions or forest fires.

Dawn and twilight are roughly equal in length on any given day at any particular latitude. The duration of both is a function of latitude, because the angle of the Sun's path above the horizon determines the thickness of the atmosphere through which the Sun's rays must pass and thus the length of dawn and twilight.

At the equator, where the Sun's rays are almost directly above the horizon throughout the year, dawn and twilight are limited to 30–45 minutes each. These times increase to 1–2 hours at 40° latitude, and at 60° latitude they range upward from 2.5 hours, with little true night in summer. The poles experience about 7 weeks of dawn and 7 weeks of twilight, leaving only 2.5 months of near darkness during the 6 months when the Sun is completely below the horizon.

**FIGURE 2-19**
**The midnight Sun.**
The midnight Sun north of the Arctic Circle captured in a series of 18 exposures on the same piece of film. Midnight is the exposure showing the Sun closest to the horizon. [Photo by Gary Braasch/Tony Stone Images.]

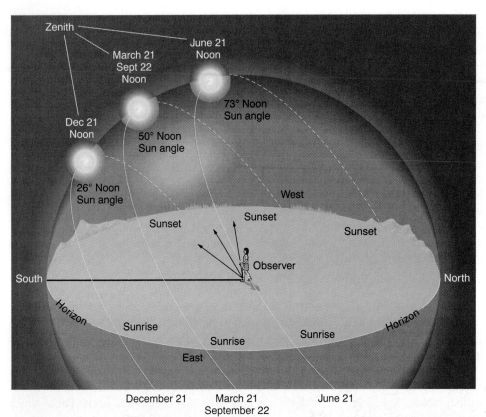

South

North

East

December 21     March 21     June 21
September 22

**FIGURE 2-20**

**Seasonal observations—sunrise, noon, and sunset through the year.**

Seasonal observations at 40° N latitude for the December solstice, March equinox, June solstice, and September equinox. The Sun's altitude increases from 26° in December to 73° above the horizon in June—a difference of 47°.

(a)

(b)

(c)

(d)

**FIGURE 2-21**

**The four seasons.**
Seasonality produces dramatic change in the leaves of an ornamental pear tree (*Pyrus calleryana*) in January, April, July, and November. [Photos by author.]

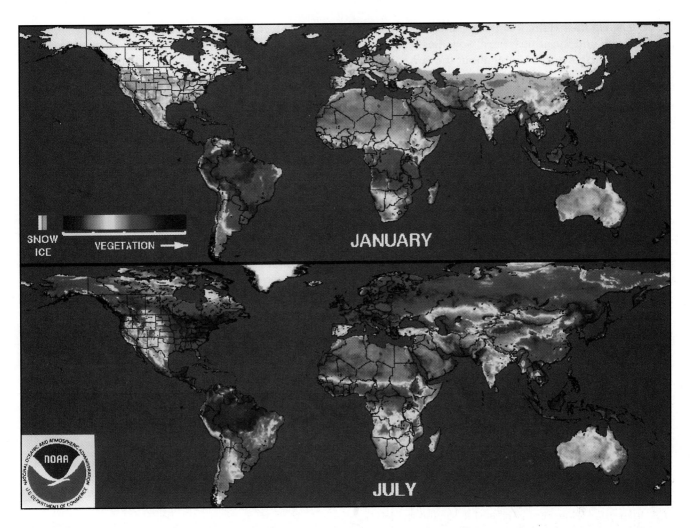

**FIGURE 2-22**
**Seasonal change from orbit.**
Seasonal change is monitored and measured by sensors aboard polar-orbiting satellites. [Images courtesy of Garik Gutman, NOAA, National Environmental Satellite, Data, and Information Service.]

# Summary and Review—Solar Energy to Earth and the Seasons

✔ *Distinguish* **among galaxies, stars, and planets and locate Earth.**

Our Solar System—Sun and nine planets—is located on a remote, trailing edge of the **Milky Way Galaxy**, a flattened, disk-shaped mass estimated to contain up to 400 billion stars. **Gravity**, the mutual attracting force exerted by the mass of an object upon all other objects, is an organizing force in the Universe. The process of suns (stars) condensing from nebular clouds with planetesimals (protoplanets) forming in orbits around their central masses is the **planetesimal hypothesis**. The development of hypotheses and theories about the Universe, Earth, and life involves the **scientific method**.

Milky Way Galaxy (p. 40)
gravity (p. 40)
planetesimal hypothesis (p. 41)
scientific method (p. 41)

1. Describe the Sun's status among stars in the Milky Way Galaxy. Describe the Sun's location, size, and relationship to its planets.

2. If you have seen the Milky Way at night, briefly describe it. Use specifics from the text in your description.

3. Briefly describe Earth's origin as part of the Solar System.

4. Define the scientific method, and give an example of its application.

✔ *Overview* **the origin, formation, and development of Earth and the atmosphere. Specifically, *construct* Earth's annual orbit about the Sun.**

The Solar System, planets, and Earth began to condense from a nebular cloud of dust, gas, debris, and icy comets approximately 4.6 billion years ago. Distances in space are so vast that the **speed of light** (300,000 kmps, or 186,000 mps, which is about 9.5 trillion kilometers, or nearly 6 trillion miles, per year) is used to express distance.

In its orbit, Earth is at **perihelion** (its closest position to the Sun) during our Northern Hemisphere winter (January 3 at 147,255,000 km, or 91,500,000 mi). It is at **aphelion** (its farthest position from the Sun) during our Northern Hemisphere summer (July 4 at 152,083,000 km, or 94,500,000 mi). Earth's average distance from the Sun is approximately 8 minutes and 20 seconds from the Sun in terms of light speed. In the Solar System, an imaginary plane touching all points of Earth's orbit is termed the **plane of the ecliptic**. Over the span of Earth's history the atmosphere evolved in four transitional stages—**primordial**, **evolutionary**, **living**, and the **modern** atmosphere we all share.

> speed of light (p. 41)
> perihelion (p. 42)
> aphelion (p. 42)
> plane of the ecliptic (p. 43)
> primordial atmosphere (p. 44)
> evolutionary atmosphere (p. 44)
> living atmosphere (p. 44)
> modern atmosphere (p. 44)

5. How far is Earth from the Sun in terms of light speed? In terms of kilometers and miles? Relate this distance to the shape of Earth's orbit during the year.

6. How many distinct atmospheres have there been on Earth? How does Earth's present atmosphere compare with past atmospheres?

7. Within which of Earth's atmospheres, past or present, did photosynthesis begin?

---

✔ *Describe* **the Sun's operation and *explain* the characteristics of the solar wind and the electromagnetic spectrum of radiant energy.**

The **fusion** process—hydrogen atoms forced together under tremendous temperature and pressure in the Sun's interior—generates incredible quantities of energy. The Sun's most conspicuous features are large **sunspots**, caused by magnetic disturbances. Solar energy in the form of charged particles of **solar wind** travels out in all directions from disturbances on the Sun. Solar wind is deflected by Earth's **magnetosphere**, producing various effects in the upper atmosphere, including spectacular **auroras**, the northern and southern lights, that surge across the skies at higher latitudes. Another effect of the solar wind in the atmosphere is its possible influence on weather.

The **electromagnetic spectrum** of radiant energy travels outward in all directions from the Sun. The total spectrum of this radiant energy is made up of different **wavelengths**—the distance between corresponding points on any two successive waves of radiant energy. Electromagnetic radiation from the Sun passes through Earth's magnetic field to the top of the atmosphere—the **thermopause**, at approximately 500 km (300 mi) altitude. Eventually some of this radiant energy reaches Earth's surface.

> fusion (p. 46)
> sunspots (p. 46)
> solar wind (p. 46)
> magnetosphere (p. 47)
> auroras (p. 48)
> electromagnetic spectrum (p. 49)
> wavelength (p. 49)
> thermopause (p. 51)

8. How does the Sun produce such tremendous quantities of energy? Write out a simple fusion reaction, using the quantities (in tons) of hydrogen involved.

9. What is the sunspot cycle? At what stage in the cycle is 1997?

10. Describe Earth's magnetosphere and its effects on the solar wind and the electromagnetic spectrum.

11. Summarize the presently known effects of the solar wind relative to Earth's environment.

12. Describe the various segments of the electromagnetic spectrum, from shortest to longest wavelength. What wavelengths are mainly produced by the Sun? Which are principally radiated by Earth to space?

---

✔ *Portray* **the intercepted solar energy and its uneven distribution at the top of the atmosphere.**

Solar radiation that reaches a horizontal plane at Earth is called **insolation**, a term specifically applied to radiation arriving at Earth's surface and atmosphere. Insolation at the top of the atmosphere is expressed as the **solar constant**: the average insolation received at the thermopause when Earth is at its average distance from the Sun. The solar constant is measured as 1372 W/m² (2.0 cal/cm²/min; 2 langleys/min). The place receiving maximum insolation is the **subsolar point**, where solar rays are perpendicular to the surface (radiating from directly overhead). All other locations away from the subsolar point receive slanting rays and more diffuse energy.

> insolation (p. 51)
> solar constant (p. 51)
> subsolar point (p. 51)

13. What is the solar constant? Why is it important to know?

14. Select 40° or 50° north latitude on Figure 2-13 and plot the amount of energy in watts per square meter (W/m²) per day on a graph for each month throughout the

year. Compare this with the amount at the North Pole and at the equator.

15. If Earth were flat and oriented perpendicularly to incoming solar radiation (insolation), what would be the latitudinal distribution of solar energy at the top of the atmosphere?

---

✔ *Define* **solar altitude, solar declination, and daylength** and *describe* **the annual variability of each—Earth's seasonality.**

The angle between the Sun and the horizon is the Sun's **altitude**. The Sun's **declination** is the latitude of the subsolar point. Declination annually migrates through 47° of latitude, moving between the *Tropic of Cancer* at 23.5° N (June) and the *Tropic of Capricorn* at 23.5° S latitude (December). Seasonality means an annual change in the Sun's altitude and changing **daylength,** or duration of exposure.

Earth's distinct seasons are produced by interactions of **revolution** (annual orbit about the Sun), **rotation** (turning on the axis), **axial tilt** (23.5° from a perpendicular to the plane of the ecliptic), **axial parallelism** (the parallel alignment of the axis throughout the year), and *sphericity.* Earth rotates about its **axis,** an imaginary line extending through the planet from the geographic North Pole to the South Pole. As it rotates, the traveling boundary that divides daylight and darkness is called the **circle of illumination**. Daylength is the interval between **sunrise,** the moment when the disk of the Sun first appears above the horizon in the east, and **sunset,** that moment when it totally disappears below the horizon in the west.

On December 21 or 22, at the moment of the **winter solstice** ("winter Sun stance"), or **December solstice,** the circle of illumination excludes the North Pole but includes the South Pole. The subsolar point is at 23.5° S latitude, the parallel called the **Tropic of Capricorn**. The moment of the **vernal equinox,** or **March equinox,** occurs on March 20 or 21. At that time, the circle of illumination passes through both poles so that all locations on Earth experience a 12-hour day and a 12-hour night.

June 20 or 21 is the moment of the **summer solstice,** or **June solstice**. The subsolar point now has shifted from the equator to 23.5° N latitude, the **Tropic of Cancer**. Because the circle of illumination now includes the North Polar region, everything north of the Arctic Circle receives 24

hours of daylight—the "midnight Sun." September 22 or 23 is the time of the **autumnal equinox,** or **September equinox,** when Earth's orientation is such that the circle of illumination again passes through both poles so that all parts of the globe experience a 12-hour day and a 12-hour night.

altitude (p. 54)
declination (p. 54)
daylength (p. 54)
revolution (p. 55)
rotation (p. 55)
axial tilt (p. 56)
axial parallelism (p. 56)
axis (p. 55)
circle of illumination (p. 56)
sunrise (p. 57)
sunset (p. 57)
winter solstice, December solstice (p. 58)
Tropic of Capricorn (p. 58)
vernal equinox, March equinox (p. 58)
summer solstice, June solstice (p. 58)
Tropic of Cancer (p. 58)
autumnal equinox, September equinox (p. 59)

16. How is the Stonehenge monument related to seasons?

17. The concept of seasonality refers to what specific phenomena? How do these two aspects of seasonality change during a year at 0° latitude? At 40°? At 90°?

18. Differentiate between the Sun's altitude and its declination at Earth's surface.

19. For the latitude at which you live, how does daylength vary during the year? How does the Sun's altitude vary? Does your local newspaper publish a weather calendar containing such information?

20. List the five physical factors that operate together to produce seasons.

21. Describe Earth's revolution and rotation, and differentiate between them.

22. Define Earth's present tilt relative to its orbit about the Sun.

23. Describe seasonal conditions at each of the four key seasonal anniversary dates during the year. What are the solstices and equinoxes, and what is the Sun's declination at these times?

---

## NetWork

The *Geosystems Home Page* provides on-line resources for this chapter on the World Wide Web. You will find review exercises, specific updates for items in the chapter, suggested readings, and links to interesting related pathways on the Internet (click on the Table of Contents link and select this chapter). *Geosystems* is at: **http://www.prenhall.com/geosystm**

# 3

# Earth's Modern Atmosphere

**Atmospheric Structure and Function**

**Variable Atmospheric Components**

**Summary and Review**

## Key Learning Concepts

After reading the chapter, you should be able to:

- *Construct* a general model of the atmosphere based on composition, temperature, and function and *diagram* this model in a simple sketch.

- *List* the stable components of the modern atmosphere and their relative percentage contributions by volume and *describe* each.

- *Describe* conditions within the stratosphere; specifically, *review* the function and status of the ozonosphere (ozone layer).

- *Distinguish* between natural and anthropogenic variable gases and materials in the lower atmosphere.

- *Describe* the sources and effects of carbon monoxide, nitrogen dioxide, and sulfur dioxide, and *construct* a simple equation that illustrates photochemical reactions that produce ozone, peroxyacetyl nitrates, nitric acid, and sulfuric acid.

*Rainbow over Mt. Rundle, Banff, Alberta.* [Photo by author.]

Earth's atmosphere is a unique reservoir of gases, the product of nearly 5 billion years of development. We all participate in the atmosphere with each breath we take. This chapter examines the modern atmosphere's structure, composition, and function. Our consideration of the modern atmosphere also must include the spatial aspects of human-induced problems that affect it, such as air pollution, the stratospheric ozone predicament, and the blight of acid deposition. We consider these critical topics carefully, for we are participating in producing the atmosphere of the future.

## Atmospheric Structure and Function

As explained in Chapter 2, the modern atmosphere probably is the fourth general atmosphere in Earth's history (the other three were the *primordial, evolutionary*, and *living* atmospheres). This modern atmosphere is a gaseous mixture of ancient origin, the sum of all the exhalations and inhalations of life on Earth throughout time. The principal substance of this atmosphere is air, the medium of life as well as a major industrial and chemical raw material. *Air* is a simple mixture of gases that is naturally odorless, colorless, tasteless, and formless, blended so thoroughly that it behaves as if it were a single gas.

In his book *The Lives of a Cell*, physician and self-styled "biology-watcher" Lewis Thomas compared the atmosphere of Earth to an enormous cell membrane. The membrane around a cell regulates the interactions between the cell's delicate inner workings and the potentially disruptive outer environment. Each cell membrane is very selective as to what it will allow to pass through. The modern atmosphere acts as Earth's protective membrane, as Thomas described so vividly (Figure 3-1).

Earth's modern atmosphere is arranged in a series of imperfectly shaped concentric "shells" or "spheres" that grade into one another, all bound to the planet by gravity. As critical as the atmosphere is to us, it represents only the thinnest envelope, amounting to less than one-millionth of Earth's total mass. We study the atmosphere by viewing it in layers that have distinctive compositions, temperatures, and functions.

### Atmospheric Composition, Temperature, and Function

Figure 3-2 charts essential aspects of the atmosphere in a vertical cross section, or side view, and it is keyed to the following discussion. Our discussion is simplified by looking at the atmosphere in three ways: by composition, by temperature, and by function. These categories are noted along the left side of Figure 3-2.

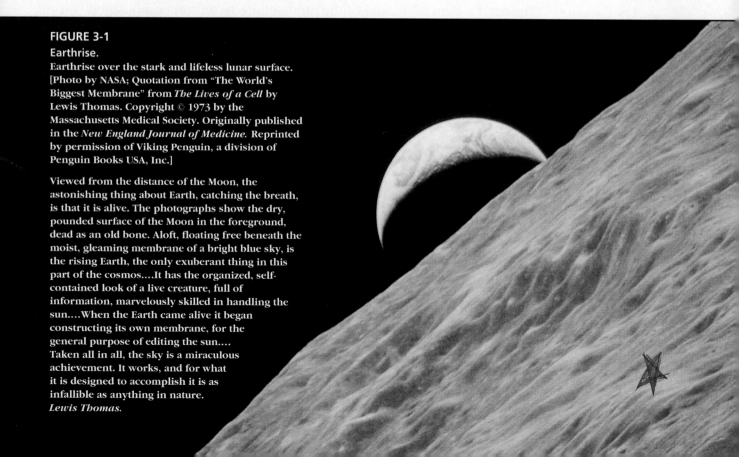

**FIGURE 3-1**
Earthrise.
Earthrise over the stark and lifeless lunar surface. [Photo by NASA; Quotation from "The World's Biggest Membrane" from *The Lives of a Cell* by Lewis Thomas. Copyright © 1973 by the Massachusetts Medical Society. Originally published in the *New England Journal of Medicine*. Reprinted by permission of Viking Penguin, a division of Penguin Books USA, Inc.]

Viewed from the distance of the Moon, the astonishing thing about Earth, catching the breath, is that it is alive. The photographs show the dry, pounded surface of the Moon in the foreground, dead as an old bone. Aloft, floating free beneath the moist, gleaming membrane of a bright blue sky, is the rising Earth, the only exuberant thing in this part of the cosmos....It has the organized, self-contained look of a live creature, full of information, marvelously skilled in handling the sun....When the Earth came alive it began constructing its own membrane, for the general purpose of editing the sun.... Taken all in all, the sky is a miraculous achievement. It works, and for what it is designed to accomplish it is as infallible as anything in nature.
*Lewis Thomas.*

## News Report 1

## Earth's Atmosphere—Unique among Planets

Scientists have determined that in our Solar System, Earth's life-supporting atmosphere is unique, formed by gases from the crust and interior and the breath of all life over time. An array of satellites and remote-sensing techniques are determining that the atmospheres of the other three inner planets (Mercury, Venus, and Mars) bear no resemblance to Earth's.

Mars has a cold, thin atmosphere of carbon dioxide, equivalent in its low pressure to Earth's atmosphere 32 km (20 mi) above the surface. Venus has a hot, dense atmosphere dominated by carbon dioxide, and it is shrouded in clouds of sulfuric acid; its surface pressures are about 90 times those on Earth, and surface temperatures average 500°C (930°F). Mercury has no appreciable atmosphere and

appears similar to the Moon, pockmarked by innumerable craters.

The five outer planets (giants Jupiter, Saturn, Uranus, Neptune, and little Pluto) are starkly different from Earth. These planets are frozen worlds with atmospheres of methane, ammonia, hydrocarbons, and other gases. Earth stands alone among all nine planets in possessing the only biosphere.

---

By chemical *composition*, the atmosphere divides into two broad regions, the *heterosphere* and the *homosphere*. If we use temperature as a criterion, the atmosphere has four distinct zones—the *thermosphere, mesosphere, stratosphere*, and *troposphere*. By *function*, the atmosphere has two specific zones that remove most of the harmful wavelengths of incoming solar radiation: the *ionosphere* and the *ozonosphere* (*ozone layer*).

We divide the following discussion by composition, examining the heterosphere and homosphere. As you read, note that we follow the same path that incoming solar radiation travels through the atmosphere to Earth's surface.

### Heterosphere

By composition, the **heterosphere** is defined as the outer atmosphere. It begins at about 80 km (50 mi) altitude and extends outward to the transition to interplanetary space (Figure 3-2). As the prefix *hetero-* implies, this region is not uniform—its gases are *not evenly mixed*. This distribution is quite different from the nicely blended gases we breathe near Earth's surface, in the homosphere. Gases in the heterosphere are distributed in distinct layers that are sorted by gravity according to their atomic weight, with the lightest elements (hydrogen and helium) at the margins of outer space and the heavier elements (oxygen and nitrogen) dominant in the lower heterosphere.

As a practical matter, we consider the top of our atmosphere to be around 480 km (300 mi), the same altitude we use for measuring the solar constant (Chapter 2). Beyond that altitude, the atmosphere is rarefied (nearly a vacuum) and is called the **exosphere**, which means "outer sphere." It contains scarce lightweight hydrogen and helium atoms, weakly bound by gravity as far as 32,000 km (20,000 mi) from Earth.

In addition to its distinctive layered composition, the heterosphere has thermal properties (in what is called the

thermosphere) and also is the location of a functional layer (called the ionosphere).

***Thermosphere.*** Using temperature, we define the **thermosphere** ("heat sphere"), which roughly corresponds to the heterosphere (80 km out to 480 km, or 50–300 mi). The upper limit of the thermosphere is called the **thermopause** (the suffix *pause* means "to change"). During periods of a less active Sun (fewer sunspots and coronal bursts), the thermopause may lower in altitude from the average 480 km (300 mi) to only 250 km altitude (155 mi). An active Sun will cause the outer atmosphere to swell to an altitude of 550 km (340 mi).

The temperature profile in Figure 3-2 (red curve) shows that temperatures rise sharply in the thermosphere, to 1200°C (2200°F) and higher. Despite such high temperatures, the thermosphere is not "hot" in the way you might expect. Temperature and heat are different concepts. The intense solar radiation in this portion of the atmosphere excites individual molecules (principally nitrogen and oxygen) and atoms (oxygen) to high levels of vibration. This **kinetic energy**, the energy of motion, is the vibrational energy that we measure as *temperature*.

However, the actual heat involved is very small. The reason is that the density of the molecules is so low that little actual **heat**, the flow of kinetic energy from one body to another because of a temperature difference between them, is produced. Heating in the atmosphere near Earth's surface is different because the greater number of molecules in the denser atmosphere transmit their kinetic energy as **sensible heat**, meaning that we can measure it. (Density, temperature, and heat capacity determine the sensible heat of a substance.)

***Ionosphere.*** On the basis of function, we identify two layers that filter harmful wavelengths of solar radiation, protecting Earth's surface. The outer functional layer, the **ionosphere**, extends throughout the thermosphere

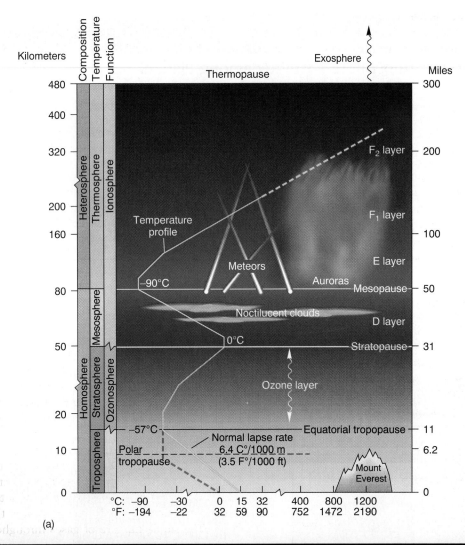

**FIGURE 3-2**

**Profile of the modern atmosphere.**

(a) An integrated chart of our modern atmosphere. The columns along the left side show the division of the atmosphere by composition, temperature, and function. The chart spans the atmosphere from Earth's surface to the thermopause at 480 km. The plot of temperature and the scale along the bottom permit you to tell the temperature at any altitude. The illustration pictures the approximate locations of auroras, meteors, and noctilucent clouds. (b) Space Shuttle astronauts captured a dramatic sunset through various atmospheric layers across the "edge" of our planet—called Earth's limb. A silhouetted cumulonimbus thunderhead cloud is seen rising to the tropopause. [Space Shuttle photo from NASA.]

(Figure 3-2). The ionosphere absorbs cosmic rays, gamma rays, X-rays, and shorter wavelengths of ultraviolet radiation, all of which are absorbed, changing atoms to positively charged *ions* and giving the ionosphere its name. Radiation bombards the ionosphere continuously, producing a constant flow of electrons. Figure 3-3 depicts in a general way the absorption of radiation by the various functional layers of the atmosphere.

Figure 3-2 shows the average daytime altitudes of four regions within the ionosphere, known as the D, E, $F_1$, and $F_2$ layers. These are important to broadcast communications. At night, the D layer virtually disappears, and the $F_1$ and $F_2$ layers usually merge and reflect certain radio wavelengths, including AM radio and other shortwave broadcasts, returning them at predictable angles to receivers hundreds of kilometers from the transmitters. However, during the day, the D and E regions in the ionosphere actively absorb arriving radio signals, preventing such distant reception. Normally unaffected are FM or television broadcast wavelengths, which pass through to space, thus necessitating use of communication satellites to distribute these signals to the surface. Because they are not affected by the atmosphere, such wavelengths are used for deep-space satellite transmissions.

During times of intense solar wind and auroral activity, the ionosphere becomes so active that radio communications can be disrupted. The glowing auroral lights discussed in Chapter 2 occur principally within the ionosphere.

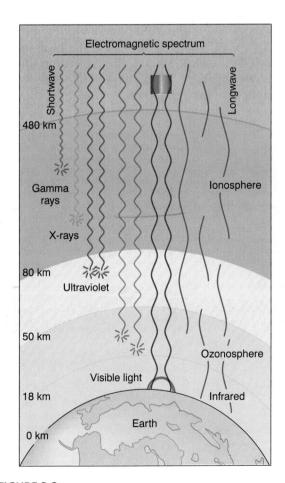

**FIGURE 3-3**
**The atmosphere protects Earth's surface.**
As solar energy passes through the atmosphere the shortest wavelengths are absorbed. Only a fraction of the ultraviolet, most of the visible light, and infrared reach Earth's surface.

**TABLE 3-1**

| Stable Components of the Modern Homosphere | | |
|---|---|---|
| **Gas (Symbol)** | **Percentage by Volume** | **Parts per Million (ppm)** |
| Nitrogen ($N_2$) | 78.084 | 780,840 |
| Oxygen ($O_2$) | 20.946 | 209,460 |
| Argon (Ar) | 0.934 | 9,340 |
| Carbon dioxide ($CO_2$) | 0.036 | 360 |
| Neon (Ne) | 0.001818 | 18 |
| Helium (He) | 0.000525 | 5 |
| Methane ($CH_4$) | 0.00014 | 1.4 |
| Krypton (Kr) | 0.00010 | 1.0 |
| Ozone ($O_3$) | variable | |
| Nitrous oxide ($N_2O$) | trace | |
| Hydrogen (H) | trace | |
| Xenon (Xe) | trace | |

## Homosphere

Between the heterosphere and Earth's surface is the other compositional shell of the atmosphere, the **homosphere**. This region, defined by composition, extends from an altitude of 80 km (50 mi) to the surface. Even though the atmosphere rapidly changes density in the homosphere, decreasing with increasing altitude, the blend (proportion) of gases is nearly uniform throughout the homosphere. The only exceptions are the concentration of ozone ($O_3$) in the "ozone layer," from 19 to 50 km (12 to 31 mi), and the variations in water vapor, pollutants, and some trace chemicals in the lowest portion of the atmosphere.

The stable mixture of gases throughout the homosphere has evolved slowly. The present proportion, which includes oxygen, was attained approximately 500 million years ago. Table 3-1 lists by volume and Figure 3-4 illustrates the stable ingredients that constitute the homosphere.

The homosphere is a vast reservoir of relatively inert nitrogen, originating principally from volcanic sources. *Nitrogen* is a key element of life, yet we exhale all the nitrogen that we inhale. This apparent contradiction is explained by the fact that nitrogen comes into our bodies not with the air we breathe but through compounds in food. In the soil, nitrogen is bound to these compounds by nitrogen-fixing bacteria, and it is returned to the atmosphere by denitrifying bacteria that remove nitrogen from organic materials.

*Oxygen*, a by-product of photosynthesis, also is essential for life processes. Slight spatial variations occur in the percentage of oxygen in the atmosphere because of variations in photosynthetic rates with latitude, seasonal changes, and the lag time as atmospheric circulation slowly mixes the air. Although it forms about one-fifth of the atmosphere, oxygen forms compounds that compose about half of Earth's crust. Oxygen readily reacts

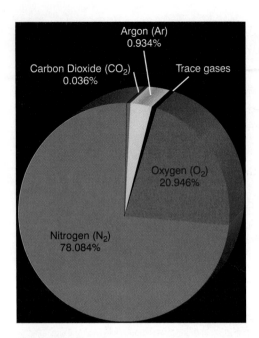

**FIGURE 3-4**

**Composition of the homosphere.**

Stable components of the atmosphere illustrated.

*Mesosphere.* The **mesosphere** is the area from 50 out to 80 km (30 to 50 mi). As Figure 3-2 shows, the mesosphere's outer boundary, the *mesopause*, is the coldest portion of the atmosphere, averaging –90°C (–130°F), although that temperature may vary considerably (25 to 30 C°, or 45 to 54 F°). Very low pressures (low density of molecules) exist in the mesosphere.

The mesosphere sometimes receives cosmic or meteoric dust, acting as nuclei around which fine ice crystals form. At high latitudes, an observer may see these bands of crystals glow in rare and unusual night clouds called **noctilucent clouds**.

The *Upper Atmosphere Research Satellite* (*UARS*), launched in 1991, is detecting large, continent-sized windstorms in the mesosphere. The very rarefied air is moving in vast waves at speeds exceeding 320 kmph (200 mph). The significance of this discovery to surface weather patterns and the Earth-atmosphere energy budget is being studied.

*Stratosphere and Ozonosphere.* The **stratosphere** extends from 18 to 50 km (11 to 31 mi) from Earth's surface. Temperatures increase with altitude throughout the stratosphere, from –57°C (–70°F) at 18 km (tropopause), warming to 0°C (32°F) at 50 km (stratopause). The heat source throughout the stratosphere is the other functional layer, called the **ozonosphere**, or **ozone layer**. Thus, the stratosphere corresponds with a functional layer, the ozonosphere.

with many elements to form these materials. Both nitrogen and oxygen reserves in the atmosphere are so extensive that, at present, they far exceed human capabilities to disrupt or deplete them.

The gas *argon*, constituting about 1% of the homosphere, is completely inert (an unreactive "noble" gas) and therefore is unusable in life processes. Argon is a residue from the radioactive decay of an isotope (form) of potassium called potassium-40 (symbolized $^{40}K$). Slow accumulation over millions of years accounts for all the argon present in the modern atmosphere. Because industry has found uses for inert argon (in light bulbs, welding, and some lasers), it is extracted or "mined" from the atmosphere, in addition to nitrogen and oxygen, for commercial and industrial uses (Figure 3-5).

*Carbon dioxide* is a natural by-product of life processes. It is essentially a stable atmospheric component, qualifying it for inclusion in Table 3-1. Although its present percentage in the atmosphere is small (0.036%), it is important in maintaining global temperatures. Its percentage has been increasing over the past 200 years as a result of human activities. The implications of this increase to global warming and our future are discussed in Chapter 10.

Shifting our view of the homosphere to temperature, we can subdivide the homosphere into three layers: the *mesosphere*, the *stratosphere*, and the one closest to Earth's surface, the *troposphere*.

**FIGURE 3-5**

**An air-mining operation.**

Air is a major industrial and chemical raw material that is extracted from the atmosphere by air-mining companies. Nitrogen, oxygen, and argon are extracted using a cryogenic process (at very low temperatures). [Photo by author.]

In addition to the usual mixture of gases in the homosphere, the ozone layer contains an increased level of ozone, a highly reactive oxygen molecule made up of three oxygen atoms ($O_3$) instead of the usual two atoms ($O_2$) that make up oxygen gas. Ozone absorbs wavelengths of ultraviolet light (0.1–0.3μm) and subsequently reradiates this energy at longer wavelengths, as infrared energy. This process converts most harmful ultraviolet radiation, effectively "filtering" it and safeguarding life at Earth's surface.

The ozone layer is presumed to have been relatively stable over the past several hundred million years (allowing, of course, for daily and seasonal fluctuations). Today, however, it is in a state of continuous change. Focus Study 3-1 presents a background analysis of the developing crisis in this critical portion of our atmosphere. The spatial implications for humanity of losses in stratospheric ozone make this an important topic for applied studies in physical geography and atmospheric sciences.

***Troposphere.*** The **troposphere** is the final layer encountered by incoming solar radiation as it surges through the atmosphere to the surface. It is the home of the biosphere, the atmospheric layer that supports life, and the region of principal weather activity.

Approximately 90% of the total mass of the atmosphere and the bulk of all water vapor, clouds, weather, air pollution, and life forms are contained within the troposphere. The **tropopause**, its upper limit, is defined by an average temperature of –57°C (–70°F), but its exact elevation varies with the season, latitude, and surface temperatures and pressures. Near the equator, because of intense heating from the surface, the tropopause occurs at 18 km (11 mi); in the middle latitudes, it occurs at 13 km (8 mi); and at the North and South Poles it is only 8.0 km (5.0 mi) or less above Earth's surface.

Figure 3-6 illustrates the normal temperature profile within the troposphere during daytime. As the graph shows, temperatures decrease rapidly with increasing altitude at an average of 6.4 C° per kilometer (3.5 F° per 1000 feet), a rate known as the **normal lapse rate**. This temperature plot also appears in Figure 3-2.

The normal lapse rate is an average. The actual lapse rate at any particular time and place, which may deviate considerably because of local weather conditions, is called the **environmental lapse rate**. This variation in temperature gradient in the lower troposphere is central to our discussion of weather processes (Chapter 8).

---

## News Report 2

### New UV Index Announced to Help Save Your Skin

The television weather reporter says, "The temperature is presently 25°C (78°F), relative humidity is 55%, light winds are from the south, and the *UV Index is 9*." This last item is what you are hearing in many cities where the National Weather Service (NWS) and the Environmental Protection Agency (EPA) are providing an ultraviolet radiation index. Table 1 presents a sampling of the ultraviolet index numbers and their application to two skin types. As an example, a moderate 6 on the index means a burn in 10 minutes for the most susceptible skin to a burn in 50 minutes for the least susceptible skin.

Environment Canada has been reporting a UV index on a regular basis since 1992, and the NWS and EPA in the United States have been doing so since 1994. An international standard index will eventually emerge. The Arctic ozone depletion affects concentrations at lower latitudes thus drawing the United States into further cooperation.

As stratospheric ozone levels continue to thin—midlatitude ozone losses of 21% during 1992–1993 set a record—surface exposure to cancer-causing radiation continues to climb. The public now is alerted when to take extra precautions in the form of sunscreens, hats, and sunglasses. Remember, damage is *cumulative* and it may be decades before you experience the ill effects triggered by last summer's sunburn. (For more information call the American Cancer Society, 800-227-2345; or write the American Academy of Dermatology, P.O. Box 681069, Shaumburg, Illinois 60168-1069.)

**TABLE 1**

| UV Index (EPA) | | | |
| --- | --- | --- | --- |
| *Exposure Category/ Index Value* | | *Minutes to Burn for "Never Tans" (most susceptible)* | *Minutes to burn for "Rarely Burns" (least susceptible)* |
| Minimal | 0–2 | 30 minutes | >120 minutes |
| Low | 4 | 15 minutes | 75 minutes |
| Moderate | 6 | 10 minutes | 50 minutes |
| High | 8 | 7.5 minutes | 35 minutes |
| Very high | 10 | 6 minutes | 30 minutes |
| | 15 | <4 minutes | 20 minutes |

# Focus Study 3-1

## Stratospheric Ozone Losses: A Worldwide Health Hazard

Consider these news reports and conditions:

- Three scientists who developed, confirmed, and promoted the stratospheric ozone depletion theory were awarded the 1995 Nobel Prize for Chemistry (see News Report 3).
- In 1994 an international scientific consensus confirmed previous assessments of the anthropogenic disruption of the ozone layer. The report, *Scientific Assessment of Ozone Depletion*, was prepared by NASA, NOAA, United Nations Environment Programme, and World Meteorological Organization.
- Measurements confirm that chlorine atoms and chlorine monoxide molecules in the stratosphere are of human origin and that ozone thinning cannot be explained by volcanic activity. The recently measured concentration of chlorofluorocarbons is about equal to that produced by industry and emitted to the atmosphere.
- During the months of Antarctic spring (September to November), ozone concentrations approach total depletion in the stratosphere over an area twice the size of the Antarctic continent. The protective ozone layer is being depleted over southern South America, southern Africa, Australia, and New Zealand.
- Antarctic researchers have recorded steady losses in the ozone layer since 1970, noting a huge temporary decrease every October. Each year, the "ozone hole" (actually a thinning) widens and deepens. In addition to ground-based measurements, NOAA satellites also record this alarming situation. The 1993, 1994, and 1995, holes were the worst in the past two decades (Figure 1).
- At Earth's opposite pole, a similar ozone depletion over the Arctic exceeded 35% in 1989. Losses have increased in each subsequent year. Since 1992, the Canadian government regularly reports an "ultraviolet index" to help the public protect themselves. A similar program began in the United States in 1994.
- Environment Canada measured an 8% loss in normal ozone levels in

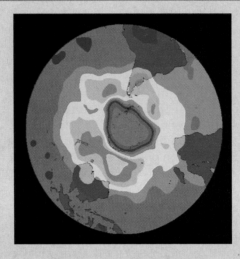

**FIGURE 1**
**The Antarctic ozone hole.**
*Tiros* TOVS (*Tiros* ozone vertical sounder) measurements made during October 1995. The color scale represents ozone concentrations in Dobson units, with purples for amounts less than 180. (One Dobson unit equals 2.69 x 10$^{16}$ molecules of O$_3$ per cm$^2$.) Measurements have dropped below 100 Dobson units. The ozone "hole" has grown larger since 1979, covering a record 25 million km$_2$, (9 million mi$_2$), or more than double the surface area of Antarctica—an area larger than Canada, the United States, and Mexico combined. [NOAA/Science Photo Library/Photo Researchers, Inc.]

1993, with greatest thinning between January and April, when it dropped to 14% below normal. Extreme lows of 22% below normal occurred in March 1993 over Toronto and Edmonton; this drastic reduction produced increases in ultraviolet (UV) radiation at the surface.
- In 1994, Environment Canada opened a new scientific observatory to monitor ozone losses over Canada. This high-Arctic facility is at a remote weather station on Ellesmere Island, Northwest Territories, about 1000 km from the North Pole.
- Overall ozone losses in the midlatitudes are continuing at 6% to 8% per decade. In North America, related skin cancers are increasing at an alarming rate, now totaling more than 700,000 cases annually, with over 34,000 malignant melanomas resulting in 10,000 deaths—now increasing at 4.2% per year. The American Academy of Dermatology reports a 500% increase in malignant melanomas from 1950 to 1985. Most affected are light-skinned persons who live at higher elevations and those who work principally outdoors.
- Increased ultraviolet is affecting atmospheric chemistry, biological systems, oceanic phytoplankton (small photosynthetic organisms that form the basis of the ocean's primary food production) and fisheries, crop yields, and human skin, eye tissues, and

immunity. Specifically, in Antarctic water, losses of 6%-12% now are measured in primary production.
- Some chemicals, banned by international agreements to protect the ozone layer, have already begun a measurable decline in concentration. Others continue to increase, but should begin a predicted downturn in about a decade.

Given this deluge of news items and facts, it is clear that more ultraviolet radiation from the Sun is breaking through Earth's protective ozone layer. What is happening in the stratosphere everywhere and in the polar regions specifically? How is this happening, and how are people and their governments responding?

### Monitoring Earth's Fragile Safety Screen

A sample of ozone from the layer's densest part (at 29 km, or 18 mi altitude) contains only 1 part ozone per 4 million parts of air—compressed to surface pressure, it would be only 3 mm thick. Yet, this rarefied zone was in steady-state equilibrium for at least several hundred million years, absorbing intense UV radiation and permitting life to proceed safely on Earth.

The ozone layer has been monitored since the 1920s, measured with instrumented balloons, aircraft, orbiting satellites, and a 30-station ozone-monitoring network (mostly in North America) established in 1957. The *total ozone mapping spectrometer* (*TOMS*) has been operating since 1978 aboard *Nimbus*, and the *Upper Atmosphere Research Satellite* (*UARS*), among other satellites and aircraft, is now monitoring ozone losses and the invasion of chlorine and bromine compounds in detail.

## Ozone Losses Explained

What is causing the decline in stratospheric ozone? Paul Crutzen, a scientist at the Max Planck Institute in Germany, determined in 1970 that nitrogen oxides were acting as catalysts in ozone-reducing reactions. In 1974, two atmospheric chemists, F. Sherwood Rowland and Mario Molina, hypothesized that some synthetic chemicals were releasing chlorine atoms that decompose ozone. These **chlorofluorocarbon compounds**, or **CFCs**, are complex synthetic molecules of chlorine, fluorine, and carbon. (See their report: "Stratospheric sink for chlorofluoromethanes: Chlorine atom catalyzed destruction of ozone," *Nature*, vol. 249 (1974): 810.

CFCs are very stable (inert) under conditions at Earth's surface, and they possess remarkable heat properties. Both qualities make them valuable as propellants in aerosol sprays and as refrigerants. Also, some 45% of CFCs are used as solvents in the electronics industry and as foaming agents. Being inert, CFC molecules do not dissolve in water and do not break down in biological processes. (In contrast, chlorine compounds derived from volcanic eruptions and the ocean are water-soluble and rarely reach the stratosphere.)

Researchers Rowland and Molina hypothesized that stable CFC molecules slowly migrate into the stratosphere, where intense UV radiation splits them, freeing chlorine (Cl) atoms. This process produces a complex set of reactions that breaks up ozone molecules ($O_3$) and leaves oxygen gas molecules ($O_2$) in their place. The effect is severe, for a single chlorine atom decomposes more than 100,000 ozone molecules. The long residence time of chlorine atoms in the ozone layer (40 to 100 years) is likely to produce long-term consequences through the next century from the chlorine already in place. By 1991, 18 million metric tons (19.8 million tons) of CFCs

had been sold worldwide and subsequently released into the atmosphere.

Of many identified reactions, the following two simplified equations demonstrate chlorine monoxide–produced (ClO) and chlorine-produced ozone losses.

Ozone depletion by ClO and by Cl:
$$ClO + O_3 \rightarrow Cl + 2\,O_2$$
$$Cl + O_3 \rightarrow ClO + O_2$$

Note that chlorine monoxide (ClO) reacts with ozone ($O_3$) to produce a free chlorine atom (Cl) and two oxygen gas molecules ($O_2$). The free chlorine atom then reacts with an ozone molecule ($O_3$) to produce more chlorine monoxide (ClO) and oxygen gas ($O_2$). The chlorine monoxide is then available to do it all over again. The net reaction, with the chlorine catalyst removed, is a loss of ozone ($O_3$) and its conversion to oxygen gas molecules. The problem then is that $O_2$, unlike $O_3$, is transparent to ultraviolet radiation.

The hypothesis stated that the increase in Cl and ClO molecules in the ozonosphere was producing a corresponding decline of $O_3$ molecules. Subsequent evidence confirms this hypothesis. Initially, scientists underestimated the risk to the ozonosphere, but today, this model is accepted by all principal research scientists.

## Political Realities—Action Begins

Between 1976 and 1979, Canada, Sweden, Norway, and the state of Oregon banned CFC propellants in aerosols, and a U.S. federal ban began in 1978. However, more than half of U.S. production was exempted, including CFCs used as air-conditioning refrigerants and to make polyurethane foam. CFC sales initially dropped in the late 1970s, but they rose again under a 1981 presidential order that permitted the export of banned products. Sales increased and hit a new peak in 1987 at 1.2 million metric tons (1.32 million tons).

Chemical manufacturers once claimed that no hard evidence existed to prove the ozone-depletion model, and they successfully delayed remedial action for 15 years. With such political delays in response to actual measurements, Rowland expressed his frustration:

What's the use of having developed a science well enough to make predictions, if in the end all we are willing to

do is stand around and wait for them to come true….Unfortunately, this means that if there is a disaster in the making in the stratosphere, we are probably not going to avoid it.[*]

Today, with extensive scientific evidence and verification of losses, even the CFC manufacturers admit that the problem is serious. Remaining critics are outside of the scientific community.

## An International Response and the Future

In 1985, 31 nations agreed to a short-term CFC production freeze and a long-term phaseout. This agreement led to the Montreal Protocol, which by 1995 had been endorsed by 140 countries, responsible for over 90% of CFC production. These countries initially agreed to freeze production at 1986 levels. The Montreal Protocol, and subsequent agreements called Handbooks, currently call for an accelerated complete ban of CFCs by 1996. The Protocol also bans other chemicals, such as bromine, that are damaging to the atmosphere, using different target dates. Bromine sources include halon-charged fire extinguishers and methyl bromide pesticides. These bromine compounds, like CFCs, are inert in the lower atmosphere and reach the stratosphere by slow transport. They are 30 to 100 times more effective than chlorine as catalysts, causing reactions that destroy ozone.

## Ozone Losses over the Poles

How do Northern Hemisphere CFCs become concentrated over the South Pole? Evidently, chlorine freed in the Northern Hemisphere midlatitudes is redistributed by atmospheric winds, concentrating over Antarctica. Persistent cold temperatures over the South Pole, and the presence of thin, icy clouds in the stratosphere, promote development of the ozone hole. Over the North Pole, conditions are more changeable, so the hole is smaller, although growing each year.

The thin clouds that are important catalysts in the release of chlorine for ozone-depleting reactions are *polar stratospheric clouds* (PSCs). PSCs form over the poles during the long, cold winter months as a tight circulation pattern

[*]R. B. Barry, "The annals of chemistry," *The New Yorker*, June 9, 1986, p. 83.

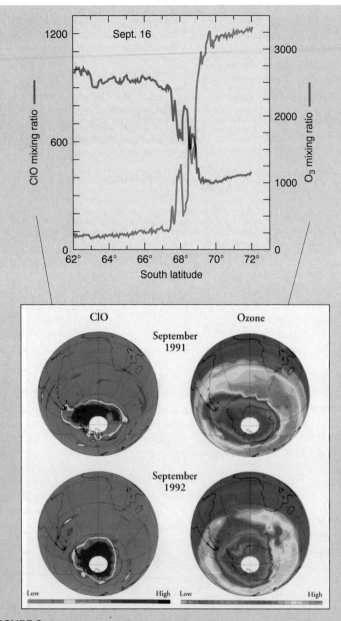

In 1990, scientists at the University of California and at Moss Landing Marine Labs, measured a 6% to 12% reduction in primary production by phytoplankton in Antarctic waters due to increased ultraviolet radiation reaching the surface. The scientists concluded:

> We find that the $O_3$-induced loss to a natural community of phytoplankton in the marginal ice zone is measurable, and the subsequent ecological consequence...of this early spring loss [August to October in the Southern Hemisphere] remains to be determined.[*]

## The Battle Is not Over

In a strange turn of events, members of the U.S. House of Representative proposed legislation to remove the United States from international treaties that protect the ozone layer. At a hearing in 1995, an environmental expert for Du Pont, formerly the largest manufacturer of CFCs and now of ozone-safe substitutes, criticized this political posture. He testified to the committee, "We know about as much about the scientific aspects of this issue as anyone. We've concluded that turning back the [CFC] phaseout at this point would be totally counterproductive."

Despite these counterefforts, CFC sales now are declining as many countries reduce demand and as worldwide industry phases in alternative chemicals. By the end of 1995, production of problem CFCs is estimated to have declined below 400,000 metric tons (440,000 tons)—about one-third that in the record-sales year of 1987. Action on an international scale produced these positive results.

Holes in the polar ozone layer and general depletion across the globe provide an early warning to civilization about this long-term worldwide health problem. It is hoped that, with the international actions taken, the hazard to life in the next century will be reduced, and science will have scored a significant victory.

**FIGURE 2**

**Chemical evidence of ozone damage by humans.**

(a) The negative correlation between ClO and $O_3$ over the Antarctic continent poleward of 68° S latitude. The data were collected from flights during September 1987 at stratospheric altitudes. (b) Chlorine monoxide (ClO) and ozone ($O_3$) concentrations above 20 km (12.5 mi) measured by the *Upper Atmosphere Research Satellite* (*UARS*) from September 1991 and 1992—the beginning of the Antarctic spring. Such evidence confirmed the CFC-related catalyzed chemistry that is devastating the ozonosphere. [(a) Data from NASA. (b) Microwave Limb Sounder, developed and operated by Jet Propulsion Laboratory aboard Goddard Space flight Center's *UARS*.]

forms over the Antarctic continent—the *polar vortex*. Chlorine that is freed from droplets in PSCs triggers the breakdown of the otherwise inert molecules and frees the chlorine for the reactions discussed. Figure 2 graphs the negative correlation between ClO and $O_3$ over the Antarctic continent within the polar vortex. You can see that concentrations of ClO increase toward the pole as $O_3$ levels decline. The ozone hole usually fills by early December with stratospheric ozone from lower latitudes, thus producing ozone thinning in middle and high latitudes.

[*]R. C. Smith, B. B. Prézelin, K. S. Baker, "Ozone depletion: Ultraviolet radiation and phytoplankton biology in Antarctic waters," *Science* 255 (February 12, 1992): 952–58.

## News Report 3

### 1995 Nobel Chemistry Prize for Ozone Depletion Researchers

For the first time, a Nobel Prize has been awarded to atmospheric chemists. The three recipients were credited with discovering that human-produced chemicals were destroying the stratospheric ozone layer. Along the way they pushed the frontier of atmospheric chemistry as a science. In making the award, the Royal Swedish Academy of Sciences said, "By explaining the chemical mechanism that affects the thickness of the ozone layer, the three researchers have contributed to our salvation from a global environmental problem that could have catastrophic consequences. It has been possible to make far-reaching decisions on prohibiting the release of the gases that destroy ozone."

Paul Crutzen is a native of Amsterdam and a researcher at the Max Plank Institute for Chemistry in Mainz, Germany. In 1970 he described the catalytic effects of nitric oxides and nitrogen dioxide in reducing stratospheric ozone. And in 1987 he discovered the role of polar stratospheric clouds and how they enhance the action of chlorine. In contrast to homogenous reactions involving gases only, he described heterogeneous chemical reactions, or which involve particles such as sulfate aerosols, ice crystals, and chlorine atoms—a new field for atmospheric chemistry.

F. Sherwood Rowland, University of California at Irvine, and Mario

Molina, Massachusetts Institute of Technology, told the world that chlorofluorocarbons (CFCs) were migrating into the stratosphere and disrupting Earth's protective shield in a self-sustaining chain reaction.

Part of the prize is obviously for the persistence of these scientists who endured years of criticism from the affected industries. All three scientists pressed for international accords which began with the Montreal Protocols in 1987. Subsequent strengthening of the treaty, as more serious evidence accumulated, led to the total ban of the most dangerous gases by 1996.

---

In the stratosphere, the marked warming with increasing altitude causes the tropopause to act like a lid, essentially preventing whatever is in the cooler (denser) air below from mixing into the warmer (less dense) stratosphere. However, the tropopause is disrupted above the midlatitudes wherever jet streams produce vertical turbulence and interchange between the troposphere and the stratosphere (Chapter 6). Also, hurricanes occasionally inject moisture above this inverted temperature layer at the tropopause, and powerful volcanic eruptions may loft ash and sulfuric acid mists into the stratosphere, as did the Mount Pinatubo eruptions in 1991.

## Variable Atmospheric Components

Within the troposphere, both natural and human-caused variable gases, particles, and other chemicals are part of the atmosphere. The spatial aspects of these variable pollutants are an important applied topic in physical geography.

Air pollution is not a new problem. Romans complained over 2000 years ago about the foul air of their cities. Filling Roman air was the stench of open sewers, smoke from fires, and fumes from ceramic-making kilns and smelters (ore furnaces) that converted ores into metals. In human experience, cities are always the place where the environment's natural ability to process and recycle waste is most taxed. English diarist John Evelyn recorded in 1684 that the air in London was so filled with

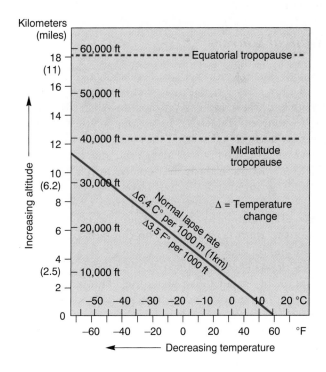

**FIGURE 3-6**

**The temperature profile of the troposphere.**

During daytime, temperatures decreases at a rate known as the *normal lapse rate*. Scientists use a concept called the *standard atmosphere* as an accepted description of air temperature and pressure changes with altitude. The values in this profile are part of the standard atmosphere. The approximate locations of the equatorial and midlatitude tropopauses are noted.

**TABLE 3-2**

| Sources of Natural Variable Gases and Materials | |
| --- | --- |
| **Source** | **Contribution** |
| Volcanoes | Sulfur oxides, particulates |
| Forest fires | Carbon monoxide and dioxide, nitrogen oxides, particulates |
| Plants | Hydrocarbons, pollens |
| Decaying plants | Methane, hydrogen sulfides |
| Soil | Dust and viruses |
| Ocean | Salt spray and particulates |

smoke that "one could hardly see across the streets, and this filling the lungs with its gross particles exceedingly obstructed the breast so as one could hardly breathe."[*]

Air pollution is closely related to our production and consumption of energy, and as human occupation of a region grows longer in duration and complexity, and as our consumption of resources increases, impact on the troposphere intensifies. The loss of clear air in our national parks, which has converted picture-postcard vistas into muted, dull landscapes, points to the difficulty posed by air pollution. Solutions require regional strategies, national strategies, and government coordination because the pollution sources oftentimes are distant from the observed impact.

## Natural Sources

Natural sources produce a greater quantity of pollutants—nitrogen oxides, carbon monoxide, hydrocarbons from plants and trees, and carbon dioxide—than do human-made sources. Table 3-2 lists some of these natural sources and the substances they contribute to the air. However, any attempt to diminish the impact of human-made air pollution through a comparison with natural sources is irrelevant, for we have evolved with and adapted to the natural ingredients in the air. We have *not* evolved in relation to the comparatively recent concentrations of *anthropogenic* (human-caused) contaminants in our metropolitan areas.

A dramatic natural source of pollution was the 1991 eruption of Mount Pinatubo in the Philippines (15° N 120° E). This event injected between 15 and 20 million tons of sulfur dioxide ($SO_2$) into the stratosphere. The spread of these emissions is shown in a sequence of satellite images presented at the beginning of Chapter 6. This event has given scientists an opportunity to study naturally produced sulfuric acid mists in the atmosphere. Evidently, the Antarctic ozone hole of 1992 was slightly worsened at lower altitudes (13–16 km, or 8–10 mi) by the debris from Pinatubo.

[*]W. Wise, *Killer Smog: The World's Worst Air Pollution Disaster* (Chicago: Rand McNally, 1968).

## Natural Factors That Affect Air Pollution

The problems resulting from both natural and human-made atmospheric contaminants are made worse by several important natural factors. Among these are wind, local and regional landscape characteristics, and temperature inversions in the troposphere.

Winds gather and move pollutants from one area to another, sometimes reducing the concentration of pollution in one location while increasing it in another. Such air movements make the atmosphere's condition an international issue. Indeed, prevailing winds transport air pollution from the United States to Canada, causing much complaint and negotiation between the two governments. In Europe, the cross-boundary drift of pollution is a major issue because of the proximity of nations. This issue has led in part to Europe's unification.

Wind can produce dramatic episodes of dust movement (Figure 3-7). (Dust is defined as particles less than 62 µm, or 0.0025 in.). In 1977, a *GOES* weather satellite followed a 1,000,000-km² (400,000-mi²) dust cloud from Colorado and New Mexico eastward. The storm lofted former soil 5000 m (3 mi) into the atmosphere. Traveling on prevailing winds, dust from Africa contributes to the soils of South America, and Texas dust ends up in Europe. Such movement is confirmed by chemical analysis, frequently employed by scientists to track dust to its source area.

An example of a point-source dust storm is shown in Figure 3-7b. Silt and alkaline dust are picked up by high-altitude winds in the Andes of South America and eventually carried out over the Atlantic. Dust storms originating in Australia's desert lands are blown many kilometers to devastate farms and cities (Figure 3-7c).

Local and regional landscapes are another important factor in air pollution. Surrounding mountains and hills can form barriers to air movement or can direct pollutants from one area to another. Some of the worst incidents have resulted when local landscapes have trapped and concentrated air pollution.

Places such as Iceland and Hawaii have their own natural pollution with which to deal. The volcanic activity in Hawaii at Kilauea produces 1000 tons of sulfur dioxide a day. Concentrations are sometimes high enough to produce health concerns, losses to agriculture, and other economic impacts. Negative effects are localized near the source, however, because the fumes disperse through chemical reactions. Hawaiians coined the word *vog* to describe their volcanic smog.

***Temperature Inversion.*** Vertical temperature and atmospheric density distribution in the troposphere also can worsen pollution conditions. A **temperature inversion** occurs when the normal temperature decrease with altitude (normal lapse rate) begins to *increase* at some altitude.

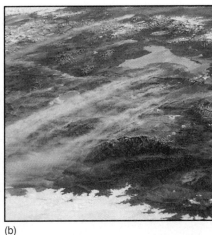

(a)                                    (b)                                    (c)

**FIGURE 3-7**
**Natural variable dust in the atmosphere.**
(a) A dust storm in central Nevada. (b) An orbital view of wind-blown alkali dust rising from high interior-drainage basins in the Andes Mountains of Chile and Argentina and blowing far over the Atlantic Ocean. (c) A wall of dust blocks the sky as it moves over Melbourne, Australia. [(a) Photo by author; (b) Space Shuttle photo from NASA; (c) photo by Bill Bachman/Photo Researchers, Inc.]

This can happen at any point from ground level to several thousand meters. Figure 3-8 compares a normal temperature profile with that of a temperature inversion. The normal profile (Figure 3-8a) permits warmer (less dense) air at the surface to rise, ventilating the valley and moderating surface pollution. But the warm air inversion (Figure 3-8b) prevents the rise of cooler (denser) air beneath, halting the vertical mixing of pollutants with other atmospheric gases. Thus, instead of being carried away, pollutants are trapped under the *inversion layer* (Figure 3-8 photo).

Inversions most often result from certain weather conditions, such as when the air near the ground is radiatively cooled on clear nights, or from topographic situations that produce cold-air drainage into valleys. In addition, the air above snow-covered surfaces or beneath subsiding air in a high-pressure system may cause a temperature inversion. As an example, in winter months in midwestern and eastern North America, high-pressure areas created by subsiding cold air masses produce inversion conditions that trap air pollution. In the western United States, summer subtropical high-pressure systems also cause inversions that trap air pollution and produce air stagnation. (The concepts of high- and low-pressure systems are discussed in Chapter 6.)

### Anthropogenic Pollution

Anthropogenic, or human-caused, air pollution remains most prevalent in urbanized regions. The human population is increasingly urbanized and susceptible to air pollution. Approximately 1.3 billion people live in metropolitan regions with unacceptable levels of suspended particles in their air; 700 million live in urban settings that exceed World Health Organization guidelines for sulfur dioxide levels. In 1992 the United Nations issued a report on major metropolitan pollution. Every city exceeded recommended levels of at least one pollutant, 14 cities were in excess for two, and 7 cities failed for three pollutants studied.

Table 3-3 lists the names, chemical symbols, and principal sources of variable anthropogenic components in the air. The first seven pollutants in the table result from combustion of fossil fuels in transportation (specifically automobiles) and at stationary sources such as power plants and factories. Overall, automobiles contribute about 60% of U.S. and 50% of Canadian human-caused air pollution. In the United States, transportation produces 70% of the carbon monoxide, 60% of the hydrocarbons, and 50% of the nitrogen oxides. Environment Canada, an agency similar to the U.S. Environmental Protection Agency, reports that, with Canada's smaller transportation fleet, Canadian automobile emissions contribute 45% of the carbon monoxide, 24% of the hydrocarbons, and 20% of the nitrogen oxides.

The apparent manageability of this transportation-pollution problem is interesting. In one California study, only 7% of the vehicles contributed half of the carbon monoxide and only 10% contributed half of the hydrocarbon pollution. These "gross polluting" vehicles are not old cars—a common misconception—but include new cars. In random highway checks, 41% of vehicles had pollution equipment that was deliberately tampered with and 25% had defective or missing emission controls. Resolving sources of air pollution from the transportation sector does not pose many mysteries.

Stationary sources, such as electric power plants and industrial plants that use fossil fuels, contribute the most sulfur oxides and particulates. For this reason, concentrations are focused in the Northern Hemisphere and the industrial, developed countries.

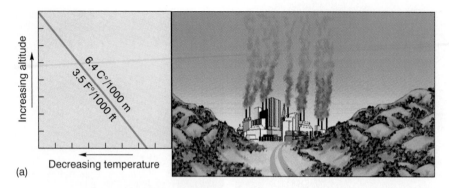

(a)

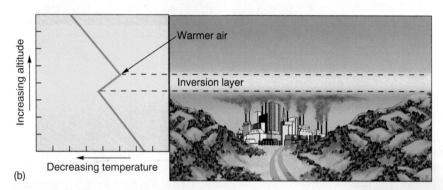

(b)

**FIGURE 3-8**
**Normal and inverted temperature profiles.**
A comparison of a normal temperature profile in the atmosphere (a) with a temperature inversion in the lower atmosphere (b). Note how the warmer air layer prevents mixing of the denser (cooler) air below the inversion, thereby trapping pollution. An inversion layer is visible in the morning hours over a valley as shown in the photograph. [Inset photo by author.]

The last three gases shown in Table 3-3 are discussed elsewhere in this text: water vapor is examined with water and weather (Chapters 7 and 8); carbon dioxide and methane with greenhouse gases and climate (Chapters 4, 5, and 10).

Figure 3-9 identifies major human-caused pollutants and their proportional sources in the United States. Some of these pollutants are described in the following sections.

***Carbon Monoxide Pollution.*** Odorless, colorless, and tasteless, **carbon monoxide (CO)** is a combination of one atom of carbon and one of oxygen. Carbon monoxide is produced by incomplete combustion (burning with limited oxygen) of fuels or other carbon-containing substances. A log decaying in the woods produces carbon monoxide, as does a forest fire or other organic decomposition. Natural sources produce up to 90% of existing carbon monoxide, whereas anthropogenic sources, principally transportation, produce the other 10%. A dangerous point source of carbon monoxide for individuals is from primary and secondary tobacco smoke.

In the tropics of south-central Africa, an interesting source of carbon monoxide is the widespread burning of biomass (trees, grasses, brush) during the dry season from August to October. This carbon monoxide spreads throughout the Southern Hemisphere—Africa, Australia, Antarctica, and the south Atlantic. Between 1988 and 1993 global carbon monoxide levels have shown a slight decline, owing to a small reduction in biomass burning in the

Southern Hemisphere. Generally, though, the carbon monoxide from human sources is concentrated in urban areas, where it directly affects human health.

The toxicity of carbon monoxide is due to its affinity for blood hemoglobin, which is the oxygen-carrying pigment in red blood cells. In the presence of carbon monoxide, the oxygen is displaced and the blood becomes deoxygenated. Elevated carbon monoxide levels (50–100 ppm) in cities are dangerous because of the resultant reduction of oxygen in the blood and the health symptoms experienced, including headaches, some vision and judgment loss, and impaired performance of simple tasks.

Carbon monoxide pollution did not become a problem until many fossil fuel-burning cars, trucks, and buses became concentrated in urban areas. Normal levels range from 10 to 30 parts per million (ppm) in metropolitan areas, but freeways and parking garages can reach 50

**TABLE 3-3**

| Anthropogenic Gases and Materials in the Lower Atmosphere | | |
|---|---|---|
| *Name* | *Symbol* | *Source* |
| Carbon monoxide | CO | Incomplete combustion of fuels |
| Nitrogen oxides | $NO_x(NO, NO_2)$ | High temperature/pressure combustion |
| Hydrocarbons | HC | Incomplete combustion of fuels |
| Ozone | $O_3$ | Photochemical reactions |
| Peroxyacetyl nitrates | PAN | Photochemical reactions |
| Sulfur oxides | $SO_x(SO_2, SO_3)$ | Combustion of sulfur-containing fuels |
| Particulates | — | Dust, dirt, soot, salt, metals, organics |
| Exotics | — | Fission products, other toxics |
| Carbon dioxide | $CO_2$ | Complete combustion |
| Water vapor | $H_2O$ vapor | Combustion processes, steam |
| Methane | $CH_4$ | Organic processes |

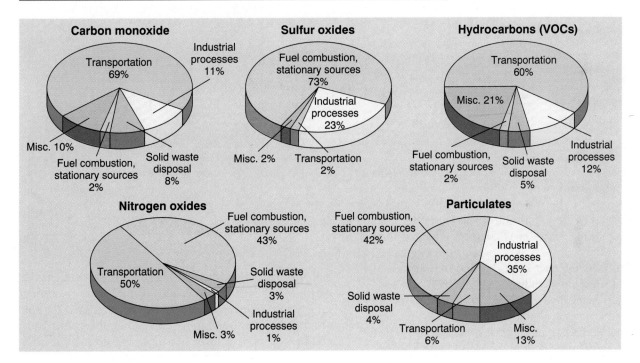

**FIGURE 3-9**

**Human-caused air pollution and sources.**

Major types of human-caused air pollution and their sources in the United States. These proportions are typical of developed, industrialized countries. (HC is also known as volatile organic compounds, or VOCs.)

ppm CO, with an additional 30 ppm possible under a temperature inversion. On several occasions, street-level concentrations in cities have reached levels high enough to set off carbon monoxide detectors.

The death in September 1994 of tennis great Vitas Gerulaitis brought widespread attention to the toxic potential of carbon monoxide. He was staying in a guest cottage on Long Island when a gas-heating system malfunctioned and filled the house with this dangerous gas. Carbon monoxide poisoning kills about 300 people a year in North America; 6000 more are injured. Media stories are common of death from carbon monoxide poisoning in an enclosed space with a running automobile engine or in a tent with an unvented heater.

***Photochemical Smog Pollution.*** Photochemical smog is another type of pollution that was not generally experienced in the past but developed with the advent of the automobile. Today it is the major component of anthropogenic air pollution. **Photochemical smog** results from the interaction of sunlight and the combustion products in automobile exhaust (nitrogen oxides and hydrocarbons). Although the term smog—a combination of the words smoke and fog—is a misnomer, it is generally used to

(a)

(b)

**FIGURE 3-10**

**Contrasts in the air over Mexico City.**

A rare day of clear air (a) in contrast to a more typical day of photochemical smog pollution (b) in the skies over Mexico City. [Photo by Larry Reider/PIX/Sipa Press.]

describe this phenomenon. Smog is responsible for the hazy sky and reduced sunlight in many of our cities, as shown occurring in Figure 3-8 photo.

Mexico City is notorious for poor air quality as its 22 million inhabitants work, commute, and live in the world's second largest metropolitan region. Conditions are worsened by frequent subtropical high-pressure systems (descending, stable air) that act as effective air traps over the Valley of Mexico, in which the city lies. The contrast between a rare, clear day and frequent polluted days in Mexico city is dramatic (Figure 3-10).

Mexico enacted new laws in 1990 to reduce unhealthy photochemical smog conditions: more public rapid transportation, controls on automobiles, limitations on factory operations, and more pollution-absorbing park space and trees. The government has spent nearly $2 billion on pollution controls. An interesting source of hydrocarbons for smog production was recently discovered: leaking liquid petroleum gas (LPG) pipes throughout the city, which carry fuel for cooking and heating. Estimates place 25% of the air pollution problem on this one controllable source.

The connection between automobile exhaust and smog was not determined until 1953, long after society had established its dependence upon individualized transportation. Despite this discovery, widespread mass transit has declined, the railroads have dwindled, and the polluting individual automobile remains America's preferred transportation.

Figure 3-11 summarizes how car exhaust is converted into major air pollutants—ozone, **peroxyacetyl nitrates (PAN)**, and nitric acid. PAN produces no known health effect in humans, but it is particularly damaging to plants, including both agricultural crops and forests. Damage in California is estimated to exceed $1 billion a year in the farming sector.

**Nitrogen dioxide** ($NO_2$) is a reddish brown gas that damages and inflames human respiratory systems, destroys lung tissue, and damages plants. Concentrations of 3 ppm are dangerous enough to require alerting parents to keep children indoors; a level of 5 ppm is extremely serious. Worldwide, the problem with nitrogen dioxide production is its concentration in metropolitan regions. North American urban areas may have from 10 to 100 times higher nitrogen dioxide concentrations than nonurban areas.

Nitrogen dioxide interacts with water vapor to form nitric acid ($HNO_3$), a contributor to acid deposition by precipitation, the subject of Focus Study 3-2. Emission levels of nitrogen oxides and their deposition by rain and snow during 1991 are shown in Figure 3-12.

***Ozone Pollution.*** Highly reactive *ozone* causes paint to oxidize, painted surfaces to peel, elastic and rubber to dry and crack, paper to dry and yellow, and plants to sustain damage—at levels of only 0.01–0.09 ppm. Exposure to 0.1 ppm for a few hours a day for a span of several days to several weeks reduces yields of a wide variety of agricultural crops by up to 50%. This damage to plants is proving significant. An example is the loblolly pine, an important commercial tree in the southeastern states, which has experienced considerable damage.

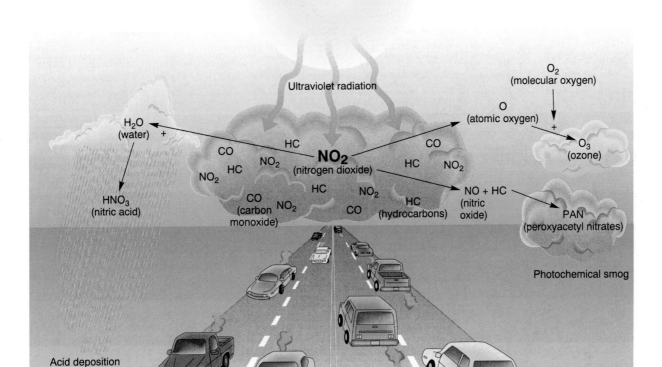

**FIGURE 3-11**

**Photochemical reactions.**

Photochemical reactions are produced through the interaction of automobile exhaust (NO$_2$, HC, CO) and ultraviolet radiation in sunlight. The high temperatures in modern automobile engines cause reactions that produce nitrogen dioxide (NO$_2$). This nitrogen dioxide, derived from automobiles and to a lesser extent from power plants, is highly reactive with ultraviolet light. The reaction liberates atomic oxygen (O) and a nitric oxide (NO) molecule from the NO$_2$. The free oxygen atom combines with an oxygen molecule (O$_2$) to form the oxidant ozone (O$_3$); this same gas that is so beneficial in the stratosphere is an air-pollution hazard at Earth's surface. In addition, the nitric oxide (NO) molecule reacts with hydrocarbons (HC) to produce a family of chemicals called peroxyacetyl nitrates (PAN).

At 0.3 ppm ozone, irritation of the eyes, nose, and throat becomes noticeable. As concentration increases to 1.0–3.0 ppm, extreme fatigue and poor coordination can appear in humans. One in four children in U.S. cities are at risk of developing health problems from ozone pollution. That ratio is significant; it means that over 12 million children are vulnerable in those cities with the worst polluted air (Los Angeles, New York City, Atlanta, Houston, and Detroit).

The hydrocarbons (HC), also known as *volatile organic compounds* (VOCs), shown in Figure 3-11 are important factors in the ozone-forming potential of these reactions. States such as California base their standards for control of ozone pollution on hydrocarbon emission controls—a scientifically accurate emphasis.

***Industrial Smog and Sulfur Oxides.*** Over the past 300 years, except in some developing countries, coal has slowly replaced wood as the basic fuel used by society. The Industrial Revolution required high-grade energy to run machines. The changes involved conversion from *animate* energy (energy from animal sources, such as animal-powered farm equipment) to *inanimate* energy (energy from nonliving sources, such as coal, steam, and water). The air pollution associated with coal-burning industries is known as **industrial smog** (Figure 3-13). The term *smog* was coined by a London physician at the turn of this century to describe the combination of fog and smoke containing sulfur gases (sulfur is an impurity in fossil fuels).

## Focus Study 3-2

## Acid Deposition: A Blight on the Landscape

Acid deposition is a major environmental problem in some areas of the United States, Canada, Europe, and Asia. Such deposition is most familiar as "acid rain," but it also occurs as "acid snow" and in a dry form as dust or aerosols. (Recall that aerosols are tiny liquid droplets or solid particles.) In addition, winds can carry the acid-producing chemicals many kilometers from their sources before they settle on the landscape, where they enter streams and lakes as runoff and groundwater flows.

Acid deposition is causally linked to serious problems: declining fish populations and fish kills in the northeastern United States, southeastern Canada, Sweden, and Norway; widespread forest damage in these same places and Germany; widespread changes in soil chemistry; and damage to buildings, sculptures, and historic artifacts. Despite scientific agreement as to the problem, corrective action has been delayed by its complexity, which the U.S. General Accounting Office calls a "combination of meteorological, chemical, and biological phenomena."

The acidity of precipitation is measured on the pH scale, which expresses the relative abundance of free hydrogen ions (H+) in a solution. Free hydrogen ions in a solution are what make an acid corrosive, for they easily combine with other ions. The pH scale is logarithmic: Each whole number represents a 10-fold change. A pH of 7.0 is neutral (neither acidic nor basic). Values less than 7.0 are increasingly *acidic*, and values greater than 7.0 are increasingly *basic*, or *alkaline*. (A pH scale for soil acidity and alkalinity is graphically portrayed in Chapter 18.)

Natural precipitation dissolves carbon dioxide from the atmosphere to form carbonic acid. This process releases hydrogen ions and produces an average pH reading for precipitation of 5.65. The normal range for precipitation is 5.3–6.0. Thus, normal precipitation is always slightly acidic.

Some anthropogenic gases are converted to acids in the atmosphere. They then are removed by wet and dry deposition processes. Specifically, nitrogen and sulfur oxides ($NO_x$ and $SO_x$) released in the combustion of fossil fuels can produce nitric acid ($HNO_3$) and sulfuric acid ($H_2SO_4$) in the atmosphere. (Figures 3-12 and 3-14 depict the patterns of $NO_x$ and $SO_x$ emissions and wet deposition in the United States and Canada.)

### Acid Precipitation Damage

Precipitation as acidic as pH 2.0 has fallen in the eastern United States, Scandinavia, and Europe. By comparison, vinegar and lemon juice register slightly less than 3.0. Aquatic plant and animal life perishes when lakes drop below pH 4.8.

More than 50,000 lakes and some 100,000 km (62,000 mi) of streams in the United States and Canada are at a pH level below normal (i.e., below 5.3), with several hundred lakes incapable of supporting any aquatic life. Acid deposition causes the release of aluminum and magnesium from clay minerals in the soil, and both of these are harmful to fish and plant communities. In some areas, acidic soils have killed snail populations, reducing the quantity of snail shells in the soils. This loss of calcium (in their shells) has affected the diets of birds, so that their eggs, in turn, have porous, weak shells, thus diminishing the number of successful hatchlings.

Also, relatively harmless mercury deposits in lake-bottom sediments are converted by acidified lake waters into highly toxic *methylmercury*, which is deadly to aquatic life. Local health advisories in two provinces and 22 U.S. states are regularly issued to warn those who fish of the methylmercury problem. Mercury atoms rapidly bond with carbon and move through biological systems as an *organometallic compound*.

Damage to forests results from the rearrangement of soil nutrients, the death of soil microorganisms, and an aluminum-induced calcium deficiency that is currently under investigation. Regional-scale decline in forest cover is significant. The most advanced impact is being seen in forests in Europe, principally because of its long history of burning coal and the density of industrial activity. In Germany, up to 50% of the forests are dead or damaged; in Switzerland 30% are afflicted.

The percentage of forest loss is highest in the nations of eastern Europe. To illustrate, Earth Satellite Corporation analyzed a small area of Poland that is heavily industrialized. The region produces and burns about 98% of Poland's coal. Two *Landsat* images allow you to do a GIS-type comparison of changes between 1981 and 1989 (Figure 1). Over 50% of Poland's forests are dead or dying. A high correlation exists between these devastated areas and acid rain-producing industrial activity.

In the United States, trees at higher elevations in the Appalachians are being injured by acid-laden cloud cover. In New England, some stands of spruce are as much as 75% affected, as evidenced through analysis of tree growth rings,

**Sulfur dioxide ($SO_2$)**, which forms during combustion, is colorless but pungent (has an irritating smell). At low concentrations between 0.1 and 1 ppm, sulfur dioxide-polluted air impairs breathing; at 0.3 ppm the air takes on a definite metallic taste. Human respiratory systems can become irritated, especially in individuals with asthma, chronic bronchitis, and emphysema.

Once in the atmosphere, $SO_2$ reacts with oxygen (O) to form sulfur trioxide ($SO_3$), which is highly reactive and, in the presence of water or water vapor, forms sulfuric acid ($H_2SO_4$). Sulfuric acid can form even in moderately polluted air, at normal temperatures. Sulfur dioxide-laden atmosphere is dangerous to health, corrodes metals, and deteriorates stone building materials at accelerated rates. Sulfuric acid deposition has increased in severity since it was first described in the 1970s and is a growing blight in the biosphere. Focus Study 3-2 discusses this vital atmospheric issue.

### FIGURE 1

**The blight of acid deposition.**

The harm done to forests and crops by acid deposition is well established, especially in Europe. A heavily industrialized area of Poland was scanned by a *Landsat* satellite in 1981 and again in 1989 for a GIS comparison of the forest cover. False coloration helps in the analysis: Brighter colors on the 1981 image (left) depict healthier vegetation, whereas the predominately blue patterns on the 1989 image denote over 50% of the forests in decline and death. [Images courtesy of Earth Satellite Corporation, Rockville, MD.]

which become narrower in adverse growing years. Another possible indicator of forest damage is the reduction by almost half of the annual production of U.S. and Canadian maple sugar.

### Cause and Effect and International Solutions

The National Academy of Sciences (NAS) and the National Research Council have identified causal relationships between fossil fuel combustion and acidic deposition, citing evidence that is "overwhelming." Government estimates of damage in the United States, Canada,

and Europe exceed $50 billion annually. Despite this consensus and annual losses, politics and special interests have halted preventive measures. Because wind and weather patterns are international, efforts at reducing acidic deposition also must be international in scope. The NAS stated that a reduction of 50% in combustion emissions would reduce acid deposition by 50%. However, this relation has turned out not to be true, and recent evidence is troublesome. The modest decline of sulfur dioxide between 1980 and 1991 as a result of the Clean Air Act has so far not

resulted in a reduction of acidity (increases in pH factor) in vulnerable soils for as yet undetermined reasons.

*Climatic Perspectives*, a publication from Environment Canada, reports acid rain data for various sites, including pH measurement and "air path to site." Not until March 1991 did the United States finally agree to a landmark agreement with Canada to lower sulfur dioxide and nitrogen oxide—a first. At best, acid deposition is an unfolding international political issue of global spatial significance for which science is providing strong incentives for action.

In the United States, coal-burning electric utilities and steel manufacturing are the main sources of sulfur dioxide. And, because of the prevailing movement of air masses, they are the main sources of sulfur dioxide in Canada, too. As much as 70% of Canadian sulfur dioxide is initiated within the United States.

Sulfur dioxide in the atmosphere reacts with other chemicals under clear-sky conditions and with cloud droplets on cloudy days to produce **sulfate aerosols**, tiny particles about 0.1 to 1 μm in diameter. These aerosols are associated with industry and electrical power generation, which produce higher concentrations in specific regions of the Northern Hemisphere. Figure 3-14 portrays representative production levels of sulfur dioxide in the 1980s and rain and snow deposition of sulfur oxides during 1991.

In 1988 alone, 20,000 industrial facilities poured 1.1 billion kilograms (2.4 billion pounds) of sulfur chemicals into the atmosphere in the United States. Concentrations have declined over the past decade as a result of Clean Air Act controls.

Certainly, society cannot simply halt two centuries of industrialization to halt air pollution; the resulting economic chaos would be devastating. But neither can it permit pollution production to continue unabated, for catastrophic environmental and human health damage inevitably will result. People have been on Earth for only a brief fragment of Earth's history, yet our activities can be a major influence on the future of the atmosphere. We are now contributing significantly to the creation of the **anthropogenic atmosphere**, a tentative label for Earth's next (fifth) atmosphere. The urban air we breathe today may be just a preview.

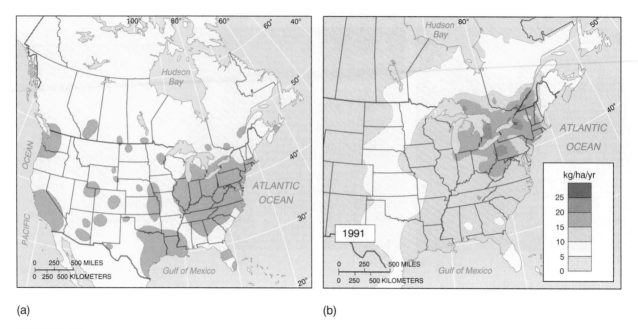

(a)

(b)

**FIGURE 3-12**

**Pollution by nitrogen oxides.**

Spatial portrayal of (a) representative emissions of nitrogen oxides in the 1980s and (b) annual wet deposition (rain and snow) of nitrogen oxides (principally $NO_3$) on the landscape for the year 1991. [Data courtesy of National Atmospheric Chemistry Data Base (NAtChem), Air Quality Measurements and Analysis Research Division, Environment Canada and the U.S. Environment Protection Agency, National Atmospheric Deposition Program/National trends Network.]

**FIGURE 3-13**

**Typical industrial smog.**

Pollution generated by industry differs from that produced by transportation. Industrial pollution has high concentrations of sulfur oxides, particulates, and carbon dioxide. [Photo by author.]

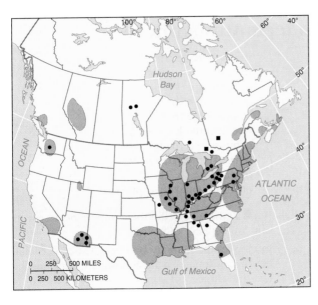

(a)

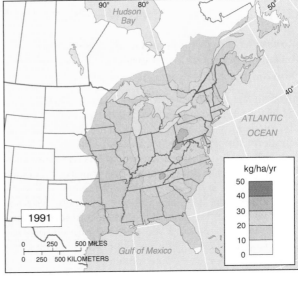

(b)

**FIGURE 3-14**

**Pollution by sulfur oxides.**

Spatial portrayal of (a) representative emissions of sulfur dioxide in the 1980s and (b) wet deposition (rain and snow) of sulfur oxides (principally $SO_4$) on the landscape in 1991. In (a), the dots locate sources producing 100–500 metric kilotons (110,000–550,000 tons) per year; the squares locate sources producing 500 metric kilotons (500,000 tons) or more per year. [Data courtesy of National Atmospheric Chemistry Data Base (NAtChem), Air Quality Measurements and Analysis Research Division, Environment Canada and the U.S. Environment Protection Agency, National Atmospheric Deposition Program/National trends Network.]

# News Report 4

## Air Pollution Abatement—Costs vs. Benefits

Are pollution controls too expensive? Much political debate swirls around this question. To be justified, abatement costs must not exceed the financial benefits derived from reducing pollution damage. Cost studies abound: the economics of ozone damage to plants (by the U.S. National Crop Loss Assessment Network, NCLAN); the cost of photochemical smog damage to human health and property, mortality rate, hospital usage, and indirect health consequences (various studies); and an 11-year study that found significant deterioration of lung function after chronic exposure to oxidants and air pollution (UCLA).

In Utah County, Utah researchers correlated the smallest particulates (10 μm or less, designated $PM_{10}$) with increased rates of hospitalization for bronchitis, asthma, pneumonia, and pleurisy (all especially in children), and therefore greatly increased medical costs. Similar studies of children affected by related illnesses in seven other cities revealed sickness rates twice as high for the city having the dirtiest air than for the city having the cleanest air. Nationally, a stud of major cities disclosed a 26% greater risk of premature death due to respirable particulate pollution as compared with nonpolluted air—further driving up medical costs and related expenses.

It is clear that the benefits of air-pollution-abatement solutions far outweigh their costs. The South Coast Air Quality Management District, which oversees the Los Angeles Basin, estimated in 1991 that full implementation of their clean air plan would cost about $6.1 billion, but it would generate at least $9.4 billion in benefits from ozone reduction alone. In transportation, repairs to emission control systems on the worst 20% of vehicles would cost one-fifth the scrappage fees to junk them. Also, these repairs would cut HC emissions by 50% and carbon monoxide by 61%—all at a savings. Thus, taking action is cheaper than allowing the situation to further worsen. Similar money savings occur in most cost-benefit studies of air, water, and toxic pollution abatement.

For an analysis of the economic benefits of the Clean Air Act, see "The Clean Air Act Brings a Windfall" in Chapter 21. Net benefits were found to exceed $6.4 trillion.

# Summary and Review—Earth's Modern Atmosphere

✔ *Construct* **a general model of the atmosphere based on composition, temperature, and function** and *diagram* **this model in a simple sketch.**

Each breath we take is a reminder of Earth's physical and natural history, for the atmosphere is a product of the planet and the inhalations and exhalations of all life that has existed. Our modern atmosphere is a gaseous mixture so evenly mixed it behaves as if it were a single gas. It is naturally odorless, colorless, tasteless, and formless. The principal substance of this atmosphere is *air*—the medium of life.

Our atmosphere is unique among the planets. By *composition*, we divide the atmosphere into the **heterosphere** and the *homosphere*. Above 480 km (300 mi) altitude, the atmosphere is rarefied (nearly a vacuum) and is called the **exosphere**, which means "outer sphere."

Using *temperature* as a criterion, we identify the **thermosphere**. Its upper limit is called the **thermopause** at approximately 480 km altitude. **Kinetic energy**, the energy of motion, is the vibrational energy that we measure and call *temperature*. However, the actual heat produced in the thermosphere is very small. The density of the molecules is so low that little actual **heat**, the flow of kinetic energy from one body to another because of a temperature difference between them, is produced. Nearer Earth's surface the greater number of molecules in the denser atmosphere transmit their kinetic energy as **sensible heat**, meaning that we can feel it. We distinguish a region in the heterosphere by *function*: The **ionosphere** absorbs cosmic rays, gamma rays, X-rays, and shorter wavelengths of ultraviolet radiation and converts them into kinetic energy.

Between the heterosphere and Earth's surface is the other compositional shell of the atmosphere, the **homosphere**, extending to 80 km above Earth's surface. It includes the **mesosphere, stratosphere**, and **troposphere**. Within the mesosphere, cosmic or meteoric dust acts as nuclei around which fine ice crystals form to produce rare and unusual night clouds called **noctilucent clouds**. By *function*, a second region within the stratosphere is the **ozonosphere** (or **ozone layer**), which absorbs life-threatening ultraviolet radiation, subsequently raising the temperature of the stratosphere.

The normal temperature profile within the troposphere during the daytime decreases rapidly with increasing altitude at an average of 6.4 C° per kilometer (3.5 F° per 1000 ft), a rate known as the **normal lapse rate**. The top of the troposphere is wherever a temperature of –57°C (–70°F) is recorded, a transition known as the **tropopause**. The actual lapse rate at any particular time and place may deviate considerably because of local weather conditions and is called the **environmental lapse rate**.

heterosphere (p. 67)
exosphere (p. 67)
thermosphere (p. 67)
thermopause (p. 67)
kinetic energy (p. 67)
heat (p. 67)
sensible heat (p. 67)
ionosphere (p. 67)
homosphere (p. 69)
mesosphere (p. 70)
stratosphere (p. 70)
troposphere (p. 71)
noctilucent clouds (p. 70)
ozonosphere, ozone layer (p. 71)
normal lapse rate (p. 71)
tropopause (p. 71)
environmental lapse rate (p. 71)

1. What is air? Where did the components in Earth's present atmosphere originate?

2. In view of the analogy by Lewis Thomas, characterize the various functions the atmosphere performs that protect the surface environment.

3. What three distinct criteria are employed in dividing the atmosphere for study?

4. Describe the overall temperature profile of the atmosphere, and list the four layers defined by temperature.

5. Describe the two divisions of the atmosphere on the basis of composition.

6. What are the two primary functional layers of the atmosphere and what does each do?

---

✔ *List* **the stable components of the modern atmosphere and their relative percentage contributions by volume and** *describe* **each.**

Even though the atmosphere's density decreases with increasing altitude in the homosphere, the blend (proportion) of gases is nearly uniform. This stable mixture of gases has evolved slowly.

The homosphere is a vast reservoir of relatively inert *nitrogen*, originating principally from volcanic sources and from bacterial action in the soil; *oxygen*, a by-product of photosynthesis; *argon*, constituting about 1% of the homosphere, is completely inert; *carbon dioxide*, a natural by-product of life processes and fuel combustion.

7. Name the four most prevalent stable gases in the homosphere. Where did each originate? Is the prevalence of any of these changing at this time?

---

✔ *Describe* **conditions within the stratosphere; specifically,** *review* **the function and status of the ozonosphere (ozone layer).**

The overall reduction of the stratospheric ozonosphere, or ozone layer, during the past several decades represents a hazard for society and many natural systems and is caused by chemicals introduced into the atmosphere by humans. Since World War II quantities of human-made **chlorofluorocarbons (CFCs)** and bromine-containing compounds have made their way into the stratosphere. The increased ultraviolet light at those altitudes breaks down these stable chemical compounds, thus freeing chlorine and bromine atoms. These atoms act as catalysts in reactions that destroy ozone molecules.

**chlorofluorocarbons (CFCs) (p. 73)**

8. Why is stratospheric ozone ($O_3$) so important? Describe the effects created by increases in ultraviolet light reaching the surface.

9. Summarize the ozone predicament and present trends and any treaties that intend to protect the ozone layer.

10. Evaluate Crutzen, Rowland, and Molina's use of the scientific method in investigating stratospheric ozone depletion.

✔ *Distinguish* **between natural and anthropogenic variable gases and materials in the lower atmosphere.**

Within the troposphere, both natural and human-caused variable gases, particles, and other chemicals are part of the atmosphere. We coevolved with natural "pollution" and thus are adapted to it. But we are not adapted to cope with our own anthropogenic pollution. It constitutes a major health threat, particularly where people are concentrated in cities.

Vertical temperature and atmospheric density distribution in the troposphere also can worsen pollution conditions. A **temperature inversion** occurs when the normal temperature decrease with altitude (normal lapse rate) reverses. In other words, temperature begins to *increase* at some altitude.

**temperature inversion (p. 76)**

11. Why are anthropogenic gases more significant to human health than are those produced from natural sources?

12. In what ways does a temperature inversion worsen an air pollution episode? Why?

✔ *Describe* **the sources and effects of carbon monoxide, nitrogen dioxide, and sulfur dioxide, and** *construct* **a simple equation that illustrates photochemical reactions that produce ozone, peroxyacetyl nitrates, nitric acid, and sulfuric acid.**

Odorless, colorless, and tasteless, **carbon monoxide (CO)** is produced by incomplete combustion (burning with limited oxygen) of fuels or other carbon-containing substances. Transportation is the major human-caused source for carbon monoxide. The toxicity of carbon monoxide is due to its affinity for blood hemoglobin, which is the oxygen-carrying pigment in red blood cells. In the presence of carbon monoxide, the oxygen is displaced and the blood becomes deoxygenated.

**Photochemical smog** results from the interaction of sunlight and the products of automobile exhaust, the single largest contributor of pollution that produces smog. Car exhaust, containing *nitrogen dioxide* and *hydrocarbons*, in the presence of ultraviolet light in sunlight converts into major air pollutants—*ozone, peroxyacetyl nitrates* (*PAN*), and *nitric acid*. **Nitrogen dioxide** ($NO_2$) damages and inflames human respiratory systems, destroys lung tissue, and damages plants. The principal photochemical by-products include *ozone* ($O_3$), which causes negative health effects, oxidizes surfaces, and kills or damages plants; and **peroxyacetyl nitrates** (**PAN**) which produces no known health effect in humans but is particularly damaging to plants, including both agricultural crops and forests. Nitric oxides participate in reactions that form nitric acid ($HNO_3$) in the atmosphere, forming acid rain and snow deposition.

The distribution of human-produced **sulfur dioxide** over North America, Europe, and Asia is related to transportation and electrical production. Such characteristic pollution is termed **industrial smog**. Sulfur dioxide in the atmosphere reacts to produce **sulfate aerosols**, that produce sulfuric acid ($H_2SO_4$) deposition and affect the Earth energy budget by scattering and reflecting solar energy. Energy conservation and efficiency and cleaning of emissions are essential strategies for abating air pollution. Earth's next atmosphere most accurately may be described as the **anthropogenic atmosphere** (human-influenced atmosphere).

> **carbon monoxide (CO) (p. 78)**
> **photochemical smog (p. 79)**
> **nitrogen dioxide (p. 80)**
> **peroxyacetyl nitrates (PAN) (p. 80)**
> **sulfur dioxide (p. 82)**
> **industrial smog (p. 81)**
> **sulfate aerosols (p. 83)**
> **anthropogenic atmosphere (p. 83)**

13. What is the difference between industrial smog and photochemical smog?

14. Describe the relationship between automobiles and the production of ozone and PAN in city air. What are the principal negative impacts of these gases?

15. How are sulfur impurities in fossil fuels related to the formation of acid in the atmosphere and acid deposition on the land?

16. In your opinion, what are the solutions to the increasing problems in the anthropogenic atmosphere?

---

 **NetWork**

The *Geosystems Home Page* provides on-line resources for this chapter on the World Wide Web. You will find review exercises, specific updates for items in the chapter, suggested readings, and links to interesting related pathways on the Internet (click on the Table of Contents link and select this chapter). *Geosystems* is at: **http://www.prenhall.com/geosystm**

# 4

# Atmosphere and Surface Energy Balances

**Energy Balance in the Troposphere**

**Energy Balance at Earth's Surface**

**Summary and Review**

## Key Learning Concepts

After reading the chapter, you should be able to:

- *Identify* the pathways of solar energy through the troposphere to Earth's surface: transmission, refraction, albedo (reflectivity), scattering, diffuse radiation, conduction, convection, and advection.

- *Analyze* the effect of clouds and air pollution on solar radiation received at ground level and *describe* what happens to insolation when clouds are in the atmosphere.

- *Review* the energy pathways in the Earth-atmosphere system, the greenhouse effect, and the patterns of global net radiation.

- *Plot* the daily radiation curves for Earth's surface and *label* the key aspects of incoming radiation, air temperature, and the daily temperature lag.

- *Portray* typical urban heat island conditions and *contrast* the microclimatology of urban areas with that of surrounding rural environments.

*Saguaros, spring storm, and sunset in Saguaro National Monument, in the Sonoran desert of Arizona.* [Photo by George Ranalli/Photo Researchers, Inc.]

arth's biosphere pulses with flows of solar energy that affect our activities by producing the seasons, climate, and daily weather. We examined the shifting seasonal rhythms of these energy patterns in Chapter 2. This chapter follows the cascade of solar energy through the troposphere to Earth's surface.

The *input* of insolation is countered by the *outputs* of reflected light and emitted infrared energy from the atmosphere and surface environment. Together, this input and output determine the net energy available to perform work. We examine surface energy budgets and analyze how net radiation is expended. Energy and moisture exchanges between Earth's surface and the atmosphere are essential elements of climate.

The climate of our urban areas differs measurably from that of surrounding rural areas. The chapter concludes with a look at the unique energy and moisture environment in our cities. This chapter features a Focus Study on solar energy, a renewable resource of great potential, as an application of surface energy.

# Energy Balance in the Troposphere

Earth's atmosphere and surface are heated by solar energy, which is unevenly distributed by latitude and which fluctuates seasonally. Our budget of atmospheric energy comprises shortwave radiation *inputs* (ultraviolet light, visible light, and near-infrared wavelengths) and longwave radiation *outputs* (thermal infrared). **Transmission** refers to

the passage of shortwave and longwave energy through either the atmosphere or water. The atmosphere and surface eventually radiate infrared energy back to space, and this energy, together with reflected energy, equals the initial solar input. Thus, a natural balance of energy input and output exists in our atmosphere.

## *Energy in the Atmosphere: Some Basics*

When you look at a photograph of Earth taken from space, you can clearly see the pattern of surface response to incoming insolation (see the back cover of this book). Solar energy is intercepted by land and water surfaces, clouds, and atmospheric gases and dust. The flows of energy are manifest in swirling weather patterns, powerful oceanic currents, and the varied distribution of vegetation. Specific energy patterns differ for deserts, oceans, mountaintops, plains, rain forests, and ice-covered landscapes. In addition, the presence or absence of clouds might make a 75% difference in the amount of energy that reaches the surface, because clouds are reflective.

Figure 4-1 is a simplified flow diagram of shortwave and longwave radiation in the Earth-atmosphere system that is discussed in the pages that follow. You will find it helpful to refer to that figure, and the more-detailed energy balance illustration in Figure 4-12, as you read through the following section.

***Refraction.*** When insolation enters the atmosphere, it passes from one medium to another (from virtually empty space into atmospheric gases). This transition subjects the insolation to a change of speed, which also shifts

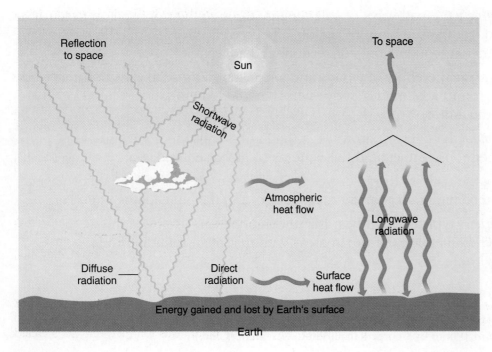

**FIGURE 4-1**

**Energy gained and lost by Earth's surface and atmosphere.** Simplified view of the Earth-atmosphere energy system—circuits include incoming shortwave insolation, reflected shortwave radiation, and outgoing longwave radiation.

its direction, a bending action called **refraction**. In the same way, a crystal or prism refracts light passing through it, bending different wavelengths to different angles, separating the light into its component colors to display the spectrum. A rainbow is created when visible light passes through myriad raindrops and is refracted and reflected toward the observer at a precise angle (see Figure 4-2).

Another example of refraction is a *mirage*, an image that appears near the horizon where light waves are refracted by layers of air of different temperatures (and consequently of different densities) on a hot day. The atmospheric distortion of the setting Sun in Figure 4-3 is also a product of refraction—light from the Sun low in the sky must penetrate more air than when the sun is high; it is refracted through air layers of different densities on its way to the viewer.

An interesting function of refraction is that it adds approximately 8 minutes of daylight that we would lack if Earth had no atmosphere. Sunlight is refracted in its passage from space through the atmosphere, and so, at sunrise, we see the Sun's image about 4 minutes before the Sun actually peeks over the horizon. Similarly, the Sun actually sets at sunset, but its image is refracted from over the horizon for about 4 minutes afterward, so we see sunset later than it truly happens. To this day, modern science cannot predict the exact time of visible sunrise or sunset within these 4 minutes because the degree of refraction continually varies with temperature, moisture, and pollutants.

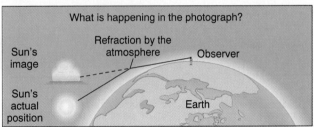

**FIGURE 4-3**
**Sun refraction.**
The distorted appearance of the Sun nearing sunset over the ocean is produced by refraction of the Sun's image in the atmosphere. [Photo by author.]

**FIGURE 4-2**
**A rainbow.**
Raindrops refract and reflect light to produce a primary rainbow. Note that the color order in the rainbow is distributed with the shortest wavelengths inside the bow and the longest wavelengths outside the bow. [Photo by author.]

***Insolation Input.*** *Insolation* is the single energy input driving the Earth-atmosphere system. The world map in Figure 4-4 shows the distribution of average annual solar energy received at Earth's surface. It includes all the radiation that arrives at Earth's surface, both direct and diffuse (scattered by the atmosphere).

Several patterns are notable on the map. Insolation decreases poleward from about 25° latitude in both the Northern and Southern Hemispheres. Consistent daylength and high Sun altitude produce average annual values of 180–220 W/m² throughout the equatorial and tropical latitudes. In general, greater insolation of 240–280 W/m² occurs in low-latitude deserts worldwide because of frequently cloudless skies. Note this energy pattern in the cloudless subtropical deserts in both hemispheres (for example, the Sonoran, Saharan, Arabian, Gobi, Atacama, Namib, Kalahari, and Australian deserts).

***Albedo and Reflection.*** A portion of arriving energy bounces directly back into space without being absorbed or performing any work. This returned energy is called

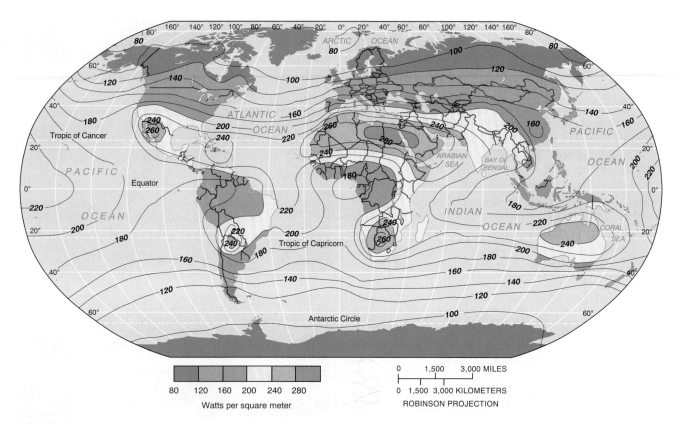

**FIGURE 4-4**
**Insolation at Earth's surface.**
Average annual solar radiation received on a horizontal surface at ground level in watts per square meter (100 W/m² = 75 kcal/cm²/yr). [After M. I. Budyko, 1958, *The Heat Balance of the Earth's Surface*, Washington, DC, U.S. Department of Commerce, p. 99, and *Atlas Teplovogo Balansa*, Moscow, 1963.]

**reflection**. <u>**Albedo**</u> is the reflective quality (intrinsic <u>brightness</u>) of a surface. It is an important control over the amount of insolation that is available for absorption by a surface. We state albedo as the percentage of insolation that is reflected, as shown in Figure 4-5.

In the visible wavelengths, darker colors have lower albedos, and lighter colors have higher albedos. On water surfaces, the angle of the solar rays also affects albedo values; lower angles produce a greater reflection than do higher angles (Figure 4-6). In addition, smooth surfaces increase albedo, whereas rougher surfaces reduce it.

Specific locations experience highly variable albedo values during the year in response to changes in cloud and ground cover. Earth Radiation Budget (ERB) sensors aboard the *Nimbus-7* satellite measured average albedos of 19%–38% between the tropics (23.5° N to 23.5° S) to as high as 80% in the polar regions.

Earth and its atmosphere reflect 31% of all insolation when averaged over a year. Looking ahead to Figure 4-12, you can see that Earth's average albedo is a combination of 21% reflected by clouds, 3% reflected by the ground (combined land and ocean surfaces), and 7% reflected and scattered by the atmosphere. By comparison, a full Moon, which is bright enough to read by

under clear skies, has only a 6%–8% albedo value. Thus, with *earthshine* being four times brighter than moonlight (four times the albedo), and with Earth four times greater in diameter than the Moon, it is no surprise that astronauts report how startling our planet looks from space. News Report 1 presents more information on earthshine.

Figure 4-7 portrays total albedos for July 1985 and January 1986 as measured by the Earth Radiation Budget experiments aboard several satellites. These patterns are typical of most years. As compared with July albedos, January albedos are higher poleward of 40° N, because of the snow and ice that cover the ground. Tropical forests are characteristically low in albedo (15%), whereas generally cloudless deserts have high albedos (35%). The southward-shifting cloud cover over equatorial Africa is quite apparent on the January map.

***Clouds and the Atmosphere's Albedo.*** A major uncertainty in the tropospheric energy budget, and therefore in refining climatic models, is the role of clouds. Clouds reflect insolation and thus cool Earth's surface. An increase in albedo caused by clouds is described by the term **cloud-albedo forcing**. Yet clouds act as insulation, trapping long-wave radiation and raising minimum temperatures. An

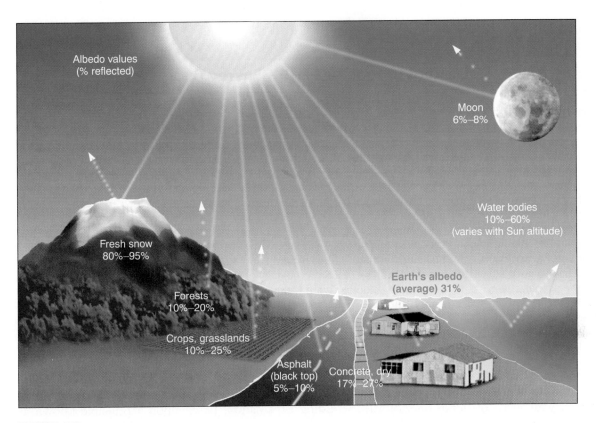

**FIGURE 4-5**
**Various albedo values.**
Different surfaces have different albedo values. In general, light surfaces are more reflective than dark surfaces and thus have higher albedo values. [Data from M.I. Budyko, 1958, *The Heat Balance of the Earth's Surface*, Washington, DC, U.S. Department of Commerce, p. 36.]

**FIGURE 4-6**
**Sunlight reflected off the ocean.**
An astronaut's view of reflected sunlight off the Mozambique Channel, between the east coast of Africa and the island country of Madagascar. Ocean surface albedo values increase with lower sun angles and calmer seas. [Space Shuttle photo from NASA.]

## News Report 1

### Earthshine—A Diagnostic Tool for Earth's Energy Budget?

Earth's reflection of light to space might prove to be a useful tool in analyzing energy budgets and climate. Earth's average albedo value is 31%—that is, 31% of all arriving insolation is returned to space without being absorbed. The Moon reflects only 6% to 8% of the sunlight that hits its surface, producing the "moonlight" familiar to everyone on Earth. Despite the bright, white appearance of a full Moon, the lunar surface albedo value is about the same as that of an asphalt street.

Because Earth reflects about 400% as much light as the Moon and is about 400% as great in diameter, it is a startling presence in space. We have all seen the effect of *earthshine* as a faint illumination of the dark portion of the Moon, especially in a crescent phase, when the dark portion of the lunar disk is barely visible.

Earthshine, if properly analyzed, is a measure of Earth's albedo. Scientists from Arizona State University and the California Institute of Technology now measure earthshine twice a month during the Moon's crescent phase. Such earthshine studies were begun by André Danjon of France in 1925. The significance of measuring earthshine is this: An approximate 1% change in albedo will produce about a 1 C° (2 F°) change in average atmospheric temperature. Thus, an albedo change can produce a temperature change, which can produce further albedo change and further alter climate.

A lot remains to be learned about this yardstick for Earth's energy balance. For example, equatorial reflection is dominant in earthshine, whereas polar reflection is scattered. Earthshine is yet another tool as society struggles to understand the Home Planet—a lot to think about the next time you take a moonlight walk.

---

increase in greenhouse warming caused by clouds is described by the term **cloud-greenhouse forcing**. Figure 4-8 illustrates the general effects of clouds on shortwave radiation and longwave radiation. More on clouds is presented later in this chapter and in a detailed section in Chapter 8.

Other mechanisms affect atmospheric albedo and therefore temperature patterns. Industrialization is producing a haze of pollution that is increasing the reflectivity of the atmosphere. Emissions of sulfur dioxide and the subsequent chemical reactions in the atmosphere form *sulfate aerosols*. These aerosols act as an insolation-reflecting haze in clear-sky conditions, or they act as a stimulus to condensation in clouds that increases their reflectivity.

The eruption of Mount Pinatubo in the Philippines, beginning explosively during June 1991, illustrates how Earth's internal processes can affect the atmosphere. Approximately 15–20 megatons of sulfur dioxide were injected into the stratosphere; winds rapidly spread this aerosol (tiny droplets) worldwide (see Figure 6-1). As a result, atmospheric albedo increased and produced a temporary worldwide average cooling of 0.5 C° (0.9 F°).

**Scattering (Diffuse Radiation).** Insolation encounters an increasing density of atmospheric gases as it travels toward the surface. The gas molecules absorb and reemit the radiation, changing the direction of the light's movement *without altering its wavelengths.* This phenomenon is known as **scattering** and represents 7% of Earth's reflectivity, or albedo (see Figure 4-12). Dust particles, pollutants, ice, cloud droplets, and water vapor produce further scattering.

Why is Earth's sky blue? And why are sunsets and sunrises often red? These simple questions have an interesting explanation, based upon a principle known as Rayleigh scattering (named for English physicist Lord Rayleigh, who stated the principle in 1881). This principle relates wavelength to the size of molecules or particles that cause the scattering. The general rule is: *The shorter the wavelength, the greater the scattering, and the longer the wavelength, the less the scattering.* Shorter wavelengths of light are scattered by small gas molecules in the air. Thus, the shorter wavelengths of visible light—the blues and violets—are scattered the most and dominate the lower atmosphere. And, because there are more blue than violet wavelengths in sunlight, a blue sky prevails (Figure 4-9). A sky filled with smog and haze appears almost white because the larger particles associated with air pollution act to scatter all wavelengths of the visible light.

The angle of the Sun's rays determines the thickness of atmosphere they must pass through to reach the surface. Therefore, direct rays (from more overhead) experience less scattering and absorption than do low, oblique-angle rays that must travel farther through the atmosphere. Insolation from the low-altitude Sun undergoes more scattering of shorter wavelengths, leaving only the residual oranges and reds to reach the observer at sunset or sunrise.

Some incoming insolation is diffused by clouds and atmosphere and is transmitted to Earth as **diffuse radiation**, the downward component of scattered light (labeled in Figure 4-12). This light is multidirectional and thus casts shadowless light on the ground.

**Absorption.** **Absorption** is the assimilation of radiation by molecules of matter and its conversion from one form

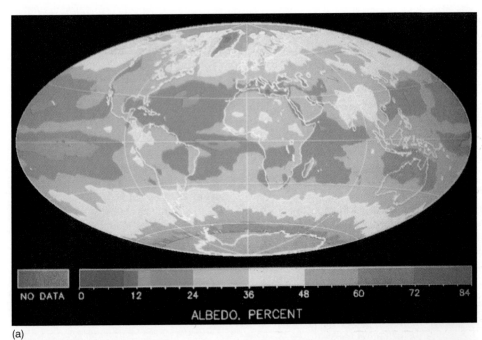

(a)

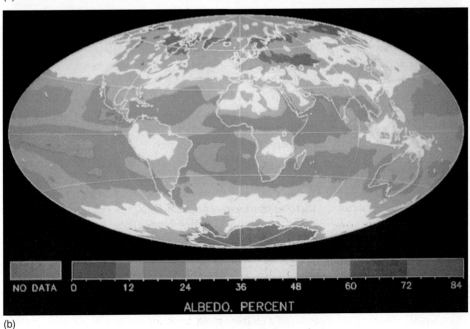

(b)

**FIGURE 4-7**
**Total albedos for July and January.**
Total albedos for July 1985 (a) and January 1986 (b) as measured by the Earth Radiation Budget experiments (ERB) aboard satellites *Nimbus-7, NOAA-9,* and *ERBS* (ERB Satellite). Each color represents a 12% interval in albedo values. (A modified elliptical equal-area projection is used.) [Courtesy of Radiation Services Branch, Langley Research Center, NASA.]

of energy to another. Insolation (both direct and diffuse) that is not part of the 31% reflected from Earth's surface and atmosphere is absorbed. It is converted into either infrared radiation or chemical energy (by plants in photosynthesis). The temperature of the absorbing surface is raised in the process, and that warmer surface radiates more total energy at shorter wavelengths—thus, the hotter the surface the shorter the wavelengths that are emitted. In addition to absorption by land and water surfaces (about 45% of incoming), absorption also occurs in atmospheric gases, dust, clouds, and stratospheric ozone (about 24% of incoming insolation). Figure 4-12 summarizes the pathways of insolation and the flow of heat in the atmosphere and at the surface.

***Conduction, Convection, and Advection.*** Heat energy is transferred through several means. **Conduction** is the molecule-to-molecule transfer of heat energy as it diffuses through a substance. As molecules warm, their vibration increases, causing collisions that produce motion in neighboring molecules, thus transferring heat from warmer to cooler materials.

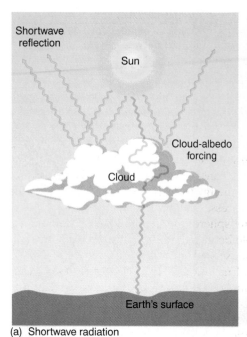

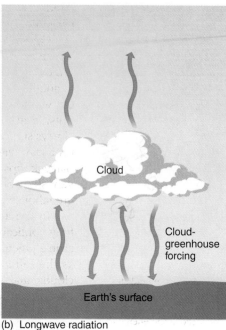

(a) Shortwave radiation

(b) Longwave radiation

**FIGURE 4-8**

**The effects of clouds on shortwave and longwave radiation.**
(a) Shortwave radiation is reflected and scattered by clouds; a high percentage is returned to space.
(b) Longwave radiation emitted by Earth is absorbed and reradiated by clouds; some infrared energy is radiated to space and some back toward the surface.

Different materials (gases, liquids, and solids) conduct sensible heat directionally from areas of higher temperature to those of lower temperature. This heat flow transfers energy through matter at varying rates, depending on the conductivity of the material—Earth's land surface is a better conductor than air; moist air is a slightly better conductor than dry air.

Energy also is transferred through gases and liquids by movements called **convection**, when the physical mixing involves a strong vertical motion. When a lateral (horizontal) motion is dominant the term **advection** is used. In the atmosphere or bodies of water, warmer (less dense) masses tend to rise and cooler (denser) masses tend to sink, establishing patterns of convection. Sensible heat is transported through the medium in this way.

You commonly experience such of energy flows in the kitchen: Energy is conducted through the handle of a pan, or boiling water bubbles in the saucepan in convective motions (Figure 4-10). Also in the kitchen you may use a convection oven that uses a fan to circulate the heated air to uniformly cook food.

In physical geography, we find many examples of *conduction* (surface energy budgets, temperature differences between land and water bodies, the heating of surfaces

**FIGURE 4-9**

**The color of the sky.**
Scattering produces the blue color of the sky. Residual reds and oranges dominate sunrises and sunsets, after other wavelengths have been scattered. [Photo by author.]

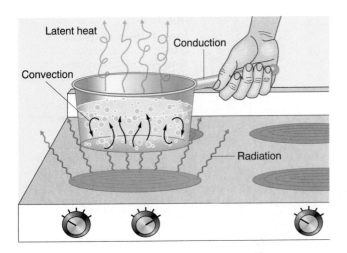

**FIGURE 4-10**

**Heat energy transfer processes.**

A pan on the stove illustrates heat transfer. Infrared energy radiates from the burner to the saucepan and the air. Energy is conducted through the molecules of the pan and the handle. The water physically mixes, carrying heat energy by convection. The energy in the water and handle is measurable as sensible heat. The vapor leaving the surface of the water contains the latent heat absorbed in the change of water to a vapor.

and overlying air, soil temperatures), *convection* (atmospheric and oceanic circulation, air mass movements and weather systems, internal motions deep within Planet Earth that produce a magnetic field and movements in the crust), and *advection* (horizontal movement of winds from land to sea and back, fog that forms and moves to another area).

### Earth Reradiation and the Greenhouse Effect.

Previously, we characterized Earth as a cool-body radiator, emitting energy in infrared wavelengths from its surface and atmosphere toward space. However, some of this infrared radiation is absorbed by carbon dioxide, water vapor, methane, chlorofluorocarbons (CFCs), and other gases in the lower atmosphere and then reradiated back to Earth, thus delaying energy loss to space. This counterradiation process is an important factor in warming the troposphere. The rough similarity between this process and the way a greenhouse operates gives the process its name—the **greenhouse effect**.

In a greenhouse, the glass is transparent to short-wave insolation, allowing light to pass through to the soil, plants, and materials inside, where absorption and conduction take place. The absorbed energy is then radiated as infrared energy back toward the glass, but the glass effectively traps both the longer infrared wavelengths and the warmed air inside the greenhouse. Thus, the glass acts as a one-way filter, allowing the light energy in but not allowing the heat energy out, except through

conduction. The same process also is quite evident in a car parked in direct sunlight.

Opening a roof vent allows the air inside the greenhouse to mix with the outside environment, thereby removing heat by moving air physically from one place to another—the process of convection. Rolling down your car window accomplishes the same thing, allowing the outside air to physically mix with the inside air. It is always surprising how hot the interior of a car gets, even on a day with mild temperatures outside. Many people place an opaque sunscreen across the windshield to prevent shortwave energy from entering the car to begin the greenhouse process.

In the atmosphere, the greenhouse analogy is not fully applicable because infrared radiation is not trapped as in a greenhouse. Rather, its passage to space is *delayed* as the infrared radiation is absorbed by certain gases, clouds, and dust in the atmosphere and is reradiated to Earth's surface. According to many scientists, today's increasing carbon dioxide concentration is causing more infrared radiation absorption in the lower atmosphere, thus forcing a warming trend and changes in the Earth-atmosphere energy system.

### Clouds and Earth's "Greenhouse."

Clouds affect the heating of the lower atmosphere in several ways, depending on cloud type. Not only is the percentage of cloud cover important but also the cloud type, height, and thickness (water content and density). High-altitude, ice-crystal clouds reflect insolation with albedos of about 50%, whereas thick, lower cloud cover reflects about 90%.

To understand the actual effects on the atmosphere's energy budget, however, we must consider both transmission of shortwave and longwave radiation and cloud type. Figure 4-11a portrays the *cloud-greenhouse forcing* caused by high clouds (warming, because their greenhouse effects exceed their albedo effects); and Figure 4-11b portrays the *cloud-albedo forcing* produced by lower, thicker clouds (cooling, because albedo effects exceed greenhouse effects). Understanding the nature of global cloud cover is crucial in refining computer models that forecast global climate change.

## Earth-Atmosphere Radiation Balance

The Earth-atmosphere energy system naturally balances itself in a steady-state equilibrium. If Earth's surface and its atmosphere are considered separately, neither exhibits a balanced radiation budget. The average annual energy distribution is positive (an energy surplus) for Earth's surface and negative (an energy deficit) for the atmosphere. Considered together, however, these two balance each other, making it possible for us to construct an overall energy balance.

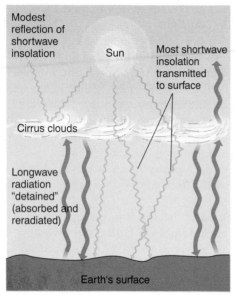

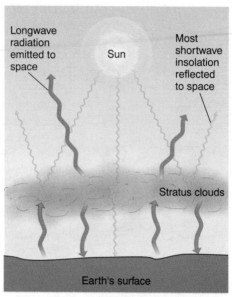

**FIGURE 4-11**

**Energy effects of two cloud types.**

Cloud effects vary depending on cloud type. (a) High, ice-crystal clouds (called *cirrus*) transmit most of the insolation. However, they absorb and delay losses of outgoing longwave infrared, producing a greater greenhouse forcing and a net warming of Earth. (b) Low, thick clouds (*stratus*) reflect most of the incoming insolation and radiate longwave infrared to space, producing a greater albedo forcing and a net cooling of Earth.

(a) High clouds: net greenhouse forcing and atmospheric warming

(b) Low clouds: net albedo forcing and atmospheric cooling

The natural energy balance occurs through energy transfers that are both *nonradiative* (physical motion) and *radiative*. Radiative transfer is by infrared radiation between the surface, the atmosphere, and space. Nonradiative transfers include convection, conduction, and the latent heat of evaporation (energy that is absorbed and dissipated by water as it evaporates).

Figure 4-12 summarizes the Earth-atmosphere radiation balance. It brings together all the elements discussed to this point in the chapter by following 100% of arriving insolation through the troposphere. The shortwave portion of the budget is on the left; the longwave part of the budget is on the right. Of 100% of solar energy arriving, Earth's average albedo involves 31% energy reflected to space. Absorption by atmospheric clouds, dust, and gases involves another 21% and accounts for the atmospheric heat input. Stratospheric ozone absorption and radiation accounts for another 3% of the atmospheric budget. About 45% of the incoming insolation is left to actually reach Earth's surface as direct and diffuse radiation. Earth eventually reradiates the remaining 69% into space as infrared radiation: 21% (atmosphere heating) + 45% (surface heating) + 3% (ozone emission) = 69%.

Earlier, in Figure 4-7, we saw average global albedos for July and January. A representative pattern of longwave radiation from the Earth-atmosphere system (net outgoing longwave) is shown in Figure 4-13. Data gathered from ERB instruments aboard several satellites show that variations in longwave radiation are generally latitudinal (zonal), with higher values at lower latitudes. An exception to distribution is the frequently cloud-covered region of the

tropics (Amazon, equatorial Africa, and Indonesia)—here, lower values are at lower latitudes. Major desert areas have greater longwave radiation emissions owing to the presence of little cloud cover and greater radiative energy losses from surfaces that have absorbed a lot of energy.

***Global Net Radiation.*** Figure 4-14 summarizes the radiation balance for all shortwave and longwave energy by latitude:

- Between the tropics, the angle of incoming insolation is high and daylength is consistent, with little seasonal variation, so more energy is gained than lost—energy surpluses dominate.

- In the polar regions, the Sun is extremely low in the sky, surfaces are light (ice and snow) and reflective, and for up to 6 months during the year no insolation is received, so more energy is lost than gained—energy deficits prevail.

- At around 36° latitude, a balance exists between energy gains and losses for the Earth-atmosphere system, according to *Nimbus-7* measurements.

This pattern of energy surpluses and deficits is established by the input of insolation minus the output of longwave radiation and shortwave reflection.

The imbalance of net radiation from tropical surpluses to the polar deficits drives a vast global circulation of both energy and mass. The meridional (north-south) transfer agents are global winds, ocean currents, dynamic weather

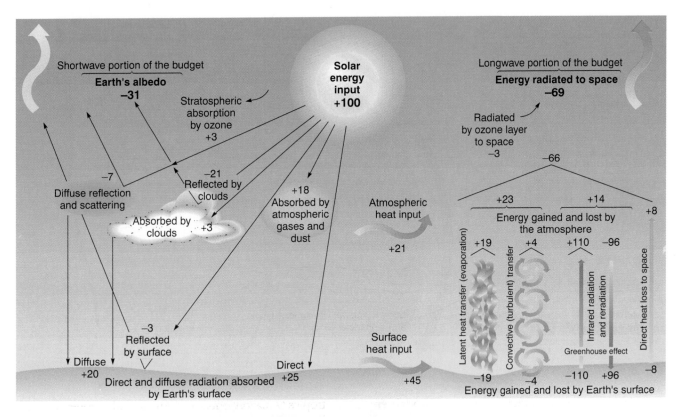

**FIGURE 4-12**

**Detail of the Earth-atmosphere energy balance.**

Solar energy cascades through the lower atmosphere (left-hand portion of the illustration), where it is absorbed, reflected, and scattered. Clouds, atmosphere, and the surface reflect 31% of this insolation back to space. Atmospheric gases and dust and Earth's surface absorb energy and radiate infrared radiation. Earth and atmosphere exchange energy through latent heat transfer in water vapor, convective transfer (moving air), and infrared radiation (right-hand portion of the illustration). Over time, Earth reradiates, on average, 69% of incoming energy to space. When added to Earth's average albedo (31%, reflected energy), this equals the total energy input from the Sun.

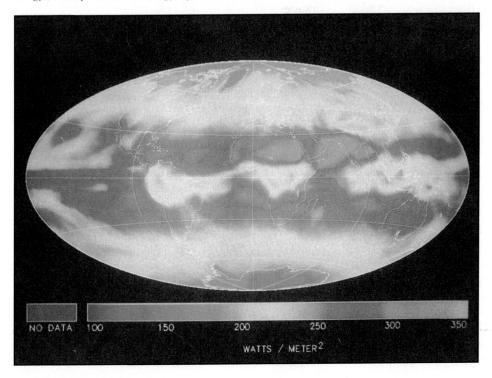

**FIGURE 4-13**

**Longwave radiation from Earth.**

Longwave radiation leaving the Earth-atmosphere system (net outgoing longwave) in average monthly values for April 1985. Data are in watts per square meter. Latitudinal distribution of longwave radiation is apparent, with cooler regions toward the poles and warmer regions in the tropics. [Courtesy of Radiation Services Branch, Langley Research Center, NASA.]

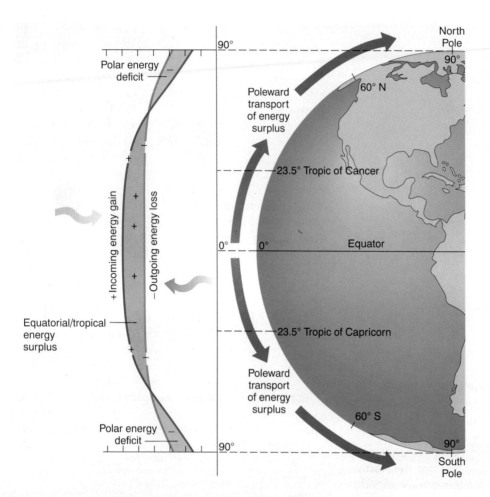

**FIGURE 4-14**
**Energy budget by latitude.**
Earth's energy surpluses and deficits
by latitude produce poleward
transport of energy and mass in each
hemisphere.

systems, and related phenomena. Dramatic examples of such energy and mass transfers are tropical cyclones—hurricanes and typhoons. Forming in the tropics, they mature and migrate to higher latitudes, carrying with them energy, water, and water vapor.

Having examined the Earth-atmosphere radiation balance, let us now focus on energy characteristics at Earth's surface.

# Energy Balance at Earth's Surface

Solar energy is the principal heat source at Earth's surface. The direct and diffuse radiation and infrared radiation arriving at the ground surface are shown in Figure 4-12 (please note the terms along the ground surface in the figure). These radiation patterns at Earth's surface are of great interest to geographers.

## *Daily Radiation Patterns*

The fluctuating daily pattern of incoming shortwave energy absorbed and the resultant air temperature are shown in Figure 4-15. This graph represents idealized conditions for

bare soil on a cloudless day in the middle latitudes. Incoming energy arrives during daylight, beginning at sunrise, peaking at noon, and ending at sunset.

The shape and height of this insolation curve vary with season and latitude. The highest trend for such a curve occurs at the time of the summer solstice (around June 21 in the Northern Hemisphere and December 21 in the Southern Hemisphere). The air temperature plot also responds to seasons and variations in insolation input. Within a 24-hour day, air temperature peaks at around 3:00–4:00 P.M. and dips to its lowest point right at or slightly after sunrise.

The relationship between the insolation curve and the air temperature curve on the graph is interesting—they do not align; there is a *lag*. As long as the incoming energy exceeds the outgoing energy, air temperature continues to increase, not peaking until the incoming energy begins to diminish in the afternoon as the Sun's altitude decreases.

The warmest time of day occurs not at the moment of maximum insolation but at the moment when a *maximum of insolation is absorbed*. Thus, this temperature lag places the warmest time of day 3–4 hours after solar noon as absorbed energy is supplied to the atmosphere from the ground. Then, as the insolation input decreases toward sunset, the amount of energy lost exceeds the

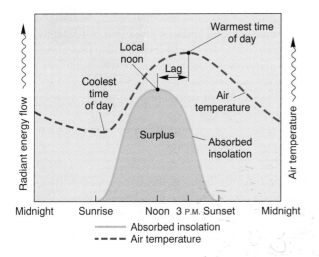

**FIGURE 4-15**
**Daily radiation curves.**
Sample radiation plot for a typical day shows the changes in insolation (orange line) and air temperature (dashed line). Comparing the curves demonstrates a lag between local noon (the insolation peak for the day) and the warmest time of day.

input, and air temperatures begin to drop until the surface has radiated away the maximum amount of energy, just at dawn. If you have ever gone camping you no doubt experienced this wake-up chill at sunrise!

Like the daily pattern, the annual pattern of insolation and air temperature exhibits a similar lag. For the Northern Hemisphere, January is usually the coldest month, occurring after the December solstice and the shortest days. Similarly, the warmest months of July and August occur after the June solstice and the longest days.

## Simplified Surface Energy Balance

Earth's surface is supplied with energy that varies daily and seasonally. Energy and moisture are continually exchanged at the surface, creating worldwide "boundary layer climates" of great variety. Physical conditions at or near Earth's surface are studied in **microclimatology**— the science of this lowest portion of the atmosphere. The following discussion is more meaningful if you visualize an actual surface—perhaps a park, a front yard, or a place on campus.

The surface receives visible light and infrared radiation, and it reflects light and radiates infrared according to the following basic scheme:*

$$+\text{SW}\!\downarrow \quad -\text{SW}\!\uparrow \quad +\text{LW}\!\downarrow \quad -\text{LW}\!\uparrow \quad = \text{NET R}$$
(Insolation) (Reflection) (Infrared) (Infrared) (Net Radiation)

* We use SW for shortwave, LW for longwave for simplicity. You may come across other symbols in the microclimatology literature, such as Q* for NET R, K for shortwave, and L for longwave.

Figure 4-16 shows the components of a surface energy balance. The soil column shown continues to a depth at which energy exchange with surrounding materials or with the surface becomes negligible, usually less than a meter. Energy from the atmosphere that is moving toward the surface is regarded as positive (a gain), and energy that is moving away from the surface toward the atmosphere is considered negative (a loss) to the surface account. In the soil column, energy that is moving out of the column or surrounding soil is negative, and energy that is moving into the column is positive.

Adding and subtracting the energy flow at the surface completes the calculation of **net radiation** (**NET R**), or the balance of all radiation at Earth's surface. As the components of this simple equation vary with daylength through the seasons, cloudiness, and latitude, so too does the resultant NET R. Figure 4-17 illustrates the components of a surface energy balance for a typical summer day at a midlatitude location. The figure shows how net radiation is derived during the day. The items plotted are direct and diffuse insolation (+SW), reflected energy (surface albedo value, −SW), and infrared radiation arriving at (+LW) and leaving from (−LW) the surface.

Surface albedo values (−SW↑) dictate the amount of insolation reflected, and therefore not absorbed, at the surface: Darker and rougher surfaces (such as a plowed field or parking lot) reflect less shortwave energy, absorb more energy, and therefore produce more net radiation; lighter and smoother surfaces (concrete, light-colored clothing) reflect more shortwave energy, producing less net radiation. Snow-covered surfaces reflect most of the

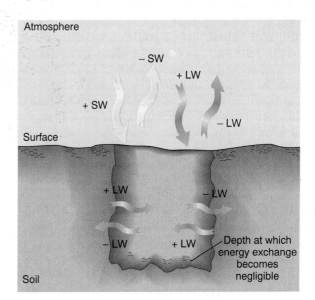

**FIGURE 4-16**
**Surface energy budget.**
Idealized input and output energy budget components for a surface and soil column. (SW = shortwave, LW = longwave.)

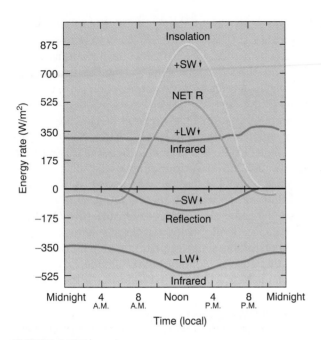

**FIGURE 4-17**
**A day's radiation budget.**
Radiation budget components and resulting net radiation (NET R) on a typical summer day for a midlatitude location (Matador in southern Saskatchewan, about 51° N, on July 30, 1971). [Adapted by permission from T. R. Oke, *Boundary Layer Climates*. New York: Methuen & Co., 1978, p. 21.]

light reaching them, whereas darker urban areas absorb more insolation and radiate more infrared radiation (review Figure 4-5).

As an example, imagine a snow-covered landscape as a surface energy system. Snow has high reflectivity, sending sunlight back to space and reducing the amount of insolation absorbed. Consequently, surfaces and air temperatures are lower, and thus the snow cover does not melt. And snow has a higher albedo value than bare ground, so it reflects insolation—and so on. This is a system with *positive feedback*. (Remember from Chapter 1 that positive feedback increases response in the system: more cooling–more snow–increasing albedo–more cooling–then colder, drier air–then less snow–and so on.)

At night, the net radiation value becomes negative because the SW component ceases at sunset and the surface continues to lose infrared energy to the atmosphere. The surface rarely reaches a zero net radiation value—a perfect balance—at any one moment, but over time, Earth's surface generally balances incoming and outgoing energies.

**Net Radiation.** The net radiation (NET R of all wavelengths) available at Earth's surface is the final outcome of the entire radiation-balance process discussed in this chapter. Figure 4-18 displays the mean annual net radiation at ground level. The abrupt change in radiation balance from ocean to land surfaces is quite evident on the map. Note

that all values are positive; negative values probably occur only over ice-covered surfaces poleward of 70° latitude in both hemispheres. The highest net radiation occurs north of the equator in the Arabian Sea at 185 W/m² per year. Aside from the obvious interruption caused by landmasses, the pattern of values appears generally zonal, or parallel to the equator.

Net radiation is expended from a nonvegetated surface through three pathways:

- *H*, or sensible heat, is the back-and-forth transfer between air and surface in turbulent eddies, through convection and through conduction within materials. This activity depends on surface and boundary-layer temperatures and on the intensity of convective motion in the atmosphere. About 18% of Earth's entire NET R is mechanically radiated as sensible heat from the surface, especially over land.

- *LE*, or latent heat of evaporation, is the energy that is stored in water vapor as water evaporates. Large quantities of heat are absorbed into water vapor, becoming *latent heat*, during water's change of state from liquid to gas. This heat energy is thereby removed from the surface. Conversely, this heat energy is released to the environment when water vapor changes state back to a liquid (Chapter 7). Latent heat is the dominant expenditure of Earth's entire NET R, especially over water surfaces. Latent heat links Earth's energy, water (hydrologic), and biological systems.

- *G*, or ground heating and cooling, is the energy that flows into and out of the ground surface (land or water) by conduction.

On land, the highest annual values for latent heat of evaporation (LE) occur in the tropics and decrease toward the poles (Figure 4-19). Over the oceans, the highest LE values are over subtropical latitudes where hot, dry air comes into contact with warm ocean water.

The values for sensible heat (H) are distributed differently, being highest in the subtropics (Figure 4-20). Here, nearly waterless and cloudless expanses of subtropical deserts stretch almost vegetation-free across vast regions. The bulk of NET R is expended as sensible heat in these dry regions. Moist and vegetated surfaces that are higher in LE have lower NET R expenditures in H, as you can see by comparing the maps in Figures 4-19 and 4-20.

During a year, the overall G value (ground heating and cooling) is zero because the stored energy from spring and summer is equaled by losses in fall and winter. The input for G is absorption of insolation and infrared. The output for G is energy lost by radiation to the atmosphere, sensible heat transfer when the air is cooler than the ground, and energy absorbed for evaporation. Of course, moist surfaces covered with vegetation modify ground

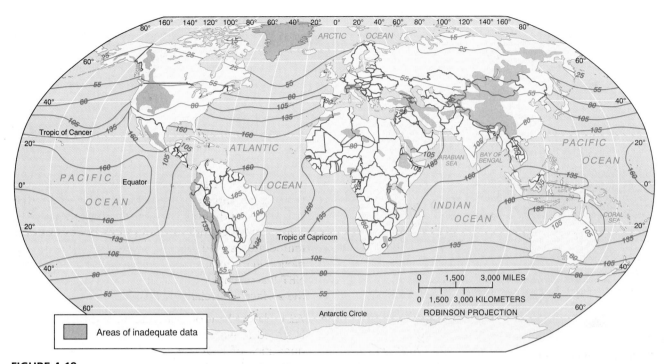

**FIGURE 4-18**
**Global net radiation.**
Distribution of global mean annual net radiation (NET R) at surface level in watts per square meter (100 W/m$^2$ = 75 kcal/cm$^2$/year). [After M. I. Budyko, 1958, *The Heat Balance of the Earth's Surface*, Washington, DC, U.S. Department of Commerce, p. 106; and *Atlas Teplovogo Balansa*, Moscow, 1963.]

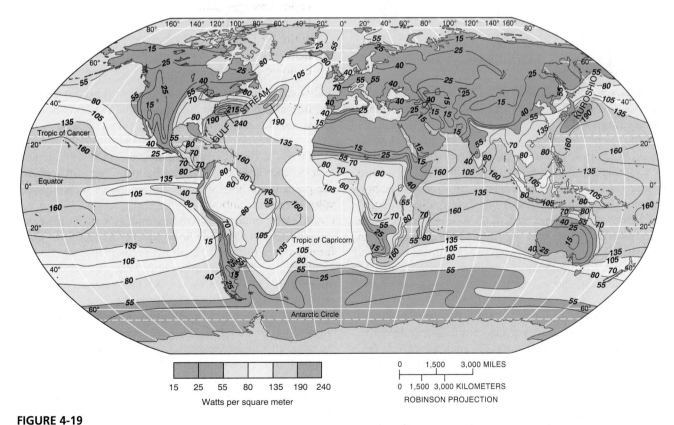

**FIGURE 4-19**
**Global latent heat of evaporation.**
Distribution of annual energy expenditure as the latent heat of evaporation (LE) at surface level in watts per square meter (100 W/m$^2$ = 75 kcal/cm$^2$/year). Note the high values associated with high sea-surface temperatures in the area of the Gulf Stream and Kuroshio currents. [Adapted by permission from M.I. Budyko, *The Earth's Climate Past and Future*, New York: Academic Press, 1982, p. 56.]

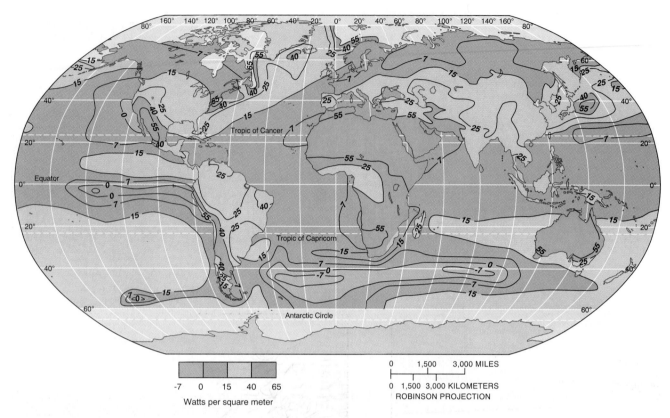

**FIGURE 4-20**

**Global sensible heat.**

Distribution of annual energy expenditure as sensible heat (H) at surface level in watts per square meter (100 W/m² = 75 kcal/cm²/year). [Adapted by permission from M.I. Budyko, *The Earth's Climate Past and Future*, New York: Academic Press, 1982, p. 59.]

heating. Net radiation available at the surface is decreased because incoming shortwave energy is intercepted by a forest canopy of leaves.

Another factor in ground heating is energy absorbed at the surface to melt snow or ice. In snow- or ice-covered landscapes, almost all available energy is absorbed as sensible and latent heat used in the melting and warming process.

Understanding net radiation is essential to solar energy technologies that concentrate shortwave energy for use. Solar energy offers great potential worldwide and will be used more in the near future. Focus Study 4-1 briefly reviews solar energy issues.

***Sample Stations.*** A couple of real locations can bring all of what you have just read into meaningful perspective. Variation in the expenditure of NET R among sensible heat (H, energy we can feel), latent heat (LE, energy for evaporation), and ground heating and cooling (G) produces the variety of environments we experience in nature. Let us examine the daily energy balance at two locations, El Mirage in California and Pitt Meadows in British Columbia.

El Mirage, at 35° N, is a hot desert location characterized by bare, dry soil with sparse vegetation (Figure 4-21a, b).

Our sample is a clear summer day, with a light wind in the late afternoon. The NET R value is lower than might be expected, considering the Sun's position close to zenith (June solstice) and the absence of clouds. But the income of energy at this site is countered by surfaces of higher albedo than forest or cropland and by hot soil surfaces that radiate infrared back to the atmosphere throughout the afternoon.

El Mirage has little or no energy expenditure for evaporation (LE). With little water and sparse vegetation, most of the available radiant energy is dissipated by the turbulent transfer of sensible heat (H), warming air and soil to high temperatures. Over a 24-hour period, H is 90% of NET R; the remaining 10% goes to ground heating (G). The G component is greatest in the morning, when winds are light and turbulent transfers are lowest. In the afternoon, heated air rises off the hot ground, and convective heat expenditures are accelerated as winds increase.

Compare El Mirage (Figure 14-21a) and Pitt Meadows (Figure 14-21c). Note that the NET R at the El Mirage desert location is quite similar to that of Pitt Meadows, British Columbia. But, Pitt Meadows is midlatitude (49° N), vegetated, and moist, and its energy expenditures differ greatly

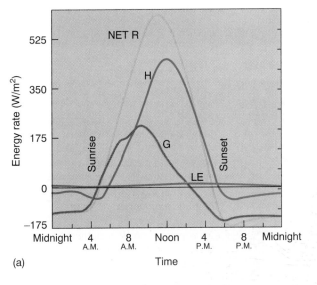

(a)

(b)

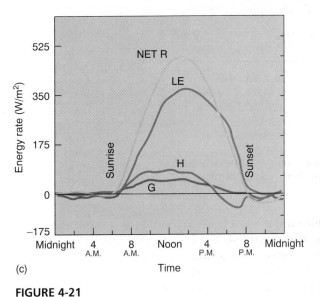

(c)

(d)

**FIGURE 4-21**

**Radiation budget comparison for two stations.**

(a) Graph shows daily net radiation expenditure for El Mirage, California, east of Los Angeles, at about 35° N. (b) Photo of the typical desert landscape near the site. (c) Graph of the daily net radiation expenditure for Pitt Meadows in southern British Columbia, at about 49° N. (d) Photo of irrigated blueberry orchards characteristic of the agricultural activity in this moist environment of moderate temperatures. (H = turbulent sensible heat transfer; LE = latent heat of evaporation; G = ground heating and cooling.) [(a) Adapted by permission from William D. Sellers, *Physical Climatology*, fig. 33, copyright © 1965 by The University of Chicago. All rights reserved. (c) Adapted by permission from T. R. Oke, *Boundary Layer Climates*. New York: Methuen & Co., 1978, p. 23. Photos by author.]

from those at El Mirage. The Pitt Meadows landscape is able to retain much more of its energy because of lower albedo values (less reflection), the presence of more water and plants, and lower surface temperatures than those of El Mirage.

The energy balance data for Pitt Meadows are plotted for a cloudless summer day. The higher LE values are attributed to the moist environment of rye grass and irrigated mixed-orchard ground cover for the sample area (Figure 14-21d), contributing to the more moderate sensible heat (H) levels during the day.

## The Urban Environment

For most of you reading this book, an urban landscape affects the temperatures you feel each day. Urban microclimates generally differ from those of nearby nonurban areas. In fact, the surface energy characteristics of urban areas are similar to energy balance traits of desert locations. Because more than 60% of the world's population will be living in cities by the year 2000, urban microclimatology and other specific environmental effects related to cities are important topics for physical geographers.

# Focus Study 4-1

## Solar Energy Collection and Concentration

Consider the following:

• Turn-of-the-century photographs of residences in southern California and Florida show solar (flat-plate) water heaters on many rooftops. Early twentieth-century ads in newspapers and merchandise catalogues featured the Climax solar water heater (1905) and the Day and Night water heater (1909). Applied solar energy principles and technologies are old! However, low-priced natural gas and oil displaced many of these early applications.

• The insolation receipt in just 35 minutes at the surface of the United States exceeds the amount of energy derived from the burning of fossil fuels (coal, oil, natural gas) in a year.

• An average building in the United States receives 6 to 10 times more energy from the Sun hitting its exterior than is required to heat the inside!

• A power line costs anywhere from $15,000 to $35,000 per kilometer to install into a rural site. A 500-W photovoltaic system that converts sunlight directly into electricity costs $15,000 (including batteries, at 1994 prices), and prices are dropping. Such a system supplies more than enough electricity for lights, television, computer, a water pump, and some appliances.

• Simple strategies for collecting and concentrating solar energy are economically competitive in many parts of the world Figure 1. Yet they are ignored by oil companies and politicians and receive few subsidies or incentives for research, investment, and installation.

Insolation not only warms Earth's surface; it provides an inexhaustible supply of energy far into the future for humanity. Sunlight is direct, pervasive, and renewable and has been collected for centuries through various technologies. Yet it is essentially untapped.

By the early 1970s, the depletion of fossil fuels, their rising prices, and our growing dependence upon foreign sources spurred a reintroduction of tried-and-true solar technologies and energy-efficiency strategies. Cities such as Davis,

California, passed energy-conservation and solar-utilization ordinances to guide construction and to encourage solar-energy installations. The country seemed to be rediscovering these proven techniques. Unfortunately, during the early 1980s, much of this progress was halted by political decisions and severe budget cuts made by the federal government. A new national energy plan promises to renew the push for solar applications.

Rural villages in developing countries could benefit greatly from the simplest, most cost-effective solar application—the *solar-box cooker*, or *solar-panel cooker*. For example, people in Kenya walk many kilometers collecting fuelwood for cooking fires (Figure 1a). Each village and refugee camp is surrounded by impoverished land, stripped bare of wood. Using solar cookers, villagers are able to cook meals and sanitize their drinking water without scavenging for wood (Figure 1b). Villagers in Guatemala have learned to make their own solar-box cookers (Figure 1c).

In developing countries, the money for electrification (a *centralized* technology) is not available despite the push from developed countries and energy corporations for large capital-intensive power projects. In such countries, the pressing need is for *decentralized* energy sources, appropriate in scale to everyday needs, such as cooking, heating water, and pasteurization. Net per capita (per person) cost for solar cookers is far less than for centralized electrical production, regardless of fuel source.

### Collecting and Concentrating Solar Energy

Any surface that receives light from the Sun is a *solar collector*. But the diffuse nature of solar energy received at the surface requires that it be collected, concentrated, transformed, and stored to be useful. Space heating is the simplest application. Windows that are carefully designed and placed allow sunlight to shine into a building, where it is absorbed and converted into sensible heat. Here we have an everyday application of the greenhouse effect.

A *passive solar system* captures heat energy and stores it in a "thermal mass," such as water-filled tanks, adobe, tile, or concrete. Three key features of passive solar design are (1) large areas of glass facing south toward the Sun (facing north in the Southern Hemisphere), (2) a thermal storage medium, and (3) shades or screens to prevent light entry in the warm summer months. At night, shades or other devices can cover glass areas to prevent energy loss from the structure. An *active solar system* involves heating water or air in a collector and then pumping it through a plumbing system to a tank where it can provide hot water for direct use or for space heating.

Solar energy systems can generate heat energy of an appropriate scale for approximately half the present domestic applications in the United States (space heating and water heating). In marginal climates, solar-assisted water and space heating is feasible as a backup; even in New England and the Northern Plains states, solar-efficient collection systems prove effective.

Focusing (concentrating) mirrors, such as Fresnel lenses, or parabolic (curved surface) troughs and dishes can be used to attain very high temperatures to heat water or other heat-storing fluids. Kramer Junction, California, about 225 km (140 mi) northeast of Los Angeles, has the world's largest operating solar electric-generating facility, with a capacity of 354 MW (megawatts; 354 million watts). This is a moderate-sized power plant. Long troughs of computer-guided curved mirrors concentrate sunlight to create temperatures of 390°C (735°F) in vacuum-sealed tubes filled with synthetic oil. The heated oil is then used to heat water; the heated water produces steam that rotates turbines to generate cost-effective electricity. The facility converts 23% of the sunlight it receives into electricity during peak hours.

Thirty years of contracts to buy this electrical power already are signed, and utility companies are seeking the construction of more installed capacity (Figure 2). But the success of this private

**FIGURE 1**

**The solar-cooking solution.**
(a) Five women haul firewood many miles to the Dadaab refugee camp in northeastern Kenya.

(b) Kenyan women in training to use their solar panel cookers, which do not require scavenging the countryside for scarce fuelwood.

(c) A woman in Guatemala cooking with her homemade solar-box cooker. These simple cookers collect direct and diffuse insolation through transparent glass or plastic and trap infrared radiation in an enclosed box or cooking bag. (This is a small-scale, efficient application of the greenhouse effect.) Construction is easy, using cardboard components. Temperatures easily exceed 105°C (220°F) for baking, boiling, purifying water, and sterilizing instruments. [Photos by Solar Cookers International, Sacramento, CA.]

venture appears to be tied to the political arena, for without tax incentives and formal encouragement, in amounts at least equal to those given the fossil fuel industry, it is difficult to operate such a plant. The original operators declared bankruptcy in 1992. The plant is now operating under the Belgian-owned, Israel-based, Solel Corporation.

**Electricity Directly from Sunlight**

Producing electricity by photovoltaic cells (PVs) is a technology that has been used in spacecraft since 1958. Familiar to us all are the solar cells in pocket calculators (some 100 million units now in use). They represent about 4 MW of electrical generation, or 8% of the global market for PVs. When sunlight shines upon a semiconductor material in these cells, it stimulates a flow of electrons (an electrical current). The efficiency of these cells has gradually improved, and they now are generally cost-competitive.

Obvious drawbacks of both solar-heating and solar-electric systems are periods of cloudiness and night, which inhibit operations. Research is under way to enhance energy-storage technologies, such as hydrogen fuel production (using energy to extract hydrogen from water for later use in producing more energy) and to improve battery technology.

Rooftop photovoltaic electrical generation is now cheaper than power line construction to rural sites. As of 1995, some 200,000 homes in Mexico, Indonesia, South Africa, India, and elsewhere have PV roof systems. Norway, at the same high latitudes as Alaska, has 50,000 units operating and is adding about 8,000 new PV systems a year.

**The Promise of Solar Energy**

Solar energy is a wise choice for the future. It is directly available to the consumer; it is based on a renewable energy source of an appropriate scale for end-use needs (it matches them well); and most solar strategies are labor-intensive (rather than capital-intensive). Solar seems preferable to further development of our decreasing fossil fuel reserves, further increases in oil imports, investment in foreign military incursions, or more troubled nuclear power.

Continued dependence on large centralized power production is undesirable because it is indirect from the consumer (fuel is burned elsewhere), nonrenewable, costly, and capital-intensive. Whether the alternative path of solar energy is followed comes down to political control. Much of the technology is ready for installation and is cost-effective when all direct and indirect costs are considered.

**FIGURE 2**
**Solar thermal energy production.**
Kramer International solar thermal energy installation in southern California. [Photos courtesy of Kramer International Limited, Los Angeles.]

At least six factors contribute to urban microclimates:

1. *Urban surfaces typically are metal, glass, asphalt, concrete, or stone.* These surfaces conduct up to three times more energy than a wet sandy soil and thus are warmer. The energy-storage capacity of these materials also exceeds that of most natural surfaces, making cities **urban heat islands**. Consequently, during the day and evening, temperatures above urban surfaces are higher than those above natural areas. At night urban surfaces rapidly radiate this stored energy to the atmosphere, producing minimum temperatures some 5–8 C° (9–14 F°) warmer than the surrounding countryside, especially on calm, clear nights. As a result, both maximum and minimum temperatures are higher in urban areas, although the

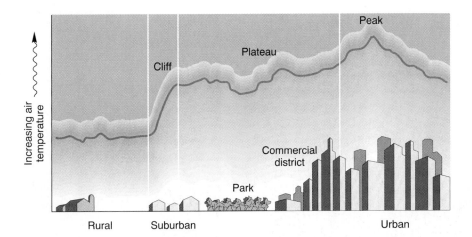

**FIGURE 4-22**
**Urban heat island profile.**
Generalized cross section of a typical urban heat island. The trend of the temperature gradient is described by the terms cliff at the edge of the city, where temperatures steeply rise; plateau over the suburban built-up area, and peak where temperature is highest at the urban core. Note the slight cooling over the park area. [Reprinted by permission from T. R. Oke, *Boundary Layer Climates.* New York: Methuen & Co., 1978, p. 254.]

effect of heat island characteristics is more profound during nighttime.

2. *Urban surfaces behave differently from natural surfaces in energy characteristics.* Albedo values of urban surfaces are lower, leading to a higher net radiation value. However, urban surfaces expend more of that energy as sensible heat than do nonurban areas; more than 70% of the net radiation in an urban setting is spent in this way. Figure 4-22 illustrates a generalized cross section of a typical urban heat island, showing increasing temperatures toward the center of the city.

3. *The irregular geometric shapes of a modern city affect radiation patterns and winds.* Incoming insolation is caught in mazelike reflection and radiation "canyons," which tend to trap energy that is conducted into surface materials, thus increasing temperatures (Figure 4-23). In natural settings, insolation is more readily reflected, stored in plants, or converted into latent heat through evaporation. Buildings also interrupt wind flows, diminishing heat loss through advective (horizontal) movement. Winds average 25% lower velocity in cities than in rural areas (although buildings create local turbulence and funneling effects). Wind flows can significantly reduce heat island effects by increasing convective mixing and sensible heat loss. Thus, maximum heat island effects in the most built-up areas of a city occur on calm, clear days and nights.

4. *Human activity alters the heat characteristics of cities.* For example, during summer in New York City, electricity production and use of fossil fuels release energy equivalent to 25–50% of the arriving insolation. In the winter, urban-generated sensible heat averages 250% greater than arriving insolation! In fact, for many northern cities in North America, winter heating requirements are actually reduced by this enhanced community heat island effect.

5. *Many urban surfaces are sealed (built on and paved), so water cannot reach the soil.* Central business district surfaces are on the average 50% sealed, whereas suburbs are about 20% sealed. Because urban precipitation cannot penetrate the soil, there is more water runoff. Urban areas respond much as a desert landscape: A storm may cause a flash flood over the hard, sparsely vegetated surfaces, only to be followed by a return to dry conditions a few hours later. Little of the net radiation is expended for evaporation on such a surface.

6. *Air pollution, including gases and aerosols, is greater in urban areas than in comparable natural settings.* Pollution increases the atmosphere's reflectivity above a city, thus reducing insolation to the ground. As a countering action, this pollution blanket also absorbs infrared radiation, reradiating infrared downward. Every major city produces its own **dust dome** of airborne pollution, which can be blown from the city in elongated plumes (Figure 4-23). The increased particulates in pollution are condensation nuclei for water vapor, producing increased cloud formation and the potential for increased precipitation. Although these precipitation effects of the urban heat island are difficult to isolate, research suggests that urban-stimulated precipitation occurs downwind from cities (Figure 4-24).

Table 4-1 compares climatic factors of rural and urban environments. The worldwide trend toward greater urbanization is placing more and more people on urban heat islands.

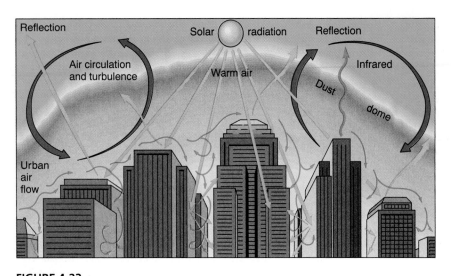

**FIGURE 4-23**
**The urban environment.**
Insolation, wind movements, and dust dome in city environments. The photograph shows an all too common sight on an urban afternoon. [Photo inset by author.]

**FIGURE 4-24**
**An urbanized landscape.**
Urbanized areas such as Chicago possess physical attributes different from surrounding natural areas. Studies in the 1970s established the effects of the Chicago metropolitan region on weather downwind. [Photo by author.]

# Summary and Review—Atmosphere and Surface Energy Balances

✔ *Identify* **the pathways of solar energy through the troposphere to Earth's surface: transmission, refraction, albedo (reflectivity), scattering, diffuse radiation, conduction, convection, and advection.**

Earth's biosphere is powered by radiant energy from the Sun that cascades through complex circuits to the surface. Our budget of atmospheric energy comprises short-wave radiation *inputs* (ultraviolet light, visible light, and

**TABLE 4-1**

| Average Differences in Climatic Elements between Urban and Rural Environments | |
|---|---|
| **Element** | **Urban Compared with Rural Environs** |
| **Contaminants** | |
| Condensation nuclei | 10 time more |
| Particulates | 10 times more |
| Gaseous admixtures | 5–25 times more |
| **Radiation** | |
| Total on horizontal surface | 0%–20% less |
| Ultraviolet, winter | 30% less |
| Ultraviolet, summer | 5% less |
| Sunshine duration | 5%–15% less |
| **Cloudiness** | |
| Clouds | 5%–10% more |
| Fog, winter | 100% more |
| Fog, summer | 30% more |
| **Precipitation** | |
| Amounts | 5%–15% more |
| Days with <5 mm (0.2 in.) | 10% more |
| Snowfall, inner city | 5%–10% less |
| Snowfall, downwind (lee) of city | 10% more |
| Thunderstorms | 10%–15% more |
| **Temperature** | |
| Annual mean | 0.5–3.0 C° (0.9–5.4 F°) more |
| Winter minima (average) | 1.0–2 C° (1.8–3.6 F°) more |
| Summer maxima | 1.0–3 C° (1.8–3.0 F°) more |
| Heating degree days | 10% less |
| **Relative Humidity** | |
| Annual mean | 6% less |
| Winter | 2% less |
| Summer | 8% less |
| **Wind Speed** | |
| Annual mean | 20%–30% less |
| Extreme gusts | 10%–20% less |
| Calm | 5%–20% more |

*Source:* H. E. Landsberg, *The Urban Climate*, International Geophysics Series, vol. 28 (1981), p. 258. Reprinted by permission from Academic Press.

near-infrared wavelengths) and longwave radiation *outputs* (thermal infrared).

**Transmission** refers to the passage of shortwave and longwave energy through either the atmosphere or water. The speed of insolation entering the atmosphere changes as it passes from one medium to another; the change of speed causes a bending action called **refraction**.

A portion of arriving energy bounces directly back into space without being converted into heat or performing any work. This returned energy is called **reflection**. **Albedo** is the reflective quality (intrinsic brightness) of a surface. It is an important control over the amount of insolation that is available for absorption by a surface. We state

albedo as the percentage of insolation that is reflected. Earth and its atmosphere reflect 31% of all insolation when averaged over a year.

An increase in albedo and reflection of shortwave radiation caused by clouds is described by the term **cloud-albedo forcing**. Also, clouds can act as insulation, thus trapping longwave radiation and raising minimum temperatures. An increase in greenhouse warming caused by clouds is described by the term **cloud-greenhouse forcing**.

Insolation encounters an increasing density of atmospheric gases that absorb and reemit it *without altering its wavelength*s. This **scattering** represents 7% of Earth's reflectivity, or albedo. Some incoming insolation is diffused by clouds and atmosphere and is transmitted to Earth as **diffuse radiation**, the downward component of scattered light.

**Absorption** is the assimilation of radiation by molecules of a substance and its conversion from one form to another—for example, visible light to infrared radiation. **Conduction** is the molecule-to-molecule transfer of energy as it diffuses through a substance.

Energy also is transferred by gases and liquids by **convection** (when the physical mixing involves a strong vertical motion) or **advection** (when the dominant motion is horizontal). In the atmosphere or bodies of water, warmer portions tend to rise (they are less dense) and cooler portions tend to sink (they are more dense), establishing patterns of convection.

transmission (p. 90)
refraction (p. 91)
reflection (p. 91-92)
albedo (p. 92)
cloud-albedo forcing (p. 92)
cloud-greenhouse forcing (p. 94)
scattering (p. 94)
diffuse radiation (p. 94)
absorption (p. 94)
conduction (p. 95)
convection (p. 96)
advection (p. 96)

1. Diagram a simple energy balance for the troposphere. Label each shortwave and longwave component and the directional aspects of related flows.

2. Define refraction. How is it related to daylength? To a rainbow? To the beautiful colors of a sunset?

3. List several types of surfaces and their albedo values. Explain the differences among these surfaces. What determines the reflectivity of a surface?

4. Using Figure 4-7, explain the seasonal differences in albedo values for each hemisphere. Be specific, using the albedo values given in Figure 4-5 where appropriate.

5. Why is the lower atmosphere blue? What would you expect the sky color to be at 50 km (30 mi) altitude? Why?

6. Define the concepts transmission, absorption, diffuse radiation, conduction, and convection.

✔ *Analyze* **the effect of clouds and air pollution on solar radiation received at ground level and *describe* what happens to insolation when clouds are in the atmosphere.**

Clouds reflect *insolation*, thus cooling Earth's surface—*cloud-albedo forcing*. Yet clouds also act as insulation, thus trapping longwave radiation and raising minimum temperatures—*cloud-greenhouse forcing*. Clouds affect the heating of the lower atmosphere depending on cloud type, height, and thickness (water content and density). High-altitude, ice-crystal clouds reflect insolation with albedos of about 50%, producing a net *cloud-greenhouse forcing* (warming); thick, lower cloud cover reflects about 90%, producing a net *cloud-albedo forcing* (cooling).

Emissions of sulfur dioxide, and the subsequent chemical reactions in the atmosphere form *sulfate aerosols*, which act either as insolation-reflecting haze in clear-sky conditions or as a stimulus to condensation in clouds that increases reflectivity.

7.   What role do clouds play in the Earth-atmosphere radiation balance? Is cloud type important? Compare high, thin cirrus clouds and lower, thick stratus clouds.

8.   In what way does the presence of sulfate aerosols affect solar radiation received at ground level? How does it affect cloud formation?

---

✔ *Review* **the energy pathways in the Earth-atmosphere system, the greenhouse affect, and the patterns of global net radiation.**

The Earth-atmosphere energy system naturally balances itself in a steady-state equilibrium. It does so through energy transfers that are *nonradiative* (convection, conduction, and the latent heat of evaporation) and *radiative* (by infrared radiation between the surface, the atmosphere, and space).

Some infrared radiation is absorbed by carbon dioxide, water vapor, methane, CFCs (chlorofluorocarbons), and other gases in the lower atmosphere and is then reradiated to Earth, thus delaying energy loss to space. This process is the **greenhouse effect**. In the atmosphere, infrared radiation is not actually trapped, as it would be in a greenhouse, but its passage to space is delayed (heat energy is detained in the atmosphere) as it is absorbed and reradiated.

Between the tropics, a high insolation angle and consistent daylength cause more energy to be gained than lost (there are energy surpluses). In the polar regions, an extremely low insolation angle, highly reflective surfaces, and up to 6 months of no insolation annually cause more energy to be lost (there are energy deficits). This imbalance of net radiation from tropical surpluses to the polar deficits drives a vast global circulation of both energy and mass.

Surface energy balances are used to summarize the energy expenditure for any location. Surface energy measurements are used as an analytical tool of **microclimatology**. Adding and subtracting the energy flow at the surface lets us calculate **net radiation (NET R)**, or the balance of all radiation at Earth's surface—shortwave (SW) and longwave (LW).

> greenhouse effect (p. 97)
> microclimatology (p. 101)
> net radiation (NET R) (p. 101)

9.   What are the similarities and differences between an actual greenhouse and the gaseous atmospheric greenhouse? Why is Earth's greenhouse changing?

10.   In terms of energy expenditures for latent heat of evaporation, describe the annual pattern as mapped in Figure 4-19.

11.   Generalize the pattern of global net radiation. How might this pattern drive the atmospheric weather machine? (See Figures 4-14 and 4-18.)

12.   In terms of surface energy balance, explain the term net radiation (NET R).

13.   What are the expenditure pathways for surface net radiation? What kind of work is accomplished?

14.   What is the role played by latent heat in surface energy budgets?

15.   Compare the daily surface energy balances of El Mirage and Pitt Meadows. Explain the differences.

---

✔ *Plot* **the daily radiation curves for Earth's surface and *label* the key aspects of incoming radiation, air temperature, and the daily temperature lag.**

The greatest insolation input occurs at the time of the summer solstice in each hemisphere. Air temperature responds to seasons and variations in insolation input. Within a 24-hour day, air temperature peaks at around 3:00–4:00 P.M. and dips to its lowest point right at or slightly after sunrise.

Air temperature lags behind each day's peak insolation. The warmest time of day occurs not at the moment of maximum insolation but at that moment when a maximum of insolation is absorbed.

16.   Why is there a temperature lag between the highest Sun altitude and the warmest time of day? Relate your answer to the insolation and temperature patterns during the day.

---

✔ *Portray* **typical urban heat island conditions and *contrast* the microclimatology of urban areas with that of surrounding rural environments.**

A growing percentage of Earth's people live in cities and experience their unique set of altered microclimatic effects:

increased conduction, lower albedos, higher NET R values, increased runoff, complex radiation and reflection patterns, anthropogenic heating, and the gases, dusts, and aerosols of urban pollution. Urban surfaces of metal, glass, asphalt, concrete, and stone conduct up to three times more energy than wet sandy soil and thus are warmed as described by the term **urban heat island**. Air pollution, including gases and aerosols, is greater in urban areas than in rural ones. Every major city produces its own **dust dome** of airborne pollution.

urban heat island (p. 108)
dust dome (p. 109)

17. What is the basis for the urban heat island concept? Describe the climatic effects attributable to urban as compared with nonurban environments.

18. Which of the items in Table 4-1 have you yourself experienced? Explain.

19. Assess the potential for solar energy applications in our society. What are some negatives? What are some positives?

---

 ## NetWork

The *Geosystems Home Page* provides on-line resources for this chapter on the World Wide Web. You will find review exercises, specific updates for items in the chapter, suggested readings, and links to interesting related pathways on the Internet (click on the Table of Contents link and select this chapter). *Geosystems* is at: **http://www.prenhall.com/geosystm**

# 5

# Global Temperatures

**Principal Temperature Controls**

**Earth's Temperature Patterns**

**Air Temperature and the Human Body**

**Summary and Review**

## Key Learning Concepts

After reading the chapter, you should be able to:

- *Define* the concepts of temperature, kinetic energy, sensible heat and *distinguish* among Kelvin, Celsius, and Fahrenheit scales and how they are measured.

- *List* and *review* the principal controls and influences that produce global temperature patterns.

- *Review* the factors that produce marine and continentality differences as they influence temperatures and *utilize* several pairs of stations to illustrate these differences.

- *Interpret* the pattern of Earth's temperatures from their portrayal on January and July temperature maps and on a map of annual temperature ranges.

- *Define* apparent temperature, or sensible temperature, and *relate* specific physiological effects of low temperature and high temperature on the human body; *utilize* the wind-chill and heat-index charts to determine some apparent temperatures.

*People and sheep seeking shade in Meroë, Sudan.* [Photo by Douglas Waugh/Peter Arnold, Inc.]

Wat is the temperature now—both indoors and outdoors—as you read these words? How is it measured, and what does the value mean? How is air temperature influencing your plans for the day? Air temperature plays a remarkable role in our lives, both at the microlevel and at the macrolevel. We read of heat waves and cold spells affecting people, crops, events, and energy consumption. For example, in 1995 devastating summer heat and high humidity killed over 1000 people in the Midwest and East. In 1986 the Space Shuttle *Challenger* exploded when below-freezing temperatures caused a gasket to crack and a fuel leak. Our bodies subjectively sense temperature and judge comfort, and they react to changing temperatures.

A variety of temperature regimes affect cultures, decision making, and resources consumed across the globe. Global temperature patterns presently are changing slowly in a warming trend that is the subject of much scientific, geographic, and political interest. International cooperation and research are progressing toward a better understanding of such short-term global change.

Of course, Earth's temperatures have not always been as they are today. In fact, humans have too recently populated Earth to have experienced the planet's "normal" climate—those climate patterns characteristic of Earth's first 4 billion years. Chapter 17 examines Earth's past temperatures and some of the temperature controls that influence long-term changes. Focus Study 5-1 introduces some important temperature concepts to help you with our study of global temperatures.

# Principal Temperature Controls

**Temperature** is a measure of the average kinetic energy (motion) of individual molecules in matter. We feel the effect of temperature as the *sensible heat* transfer from warmer objects to cooler objects. For instance, when you jump into a cool lake you can sense the heat transfer from your skin to the water as kinetic energy leaves your body and flows to the water; a chill develops.

Principal controls and influences upon temperature patterns include latitude, altitude, cloud cover, and land-water heating differences. Let us begin by examining each of the principal temperature controls.

## *Latitude*

*Insolation* is the single most important influence on temperature variations. Figures 2-13 and 2-14 show how insolation intensity decreases as one moves away from the subsolar point, a point that migrates annually between the Tropic of Cancer and Tropic of Capricorn—that is, between 23.5° N and 23.5° S. In addition, daylength and Sun angle change throughout the year, increasing in seasonal effect with increasing latitude. From equator to poles, Earth ranges from continually warm, to seasonally variable, to continually cold (Figure 5-1).

(a)

(b)

**FIGURE 5-1**
**Cold and warm conditions.**
Within its moderate climate range, Earth's surface experiences a variety of temperature regimes, from the cold of a Canadian winter near Igoolik, N.W.T. Canada (a) to the hot and humid summers along the Gulf Coast in Florida (b).
[Photos by (a) Robert Semeniuk/Stock Market; (b) author.]

# Focus Study 5-1

## Temperature Concepts, Terms, and Measurements

Heat and temperature are not the same. They are two different but related concepts. *Heat* is a form of energy that flows from one system or object to another because the two are at different temperatures. *Temperature* is a measurement—a measure of molecular motion in a substance caused by the kinetic energy present; the relative hotness or coldness of something. Temperature and heat are related because changes in temperature are caused by the absorption or emission (gain or loss) of heat energy. The term *heat energy* is frequently used to describe energy that is added to or removed from a system or substance.

### Temperature Scales

The temperature at which all atomic and molecular motion in a substance completely stops is called 0° *absolute temperature* (commonly, "absolute zero"). Its value on the different temperature-measuring scales is −273° Celsius (C), −459.4° Fahrenheit (F), and 0 Kelvin (K). Figure 1 compares these three scales. Formulas for converting between Celsius, SI (Système International), and English units are on the inside back cover of this text.

The Fahrenheit scale places the melting point of ice at 32°F and the boiling point of water at 212°F. The scale is named for Daniel G. Fahrenheit, a German physicist (1686–1736), who used these odd values based on the coldest temperature he could achieve in his laboratory (which he called 0°F) and the approximate temperature of the human body, thought to be about 100°F. His estimates placed the melting point of ice at 32°F, with 180 subdivisions in his scale to the boiling point of water at 212°F. (Note that there is only one melting point for ice, but there are many freezing points for water ranging from 32°F down to −40°F, depending on its purity, volume, and certain conditions in the atmosphere.)

About a year after the adoption of the Fahrenheit scale, the Celsius scale (formerly called centigrade) was developed by Swedish astronomer Anders Celsius (1701–1744). He placed the melting point of ice at 0° and divided his scale into 100 degrees, using a decimal scale, to the 100° boiling temperature of water at sea level.

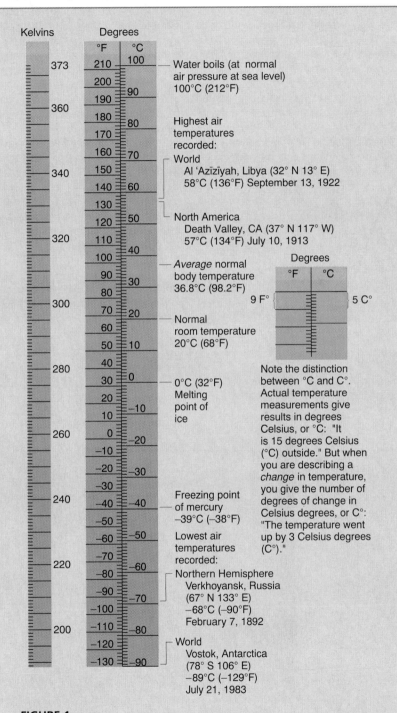

**FIGURE 1**
**Temperature scales.**
Scales for expressing temperature in Kelvins (K) and degrees Celsius and Fahrenheit.

The Kelvin scale is named after British physicist Lord Kelvin (born William Thomson, 1824–1907). He proposed the scale in 1848. It is used in science because temperature readings start at absolute zero and thus are proportional to the actual kinetic energy in a material. The Kelvin scale's melting point of ice is 273K, and its boiling point of water is 373K.

Most countries use the Celsius scale to express temperature. The United States remains the only major country still using the Fahrenheit scale. This textbook presents Celsius (with Fahrenheit equivalents in parentheses) throughout, to help bridge this transitional era in the United States. The continuing pressure of the scientific community and corporate interests, not to mention the rest of the world, makes adoption of Celsius and SI units inevitable for the United States.

### Measuring Temperature

Outdoor temperature usually is measured with a mercury thermometer or alcohol thermometer in a sealed glass tube.

(Fahrenheit invented the alcohol and mercury thermometers.) Alcohol is preferred in cold climates because it freezes at –112°C (–170°F), whereas mercury freezes at –39°C (–38.2°F). The principle of these thermometers is simple: When these fluids are heated, they expand; upon cooling, they contract. A thermometer stores fluid in a small reservoir at one end and is marked with calibrations to measure the expansion or contraction of the fluid, which reflects the temperature of the thermometer's environment.

Figure 2 shows a mercury minimum-maximum thermometer. It preserves readings of the day's highest and lowest temperatures until it is reset (by moving the markers with a magnet). Another type, the recording thermometer, creates an inked record on a turning drum; it is usually set for one full rotation every 24 hours or every 7 days.

Thermometers for standardized official readings are placed outdoors, in shelters that are white (for high albedo) and louvered (for ventilation) to avoid

overheating of the instruments (Figure 3). They are placed at least 1.2–1.8 m (4–6 ft) above a surface, usually on turf; in the United States 1.2 m (4 ft) is standard. Official temperatures are measured in the shade to prevent the effect of direct insolation. Temperature readings are taken daily, sometimes hourly, at more than 10,000 weather stations worldwide. Some stations with recording equipment also report the duration of temperatures, rates of rise or fall, and variation over time throughout the day and night.

Daily minimum-maximum readings are averaged to get the *daily mean temperature*. The *monthly mean temperature* is the total of daily mean temperatures for the month divided by the number of days in the month. An *annual temperature range* expresses the difference between the lowest and highest monthly mean temperatures for a given year. If you install a thermometer for outdoor temperature reading, be sure to avoid direct sunlight on the instrument and place it in an area of good ventilation.

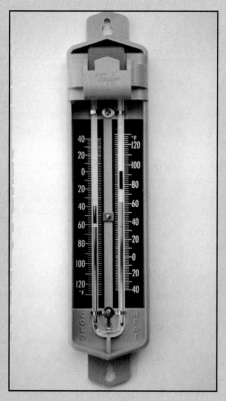

**FIGURE 2**
**A minimum-maximum thermometer.**
[Photo by Bobbé Christopherson.]

**FIGURE 3**
**Instrument shelter.**
This standard weather instrument shelter is white (for high albedo) and louvered (for ventilation).
[Photo by Mettee Photography/Belfort Instruments, Inc.]

## Altitude

Within the troposphere, temperatures decrease with increasing altitude above Earth's surface. (Recall that the *normal lapse rate* of temperature change with altitude is 6.4 C°/1000 m, or 3.5 F°/1000 ft; see Figure 3-6). Thus, worldwide, mountainous areas experience lower temperatures than do regions nearer sea level, even at similar latitudes. The density of the atmosphere also diminishes with increasing altitude. In fact, the density of the atmosphere at an elevation of 5500 m (18,000 ft) is about half of that at sea level. As the atmosphere thins, its ability to absorb and radiate sensible heat is reduced.

The consequences are that, at high elevations, average air temperatures are lower, nighttime cooling is greater, and the temperature range between day and night is greater than at low elevations. Also, the temperature difference between areas of sunlight and shadow is greater than at sea level. Temperatures may decrease noticeably in the shadows and shortly after sunset. Surfaces both gain energy rapidly and lose energy rapidly to the thinner atmosphere. Also, at higher elevations, the insolation received is more intense because of the reduced mass of atmospheric gases. As a result of this intensity, the ultraviolet energy component makes sunburn a distinct hazard.

The snowline seen in mountain areas indicates where winter snowfall exceeds the amount of snow lost through summer melting and evaporation. The snowline's location is a function both of elevation and latitude, so glaciers can exist even at equatorial latitudes if the elevation is great enough. In equatorial mountains, the snowline occurs at approximately 5000 m (16,400 ft) because of the latitude (more insolation in the tropics produces temperatures that place the snowline at higher altitude). Permanent ice fields and glaciers exist on equatorial mountain summits in the Andes and East Africa. With increasing latitude toward the poles, snowlines gradually descend in elevation from 2700 m (8850 ft) in the midlatitudes to lower than 900 m (2950 ft) in southern Greenland.

Two cities in Bolivia illustrate the interaction of the two temperature controls, latitude and altitude. Figure 5-2 displays temperature data for the cities of Concepción and La Paz, which are near the same latitude (about 16° S). Note the elevation, average annual temperature, and precipitation for each location noted on the figure.

The hot, humid climate of Concepción at its much lower elevation stands in marked contrast to the cool, dry climate of highland La Paz. People living around La Paz actually grow wheat, barley, and potatoes—crops characteristic of the cooler midlatitudes—despite the fact that La Paz is 4103 m (13,461 ft) above sea level (Figure 5-3). (For comparison, the summit of Pikes Peak in Colorado is at 4301 m or 14,111 ft, and Mount Rainier, Washington, is at 4392 m or 14,410 ft.)

The combination of elevation and low-latitude location guarantee La Paz nearly constant daylength and moderate temperatures, averaging about 9°C (48°F) every month. Such moderate temperature and moisture conditions lead to the formation of more fertile soils than those found in the warmer, wetter climate of Concepción.

## Cloud Cover

Orbital surveys reveal that approximately 50% of Earth is cloud-covered at any one time. Clouds moderate temperature, and their effect varies with cloud type, height, and density. Because their moisture reflects, absorbs, and liberates large amounts of energy, clouds reduce the insolation that reaches the surface. In general, they lower daily maximum temperatures and raise nighttime minimum temperatures. Clouds also reduce latitudinal and seasonal temperature differences.

At night, clouds act as insulation and radiate longwave energy, preventing rapid energy loss. During the day, clouds reflect insolation as a result of their high albedo values. In the last chapter (Figures 4-8 and 4-11) we discussed cloud-albedo forcing and cloud-greenhouse forcing as they are related to the presence of clouds and cloud types.

Clouds are the most variable factor influencing Earth's radiation budget, so they are the subject of much investigation and are critical to computer models of atmospheric behavior. The International Satellite Cloud Climatology Project, part of the World Climate Research Program, is presently in the midst of such research. The Earth Radiation Budget experiments mentioned in Chapters 2 and 4 are assessing cloud effects on longwave, shortwave, and net radiation patterns as never before possible.

## Land-Water Heating Differences

Another major control over temperature is the pronounced difference in the heating of land and water by insolation. Earth presents these two surfaces in an irregular arrangement of continents and oceans. Each absorbs and stores energy differently than the other and therefore contributes to the global pattern of temperature. Moderate temperature patterns are associated with water bodies; extreme temperatures occur inland.

The physical nature of land (rock and soil) and water (oceans, seas, and lakes) is the reason for these **land-water heating differences**. Figure 5-4 visually summarizes the following discussion of the five land-water temperature controls: evaporation, transparency, specific heat, movement, and ocean currents and sea-surface temperatures.

***Evaporation.*** More of the energy arriving at the ocean's surface is expended for evaporation than is expended over a comparable area of land, simply because so much

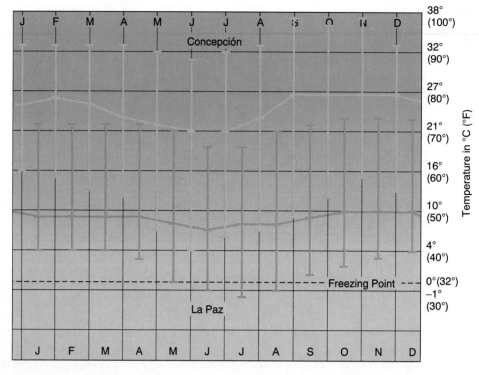

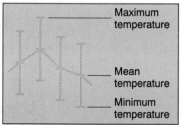

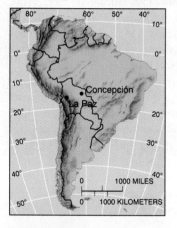

| | Station | |
|---|---|---|
| | **Concepción, Bolivia** | **La Paz, Bolivia** |
| Latitude/longitude | 16° 15′ S 62° 03′ W | 16° 30′ S 68° 10′ W |
| Elevation | 490 m (1607.6 ft) | 4103 m (13,461 ft) |
| Avg. ann. temperature | 24°C (75.2°F) | 9°C (48.2°F) |
| Ann. temperature range | 5 C° (9 F°) | 3 C° (5.4 F°) |
| Ann. precipitation | 121.2 cm (47.7 in.) | 55.5 cm (21.9 in.) |
| Population | 768,000 | 993,000 |

**FIGURE 5-2**

**Effects of latitude and altitude.**

Comparison of temperature patterns in two Bolivian cities: La Paz (4103 m, 13,460 ft) and Concepción (490 m, 1608 ft).

water is available. An estimated 84% of all evaporation on Earth is from the oceans. When water evaporates and thus changes to water vapor, *heat energy is absorbed in the process and is stored in the water vapor as latent heat.* We saw this process in Figure 4-19, the map of energy expended for the latent heat of evaporation. Latent heat is discussed fully in Chapter 7.

You can experience this evaporative heat loss (cooling) by wetting the back of your hand and then blowing on the moist skin. Sensible heat energy is drawn from your skin to supply some of the energy for evaporation, and you feel the cooling. Similarly, as surface water evaporates, substantial energy is absorbed from the immediate environment, resulting in a lowering of temperatures. Temperatures over land, which has far less water, are not as moderated by evaporative cooling as are marine locations.

***Transparency.*** The transmission of light obviously differs between soil and water: Solid ground is opaque, water is transparent. Consequently, light striking a soil surface does not penetrate but is absorbed, heating the ground surface. That energy is accumulated during times of exposure and is rapidly lost at night or in shadows.

Figure 5-5 shows the profile of diurnal (daily) temperatures for a column of soil and the atmosphere above it at a midlatitude location. You can see that maximum and minimum temperatures generally are experienced right at ground level. Below the surface, even at shallow depths, temperatures remain about the same throughout the day. This situation often exists at a beach, where surface sand may be painfully hot to your feet, but dig in your toes and you will feel the sand a few centimeters below the surface feels cooler and offers relief.

**FIGURE 5-3**

**High-elevation farming.**

People in these high-elevation villages grow potatoes and wheat in view of the permanent ice-covered peaks of the Andes Mountains in the background. The combination of low latitude and high elevation create consistent, but moderate, temperatures throughout the year, averaging about 9°C (48°F). [Photo by Comstock.]

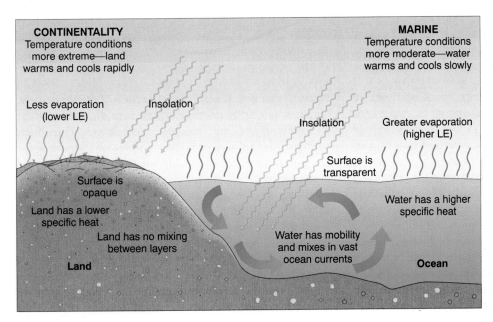

**CONTINENTALITY**
Temperature conditions more extreme—land warms and cools rapidly

Less evaporation (lower LE)

Insolation

Surface is opaque

Land has a lower specific heat

Land has no mixing between layers

**Land**

**MARINE**
Temperature conditions more moderate—water warms and cools slowly

Insolation

Greater evaporation (higher LE)

Surface is transparent

Water has a higher specific heat

Water has mobility and mixes in vast ocean currents

**Ocean**

**FIGURE 5-4**

**Land-water heating differences.**

The differential heating of land and water produces contrasting marine (more moderate) and continental (more extreme) temperature regimes. (LE, latent heat of evaporation, is the energy stored in water vapor.)

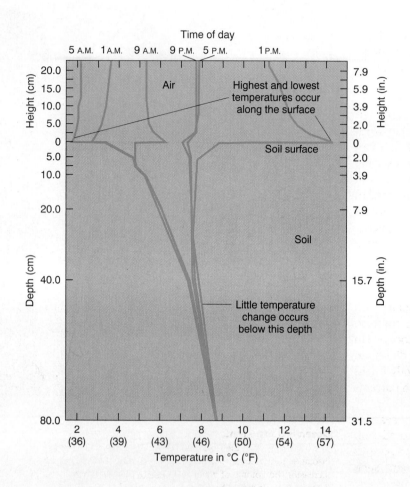

**FIGURE 5-5**

**Land is opaque.**

Profile of air and soil temperatures in Seabrook, New Jersey. Note that little temperature change occurs throughout the year at depths beyond a meter, whereas the daily extremes of temperature register along the surface, where insolation is absorbed. [Adapted from John R. Mather, *Climatology: Fundamentals and Applications.* New York: McGraw-Hill, 1974, p. 36. Adapted by permission.]

In contrast, when light reaches a body of water, it penetrates the surface because of water's **transparency**, transmitting light to an average depth of 60 m (200 ft) in the ocean (Figure 5-6). This illuminated zone is known as the *photic layer* and has been recorded in some ocean waters to depths of 300 m (1000 ft). *This characteristic of water results in the distribution of available heat energy over a much greater depth and volume*, forming a larger energy reservoir than that of the surface layers of the land.

***Specific Heat.*** When equal volumes of water and land are compared, water requires far more energy to increase its temperature than does land. In other words, *water can hold more heat than can soil or rock*, and therefore water is said to have a higher **specific heat**. On the average, the specific heat of water is about four times that of soil. A given volume of water represents a more substantial energy reservoir than an equal volume of land, so changing the temperature of the oceanic energy reservoir is a slower process than changing the temperature of land. Likewise, for that oceanic heat reservoir to lose heat energy requires more time than would a similar volume of land. The temperature response of water bodies is "sluggish" in comparison with land surfaces. For this reason, day-to-day temperatures near a substantial body of water tend to be moderated.

***Movement.*** Land is a rigid, solid material, whereas water is a fluid and is capable of movement. Differing temperatures and currents result in a mixing of cooler and warmer waters, and that *mixing spreads the available energy over an even greater volume* than if the water were still. Surface water and deeper waters mix, redistributing energy. Both ocean and land surfaces radiate longwave radiation at night, but land loses its energy more rapidly than does the greater mass of the moving oceanic energy reservoir.

***Ocean Currents and Sea-Surface Temperatures.*** Warm water adds energy to overlying air through high evaporation rates and transfers of latent heat. Thus, in an air mass, the amount of water vapor is affected by ocean temperatures in an interesting *negative feedback* mechanism. Across the globe, ocean water is rarely found warmer than 31°C (88°F). The higher the ocean temperature, the higher the evaporation rate will be with increasing amounts of energy lost from the ocean as latent heat. As water vapor content of the overlying air mass increases, the ability of the air to absorb longwave radiation also increases. Therefore, the air mass becomes warmer, enhancing the greenhouse effect. The warmer the air and the ocean become, the more evaporation will occur and more water vapor will enter the air mass. More water vapor leads to cloud formation, which reflects insolation and produces lower temperatures. Lower temperatures of air and ocean reduce evaporation rates and the ability of the air mass to absorb water vapor.

As a specific example, the **Gulf Stream** (described in Chapter 6) moves northward off the east coast of North America, carrying warm water far into the North Atlantic (Figure 5-7). As a result, the southern third of Iceland experiences much milder temperatures than would be expected for a latitude of 65° N, just below the Arctic Circle (66.5°). In Reykjavík, on the southwestern coast of Iceland, monthly temperatures average above freezing during all months of the year. Similarly, the Gulf Stream moderates temperatures in coastal Scandinavia and northwestern Europe.

In the western Pacific Ocean, the *Kuroshio*, or *Japan Current*, similar to the Gulf Stream, functions much the same in its warming effect on Japan, the Aleutians, and the northwestern margin of North America. In contrast, along midlatitude and subtropical west coasts, cool ocean currents influence air temperatures. When conditions in these regions are warm and moist, fog frequently forms in the chilled air over the cooler currents. Ocean currents are detailed further in Chapter 6.

Remote sensing from satellites is providing sea-surface temperature (SST) data that are well-correlated with actual sea-surface measurements. This correlation permits a thorough global assessment of SSTs in programs such

**FIGURE 5-6**
**Ocean is transparent.**
The transparency of the ocean at Taveuni Reef near Fiji permits insolation to penetrate to average depths of 60 m (200 ft), greatly increasing the volume of water that absorbs energy. [Photo by Copr. F. Stuart Westmorland/Photo Researchers, Inc.]

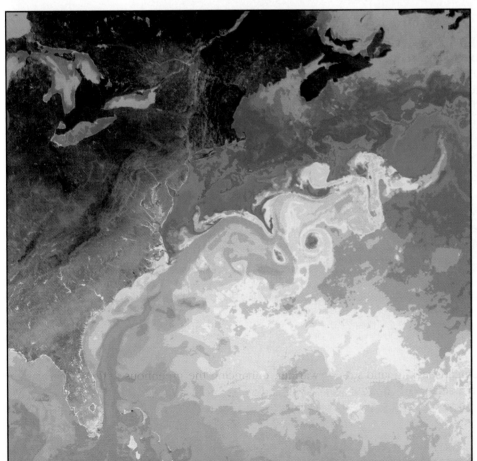

**FIGURE 5-7**
**The Gulf Stream.**
Satellite image of the warm Gulf Stream as it flows northward along the North American eastern coast. It is the streamlike flow in red, orange, and yellow. Instruments sensitive to infrared wavelengths produced this remote-sensing image, which covers approximately 11.4 million square kilometers (4.4 million square miles). Temperature differences are distinguished by computer-enhanced coloration: reds/oranges = 25°–29°C (76°–84°F), yellows/greens = 17°–24°C (63°–75°F); blues = 10°–16°C (50°–61°F); and purples = 2°–9°C (36°–48°F). [Imagery by NASA and NOAA.]

as Tropical Ocean Global Atmosphere (TOGA) and Coupled Ocean-Atmosphere Response Experiment (COARE).

Figure 5-8 displays data from *TIROS-N* and *NOAA* satellites, which provided University of Delaware scientists with a 10-year record of sea-surface temperatures. The red and orange area in the southwestern Pacific Ocean with temperatures above 28°C (82.4°F), occupying a region larger than the United States, is called the *Western Pacific Warm Pool*. Mean annual SSTs increased steadily from 1982 to 1991, with some fluctuations after 1987.

The Western Pacific Warm Pool has the highest average ocean temperatures in the world. This fact appears to be important in the early stages of El Niño–Southern Oscillation development (see the Focus Study in Chapter 10). Sea-surface temperatures are affected by these recurring weather events. Understanding the linkage between ocean and atmosphere systems is important in establishing accurate general circulation models for forecasting weather and long-term climate conditions.

***Summary of Marine vs. Continental Conditions.*** As noted, Figure 5-4 summarizes the operation of the five land-water temperature controls presented: evaporation, transparency, specific heat, movement, and ocean currents and sea-surface temperatures. The term **marine**, or *maritime*, is used to describe locations that exhibit the moderating influences of the ocean, usually along coastlines or on islands. **Continentality** refers to the condition of areas that are less affected by the sea and therefore have a greater range between maximum and minimum temperatures, daily and yearly.

The Canadian cities of Vancouver, British Columbia, and Winnipeg, Manitoba, exemplify marine and continental conditions (Figure 5-9). Both cities are at approximately 49° N latitude. Vancouver is at sea level and Winnipeg is at 248 m (814 ft) elevation. However, Vancouver has a more moderate pattern of average maximum and minimum temperatures. Vancouver's annual range of 16.0 C° (28.8 F°) is far less than Winnipeg's 38.0 C° (68.4 F°) range. In fact, Winnipeg's continental temperature pattern is more extreme in every aspect than that of maritime Vancouver.

A similar comparison of San Francisco and Wichita, Kansas, provides us another comparison of marine and continental conditions (Figure 5-10). Both cities are at approximately 37° 40′ N latitude. San Francisco is at about 5 m (16 ft), and Wichita is at 400 m (1300 ft) elevation. Summer fog helps delay until September the warmest

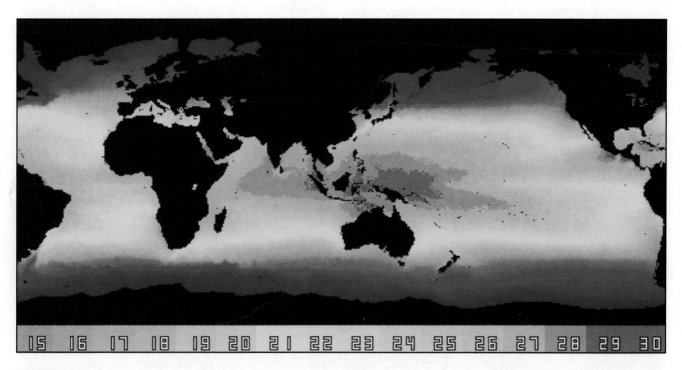

**FIGURE 5-8**

**Sea-surface temperatures.**

Average annual sea-surface temperatures for the period 1982–1991 as measured by *TIROS-N* and *NOAA* satellites. The Western Pacific Warm Pool, the warmest area of all oceans, is well defined. These remotely sensed data are closely correlated with actual measurements of the ocean's surface temperature. During an El Niño event, discussed in Chapter 10, these warm regions of the Pacific Ocean spread to the east toward South America. [Satellite image processed by Dr. Xiao-Hai Yan, Center for Remote Sensing, University of Delaware. Reprinted by permission.]

**FIGURE 5-9**

**Marine and continental cities—Canada.**

Comparison of temperatures in coastal Vancouver, British Columbia, and continental Winnipeg, Manitoba. [Photos by: Vancouver–author; Winnipeg–John Eastcott/ Yva Momatiak/The Image Works.]

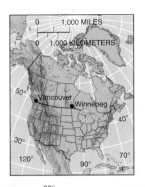

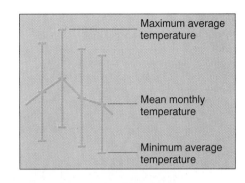

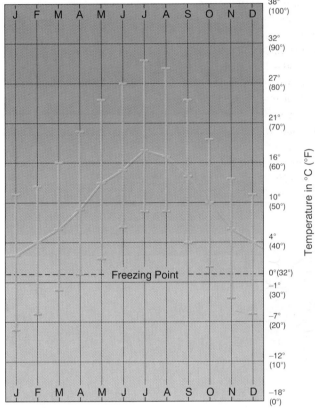

**Station:** Vancouver, British Columbia
**Lat/long:** 49° 11′ N 123° 10′ W
**Avg. ann. temp.:** 10°C (50°F)
**Total ann. precip.:**
    104.8 cm (41.3 in.)

**Elevation:** sea level
**Population:** 431,000
**Ann. temp. range:**
    16 C° (28.8 F°)

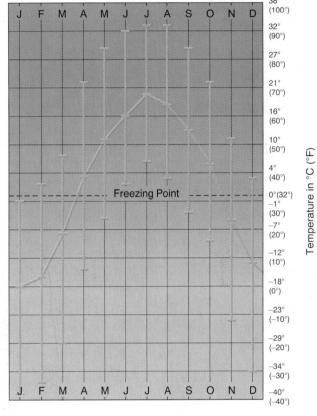

**Station:** Winnipeg, Manitoba
**Lat/long:** 49° 54′ N 97° 14′ W
**Avg. ann. temp.:** 2°C (35.6°F)
**Total ann. precip.:**
    51.7 cm (20.3 in.)

**Elevation:** 248 m (813.6 ft)
**Population:** 561,000
**Ann. temp. range:**
    38 C° (68.4 F°)

(a)

(b)

## FIGURE 5-10
### Marine and continental cities—United States.

Comparison of temperatures in coastal San Francisco, California, and continental Wichita, Kansas. [San Francisco photo by author; Wichita, photo by Garry D. McMichael/Photo Researchers, Inc.]

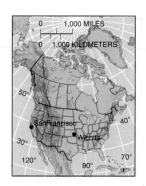

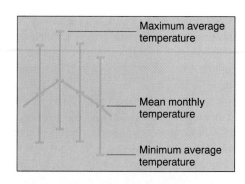

Maximum average temperature

Mean monthly temperature

Minimum average temperature

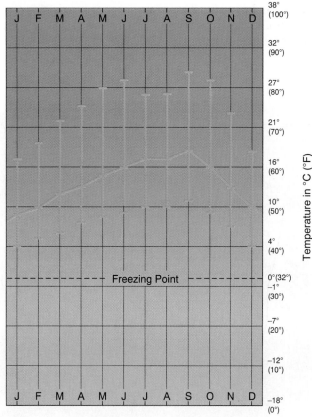

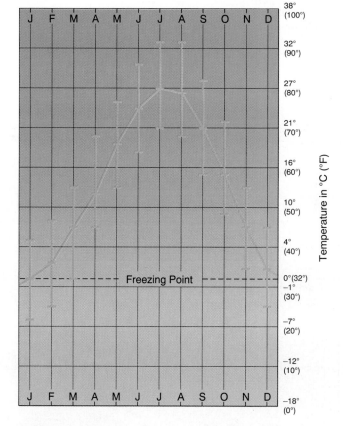

**Station:** San Francisco, CA
**Lat/long:** 37° 37′ N 122° 23′ W
**Avg. ann. temp.:** 14°C (57.2°F)
**Total ann. precip.:** 47.5 cm (18.7 in.)

**Elevation:** 5 m (16.4 ft)
**Population:** 750,000
**Ann. temp. range:** 9 C° (16.2 F°)

**Station:** Wichita, KS
**Lat/long:** 37° 39′ N 97° 25′ W
**Avg. ann. temp.:** 13.7°C (56.6°F)
**Total ann. precip.:** 72.2 cm (28.4 in.)

**Elevation:** 402.6 m (1321 ft)
**Population:** 280,000
**Ann. temp. range:** 27 C° (48.6 F°)

(a)

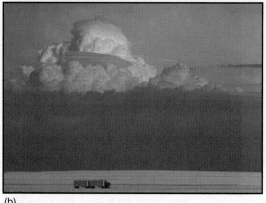

(b)

summer month in San Francisco. In 90 years of weather records there, the ocean has moderated temperatures so that, on average, only once a year has the summer maximum exceeded 32.2°C (90°F) or the winter minimum dropped below freezing. Interestingly, during the past 15 years, the number of days above 32°C has occurred at more than double the 90-year average.

In contrast, Wichita is susceptible to freezes from late October to mid-April, with daily variations slightly increased by its elevation. Wichita has temperatures of 32.2°C (90°F) or higher at least 65 days each year. West of Wichita, winters increase in severity with increasing distance from the moderating influences of invading air masses from the Gulf of Mexico.

## Earth's Temperature Patterns

Earth's temperature patterns result from the combined effect of the controlling factors just discussed. Let us now look at temperatures portrayed on maps that show worldwide mean air temperatures for January (Figure 5-11) and July (Figure 5-13). To complete the analysis, Figure 5-14 presents the temperature range differences between the January and July maps, or the difference between averages of the coolest and warmest months. The temperature range map helps identify areas that experience the greatest annual extremes (continental locations) and the most moderate temperature regimes (marine locations).

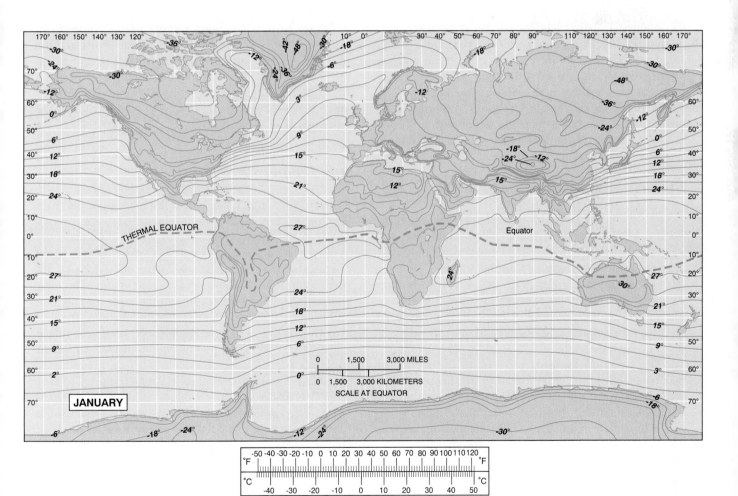

**FIGURE 5-11**

**Global temperatures for January.**

Worldwide mean air temperatures for January. Temperatures are in Celsius (convertible to Fahrenheit by means of the scale) as taken from separate air temperature data bases for ocean and land. (Compare with Figure 5-13.) [Adapted from National Climatic Data Center, *Monthly Climatic Data for the World*, 47, no. 1, January 1994. Prepared in cooperation with the World Meteorological Organization. Washington, DC: National Oceanic and Atmospheric Administration.]

We use maps for January and July instead of the solstice months of December and June because, as explained earlier in Chapter 4, a lag occurs between insolation received and maximum or minimum temperatures experienced. The data used for preparation of our maps were provided by the U.S. National Climate Data Center. Some ship reports go back to 1850 and land reports to 1890, although the bulk of the record is representative of conditions since 1950. An important consideration in using these maps is to remember that their small scale permits only generalizations about actual temperatures.

The lines on temperature maps are known as isotherms. An **isotherm** is an *isoline* that connects points of equal temperature and portrays the temperature pattern, just as a contour line on a topographic map illustrates points of equal elevation. Geographers are concerned with spatial analysis of temperatures, and isotherms help with this analysis.

With each map, begin by finding your own city or town and noting the temperatures indicated by the isotherms. Record the information from these maps in your notebook. As you work through the different maps throughout this text, note atmospheric pressure and winds, precipitation, climate, landforms, soil orders, vegetation, and terrestrial biomes. By the end of the course you will have recorded a complete physical geography profile for your regional environment.

## January Temperature Map

Figure 5-11 maps January's mean temperatures. In the Southern Hemisphere, the higher Sun altitude causes longer days and summer weather conditions; in the Northern Hemisphere, the lower Sun angle causes the short days of winter. Isotherms generally are zonal, trending east-west, parallel to the equator, and they appear to be interrupted by the presence of landmasses. They mark the general decrease in insolation and net radiation with distance from the equator.

The **thermal equator** (an isoline connecting all points of highest mean temperature) trends southward into the interior of South America and Africa, indicating higher temperatures over landmasses. In the Northern Hemisphere, isotherms shift equatorward as cold air chills the continental interiors. The oceans, on the other hand, are more moderate, with warmer conditions extending farther north than over land at comparable latitudes.

As an example, follow along 50° north latitude (the 50th parallel) and compare isotherms: 2° to 4°C in the North Pacific and 4° to 10°C in the North Atlantic, as contrasted to –18°C in the interior of North America and –24° to –30°C in central Asia. Also, note on the maps the effects of altitude as mountain ranges influence the orientation of isotherms.

The coldest area on the map is in Russia, specifically northeastern Siberia. The intense cold experienced there results from winter conditions of consistent clear, dry, calm air, small insolation input, and an inland location far from any moderating maritime effects. Verkhoyansk, Russia, (located within the –48°C isotherm on the map) has experienced a minimum temperature of –68°C (–90°F) and a daily average of –50.5°C (–58.9°F) for January. Verkhoyansk (Figure 5-12) experiences at least 7 months of temperatures below freezing, including at least 4 months below –34°C (–30°F)! In contrast, these same towns experience maximum temperatures of +37°C (+98°F) in July. People do live and work in Verkhoyansk, which has a population of 1400; it has been continuously occupied since 1638 and is today a minor mining district.

Trondheim, Norway, is near the latitude of Verkhoyansk and at a similar elevation. But, Trondheim's coastal location moderates its annual temperature regime (Figure 5-12): January minimum and maximum temperatures range between –17° and +8°C (+1.4° and +46°F), and the minimum-maximum range for July is from +5° to +27°C (+41° to +81°F). The most extreme minimum and maximum temperatures ever recorded in Trondheim are –30° and +35°C (–22° and +95°F)—quite a difference from the extremes at Verkhoyansk.

## July Temperature Map

Average July temperatures are presented in Figure 5-13. The longer days of summer and higher Sun altitude now are in the Northern Hemisphere. Winter dominates the Southern Hemisphere, although it is milder there because continental landmasses are smaller and the dominant oceans and seas store and release more energy. The *thermal equator* shifts northward with the high summer Sun and reaches the Persian Gulf–Pakistan–Iran area. The Persian Gulf is the site of the highest recorded sea-surface temperature of 36°C (96°F), difficult to imagine for a large water body.

July is a time when nights in Antarctica are 24 hours long. This lack of insolation caused the lowest natural temperature reported on Earth, a frigid –89.2°C (–128.6°F). It was recorded on July 21, 1983, at the Russian research base at Vostok, Antarctica (78° 27′ S; elevation 3420 m, or 11,220 ft): Such a temperature is 11 C° (19.8 F°) colder than the freezing point of dry ice (solid carbon dioxide)! If the concentration of carbon dioxide were large enough such a cold temperature would theoretically freeze tiny carbon dioxide dry-ice particles from the sky.

During July in the Northern Hemisphere, isotherms shift poleward over land, as higher temperatures dominate continental interiors. July temperatures in Verkhoyansk average more than 13°C (56°F), which represents a 63 C° (113 F°) seasonal range between winter and summer averages.

**FIGURE 5-12**

**Marine and continental cities—Eurasia.**

Comparison of temperatures in coastal Trondheim, Norway, and continental Siberian Russia. [Photos by (a) Norman Benton/Peter Arnold, Inc., (b) Sovfoto/Eastfoto.]

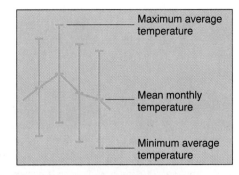

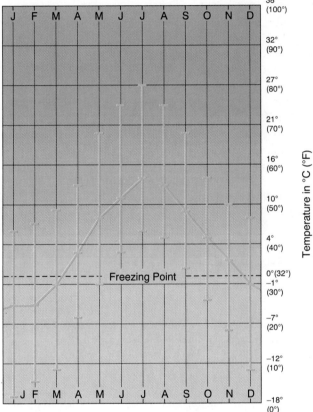

**Station:** Trondheim, Norway
**Lat/long:** 63° 25′ N 10° 27′ E
**Avg. ann. temp.:** 5°C (41°F)
**Total ann. precip.:**
  85.7 cm (33.7 in.)

**Elevation:** 115 m (377.3 ft)
**Population:** 135,000
**Ann. temp. range:**
  17 C° (30.6 F°)

**Station:** Verkhoyansk, USSR
**Lat/long:** 67° 35′ N 135° 23′ E
**Avg. ann. temp.:** −15°C (5°F)
**Total ann. precip.:**
  15.5 cm (6.1 in.)

**Elevation:** 137 m (449.5 ft)
**Population:** 1400
**Ann. temp. range:**
  63 C° (113.4 F°)

(a)

(b)

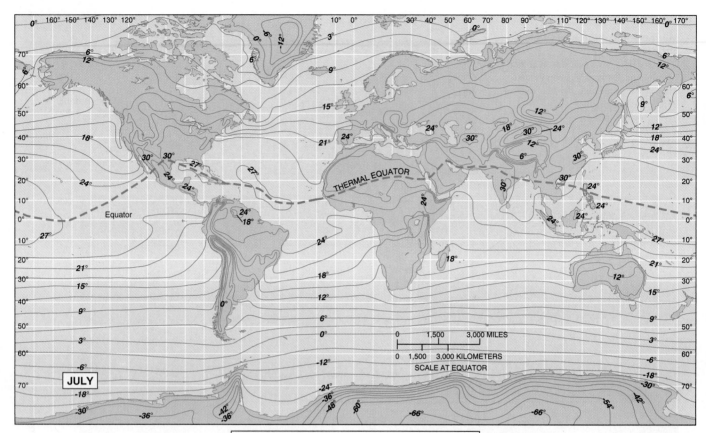

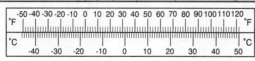

**FIGURE 5-13**

**Global temperatures for July.**

Worldwide mean air temperatures for July. Temperatures are in Celsius (convertible to Fahrenheit by means of the scale) as taken from separate air temperature data bases for ocean and land. (Compare with Figure 5-11.) [Adapted from National Climatic Data Center, *Monthly Climatic Data for the World*, 47, no. 7, July 1994. Prepared in cooperation with the World Meteorological Organization. Washington, DC: National Oceanic and Atmospheric Administration.]

The Verkhoyansk region of Siberia is probably the greatest example of continentality on Earth.

The hottest places on Earth occur in Northern Hemisphere deserts during July. The reasons are simple: clear skies, strong surface heating, virtually no surface water, and few plants. Prime examples are portions of the Sonoran Desert of North America and the Sahara of Africa. Africa recorded a shade temperature over 58°C (136°F), a record set on September 13, 1922 at Al 'Azīzīyah,' Libya (32° 32′ N; 112 m, or 367 ft elevation).

The highest maximum and annual average temperatures in North America occurred in Death Valley, California, where the Greenland Ranch Station reached 57°C (134°F) in 1913. (The station is at 37° N and is −54.3 m, or −178 ft, below sea level). Such hot, arid lands are discussed further in Chapter 15.

## Annual Temperature Range Map

The pattern of marine versus continental influence emerges dramatically when the temperature ranges between the coolest and warmest months are mapped for a full year (Figure 5-14). As you might expect, the largest temperature ranges occur in subpolar locations in North America and Asia, where average ranges of 64 C° (115 F°) are recorded. The Southern Hemisphere, on the other hand, has little seasonal variation in mean temperatures, owing to the lack of large landmasses and the moderating influence of vast expanses of water.

For example, in January (Figure 5-11), Australia is dominated by isotherms of 20° to 30°C (68° to 86°F), whereas in July (Figure 5-13), Australia is crossed by the 12°C (54°F) isotherm. In general, Southern Hemisphere temperature

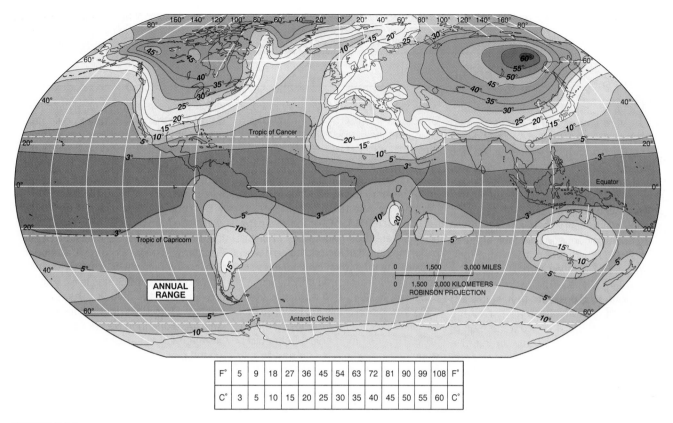

**FIGURE 5-14**
**Global annual temperature ranges.**
A generalized portrait of the annual range of global temperatures in Celsius (conversions to Fahrenheit shown on scale).

| F° | 5 | 9 | 18 | 27 | 36 | 45 | 54 | 63 | 72 | 81 | 90 | 99 | 108 | F° |
|----|---|---|----|----|----|----|----|----|----|----|----|----|-----|----|
| C° | 3 | 5 | 10 | 15 | 20 | 25 | 30 | 35 | 40 | 45 | 50 | 55 | 60 | C° |

patterns are mostly marine, and Northern Hemisphere patterns feature continentality. The Northern Hemisphere, with greater land area overall, registers a slightly higher average surface temperature than does the Southern Hemisphere: 15°C (59°F) as opposed to 12.7°C (55°F).

# Air Temperature and the Human Body

We all sense small temperature changes in our environment, although our sense of warm and cold is subjective; it is difficult for the human body to accurately determine precise temperature. We describe our perception of temperature with the term **apparent temperature**, or *sensible temperature*. This perception varies among individuals and cultures. Through complex mechanisms, our bodies maintain an average internal temperature within a degree of 36.8°C (98.2°F), slightly lower in the morning or in cold weather and slightly higher at emotional times or during exercise and work.*

The water vapor content of air, wind speed, and air temperature taken together affect each individual's sense of comfort. The most discomfort comes with high temperatures, high humidity, and low winds. More comfort comes with low humidity and stronger winds.

Although modern heating and cooling systems, where available, reduce the impact of extreme temperatures indoors, the danger to human life from excessive heat or cold persists. A notable example was the death of nearly 1000 people in the Midwest and East during a summer heat wave in 1995 (see News Report 1).

---

*The traditional value for "normal" body temperature, 37°C (98.6°F), was set in 1868 using old methods of measurement. According to Dr. Philip Mackowiak of the University of Maryland School of Medicine, a more accurate modern assessment places the normal value at 36.8°C (98.2°F), with a range of 2.7 C° (4.8 F°), for the human population overall or about one degree of a specific individual's normal temperature (*Journal of the American Medical Association*, September 23–30, 1992).

# News Report 1

## The Heat Index Can Kill

The summer of 1995 will be remembered as a killer season when nearly 1000 people died in the Midwest and East from high temperatures and humidity. This combination of heatwave and water vapor produced stifling conditions. The sick and elderly, particularly, were affected. On July 13 in Chicago, some apartments without air conditioning exceeded indoor sensible temperatures of 54°C (130°F) when the official temperature at Midway Airport reached a record 41°C (106°F).

With high pressure (hot, stable air) dominating from the Midwest to the Atlantic and moist air from the Gulf of Mexico, all the ingredients for this disaster were present. Afternoon temperatures were above 32°C (90°F) for a week in Chicago, 38°C (100°F) in South Dakota, and 40°C (104°F) in Toledo, Ohio, and New York City; the temperature hit 43°C (109°F) in Omaha, Nebraska. Cities in New England that had reached 37.7°C (100°F) only twice in their history exceeded that for 4 or 5 days during July 1995!

The death toll in Chicago was especially severe; 700 people died. For this major metropolitan region, the heat wave became the third worst disaster in its colorful history—exceeded only by a ship's capsizing in 1915 and a theater fire in 1903. The famous Chicago fire of 1871 is far overshadowed by the loss of life in 1995. The rush for medical help quickly swamped hospitals, which had to turn away hundreds of patients. The rate at which the heat claimed its victims was far greater than the coroner could handle, so the dead were stored in many refrigerated trucks parked outside the morgue. We can only wonder how many lives would have been saved if refrigeration for the living was available.

Although thousands have died worldwide from heat-related causes over the past several years—Pakistan, Spain, Germany, Great Britain, Cambodia, Greece, Albania—it was a shock to residents of the United States to experience the magnitude of this "heat-index" tragedy. More died during July 1995 in Chicago alone than were killed in all the hurricanes that struck U.S. shores during the past 20 years.

**TABLE 5-1**

| The Human Body's Response to Temperature Stress | |
| --- | --- |
| *At Low Temperatures* | *At High Temperatures* |
| **Temperature Regulation Methods** | |
| Heat-gaining mechanism | Heat-dissipation mechanism |
| Constriction of surface blood vessels | Dilation of surface blood vessels |
| Concentration of blood | Dilution of blood |
| Flexing to reduce surface exposure | Extending to increase exposure |
| Increased muscle tone | Decreased muscle tone |
| Decrease in sweating | Sweating |
| Inclination to increase activity | Inclination to decrease activity |
| Shivering | |
| Increased cell metabolism | |
| **Consequent Disturbances** | |
| Increased urine volume | Decreased urine volume |
| Danger of inadequate blood supply to exposed parts; frostbite | Reduced blood supply to brain; dizziness, nausea, fainting |
| Discomfort leading to neuroses | Discomfort leading to neuroses |
| Increased appetite | Decreased appetite |
| | Mobilization of tissue fluid |
| | Thirst and dehydration |
| | Reduced chloride balance; heat cramps |
| **Failure of Regulation** | |
| Falling body temperature | Rising body temperature |
| Drowsiness | Impaired heat-regulating center |
| Cessation of heartbeat and respiration | Failure of nervous regulation; cessation of breathing |

*Source: Climatology*—Arid Zone Research X. © UNESCO 1958. Reproduced by permission of UNESCO.

When changes occur in the surrounding air, the human body reacts to maintain its core temperature and to protect the brain at all cost. Table 5-1 summarizes the human body's response to stress induced by exposure to low temperature (*hypothermia*) and high temperature (*hyperthermia*). Carefully review these two lists of responses for future reference.

## Wind Chill

The wind-chill index (Figure 5-15) is important to people who experience winters with freezing temperatures. The **wind-chill factor** indicates the enhanced rate at which body heat is lost to the air and at best represents an estimate of heat energy loss. As wind speeds increase, heat loss from the skin increases. For example, if the air temperature is −1°C (30°F) and the wind is blowing at 32 kmph (20 mph), skin temperatures will be a dangerous −16°C (4°F). The lower wind-chill values present a serious freezing hazard to exposed flesh. Imagine the wind chill experienced by a downhill ski racer going 130 kmph (80 mph)–frostbite is a definite possibility.

## Heat Index

The *heat index (HI)* indicates the human body's reaction to the combination of air temperature and water vapor. The heat index, combining the effects of temperature and humidity, indicates how hot the air feels to an average person. Water vapor in air is expressed as relative humidity, a concept presented in Chapter 7. For now, know that the amount of water vapor in the air affects the evaporation rate of perspiration on the skin. The reason is simply available space: the more water vapor in the air (higher humidity), the less space the air has to absorb water vapor evaporated from perspiration.

Figure 5-16 is an abbreviated version of the heat index that the National Weather Service (NWS) now includes in its daily weather summaries during summer months. The table beneath the graph describes the effects of heat-index categories on higher-risk groups. A combination of high temperature and high humidity can severely reduce the body's natural ability to regulate internal temperature. For nearly a week during July 1995, Chicago's heat-index values went to category I in

**Actual Air Temperature in °C (°F)**

| | 10° (50°) | 4° (40°) | −1° (30°) | −7° (20°) | −12° (10°) | −18° (0°) | −23° (−10°) | −29° (−20°) | −34° (−30°) | −40° (−40°) | −46° (−50°) | −51° (−60°) |
|---|---|---|---|---|---|---|---|---|---|---|---|---|
| **Calm** | 10° (50°) | 4° (40°) | −1° (30°) | −7° (20°) | −12° (10°) | −18° (0°) | −23° (−10°) | −29° (−20°) | −34° (−30°) | −40° (−40°) | −46° (−50°) | −51° (−60°) |
| 8 (5) | 9° (48°) | 3° (37°) | −2° (28°) | −9° (16°) | −14° (6°) | −21° (−5°) | −26° (−15°) | −32° (−26°) | −38° (−36°) | −44° (−47°) | −49° (−57°) | −56° (−68°) |
| 16 (10) | 4° (40°) | −2° (28°) | −9° (16°) | −16° (4°) | −23° (−9°) | −29° (−21°) | −36° (−33°) | −43° (−46°) | −50° (−58°) | −57° (−70°) | −64° (−83°) | −71° (−95°) |
| 24 (15) | 2° (36°) | −6° (22°) | −13° (9°) | −21° (−5°) | −28° (−18°) | −38° (−36°) | −43° (−45°) | −50° (−58°) | −58° (−72°) | −65° (−85°) | −73° (−99°) | −74° (−102°) |
| 32 (20) | 0° (32°) | −8° (18°) | −16° (4°) | −23° (−10°) | −32° (−25°) | −39° (−39°) | −47° (−53°) | −55° (−67°) | −63° (−82°) | −71° (−96°) | −79° (−110°) | −87° (−124°) |
| 40 (25) | −1° (30°) | −9° (16°) | −18° (0°) | −26° (−15°) | −34° (−29°) | −42° (−44°) | −51° (−59°) | −59° (−74°) | −64° (−83°) | −76° (−104°) | −81° (−113°) | −92° (−133°) |
| 48 (30) | −2° (28°) | −11° (13°) | −19° (−2°) | −28° (−18°) | −36° (−33°) | −44° (−48°) | −53° (−63°) | −62° (−79°) | −70° (−94°) | −78° (−109°) | −87° (−125°) | −96° (−140°) |
| 56 (35) | −3° (27°) | −12° (11°) | −20° (−4°) | −29° (−20°) | −37° (−35°) | −45° (−49°) | −53° (−64°) | −63° (−82°) | −72° (−98°) | −84° (−119°) | −89° (−129°) | −98° (−145°) |
| 64 (40) | −3° (26°) | −12° (10°) | −21° (−6°) | −29° (−21°) | −38° (−37°) | −47° (−53°) | −56° (−69°) | −65° (−85°) | −74° (−102°) | −82° (−116°) | −91° (−132°) | −100° (−148°) |

Temperature Effects on Exposed Flesh

Wind speed, kmph (mph)

Low danger with proper clothing | Exposed flesh in danger of freezing

**FIGURE 5-15**

**Wind chill.**

Wind-chill factor for various temperatures and wind speeds. A wind chill of −20°C results in a loss of 1420 W/m²; −40°C, a loss of 1955 W/m²; −59°C, a loss of 2490 W/m² from exposed flesh. [Adapted from the National Weather Service and U.S. Army.]

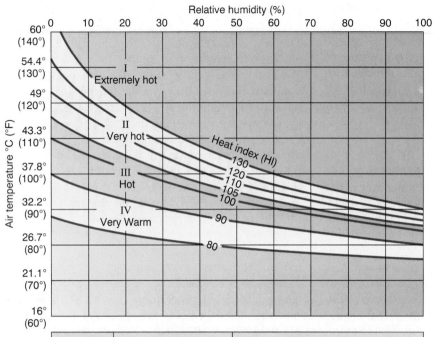

**FIGURE 5-16**
**Heat index.**
Heat index graph for various temperatures and relative humidity levels. [Courtesy of the National Weather Service.]

| Category | Heat Index Apparent Temperature | General Effect of Heat Index on People in High-Risk Groups |
|---|---|---|
| I | 54°C (130°F) or higher | Heat/sunstroke highly likely with continued exposure |
| II | 41° – 54°C (105° – 130°F) | Sunstroke, heat cramps, or heat exhaustion likely and heatstroke possible with prolonged exposure and/or physical activity |
| III | 32° – 41°C (90° – 105°F) | Sunstroke, heat cramps, and heat exhaustion possible with prolonged exposure and/or physical activity |
| IV | 27° – 32°C (80° – 90°F) | Fatigue possible with prolonged exposure and/or physical activity |

**FIGURE 5-17**
**Chicago heat wave.**
The Midwest and eastern United States were hit with a devastating heat wave during July 1995. City emergency, medical, and coroner services were overwhelmed by the tragedy. [Photo by Dave Weaver/Gamma Liaison.]

dwellings that lacked air conditioning; that is an apparent temperature of over 54°C (130°F) (Figure 5-17).

There is a distinct possibility that future humans may experience greater temperature-related challenges owing to complex changes now under way in the lower atmosphere. See News Report 2 for an introduction to some of these global changes. Possible global warming and cooling are discussed in Chapter 10. Meanwhile the heat-wave deaths in the 1995 summer season remain a powerful reminder of the role of temperatures in our lives.

## News Report 2

### Record Temperatures Suggest a Greenhouse Warming

Scientists agree that human activities are enhancing Earth's natural greenhouse. Certain radiatively active gases are absorbing longwave radiation and delaying losses of heat energy to space, thus throwing off the natural Earth-atmosphere energy equilibrium and producing a dynamic warming trend.

The Intergovernmental Panel on Climate Change (IPCC), through major reports in 1990, 1992, 1994, and 1995, confirmed that global warming is occurring and that the changes are caused by human activity, principally the burning of fossil fuels. Uncertainty remains in forecasting the severity of such global change and the implications this warming will have on Earth systems.

The period 1980 to 1995 registered 11 of the warmest years in the history of instrumental measurements. Effects of this warming are the subject of many scientific studies by many scientists: melting and retreat of mountain glaciers worldwide, disintegration of coastal ice around Antarctica, rising sea level at a faster rate than previously observed, more-intense thunderstorms and weather extremes, changes in vegetation, and heat-wave effects on grain production. The snowline in portions of the European Alps has risen 100 m (330 ft) just since 1980.

With the blame resting squarely on fossil fuel consumption and our modern technological society, the political fallout and debate is intense. Often lacking from the media and popular discussion are the many scientific measurements and discoveries concerning the Earth-atmosphere system. Scientists are equipped with an array of sophisticated satellites, remote sensing capabilities, and powerful computers on which they run global circulation models.

In Chapter 10 you will find a full discussion of global warming and climate change: what is known and unknown, how global models work, the radiatively active gases causing the problem, the consequences for natural systems, and most important, the solutions and a viable action plan as well as its alternative, a nonaction plan.

## Summary and Review —Global Temperatures

✔ **Define the concepts of temperature, kinetic energy, sensible heat and distinguish among Kelvin, Celsius, and Fahrenheit scales and how they are measured.**

**Temperature** is a measure of the average kinetic energy (motion) of individual molecules in matter. We feel the effect of temperature as the *sensible heat* transfer from warmer objects to cooler objects when these objects are touching. Temperatures scales include:

- Kelvin scale: 100 units between ice's melting point (273K) and water's boiling point (373K).

- Celsius scale: 100 degrees between ice's melting point (0°C) and water's boiling point (100°C).

- Fahrenheit scale: 180 degrees between ice's melting point (32°F) and water's boiling point (212°F).

The Kelvin scale is used in scientific research because temperature readings start at absolute zero and thus are proportional to the actual kinetic energy in a material.

temperature (p. 117)

1. Distinguish between sensible heat and sensible temperature.

2. What does air temperature indicate about energy in the atmosphere?

3. Compare the three scales that express temperature. What is the basic assumption for each?

4. What is your source of daily temperature information? Describe the highest temperature you have experienced and the lowest temperature. From what we have discussed in this chapter, can you identify the factors that may have contributed to these temperatures?

---

✔ **List and review the principal controls and influences that produce global temperature patterns.**

Principal controls and influences upon temperature patterns include latitude (the distance north or south of the equator), altitude (location above sea level), cloud cover (reflect, absorb, and reradiate energy), land-water heating differences (the nature of evaporation, transparency, specific heat, movement, and ocean currents and sea-surface temperatures).

5. Explain the effect of altitude on air temperature. Why is air at higher altitudes lower in temperature? Why does it feel cooler standing in shadows at higher altitude than at lower altitude?

6. What noticeable effect does air density have on the absorption and radiation of energy? What role does altitude play in that process?

7. How is it possible to grow moderate-climate type crops such as wheat, barley, and potatoes at an elevation of 4103 m (13,460 ft) near La Paz, Bolivia, so near the equator?

8. Describe the effect of cloud cover with regard to Earth's temperature patterns. From the last chapter, review the cloud-albedo forcing and cloud-greenhouse forcing of different cloud types.

---

✔ *Review* the factors that produce marine and continentality differences as they influence temperatures and *utilize* several pairs of stations to illustrate these differences.

The physical nature of land (rock and soil) and water (oceans, seas, and lakes) is the reason for **land-water heating differences**. Moderate temperature patterns are associated with water bodies, and extreme temperatures occur inland. The five controls that differ between land and water surfaces include: evaporation, transparency, specific heat, movement, and ocean currents and sea-surface temperatures.

Light penetrates water because of its **transparency**, to an average depth of 60 m (200 ft) in the ocean. This penetration distributes available heat energy over a much greater volume that could occur on opaque land, thus a larger energy reservoir is formed. When equal volumes of water and land are compared, water requires far more energy to increase its temperature than does land. In other words, water can hold more energy than can soil or rock, so water has a higher **specific heat**, averaging about four times that of soil.

Ocean currents affect temperature. An example of the effect of ocean currents is the **Gulf Stream**, which moves northward off the east coast of North America, carrying warm water far into the North Atlantic. As a result, the southern third of Iceland experiences much milder temperatures than would be expected for a latitude of 65° N, just below the Arctic Circle (66.5°).

**Marine**, or *maritime*, describes locations that exhibit the moderating influences of the ocean, usually along coastlines or on islands. **Continentality** refers to the condition of areas that are less affected by the sea and therefore have a greater range between maximum and minimum temperatures diurnally and yearly.

land-water heating differences (p. 119)
transparency (p. 122)
specific heat (p. 122)
Gulf Stream (p. 123)
marine (p. 124)
continentality (p. 124)

9. List the physical aspects of land and water that produce their different responses to heating from absorption of insolation. What is the specific effect of transparency in a medium?

10. What is specific heat? Compare the specific heat of water and soil.

11. Describe the pattern of sea-surface temperatures (SSTs) as determined by satellite remote sensing. Where is the warmest ocean region on Earth?

12. What effect does sea-surface temperature have on air temperature? Describe the negative feedback mechanism created by higher sea-surface temperatures and evaporation rates.

13. Differentiate between marine and continental temperatures. Give geographic examples of each from the text: in Canada, the United States, Norway, and Russia.

---

✔ *Interpret* the pattern of Earth's temperatures from their portrayal on January and July temperature maps and on a map of annual temperature ranges.

Maps for January and July instead of the solstice months of December and June are used for temperature comparison because of the natural lag that occurs between insolation received and maximum or minimum temperatures experienced. Each line on these temperature maps is an **isotherm**, an isoline that connects points of equal temperature. Isotherms portray temperature patterns.

Isotherms generally are zonal, trending east-west, parallel to the equator. They mark the general decrease in insolation and net radiation with distance from the equator. The **thermal equator** (isoline connecting all points of highest mean temperature) trends southward in January and shifts northward with the high summer Sun in July. In January it extends farther south into the interior of South America and Africa, indicating higher temperatures over landmasses.

In the Northern Hemisphere in January, isotherms shift equatorward as cold air chills the continental interiors. The coldest area on the map is in Russia, specifically northeastern Siberia. The intense cold experienced there results from winter conditions of consistent clear, dry, calm air, small insolation input, and an inland location far from any moderating maritime effects.

isotherm (p. 128)
thermal equator (p. 128)

14. What is the thermal equator? Describe its location in January and in July. Explain why it shifts position annually.

15. Observe trends in the pattern of isolines over North America and compare the January average temperature map with the July map. Why do the patterns shift locations?

16. Explain the extreme temperature range experienced in north-central Siberia between January and July.

17. Where are the hottest places on Earth? Are they near the equator or elsewhere? Explain. Where is the coldest place on Earth?

18. From the maps in Figures 5-11, 5-13, and 5-14, determine the average temperature values and annual range of temperatures for your present location.

---

✔ *Define* **apparent temperature, or sensible temperature,** and *relate* **specific physiological effects of low temperature and high temperature on the human body;** *utilize* **the wind-chill and heat-index charts to determine some apparent temperatures.**

We describe our perception of temperature with the term **apparent temperature**, or *sensible temperature*. This perception varies among individuals and cultures. The **wind-chill factor** indicates the enhanced rate at which body heat energy is lost to the air and at best is an estimate. As wind speeds increase, heat loss from the skin increases and flesh temperatures decrease. The *heat index (HI)* indicates the human body's reaction to air temperature and water vapor. The heat index, combining the effects of temperature and humidity, indicates how hot the air feels to an average person. The amount of water vapor in the air affects the evaporation rate of perspiration on the skin—the body's natural cooling mechanism. The heat-index and wind-chill charts are two useful tools for determining apparent temperature.

apparent temperature (p. 131)
wind-chill factor (p. 133)

19. Identify the different responses of the human body to low-temperature and high-temperature stress.

20. Describe the interaction between air temperature and wind speed and their affect on skin chilling. Select several temperatures and wind speeds from Figure 5-15 and determine the wind-chill factor.

21. Define the heat index. What characteristic patterns are experienced throughout the year where you live, relative to the graph in Figure 5-16?

---

 ## NetWork

The *Geosystems Home Page* provides on-line resources for this chapter on the World Wide Web. You will find review exercises, specific updates for items in the chapter, suggested readings, and links to interesting related pathways on the Internet (click on the Table of Contents link and select this chapter). *Geosystems* is at: **http://www.prenhall.com/geosystm**

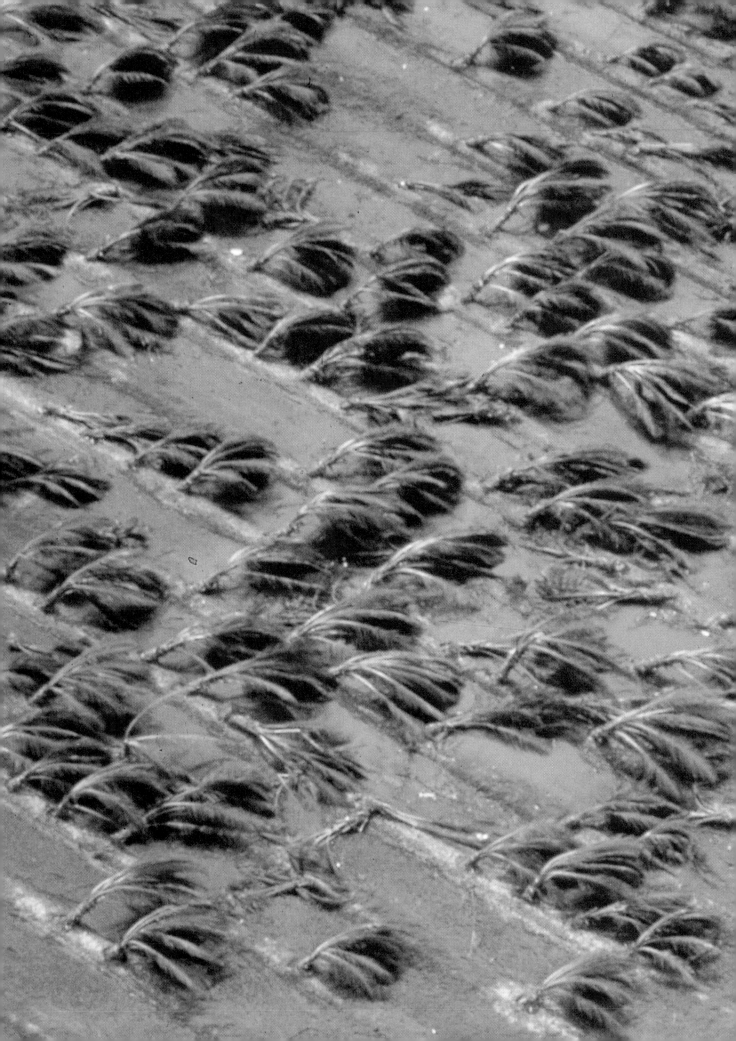

# 6

# Atmospheric and Oceanic Circulations

**Wind Essentials**

**Driving Forces Within the Atmosphere**

**Atmospheric Patterns of Motion**

**Oceanic Currents**

**Summary and Review**

### Key Learning Concepts

After reading the chapter, you should be able to:

- *Define* wind and *describe* how wind is measured, how wind-direction is determined, and how winds are named.

- *Define* the concept of air pressure and *portray* the pattern of global pressure systems on isobaric maps.

- *Explain* the four driving forces within the atmosphere—gravity, pressure gradient force, Coriolis force—and friction force, and *describe* the primary high- and low-pressure areas and principal winds.

- *Describe* upper-air circulation and its support role for surface systems and *define* the jet streams.

- *Explain* several types of local winds: land-sea breezes, mountain-valley breezes, katabatic winds, and the regional monsoons.

- *Discern* the basic pattern of Earth's major surface and deep ocean currents.

*The powerful winds of Hurricane Andrew blew down thousands of trees in southern Florida.* [Photo by John Lopinot/Palm Beach Post/Sygma.]

Early in April 1815, on an island named Sumbawa in present-day Indonesia, the volcano Tambora erupted violently. It spewed an estimated 150 km³ (36 mi³) of material, 25 times the volume produced by the 1980 Mount St. Helens eruption in Washington State. Some material from Tambora was carried worldwide by global atmospheric circulation, creating a stratospheric veil of dust and acid mist. The result was both a higher atmospheric albedo and absorption of energy by the particulate materials injected into the stratosphere. Remarkable optical and meteorological effects resulting from the spreading dust were noted worldwide for years after the eruption—beautiful sunrises and sunsets and a temporary lowering of temperatures.

Scientists in 1815 lacked the remote-sensing capability of satellite technology, and they had no way of knowing the global impact of Tambora's eruption. Today, technology permits a depth of analysis unknown in the past—satellites now track the atmospheric effects from dust storms, forest fires, industrial haze, warfare, and the dispersal of volcanic explosions, among many other things.

After 635 years of dormancy, Mount Pinatubo in the Philippines erupted in 1991. This event had tremendous atmospheric impact, lofting 15–20 million tons of ash, dust, and sulfur dioxide ($SO_2$) into the atmosphere. As the sulfur dioxide rose into the stratosphere, it quickly formed sulfuric acid ($H_2SO_4$) aerosols, which became concentrated at 16–25 km (10–15.5 mi) altitude. This debris increased atmospheric albedo about 1.3% giving scientists an estimate of the aerosol volume generated by Mount Pinatubo. The eruption provided a unique insight into the dynamics of atmospheric circulation.

The AVHRR (advanced very high resolution radiometer) instrument aboard *NOAA-11* monitored the reflected solar radiation from Mount Pinatubo's aerosols as they were swept around Earth by global winds. Figure 6-1 shows images made at 2-week intervals that clearly track this handiwork of our atmospheric circulation. The earliest image (Figure 6-1a), near the time of the eruption, shows winds blowing dust westward from Africa, smoke from Kuwaiti oil well fires set during the Persian Gulf War, and some haze off the east coast of the United States. Also visible is the Pinatubo aerosol layer beginning to emerge north of Indonesia.

Atmospheric winds spread the debris worldwide in just 21 days (Figure 6-1a, b), dusting skies in a band from 15° S to 25° N in width. By the last image in the sequence (Figure 6-1d), some 60 days after the eruption, the aerosol cloud spanned from 20° S to 30° N and covered about 42% of the globe. Most of the world's population experienced spectacular sunrises and sunsets and a small lowering of average temperatures during the 2 years that followed.

In this chapter, we examine the dynamic circulation of Earth's atmosphere. The general atmospheric circulation carried Tambora's debris, Mount Pinatubo's aerosols, and society's pollution, along with oxygen, carbon dioxide, and water vapor around the globe. We also consider Earth's wind-driven oceanic currents. The energy driving all this movement comes from one source: the Sun.

# Wind Essentials

Earth's atmospheric circulation transfers both energy and mass on a grand scale. In the process, the imbalance between equatorial energy surpluses and polar energy deficits (Chapter 4) is partly resolved, Earth's weather patterns are generated, and ocean currents are produced. Air pollutants, whether natural or human-caused, also are spread worldwide by atmospheric circulation, far from their point of origin.

Global winds are certainly an important reason why the United States, the former Soviet Union, and Great Britain signed the 1963 Limited Test Ban Treaty. That treaty banned aboveground testing of nuclear weapons because atmospheric circulation was spreading radioactive contamination worldwide. Such agreements illustrate how the fluid movement of the atmosphere socializes humanity more than any other natural or cultural factor. Our atmosphere makes all the world a spatially linked society—one person's or country's exhalation is another's inhalation.

Atmospheric circulation is generally categorized at three levels: *primary circulation* (general worldwide circulation), *secondary circulation* of migratory high-pressure and low-pressure systems, and *tertiary circulation* that includes local winds and temporal weather patterns. Winds that move principally north or south along meridians are known as *meridional flows*. Winds moving east or west along parallels of latitude are called *zonal flows*.

## Wind: Description and Measurement

Simply stated, **wind** is the horizontal motion of air across Earth's surface. *It is produced by differences in air pressure from one location to another*. Wind's two principal properties are speed and direction, and instruments are used to measure each. An **anemometer** measures wind speed in kilometers per hour (kmph), miles per hour (mph), meters per second (mps), or knots. (A *knot* is a nautical mile per hour, covering 1 minute of Earth's arc in an hour, equivalent to 1.85 kmph, or 1.15 mph.) A **wind vane** determines wind direction; the standard measurement is taken 10 m (33 ft) above the ground to reduce the effects of local topography on wind direction (Figure 6-2).

Winds are named for the direction *from which they originate*. For example, a wind from the west is a *westerly* wind (it blows eastward); a wind out of the south is a *southerly* wind (it blows northward). Figure 6-3

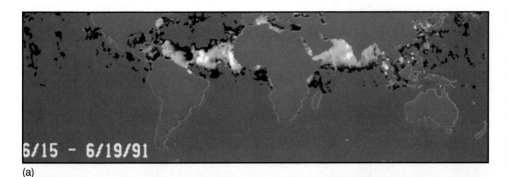

(a)

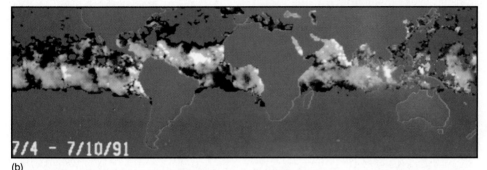

(b)

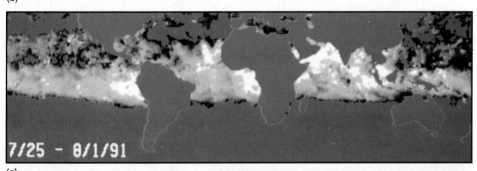

(c)

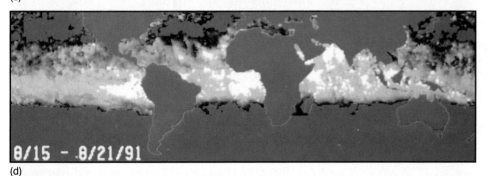

(d)

(e)

**FIGURE 6-1**

**Volcanic eruption effects spread worldwide by winds.**

False-color images (a through d) show aerosols from Mount Pinatubo, smoke from fires, and dust storms, all swept about the globe by the general atmospheric circulation. The dramatic increase in aerosols over the oceans from mid-June 1991 (a) to mid-August (d) is due to the Mount Pinatubo eruption on June 15. The false color shows aerosol concentration, measured by the atmosphere's aerosol optical thickness (AOT): White is densest, dull yellow indicates medium values, and brown areas have the lowest aerosol concentration. In (d), also note the wind-blown smoke from forest fires in Siberia (some white and dull yellow at extreme upper right). Satellite image (e) shows Mount Pinatubo as it erupted. [AVHRR satellite images of aerosols courtesy of Dr. Larry L. Stowe, National Environmental Satellite, Data, and Information Service, National Oceanic and Atmospheric Administration. Used by permission. (e) AVHRR satellite eruption image courtesy of EROS Data Center.]

**FIGURE 6-2**

**Wind vane and anemometer.**

Instruments used to measure wind direction (wind vane, right) and wind speed (anemometer, left) at a weather station installation. [Photo by Belfort Instruments.]

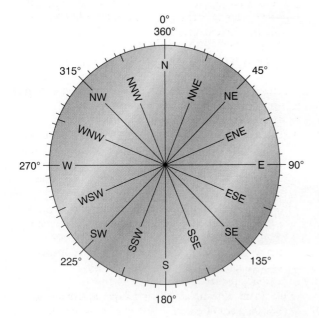

**FIGURE 6-3**

**A wind compass.**

Sixteen wind directions identified on a wind compass.

illustrates a simple wind compass, naming 16 principal wind directions used by meteorologists.

The traditional **Beaufort wind scale** is a descriptive scale useful in visually estimating wind speed. In 1806,

Admiral Beaufort of the British Navy introduced his wind scale. In 1926, G. C. Simpson expanded Beaufort's scale to include wind speeds on land. The scale was standardized by the National Weather Service (Weather Bureau) in 1955. It is still referenced on ocean charts, enabling estimation of wind speed without instruments. Table 6-1 is a modernized wind scale, incorporating the earlier ones. (Try using the Beaufort scale to estimate wind speed as you walk across campus today.) Most modern ships use sophisticated equipment to perform such measurements.

## Global Winds

The primary circulation of winds across Earth has fascinated travelers, sailors, and scientists for centuries, although only in the modern era is a true picture emerging of the pattern and causes of global winds. Breakthroughs in space-based observations and Earth-bound computer technology are refining models that simulate total atmospheric and oceanic circulation.

A remarkable portrait of surface winds across the Pacific Ocean was painstakingly assembled by scientists at the Jet Propulsion Laboratory and the University of California–Los Angeles (Figure 6-4). The image was produced indirectly, using satellite-borne radar to measure the motion and direction of ocean waves. Because wind drives waves on the ocean surface, the waves reflect wind patterns. The researchers used 150,000 measurements made during a single day by the *Seasat* satellite.

The patterns in the figure are the result of specific forces at work in the atmosphere: *pressure gradient force*, *Coriolis force*, *friction force*, and gravity. These forces are our next topic. As we progress through this chapter, you may want to refer to this *Seasat* image to identify the winds, eddies, and vortexes it portrays. Let's begin our study of atmospheric motion by analyzing the concept of air pressure, key to forming pressure gradients.

## Air Pressure

Important to an understanding of wind is the concept of air pressure, its measurement and expression. The molecules that constitute air create air pressure through their motion, size, and number—the factors that determine the temperature and density of the air. Pressure is exerted on all surfaces in contact with the air. The weight (force over a unit area) of the atmosphere, or **air pressure**, crushes in on all of us. Fortunately, that same pressure also exists inside us, pushing outward; otherwise we would be crushed by the mass of air around us.

The atmosphere exerts an average force of approximately $1 \text{ kg/cm}^2$ ($14.7 \text{ lb/in.}^2$) at sea level. Under the influence of gravity, air is compressed and therefore denser

**TABLE 6-1**

## Beaufort Wind Scale

| Wind Speed | | | Beaufort Wind Scale | | | |
|---|---|---|---|---|---|---|
| *kmpb* | *mpb* | *knots* | *Beaufort Number* | *Wind Description* | *Observed Effects at Sea* | *Observed Effects on Land* |
| <1 | <1 | <1 | 0 | Calm | Glassy calm, like a mirror | Calm, no movement of leaves |
| 1–5 | 1–3 | 1–3 | 1 | Light air | Small ripples; wavlet scales; no foam on crests | Slight leaf movement; smoke drifts; wind vanes still |
| 6–11 | 4–7 | 4–6 | 2 | Light breeze | Small wavelets; glassy look to crests, which do not break | Leaves rustling; wind felt; wind vanes moving |
| 12–19 | 8–12 | 7–10 | 3 | Gentle breeze | Large wavelets; dispersed whitecaps as crests break | Leaves and twigs in motion; small flags and banners extended |
| 20–29 | 13–18 | 11–16 | 4 | Moderate breeze | Small longer waves; numerous whitecaps | Small branches moving; raising dust, paper, litter, and dry leaves |
| 30–38 | 19–24 | 17–21 | 5 | Fresh breeze | Moderate, pronounced waves; many whitecaps; some spray | Small trees and branches swaying; wavelets forming on inland waterways |
| 39–49 | 25–31 | 22–27 | 6 | Strong breeze | Large waves, white foam crests everywhere; some spray | Large branches swaying; overhead wires whistling; difficult to control an umbrella |
| 50–61 | 32–38 | 28–33 | 7 | Moderate (near) gale | Sea mounding up; foam and sea spray blown in streaks in the direction of the wind | Entire trees moving; difficult to walk into wind |
| 62–74 | 39–46 | 34–40 | 8 | Fresh gale (or gale) | Moderately high waves of greater length; breaking crests forming sea spray; well-marked foam streaks | Small branches breaking; difficult to walk; moving automobiles drifting and veering |
| 75–87 | 47–54 | 41–47 | 9 | Strong gale | High waves; wave crests tumbling and the sea beginning to roll; visibility reduced by blowing spray | Roof shingles blown away; slight damage to structures; broken branches littering the ground |
| 88–101 | 55–63 | 48–55 | 10 | Whole gale (or storm) | Very high waves and heavy, rolling seas; white appearance to foam-covered sea; overhanging waves; visibility reduced | Uprooted and broken trees; structural damage; considerable destruction; seldom occurring |
| 102–116 | 64–73 | 56–63 | 11 | Storm (or violent storm) | White foam covering a breaking sea of exceptionally high waves; small- and medium-sized ships lost from view in wave troughs; wave crests frothy | Widespread damage to structures and trees, a rare occurrence |
| >117 | >74 | >64 | 12–17 | Hurricane | Driving foam and spray filling the air; white sea; visibility poor to nonexistent | Severe to catastrophic damage; devastation to affected society |

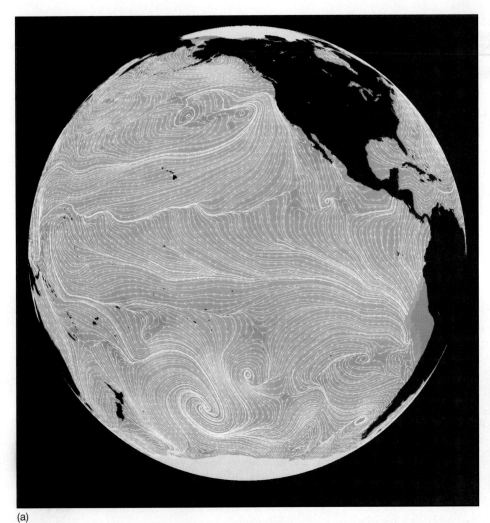

(a)

**FIGURE 6-4**

**Wind portrait of the Pacific Ocean.**

Surface wind measured by radar scatterometer aboard the *Seasat* satellite on September 14, 1978. Scientists analyzed 150,000 measurements to produce this image (a). Colors are correlated with wind speeds, and the white arrows note wind direction. Try comparing the wind patterns with a visible-light image of the same region (b). Can you identify the pattern of trade winds, westerlies, high-pressure cells, and low-pressure cells from the cloud patterns on the image? [Wind portrait courtesy of Dr. Peter Woiceshyn, Jet Propulsion Laboratory, Pasadena, California. Satellite image inset from Laboratory of Planetary Studies, Cornell University. Used by permission.]

(b)

near Earth's surface; it thins rapidly with increasing altitude (Figure 6-5). Consequently, over half the total mass of the atmosphere is compressed below 5500 m (18,000 ft), 75% is compressed below 10,700 m (35,100 ft), and 90% is below the tropopause at 16,000 m (52,500 ft). All but 0.1% of the atmosphere exists within an altitude of 50 km (31 mi), as shown in the pressure profile in Figure 6-6.

In A.D. 1643, Galileo's pupil Evangelista Torricelli was working on a mine-drainage problem. His work led him to discover a method for measuring air pressure (Figure 6-7a). He knew that pumps in the mine were able to "pull" water upward about 10 m (33 ft), but no higher, and he did not know why. Careful observation led him to discover that this limitation was caused not by weak pumps but by the atmosphere itself.

A pump of the type shown in Figure 6-7a works by creating a vacuum above the water. This vacuum allows the pressure of the atmosphere that is pushing on the water in the mine to force water up the pipe. It is the same principle you use with a drinking straw; you lower the pressure inside the straw, thereby allowing atmospheric

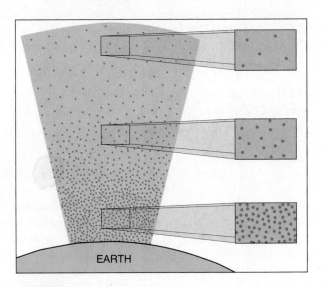

**FIGURE 6-5**

**Density decreases with altitude.**

The atmosphere, denser near Earth's surface, rapidly decreases in density with altitude. This difference in density is easily measured because air exerts its weight as pressure.

pressure against the drink to force liquid up the straw to your mouth. Torricelli noted that the water level in the vertical pipe fluctuated from day to day. He figured out that air pressure was varying with changing weather conditions.

To simulate the problem at the mine, Torricelli devised an instrument at Galileo's suggestion. Instead of water, which required a tube 10 m (33 ft) long, he used a much denser fluid, mercury (Hg). He reasoned that atmospheric pressure could not push a denser fluid as high and that he could use a shorter glass laboratory tube, only 1 m high.

Torricelli sealed the glass tube at one end, filled it with mercury, and inverted it into a dish of mercury (Figure 6-7b). He determined that the average height of the column of mercury in the tube was 760 mm (29.92 in.), and that it did vary day to day as the weather changed. He concluded that the column of mercury was counterbalanced by the mass of surrounding air exerting an equivalent pressure on the mercury in the dish. Thus, the 10 m (33 ft) limit to which the mine pumps could draw water actually was controlled by the mass of the atmosphere and the density of the water. (Had the mine been flooded with mercury, the pumps could have raised the mercury less than a meter!)

Using similar instruments today, scientists have determined that *normal sea-level pressure* is 1013.2 mb (millibar, mb, expresses force per square meter of surface area), or 29.92 in. mercury. In Canada and other countries normal is expressed as 101.32 kPa (kilopascal; 1 kPa = 10 mb).

Any instrument that measures air pressure is called a barometer (from the Greek *baros*, meaning "weight"). Torricelli developed a **mercury barometer**. A more compact barometer design, which works without a meter-long tube of mercury, is the aneroid barometer shown in Figure 6-7c. *Aneroid* means "using no liquid." The **aneroid barometer** principle is simple: Imagine a small chamber, partially emptied of air, that is sealed and connected to a mechanism attached to a needle on a dial. As air pressure increases, it presses on the chamber; as the air pressure decreases it relieves pressure on the chamber. The chamber responds to these changes in air pressure and moves the needle. An aircraft altimeter is a type of aneroid barometer. It accurately measures altitude because air pressure diminishes with elevation above sea level. For accuracy the altimeter instrument must be adjusted for temperature changes.

Figure 6-8 illustrates comparative scales in millibars and inches used to express air pressure and its relative force. The normal range of Earth's atmospheric pressure is from strong high pressure to deep low pressure, about 1050 to 980 mb. The figure also indicates the extreme highest and lowest pressures ever recorded in the United States, Canada, and on Earth. The concept of air pressure as a driving force in atmospheric circulation is our next topic.

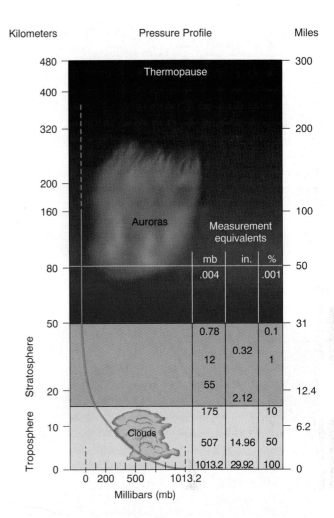

**FIGURE 6-6**
**Atmospheric pressure profile.**
Pressure profile of the atmosphere. Note that about 90% of the atmospheric mass resides in the troposphere. Clouds in the troposphere and auroras in the thermosphere are shown to help you relate this profile to the discussion of the atmosphere in Chapter 3 and Figure 3-2a.

# Driving Forces Within the Atmosphere

Four forces determine both speed and direction of winds:

- Earth's *gravitational force* on the atmosphere is virtually uniform. Gravity equally compresses the atmosphere worldwide, with the density decreasing as altitude increases. Without gravity, there would be no atmospheric pressure—or atmosphere, for that matter.

- The **pressure gradient force** drives air from areas of higher barometric pressure to areas of lower barometric pressure, thereby causing winds. Without a pressure gradient force, there would be no wind.

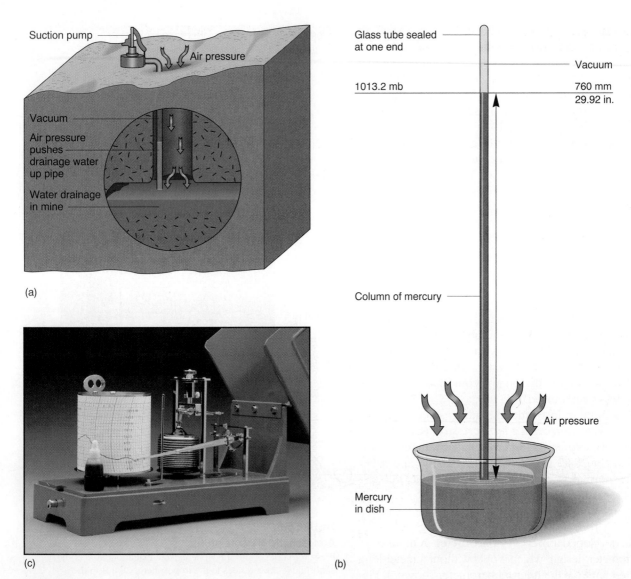

(a)

(c)

(b)

**FIGURE 6-7**

**Developing the barometer.**

Evangelista Torricelli developed the barometer to measure air pressure as a by-product of trying to solve a mine-drainage problem (a). Two types of instruments are used to measure atmospheric pressure: (b) an idealized sketch of a mercury barometer, (c) an aneroid barometer. [(c) Courtesy of Qualimetrics, Inc., Sacramento, CA.]

- The **Coriolis force**, a deflective force, makes wind that travels in a straight path appear to be deflected in relation to Earth's rotating surface. Coriolis force deflects wind to the right in the Northern Hemisphere and to the left in the Southern Hemisphere. Without Coriolis force, winds would move along straight paths between high and low pressure areas.

- The **friction force** drags on the wind as it moves across surfaces; it decreases with height above the surface. Without friction, winds would simply move in paths parallel to isobars and at high rates of speed.

All four of these forces operate on moving air and water at Earth's surface and affect the circulation patterns of

global winds. The following sections describe the actions of the pressure gradient, Coriolis, and friction forces. (The gravitational force operates uniformly worldwide.)

## Pressure Gradient Force

High- and low-pressure areas exist in the atmosphere principally because of unequal heating at Earth's surface and certain dynamic forces in the atmosphere. These pressure differences establish a *pressure gradient force*. For example, cold, dense air at the poles exerts greater pressure than warm, less-dense air along the equator.

An **isobar** is an isoline (a line on a map along which there is a constant value) plotted on a weather map to

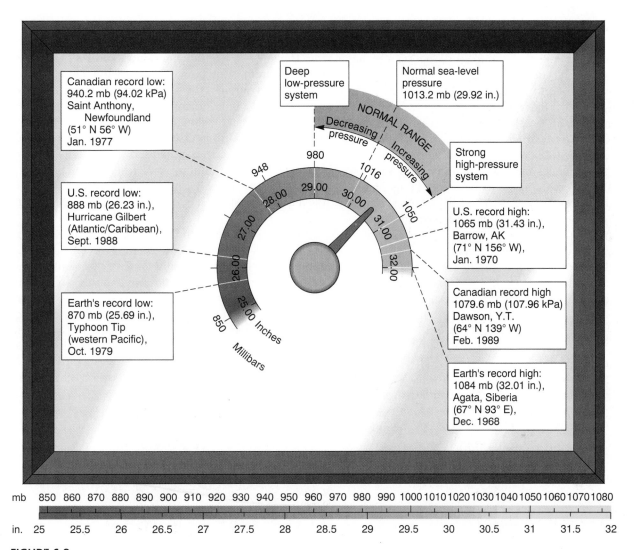

**FIGURE 6-8**

**Air pressure readings and conversions.**

Scales for expressing barometric air pressure in millibars and inches, with average air pressure values and recorded pressure extremes. Canadian values include kilopascal equivalents (10 mb = 1 kPa).

connect points of equal pressure. Thus, a pressure gradient between an area of higher pressure and one of lower pressure is portrayed by isobars. The spacing between isobars indicates the intensity of the pressure difference, or pressure gradient.

Just as closer contour lines on a topographic map indicate a steeper slope on land, so do closer isobars denote steepness in the pressure gradient. In Figure 6-9a, note the spacing of the isobars (purple lines). A steep gradient causes faster air movement from a high-pressure area to a low-pressure area. Isobars spaced at greater distances from one another mark a more gradual pressure gradient, one that creates a slower air flow. Along a horizontal surface, the pressure gradient force acts at right angles to the isobars, so wind blows across them at right angles. Note the location of steep and gradual pressure gradients and their relationship to wind intensity on the map in Figure 6-9b.

Figure 6-10 illustrates the forces that direct the wind. Figure 6-10a shows the pressure gradient force acting alone. In a high-pressure area, as air *descends*, a field of subsiding, or sinking, air develops. Air *diverges* out of the high-pressure area, moving outward in all directions. On the other hand, in a low-pressure area, as air *rises*, it *converges* from all directions into the area of lower pressure. For instance on a warm day, the temperature of the air increases, the air is less dense and more buoyant, so it rises. In contrast on a cold day, the temperature of the air is lower, the air is denser, less buoyant, so it descends. This behavior of a parcel of air is discussed further in Chapter 7.

## *Coriolis Force*

You might expect surface winds to move in a straight line from areas of higher pressure to areas of lower pressure. On a nonrotating Earth, they would. But on our rotating planet, anything that flies or flows across Earth's surface—wind, an airplane, or ocean currents—is deflected from a straight path by the *Coriolis force*, which is an effect of Earth's rotation.

Earth's rotational speed varies with latitude, increasing from 0 kmph at the poles to 1675 kmph (1041 mph) at the equator (see Table 2-3). The Coriolis force is zero along the equator, increases to half the maximum deflection at 30° N and 30° S latitudes, and reaches maximum deflection flowing away from the poles.

The deflection occurs regardless of the direction in which the object is moving. Because Earth rotates eastward, objects that move in an absolute straight line over a distance (such as winds and ocean currents) appear to curve to the *right* in the Northern Hemisphere and to the *left* in the Southern Hemisphere (Figure 6-11a). The effect of the Coriolis force increases as the speed of the moving object increases; thus, the faster the wind speed, the greater its apparent deflection. The Coriolis force does not affect small scale motions that cover insignificant distance and time (see News Report 1).

The key to understanding Coriolis force is one's viewpoint. From the viewpoint of an airplane that is passing over Earth's surface, the surface can be seen to rotate slowly below. But, looking from the surface at the airplane, the surface seems stationary, and the airplane appears to curve off course. The airplane does not actually deviate from a straight path, but it *appears to* do so because we are standing on Earth's rotating surface beneath the airplane. Because of this apparent deflection the airplane must make constant corrections in flight path to maintain its "straight" heading.

As an example of the effect of this force, see Figure 6-11b. A pilot leaves the North Pole and flies due south toward Quito, Ecuador. If Earth were not rotating, the aircraft would simply travel along a meridian of longitude and arrive at Quito. But Earth is rotating eastward beneath the aircraft's flight path. If the pilot does not allow for this rotation, the plane will reach the equator over the ocean along an apparently curved path far to the west of the intended destination. Pilots must correct for this Coriolis deflection in their navigational calculations.

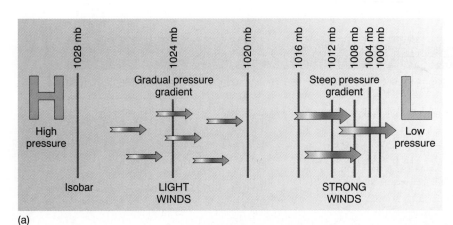

(a)

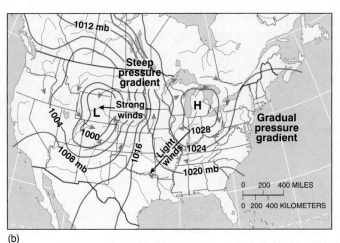

(b)

**FIGURE 6-9**

**Pressure gradient determines wind speed.**

Pressure gradient (a). On a weather map (b), the closer spacing of isobars represents a steeper pressure gradient that produces stronger winds; wider spacing of isobars denotes a gradual pressure gradient that leads to lighter winds.

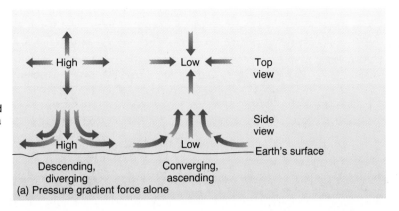

Top view and side view of air movement in an idealized high-pressure area and low-pressure area on a nonrotating Earth.

(a) Pressure gradient force alone

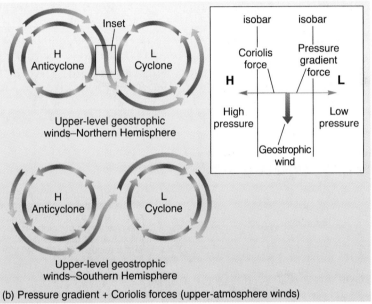

Earth's rotation adds the Coriolis force and a "twist" to air movements. High-pressure and low-pressure areas develop a rotary motion, and wind flowing between highs and lows flows parallel to isobars.

(b) Pressure gradient + Coriolis forces (upper-atmosphere winds)

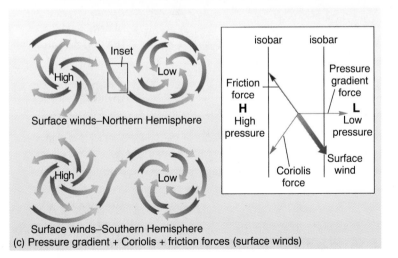

Surface friction adds a countering force to Coriolis, producing winds that spiral out of a high-pressure area and into a low-pressure area. Surface winds cross isobars at an angle.

(c) Pressure gradient + Coriolis + friction forces (surface winds)

**FIGURE 6-10**

**Three physical forces that produce winds.**

Three physical forces integrate to produce wind patterns at the surface and aloft. (a) the pressure gradient force; (b) the Coriolis force counters the pressure gradient force, producing a geostrophic wind flow in the upper atmosphere; (c) the friction force, which, combined with the other two forces, produces characteristic surface winds. The gravitational force is assumed. Note the two inset diagrams showing the interaction of forces that form prevailing geostrophic and surface winds. In (b) and (c), note the reverse circulation pattern in the Southern Hemisphere.

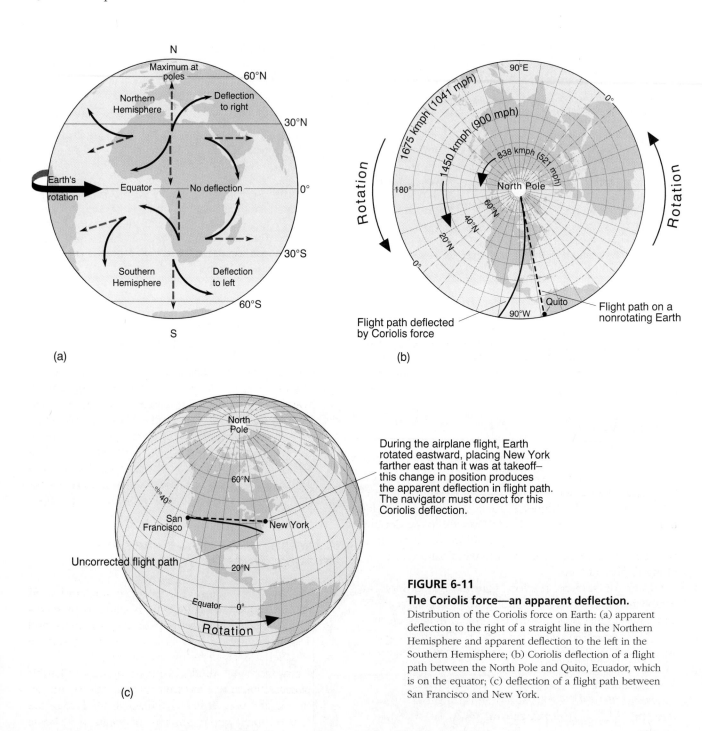

(a)

(b)

(c)

During the airplane flight, Earth rotated eastward, placing New York farther east than it was at takeoff—this change in position produces the apparent deflection in flight path. The navigator must correct for this Coriolis deflection.

**FIGURE 6-11**

**The Coriolis force—an apparent deflection.**

Distribution of the Coriolis force on Earth: (a) apparent deflection to the right of a straight line in the Northern Hemisphere and apparent deflection to the left in the Southern Hemisphere; (b) Coriolis deflection of a flight path between the North Pole and Quito, Ecuador, which is on the equator; (c) deflection of a flight path between San Francisco and New York.

These effects are in force regardless of the direction of the moving object. A flight from California to New York is shown in Figure 6-11c. Note the Coriolis deflection occurs because, as the airplane flew to New York, Earth continued to rotate eastward, so the destination moved farther to the east. Unless the pilot corrected for Earth's rotational motion, the flight would end up somewhere in North Carolina.

How does the Coriolis force affect wind? As air rises from the surface through the lowest levels of the atmosphere, leaving the drag of surface friction behind, its speed increases. This increase in speed increases the

Coriolis force, spiraling the winds to the right in the Northern Hemisphere or to the left in the Southern Hemisphere. In the upper troposphere, the Coriolis force just balances the pressure gradient force. Consequently, the winds between higher-pressure and lower-pressure areas aloft flow parallel to the isobars.

Figure 6-10b illustrates the combined effect of the pressure gradient force and the Coriolis force on air currents aloft. Together, they produce winds that do not flow directly from high to low, but *around* the pressure areas instead, remaining parallel to the isobars. Such winds are called **geostrophic winds** and are characteristic of upper

## News Report 1

### Coriolis, a Forceful Effect on Drains?

A common misconception about the Coriolis force is that it affects water draining out of a sink, tub, or toilet. Can the Coriolis force cause this twist? When a ship crosses the equator, does the direction of a draining spiral of water in a sink suddenly reverse?

Moving water or air must cover some distance across space and time before it is noticeably deflected by the Coriolis force. Long-range artillery shells and guided missiles do exhibit small amounts of deflection that must be corrected for accuracy. But water movements down a drain are too small to be noticeably affected by this force.

Note that we call this a force. The label *force* is appropriate because, as the physicist Sir Isaac Newton (1643–1727) stated, when something is accelerating over a space, *a force is in operation*. This *apparent force* (in classical mechanics, an inertial force) *acts as an effect* on moving objects. It is named for Gaspard Coriolis, a French mathematician and researcher of applied mechanics, who first described the phenomenon in 1831.

---

tropospheric circulation. (The suffix *-strophic* means "to turn.") Geostrophic winds produce the characteristic pattern shown on the upper-air weather chart in Figure 6-12. Note the inset illustration showing the effects of the pressure gradient and Coriolis forces that produce a geostrophic flow of air.

### Friction Force

The effect of surface friction extends to a height of about 500 m (around 1600 ft) and varies with surface texture, wind speed, time of day and year, and atmospheric conditions. In general, rougher surfaces produce more friction. Figure 6-10c adds the effect of friction to the Coriolis and pressure gradient forces on wind movements; combining all three forces produces the wind patterns we see along Earth's surface.

Near the surface, friction disrupts the equilibrium established in geostrophic wind flows between the pressure gradient and Coriolis forces—note the inset illustration in Figure 6-10c. Because surface friction decreases wind speed, it reduces the effect of the Coriolis force and causes winds to move across isobars at an angle.

In Figure 6-10c, you can see that the Northern Hemisphere winds spiral out from a high-pressure area *clockwise* to form an **anticyclone** and spiral into a low-pressure area *counterclockwise* to form a **cyclone**. (In the Southern Hemisphere these circulation patterns are reversed, flowing out of high-pressure cells counterclockwise and into low-pressure cells clockwise.)

## Atmospheric Patterns of Motion

With these forces and motions in mind, we are ready to understand the *Seasat* image of winds (see Figure 6-4) and to build a general model of total atmospheric circulation. The warmer, less-dense air along the equator rises, creating low pressure at the surface, and the colder, denser air at the poles sinks, creating high pressure. If Earth did not rotate, the result would be a simple wind flow from the poles to the equator (a meridional flow), established purely by pressure gradient.

But, because Earth does rotate, a much more complex system exists. This system transfers thermal energy and air and water masses from equatorial surpluses to polar deficits, using waves, streams, and eddies on a planetary scale. Instead of the nonrotating-Earth, poles-to-equator flow, the flow is predominantly zonal (latitudinal), both at the surface and aloft. Winds are westerly (eastward-moving) in the middle and high latitudes and easterly (westward-moving) in the low latitudes of both hemispheres.

### Primary High-Pressure and Low-Pressure Areas

The following discussion of Earth's pressure and wind patterns refers often to Figure 6-13, isobaric maps showing average surface barometric pressure in January and July. Indirectly, these maps indicate prevailing surface winds, which are suggested by the isobars.

The primary high- and low-pressure areas of Earth's general circulation appear on these maps as cells or uneven belts of similar pressure that stretch across the face of the planet, interrupted by landmasses. Between these areas flow the primary winds, which have been noted in adventure stories and myths throughout human experience.

Secondary highs and lows, from a few hundred to a few thousand kilometers in diameter and hundreds to thousands of meters high, are formed within these primary pressure areas. The secondary systems seasonally migrate to produce changing weather patterns in the regions over which they pass.

Four broad pressure areas cover the Northern Hemisphere, and a similar set exists in the Southern Hemisphere. In each hemisphere, two of the pressure areas are stimulated by thermal (temperature) factors: the **equatorial low-pressure trough** (marked by the ITCZ line on

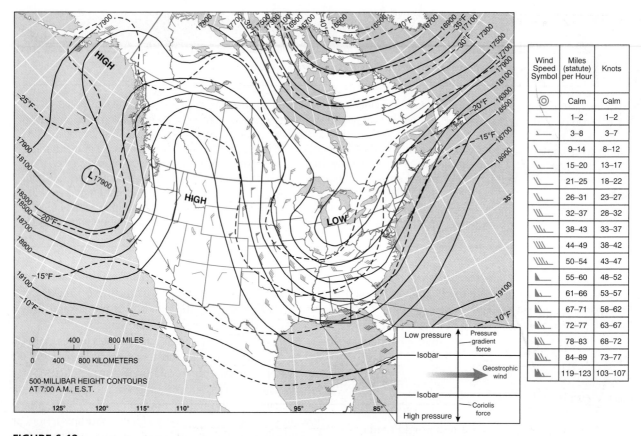

**FIGURE 6-12**

**A 500 mb pressure map and geostrophic winds aloft.**

Isobaric chart for April 25, 1977. Contours show elevation (in feet) at which 500 mb pressure occurs. The pattern of contours reveals geostrophic wind patterns in the troposphere at approximately 5500 m (18,000 ft) altitude. Note the "ridge" of high pressure over the Intermountain West (through the Rockies) and the "trough" of low pressure over the Great Lakes region. Note the inset diagram showing the interaction of forces that form prevailing geostrophic winds. [Data provided by National Weather Service, NOAA.]

the maps) and the weak **polar high-pressure cells** at the North and South Poles (not shown, as the maps are cut off at 80° N and 80° S). The other two pressure areas are formed by dynamic factors: the **subtropical high-pressure cells** (H) and **subpolar low-pressure cells** (L). Table 6-2 summarizes the characteristics of these pressure areas. We now examine each principal pressure region.

***Equatorial Low-Pressure Trough (ITCZ): Clouds and Rain.*** Figure 6-14 is a satellite image showing the equatorial low-pressure trough. Its presence is revealed by the broken band of clouds that straddles the equator across the middle of the image. The equatorial low-pressure trough is an elongated, undulating narrow band of low pressure (converging, ascending air flow) that nearly encircles the planet.

Constant high Sun altitude and consistent daylength (12 hours a day, year-round) make large amounts of energy available in this region throughout the year. The warming creates lighter, less-dense, ascending air, with

surface winds converging along the entire extent of the low-pressure "trough." This converging air is extremely moist and full of latent heat energy. As it rises, the air expands and cools, producing condensation; consequently, precipitation is heavy throughout this zone. Vertical cloud columns frequently reach the tropopause.

The combination of heating and convergence forces air aloft and forms the **intertropical convergence zone (ITCZ)**. The ITCZ is identified by bands of clouds associated with the convergence of winds along the equator and is noted on the pressure maps (Figure 6-13). You easily can identify this zone on the photograph on this book's back cover.

During summer, a marked wet season accompanies the shifting ITCZ over various regions. The maps in Figure 6-13 show the ITCZ as a dashed line in January and July. In January, the zone crosses northern Australia and dips southward in eastern Africa and South America. By July the zone shifts northward with the Sun, as far north as Pakistan and southern Asia.

**FIGURE 6-13**

**Global barometric pressures for January and July.**

Average surface barometric pressure (millibars) for (a) January and (b) July. Dashed line marks the intertropical convergence zone (ITCZ). Compare specific regions for January and July—for instance, the North Pacific, the North Atlantic, and the central Asian landmass. [Adapted from National Climatic Data Center, *Monthly Climatic Data for the World*, 46, no. 1, January and July 1993. Prepared in cooperation with the World Meteorological Organization. Washington, DC: National Oceanic and Atmospheric Administration.]

Figure 6-15 shows two views of Earth's general circulation. Both parts of the figure show Hadley cells, named for the eighteenth-century English scientist who described the trade winds. Each **Hadley cell** is generated by the equatorial low-pressure system of converging, ascending air. The winds converging on the equatorial low-pressure trough are known generally as the **trade winds**, or *trades*. *Northeast trade winds* blow in the

Northern Hemisphere and *southeast trade winds* in the Southern Hemisphere. These are labeled in Figure 6-15a. The trade winds were named during the era when sailing ships dominated the seas and world commerce was propelled by these winds.

The trade winds pick up large quantities of moisture as they return through the Hadley circulation cell for another cycle of uplift and condensation (shown in cross section in Figure 6-15b). Within the ITCZ, winds are calm or mildly variable because of the even pressure gradient and the vertical ascent of air. These equatorial calms are the **doldrums**, a name formed from an older English word meaning foolish, because of the difficulty sailing ships encountered when attempting to move through this zone.

The rising air from the equatorial low-pressure area spirals upward into a geostrophic flow to the north and

south. These upper-air winds turn eastward, flowing from west to east, beginning at about 20° N and 20° S, forming descending anticyclonic flows above the subtropics. In each hemisphere, this circulation pattern appears most vertically symmetrical near the equinoxes.

### Subtropical High-Pressure Cells: Hot, Dry, Desert Air.

Between 20° and 35° latitude in both hemispheres, a broad high-pressure zone of hot, dry air is evident across the globe (Figures 6-13, 6-15b). It is represented by the clear, frequently cloudless skies over the Sahara and Arabian Desert and portions of the Indian Ocean on the satellite imagery in Figure 6-14. The dynamic cause of these subtropical anticyclones is too complex to detail here, but they generally form as air in the Hadley cell

**TABLE 6-2**

| **Four Hemispheric Pressure Areas** | | | |
|---|---|---|---|
| *Name* | *Cause* | *Location* | *Air Temperature/ Moisture* |
| Polar high-pressure cells | Thermal | 90° N 90° S | Cold/dry |
| Subpolar low-pressure cells | Dynamic | 60° N 60° S | Cool/wet |
| Subtropical high-pressure cells | Dynamic | 20° to 35° N and S | Hot/dry |
| Equatorial low-pressure trough | Thermal | 10° N to 10° S | Warm/wet |

**FIGURE 6-14**

**Clouds portray equatorial and subtropical circulation patterns on a satellite image.**

An interrupted band of clouds along the equator denotes the intertropical convergence zone (ITCZ), flanked to the north and south by several subtropical high-pressure systems and clear skies. This natural-color image was taken by the solid–state imaging instrument aboard the *Galileo* spacecraft during its December 8, 1990, flyby of Earth on its successful voyage to the planet Jupiter. [The Solid State Imaging instrument (violet, green, and red filters) courtesy of Dr. W. Reid Thompson, Laboratory of Planetary Studies, Cornell University. Used by permission.]

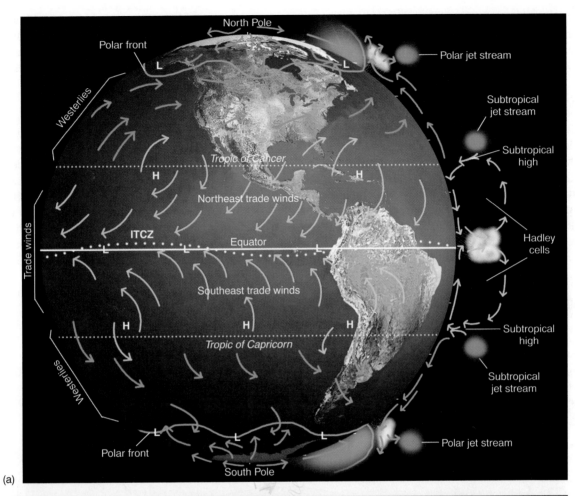

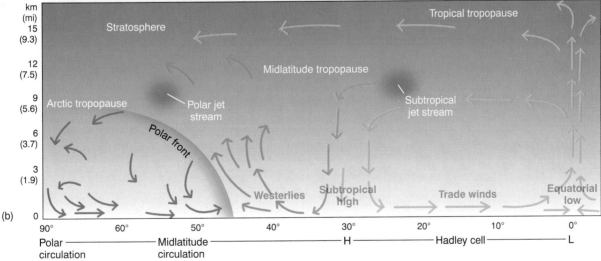

**FIGURE 6-15**

**General atmospheric circulation model.**

Two views of the general atmospheric circulation: (a) general circulation schematic; (b) equator-to-pole cross section of the Northern Hemisphere. Both views show Hadley cells, subtropical highs, polar front, the subpolar low-pressure cells, and approximate locations of the subtropical and polar jet streams.

descends in these latitudes, constantly supported by a dynamic mechanism in the upper-air circulation.

Air above the subtropics is mechanically pushed downward and is heated as it is compressed on its descent to the surface, as illustrated in Figure 6-15. Warmer air has a greater capacity to hold water vapor than does cooler air, making this descending warm air relatively dry. The air is also dry because heavy precipitation along the equator removed moisture.

Surface air diverging from the subtropical high-pressure cells generates Earth's principal surface winds: the westerlies and the trade winds. The **westerlies** are the dominant surface winds from the subtropics to high latitudes. They diminish somewhat in summer and are stronger in winter.

As you examine the global pressure maps in Figure 6-13, you will find several high-pressure areas. In the Northern Hemisphere, the Atlantic subtropical high-pressure cell is called the **Bermuda high** (in the western Atlantic) or the **Azores high** (when it migrates to the eastern Atlantic in winter). The area in the Atlantic under this subtropical high features clear, warm waters and large quantities of *Sargassum* (a seaweed), which gives the area its name—the Sargasso Sea.

The **Pacific high**, or Hawaiian high, dominates the Pacific in July, retreating southward in January. In the Southern Hemisphere, three large high-pressure centers dominate the Pacific, Atlantic, and Indian Oceans, especially in January, and tend to move along parallels of latitude in shifting zonal positions.

The entire high-pressure system migrates with the summer high Sun, fluctuating about 5°–10° in latitude. The *eastern sides* (right-hand side) of these anticyclonic systems are drier and more stable (less convective activity) and feature cooler ocean currents than do the *western sides* (left-hand side). Thus, subtropical and midlatitude west coasts are influenced by the drier eastern side of these systems and therefore experience dry-summer conditions (Figure 6-16). In fact, Earth's major deserts generally occur within the subtropical belt and extend to the west coast of each continent except Antarctica. In the figure the desert region of Africa comes right to the shore, with the cool, southward flowing *Canaries Current* offshore.

The western sides of subtropical high-pressure cells tend to be moist and unstable, as characterized in Figure 6-16. These conditions cause warm, moist weather in Hawaii, Japan, southeastern China, and the southeastern United States (shown in the figure).

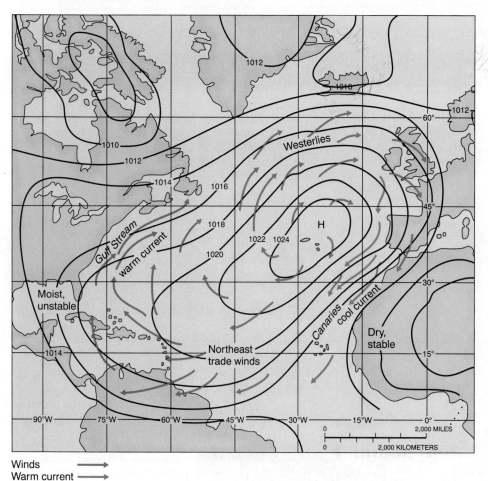

**FIGURE 6-16**
**Subtropical high-pressure system in the Atlantic.**
Characteristic circulation and climate conditions related to the Atlantic subtropical high-pressure anticyclone in the Northern Hemisphere.

Because the location of the subtropical belts are near 25° N and 25° S latitudes, these areas sometimes are known as the *calms of Cancer* and the *calms of Capricorn*. The name **horse latitudes** also is used for these zones of windless, hot, dry air, so deadly in the era of sailing ships. This label is popularly attributed to becalmed and stranded sailing crews of past centuries, who destroyed the horses on board, not wanting to share food or water with the livestock. The term's true origin may never be known; the *Oxford English Dictionary* calls it "uncertain."

***Subpolar Low-Pressure Cells: Cool and Moist.*** In January, two low-pressure cyclonic cells exist over the oceans around 60° N latitude, near their namesakes: the North Pacific **Aleutian low** and the North Atlantic **Icelandic low** (Figure 6-13a, January map). Both cells are dominant in winter and weaken or disappear in summer with the strengthening of high-pressure systems in the subtropics. The area of contrast between cold and warm air masses forms the **polar front**, a contact zone that encircles Earth at this latitude and is focused in these low-pressure areas.

Figure 6-15 illustrates this confrontation between warm, moist air from the westerlies and cold, dry air from the polar and Arctic regions. The upward displacement of the warm air by the heavier cold air forces cooling and condensation in the lifted air. Low-pressure cyclonic storms migrate out of the Aleutian and Icelandic frontal areas and may produce precipitation in North America and Europe, respectively. Northwestern sections of North America and Europe generally are cool and moist as a result of the passage of these cyclonic systems onshore—consider the weather in Washington State and Ireland.

In the Southern Hemisphere, a noncontinuous belt of subpolar cyclonic pressure systems surrounds Antarctica. The spiraling cloud patterns produced by these cyclonic systems are visible on the satellite image in Figure 6-17. Severe cyclonic storms can cross Antarctica, producing strong winds and new snowfall. How many cyclonic systems can you identify on the satellite image?

***Polar High-Pressure Cells: Frigid, Dry Deserts.*** Polar high-pressure cells are weak. The polar atmospheric mass is small, receiving little energy to put it into motion. Variable winds, cold and dry, move away from the polar region in an anticyclonic direction. They descend and diverge clockwise in the Northern Hemisphere (counterclockwise in the Southern Hemisphere) and form weak, variable **polar easterlies**.

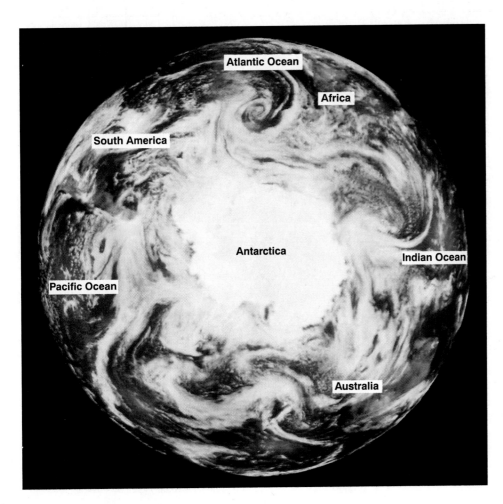

**FIGURE 6-17**
**Clouds portray subpolar and polar circulation patterns.**
Centered on Antarctica, this satellite image shows a series of subpolar low-pressure cyclones in the Southern Hemisphere. Antarctica is fully illuminated by a midsummer Sun as the continent approaches the December solstice. This natural-color image was taken by the solid–state imaging instrument aboard the *Galileo* spacecraft during its December 8, 1990, flyby of Earth on its way to a successful rendevous with the planet Jupiter in 1995. [The Solid State Imaging instrument (violet, green, and red filters) courtesy of Dr. W. Reid Thompson, Laboratory of Planetary Studies, Cornell University. Used by permission.]

Of the two polar regions, the **Antarctic high** is stronger and more persistent forming over the Antarctic landmass. Over the Arctic Ocean, a polar high-pressure cell is less pronounced. When it does form, it tends to locate over the colder northern continental areas in winter (Canadian and Siberian highs) rather than directly over the relatively warmer Arctic Ocean.

## Upper Atmospheric Circulation

Circulation in the middle and upper troposphere is an important component of the atmosphere's general circulation. As indicated, these upper-atmosphere winds blow eastward (west to east) from the subtropics to the poles.

Just as we use sea level as a reference datum for evaluating air pressure at the surface, we use a pressure level such as 500 mb as a **constant isobaric surface**, for a pressure-reference datum. On upper-air pressure maps, we plot the height above sea level at which an air pressure of 500 mb occurs. In contrast, on surface weather maps, we plot different pressures at the fixed elevation of sea level—a *constant height surface*. Using the isobaric chart for April 25, 1977 (Figure 6-12), Figure 6-18 illustrates such an isobaric surface, upon which all points have the same pressure.

We use this 500 mb level to analyze upper-air winds and possible support for surface weather conditions. As on surface maps, closer spacing of the isobars indicates

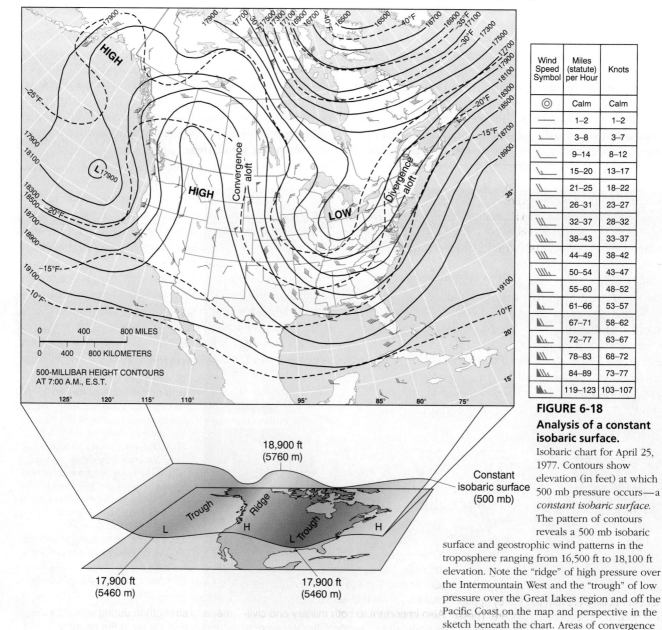

| Wind Speed Symbol | Miles (statute) per Hour | Knots |
|---|---|---|
| ◎ | Calm | Calm |
| — | 1–2 | 1–2 |
| ⌐ | 3–8 | 3–7 |
| ⌐ | 9–14 | 8–12 |
| ⌐ | 15–20 | 13–17 |
| ⌐ | 21–25 | 18–22 |
| ⌐ | 26–31 | 23–27 |
| ⌐ | 32–37 | 28–32 |
| ⌐ | 38–43 | 33–37 |
| ⌐ | 44–49 | 38–42 |
| ⌐ | 50–54 | 43–47 |
| ⌐ | 55–60 | 48–52 |
| ⌐ | 61–66 | 53–57 |
| ⌐ | 67–71 | 58–62 |
| ⌐ | 72–77 | 63–67 |
| ⌐ | 78–83 | 68–72 |
| ⌐ | 84–89 | 73–77 |
| ⌐ | 119–123 | 103–107 |

**FIGURE 6-18**

**Analysis of a constant isobaric surface.**

Isobaric chart for April 25, 1977. Contours show elevation (in feet) at which 500 mb pressure occurs—a *constant isobaric surface*. The pattern of contours reveals a 500 mb isobaric surface and geostrophic wind patterns in the troposphere ranging from 16,500 ft to 18,100 ft elevation. Note the "ridge" of high pressure over the Intermountain West and the "trough" of low pressure over the Great Lakes region and off the Pacific Coast on the map and perspective in the sketch beneath the chart. Areas of convergence and divergence aloft are noted. Data for map from the National Weather Service, NOAA.

faster winds; wider spacing indicates slower winds. On this isobaric pressure surface, altitude variations from the reference datum are called *ridges* for high pressure (with isobars on the map bending poleward) and *troughs* for low pressure (with isobars on the map bending equatorward). Looking at the figure, can you identify such ridges and troughs in the isobaric surface?

The pattern of ridges and troughs in the upper-air wind flow is important in sustaining surface cyclonic (low) and anticylonic (high) circulation. Frequently, surface pressure systems are generated by the upper-air wind flow. Near ridges in the isobaric surface, winds slow and converge (pile-up). Whereas, winds near the area of maximum wind speeds in the isobaric surface, accelerate and diverge (spread out). Note the wind speed indicators in Figure 6-18 near the ridge (over Alberta, Saskatchewan, Montana, and Wyoming); now compare these with the wind-speed indicators around the trough (over Kentucky, West Virginia, the New England states, and Maritimes). Also, note the wind relationships off the Pacific Coast.

Think for a moment of traffic along a busy three-lane street, interrupted every few kilometers by a section reduced to two-lanes, then increasing back to three-lanes. When traffic approaches the slowdown it jams up and converges. When the road widens and maximum speeds are reached, the cars spread out and diverge. Likewise, as wind moves along in geostrophic flow it is constantly experiencing horizontal convergence (piling up) and divergence (spreading out). This divergence in the upper-air flow is important to cyclonic circulation at the surface because it creates an outflow of air aloft that stimulates an inflow of air into the low-pressure cyclone.

Within the westerly flow of geostrophic winds are great waving undulations called **Rossby waves**, named for meteorologist Carl G. Rossby, who, in 1938, first described them mathematically. The polar front is the contact between colder air to the north and warmer air to the south (Figure 6-19). Rossby waves bring tongues of cold air southward (troughs), with warmer tropical air moving northward (ridges). The development of Rossby waves in the upper-air circulation is shown in the three-part figure. As these disturbances mature, distinct cyclonic circulations form, with warmer air and colder air mixing along distinct fronts. The development of cyclonic storm systems at the surface is supported by these wave-and-eddy formations and upper-air divergence. These Rossby waves develop along the flow axis of a jet stream.

***Jet Streams.*** The most prominent movement in these upper-level westerly wind flows is the **jet stream**, an irregular, concentrated band of wind occurring at several different locations. (Figure 6-15a shows the location of four jet streams.) Rather flattened in vertical cross section, the jet streams normally are 160–480 km (100–300 mi) wide by 900–2150 m (3000–7000 ft) thick, with core speeds that can exceed 300 kmph (190 mph). Jet streams in each hemisphere tend to weaken during the hemisphere's summer and strengthen during winter as the streams shift closer to the equator. The patterns of ridges and troughs cause variations in wind speeds (convergence and divergence). These upper-level westerly wind systems also affect transportation as discussed in News Report 2.

The **polar jet stream** is located at the tropopause along the polar front, at altitudes between 7600 and 10,700 m (24,900 and 35,100 ft), meandering between 30° and 70° N latitude (Figure 6-15). The polar jet stream can migrate as far south as Texas, steering colder air masses into North America and influencing surface storm paths traveling eastward. In the summer, the polar jet stream exerts less influence on storms by staying far poleward. The polar jet stream is commonly shown on television weather broadcasts—shifting daily and supporting surface weather patterns. Figure 6-20 shows a map and a stylized view of a polar jet stream flow.

In subtropical latitudes, near the boundary between tropical and midlatitude air, another jet stream flows near the tropopause (Figure 6-15). This **subtropical jet stream** ranges from 9100 to 14,000 m (30,900 to 45,000 ft) in altitude, and, although it is generally weaker, it can

## News Report 2

### Jet Streams Affect Flight Times

As airplane travel increased during World War II, flight crews reported strong headwinds on routes to the west. In some cases, planes leaving San Francisco for the Pacific were turned back by opposing wind in the upper troposphere. These pilots had encountered the jet streams.

Next time you plan a flight, note that airline schedules reflect the presence of these upper-level westerly winds, for they allot shorter flight times from west to east and longer flight times from east to west. Also important to both military and civilian aircraft is the effect the jet streams have on fuel consumption and the existence of air turbulence. Seasonal adjustments are necessary because the jet streams in both hemispheres tend to weaken during each hemisphere's summer and strengthen during winter as the streams shift closer to the equator.

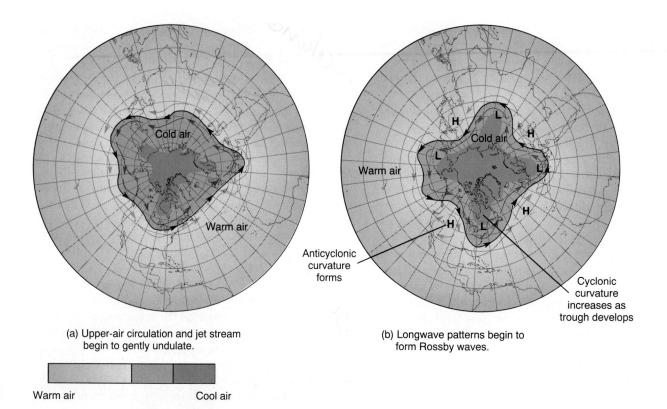

(a) Upper-air circulation and jet stream begin to gently undulate.

Warm air — Cool air

(b) Longwave patterns begin to form Rossby waves.

Anticyclonic curvature forms

Cyclonic curvature increases as trough develops

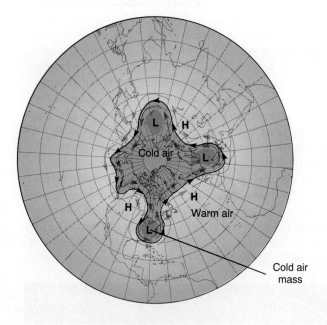

Cold air mass

(c) Strong development of waves produces cells of cold and warm air—high-pressure ridges and low-pressure troughs.

**FIGURE 6-19**

**Rossby upper-atmosphere waves.**

Development of waves in the upper-air circulation first described by C. G. Rossby in 1938 and expanded by J. Namias in 1952.

reach greater speeds than those of the polar jet stream. The subtropical jet stream meanders from 20° to 50° latitude and may occur over North America simultaneously with the polar jet stream—sometimes the two will actually merge for brief episodes.

Again, these atmospheric patterns of fronts, highs, and lows are summarized in Figure 6-15. Use this figure as a convenient reference for review of global circulation. Now, let us move on to some important local winds.

**FIGURE 6-20**

**Jet streams.**
(a) Average locations of the two jet streams over North America. (b) Stylized portrait of a polar jet stream.

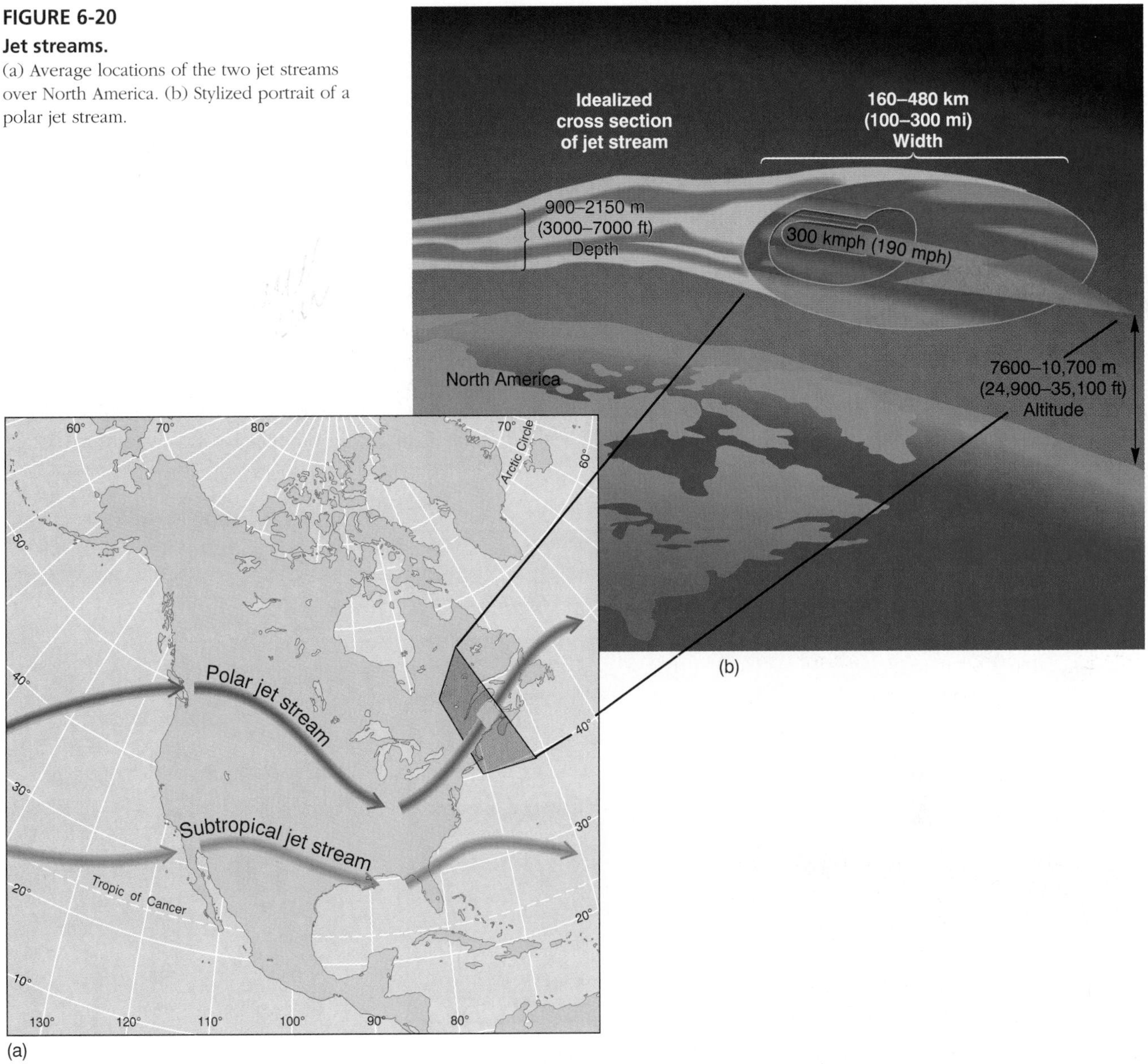

Idealized
cross section
of jet stream

160–480 km
(100–300 mi)
Width

900–2150 m
(3000–7000 ft)
Depth

300 kmph (190 mph)

North America

7600–10,700 m
(24,900–35,100 ft)
Altitude

(b)

Polar jet stream

Subtropical jet stream

Tropic of Cancer

Arctic Circle

(a)

## Local Winds

Several winds form in response to local terrain. These local effects can, of course, be overwhelmed by weather systems passing through an area.

**Land-sea breezes** occur on most coastlines (Figure 6-21). The breezes are created by different heating characteristics of land and water surfaces. During the day, land gains heat energy faster and becomes warmer than the water offshore. Because warm air is less dense, it rises and triggers an onshore flow of cooler marine air to replace the rising warm air—the flow is usually strongest in the afternoon. At night, inland areas cool (radiate heat energy) faster than offshore waters. As a result, the cooler air over the land subsides and flows offshore over the

warmer water, where the air is lifted. This night pattern reverses the process that developed during the day.

Sacramento, California, which is well inland from the Pacific Ocean—160 km (100 mi)—demonstrates the sea-breeze effect. The city is at 39° N and 5 m (17 ft) elevation. The average July maximum and minimum temperatures are 34°C and 14°C (93°F and 57°F), respectively. The evening cooling by the natural flow of marine air establishes a monthly mean of only 24°C (75°F), despite high daytime temperatures. In some similar locations (for example, Hyderabad, Pakistan), large deflectors are placed over roof openings to direct onshore sea-to-land breezes into dwellings to take advantage of the natural cooling.

**Mountain-valley breezes** are created in a somewhat similar exchange. Mountain air cools rapidly at night, and valley air gains heat energy rapidly during the day (Figure 6-22). Thus, warm air rises upslope during the day, particularly in the afternoon; at night, cooler air subsides downslope into the valleys. Other downslope winds are discussed in Chapter 8.

**Katabatic winds**, or gravity drainage winds, are of larger regional scale and usually stronger than mountain-valley breezes, under certain conditions. An elevated plateau or highland is essential, where layers of air at the surface cool, become denser, and flow downslope. The ferocious winds that can blow off the ice sheets of Antarctica and Greenland are katabatic in nature.

Worldwide, a variety of terrains produce such winds and bear many local names. The *mistral* of the Rhône Valley in southern France can cause frost damage to vineyards as the cold north winds move over the region to the Gulf of Lions and the Mediterranean Sea. The frequently stronger *bora*, driven by the cold air of winter high-pressure systems inland, flows across the Adriatic Coast to the west and south. In Alaska such winds are called the *taku*.

*Santa Ana winds* are another local wind type, generated by high pressure over the Great Basin of the western United States. A strong, dry wind is produced that flows out across the desert to southern California coastal areas. The air is heated by compression as it flows from higher to lower elevations and with increasing speed it moves through constricting valleys to the southwest. These winds irritate the population with their dust, dryness, and heat.

Locally, wind represents a source of renewable energy of great promise. Focus Study 6-1 briefly explores the potential for development of wind resources.

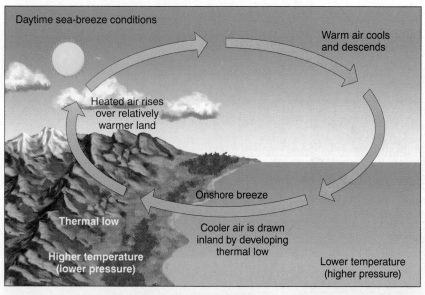

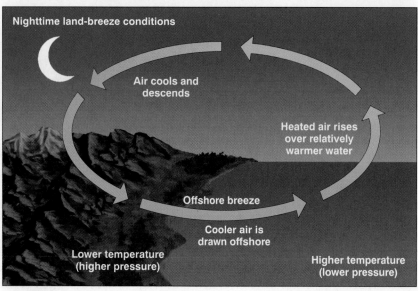

**FIGURE 6-21**

**Land-sea breezes characteristic of day and night.**

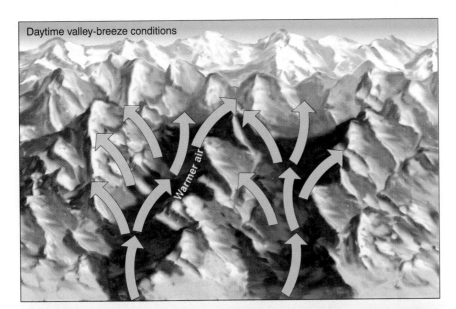

Daytime valley-breeze conditions

Warmer air

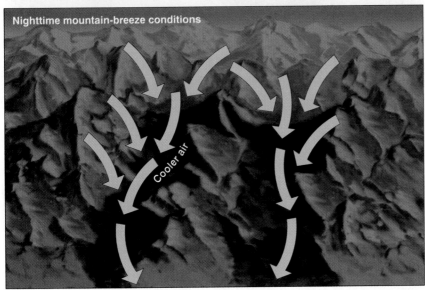

Nighttime mountain-breeze conditions

Cooler air

**FIGURE 6-22**
**Pattern of mountain-valley breezes during day and night.**

***Monsoonal Winds.*** Some regional wind systems seasonally change direction. Intense, seasonally shifting wind systems occur in the tropics over Southeast Asia, Indonesia, India, northern Australia, and equatorial Africa. These winds involve an annual cycle of returning precipitation with the summer Sun and are named after the Arabic word for season, *mausim*, or **monsoon**. (Specific monsoonal weather, associated climate types, and vegetation regions are discussed in Chapters 8, 10, and 20.)

The monsoons of southern and eastern Asia (Figure 6-23) are driven by the location and size of the Asian landmass and its proximity to the Indian Ocean. Also important to the generation of monsoonal flows are wind and pressure patterns in the upper-air circulation.

The extreme temperature range from summer to winter over the Asian landmass is due to its continentality and isolation from the modifying effects of the ocean. This continental landmass is dominated by an intense high-pressure anticyclone in winter (Figures 6-13a and 6-23a), whereas the central area of the Indian Ocean is dominated by the equatorial low-pressure trough (ITCZ). Resulting cold, dry winds blow from the Asian interior over the Himalayas, downslope, and across India, producing average temperatures between 15° and 20°C (60° and 68°F) at lower elevations. These dry winds desiccate (dry out) the landscape and then give way to hot weather from March through May.

During the June–September wet period, the subsolar point shifts northward to the Tropic of Cancer, near the mouths of the Indus and Ganges Rivers. The intertropical convergence zone shifts northward over southern Asia, and the Asian continental interior develops a thermal low pressure, associated with high average temperatures.

## Focus Study 6-1

## Wind Power: An Energy Resource for the Present and Future

The principles of wind power are ancient, but the technology is modern and the benefits are worth pursuing. In developed nations, energy sources are dominated by the use of nonrenewable fuels—coal, gas, oil, nuclear—and centralized energy production. In developing countries, however, more than half of the world's population relies on renewable energy—small hydroelectric plants, wind-power systems, diminishing wood supplies, solar energy—for cooking, heating, and pumping water. These resources are considered renewable because they are not depleted in the span of a human lifetime. The developed countries are fast approaching a reality of scarcity and need to install renewable energy technology.

In developed countries, some electricity now is generated at wind farms, where groups of wind machines are massed. More than 50,000 wind turbines have been installed worldwide since 1974, with 15,000 of them in California alone. Denmark, with 3600 turbines, and California lead the world in this technology, with Denmark moving forward under its own official National Wind Strategy (Figure 1).

Both California and Denmark have a goal to provide 10% of their electrical generating capacity with wind energy by A.D. 2000 when more than 20,000 turbines should be operating—producing approximately 3000 MW (megawatts) of electrical capacity. Sweden is implementing plans to dismantle nuclear power plants and no doubt will increase use of its appreciable wind resources. Other countries are putting forward serious wind initiatives: Great Britain, Ireland, Greece, Spain, Netherlands, China, and New Zealand. The countries of the European Union together have established a goal of 8000 MW of electrical capacity by the year 2005.

### The Nature of Wind Energy

Power generation from wind is site-specific, because conditions that produce adequate winds are limited to certain areas. Wind resources are greatest in three basic settings: (1) along coastlines

(a)

(b)

**FIGURE 1**
**Wind farms in California and Denmark.**
(a) Wind farm in the Tehachapi Mountains of southern California. (b) Wind turbines along the Jutland coastline of Denmark. [Photos by (a) Steve Mulligan. (b) Tony Craddock/Science Photo Library/Photo Researchers, Inc.]

that are influenced by trade winds and westerly winds; (2) where mountain passes constrict air flow and interior valleys develop thermal low-pressure areas, thus drawing air across the landscape; or (3) where localized winds occur, such as katabatic or monsoonal flows. Cash-short developing countries are generally located in areas blessed by such steady winds.

The mountains of North America, northwestern Europe, southwestern

Australia, and the Arctic coastlands of Russia all offer favorable sites. Where wind is reliable less than 25%–30% of the time, only small-scale uses are economically feasible.

Worldwide, the global wind-energy potential is estmated to be about five times the present total electrical demand! In the United States, the winds of North and South Dakota and Texas could meet all the U.S. electrical needs. Of course, transmitting this electricity from areas of production to areas of consumption is an unresolved problem, although existing grids can serve some transmission needs.

One study done in Wyoming found that land presently selling for $100 per hectare (2.47 acres) would yield about $25,000 worth of wind-generated electricity annually. With this carbon-free, sulfur-free, radiation-free energy, the land used for a wind farm still is available for multiple use such as livestock ranching.

## Why Use Wind When We Have Fossil Fuels?

Can wind energy replace fossil fuel or nuclear electrical power plants? In some cases, yes. Let us look at some realities of the fossil fuel supply situation. The United States, with less than 5% of the world's population, consumes nearly 30% of the crude oil extracted worldwide to supply a domestic consumption exceeding 6.2 billion barrels per year (17.1 million barrels a day).

The U.S. domestic reserve will be depleted before the year 2020, and in even less time if today's consumption rate increases (this projection is based on optimistic assessments by the USGS in 1995 of domestic reserves totaling 110 billion barrels). Imports at the level of 50% or greater will stretch this reserve, but with unknown economic, environmental, and military consequences. Of course, imports will be an increasingly difficult problem because, at global consumption rates, the planet will be out of crude oil reserves by 2040. The domestic natural gas supply is a little better, with 35 years projected until depletion of domestic reserves. Present fossil fuel prices do not seem to be related to long-term supplies but rather to short-term exploitation.

Coal is the energy source for 55% of electrical power in the U.S. Burning coal in power plants consumes approximately 70% of annual U.S. production. Several hundred years of coal reserves remain in the United States and Canada if extraction and consumption rates continue at the 1980s level, about 1 billion tons a year. However, if annual production grows 3% to 5% per year, the reserve life will drop to less than a century.

Coal is a valuable source of chemicals, including medicines, so its depletion for the production of electricity could deprive future generations. Expanded use of coal will cause more air pollution, emission of carbon dioxide (a greenhouse gas) and sulfur dioxide (a component of acid deposition), and more surface mining to alter the landscape, and more underground mining to cause land subsidence (sinking).

Despite dwindling fossil fuel supplies, proliferation of wind farms slowed in the United States because of political decisions made in the 1980s. All alternative energy programs of the federal government, including wind energy, were reduced or canceled. Tax incentives vanished—at the same time that tax incentives and research for nonrenewable fossil fuels and nuclear power were increased.

Problems of wind-power engineering are resolved now and presently available turbines range from 300 to 750 kW (kilowatts). Power costs with new technology are down to 4 to 5 cents per kilowatt-hour, whereas coal and natural gas are at 4 to 6 cents per kilowatt-hour. And compare installation costs: Mass production brought wind energy down to $800 per installed kilowatt, whereas, in 1985 dollars, the Diablo Canyon nuclear power plants in California cost $2460 per installed kilowatt.

## The Benefits of the Wind Resource

Economic reality should override further delaying actions, especially where peak winds are in concert with peak electrical demand for air cooling, space heating, or agricultural water pumping. *Long-term* and *marginal costs* also favor wind-energy deployment, for it does not produce carbon dioxide, radioactivity, sulfur dioxide, toxic wastes, or ash, nor does it require mining or the importation of foreign fuels—all of which represent significant cost with nonrenewable fuels. Wind energy is competitive with coal and nuclear power even without considering resource-depletion costs or the environment.

History is rich with the accomplishments of wind-powered sails. Experiments are going forward today to see whether the wind can still drive ships across the sea as in days past—an obvious answer awaits! In an experiment more than 10 years ago, the Japanese oil tanker *Shin Aitoku Maru* installed two computer-guided sails of 300 m$^2$ (about 3200 ft$^2$) each. Fuel consumption was reduced by 50% in early tests. If we consider the world's cargo fleet of more than 25,000 vessels, consuming 4–5 million barrels of oil a day, a savings of just 10%–15% would be significant.

Wind is a renewable resource, waiting to be rediscovered and used as it was for sailing ships and windmills of old. Needless to say, whether or not governments and transnational corporations begin full-scale planning and implementation of renewable resources now, the energy realities of the near future will leave no alternative. By the middle of the next century, wind-generated electricity will be a necessity, along with other renewable energy sources, conservation, and energy efficiency strategies.

---

Meanwhile, the Indian Ocean, with a surface temperature of 30°C (86°F), is under subtropical high pressure. As a result, hot, dry subtropical air sweeps over the warm ocean, producing extremely high evaporation rates (Figure 6-23b).

By the time this air mass reaches India and the convergence zone, it is laden with moisture in thunderous, dark clouds. The warmth of the land lends additional lifting to the incoming air, as do the Himalayas, which force the air mass to higher altitudes. These conditions produce the wet monsoon of India, where world-record rainfalls have occurred. When the monsoonal rains arrive from June to September, they are welcome relief from the dust,

heat, and parched land of Asia's springtime. Both the second-highest average rainfall (1143 cm, or 450 in.) and the highest single-year rainfall (2647 cm, or 1042 in.) were measured at Cherrapunji, India. This record rainfall is discussed further in Chapter 8.

The annual monsoon is an integral part of Indian culture and life, as writer Khushwant Singh described:

> To know India and her peoples, one has to know the monsoon....It has to be a personal experience because nothing short of living through it can fully convey all it means....What the four seasons of the year mean to the European, the one season of the monsoon means to the

Indian. It is preceded by desolation; it brings with it the hopes of spring; it has the fullness of summer and the fulfillment of autumn all in one....Much of India's art, music, and literature is concerned with the monsoon....First it falls in fat drops; the earth rises to meet them. She laps them up thirstily and is filled with fragrance. It brings the odor of the earth and of green vegetation to the nostrils.*

# Oceanic Currents

Earth's atmospheric and oceanic circulations are intimately related. The driving force for ocean currents is the frictional drag of the winds. Also important in shaping these currents is the interplay of the Coriolis force, density differences caused by temperature and salinity, the configuration of the continents and ocean floor, and astronomical forces (the tides).

* K. Singh, *I Hear the Nightingale* (New York: Grove Press, 1959), pp. 101–102. Reprinted by permission.

## Surface Currents

The general patterns of major ocean currents are shown in Figure 6-24. Because ocean currents flow over distance and through time, they are deflected by the Coriolis force. However, their pattern of deflection is not as tightly circular as that of the atmosphere. Compare this ocean-current map with the map showing Earth's pressure systems (Figure 6-13) and you can see that ocean currents are driven by the circulation around subtropical high-pressure cells in both hemispheres. These circulation systems are known as **gyres** and generally appear to be offset toward the western side of each ocean basin. (Remember, in the Northern Hemisphere, winds and ocean currents move *clockwise* about high pressure cells—note the currents in the North Pacific and North Atlantic on the map. In the Southern Hemisphere, the *counterclockwise* circulation about a high is evident on the map.)

In sailing days the Spanish galleons (the *Manila Galleons*) would leave San Blas and Acapulco, Mexico (16.5° N), and sail southwest, catching the northeast trade winds across the Pacific to the Philipines and Manila (14° N). Goods would be traded, and new commodities

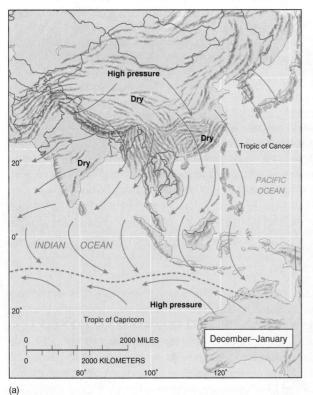 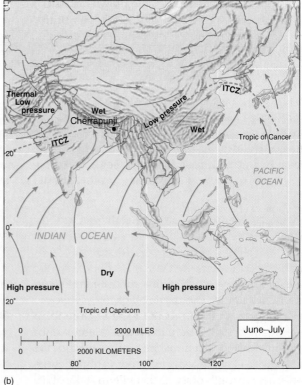

(a)    (b)

**FIGURE 6-23**
**The Asian monsoons.**
Asian monsoon pressure and wind patterns: (a) winter, (b) summer. Note the shifting location of the ITCZ, the changing pressures over the Indian Ocean, and the different conditions over the Asian landmass. [Adapted from Joseph E. Van Riper, *Man's Physical World*, p. 215. Copyright 1971 by McGraw-Hill. Adapted by permission.]

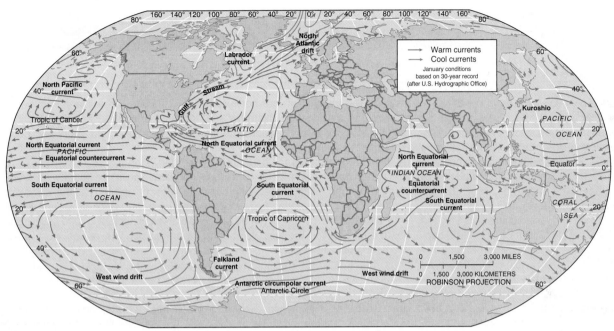

**FIGURE 6-24**
**Major ocean currents.**
[After the U. S. Navy Hydrographic Office.]

obtained and the ships would move northward to catch the westerlies in the midlatitudes to be blown across the ocean to the shores of present-day Alaska or British Columbia down to Northern California. The galleons would make their way south, fighting the rain, frequent fog, and the right-hand Coriolis deflection pushing them away from the coast. The journey would end back in Mexico with cargo from the Orient. An interesting footnote to the sailing difficulties southward along the Pacific Coast is that these ships missed sighting the entrance to San Francisco Bay for over 250 years; finally a land party discovered it in 1769. Thus the history of ships called the *Manilla Galleons* is a lesson in oceanic currents around the Pacific gyre.

***Equatorial Currents.*** Gyres feature slightly stronger currents along their western margins. These are the result of a phenomenon called **western intensification**.

In Figure 6-24, you can see that trade winds drive the ocean surface waters westward in a concentrated channel along the equator. These currents, called *equatorial currents*, are kept near the equator by the Coriolis force that diminishes to zero at the equator. As these surface currents approach the western margins of the oceans, the water actually piles up against the eastern shores of the continents. The average height of this pileup is 15 cm (6 in.). The piled-up ocean water then goes where it can, spilling northward and southward in strong currents, flowing in tight channels along the eastern shorelines. A fine example is the splitting of the South Equatorial current against Brazil in Figure 6-24.

In addition, a strong countercurrent is generated. It flows through the full extent of the Pacific, Atlantic, and Indian Oceans. This occasionally strong, somewhat sporadic eastward **equatorial countercurrent** may flow alongside or just beneath the surface current, at depths of 100 m (300 ft).

In the Northern Hemisphere, the *Gulf Stream* and the *Kuroshio* (current east of Japan) move forcefully northward as a result of western intensification. Their speed and depth are increased by the constriction of the area they occupy. The warm, deep, clear water of the ribbon-like Gulf Stream (Figure 5-7) usually is 50–80 km (30–50 mi) wide and 1.5–2.0 km (0.9–1.2 mi) deep, moving at 3–10 kmph (1.8–6.2 mph). In 24 hours, ocean water can move 70–240 km (40–150 mi) in the Gulf Stream. It may take a year for water to make a complete circuit around an entire gyre. (See News Report 3 for two interesting examples of ocean currents.)

## Deep Currents

Where surface water is swept away from a coast, either by surface divergence (induced by the Coriolis force) or by offshore winds, an **upwelling current** occurs. This cool water generally is nutrient-rich and rises from great depths to replace the vacating water. Such cold upwelling currents exist off the Pacific coasts of North and South America and the subtropical and midlatitude west coast of Africa. These areas are some of Earth's prime fishing regions.

# News Report 3

## A Message in a Bottle and Rubber Duckies

To give you an idea of the dynamic circulation of the ocean, let us briefly look at two recent examples. A 9-year-old child at Dana Point, California (33.5° N), a small seaside community south of Los Angeles, placed a letter in a glass juice bottle in July 1992 and tossed it into the waves. Thoughts of distant lands and fabled characters filled the child's imagination as the bottle disappeared. The vast circulation around the Pacific high, clockwise-circulating gyre, now took command (Figure 1).

Three years passed before ocean currents carried the message in a bottle to the coral reefs and white sands of Mogmog, a small island in Micronesia (7° N). A 7-year-old child there had found a pen pal from afar and immediately sent a photo and card to the sender of the message in the bottle.

Imagine the journey of that note from California—traveling through storms and calms, clear moonlit nights and typhoons, as it floated on ancient currents as the galleons once had.

In January 1994, a large container ship from Hong Kong loaded with toys and other goods was ravaged by a powerful storm. One of the containers on board split apart in the wind, dumping nearly 30,000 rubber ducks, turtles, and frogs into the North Pacific. Westerly winds and the North Pacific current swept this floating cargo across the ocean to the coast of Alaska and Canada. Other toys, still adrift went through the Bering Sea and into the Arctic Ocean (this route is indicated by the dashed line in Figure 1). These will eventually end up in the Atlantic Ocean as they drift around the Arctic Ocean frozen into pack ice.

The bottle from Dana Point was relatively immune to wind, because it rode low in the water, but the rubber duckies rode higher and were susceptible to winds. Scientists are using these incidents to learn more about wind systems and ocean currents. The bottle and ducks offered a much better opportunity than did an earlier spill of 60,000 athletic shoes near Japan. The shoes were tracked across the Pacific Ocean until they landed in the Northwest. Because the shoes became soaked, they were less affected by wind than the ducks were, so they were a better indicator of currents. By the way, the shoes that did not make landfall headed (or footed) back around the Pacific gyre into the tropics and westward, back toward Japan!

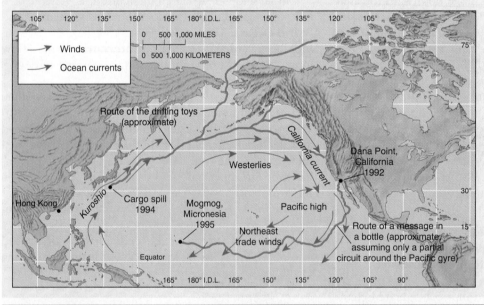

**FIGURE 1**

**Pacific Ocean currents transport human artifacts.** The approximate route of a message in a bottle from Dana Point, California, to Mogmog, Micronesia, assumes a partial circuit around the Pacific gyre. Given the 3-year travel time, it is possible that the message circumnavigated the Pacific Ocean more than once. The rubber duckies voyaged eastward across the Pacific and beyond.

In other portions of the sea where there is an accumulation of water—as at the western end of an equatorial current, or the Labrador Sea, or along the margins of Antarctica—excess water gravitates downward in a **downwelling current**. These currents flow along the ocean floor and travel the full extent of the ocean basins, carrying heat energy and salinity.

To picture such a deep current, imagine a continuous channel of water beginning with cold water downwelling

in the North Atlantic, flowing deep, and upwelling in the Indian Ocean and North Pacific (Figure 6-25). Here it warms and then is carried in surface currents back to the North Atlantic. A complete circuit of the current may require 1000 years from downwelling in the Labrador Sea off Greenland to its reemergence in the southern Indian Ocean and return. Even deeper Antarctic bottom water flows northward in the Atlantic Basin beneath these currents. These current systems appear to play a profound role in global climate.

Scientists at the Institute of Oceanographic Sciences, Wormley, England, have found that the surface warm currents returning from the Indian Ocean, flow into the Atlantic—as the aggressive *Agulhas current* that rounds South Africa—bringing large quantities of heat energy into Atlantic waters and into the Gulf Stream current. The work of this research group is beautifully portrayed in *The FRAM (Fine Resolution Antarctic Model) Atlas of the Southern Ocean* (Natural Environment Research Council, 1991).

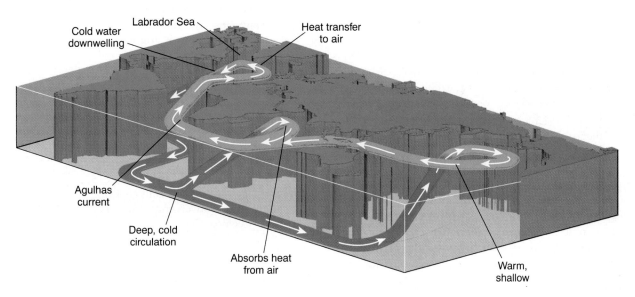

**FIGURE 6-25**
**Deep-ocean circulation.**
Centuries-long deep circulation in the oceans is being deciphered by scientists. This global circulation mimics a vast conveyor belt of water, drawing heat energy from some regions and transporting it for release in others.

# Summary and Review — Atmospheric and Oceanic Circulations

✔ *Define* **wind and** *describe* **how wind is measured, how wind-direction determined, and how winds are named.**

Volcanic eruptions such as those of Tambora in 1815 and Mount Pinatubo in 1991 dramatically demonstrate the power of global winds to disperse aerosols and pollution worldwide in a matter of weeks. Atmospheric circulation facilitates important transfers of energy and mass on Earth, thus maintaining Earth's natural balances. Earth's atmospheric and oceanic circulations represent a vast heat engine powered by the Sun.

**Wind** is the horizontal movement of air across Earth's surface. Its speed is measured with an **anemometer** (a device with cups that are pushed by the wind) and its direction with a **wind vane** (a flat blade or surface that is directed by the wind). A descriptive scale useful in visually estimating wind speed is the traditional **Beaufort wind scale**, which allows the observer to determine approximate wind speeds by the behavior of the surface of a body of water or the motions of objects on land.

> wind (p. 140)
> anemometer (p. 140)
> wind vane (p. 140)
> Beaufort wind scale (p. 142)

1. What is a possible explanation for the beautiful sunrises and sunsets during the summer of 1816 in New England? Relate your answer to global circulation.

2. Explain this statement: "The atmosphere socializes humanity, making all the world a spatially linked society." Illustrate your answer with some examples.

3. Define wind. How is it measured? How is its direction determined?

4. Distinguish among primary, secondary, and tertiary classifications of global atmospheric circulation.

5. What is the purpose of the Beaufort wind scale? Characterize winds given Beaufort numbers of 4, 8, and 12, giving effects over both water and land.

✔ *Define* **the concept of air pressure and** *portray* **the pattern of global pressure systems on isobaric maps.**

The weight (created by motion, size, and number of molecules) of the atmosphere is **air pressure**, which exerts an average force of approximately 1 kg/cm$^2$ (14.7 lb/in.$^2$). Air pressure is measured with a **mercury barometer** at the surface (mercury in a tube—closed at one end and open at the other, with the open end placed in a vessel of mercury—that changes level in response to pressure changes) or an **aneroid barometer** (a closed cell, partially evacuated of air, that detects changes in pressure). Maps portray air pressure patterns using **isobars**—the isolines that connect points of equal pressure.

> air pressure (p. 142)
> mercury barometer (p. 145)
> aneroid barometer (p. 145)
> isobars (p. 146)

6. How does air exert pressure? Describe the basic instrument used to measure air pressure. Compare the operation of two different types of instruments discussed.

7. What is normal sea-level pressure in millimeters? Millibars? Inches? Kilopascals?

8. What does an isobaric map of surface air pressure portray? Contrast pressures over North America for January and July.

---

✔ *Explain* the four driving forces within the atmosphere—gravity, pressure gradient force, Coriolis force, and friction force—and *describe* the primary high- and low-pressure areas and principal winds.

Earth's gravitational force on the atmosphere operates uniformly worldwide. Winds are directed and driven by the **pressure gradient force** (air moves from areas of high pressure to areas of low pressure), the **Coriolis force** (an apparent deflection in the path of winds or ocean currents caused by the rotation of Earth; deflecting objects to the right in the Northern Hemisphere and to the left in the Southern Hemisphere), and **friction force** (Earth's varied surfaces exert a drag on wind movements in opposition to the pressure gradient). A combination of the pressure gradient and Coriolis forces alone produces **geostrophic wind**, which move parallel to isobars, characteristic of winds above the surface frictional layer.

Winds descend and diverge, spiraling outward to form an **anticyclone** (clockwise in the Northern Hemisphere), and they converge and ascend, spiraling upward to form a **cyclone** (counterclockwise in the Northern Hempisphere). The pattern of high and low pressures on Earth in generalized belts in each hemisphere produces the distribution of specific wind systems. These primary pressure regions are the: **equatorial low-pressure trough**, the weak **polar high-pressure cells**, (at both the North and South Poles), and the **subtropical high-pressure cells** and **subpolar low-pressure cells**.

All along the equator winds converge into the equatorial low creating the **intertropical convergence zone (ITCZ)**. Air rising along the equator and descending in the subtropics, in each hemisphere, produce **Hadley cells**. The winds returning to the ITCZ from the northeast in the Northern Hemisphere and from the southeast in the Southern Hemisphere produce the **trade winds** and a region of poor sailing in the convergence zone nicknamed the **doldrums**.

Winds flowing out of the subtropics to higher latitudes produce the **westerlies** in either hemisphere. The subtropical high-pressure cells on Earth, generally between 20° and 35° in

either hemisphere, are variously named the **Bermuda high**, **Azores high**, and **Pacific high**. Sailing in these latitudes, nicknamed the **horse latitudes**, is dangerous because of the calms.

Along the polar front and the series of low-pressure cells, the **Aleutian low** and **Icelandic low** dominate the North Pacific and Atlantic, respectively. This region of contrast between colder air toward the poles and warmer air equatorward is called the **polar front**. The weak and variable **polar easterlies** diverge from the polar high-pressure cells, particularly the **Antarctic high**.

> pressure gradient force (p. 145)
> Coriolis force (p. 146)
> friction force (p. 146)
> geostrophic winds (p. 150)
> anticyclone (p. 151)
> cyclone (p. 151)
> equatorial low-pressure trough (p. 151)
> polar high-pressure cells (p. 152)
> subtropical high-pressure cells (p. 152)
> subpolar low-pressure cells (p. 152)
> intertropical convergence zone (ITCZ) (p. 152)
> Hadley cell (p. 153)
> trade winds (p. 153)
> doldrums (p. 154)
> westerlies (p. 156)
> Bermuda high (p. 156)
> Azores high (p. 156)
> Pacific high (p. 156)
> horse latitudes (p. 157)
> Aleutian low (p. 157)
> Icelandic low (p. 157)
> polar front (p. 157)
> polar easterlies (p. 157)
> Antarctic high (p. 158)

9. Describe the effect of the Coriolis force. Explain how it apparently deflects atmospheric and oceanic circulations.

10. What are geostrophic winds, and where are they encountered in the atmosphere?

11. Describe the horizontal and vertical air motions in a high-pressure anticyclone and in a low-pressure cyclone.

12. Construct a simple diagram of Earth's general circulation, including the four principal pressure belts or zones and the three principal wind systems.

13. How is the intertropical convergence zone (ITCZ) related to the equatorial low-pressure trough? How might it appear on a satellite image?

14. Characterize the belt of subtropical high pressure on Earth: Name the of specific cells. Describe the generation of westerlies and trade winds. Discuss sailing conditions.

15. What is the relation among the Aleutian low, the Icelandic low, and migratory low-pressure cyclonic storms in North America? In Europe?

---

✔ *Describe* upper-air circulation and its support role for surface systems and *define* the jet streams.

Air pressure in the middle and upper troposphere is described using a **constant isobaric surface**, or surface along which the same pressure, such as 500 mb, is recorded regardless of altitude. The height of this surface above the ground is described in ridges and troughs that support the development of and help sustain surface pressure systems; lows are sustained by divergence aloft, and highs are sustained by convergence aloft.

Vast, flowing longwave undulations in these upper-air westerlies form wave motions called **Rossby waves**. The prominent movements in upper level, westerly winds are streams of high speed winds called the **jet streams**. Depending on their latitudinal position in either hemisphere they are termed the **polar jet stream** or the **subtropical jet stream**.

> constant isobaric surface (p. 158)
> Rossby waves (p. 159)
> jet streams (p. 159)
> polar jet stream (p. 159)
> subtropical jet stream (p. 159)

16. What is the relation between wind speed and the spacing of isobars?

17. How is the constant isobaric surface (ridges and troughs) related to surface pressure systems? To divergence aloft and surface lows? To convergence aloft and surface highs?

18. Relate the jet-stream phenomenon to general upper-air circulation. How is the presence of this circulation related to airline schedules from New York to San Francisco and the return trip to New York?

---

✔ *Explain* **several types of local winds: land-sea breezes, mountain-valley breezes, katabatic winds, and the regional monsoons.**

Different heating characteristics of land and water surfaces create **land-sea breezes**. **Mountain-valley breezes** are caused by changing temperature differences during the day and evening between valleys and mountain summits. **Katabatic winds**, or gravity drainage winds, are of larger regional scale and are usually stronger than mountain-valley breezes, under certain conditions. An elevated plateau or highland is essential, where layers of air at the surface cool, become denser, and flow downslope.

Intense, seasonally shifting wind systems occur in the tropics over Southeast Asia, Indonesia, India, northern Australia, and equatorial Africa. These winds involve an annual cycle of returning precipitation with the summer Sun and are named after the Arabic word for season, *mausim*, or **monsoon**. The monsoons of southern and eastern Asia are driven by the location and size of the Asian landmass and its proximity to the Indian Ocean. Also important to the generation of monsoonal flows are wind and pressure patterns in the upper-air circulation.

> land-sea breezes (p. 161)
> mountain-valley breezes (p. 161)
> katabatic winds (p. 162)
> monsoon (p. 163)

19. People living along coastlines generally experience variations in winds from day to night. Explain the factors that produce these changing wind patterns.

20. The arrangement of mountains and nearby valleys produces local wind patterns. Explain the day and night winds that might develop.

21. Describe the seasonal pressure patterns that produce the Asian monsoonal wind and precipitation patterns. Contrast January and July conditions.

---

✔ *Discern* **the basic pattern of Earth's major surface and deep ocean currents.**

Ocean currents are primarily caused by the frictional drag of wind and occur worldwide at varying intensities, temperatures, and speeds, both along the surface and at great depths in the oceanic basins. The circulation around subtropical high-pressure cells in both hemispheres is notable on the ocean circulation map—these **gyres** are usually offset toward the western side of each ocean basin.

The trade winds converge along the ITCZ and push enormous quantities of water in a process known as the **western intensification**. Strong countercurrents flow the full extent of the Pacific, Atlantic, and Indian Oceans, in a sporadic eastward **equatorial countercurrent**. Where surface water is swept away from a coast, either by surface divergence (induced by the Coriolis force) or by offshore winds, an **upwelling current** occurs. This cool water generally is nutrient-rich and rises from great depths to replace the vacating water. In other portions of the sea where there is an accumulation of water—as at the western end of an equatorial current, or the Labrador Sea, or along the margins of Antarctica—excess water gravitates downward in a **downwelling current**. These currents generate important mixing currents that flow along the ocean floor and travel the full extent of the ocean basins, carrying heat energy and salinity.

> gyres (p. 166)
> western intensification (p. 167)
> equatorial countercurrent (p. 167)
> upwelling current (p. 167)
> downwelling current (p. 168)

22. What is the relationship between global atmospheric circulation and ocean currents? Relate oceanic gyres to patterns of subtropical high pressure.

23. Define the western intensification. How is it related to the Gulf Stream and Kuroshio currents?

24. Where on Earth are upwelling currents experienced? What is the nature of these currents?

25. What is meant by deep-ocean circulation? At what rates do these currents flow? How might this circulation be related to the Gulf Stream in the western Atlantic Ocean?

---

 ## NetWork

The *Geosystems Home Page* provides on-line resources for this chapter on the World Wide Web. You will find review exercises, specific updates for items in the chapter, suggested readings, and links to interesting related pathways on the Internet (click on the Table of Contents link and select this chapter). *Geosystems* is at: **http://www.prenhall.com/geosystm**

# *Part 2*

# The Water, Weather, and Climate Systems

*Cumulonimbus clouds and thunderstorm in eastern Washington.* [Photo by author.]

Earth is the water planet. Its surface waters are unique in the solar system. Chapter 7 explains why water exists here in such quantity, covering more than two-thirds of the globe, and describes the remarkable qualities and properties it possesses. It also discusses the daily dynamics of the atmosphere—the powerful interaction of moisture and energy and the resulting stability and instability—as a preamble to understanding weather. Chapter 8 examines weather and its causes. Topics include interpretation of cloud forms, interaction of air masses, understanding the daily weather map, and the violent phenomena of thunderstorms, tornadoes, and hurricanes.

In Chapter 9, water circulation over Earth—the hydrologic cycle—is explained. We examine the water-budget concept, which is useful in understanding soil-moisture and water-resource relationships at all levels—global, regional, and local. In Chapter 10, we see the spatial implications over time of the energy-atmosphere and water-weather systems: Earth's climate patterns are generated. We close with a discussion of global climate change and future climate trends.

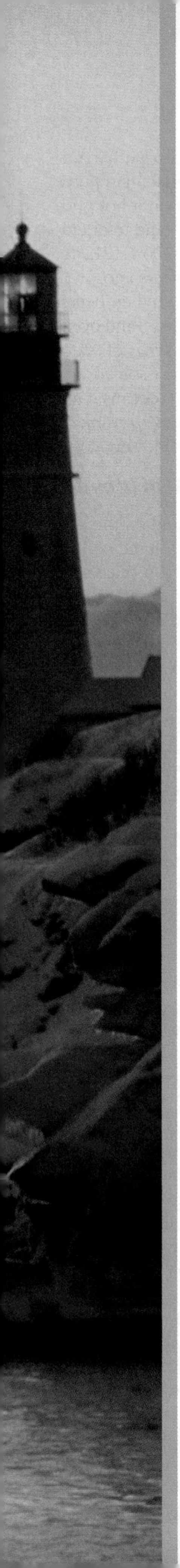

# 7

# Water and Atmospheric Moisture

**Water on Earth**

**Unique Properties of Water**

**Humidity**

**Atmospheric Stability**

**Summary and Review**

## Key Learning Concepts

After reading the chapter, you should be able to:

- *Describe* the origin of Earth's waters, *relate* the quantity of water that exists today, and *list* the locations of Earth's freshwater supply.

- *Describe* the heat properties of water and *identify* the traits of its three phases: solid, liquid, and gas.

- *Define* humidity and the expressions of the relative humidity concept; *explain* dew-point temperature and saturated conditions in the atmosphere.

- *Define* atmospheric stability and *relate* it to a parcel of air that is ascending or descending.

- *Illustrate* three atmospheric conditions—unstable, conditionally unstable, and stable—with a simple graph that relates the environmental lapse rate to the dry adiabatic rate (DAR) and moist adiabatic rate (MAR).

*Fog bank, sea smoke, and snow at Portland Head Lighthouse, Cape Elizabeth, Maine.*
[Photo by Randy Ury/Stock Market.]

Walden is blue at one time and green at another, even from the same point of view. Lying between the earth and the heavens, it partakes of the color of both....A lake is the landscape's most beautiful and expressive feature. It is earth's eye; looking into which the beholder measures the depth of his own nature....Sky water. It needs no fence. Nations come and go without defiling it. It is a mirror which no stone can crack....Nature continually repairs...a mirror in which all impurity presented to it sinks, swept and dusted by the sun's hazy brush. A field of water...is continually receiving new life and motion from above. It is intermediate in its nature between land and sky.*

Thus did Thoreau speak of the water so dear to him—Walden Pond in Massachusetts, where he lived along the shore.

Water is critical to our daily lives and is an extraordinary compound in nature. It covers 71% of Earth; in the Solar System, it occurs in such significant quantities only on our planet. Pure water is colorless, odorless, and tasteless; yet, because it is a solvent, pure water rarely occurs in nature. Water weighs 1 g/cm$^3$ (gram per cubic centimeter), or 1 kg/l (kilogram per liter). In the English system water weighs 62.3 lb/ft$^3$, or 8.337 lb/gal.

Water constitutes nearly 70% of our bodies by weight and is the major ingredient in plants, animals, and our food. A human can survive 50 to 60 days without food but only 2 or 3 days without water. The water we use must be adequate, both in quantity and quality, for its many tasks—everything from personal hygiene to vast national water projects. Water indeed occupies that place between land and sky, mediating energy and shaping both the lithosphere and atmosphere, as Thoreau revealed.

This chapter examines water and the dynamics of atmospheric moisture and stability—the essentials of weather.

# Water on Earth

Earth's hydrosphere contains about 1.36 billion cubic kilometers (specifically, 1,359,208,000 km$^3$, or 326,074,000 mi$^3$). According to scientific evidence, much of Earth's water originated from icy comets, which were a portion of the planetesimals that accreted (coalesced) to form the planet. The water within the planet then reached the surface by **outgassing**. Outgassing is a continuing process by which water and water vapor emerge from layers deep within and below the crust, 25 km (15.5 mi) or more below Earth's surface (Figure 7-1).

In the early atmosphere, massive quantities of out-

gassed water vapor condensed and fell back to Earth in torrents, only to vaporize again because of high temperatures at Earth's surface, like water drops sizzling on a hot griddle. For water to remain on Earth's surface, land temperatures had to drop below the boiling point of 100°C (212°F), something that occurred about 3.8 billion years ago.

The lowest places across the face of Earth then began to fill with water: first ponds, then lakes and seas, and eventually ocean-sized bodies of water. Massive flows of water washed over the landscape, carrying both dissolved and undissolved materials to these early seas and oceans. Outgassing of water has continued ever since and is visible in volcanic eruptions, geysers, and seepage to the surface.

## Water Quantity: In Equilibrium Worldwide

Today, water is the most common compound on the surface of Earth, having attained the present volume of 1.36 billion cubic kilometers approximately 2 billion years ago. This quantity has remained relatively constant, even though water is continuously being lost from the system. It is lost when water dissociates into hydrogen and oxygen, and the hydrogen escapes Earth's gravity to space. It is lost through breaking down and forming new compounds with other elements. Lost water is replaced by pristine water not previously present on the surface, which emerges from within Earth's crust. The net result of these water inputs and outputs is that Earth's hydrosphere is in a steady-state equilibrium.

Despite this overall net balance in water quantity, worldwide changes in sea level called **eustasy**, do occur. Eustatic sea-level changes are related to changes in the water volume of the oceans. Some of these changes are explained by the amount of water stored in glaciers and ice sheets; these are called **glacio-eustatic** factors (see Chapter 17). In cooler times, as more water is bound up in glaciers (on mountains worldwide) and in ice sheets (Greenland and Antarctica), sea level lowers. In a warmer era, less water is stored as ice, so sea level rises.

Some 18,000 years ago, during the most recent ice-age pulse, sea level was more than 100 m (330 ft) lower than it is today; 40,000 years ago it was 150 m (about 500 ft) lower. Over the past 100 years, mean sea level has steadily risen by 20–40 cm (8–16 in.) and is still rising worldwide.

Apparent changes in sea level also are related to actual physical changes in landmasses, such as continental uplift or subsidence. This change in land elevation, called *isostasy*, is discussed in Chapter 11.

**Distribution of Earth's Water Today.** From a geographic point of view, ocean and land surfaces are not evenly distributed. If you examine a globe, it is obvious that most of Earth's continental land is in the Northern Hemisphere, whereas the Southern Hemisphere is dominated by water. This observation has lead to the general terms *land hemi-*

**FIGURE 7-1**

**Water outgassing from the crust.**

Outgassing of water from Earth's crust in the geothermal area near Wairakei on the North Island of New Zealand. [Photo by Bill Bachman/Photo Researchers, Inc.]

*sphere* and *water hemisphere*. In fact, from certain perspectives, Earth appears to have a distinct *oceanic hemisphere* and a distinct *land hemisphere* (Figure 7-2).

The present location of all of Earth's liquid and frozen water—whether fresh or saline, surface or underground—is illustrated in Figure 7-3. The oceans contain 97.22% of all water, and it is saltwater (about 1.321 billion cubic kilometers, or 317 million cubic miles). The four oceans—Pacific, Atlantic, Indian, and Arctic—are identified on Figure 7-4. (The extreme southern portions of the Pacific, Atlantic, and Indian Oceans that surround the Antarctic continent are sometimes collectively called the "Southern Ocean.") Only 2.78% of all of Earth's water is freshwater (nonsaline and nonoceanic).

Table 7-1 divides freshwater into two categories—surface water and subsurface water. Ice sheets and glaciers are the greatest single repository of surface freshwater; they contain 77.14% of all of Earth's freshwater. Adding subsurface groundwater to frozen surface water accounts for 99.36% of all freshwater.

The remaining freshwater, which resides in lakes, rivers, and streams so familiar to us, actually represents less than 1%. All the world's freshwater lakes total only 125,000 km³ (30,000 mi³), with 80% of this volume in just 40 of the largest lakes and about 50% contained in just 7 lakes (noted on Figure 7-4).

The greatest single volume of freshwater resides in 25-million-year-old Lake Baykal in Siberian Russia. It contains almost as much water as all five U.S. Great Lakes combined. Africa's Lake Tanganyika contains the next largest volume, followed by the five Great Lakes. Overall, 70% of lake water is in North America, Africa, and Asia, with about a fourth of lake water worldwide in innumerable small lakes. Over 3 million lakes exist in Alaska alone, and Canada has over 750 km² of lake surface.

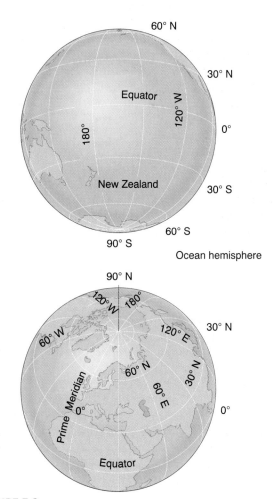

**FIGURE 7-2**

**Land and water hemispheres.**

Two perspectives that divide Earth's surface into an ocean hemisphere and a land hemisphere.

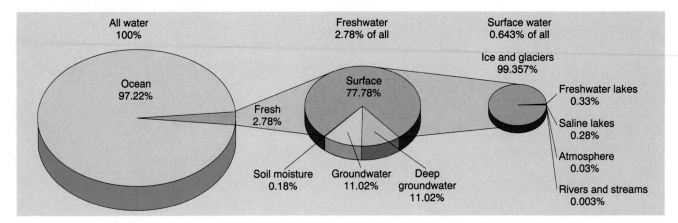

**FIGURE 7-3**
**Ocean and freshwater distribution on Earth.**

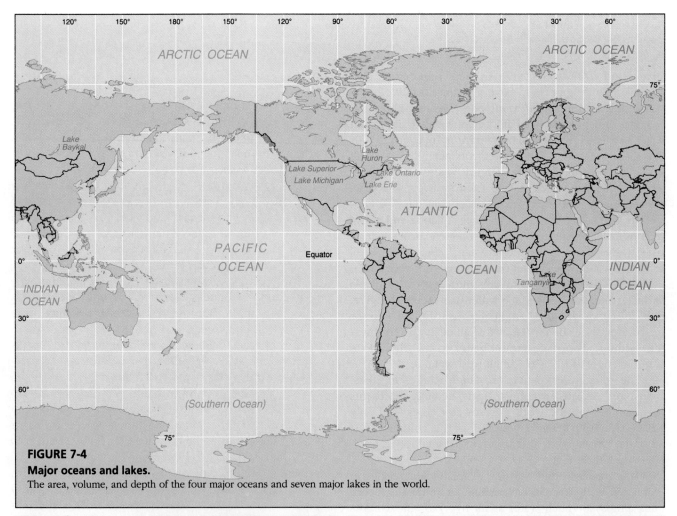

**FIGURE 7-4**
**Major oceans and lakes.**
The area, volume, and depth of the four major oceans and seven major lakes in the world.

| Major Freshwater Lake | Volume [km³ (mi³)] | SurfaceArea [km² (mi²)] | Depth [m (ft)] |
|---|---|---|---|
| Baykal (Russia) | 22,000 (5280) | 31,500 (12,160) | 1620 (5315) |
| Tanganyika (Africa) | 18,750 (4500) | 39,900 (15,405) | 1470 (4923) |
| Superior (U.S./Canada) | 12,500 (3000) | 83,290 (32,150) | 397 (1301) |
| Michigan (U.S.) | 4920 (1180) | 58,030 (22,400) | 281 (922) |
| Huron (U.S./Canada) | 3545 (850) | 60,620 (23,400) | 229 (751) |
| Ontario (U.S./Canada) | 1640 (395) | 19,570 (7550) | 237 (777) |
| Erie (U.S./Canada) | 485 (115) | 25,670 (9910) | 64 (210) |

| Ocean | Earth's Ocean Area (%) | *Area [km²(mi²)] | *Volume [km³ (mi³)] | Mean Depth of Main Basin [m (ft)] |
|---|---|---|---|---|
| Pacific | 48 | 179,670 (69,370) | 724,330 (173,700) | 4280 (14,040) |
| Atlantic | 28 | 106,450 (41,100) | 355,280 (85,200) | 3930 (12,890) |
| Indian | 20 | 74,930 (28,930) | 292,310 (70,100) | 3960 (12,900) |
| Arctic | 4 | 14,090 (5440) | 17,100 (4100) | 1205 (3950) |

* Data in thousands (,000): includes all marginal seas.

**TABLE 7-1**

| Distribution of Freshwater on Earth | | | |
|---|---|---|---|
| *Location* | *Amount (km³ [mi³])* | | *Percent of Fresh Water* | *Percent of Total Water* |
| **Surface Water** | | | | |
| Ice sheets and glaciers | 29,180,000 | (7,000,000) | 77.14 | 2.146 |
| Freshwater lakes | 125,000 | (30,000) | 0.33 | 0.009 |
| Saline lakes and inland seas | 104,000 | (25,000) | 0.28 | 0.008 |
| Atmosphere | 13,000 | (3,100) | 0.03 | 0.001 |
| Rivers and streams | 1,250 | (300) | 0.003 | 0.0001 |
| Total surface water | 29,423,250 | (7,058,400) | 77.78 | 2.164 |
| **Subsurface Water** | | | | |
| Groundwater—surface to 762 m (2500 ft) depth | 4,170,000 | (1,000,000) | 11.02 | 0.306 |
| Groundwater—762 to 3962 m (2500 to 13,000 ft) depth | 4,170,000 | (1,000,000) | 11.02 | 0.306 |
| Soil moisture storage | 67,000 | (16,000) | 0.18 | 0.005 |
| Total subsurface water | 8,407,000 | (2,016,000) | 22.22 | 0.617 |
| **Total Freshwater** (rounded) | **37,800,000** | **(9,070,000)** | **100.00%** | 2.78% |

Saline lakes and salty inland seas are not connected to the ocean. They are usually in regions of interior river drainage (no outlet to the ocean), which allows salts to become concentrated. They contain 104,000 km³ (25,000 mi³) of water. Such lakes as Utah's Great Salt Lake, California's Mono Lake, Southwest Asia's Caspian Sea and Aral Sea, and the Dead Sea between Israel and Jordan, exist as remnants of past wetter climates.

Think of Earth's thousands of flowing rivers and streams. Combined, they amount to only 14,250 km³ (3400 mi³), or only 0.033% of freshwater, or 0.0011% of all water! Yet, this small amount is very dynamic. A water molecule traveling atmospheric and surface-water paths moves through the entire hydrologic cycle (ocean-atmosphere-precipitation-runoff) in less than 2 weeks. Contrast this to a water molecule in deep-ocean circulation, groundwater, or a glacier; in those systems it moves very slowly, taking thousands of years to migrate through the system. Chapter 9 explores the hydrologic system and water resources in more detail.

# Unique Properties of Water

Earth's distance from the Sun places it within a most remarkable temperate zone, compared with the locations of the other planets. This temperate location allows all three states of water—ice, liquid, and vapor—to occur naturally on Earth and to change from one to another.

Each water molecule is composed of two atoms of hydrogen and one of oxygen that readily bond (as suggested in Figure 7-5, upper left). Once hydrogen and oxygen atoms combine, they are quite difficult to separate, thereby producing a water molecule that remains stable in Earth's environment. This water molecule is a versatile solvent and possesses extraordinary heat characteristics. As the most common compound on Earth's surface, water exhibits the most uncommon of properties.

The nature of the hydrogen-oxygen bond gives the hydrogen side of a water molecule a positive charge and the oxygen side a negative charge (see Figure 7-5). As a result of this *polarity*, water molecules are attracted to each other: The positive (hydrogen) side of one water molecule is attracted to the negative (oxygen) side of another. This bonding between water molecules is called *hydrogen bonding*.

The polarity of water molecules also explains why water "acts wet" and sticks to things and dissolves so many substances. Because of this solvent ability, pure water is rare in nature.

The effects of hydrogen bonding in water are observable in everyday life. Hydrogen bonding creates the *surface tension* that allows you to float a steel needle on the surface of water, even though steel is much denser than water. This surface tension also allows you to slightly overfill a glass with water; the water surface, which is actually above the rim of the glass, is retained by a web of millions of hydrogen bonds.

Hydrogen bonding also is the cause of *capillarity*, which you observe when you "dry" something with a paper towel. The towel draws water through its fibers because hydrogen bonds make each molecule pull on its neighbor. In chemistry laboratory classes, students observe the curved *meniscus*, or surface of the water, that forms in a cylinder or a test tube because hydrogen bonding allows the water to slightly "climb" the glass walls. *Capillary action* is an important component of soil-moisture processes, discussed in Chapters 9 and 18. It is

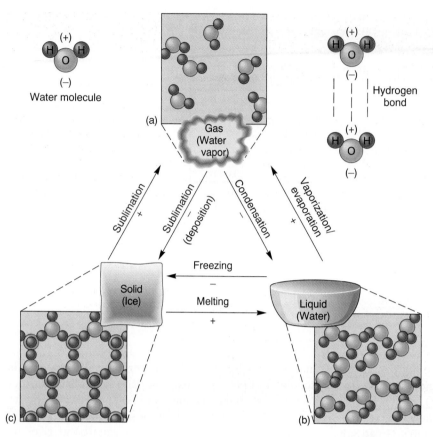

Water molecule

(a) Gas (Water vapor)

Hydrogen bond

Sublimation +

Sublimation − (deposition)

Condensation −

Vaporization/ evaporation +

Freezing −

Melting +

Solid (Ice)

Liquid (Water)

(c)

(b)

**FIGURE 7-5**

**Three states of water and water's phase changes.**

The three physical states of water: (a) water vapor, (b) water, and (c) ice. Note the molecular arrangement in each state and the terms that describe the changes from one phase to another. Also note how the polarity of water molecules bonds them to one another, loosely in the liquid state and firmly in the solid state. The plus and minus symbols denote whether heat energy is absorbed (+) or liberated (−) during the phase change.

important to note that, without hydrogen bonding to hold molecules together in water and ice, water would be a gas at normal surface temperatures.

## Heat Properties

For water to change from one state to another (solid, liquid, or vapor), *heat energy must be absorbed or liberated.* To cause a change of state, the amount of heat energy must be sufficient to affect the hydrogen bonds between molecules. This relation between water and heat energy is an important driving force in producing weather. In fact, the heat exchanged in the phase changes of water provides over 30% of the energy that powers the general circulation of the atmosphere.

Figure 7-5 presents the three states of water and the terms used to describe each **phase change**. At the bottom, *melting* and *freezing* describe the familiar phase change between solid and liquid. At the right, the terms *condensation* and *evaporation* (or *vaporization* at boiling temperature) apply to the change between liquid and vapor. At the left, the term **sublimation** refers to the direct change of water vapor to ice or ice to water vapor. Sometimes the term *deposition* is used if water vapor attaches itself directly to an ice crystal. The deposition of water vapor to ice may form *frost* on surfaces.

***Ice, the Solid Phase.*** As water cools, it behaves like most compounds and contracts. However, it reaches its greatest density not as ice, but as water at 4°C (39°F). Below that temperature, water behaves very differently from other compounds. It begins to expand as more hydrogen bonds form among the slower-moving molecules, creating the hexagonal (six-sided) structures shown in Figure 7-5c. This expansion continues to a temperature of −29°C (−20°F), with up to a 9% increase in volume possible. This expansion is important in the weathering of rocks in highway and pavement damage. News Report 1 discusses some effects of ice. Surprisingly, scientists are still learning about this structure in ice and water.

The expansion in volume that accompanies the freezing process results in a decrease in density (the same number of molecules occupy greater space). Specifically, ice has 0.91 times the density of water, so it floats. Without this change in density, much of Earth's freshwater would be bound in masses of ice on the ocean floor. Instead, we have floating icebergs, with approximately 1/11 (9%) of their mass exposed and 10/11 (91%) hidden beneath the ocean's surface (Figure 7-6b). Given the present warming conditions at higher latitudes, a record number of icebergs have broken loose from Antarctica and Greenland, posing an increased hazard to ships.

## News Report 1

## Breaking Roads and Pipes and Sinking Ships

Road crews are busy in the summer in many parts of the country repairing winter damage to streets and freeways. A major contributor to this damage is the expansive phase change from water to ice. Rainwater seeps into roadway cracks and then expands as it freezes, thus breaking up the pavement (Figure 1). Perhaps you have noticed that bridges suffer the greatest damage. The reason is that cold air can circulate beneath a bridge and so produces more freeze-thaw cycles on the bridge than in the roadbed on rock and soil.

The expansion of freezing water exerts a tremendous force—enough to crack plumbing or an automobile radiator or engine block. Wrapping water pipes with insulation is a common winter task in many places to avoid damage. People living in very cold climates use antifreeze and engine heaters to avoid auto damage.

Historically, this physical property of water was put to useful work in quarrying rock for building materials. Holes were drilled and filled with water before winter so that when cold weather arrived the water would freeze and expand, cracking the rock into manageable shapes.

A major hazard to ships in higher latitudes is posed by floating ice. Since ice has 0.91 the density of water, an iceberg sits with approximately 10/11 of its mass below water level. The irregular edges of submarine ice can slash the side of a passing ship. This is what happened to the HMS *Titanic* in 1912 on its maiden voyage.

**FIGURE 7-1**
**The power of freezing water.**
Damage to pavement is a result of the expansion that occurs as water temperature goes below 4°C (39°F). [Photo by Leonard Lessin/Peter Arnold, Inc.]

---

The freezing action of ice as an important physical weathering process is discussed in Chapter 13. Approximately 30% of Earth's surface is affected by the freeze-thaw action of surface and subsurface water, producing a variety of processes and landforms. Such periglacial landscapes are discussed in Chapter 17.

The rigid internal structure of ice dictates the six-sided appearance of all ice crystals (Figure 7-6a), which can loosely combine to form snowflakes. This six-sided preference applies to ice crystals of all shapes: plates, columns, needles, and dendrites (branching or treelike forms). Ice crystals demonstrate a unique interaction of chaos (all ice crystals are different) and the determinism of physical principles (all have a six-sided structure).

***Water, the Liquid Phase.*** As a liquid, water assumes the shape of its container and is a noncompressible fluid. For ice to change to water, heat energy must increase the motion of the water molecules to break some of the hydrogen bonds (Figure 7-5b). Despite the fact that there is no change in sensible temperature between ice at 0°C (32°F) and water at 0°C, 80 calories of heat energy must be absorbed for the phase change of 1 g of ice to 1 g of water (Figure 7-7, upper left). This heat energy involved in the phase change is called **latent heat** and is stored within the water. It becomes liberated whenever phase reverses and a gram of water freezes. These 80 calories are called the *latent heat of fusion*, or the *latent heat of melting*.

To raise the temperature of 1 g of water from freezing at 0°C (32°F) to boiling at 100°C (212°F), we must add 100 calories, gaining an increase of 1 C° (1.8 F°) for each calorie added.

***Water Vapor, the Gas Phase.*** Water vapor is an invisible and compressible gas in which each molecule moves independently of the others (Figure 7-5a). The phase change from liquid to vapor at boiling temperature, under normal sea-level pressure, requires the addition of a much greater amount of heat energy than does the phase change from solid to liquid: 540 calories are needed for each gram of water changed to water vapor (Figure 7-7). Those calories are the **latent heat of vaporization**. When water vapor condenses to a liquid, each gram gives up its hidden 540 calories as the **latent heat of condensation**. Perhaps you have felt the liberation of the latent heat of condensation on your skin from steam, as when you drain steamed vegetables or pasta or filled a hot tea kettle. News Report 2 discusses the power in clouds.

In summary, taking 1 g of ice at 0°C and changing it to water vapor at 100°C—changing it from a solid, to a liquid, to a gas—*absorbs* 720 calories (80 cal + 100 cal + 540 cal). Or, reversing the process, changing 1 gram of water vapor at 100°C to ice at 0°C *liberates* 720 calories into the surrounding environment.

***Heat Properties of Water in Nature.*** In a lake or stream or in soil water, at 20°C (68°F), every gram of

(a)

(b)

(c)

**FIGURE 7-6**
**The uniqueness of ice forms.**
(a) Computer-enhanced photo reveals ice-crystal patterns, which are dictated by the internal structure between water molecules. This open structure also explains the lower density of ice and why it floats in the denser water. (b) Icebergs (pinnacled and saddle-shaped forms) floating in Disko Bay, Greenland, calved into the ocean from the Jakobshavn glacier. (c) Frost—delicate ice crystals formed by the deposit of water vapor directly as ice. [(a) Photo enhancement © Scott Camazine/Photo Researchers, Inc., after W. A. Bentley; (b) photo by George Hunter/Allstock; (c) photo by author.]

water that breaks away from the surface through evaporation must absorb from the environment approximately 585 calories as the *latent heat of evaporation* (see the natural scene in Figure 7-7). This is slightly more energy than would be required if the water were boiling (540 cal). You can feel this absorption of latent heat as evaporative cooling on your skin when it is wet. This latent heat exchange is the dominant cooling process in Earth's energy budget.

The process reverses when air that contains water vapor is cooled. The vapor eventually condenses back into the liquid state, forming moisture droplets and thus liberating 585 calories as the *latent heat of condensation* for every gram of water.

The *latent heat of sublimation* absorbs 680 calories as a gram of ice transforms into vapor. A comparable amount of energy is released as vapor freezes directly to ice.

# Humidity

The water-vapor content of air is termed **humidity**. The capacity of air to hold water vapor is primarily a function of temperature—the temperatures of both the air and the water vapor, which are usually the same. Warmer air has a greater capacity for holding water vapor than does cooler air.

We are all aware of humidity in the air, for its relationship to air temperature determines our sense of comfort. North Americans spend billions of dollars a year to adjust humidity, either with air conditioning (extracting water vapor and cooling) or with air humidifying (adding water vapor). We discussed the relation between humidity and temperature in Chapter 5 as the heat index of apparent temperature. To determine the energy available for powering weather activities, one has to know the water-vapor content of air and relate that to the air's capacity to hold water vapor at a given temperature.

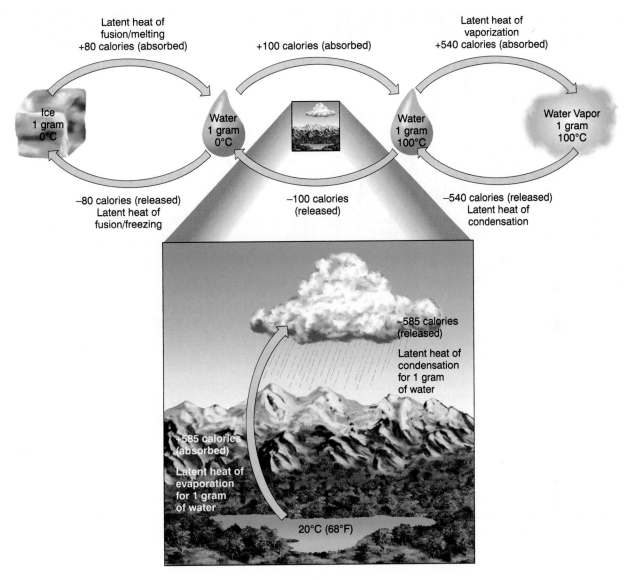

**FIGURE 7-7**

**Water's heat-energy characteristics.**

A lot of latent heat energy is involved in the phase changes of water. To transform 1 g of ice at 0°C to 1 g of water vapor at 100°C requires 720 calories (80 + 100 + 540). The landscape illustrates phase changes between water (lake at 20°C) and water vapor under typical conditions in the environment.

## *Relative Humidity*

After air temperature and barometric pressure, the most common piece of information in local weather broadcasts is relative humidity. **Relative humidity** is a ratio (expressed as a percentage) of the amount of water vapor that is *actually* in the air (content) to the maximum water vapor the air *could hold* at a given temperature (capacity)—it is content compared with capacity.

If air is relatively dry compared with its capacity, the relative humidity percentage is low; if the air is relatively moist, the percentage is higher; and if the air is saturated

with all the moisture it can hold for its temperature, the percentage is 100% (Figure 7-8). The formula to calculate relative humidity is:

$$\text{Relative humidity} = \frac{\text{Actual water-vapor content of the air}}{\text{Maximum water-vapor capacity of the air at that temperature}} \times 100$$

Relative humidity varies because of evaporation, condensation, or temperature changes. All three affect both the moisture content (the numerator) and the capacity (the denominator) of the air to hold water vapor. Relative humidity is an expression of an ongoing process

# News Report 2

## The Power in Clouds

For every gram of water that evaporates from Earth's surface, approximately 585 calories are absorbed as the latent heat of evaporation. Likewise, this heat energy is liberated upon the condensation of water vapor into a cloud. When you realize that a small, puffy, fair-weather cumulus cloud holds 500–1000 tons of moisture droplets, think of the tremendous latent heat released into the atmosphere as they condensed! You can see that weather events such as Hurricane Camille (1969), Hurricane Gilbert (1988), Hurricane Hugo (1989), and Hurricane Andrew (1992) involve a staggering amount of energy.

Government meteorologists estimate that the moisture in Andrew weighed nearly 30 trillion metric tons at its maximum power and mass. The latent heat liberated each day from these hurricane clouds approximately equals the total energy consumption in the United States for 6 months! These unique heat properties and latent heat exchanges in water, powered by solar energy, drive weather systems worldwide. The next time a beautiful puffy cumulus cloud floats by, be reminded of the power in the clouds (Figure 1).

**FIGURE 1**
**Heat energy in clouds.**
The latent heat of condensation released as these clouds form is enormous. [Photo by author.]

---

between air and moist surfaces, for condensation and evaporation operate continuously—water molecules move back and forth between the air and bodies of water or ice. It is the temperatures of air and water vapor that are the principal regulators of condensation and evaporation, and thus of relative humidity.

Think of warm air and cool air as sponges: Warm air is a large sponge that can hold a lot of water; cool air is a small sponge that can hold very little water. A cool body of air can be saturated—filled to capacity—by an amount of water vapor that will only partially fill the capacity of warm air.

Air is **saturated**, or filled to capacity, when it contains all the water vapor that it can hold at a given temperature (100% relative humidity). In saturated air, the net transfer of water molecules between a moist surface and air reaches an equilibrium. Saturation indicates that any further addition of water vapor (increase in content) or any decrease in temperature (reduction in capacity) will result

in active condensation (clouds, fog, or precipitation). Therefore, relative humidity indicates both the nearness of air to a saturated condition and when active condensation will begin.

The temperature at which a given mass of air becomes saturated is termed the **dew-point temperature**. In other words, *air is saturated when the dew-point temperature and the air temperature are the same*. A cold drink in a glass provides a common example of these conditions: The water droplets that form on the outside of the glass condense from the air because the air layer next to the glass is chilled to below its dew-point temperature and thus becomes saturated (Figure 7-9). Figure 7-9 shows two additional examples of saturated air and active condensation above a rock surface and in a fog formation overlying a cool ocean surface.

***Daily and Seasonal Relative Humidity Patterns.*** During a typical day, air temperature and relative humidity are

**FIGURE 7-8**

**Water vapor content, capacity, and relative humidity.**

The water vapor capacity of warm air is greater than the capacity of cold air, so relative humidity changes with temperature.

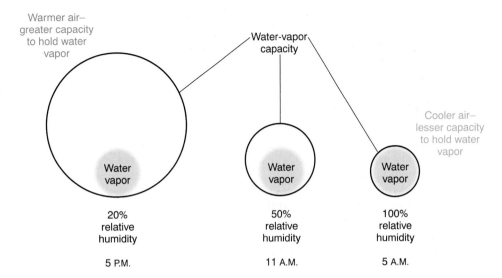

Warmer air— greater capacity to hold water vapor

Water-vapor capacity

Cooler air— lesser capacity to hold water vapor

Water vapor

Water vapor

Water vapor

20% relative humidity

50% relative humidity

100% relative humidity

5 P.M.

11 A.M.

5 A.M.

Cold glass chills the surrounding air layer to the dew-point temperature

Dew (active condensation)

Somebody forgot to put a coaster under the glass!

(a)

(b)

(c)

**FIGURE 7-9**

**Dew-point temperature examples.**

(a) The low temperature of the glass chills the surrounding air layer to the dew point and saturation. Thus, water vapor condenses out of the air and onto the glass as dew. (b) The cold air above the rain-soaked rocks is at the dew point and is saturated. Condensation shrouds the rock in a changing veil of clouds. (c) The cold ocean surface chills the moist air layer to the dew point and saturation. As water vapor condenses, a dense fog forms. In the evening when temperatures drop over coastal lands, fog forms there too, giving the appearance of moving inland. [Photos by author.]

inversely related—as temperature rises, relative humidity falls (Figure 7-10a). Relative humidity is highest at dawn, when air temperature is lower and the capacity of the air to hold water vapor is less. If you park outdoors, you know about the wetness of the dew that condenses on the car overnight. Relative humidity is lowest in the late afternoon, when higher air temperatures increase the capacity of the air to hold water vapor. The actual water-vapor content in the air may remain the same throughout the day, but because the temperature varies, relative humidity changes from morning to afternoon.

Weather records for Sacramento, California, show the seasonal variation in relative humidity by time of day, confirming the relationship of temperature and relative humidity (Figure 7-10b). January readings are higher than July readings because air temperatures are lower overall in winter. Similar relative humidity records at most weather stations demonstrate the same relation among season, temperature, and relative humidity. In your own experience, you probably have noticed this pattern—the morning dew on windows, cars, and lawns has evaporated by late morning as the capacity of the air to hold water vapor increases with air temperature.

## Expressions of Relative Humidity

There are several ways to express humidity and relative humidity. Each has its own utility and application. Here we discuss vapor pressure and specific humidity.

***Vapor Pressure.*** One way of describing humidity is related to air pressure. As free water molecules evaporate from a surface into the atmosphere, they become water vapor. Now part of the air, they also become a portion of air pressure. The share of air pressure that is made up of water vapor molecules is termed **vapor pressure**. Like air pressure, it is expressed in millibars (mb).

Water-vapor molecules continue to evaporate from a moist surface, slowly diffusing into the air, until the increasing vapor pressure in the air causes some molecules to return to the surface. As explained earlier, *saturation* is reached when the movement of water molecules between surface and air is in equilibrium. The maximum capacity of the air at a given temperature is termed the *saturation vapor pressure* and indicates the maximum pressure that water vapor molecules can exert. Any temperature increase or decrease will change the saturation vapor pressure.

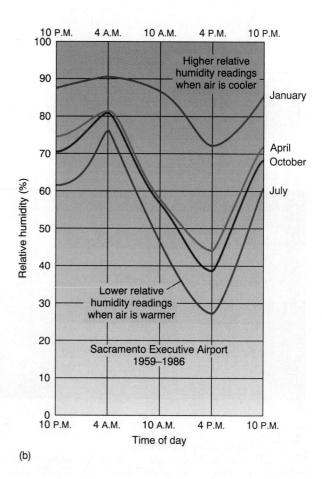

**FIGURE 7-10**

**Daily and seasonal relative humidity patterns.**

Sacramento, California: (a) typical daily variations in temperature and relative humidity, (b) seasonal variations in daily relative humidity.

Figure 7-11 graphs the saturation vapor pressure at various air temperatures. The graph illustrates that, for every temperature increase of 10 C° (18 F°), the vapor pressure capacity of air nearly doubles. This relation explains why warm tropical air over the ocean can hold so much water vapor, thus providing great latent heat to power tropical storms. It also explains why cold air is "dry" and why cold air toward the poles does not produce great storms (it holds too little water vapor).

As the graph shows, air at 20°C (68°F) has a saturation vapor pressure of 24 mb; that is, the air is saturated if the water-vapor portion of the air pressure is at 24 mb. Thus, if the water-vapor content actually present is exerting a vapor pressure of only 12 mb in 20°C air, the relative humidity is 50% (12 mb ÷ 24 mb = 0.50 × 100 = 50%). The inset in Figure 7-11 compares saturation vapor pressure over water and over ice surfaces at subfreezing temperatures. You can see that saturation vapor pressure is greater above a water surface than over an ice surface—that is, it takes more water-vapor molecules to saturate air above water than it does above ice. This fact is important to condensation processes and rain-droplet formation, both of which are discussed in the section on clouds in Chapter 8.

***Specific Humidity.*** A useful humidity measure is one that remains constant as temperature and pressure change. **Specific humidity** is the mass of water vapor (in grams) per mass of air (in kilograms) at any specified temperature. Because it is measured in mass, specific humidity is not affected by changes in temperature or pressure, such as occur when an air parcel rises to higher elevations. Specific humidity stays constant despite volume changes.

The maximum mass of water vapor that a kilogram of air can hold at any specified temperature is termed the *maximum specific humidity* and is plotted in Figure 7-12. The graph shows that a kilogram of air could hold a maximum specific humidity of 47 g of water vapor at 40°C (104°F), 15 g at 20°C (68°F), and about 4 g at 0°C (32°F). Therefore, if a kilogram of air at 40°C has a specific humidity of 12 g, its relative humidity is 25.5% (12 g ÷ 47 g = 0.255 × 100 = 25.5%). Specific humidity is useful in describing the moisture content of large air masses that are interacting in a weather system, and it is valuable information for weather forecasting.

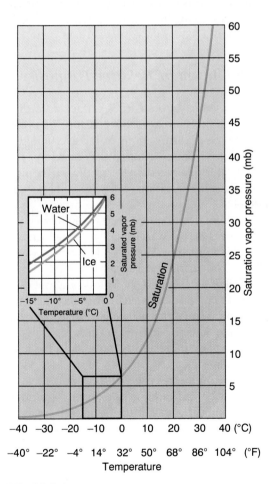

**FIGURE 7-11**

**Saturation vapor pressure.**

Saturation vapor pressure of air at various temperatures—the maximum capacity of air to hold water vapor, expressed in vapor pressure. Inset compares saturation vapor pressures over water surfaces with those over surfaces at subfreezing temperatures.

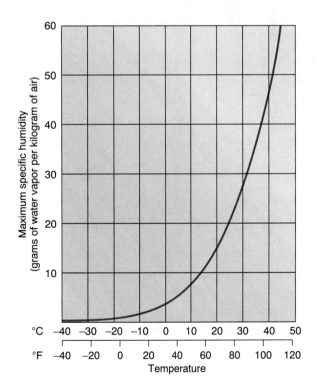

**FIGURE 7-12**

**Maximum specific humidity.**

Maximum specific humidity for a mass of air at various temperatures—the maximum capacity of air to hold water vapor.

***Instruments for Measuring Humidity.*** Relative humidity is measured with various instruments. The **hair hygrometer** uses the principle that human hair changes as much as 4% in length between 0% and 100% relative humidity. The instrument connects a standardized bundle of human hair through a mechanism to a gauge. As the

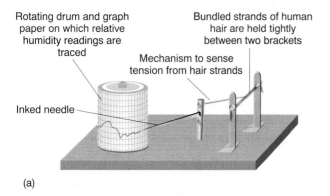

Rotating drum and graph paper on which relative humidity readings are traced

Bundled strands of human hair are held tightly between two brackets

Mechanism to sense tension from hair strands

Inked needle

(a)

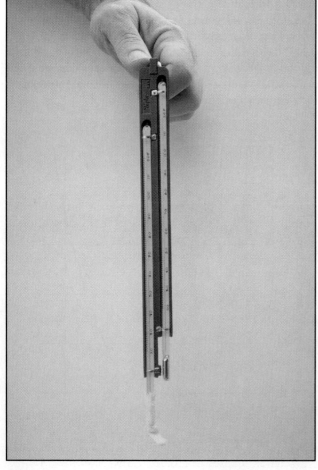

(b)

**FIGURE 7-13**

**Instruments that measure relative humidity.**

(a) The principle of a hair hygrometer. (b) Sling psychrometer with wet and dry bulbs. [Photo by Bobbé Christopherson.]

hair absorbs or loses water in the air, it changes length, indicating relative humidity (Figure 7-13a).

Another instrument used to measure relative humidity is a **sling psychrometer.** Figure 7-13b shows this device, which has two thermometers mounted side by side on a metal holder. One is called the *dry-bulb thermometer*; it simply records the ambient (surrounding) air temperature. The other thermometer is called the *wet-bulb thermometer*; it is set lower in the holder and its bulb is covered by a moistened cloth wick. The psychrometer is then spun ("slung") by its handle or placed where a fan forces air over the wet bulb.

The rate at which water evaporates from the wick depends on the relative saturation of the surrounding air. If the air is dry, water evaporates quickly, *absorbing the latent heat of evaporation from the wet-bulb thermometer*, cooling the thermometer and causing its temperature to drop (wet-bulb depression). In high humidity, little water evaporates from the wick; in low humidity, more water evaporates. After a minute or two, the temperature on each bulb is compared on a relative humidity (psychrometric) chart, from which relative humidity can be determined.

## Global Distribution of Relative Humidity

Relative humidity varies across Earth's surface according to temperature and moisture availability. Figure 7-14 plots the approximate mean distributions of relative humidity and vapor pressure by latitude for January and July. The graphs illustrate the relative dryness of the subtropical regions and the relative wetness of equatorial areas and higher latitudes.

The relative humidity (Figure 7-14a) in the subtropical regions is lower because the warmer air has a much greater capacity to hold water vapor. However, in terms of vapor pressure (Figure 7-14b), the water vapor actually in the air above subtropical deserts around 30° latitude is about double the amount at higher latitudes such as 50° (compare the dots on the graph). Thus, the relatively dry air over the generally cloudless Sahara (warmer air) actually contains more water vapor than the relatively moist air in the midlatitudes (cooler air).

Satellites now routinely sense water vapor content of the lower atmosphere. Water vapor absorbs infrared wavelengths, making it possible to distinguish areas of relatively high water-vapor content from areas of low water-vapor content, using infrared sensors. Figure 7-15 is such an image of the Western Hemisphere showing water-vapor content recorded by sensors of the "water-vapor channel." This knowledge is important to forecasting because it shows the moisture available to weather systems and therefore the energy available (latent heat) and precipitation potential of those systems.

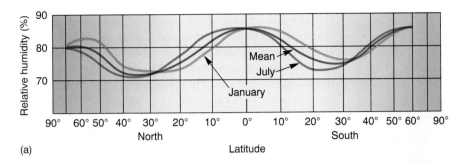

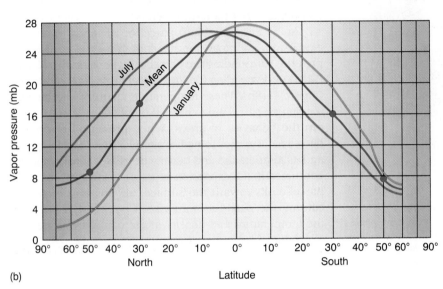

**FIGURE 7-14**
**Relative humidity and vapor pressure by latitude.**
(a) Approximate relative humidity distributed by latitude for January, July, and the mean. (b) Distribution of vapor pressure by latitude for January, July, and the mean. [After B. Haurwitz and J. Austin, *Climatology* (New York: McGraw-Hill, 1944), p. 45. Adapted by permission.]

Now that you know something about atmospheric moisture, dew point, and relative humidity, let us examine the concept of stability in the atmosphere.

## Atmospheric Stability

Meteorologists use the term *parcel* to describe a body of air that has specific temperature and humidity characteristics. Think of an air parcel as a volume of air, perhaps 300 m (1000 ft) in diameter. Differences in temperature create changes in density within the parcel. Warm air produces a lower density in a given volume of air; cold air produces a higher density. Two opposing forces work on a parcel of air: an upward buoyant force and a downward gravitational force. A parcel of lower density will rise (is more buoyant); a rising parcel expands as external pressure decreases. A parcel of higher density will descend (is less buoyant); a falling parcel compresses as external pressure increases. Figure 7-16 shows air parcels and illustrates these relationships. (In Chapter 8, we discuss air masses, which are much larger parcels; they are regional in extent.)

An indication of weather conditions is the relative stability of air parcels and air masses in the atmosphere. **Stability** refers to the tendency of an air parcel, with its water-vapor cargo, either to remain in place or to change vertical position by ascending (rising) or descending (falling). An air parcel is *stable* if it resists displacement upward or, when disturbed, it tends to return to its starting place. An air parcel is *unstable* if it continues to rise until it reaches an altitude where the surrounding air has a density (air temperature) similar to its own. A difference between an air parcel and the surrounding environment produces a buoyancy that contributes to further lifting. To visualize this, imagine a hot-air balloon launch. The air-filled balloon sits on the ground with the same temperature of air inside as in the surrounding environment, like a stable air parcel. Now, as the burner is ignited, the balloon fills with hot (less-dense) air and rises buoyantly (Figure 7-17), like an unstable parcel.

Given the concepts of stable and unstable air, let us examine the specific temperature characteristics that produce these conditions.

**FIGURE 7-15**

**Satellite image of water vapor in the atmosphere.**

Water-vapor content of the atmosphere over the Western Hemisphere as disclosed in a *GOES* infrared image. On the gray scale used for this image, the lighter tones denote higher water-vapor content. [*GOES* image courtesy of NOAA and the National Environmental Satellite, Data, and Information Service.]

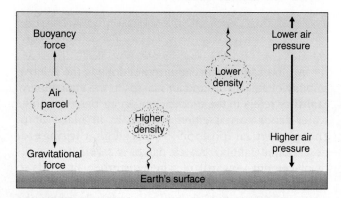

**FIGURE 7-16**

**The forces acting on an air parcel.**

Buoyancy and gravitational forces work on an air parcel. Different densities produce rising or falling parcels in response to imbalance in these forces.

## *Adiabatic Processes*

Determining the degree of stability or instability requires measuring two temperatures: the temperature in an air parcel and the temperature in the air surrounding the parcel. The contrast of these two temperatures determines

stability. Such temperature measurements are made daily with instrument packages called *radiosondes* carried aloft by helium-filled balloons at thousands of weather stations.

The *normal lapse rate*, as introduced in Chapter 3, is the average decrease in temperature with increasing altitude, a value of 6.4 C°/1000 m (3.5 F°/1000 ft). This rate of temperature change is for still, calm air, but, it can vary greatly under different weather conditions. Consequently, the actual lapse rate at a particular place and time is labeled the *environmental lapse rate*. It can vary by several degrees per thousand meters.

An ascending (rising) parcel of air tends to cool by expansion, an expansion resulting from the reduced air pressure surrounding the parcel at higher altitudes (Figure 7-18a). A descending (falling) parcel heats by compression (Figure 7-18b). These temperature changes internal to a moving air parcel are explained by physical laws that govern the behavior of gases. When pressure on a parcel decreases as it rises to higher altitudes, the parcel expands and its temperature and density decrease (sensible heat is consumed in the expansion process). Conversely, as an air parcel sinks toward the surface, air pressure increases, causing the parcel to compress. Compression increases the temperature and density in the parcel (sensible heat is produced by compression).

Temperature changes in both ascending and descending air occur *without any significant heat exchange between the surrounding environment and the vertically moving parcel of air*. The warming and cooling rates for a parcel of expanding or compressing air are termed **adiabatic**. (*Diabatic* means occurring with an exchange of heat; *adiabatic* means occurring *without* a

**FIGURE 7-17**

**Principles of air stability and balloon launches.**

Principles of stability are illustrated by hot-air balloons being launched in the Swiss Alps. As the temperature inside a balloon increases, the air in the balloon becomes less dense than the surrounding air and the buoyancy force causes the balloon to rise, like a warm air parcel. [Photo by Bap Vandystadt/Photo Researchers, Inc.]

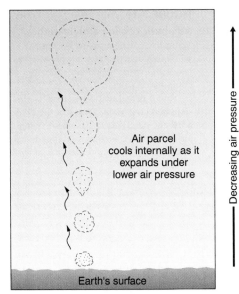

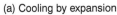

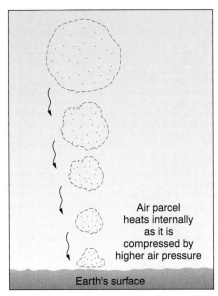

(a) Cooling by expansion

(b) Heating by compression

**FIGURE 7-18**
**Vertically moving air experiences temperature changes.**
(a) A rising air parcel cools by expansion. (b) A falling air parcel heats by compression.

loss or gain of heat energy to or from the environment.) There are two adiabatic rates, depending on moisture conditions in the vertically moving air parcel: a dry adiabatic rate (DAR) and a moist adiabatic rate (MAR). (Note that the word *lapse* is not used with adiabatic processes in proper usage. *Lapse* is correctly applied in the term *environmental lapse rate*.)

***Dry Adiabatic Rate (DAR).*** The **dry adiabatic rate** is the rate at which "dry" air cools by expansion (if ascending) or heats by compression (if descending). The term *dry* is used when air is less than saturated (relative humidity is less than 100%). The average DAR is 10 C°/1000 m (5.5 F°/1000 ft).

The principle is illustrated by the rising parcel at the left in Figure 7-19. To see how a specific example of dry air behaves, consider an unsaturated parcel of air at the surface with a temperature of 27°C (80°F), as illustrated in Figure 7-19a. It rises, expands, and cools adiabatically at the DAR as it rises to 2500 m (approximately 8000 ft). What happens to the temperature of the parcel? Calculate the temperature change in the parcel, using the dry adiabatic rate:

(10 C°/1000 m) × 2500 m = 25 C° of total cooling
(5.5 F°/1000 ft) × 8000 ft = 44 F° of total cooling

Subtracting the 25 C° (45 F°) of adiabatic cooling from the starting temperature of 27°C (81°F) gives the temperature in the air parcel at 2500 m of 2°C (36°F).

In Figure 7-19b, assume that an unsaturated air parcel with a temperature of −20°C is at 3000 m (−4°F at 9800 ft) and descends to the surface, heating adiabatically. Using the dry adiabatic lapse rate, we can deter-mine the temperature of the air parcel when it arrives at the surface:

(10 C°/1000 m) × 3000 m = 30 C° of total warming
(5.5 F°/1000 ft) × 9800 ft = 54 F° of total warming

Adding the 30 C° of adiabatic warming from the starting temperature of −20°C gives the temperature in the air parcel at the surface of 10°C (50°F).

***Moist Adiabatic Rate (MAR).*** The **moist adiabatic rate** is the rate at which an ascending air parcel that is moist (saturated) cools by expansion or that a descending parcel warms by compression. The average MAR is 6 C°/1000 m (3.3 F°/1000 ft). This is roughly 4 C° (2 F°) less than the dry adiabatic rate. From this average, the MAR varies with moisture content and temperature and can range from 4 C° to 10 C° per 1000 m (2 F° to 5.5 F° per 1000 ft). The cause of this variability, and the reason that the rate is lower than the DAR, is the latent heat of condensation. As water vapor condenses in the saturated air, sensible heat is liberated and the adiabatic rate of cooling is lowered. The release of latent heat may vary with temperature and water vapor content. The MAR is much lower than the DAR in warm air, whereas the two rates are more similar in cold air.

## Stable and Unstable Atmospheric Conditions

Now we can bring this discussion together to determine atmospheric stability. The relation among the dry adiabatic rate (DAR), moist adiabatic rate (MAR), and the environmental (actual) lapse rate at a given time and place determines the stability of the atmosphere over an area. You can see the range of possible stability relationships

(a) Air parcel *cools* adiabatically at the DAR

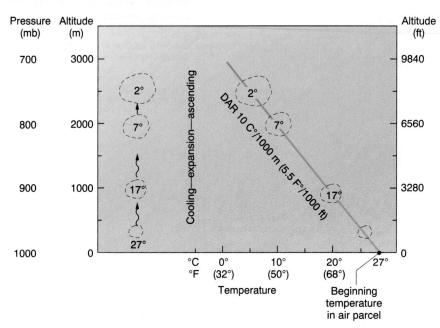

**FIGURE 7-19**
**Adiabatic cooling and heating.**
Vertically moving air parcels expand when they rise (because air pressure is less with increasing altitude) and are compressed when they descend.
(a) An air parcel that is less than saturated cools adiabatically at the dry adiabatic rate (DAR). (b) A descending air parcel that is less than saturated heats adiabatically by compression at the DAR.

(b) Air parcel *heats* adiabatically at the DAR

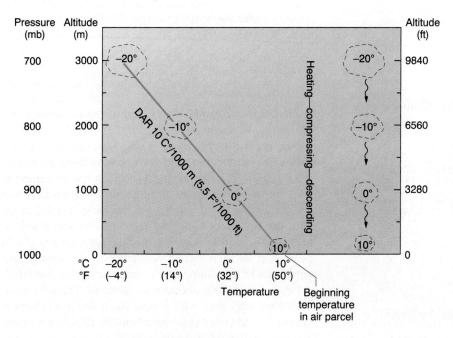

in Figure 7-20. Let's examine what produces this variation in atmospheric stability.

Temperature relationships in the atmosphere produce the three different conditions: unstable, conditionally unstable, and stable. For the sake of illustration, the three examples in Figure 7-21 begin with an air parcel at the surface at 25°C (77°F). In each example, compare the temperatures of the air parcel and the surrounding environment. Assume that a lifting mechanism is present (we will examine lifting mechanisms in Chapter 8).

Given unstable conditions in Figure 7-21a, the air parcel continues to rise through the atmosphere because it is warmer (and therefore, less dense and more buoyant) than the surrounding environment. Note that the environmental lapse rate on this occasion is measured

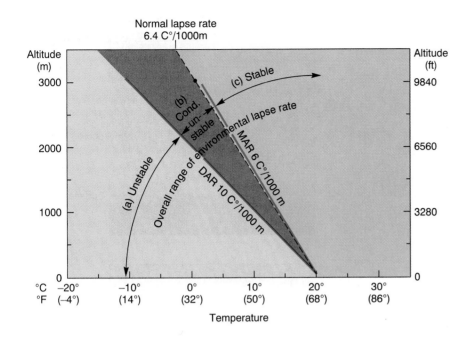

**FIGURE 7-20**
**Temperature relationships and atmospheric stability.**
The relationship between dry and moist adiabatic rates and environmental lapse rates produces three atmospheric conditions: (a) unstable (environmental lapse rate exceeds the DAR), (b) conditionally unstable (environmental lapse rate is between the DAR and MAR), and (c) stable (environmental lapse rate is less than DAR and MAR).

at 12 C°/1000 m (6.6 F°/1000 ft). That is, the air surrounding the air parcel is cooler by 12 C° for every 1000 m increase in altitude. By 1000 m (about 3300 ft), the lifting air parcel has adiabatically cooled by expansion at the DAR from 25° C to 15°C, but the surrounding air has cooled from 25°C at the surface to 13°C. By comparing the temperature in the air parcel and the surrounding environment, you can see that the temperature in the parcel is 2 C° (3.6 F°) warmer than the surrounding air. The condition is described as *unstable* because the less-dense air parcel is going to continue lifting.

Eventually, as the air parcel continues lifting and cooling, it may achieve the dew-point temperature, saturation, and active condensation. This point of saturation forms the lifting condensation level that you see in the sky as the flat bottoms on clouds.

On the other hand, if the environmental lapse rate is measured at only 5 C°/1000 m (3 F°/1000 ft) on another day, stable conditions will result, as shown in Figure 7-21c. An environmental lapse rate of 5 C°/1000 m is less than both the DAR and the MAR. This rate produces a lower temperature in the air parcel (higher density, less buoyant) than in the surrounding environment, which

forces the air parcel to settle back to its original position. This condition is described as *stable*, because the denser air parcel resists lifting and the sky remains generally cloud-free. In regions experiencing air pollution, stable conditions in the atmosphere worsen the pollution by slowing exchanges in the surface air.

You may be wondering what stability condition exists if the environmental lapse rate is somewhere between the DAR and MAR, and conditions therefore are neither unstable nor stable. In Figure 7-21b, the environmental lapse rate is measured at 7 C°/1000 m. Under these conditions, the air parcel resists upward movement if it is less than saturated. But, if the air parcel is saturated and cools at the MAR, it becomes unstable and continues to rise. One example of such conditionally unstable air occurs when stable air passes over a mountain range and thus is forcibly lifted. As the air parcel lifts and cools to the dew point, saturated conditions are reached and condensation begins. Now the MAR is in effect and the air parcel behaves in an unstable manner. The sky might be beautiful and clear, without a cloud, yet you might see huge clouds developing over a nearby mountain range.

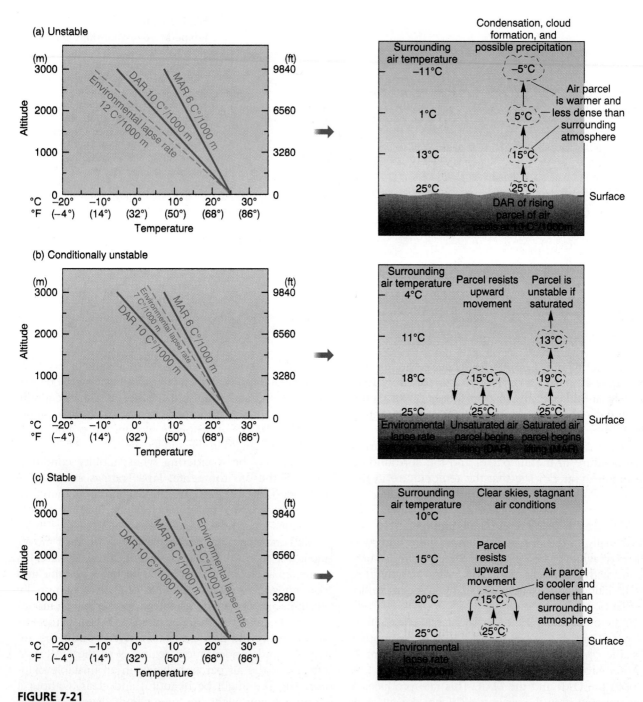

**FIGURE 7-21**

**Stability—three examples.**

Specific examples of (a) unstable, (b) conditionally unstable, and (c) stable conditions in the lower atmosphere.

# Summary and Review —Water and Atmospheric Moisture

✔ *Describe* **the origin of Earth's waters,** *relate* **the quantity of water that exists today, and** *list* **the locations of Earth's freshwater supply.**

The next time it rains where you live, pause and reflect on the journey each of those water molecules has made. Water molecules came from within Earth over billions of years, in a process called **outgassing**. The then began endless cycling through the hydrologic system of evaporation-condensation-precipitation. Water covers about 71% of Earth. Approximately 97% of it is salty seawater, and the remaining 3% is freshwater—most of it frozen.

The present volume of water on Earth is estimated to be 1.36 billion cubic kilometers (326 million cubic miles), an amount achieved roughly 2 billion years ago. This overall steady-state equilibrium might seem in conflict with the many changes in sea level that have occurred over Earth's history, but is not. Worldwide changes in sea level are called **eustasy**, are related to the change in volume of water in the oceans. Some of these changes are explained by the amount of water stored in glaciers and ice sheets, called **glacio-eustatic** factors. At present, sea level is rising because of increases in the temperature of the oceans and the record melting of glacial ice.

outgassing (p. 176)
eustasy (p. 176)
glacio-eustatic (p. 176)

1. Approximately where and when did Earth's water originate?

2. If the quantity of water on Earth has been quite constant in volume for at least 2 billion years, how can sea level have fluctuated? Explain.

3. Describe the locations of Earth's water, both oceanic and fresh. What is the largest repository of freshwater at this time? In what ways is this distribution of water significant to modern society?

4. Why might you describe Earth as the water planet? Explain.

✔ *Describe* **the heat properties of water and** *identify* **the traits of its three phases: solid, liquid, and gas.**

Water is the most common compound on the surface of Earth, and it possesses unusual solvent and heat characteristics. Part of Earth's uniqueness is that its water naturally exists in all three states—solid, liquid, and gas—owing to Earth's temperate position relative to the Sun. A change from one state to another is called a **phase change**. The change from solid to vapor is called **sublimation**; from liquid to

solid, freezing; from solid to liquid, melting; from vapor to liquid, condensation; and from liquid to vapor, vaporization, or evaporation.

The heat energy required for water to change phase is termed **latent heat**, because, once absorbed, it is hidden within the structure of the water, ice, or water vapor. For 1 g of water to become 1 g of water vapor at boiling requires addition of 540 calories, or the **latent heat of vaporization**. When this 1 g of water vapor condenses, the same amount of heat energy is liberated, as 540 calories of the **latent heat of condensation**. Weather is powered by the tremendous amount of latent heat energy involved in the phase changes between the three states of water.

phase change (p. 180)
sublimation (p. 180)
latent heat (p. 181)
latent heat of vaporization (p. 181)
latent heat of condensation (p. 181)

5. Describe the three states of matter as they apply to ice, water, and water vapor.

6. What happens to the physical structure of water as it cools below 4°C (39°F)? What are some visible indications of these physical changes?

7. What is latent heat? How is it involved in the phase changes of water?

8. Take 1 g of water at 0°C and follow it through to 1 g of water vapor at 100°C, describing what happens along the way. What amounts of energy are involved in the changes that take place?

✔ *Define* **humidity and the expressions of the relative humidity concept;** *explain* **dew-point temperature and saturated conditions in the atmosphere.**

The amount of water vapor in the atmosphere is termed **humidity**. The ability of air to hold water vapor is principally a function of the temperature of the air and of the water vapor (usually the same). **Relative humidity** is a percentage expression of the humidity content of the air compared with the capacity of the air to hold water vapor at a given temperature—content compared with capacity. Relatively dry air has a lower relative humidity value; relatively moist air a higher percentage. Air is said to be **saturated**, or filled to capacity, if it contains all the water vapor it can hold at a given temperature (100% relative humidity). The temperature at which air achieves saturation is called the **dew-point temperature**.

Among the various ways to express humidity and relative humidity are vapor pressure and specific humidity.

**Vapor pressure** is that portion of the atmospheric pressure that is produced by the presence of water vapor. A comparison of vapor pressure with the *saturation vapor pressure* at any moment produces a relative humidity percentage. **Specific humidity** is the mass of water vapor (in grams) per mass of air (in kilograms) at any specified temperature. Because it is measured as a mass, specific humidity does not change as temperature or pressure changes, making it a valuable measurement in weather forecasting. A comparison of specific humidity with the *maximum specific humidity* at any moment produces a relative humidity percentage.

Two instruments that are used to measure relative humidity, and indirectly the actual humidity content of the air, are the **hair hygrometer** and the **sling psychrometer**.

> humidity (p. 182)
> relative humidity (p. 183)
> saturated (p. 184)
> dew-point temperature (p. 184)
> vapor pressure (p. 186)
> specific humidity (p. 187)
> hair hygrometer (p. 188)
> sling psychrometer (p. 188)

9. What is humidity? How is it related to the energy present in the atmosphere? To our personal comfort and how we perceive apparent temperatures?

10. Define relative humidity. What does the concept represent? What is meant by the terms *saturation* and *dew-point temperature?*

11. Using different measures of humidity in the air given in the chapter, derive relative humidity values (vapor pressure/saturation vapor pressure; specific humidity/maximum specific humidity).

12. How do the two instruments described in this chapter measure relative humidity?

13. How does the latitudinal distribution of relative humidity compare with the actual water vapor content of air?

---

✔ *Define* **atmospheric stability and** *relate* **it to a parcel of air that is ascending or descending.**

Meteorologists use the term *parcel* to describe a body of air that has specific temperature and humidity characteristics. Think of a parcel of air as a volume of air, perhaps 300 m (1000 ft) in diameter, that is subjected to temperature differences that create changes in density within the parcel. Warm air has a lower density in a given volume of air; cold air has a higher density.

**Stability** refers to the tendency of an air parcel, with its water-vapor cargo, either to remain in place or to change vertical position by ascending (rising) or descending (falling). An air parcel is *stable* if it resists displacement upward or,

when disturbed, it tends to return to its starting place. An air parcel is *unstable* if it continues to rise until it reaches an altitude where the surrounding air has a density (air temperature) similar to its own.

> stability (p. 189)

14. Differentiate between stability and instability relative to a parcel of air rising vertically in the atmosphere.

15. What are the forces acting on a vertically moving parcel of air? How are they affected by the density of the air parcel?

---

✔ *Illustrate* **three atmospheric conditions—unstable, conditionally unstable, and stable—with a simple graph that relates the environmental lapse rate to the dry adiabatic rate (DAR) and moist adiabatic rate (MAR).**

An ascending (rising) parcel of air cools by expansion, responding to the reduced air pressure at higher altitudes. A descending (falling) parcel heats by compression. These temperature changes internal to a moving air parcel are explained by physical laws that govern the behavior of gases. Temperature changes in both ascending and descending air parcels occur *without any significant heat exchange between the surrounding environment and the vertically moving parcel of air*. The warming and cooling rates for a parcel of expanding or compressing air are termed **adiabatic.**

The **dry adiabatic rate** is the rate at which "dry" air cools by expansion (if ascending) or heats by compression (if descending). The term *dry* is used when air is less than saturated (relative humidity less than 100%). The DAR is 10 C°/1000 m (5.5 F°/1000 ft). The **moist adiabatic rate** is the average rate at which ascending air that is moist (saturated) cools by expansion, or descending air warms by compression. The average MAR is 6 C°/1000 m (3.3 F°/1000 ft). This is roughly 4 C° (2 F°) less than the dry rate. The MAR, however, varies with moisture content and temperature and can range from 4 C° to 10 C° per 1000 m (2 F° to 5.5 F° per 1000 ft).

A simple comparison of the dry adiabatic rate (DAR) and moist adiabatic rate (MAR) in a vertically moving parcel of air with that of the *environmental lapse rate* in the surrounding air determines the atmosphere's stability—whether it is unstable (lifting of air parcels continues), stable (air parcels resist vertical displacement), or conditionally unstable (air parcel behaves as though unstable if the MAR is in operation and stable otherwise).

> adiabatic (p. 190)
> dry adiabatic rate (DAR) (p. 191)
> moist adiabatic rate (MAR) (p. 191)

16. How do the adiabatic rates of heating or cooling in a vertically displaced air parcel differ from the normal lapse rate and environmental lapse rate?

17. Why is there a difference between the dry adiabatic rate (DAR) and the moist adiabatic rate (MAR)?

18. What would atmospheric temperature and moisture conditions be on a day when the weather is unstable? When it is stable? Relate in your answer what you would experience if you were outside watching.

---

 ## NetWork

The *Geosystems Home Page* provides on-line resources for this chapter on the World Wide Web. You will find review exercises, specific updates for items in the chapter, suggested readings, and links to interesting related pathways on the Internet (click on the Table of Contents link and select this chapter). *Geosystems* is at: **http://www.prenhall.com/geosystm**

# 8

# Weather

**Clouds and Fog**

**Air Masses**

**Atmospheric Lifting Mechanisms**

**Midlatitude Cyclonic Systems**

**Violent Weather**

**Summary and Review**

## Key Learning Concepts

After reading the chapter, you should be able to:

- *Identify* the requirements for cloud formation and *explain* the major cloud classes and types.

- *Identify* the basic types of fog and *explain* the conditions that lead to their formation.

- *Describe* air masses that affect North America and *relate* their qualities to source regions.

- *Identify* types of atmospheric lifting mechanisms and *describe* four principal examples.

- *List* the measurable elements that contribute to weather.

- *Analyze* various types of violent weather and the characteristics of each.

*Lightning illuminates a Florida waterspout (90 m wide by 450 m tall).*
[Photo by Fred K. Smith.]

ater has a leading role in the vast drama played out daily on Earth's stage. It affects the stability of air masses and their interactions and produces powerful and beautiful special effects in the lower atmosphere. Large air masses come into conflict; they move and shift, dominating now one region then another in response to insolation; latitude; patterns of atmospheric temperature and pressure; the arrangement of land, water, and mountains; and migrating cyclonic systems that cross the middle latitudes. Water, with its ability to absorb and release vast quantities of heat energy, drives this daily drama in the atmosphere.

Tropical cyclones travel northward and southward from tropical latitudes, moving over vulnerable coastal lowlands as hurricanes and typhoons, especially in recent years. Since record keeping began in 1870, the second-greatest year for number of Atlantic hurricanes was 1995. In that year 121 people were killed and $8 billion in property was damaged. It was exceeded only by 1933 for the number of tropical cyclones. In 1992, Hurricane Andrew hit Florida and Louisiana, Hurricane Iniki struck Hawaii, and Typhoon Omar slashed Guam, causing a combined $25 billion in destruction.

**Weather** is the short-term, day-to-day condition of the atmosphere, contrasted with *climate*, which is the long-term average (over decades) of weather conditions and extremes in a region. Weather is, at the same time, both a "snapshot" of atmospheric conditions and a technical status report of the Earth-atmosphere heat-energy budget. Important elements that contribute to the weather are temperature, air pressure, relative humidity, wind speed and direction, daylength, and Sun angle. We turn to local media for the day's weather report from the National Weather Service (in the United States) or Atmospheric Environment Service (in Canada) to see current satellite images and to hear tomorrow's forecast.

**Meteorology** is the scientific study of the atmosphere. (*Meteor* means "heavenly" or "of the atmosphere.") Meteorologists study the atmosphere's physical characteristics and motions, related chemical, physical, and geologic processes, the complex linkages of atmospheric systems, and weather forecasting. Computers handle volumes of data for accurate forecasting of near-term weather and for studying trends in long-term weather and climatic change.

New developments in supercomputers, Earth-bound instruments, and orbiting observation systems are rapidly advancing the science of the atmosphere. For example, a $5 billion weather-instrument modernization program is under way in the United States and should be completed by 2004 if everything continues as planned. The National Weather Service, in conjunction with the Federal Aviation Administration (FAA) and the Department of Defense, installed 110 new Doppler radar systems by mid-1995 as part of the NEXRAD (Next Generation Weather Radar) program (Figure 8-1). Doppler radar detects the direction of moisture droplets toward or away

**FIGURE 8-1**
**Weather installation.**
Typical Doppler radar installation at Denver's Stapleton International Airport operated by the National Weather Service. The radar antenna is sheltered within the dome structure. [Photo by National Center for Atmospheric Research.]

from the radar, indicating wind direction and air flow. This information is critical to making severe storm warnings.

Chapter 7 examined atmospheric stability and instability. That concept leads us to cloud formation, which opens this chapter. We then follow huge air masses across North America, observe powerful lifting mechanisms in the atmosphere, examine cyclonic systems, and conclude with a portrait of the violent and dramatic weather that occurs in the environment around us. The spatial implications of these weather phenomena and their relationship to human activities strongly link meteorology to physical geography.

## Clouds and Fog

Clouds are more than whimsical, beautiful decorations in the sky; they are fundamental indicators of overall atmospheric conditions: stability, moisture content, and weather. They form as air becomes saturated with water. Clouds are the subject of much scientific inquiry, especially regarding their effect on net radiation patterns, as discussed in Chapters 4 and 5. With a little knowledge and practice, you can learn to "read" the atmosphere from its signature clouds.

A **cloud** is an aggregation of tiny moisture droplets and ice crystals that are suspended in air, great enough in volume and concentration to be visible. **Fog** is simply a cloud in contact with the ground. Cloud types are too numerous to fully describe here, so the most common examples are presented in a simple classification scheme.

## Cloud Formation Processes

Clouds may contain raindrops, but not initially. At the outset, clouds are a great mass of moisture droplets, each invisible without magnification. A **moisture droplet** is approximately 20 μm (miccrometers) in diameter (0.002 cm, or 0.0008 in.). It takes a million or more such droplets to form an average raindrop of 2000 μm in diameter (0.2 cm, or 0.078 in.), as shown in Figure 8-2. After more than a century of debate, scientists have determined how these droplets form during condensation and what makes them coalesce into raindrops.

Under unstable conditions (discussed in Chapter 7), an air parcel may rise until it becomes *saturated*—that is, until the air cools to the dew-point temperature and relative humidity is 100%. (Under certain conditions, condensation may occur at slightly less or more than 100% relative humidity.) More lifting of the air parcel cools it further, producing condensation of water vapor into water. Water does not just condense among the air molecules. Condensation requires **cloud-condensation nuclei**, microscopic particles that always are present in the atmosphere.

Continental air masses average 10 billion cloud-condensation nuclei per cubic meter. These nuclei are typically derived from ordinary dust, soot, and ash from volcanoes and forest fires and particles from burned fuel, such as sulfate aerosols. The air over cities contains great concentrations of such nuclei. In maritime air masses, which average 1 billion nuclei per cubic meter, the nuclei are supplied by a high concentration of sea salts derived from ocean sprays. Salts are particularly attracted to moisture and thus are called *hygroscopic nuclei*. The lower atmosphere never lacks cloud-condensation nuclei.

***Raindrop Formation.*** Given the conditions of saturated air, availability of cloud-condensation nuclei, and the presence of cooling (lifting) mechanisms in the atmosphere, condensation occurs. Two principal processes account for the majority of the world's raindrops and snowflakes: the *collision-coalescence* and the *Bergeron ice-crystal* processes. These are summarized in Figure 8-3.

The **collision-coalescence process** predominates in clouds that form at above-freezing temperatures, principally in the *warm clouds* of the tropics. Initially, simple condensation takes place on small nuclei. Some nuclei are larger and produce larger droplets. As these larger droplets respond to gravity and fall through a cloud, they collide with and combine with smaller droplets. These gradually coalesce until their weight exceeds the strength of air circulation in the cloud to hold them aloft, at which point they fall as raindrops. Towering tropical clouds keep a growing raindrop aloft longer than do the flatter clouds of temperate regions, so tropical clouds produce more collisions and more drops of greater diameter (Figure 8-3a).

The **Bergeron ice-crystal process** also is a significant raindrop-forming mechanism, especially in the middle and high latitudes in *cold clouds*. Water in bulk freezes at 0°C (32°F), but fine, pure (distilled) water droplets do not freeze until the temperature drops much lower, to –40°C (–40°F). Consequently, *supercooled droplets* may be found in clouds at temperatures from –15° to –35°C (5° to –31°F).

Figure 7-11 showed that the saturation vapor pressure near an ice surface is less than that near a water surface; therefore, supercooled water droplets evaporate more rapidly near ice crystals than near other water droplets. The ice crystals then absorb the vapor (the vapor sublimates on the ice crystal). This ice-crystal process, proposed by meteorologist Alfred Wegener—noted in Chapter 11 for his theory of the drifting continents—was elaborated on by Swedish scientist Tor Bergeron and bears his name.

The ice crystals feed on the supercooled cloud droplets, grow in size, and eventually fall as snow or rain (Figure 8-3b). Precipitation in middle and high latitudes begins as ice crystals high in the clouds. The ice crystals melt and gather moisture through collision-coalescence as they fall through the warmer portions of the cloud.

## Cloud Types and Identification

As noted, a cloud is a collection of moisture droplets and ice crystals suspended in air in sufficient volume and concentration to be visible. In 1803, English biologist Luke Howard established a classification system for clouds and coined Latin names for them that we still use.

Clouds usually are classified by *altitude* and by *shape*. They come in three basic forms—flat, puffy, and wispy—which occur in four primary altitude classes and ten basic cloud types. Clouds that are developed horizontally—flat and layered—are called *stratiform* clouds. Clouds that are developed vertically—puffy and globular—are termed *cumuliform*. Wispy clouds usually are quite high in altitude and are composed of ice crystals; they are labeled *cirroform*.

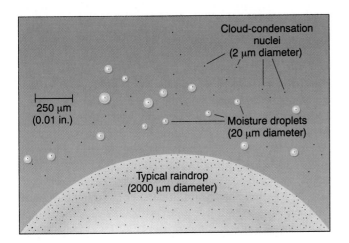

**FIGURE 8-2**
**Moisture droplets and raindrops.**
Cloud-condensation nuclei, moisture droplets, and a raindrop enlarged many times—compared at roughly the same scale.

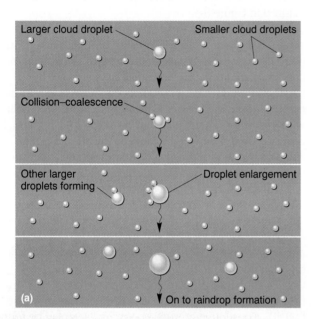 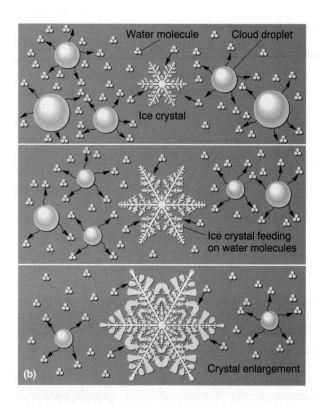

**FIGURE 8-3**
**Raindrop formation.**
Principal processes of raindrop and snowflake formation: (a) the collision-coalescence process; (b) the Bergeron ice-crystal process. [From F. K. Lutgens and E. J. Tarbuck, *The Atmosphere: An Introduction to Meteorology*, 3rd ed., copyright © 1986, p. 127. Reprinted by permission of Prentice Hall, Inc., Upper Saddle River, N.J.]

These three basic forms occur in four altitudinal classes: low, middle, high, and those that are vertically developed through the troposphere. Table 8-1 presents the basic cloud classes and types. Figure 8-4 illustrates the general appearance of each type and includes representative photographs of most of them.

***Cloud Classes.*** Low clouds, ranging from the surface up to 2000 m (6500 ft) in the middle latitudes, are simply called *stratus* or *cumulus* (Latin for "layer" and "heap," respectively). **Stratus** clouds appear dull, gray, and featureless. When they yield precipitation, they are called **nimbostratus** (*nimbo-* denotes precipitation or heavy rain), and their showers typically fall as drizzling rain (Figure 8-4e).

**Cumulus** clouds appear bright and puffy, like cotton balls. When they do not cover the sky, they float by in infinitely varied shapes. Vertically developed clouds are listed in a separate class in Table 8-1 because further vertical development can produce clouds that extend beyond low altitudes into middle and high altitudes (illustrated at the far right in Figure 8-4).

Sometimes near the end of the day, lumpy, grayish, low-level clouds called **stratocumulus** may fill the sky in patches. Near sunset, these spreading puffy stratiform remnants may catch and filter the Sun's rays, sometimes indicating clearing weather.

Middle-level clouds are denoted by the prefix *alto-* (meaning "high"). They are made of water droplets and, when cold enough, can be mixed with ice crystals. **Altocumulus** clouds, in particular, represent a broad category that occurs in many different styles: patchy rows, wave patterns, a "mackerel sky," or lens-shaped (lenticular) clouds.

Clouds occurring above 6000 m (20,000 ft) are composed principally of ice crystals in thin concentrations. These wispy filaments, usually white except when colored by sunrise or sunset, are termed **cirrus** clouds (Latin for "curl of hair"), sometimes dubbed mares'-tails. Cirrus clouds look as though an artist has taken a brush and placed delicate feathery strokes high in the sky. Often, cirrus clouds are associated with an oncoming storm, especially if they thicken and lower in elevation. The prefix *cirro-* (cirrostratus, cirrocumulus) is used for other high clouds.

A cumulus cloud can develop into a towering giant called **cumulonimbus** (again, *-nimbus* denotes precipitation; Figure 8-5). Such clouds are called *thunderheads* because of their shape and their associated lightning and thunder. Note the surface wind gusts, updrafts and downdrafts, heavy rain, and the presence of ice crystals at the top of the rising cloud column.

When the rising cloud column enters a region of temperatures well below freezing, ice crystals form and consume available moisture droplets. It is possible to witness

**TABLE 8-1**

| Cloud Classes and Types | | | | |
|---|---|---|---|---|
| *Class* | *Altitude/Composition at Midlatitudes* | *Type* | *Symbol* | *Description* |
| Low clouds ($C_L$) | Up to 2000 m (6500 ft) Water | Stratus (St) Stratocumulus (Sc) Nimbostratus (Ns) | | Uniform, featureless, gray, like high fog Soft gray globular masses in lines, groups, or waves, heavy rolls, irregular overcast patterns Gray, dark, low, with drizzling rain |
| Middle clouds ($C_M$) | 2000-6000 m (6500-20,000 ft) Ice and water | Altostratus (As) Altocumulus (Ac) | | Thin to thick, no halos, Sun's outline just visible, gray day Patches of cotton balls, dappled, arranged in lines or groups, rippling waves, the lenticular clouds associated with mountains |
| High clouds ($C_H$) | 6000-13,000 m (20,000-43,000 ft) Ice | Cirrus (Ci) Cirrostratus (Cs) Cirrocumulus (Cc) | | Mares'-tails, wispy, feathery, hairlike, delicate fibers, streaks, or plumes Veil of fused sheets of ice crystals, milky, with Sun and Moon halos Dappled, "mackerel sky," small white flakes, tufts, in lines or groups, sometimes in ripples |
| Vertically developed clouds | Near surface to 13,000 m (43,000 ft) Water below, ice above | Cumulus (Cu) Cumulonimbus (Cb) | | Sharply outlined, puffy, billowy, flat-based, swelling tops, fair weather Dense, heavy, massive, dark thunderstorms, hard showers, explosive top, great vertical development, towering, cirrus-topped plume blown into anvil-shaped head |

this transformation from a water-droplet cloud to an ice-crystal cloud, because the sharply delineated form of the cloud top softens in appearance. High-altitude winds may then shear the top of the cloud into the characteristic anvil shape of the mature thunderhead.

## Fog

By international definition, *fog* is a cloud layer on the ground, with visibility restricted to less than 1 km (3300 ft). The presence of fog tells us that the air temperature and the dew-point temperature at ground level are nearly identical, indicating saturated conditions. Fog is generally capped by an inversion layer, with as much as 30 C° (50 F°) difference in air temperature between the cool ground under the fog and the warmer clear, sunny skies above.

Almost all fog is warm—that is, its moisture droplets are above freezing. Supercooled fog, which occurs when the moisture droplets are below freezing, is special because it can be dispersed by means of artificial seeding with ice crystals or other crystals that mimic ice, following the principles of the ice-crystal formation process described earlier. Let's briefly look at several types of fog.

***Advection Fog.*** As the name implies, **advection fog** forms when air in one place *migrates* to another place where

conditions are right for saturation. For example, when warm, moist air overlays cooler ocean currents, lake surfaces, or snow masses, the layer of migrating air directly above the surface becomes chilled to the dew point, and fog develops. Off all subtropical west coasts in the world, summer fog forms in the manner just described (Figure 8-6). Some coastal desert communities actually extract usable water from such fog formations, as described in News Report 1.

Another type of advection fog forms when cold air lies over the warm water of a lake, ocean surface, or even a swimming pool. A wispy **evaporation fog**, or *steam fog*, may form as water molecules evaporate from the water surface into the cold overlying air, effectively humidifying the air, and then condense to form fog. When evaporation fog is visible at sea, it is a shipping hazard called *sea smoke* (Figure 8-7).

A type of fog forms when moist air is forced to higher elevations along a hill or mountain. This upslope lifting leads to adiabatic cooling by expansion as the air rises. The resulting **upslope fog** forms a stratus cloud at the condensation level of saturation. Along the Appalachians and the eastern slopes of the Rockies such fog is common in winter and spring. Another fog associated with topography is called **valley fog**. Because cool air is denser than warm air, it settles in low-lying areas, producing a fog in the chilled, saturated layer near the ground (Figure 8-8).

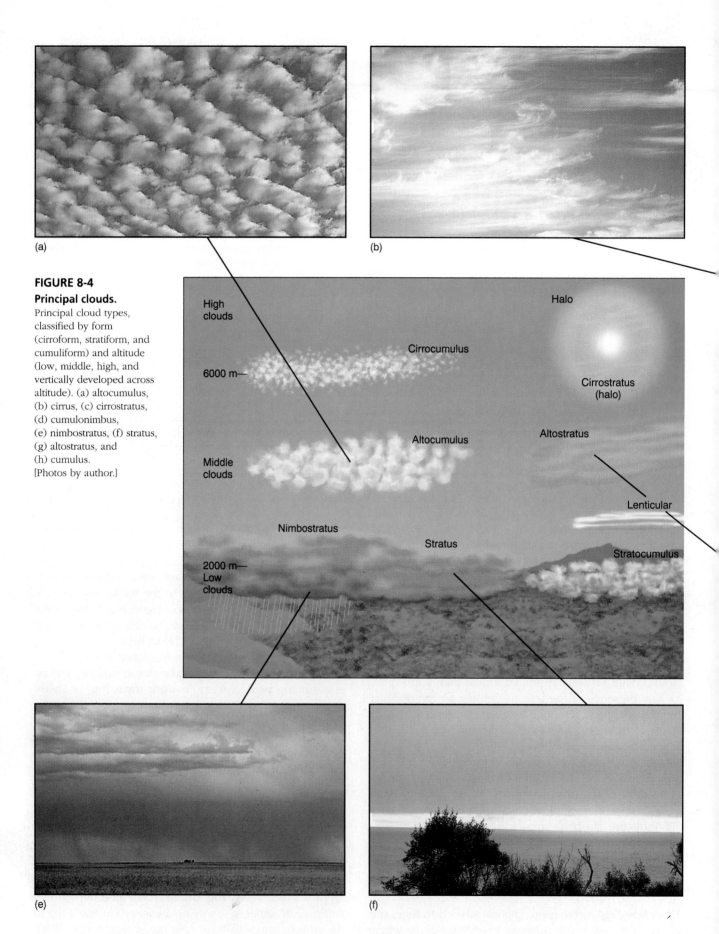

**FIGURE 8-4**
**Principal clouds.**
Principal cloud types, classified by form (cirroform, stratiform, and cumuliform) and altitude (low, middle, high, and vertically developed across altitude). (a) altocumulus, (b) cirrus, (c) cirrostratus, (d) cumulonimbus, (e) nimbostratus, (f) stratus, (g) altostratus, and (h) cumulus. [Photos by author.]

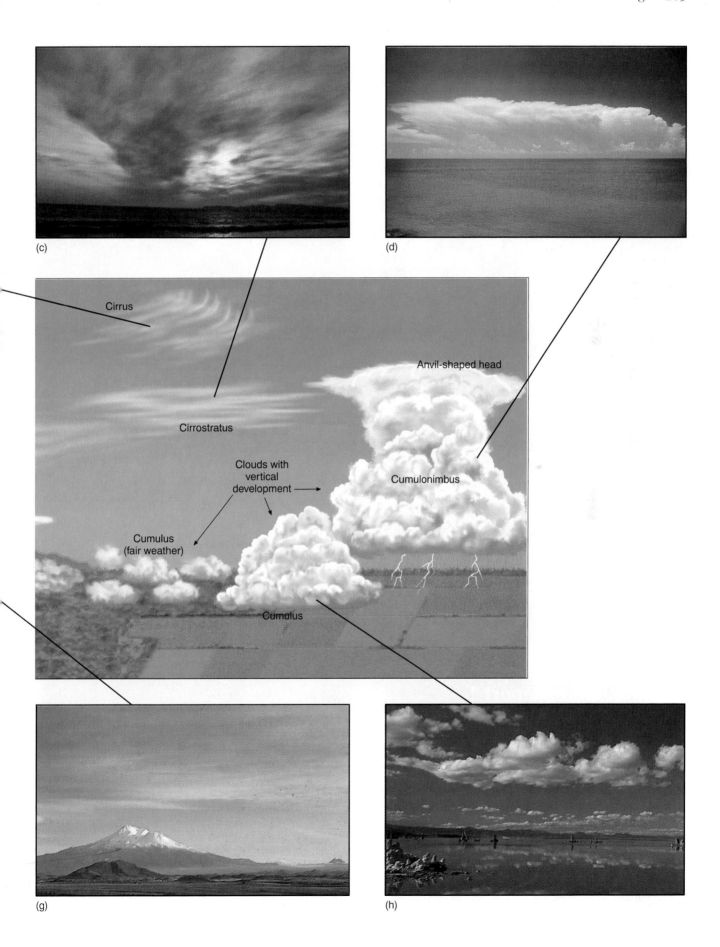

(c)

(d)

Cirrus

Cirrostratus

Anvil-shaped head

Clouds with
vertical
development →

Cumulonimbus

Cumulus
(fair weather)

Cumulus

(g)

(h)

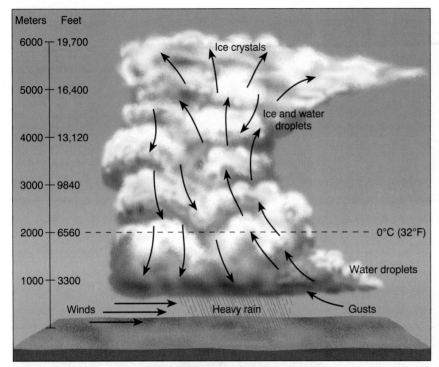

(a)

(b)

(c)

**FIGURE 8-5**
**Cumulonimbus thunderhead.**
(a) Structure and form of a cumulonimbus cloud. Violent updrafts and downdrafts mark the circulation within the cloud. Blustery wind gusts occur along the ground. (b) Shuttle astronauts capture a dramatic cumulonimbus thunderhead as it moves over Galveston Bay, Texas. (c) Few acts of nature can match the sheer power released by an intense lightning and thunder storm. [(b) Space Shuttle photo from NASA; (c) photo by Joe Towers/Stock Market.]

# News Report 1

## Harvesting Fog

Desert organisms have adapted remarkably to the presence of coastal fog along western coastlines in subtropical latitudes. For example, sand beetles in the Namib Desert in extreme southwestern Africa harvest water from the fog. They hold up their wings so condensation collects and runs down to their mouths. As the air warms during the day and the fog evaporates, the tiny beetles retreat beneath the sand. Early the next morning they emerge in the fog to harvest their next drink.

In the Atacama Desert of Chile and Peru, residents stretch large nets to intercept the fog; moisture condenses on the netting and drips into barrels. Chungungo, Chile, receives water from fog-harvesting in a pilot program developed by Canadian and Chilean interests. Large sheets of plastic mesh along a ridge of the El Tofo mountains harvest water from advection fog. The captured water flows through pipes to the fishing village below. At least 22 countries distributed on six continents experience conditions suitable for this water resource technology.

**FIGURE 8-6**
**Advection fog.**
San Francisco's Golden Gate Bridge shrouded by an invading advection fog characteristic of summer conditions along a western coast. [Photo by author.]

**FIGURE 8-7**
**Evaporation fog.**
An evaporation fog, commonly referred to as sea smoke. [Photo by author.]

***Radiation Fog.*** **Radiation fog** forms when radiative cooling of a surface chills the air layer directly above that surface to the dew-point temperature, creating saturated conditions and fog. This fog occurs over moist ground especially on clear nights; it does not occur over water, because water does not cool appreciably overnight. Slight movements of air deliver even more moisture to the cooled area for more fog formation of greater depth (Figure 8-9).

Every year the media carry stories of multicar pileups on stretches of highway that became obscured by fog. Fog is a hazard to drivers, pilots, sailors, pedestrians, and cyclists, and its conditions of formation are quite predictable. The spatial aspects of fog occurrence should be

**FIGURE 8-8**
**Valley fog.**
Cold air settles in the valleys of the Appalachian Mountains, chilling the air to the dew point and forming a valley fog. [Photo by author.]

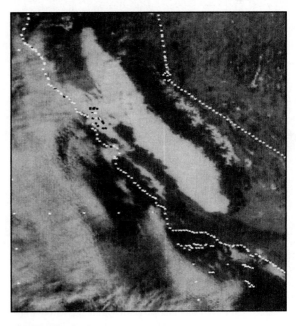

**FIGURE 8-9**
**Radiation fog.**
Satellite image of a radiation fog in Central California, locally known as a tule fog (pronounced "tooley") because of its association with the tule (bulrush) plants that line the low-elevation islands and marshes of the Sacramento River and San Joaquin River delta regions. [Image courtesy of National Environmental Satellite Data and Information Service (NESDIS).]

a key planning element for any proposed airport, harbor facility, or highway. The prevalence of fog throughout the United States and Canada is shown in Figure 8-10.

Fog can even become a health issue. Fog droplets can combine with atmospheric pollutants to form airborne acid pollution and even a "killer fog" in extreme cases, with stagnant air leading to unhealthy concentrations and possible fatalities.

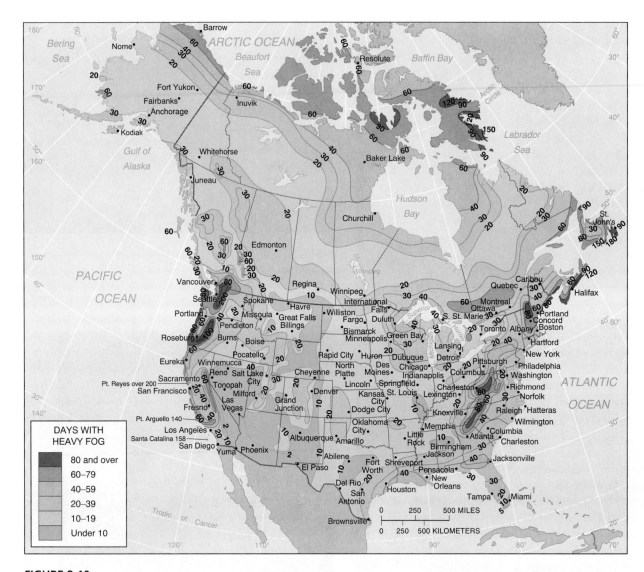

**FIGURE 8-10**

**Fog incidence map.**

Mean annual number of days with heavy fog in the United States and Canada. Officially, fog is declared if visibility is less than 1 km (3300 ft). The foggiest spot in the United States is the mouth of the Columbia River where it enters the Pacific Ocean at Cape Disappointment, Washington. One of the foggiest places in the world is Newfoundland's Avalon Peninsula, specifically Argentia and Belle Isle, which regularly exceed 200 days of fog each year. [Data courtesy of National Weather Service; Map Series 3, *Climatic Atlas of Canada*, Atmospheric Environment Service Canada, and *The Climates of Canada*, compiled by David Philips, Senior Climatologist, Environment Canada, 1990.]

# Air Masses

Each area of Earth's surface imparts its temperature and moisture characteristics to the air it touches. The effect of the surface on the air creates regional, homogeneous masses of air having specific conditions of temperature, humidity, and stability. Such masses of air possess all the physical characteristics of the atmosphere discussed so far and thus link the Earth-atmosphere energy budget and water-weather systems. These masses of air interact to produce weather patterns—in essence, they are the actors in our weather drama. Such a distinctive body of air is called an **air mass**, and it initially reflects the characteristics of

its *source region*. For example, weather reporters speak of a "cold Canadian air mass" and "moist tropical air mass."

The longer an air mass remains stationary over a region, the more definite its physical attributes become. Within each air mass there is a homogeneity of temperature and humidity that sometimes extends through the lower half of the troposphere.

## *Air Masses Affecting North America*

Air masses generally are classified according to the moisture and temperature characteristics of their source regions:

1. Moisture—designated **m** for maritime (wetter) and **c** for continental (drier).

2. Temperature (latitude)—designated **A** (arctic), **P** (polar), **T** (tropical), **E** (equatorial), and **AA** (antarctic).

The principal air masses that affect North America in winter and summer are mapped in Figure 8-11.

*Continental polar* (**cP**) air masses form only in the Northern Hemisphere and are most developed in winter when they dominate cold weather conditions. cP air masses are major players in middle- and high-latitude weather. The cold, dense cP air lifts moist, warm air in its path, producing lifting, cooling, and condensation. As we will see, the greater the temperature difference between this cold air mass and warmer air masses, the more dramatic is the weather produced. An area covered by cP air experiences cold, stable air and clear skies. The Southern Hemisphere lacks the necessary continental masses (continentality) at high latitudes to produce continental polar characteristics.

*Maritime polar* (**mP**) air masses in the Northern Hemisphere exist northwest and northeast of the North American continent over the northern oceans. Within them, cool, moist, unstable conditions prevail throughout the year. The Aleutian and Icelandic subpolar low-pressure cells reside within these mP air masses, especially in their well-developed winter pattern.

North America also is influenced by two maritime tropical (**mT**) air masses—the *mT Gulf/Atlantic* and the *mT Pacific*. The humidity experienced in the East and Midwest is created by the mT Gulf/Atlantic air mass, which is particularly unstable and active from late spring to early fall (Figure 8-12). In contrast, the mT Pacific is stable to conditionally unstable and generally much lower in moisture content and available energy. As a result, the western United States, influenced by this weaker Pacific air mass, receives lower average precipitation than the rest of the country.

Please review Figure 6-16 and the discussion of subtropical high-pressure cells off the coast of North America—the moist, unstable conditions on the western edge of the Atlantic (east coast), and the drier, stable conditions on the eastern edge of the Pacific (west coast). These conditions, coupled respectively with ocean currents that are warmer (Gulf Stream) and cooler (California current), produce the characteristics of each source region for these maritime air masses.

### Air Mass Modification

As air masses migrate from their source regions, their temperature and moisture characteristics are modified and slowly take on the characteristics of the land over which they pass. For example, a mT Gulf/Atlantic air mass may carry humidity to Chicago and on to Winnipeg, but gradually will lose its initial characteristics of high humidity and warmth with each day's passage northward.

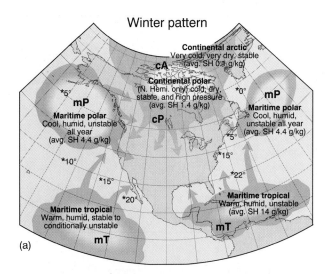

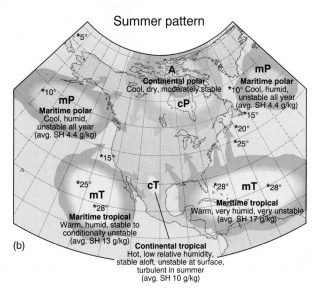

**FIGURE 8-11**

**Principal air masses.**

Air masses that influence North America in winter and summer. (*Sea-surface temperature in °C; SH=specific humidity.)

Similarly, below freezing temperatures occasionally reach into southern Texas and Florida, brought by an invading winter cP air mass from the north. However, that air mass will have warmed from the −50°C (−58°F) of its source region in central Canada. In winter, as a cP air mass moves southward over warmer land, it moderates, warming especially after it crosses the snowline.

Modification of cP air masses as they move south and east produces snowbelts that lie to the east of each of the Great Lakes. As below-freezing cP air passes over the warmer Great Lakes, it absorbs heat energy and moisture from the lake surfaces. This enhancement produces heavy lake-effect snowfall downwind into Ontario, Michigan, Pennsylvania, and New York—some areas receiving in excess of 250 cm (100 in.) in average snowfall a year (Figure 8-13).

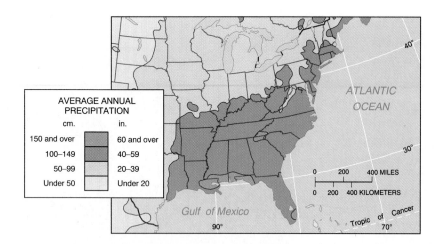

**FIGURE 8-12**

**Influence of mT Gulf/Atlantic air mass.**

The pattern of precipitation over the southeastern United States shows the influence of the warm, moist, and generally unstable mT Gulf/Atlantic air mass. Precipitation decreases with distance inland from the source region over the Gulf of Mexico and Atlantic Ocean.

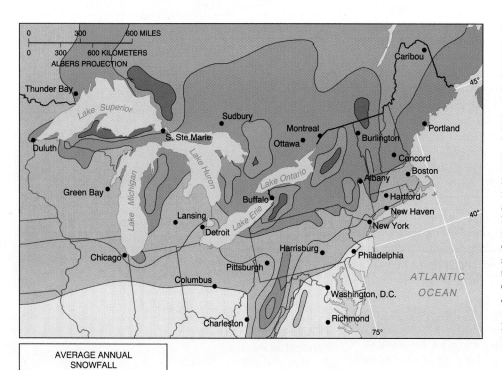

**FIGURE 8-13**

**Lake-effect snowbelts of the Great Lakes.**

Locally, heavy areas of snowfall are associated with the lee side of each of the Great Lakes. In winter, cold cP and cA air masses pass from colder land surfaces across the relatively warmer water of these lakes. The passing air masses are warmed and humidified (water vapor is added) from the lake water. The humid, now-unstable air yields heavy snowfall as it moves onshore and becomes chilled. The strongest effect is generally limited to about 50 km (30 mi) from the lake shore, although severe storms may bring these lake-enhanced snows 100 km (60 mi) or more inland. [Snowfall data from the *Climatic Atlas of the United States* (Washington DC: Department of Commerce, NOAA, 1983), p. 53.]

# Atmospheric Lifting Mechanisms

If air masses are to cool adiabatically (by expansion), and if they are to reach the dew-point temperature and saturate, condense, form clouds, and perhaps precipitate, then they must be lifted. Four principal lifting mechanisms operate in the atmosphere: *convergent lifting* (air flows toward an area of low pressure), *convectional lifting* (stimulated by local surface heating), *orographic lifting* (air is forced over a barrier such as a mountain range),

and *frontal lifting* (along the leading edges of contrasting air masses). Descriptions of all four mechanisms follow and are summarized in Figure 8-14.

## *Convergent Lifting*

Air flowing from different directions into the same low-pressure area is converging, displacing air upward in **convergent lifting**. In the tropics, convergent lifting of warm, moist air produces disturbances that can lead to the devel-

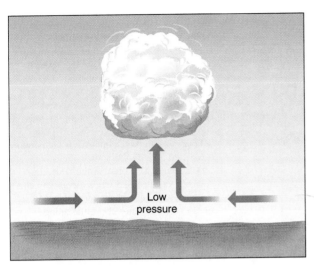

(a) Convergent

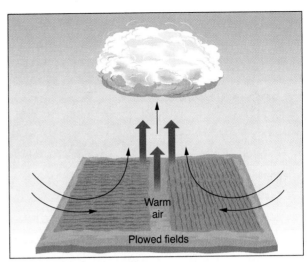

(b) Convectional (local heating)

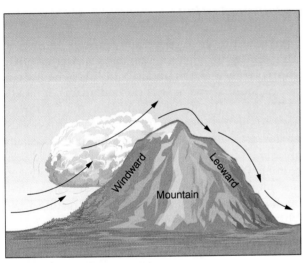

(c) Orographic (barrier)

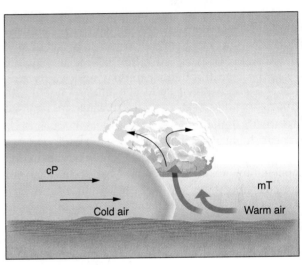

(d) Frontal (e.g. cold front)

**FIGURE 8-14**
**Four atmospheric lifting mechanisms.**
(a) Convergent lifting, (b) Convectional lifting, (c) Orographic lifting, (d) Frontal lifting.

opment of a tropical storm. We studied another example of convergence in Chapter 6, in an atmospheric circulation model. All along the equatorial region, the southeast and northeast trade winds converge, forming the intertropical convergence zone (ITCZ) and areas of extensive uplift, towering cumulonimbus cloud development, and high average annual precipitation (see Figures 6-14 and 6-15b).

## Convectional Lifting

When an air mass passes from a maritime source region to a warmer continental region, heating from the warmer land causes lifting and convection in the air mass. Other sources of surface heating might include an urbanized area (heat island) or an area of dark soil in a plowed field—the warmer surfaces produce **convectional lifting**. If conditions are unstable, initial lifting is sustained and clouds develop. Figure 8-15 illustrates convectional action stimulated by local heating, with unstable conditions present in the atmosphere. The rising parcel of air continues its ascent because it is warmer (less dense) than the surrounding environment.

In the figure, the MAR (moist adiabatic rate) is used above the lifting condensation level, which is visible at the elevation of the cloud base where the rising air parcel becomes saturated. Buoyancy is added at this level through the release of latent heat by condensation. Continued lifting above this altitude produces further cooling of the air parcel at the MAR.

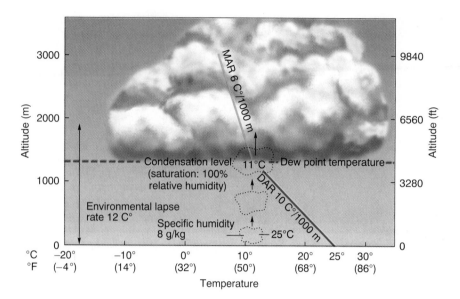

**FIGURE 8-15**

**Local heating and convection.**

Local heating and convection under unstable atmospheric conditions. Note that specific humidity is 8 g/kg and beginning temperature in the air parcel is 25°C. If you look back to Figure 7-12, you will find that air with a specific humidity of 8 g/kg must be cooled to 11°C to achieve the dew-point temperature. In the example, this temperature is reached after 14 C° of adiabatic cooling at 1400 m (4600 ft). Note that the DAR (dry adiabatic rate) is used when the air parcel is less than saturated, changing to the MAR (moist adiabatic rate) above the lifting-condensation level at 1400 m.

As an example of local heating and convectional lifting, Figure 8-16 depicts a day on which the landmass of Florida was warmer than the surrounding Gulf of Mexico and Atlantic Ocean. Because the Sun's radiation gradually heats the land throughout the day and warms the air above it, convectional showers tend to form in the afternoon and early evening, causing the highest frequency of days with thunderstorms in the United States. Florida appears highlighted and painted with clouds.

Towering cumulonimbus clouds are an expected summertime feature in the regions of North America that are affected by the mT Gulf/Atlantic air mass and, to a lesser extent, by the weaker mT Pacific air mass. Convectional precipitation also dominates along the ITCZ, over tropical islands, and anywhere that moist, unstable air is heated from below or where the inflowing trade winds produce a dynamic convergence.

## *Orographic Lifting*

The physical presence of a mountain acts as a topographic barrier to migrating air masses. **Orographic lifting** (*oro* means "mountain") occurs when air is forcibly lifted upslope as it is pushed against a mountain. It cools adiabatically. Stable air that is forced upward in this manner may produce stratiform clouds, whereas unstable or conditionally unstable air usually forms a line of cumulus and cumulonimbus clouds. An orographic barrier enhances convectional activity and causes additional lifting during the passage of weather fronts and cyclonic systems, thereby extracting more moisture from passing air masses.

Figure 8-17 illustrates the operation of orographic lifting under unstable conditions. The wetter intercepting slope is termed the *windward slope*, as opposed to the

**FIGURE 8-16**

**Convectional activity over the Florida peninsula.**

Cumulus clouds cover the land, with several cells developing into cumulonimbus thunderheads. Warm, moist air from the Gulf of Mexico is lifted by local heating as it passes over the relatively warmer land. [Project Gemini photo from NASA.]

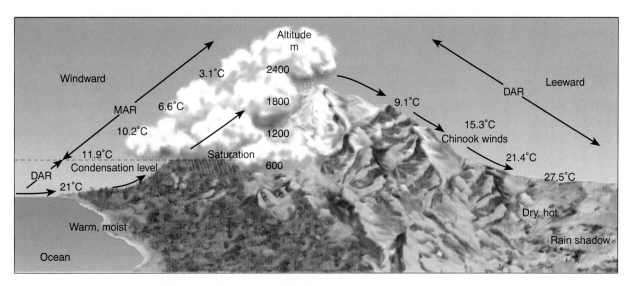

**FIGURE 8-17**
**Orographic precipitation.**
Orographic barrier and precipitation patterns—unstable conditions assumed. Prevailing winds force warm, moist air upward against a mountain range, producing adiabatic cooling, eventual saturation, and precipitation. On the leeward slope, as the "dried" air descends, compressional heating warms it, creating the hot, relatively dry rain shadow of the mountain.

drier far-side slope, known as the *leeward slope*. Moisture is condensed from the lifting air mass on the windward side of the mountain; on the leeward side, the descending air mass is heated by compression, and any remaining water in the air evaporates. Thus, air beginning its ascent up a mountain can be warm and moist, but finishing its descent on the leeward slope it has become hot and dry. In North America, **chinook winds** (called *föhn* or *foehn* winds in Europe) are the warm, downslope air flows characteristic of the leeward side of mountains. Such winds can bring a 20 C° (36 F°) jump in temperature and greatly reduced relative humidity on the lee side of the mountains.

The term **rain shadow** is applied to dry regions leeward of mountains. East of the Cascade Range, Sierra Nevada, and Rocky Mountains, such rain-shadow patterns predominate. In fact, the precipitation pattern of windward and leeward slopes persists worldwide, as confirmed by the precipitation maps for North America (Figure 9-6) and the world (Figure 10-3).

The state of Washington provides an excellent example of this concept, as shown in Figure 8-18. The Olympic Mountains and Cascade Mountains orographically lift invading mP air masses from the North Pacific Ocean, squeezing out annual precipitation of over 500 cm and 400 cm, respectively (200 in. and 160 in.). The Quinault Ranger and Rainier Paradise weather stations demonstrate windward-slope rainfall. The cities of Sequim, in the Puget Trough, and Yakima, in the Columbia Basin, are on the leeward side of these mountain ranges and are characteristically low in annual rainfall, being in the rain shadow. News Report 2 describes some record-breaking orographic effects.

## Frontal Lifting (Cold and Warm Fronts)

The leading edge of an advancing air mass is called its *front*. That term was applied by Vilhelm Bjerknes (1862–1951) and a team of meteorologists working in Norway during World War I, because it seemed to them that migrating air-mass "armies" were doing battle along their fronts. A front is a place of atmospheric discontinuity, a narrow zone that is the line of conflict between two air masses of different temperature, pressure, humidity, wind direction and speed, and cloud development. The leading edge of a cold air mass is a **cold front**, whereas the leading edge of a warm air mass is a **warm front**.

***Cold Front.*** On weather maps, such as those shown in Figure 8-24, a cold front is identified by a line of triangular spikes pointing in the direction of frontal movement along an advancing cP or mP air mass. The steep face of the cold air mass suggests its ground-hugging nature, caused by its density and uniform physical character (Figure 8-19).

Warm, moist air in advance of the cold front is lifted upward abruptly and is subjected to the same adiabatic rates and factors of stability or instability that pertain to all lifting air parcels. A day or two ahead of the cold front's passage, high cirrus clouds appear, telling observers that a lifting mechanism is on the way.

The cold front's advance is marked by a wind shift, temperature drop, and lowering barometric pressure due to the lifting. Air pressure reaches a local low as the line of most-intense lifting passes, usually just ahead of the front itself. Clouds may build along the cold front into characteristic cumulonimbus and may appear as an

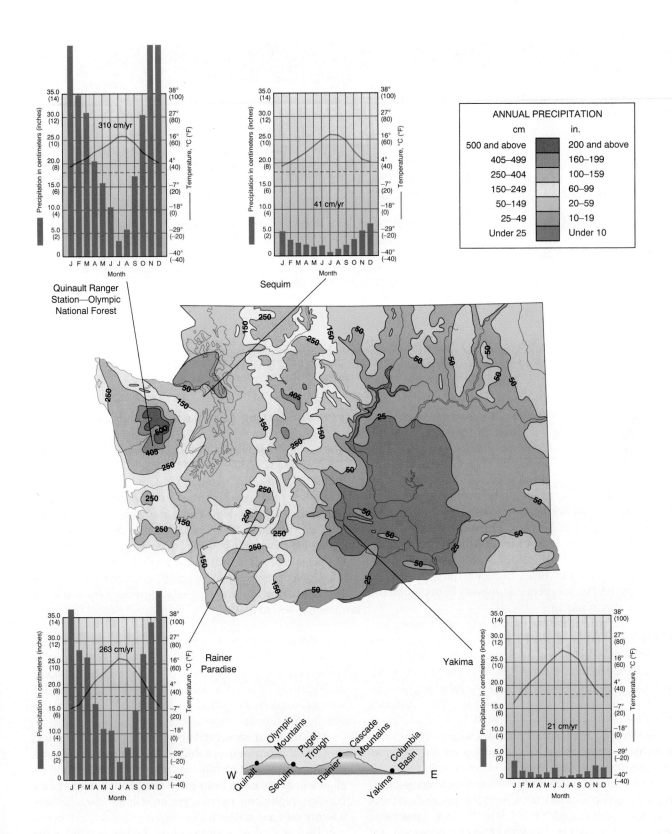

**FIGURE 8-18**

**Orographic patterns in Washington State.**

Four stations in Washington provide examples of orographic effects: windward precipitation and leeward rain shadows. Isohyets on the map indicate rainfall (in inches). [Data from J. W. Scott, and others, *Washington: A Centennial Atlas* (Bellingham, Wash.: Center for Pacific Northwest Studies, Western Washington University, 1989), p. 3.]

# News Report 2

## Mountains Set Precipitation Records

Because orographic lifting is limited in areal extent to locations where a topographic barrier exists, it is the least dominant lifting mechanism worldwide. But mountain ranges are the most consistent of all the precipitation-inducing mechanisms. Both the greatest average annual precipitation and the greatest maximum annual precipitation on Earth occur on the windward slopes of mountains that intercept moist tropical trade winds.

The world's greatest average annual precipitation occurs in the United States on Mount Waialeale, on the island of Kauai, Hawaii. This mountain rises 1569 m (5147 ft) above sea level. On its windward slope, rainfall averages 1234 cm (486 in., or 40.5 ft) a year (1941–1992). In contrast, the rain-shadow side of Kauai receives only 50 cm (20 in.) of rain annually. If no islands existed at this location, this portion of the Pacific Ocean would receive only an average 63.5 cm (25 in.) of precipitation a year.

Another place receiving world-record precipitation is Cherrapunji, India, 1313 m (4309 ft) above sea level at 25° N latitude, in the Assam Hills south of the Himalayas. It is located just north of Bangladesh, where tropical cyclones, torrential rains, and flooding in 1970 and 1991 killed hundreds of thousands of people. Because of the summer monsoons that pour in from the Indian Ocean and the Bay of Bengal, Cherrapunji has received 930 cm (366 in., or 30.5 ft) of rainfall in *one month* and a total of 2647 cm (1042 in., or 86.8 ft) in *one year*! Not surprisingly, Cherrapunji is the all-time precipitation record holder for a single year and for every other time interval from 15 days to 2 years. The average annual precipitation there is 1143 cm (450 in., 37.5 ft), placing it second only to Mount Waialeale.

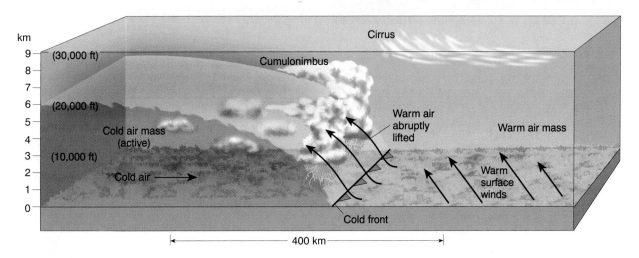

**FIGURE 8-19**

**A typical cold front.**

Denser, advancing cold air forces warm moist air to lift abruptly. As the air is lifted, it cools by expansion at the DAR, cooling to the dew-point temperature as it rises to a level of condensation and cloud formation. Cumulonimbus clouds may produce large raindrops, heavy showers, lightning and thunder, and hail.

advancing wall of clouds. Precipitation usually is heavy, containing large droplets, and can be accompanied by hail, lightning, and thunder.

The aftermath of a cold front passage usually brings northerly winds in the Northern Hemisphere (southerly winds in the Southern Hemisphere); lower temperatures; increasing air pressure from the cooler, denser air; and broken cloud cover.

The particular shape and size of the North American landmass and its latitudinal position present conditions where cP and mT air masses are best developed and have the most direct access to each other. The resulting contrast can lead to dramatic weather, particularly in late spring, with sizable temperature differences from one side of a cold front to the other.

A fast-advancing cold front can cause violent lifting and create a zone right along or slightly ahead of the front, called a **squall line**. Along a squall line, such as the one in the Gulf of Mexico shown in Figure 8-20, wind patterns are turbulent and wildly changing, and precipitation is intense. The well-defined wall of clouds in the photograph rises abruptly to almost 17,000 m (56,000 ft), with new thunderstorms forming along the front. Tornadoes also may develop along such a squall line.

**FIGURE 8-20**
**Cold front and squall line.**
Cold front and squall line are marked by a sharp line of
cumulonimbus clouds in the Gulf of Mexico. The cloud formation
rises to 17,000 m (56,000 ft). The passage of such a frontal system
over land often produces strong winds and possibly tornadoes.
[Space Shuttle photo from NASA.]

***Warm Front.*** A warm front is denoted on weather maps by
a line of semicircles facing the direction of frontal movement
(Figure 8-24). The leading edge of an advancing warm air
mass is unable to displace cooler, passive air, which is denser.
Instead, the warm air tends to push the cooler, underlying
air into a characteristic wedge shape, with the warmer air
sliding up over the cooler air. Thus, in the cooler-air region
a temperature inversion is present, sometimes causing poor
air drainage and stagnation.

Figure 8-21 illustrates a typical warm front in which
mT air is gently lifted, leading to stratiform cloud development and characteristic nimbostratus clouds and drizzly precipitation. A progression of cloud development is
associated with the warm front: High cirrus and cirrostratus clouds announce the advancing frontal system;
then clouds lower and thicken to altostratus; and finally
the clouds lower and thicken to stratus within several
hundred kilometers of the front.

## Midlatitude Cyclonic Systems

The conflict between contrasting air masses can develop
a **midlatitude cyclone**, or **wave cyclone**. Because of
the undulating nature of frontal boundaries and the steering flow of the jet streams, the term *wave* is appropriate.
This is a migrating low-pressure center with converging,
ascending air, spiraling inward counterclockwise in the
Northern Hemisphere (or converging clockwise in the
Southern Hemisphere). This cyclonic motion is generated
by the combined *pressure gradient force, Coriolis force,*

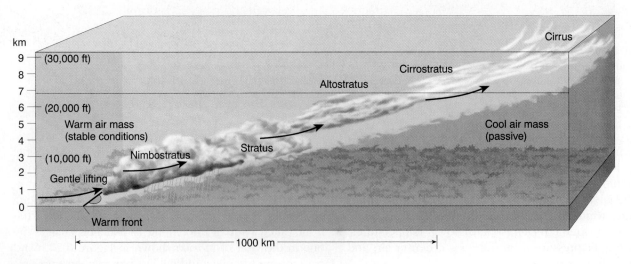

**FIGURE 8-21**
**A typical warm front.**
Note the sequence of cloud development as the warm front approaches. Warm air slides upward over a wedge of cooler,
passive air near the ground. Gentle lifting of the warm, moist air produces nimbostratus and stratus clouds and drizzly rain
showers, in contrast to the more dramatic precipitation associated with the passage of a cold front.

and *surface friction*, as discussed in Chapter 6. Before World War I, weather maps displayed only pressure and wind patterns. V. Bjerknes added the concept of fronts, and his son Jacob contributed the concept of migrating centers of cyclonic low-pressure systems.

Wave cyclones are initiated by wind-flow patterns in the upper atmosphere and can be triggered by divergence aloft—a surface low has an area of net divergence directly above it. Cyclonic systems are guided along their tracks by the intense high-speed winds in the jet stream, as portrayed in Figure 6-20.

Wave cyclones dominate the weather patterns in the middle and higher latitudes of both the Northern and Southern Hemispheres and act as a catalyst for air mass conflict. Such a midlatitude cyclone can be born along the polar front, particularly in the region of the Icelandic and Aleutian subpolar low-pressure cells in the Northern Hemisphere (see Figures 6-13a and 6-20).

## Life Cycle of a Midlatitude Cyclone

Figure 8-22 shows the birth, maturity, and death of a typical midlatitude cyclone in several stages, along with an idealized weather map. On the average, a midlatitude cyclone takes 3–10 days to progress through these stages from the area where it develops to the area where it finally dissolves.

***Cyclogenesis.*** **Cyclogenesis** is the atmospheric process in which low-pressure systems develop and strengthen. Along the polar front, cold and warm air masses converge and conflict.

The polar front is a discontinuity of temperature, moisture, and winds that establishes potentially unstable conditions. For a wave cyclone to form along the polar front, however, a point of air *convergence* at the surface must be matched by a compensating area of air *divergence* aloft. Even a slight disturbance along the polar front, perhaps a

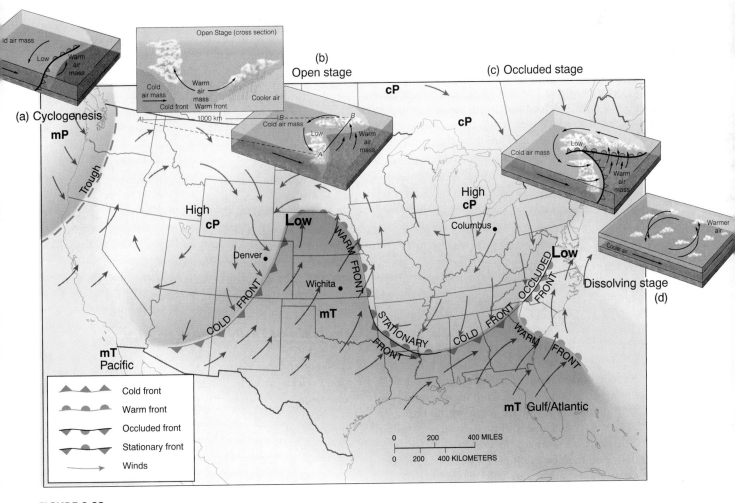

**FIGURE 8-22**
**Idealized stages of a midlatitude wave cyclone.**
(a) Cyclogenesis is noted where surface convergence and lifting begin. (b) The open stage. (c) The occluded stage. (d) The dissolving stage is reached at the end of the storm track as the cyclone spins down, no longer energized by the latent heat from condensing moisture. A system may take from 3 to 10 days to cross the continent along storm tracks that vary seasonally. After studying the text and this weather map, can you describe conditions in Denver? Wichita? Columbus?

small change in the path of the jet stream, can initiate the converging, ascending flow of air and thus a surface low-pressure system (illustrated in Figure 8-22a).

In addition to the polar front, certain other areas are associated with wave cyclone development and intensification: the eastern slope of the Rockies and other north-south mountain barriers, the Gulf Coast, and the east coasts of North America and Asia. V. Bjerknes, mentioned earlier, first proposed the polar front theory of cyclonic development.

**Open Stage.** To the east of the developing low-pressure center, warm air begins to move northward along an advancing front, while cold air advances southward to the west of the center. This movement is noted on Figure 8-22b as a trough, or area of beginning convergence. The growing circulation system then is vented into upper-level winds. As the mid-latitude cyclone matures, the counterclockwise flow (in the Northern Hemisphere) draws the cold air mass from the north and west and the warm air mass from the south. A characteristic open stage forms as shown, with a low centered over western Nebraska. In the cross section, you can see the profiles of both a cold front and a warm front and each air mass segment.

On the map, Denver has just experienced the passage of a cold front. Before the front passed, you can see on the map that winds were from the southwest, but now the winds have shifted and are from the northwest. Temperature and humidity have gone from the warm moist mT to colder conditions as the cP air mass moves over Denver. Wichita, Kansas, experienced the passage of a warm front and now is in the midst of the warm-air segment of the cyclone.

**Occluded Stage.** Because the cP air mass is more homogeneous in temperature and pressure than is the mT air mass, the cold front is the bulldozer blade for a denser, more-unified mass and therefore moves faster than the warm front. Cold fronts can travel at an average 40 kmph (25 mph), whereas warm fronts average roughly half that—16–24 kmph (10–15 mph). Thus, a cold front often overtakes the cyclonic warm front, wedging beneath it, producing an **occluded front** (*occlude* means "to close"). An occluded front stretches south from the center of low pressure in Virginia to the border between the Carolinas (Figure 8-22c). A cold front and a warm front remain active to the south. Precipitation may be moderate to heavy initially and then taper off as the warmer air wedge is lifted higher by the advancing cold air mass.

Also note the designation of a **stationary front** on the weather map (in Arkansas and Mississippi). This frontal symbol tells you that there is a stalemate between cooler and warmer air masses where air flow is almost parallel to the front, although in opposite directions on either side. Some gentle lifting is producing light to moderate precipitation. Eventually the stationary front will begin to move, as one of the air masses assumes dominance, and it will evolve into a warm or a cold front.

**Dissolving Stage.** The final, dissolving, stage of the midlatitude cyclone occurs when its lifting mechanism is completely cut off from the warm air mass, which was its source of energy and moisture. Remnants of the cyclonic system then dissipate in the atmosphere, perhaps after passage across the country (Figure 8-22d).

**Storm Tracks.** Cyclonic storms—1600 km (1000 mi) wide—and their attendant air masses move across the continent along **storm tracks**, which shift latitudinally with the Sun and seasons. Typical storm tracks that cross North America are farther northward in summer and farther southward in winter (Figure 8-23). As the storm tracks begin to shift northward in the spring, cP and mT air masses are brought into their clearest conflict. This is the time of strongest frontal activity, featuring thunderstorms and tornadoes. Storm tracks follow the path of upper-air winds, which direct storm systems across the continent.

A map of actual storm tracks for March 1991 demonstrates several areas of cyclogenesis: the northwest over the Pacific Ocean, the Gulf of Mexico, the eastern seaboard, and the Arctic. Cyclonic circulation also frequently develops on the lee side of mountain ranges, as along the Rockies from Alberta south to Colorado. As air moves downslope, the vertical axis of the air column extends, shrinking the system horizontally and intensifying wind speeds—the principle is similar to that used when ice skaters draw in their arms tightly to increase their spin speed. As the air travels downslope, its deflection in a cyclonic flow develops new cyclonic systems or intensifies existing ones. By crossing the mountains, such systems gain access to the moisture-laden, energy-rich mT air mass from the Gulf of Mexico.

## Analysis of Daily Weather Maps

The sequence in Figure 8-22 illustrates an ideal midlatitude cyclone model. The actual pattern of cyclonic passage over North America is not so tidy, it is widely varied in shape, form, and duration. Regardless, you can apply this general model, along with your understanding of warm and cold fronts, to the actual midlatitude cyclone shown in the weather-map sequence for April 1–3, 1988 (Figure 8-24).

*Synoptic analysis* is the evaluation of weather data collected at a selected time, as shown on the three synoptic weather maps in Figure 8-24. Building a data base of wind, pressure, temperature, and moisture conditions is key to *numerical* (computer-based) *weather prediction* and the development of weather-forecasting models. Development of numerical models is a great challenge because the atmosphere operates as a nonlinear (irrational) system, tending toward chaotic behavior. Slight variations in input data or slight changes in the basic assumptions of the model's behavior can produce widely varying forecasts. As our knowledge of the interactions that produce weather and our instruments and software improve so too will the accuracy of our forecasts.

Preparing a weather report and forecast requires analysis of a daily weather map and satellite images that depict atmospheric conditions. Figure 8-24 presents weather maps and satellite images for a typical 3-day period and explains the standard weather station symbols. News Report 3 describes an exciting new technology for gathering weather data. Weather data from all these sources include the following:

- Barometric pressure
- Pressure tendency (steady, rising, falling)
- Surface air temperature
- Dew-point temperature
- Wind speed and direction
- Type and movement of clouds
- Current weather
- State of the sky (current conditions)
- Precipitation since last observation

On the April 2 map in Figure 8-24b, you can identify the temperatures reported by various stations, the patterns created, and the location of warm and cold fronts. The distribution of air pressure is defined by *isobars*, lines that connect points of equal pressure on the map. Although the low-pressure center identified is not intense, it is well defined over eastern Nebraska, with winds following in a counterclockwise, cyclonic path around the low pressure. This pattern is clear on the satellite image.

The maps and images also illustrate the evolving pattern of precipitation over the 3-day period as the warm and cold fronts sweep eastward. The mT air mass, advancing northward from the Gulf/Atlantic source region, provides the moisture for afternoon and evening convectional thunder showers. The storm track for the 3 days shown in the sequence is noted as a dashed line.

# Violent Weather

Weather provides a continuous reminder that insolation is absorbed, providing energy to drive nature's atmospheric handiwork. The flow of energy across the latitudes can at times set into motion violent and destructive weather conditions. We now look at specific forms of violent weather: thunderstorms, tornadoes, and tropical cyclones.

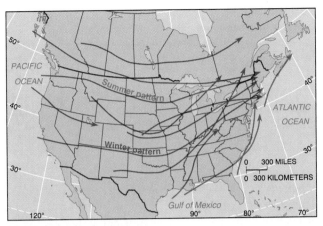

**(a) Average storm tracks**

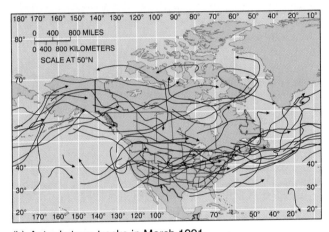

**(b) Actual storm tracks in March 1991**

**FIGURE 8-23**

**Typical and actual storm tracks.**

(a) Cyclonic storm tracks for North America vary seasonally. Several locations of cyclogenesis are indicated. (b) Actual cyclonic tracks that occurred during March 1991 over North America. [(b) From *Storm Data* 33, no. 3 (March 1991); Asheville, N.C.: NOAA, (NESDIS), National Climatic Data Center.]

## *Thunderstorms*

Tremendous energy is liberated by the condensation of large quantities of water vapor in clouds. This process locally heats the air, rapidly changing its density and buoyancy and causing violent updrafts. As rising parcels of air pull surrounding air into the column, updrafts get stronger. Raindrops form and descend through the cloud, and their frictional drag pulls air toward the ground, causing violent downdrafts. As a result, giant cumulonimbus clouds can create dramatic weather moments—heavy precipitation, lightning, thunder, hail, blustery winds, and tornadoes. Such thunderstorms may develop in three settings: within an air mass (particularly in warm, moist air), in a line along a cold front or other convergent boundary, or where mountain slopes cause orographic lifting.

**FIGURE 8-24**

**Weather map sequence.**

Weather maps and *GOES-7* satellite images for April 1, 2, and 3, 1988. The standard weather symbols used are identified and explained. Note the track of the center of low pressure from April 1 through April 3, beginning in Texas and ending in northern Michigan (dashed line on map). [Images from National Weather Service and NOAA, NESDIS, National Climatic Data Center. Image (d) is visible light; images (e) and (f) are infrared wavelengths.]

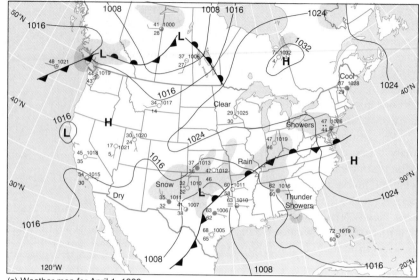

(a) Weather map for April 1, 1988

WEATHER STATION SYMBOL

WIND SPEED
TEMPERATURE °F
PRECIPITATION
DEW POINT °F

WIND DIRECTION
74   1004   PRESSURE
CLOUD COVER

COLD FRONT        WARM FRONT

STATIONARY FRONT    OCCLUDED FRONT

| PRECIPITATION TYPE | | WIND SPEED (kph/mph) | |
|---|---|---|---|
| , | Drizzle | Calm | |
| • | Rain | 5–13 | 3–8 |
| ✳ | Snow | 14–23 | 9–14 |
| ▽ | Showers | 24–32 | 15–20 |
| T̅ | Thunderstorms | 33–40 | 21–25 |
| ☰ | Fog | 89–97 | 55–60 |
| ⨀ | Dry Haze | 116–124 | 72–77 |
| ⌒ | Freezing Rain | | |
| △ | Hail | | |
| ◬ | Sleet | | |

CLOUD COVER

○ No clouds
◐ One-tenth or less
◑ Two-tenths to three-tenths
◑ Four-tenths
◑ Five tenths

◕ Six-tenths
◕ Seven-tenths to eight-tenths
◑ Nine-tenths or overcast with openings
● Completely overcast
⊗ Sky obscured

Polar stereographic projection true at 60° N

Scale at different latitudes

Surface weather map and station weather at 7:00 A.M. E.S.T.

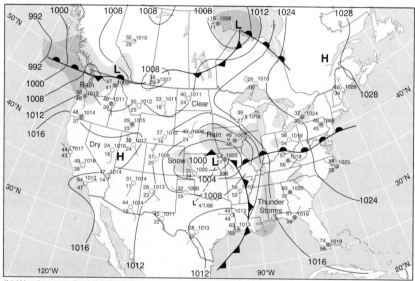

(b) Weather map for April 2, 1988

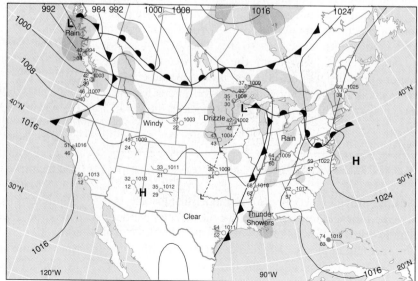

(c) Weather map for April 3, 1988

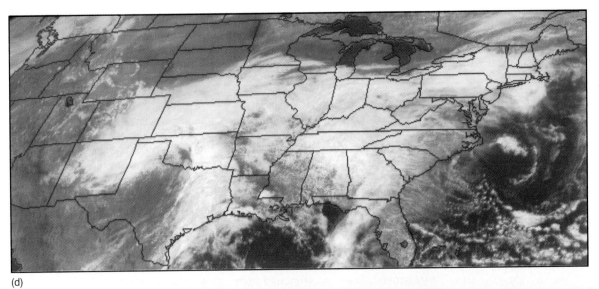

(d)

(e)

(f)

## News Report 3

### Revolutionary Use of the Global Positioning System to Forecast Weather—The GPS/MET

The Global Positioning System (GPS) of 25 orbiting satellites is accessed by hand-held receivers that calculate the latitude, longitude, and elevation of the observer within meters. In 1995, the GPS/MET receiver was placed in low-Earth orbit aboard the *MicroLab-1* satellite. Scientists at the University Corporation for Atmospheric Research (UCAR), operate the system, with support from the National Science Foundation, NASA, NOAA, and the FAA.

GPS/MET is a shoebox-sized device that uses radio signals from the GPS satellite network to gather atmospheric temperature, humidity, and pressure data worldwide. Radio signals passing through the atmosphere from any one of the GPS satellites to the GPS/MET receiver produce a *sounding*

profile of the atmosphere. Scientists monitor a comparison of the known distance between satellites and the slight delays in a signal that occur as it passes through the varying densities of the atmosphere. Temperature, humidity, and pressure measurements can be derived mathematically as a product of the "refractive effect" of the atmosphere on radio signals.

Presently, surface-launched weather balloons take soundings of conditions as they ascend. They are launched from about 1000 locations worldwide twice a day. But this single GPS/MET receiver makes 500 soundings a day, providing vertical profiles of temperature, moisture, and pressure! As procedures are refined, the number of observations will easily double from this first GPS/MET device.

Further, the GPS/MET does its sounding in *only a minute.* Radiosonde instrument packages carried aloft by balloons take over 100 minutes for the same task. The traditional balloon method covers only a 25 km (15 mi) vertical profile, but GPS/MET measures to 60 km (37 mi) and with less expected error than radiosondes.

If all goes well in fine-tuning the system, the potential is exciting. A global constellation of such receivers could provide worldwide data gathering at a fraction of the cost of the present weather-balloon system. Benefits include improved weather forecasts, better measurement of global temperature and weather changes, and more thorough atmospheric monitoring than is now possible.

---

Thousands of thunderstorms occur on Earth at any given moment. Equatorial regions and the ITCZ experience many of them, as exemplified by the city of Kampala in Uganda, East Africa (north of Lake Victoria), which sits virtually on the equator and averages a record 242 days a year of thunderstorms. Figure 8-25 shows the annual distribution of days with thunderstorms across the United States and Canada. You can see that, in North America, most thunderstorms occur in areas dominated by mT air masses.

***Lightning and Thunder.*** An estimated 8 million lightning strikes occur each day on Earth (see the chapter-opening photo). **Lightning** refers to flashes of light caused by enormous electrical discharges—tens of millions to hundreds of millions of volts—that briefly superheat the air to temperatures of 15,000°–30,000°C (27,000°–54,000°F). The violent expansion of this abruptly heated air sends shock waves through the atmosphere as the sonic bang of **thunder**.

The greater the distance a lightning stroke travels, the longer the thunder echoes as the waves of sound arrive at your ear. Lightning at great distance from the observer may be too far away for its thunder to be heard and is called "heat lightning"; a misleading term for it is no different than any other lightning.

Lightning is created by a buildup of electrical energy between areas within a cumulonimbus cloud or between the cloud and the ground. When sufficient electrical potential develops to overcome the high electrical

resistance of the atmosphere, electrons leap from one surface to the other, creating a giant spark.

The beginning of a lightning discharge, called the *leader stroke,* goes from the cloud to the surface, following the route of least resistance, usually in a succession of forked steps. The path of charged ions remaining in the air behind this relatively slow-moving stroke then becomes the lower-resistance pathway for the *main stroke,* or return stroke, which surges in an enormous rush of energy along that path a fraction of a second later. This *return stroke,* traveling at around a tenth the speed of light (30,000 kmp/s), produces most of the flashing seen as lightning. From 2 to 40 repetitions of this sequence occur in a second or two along the same path and produce *forked lightning.* A discharge of forked lightning from one portion of a cumulonimbus cloud to another, usually masked by the cloud and appearing as a broad flash of light, is termed *sheet lightning.*

Lightning poses a hazard to aircraft, people, animals, trees, and structures. Certain precautions are mandatory when a lightning discharge threatens, because lightning causes nearly 200 deaths and thousands of injuries each year in the United States and Canada. When lightning is imminent, agencies such as the National Weather Service issue *severe storm warnings* and caution people to remain indoors. (The place *not* to seek shelter is beneath a tree, for trees are good conductors of electricity and often are hit by lightning.)

The United States and Canada operate a National Lightning Detection Network of over 150 sensors to detect increases in electromagnetic fields and related lightning discharges. Atmospheric scientists at the State University of New York–Albany manage the research program.

***Hail.*** Ice pellets called **hail** generally form within a cumulonimbus cloud. Raindrops circulate repeatedly above and below the freezing level in the cloud, adding layers of ice until their weight no longer can be supported by the circulation in the cloud.

Pea-sized and marble-sized hail are common, although hail the size of golf balls and baseballs also is reported at least once or twice a year somewhere in North America. For larger hail to form, the frozen pellets must stay aloft for longer periods. The largest authenticated hailstone in the world landed in Kansas in 1970; it measured 14 cm (5.6 in.) in diameter and weighed 758 g (1.67 lb)!

Hail is common in the United States and Canada, although somewhat infrequent at any given place. It hails perhaps every one or two years in the highest-frequency areas. Annual hail damage in the United States tops $750 million (Figure 8-26). The pattern of hail occurrence across the United States and Canada is similar to that of thunderstorms shown in Figure 8-25.

## Tornadoes

The updrafts associated with a cumulonimbus cloud appear on satellite images as pulsing bubbles of clouds. Friction with the ground slows surface winds, but higher

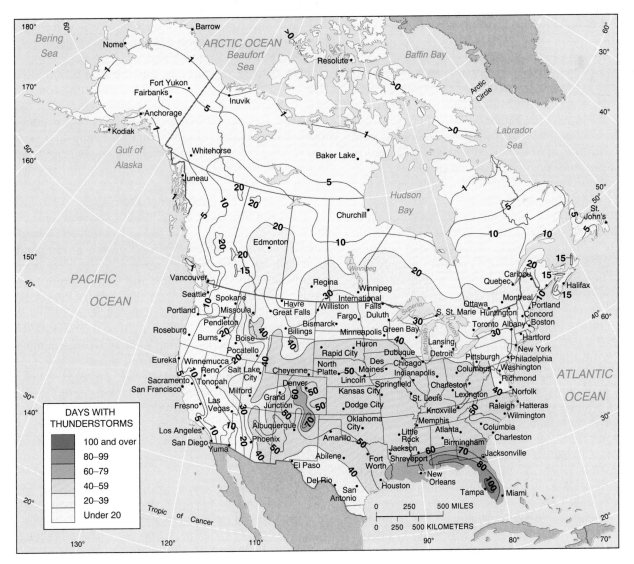

**FIGURE 8-25**

**Thunderstorm occurrence.**

Average annual number of days with thunderstorms. [Data courtesy of U.S. National Weather Service; Map Series 3, *Climatic Atlas of Canada*, Atmospheric Environment Service, Canada.]

**FIGURE 8-26**
**Hailstone damage.**
Hail can severely damage windows, automobiles, crops, livestock, and people. [Photo by *The Denver Post*, 1990.]

in the troposphere, winds blow faster. Thus, a body of air moves faster at altitude than at the surface, creating a rotation in the air along a horizontal axis parallel to the ground (picture a rolling pin; see Figure 8-27a). When that rotating air encounters the strong updrafts associated with frontal activity, the axis of rotation is shifted to a vertical alignment perpendicular to the ground (Figure 8-27b).

According to one hypothesis, this spinning, cyclonic, rising column of mid-troposphere-level air forms a **mesocyclone**. A mesocyclone can range up to 10 km (6 mi) in diameter and rotate vertically within the parent cloud to a height of thousands of meters (Figure 8-27c). As a mesocyclone extends vertically and contracts horizontally, wind speeds accelerate in an inward vortex (as ice skaters accelerate while spinning or, in miniature, as water accelerates when it spirals down a sink drain). A well-developed mesocyclone most certainly will produce heavy rain, large hail, blustery winds, and lightning; some mature mesocyclones will generate tornado activity.

As more moisture-laden air is drawn up into the circulation of a mesocyclone, more energy is liberated, and the rotation of air becomes more rapid. The narrower the mesocyclone, the faster the spin of converging parcels of air being sucked into the rotation. The swirl of the mesocyclone itself is visible, as are dark gray **funnel clouds** that pulse from the bottom side of the parent cloud. The terror of this stage of development is the lowering of such a funnel to Earth—a **tornado** (Figures 8-27c and 8-28). When tornado circulation occurs over water, a **waterspout** forms, and the sea or lake surface water is drawn up into the funnel some 3–5 m (10–16 ft). The rest of the waterspout funnel is made visible by the rapid condensation of water vapor.

***Tornado Measurement and Science.*** A tornado's diameter can range from a few meters to a few hundred meters. It can last from a few moments to tens of minutes. Pressures inside a tornado funnel usually are about 10% less than those in the surrounding air. This disparity is similar to the pressure difference between sea level and an altitude of 1000 m (3300 ft). The in-rushing convergence created by such a sharp horizontal-pressure gradient causes severe wind speeds at the funnel; they can exceed 485 kmph (300 mph), with at least a third of all tornadoes exceeding 182 kmph (113 mph).

Theodore Fujita, a noted meteorologist from the University of Chicago, designed a scale for classifying tornadoes according to wind speed and related property damage. The Fujita Scale ranks tornadoes as F0 and F1 (weak, with winds less than 180 kmph, or 112 mph; about 85% of all tornadoes); F2 and F3 (strong, with winds 181–332 kmph, or 113–206 mph; about 13%); and F4 and F5 (violent, with winds over 333 kmph, or 207 mph; about 2%, with less than 1% reaching F5).

The damage produced by such wind is catastrophic (Figure 8-29) and unpredictable. The entire 1990 tornado season caused 53 deaths. In the month of April 1991, a four-state area was devastated when over 70 tornadoes touched down from Texas to Nebraska, killing 30 people and causing millions of dollars of property damage. And in one day, March 27, 1994, 25 tornadoes hit Alabama, killing 43 people.

All 50 states have experienced tornadoes, as have all the Canadian provinces and territories. May and June are the peak months, as you can see in Figure 8-30. A small number of tornadoes are reported on other continents each year, but North America receives the greatest share because its latitudinal position and shape permit contrasting air masses to confront one another. Looking at the statistics on the map you can see the infamous "Tornado Alley" through Texas, Oklahoma, Kansas, and Nebraska—a corridor that experiences the highest average number of tornadoes on Earth. Using satellites, aircraft, and surface measurements, the National Center for Atmospheric Research and the University of Oklahoma completed a 2-year project in 1995 called VORTEX (Verification of the Origin of Rotation in Tornadoes). VORTEX is the most comprehensive effort yet to better understand these intense mesocyclones and the tornadoes they breed.

***Recent Tornado Records.*** In the United States, 31,204 tornadoes were recorded in the 37 years from 1959 through 1995, with 791 of them causing 2780 deaths. In addition, these tornadoes resulted in over 56,000 injuries and damages of over $17 billion. The yearly average of $475 million in damage is rising at about $50 million a year.

Interestingly, 1992 set a record for the most tornadoes in the United States in a single year, reaching an

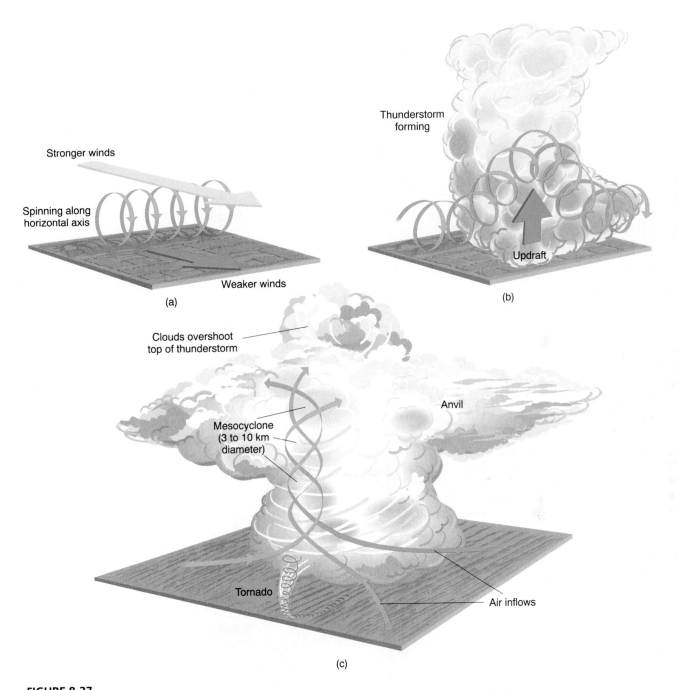

**FIGURE 8-27**
**Mesocyclone and tornado formation.**
(a) Wind is stronger aloft, establishing spinning along a horizontal axis. (b) Updraft from thunderstorm development tilts the rotating air along a vertical axis. (c) Mesocyclone forms as a rotating updraft within the thunderstorm. If a tornado forms, it will descend from the cloud base and the lower portion of the mesocyclone.

all-time high of 1297. In June alone, 421 occurred. The year 1993 ranked second, with 1173 tornadoes; 1990 ranked third, with 1133 tornadoes; 1991 ranked fourth, with 1132. The years 1994 at 1076 and 1995 at 1100 tornadoes ranked slightly lower. The annual average tornado occurrences for 1975–1989 was only 827. The reason for this marked increase in tornado frequency in North America is being researched; explanations range from global climate change to better reporting by a larger population armed with camcorders!

It is not enough to account statistically for past tornadoes; the need is for accurate forecasting. The National Severe Storms Forecast Center in Kansas City, Missouri, is the U.S. government's headquarters for such activities.

**FIGURE 8-28**
**Tornado.**
A fully developed tornado locks onto the ground in Gove County, Kansas. [Photo by Sheila Beougher/Gamma-Liaison, Inc.]

Modernization of the National Weather Service includes increasing use of *Doppler radar* that enables scientists to detect the specific flow of moisture droplets in mesocyclones and allows a warning period of 30 minutes to 1 hour in areas at risk. As a result of Doppler radar, only about 15% of tornadoes strike without some public warning. A severe storm watch sometimes is possible a few hours in advance. But at other times, this most intense phenomenon of the atmosphere can strike with destructive suddenness and little warning.

## Tropical Cyclones

A powerful manifestation of the Earth-atmosphere energy budget and moisture system is the **tropical cyclone**, which originates entirely within tropical air masses. The *tropics* extend from the Tropic of Cancer at 23.5° N to the Tropic of Capricorn at 23.5° S, containing the equatorial zone between 10° N and 10° S. Approximately 80 tropical cyclones occur annually worldwide. Some are powerful enough to be classified as hurricanes and typhoons.

Cyclonic systems forming in the tropics are quite different from midlatitude cyclones because the air of the tropics is essentially homogeneous, with no fronts or conflicting air masses. In addition, the warm air and warm seas ensure abundant water vapor and thus the necessary latent heat to fuel these storms.

What mechanism triggers the start of a tropical cyclone? Meteorologists now think that cyclonic motion begins with slow-moving easterly waves of low pressure in the trade-wind belt of the tropics, such as the Caribbean area (Figure 8-31). It is along the eastern (leeward) side of these migrating troughs of low pressure, a place of convergence and rainfall, that tropical cyclones are formed. Surface air flow then spins in to the low-

**FIGURE 8-29**
**Tornado damage.**
Aerial view of tornado devastation to homes and lives in Edmond, Oklahoma. [Photo by Chris Johns/Allstock.]

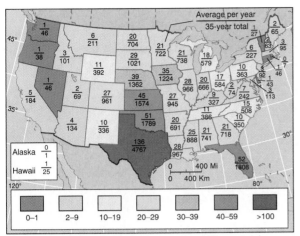

(a) Average number of tornadoes per year and 35-year totals (1959–1993)

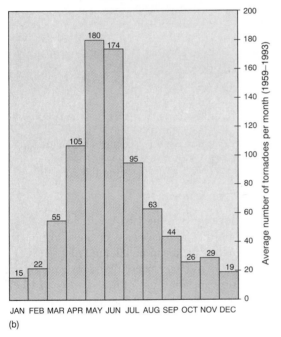

(b)

## FIGURE 8-30
**Tornado occurrence in the United States.**
(a) Average number of tornadoes per year/total number of tornadoes for 1959 to 1993. (b) Average number of tornadoes per month. During an average year, at least 830 tornadoes strike throughout the United States and Canada. [Data courtesy of the National Severe Storm Forecast Center, National Weather Service.]

pressure area (convergence), ascends, and flows outward (divergence) aloft. This important divergence aloft acts as a chimney, pulling more moisture-laden air into the developing system. Tropical cyclones also may result from extratropical cyclonic flows extending equatorward as well as along the ITCZ.

Tropical depressions intensify into tropical storms as they cross the Atlantic. A rule of thumb has emerged: If they mature early along their track, they tend to curve

northward toward the North Atlantic and miss the United States. The critical position is approximately 40° W longitude. If a tropical storm matures after it reaches the longitude of the Dominican Republic (70° W), then it has a higher probability of hitting the U.S. mainland (Figure 8-32 and Figure 1 in Focus Study 8-1, which discusses a new forecasting method.)

***Tropical Cyclone Global Patterns.*** The map in Figure 8-32 shows areas in which tropical cyclones form and some of the characteristic storm tracks traveled. It also indicates the range of months during which tropical cyclones are most likely to appear. They tend to occur when the equatorial low-pressure trough (ITCZ) is farthest from the equator, during the months following the summer solstice in each hemisphere. For example, storms that strike the southeastern United States do so mostly between August and October, following the June 21 solstice. Officially, tropical cyclone season is from June 1 to November 30 each year. (For comparison, the average storm tracks of midlatitude cyclones are plotted on the map.)

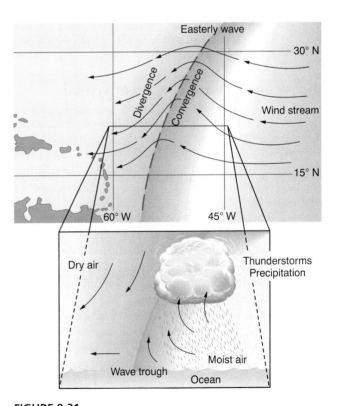

## FIGURE 8-31
**Easterly wave in the tropics.**
Development of a low-pressure center along an easterly (westward-moving) wave. Moist air rises in an area of convergence at the surface to the east of the wave trough. Windflows bend and converge before the trough and diverge downwind from the trough.

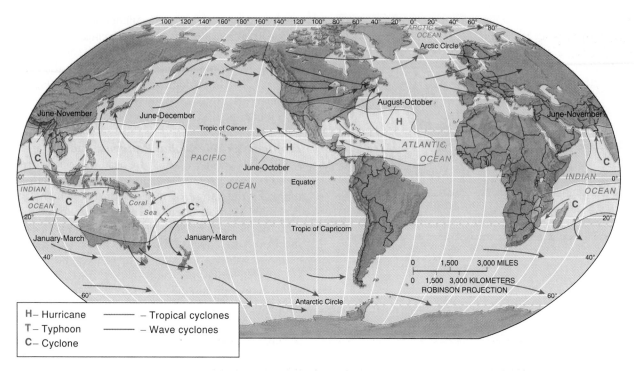

**FIGURE 8-32**
**Worldwide pattern of the most intense tropical cyclones.**
Typical tropical storm tracks with principal months of occurrence and regional names. For the North Atlantic region from 1871 to 1995, 1027 tropical cyclones (storms and hurricanes) developed. The peak day in this Atlantic region (assuming a 9-day moving average calculation) is September 10, with 63 cyclones for the 125-year period. Also shown are characteristic storm tracks of midlatitude cyclones.

## Hurricanes, Typhoons, and Cyclones

Tropical cyclones are potentially the most destructive storms experienced by humans, claiming thousands of lives each year worldwide. This is especially true when they attain wind speeds and low-pressure readings that upgrade their status to a full-fledged **hurricane** or **typhoon** (>65 knots, 74 mph). Different names are used for these storms worldwide. We call them hurricanes. They are typhoons in the western Pacific (Japan, Philippines) and cyclones in Indonesia, Bangladesh, and India. By any name they can be destructive killers. Worldwide, about 10% of all tropical disturbances have the right ingredients to become hurricanes or typhoons.

Damage depends on the degree of development at a storm's landfall site, how prepared citizens are for the blow, and the intensity of the storm. Table 8-2 presents the criteria for classification of tropical cyclones on the basis of wind speed, and it lists some of their features. When you hear meteorologists speak of a "category 4" hurricane, they are using the Saffir-Simpson Hurricane Damage Potential Scale to estimate possible damage from hurricane-force winds. This scale ranks hurricanes and typhoons in five categories, from smaller, category 1 storms to extremely dangerous (and rare) category 5 storms. These categories are defined in Table 8-3 using wind speeds and central pressure criteria, and some examples of recent storms are given.

Typhoons form in the vast Pacific, where the annual frequency of tropical cyclones is greatest. You hear about these when they make landfall in the Philippines or Japan. The magnitude and intensity of typhoons generally exceed those of hurricanes because of the vast expanses of open water in the Pacific that provide more area for development. In fact, the lowest sea-level pressure recorded on Earth was 870 mb (25.69 in.) in the center of Typhoon Tip, in October 1979, northwest of Guam.

The cyclone that struck Bangladesh in 1970 killed an estimated 300,000 people, and the one in 1991 claimed over 200,000 lives. The Galveston, Texas, hurricane of 1900 killed 6000; Hurricane Audry (1957), 400; Hurricane Gilbert (1988), 318; Hurricane Camille (1969), 256; and Hurricane Agnes (1972), 117. The record holder for property damage is 1992's Hurricane Andrew, which tallied $20 billion in damage. Had Hurricane Andrew made landfall a few kilometers farther north, it would have passed over Miami; damage estimates for that missed catastrophe range above $85 billion.

Statistically, the damage caused by tropical cyclones is increasing substantially as more and more development occurs along susceptible coastlines, whereas loss of life is decreasing in most parts of the world owing to better forecasting and understanding of these storms. Atmospheric scientists at Colorado State University, led by William Gray, successfully predicted the near-record 1995 Atlantic hurricane season (see Focus Study 8-1).

**TABLE 8-2**

| Tropical Cyclone Classification | | |
| --- | --- | --- |
| *Designation* | *Winds* | *Features* |
| Tropical disturbance | Variable, low | Definite area of surface low pressure; patches of clouds |
| Tropical depression | Up to 30 knots (63 kmph, 39 mph) | Gale force, organizing circulation; light to moderate rain |
| Tropical storm | 30–64 knots (63–117.5 kmph, 39–73 mph) | Closed isobars; definite circular organization; heavy rain; assigned a name |
| Hurricane (Atlantic and E. Pacific) | Greater than 65 knots (119 kmph, 74 mph) | Circular, closed isobars; heavy rain, storm surges; tornadoes in right-front quadrant |
| Typhoon (W. pacific) Cyclone (Indian Ocean, Australia) Baguios (Philippines) | | |

**TABLE 8-3**

| Saffir-Simpson Hurricane Damage Potential Scale | | | |
| --- | --- | --- | --- |
| *Category* | *Wind Speed* | *Central Pressure* | *Notable Atlantic Examples* |
| 1 | 64–82 knots (74–95 mph) | >979 mb | — |
| 2 | 83–95 knots (96–110 mph) | 965–979 mb | — |
| 3 | 96–113 knots (111–130 mph) | 945–964 mb | 1985 Elena; 1991 Bob; 1995 Roxanne, Marilyn |
| 4 | 114–135 knots (131–155 mph) | 920–944 mb | 1979 Frederic; 1985 Gloria; 1995 Felix, Luis, Opal |
| 5 | >135 knots (>155 mph) | <920 mb | 1935 No. 2; 1938 No. 4; 1960 Donna; 1961 Carla; 1969 Camille; 1979 David; 1988 Gilbert; 1989 Hugo; 1992 Andrew |

***Physical Structure.*** Fully organized tropical cyclones have an intriguing physical appearance (Figure 8-33). They range in diameter from a compact 160 km (100 mi), to 1000 km (600 mi), to some western Pacific typhoons that attain 1300–1600 km (800–1000 mi). Vertically, these storms dominate the full height of the troposphere.

The inward-spiraling clouds form dense rain bands, with a central area designated the *eye*. Around the eye swirls a thunderstorm cloud called the *eyewall*, which is the area of most intense precipitation. The eye remains an enigma, for in the midst of devastating winds and torrential rains the eye is quiet and warm with even a glimpse of blue sky or stars possible. The structure of the rain bands, central eye, and eyewall is clearly visible for Hurricane Gilbert in Figure 8-33a (top view), 8-33b (oblique view from an artist's perspective), and 8-33c (side view).

The entire storm lumbers along at 16–40 kmph (10–25 mph). When it makes **landfall** (moves ashore), it pushes dangerous **storm surges** of seawater inland, often rising several meters. Storm surges often catch people by surprise and cause the majority of hurricane drownings.

The strongest winds of a tropical cyclone are usually recorded in its right-front quadrant (relative to the storm's directional path), where dozens of fully developed tornadoes may be embedded at the time of landfall. For example, Hurricane Camille in 1969 had up to 100 tornadoes embedded in that quadrant.

For the Western Hemisphere, Hurricane Gilbert (September 1988) attained the record size (1600 km, or 1000 mi, in diameter) and record low barometric pressure (888 mb, or 26.22 in., of mercury). Gilbert had sustained winds of 298 kmph (185 mph) with peaks exceeding 320 kmph (200 mph).

# Focus Study 8-1

## Forecasting the 1995 Atlantic Hurricane Season

As Atlantic hurricane seasons go, 1995 was the second-most active since record keeping began in 1870. Of 19 tropical storms, 11 became hurricanes, and five of these achieved category 3 status or higher. Only 1933 (21 tropical storms), and 1969 (12 hurricanes), exceeded this total. During the Atlantic hurricane season from June 1 to November 30, the year 1995 experienced tropical storms 52% of the season and hurricanes 27% of the time—both records.

The 1995 season killed 121 people (36 in the United States) and damaged an estimated $7.7 billion worth of property ($5 billion in the United States,

matching the typical annual average for property damage). Yet this is only one-fourth the damage wrought by Hurricane Andrew in 1992.

Tropical storms Dean and Jerry made landfall in the United States, killing 6 people and causing $21.5 million in damage. Hurricanes Allison, Erin, and Opal together killed 66 and damaged $3.7 billion in U.S. property.

Figure 1 maps all 19 tropical storms, including 11 hurricanes (colored tracks). The season began with Allison on June 3–11 and ended with Tanya from October 27 through November 1. Each storm went through stages of tropical depression,

tropical storm, hurricane, and extratropical depression. These four stages are based on wind speed (depression, storm, hurricane) and location of the low-pressure center (tropical or extratropical). Top wind speed and lowest central pressure are noted for each hurricane on the map.

Figure 2 is a satellite image from August 30 that shows tropical storm Karen and Hurricanes Humberto, Iris, and Luis, and the remnants of tropical storm Jerry off the image to the far left. In addition, the Space Shuttle *Endeavour* captured a photo of Luis 1000 km (600 mi) east of the Carolina coast, near Bermuda (Figure 3).

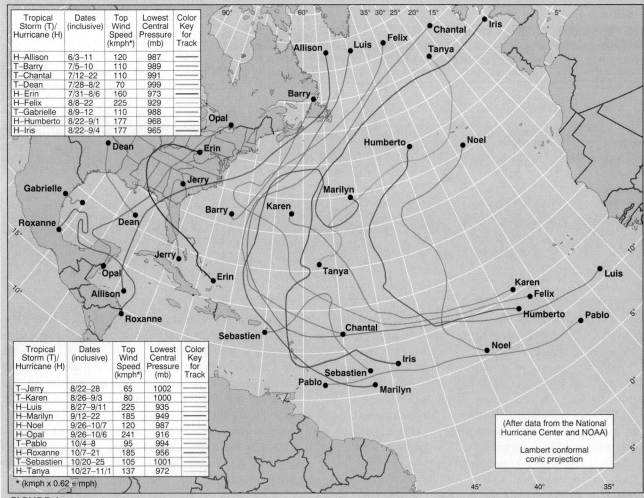

| Tropical Storm (T)/ Hurricane (H) | Dates (inclusive) | Top Wind Speed (kmph*) | Lowest Central Pressure (mb) | Color Key for Track |
|---|---|---|---|---|
| H–Allison | 6/3–11 | 120 | 987 | |
| T–Barry | 7/5–10 | 110 | 989 | |
| T–Chantal | 7/12–22 | 110 | 991 | |
| T–Dean | 7/28–8/2 | 70 | 999 | |
| H–Erin | 7/31–8/6 | 160 | 973 | |
| H–Felix | 8/8–22 | 225 | 929 | |
| T–Gabrielle | 8/9–12 | 110 | 988 | |
| H–Humberto | 8/22–9/1 | 177 | 968 | |
| H–Iris | 8/22–9/4 | 177 | 965 | |

| Tropical Storm (T)/ Hurricane (H) | Dates (inclusive) | Top Wind Speed (kmph*) | Lowest Central Pressure (mb) | Color Key for Track |
|---|---|---|---|---|
| T–Jerry | 8/22–28 | 65 | 1002 | |
| T–Karen | 8/26–9/3 | 80 | 1000 | |
| H–Luis | 8/27–9/11 | 225 | 935 | |
| H–Marilyn | 9/12–22 | 185 | 949 | |
| H–Noel | 9/26–10/7 | 120 | 987 | |
| H–Opal | 9/26–10/6 | 241 | 916 | |
| T–Pablo | 10/4–8 | 95 | 994 | |
| H–Roxanne | 10/7–21 | 185 | 956 | |
| T–Sebastien | 10/20–25 | 105 | 1001 | |
| H–Tanya | 10/27–11/1 | 137 | 972 | |

* (kmph x 0.62 = mph)

(After data from the National Hurricane Center and NOAA)

Lambert conformal conic projection

**FIGURE 1**

**1995 Atlantic storm season.**

Atlantic, Caribbean, and Gulf of Mexico hurricane season, 1995. Noted on the map are inclusive dates from each storm's birth as a tropical depression to dissipation, highest wind speed in kmph (kmph × 0.62 = mph), and lowest pressure (mb). [Data courtesy of NOAA and the National Hurricane Center, Miami.]

**FIGURE 2**
**Satellite observes four storms in the frame.**
Tropical storm Karen and Hurricanes Humberto, Iris, and Luis march westward across the Atlantic Ocean in fairly tight formation during late August 1995. The remnants of tropical storm Jerry can be seen at the far left. [Image from National Weather Service and NOAA, (NESDIS), National Climatic Data Center.]

Several areas were particularly hard hit. The Florida panhandle was crossed by three hurricanes within 160 km (100 mi) of each other in June, July, and September. Also, the eastern Caribbean was savaged, especially the islands of Dominica, Saint Thomas, and Saint Maarten, where Luis and Marilyn hit within a few days of each other in September. As one researcher noted, "maybe lightning does strike the same place twice!" It was a lesson to those who ignored warnings to evacuate.

Forecasters at the National Hurricane Center in Miami were busy throughout the season. An important innovation was a new forecasting method developed by a team at Colorado State University led by William Gray. Their analysis of weather records for the past 92 years disclosed a significant causal relationship between the timing and severity of the western African rainy season, several other variables, and the intensity of hurricanes that make landfall along the U.S. Gulf and East Coasts. The six forecast factors used by the Colorado scientists were:

1. The presence or absence of warm water off the coast of Peru. Its presence causes the El Niño phenomenon discussed in Chapter 10. A strong El Niño in the Pacific Ocean tends to dampen the development of Atlantic hurricanes. During the 1995 season, the Pacific El Niño had weakened after three strong years.

2. Rainfall patterns in West Africa, specifically the Sahel along the southern margins of the Sahara (10°–20° N). Heavy rainfall stimulates storm formation off the Africa coast.

3. Temperature and pressure patterns in the Sahelian region from February through May. Instability favors storm formation off the coast.

4. Directional flows of equatorial stratospheric winds (20,000 to 23,000 m altitude, 68,000 to 75,000 ft), which tend to shift and reverse roughly every 12 to 16 months. Winds from the west tend to double storm probability and intensity.

5. The state of tropospheric winds below 12,000 m altitude (40,000 ft). Stronger than normal westerly winds reduce hurricane activity; weaker winds enhance activity.

6. Sea-surface temperatures in the Atlantic Ocean between Africa and the Caribbean. Warmer surface waters favor tropical storm development.

As this was being written in early 1996, the forecast for the 1996 Atlantic tropical storm season was for 85% of expected Net Tropical Cyclone Activity, which translates to eight named storms, including five hurricanes. For comparison, tropical storm activity for 1995 was 235% of an average season—19 named storms, 11 hurricanes! 1996 was predicted to have 20 hurricane days within an overall 40 tropical storm days (1995 experienced 62 hurricane days out of 121 storm days). The Colorado State team will update this forecast in mid-1996 and again at year's end. (To check the accuracy of this 1996 forecast, see the *Geosystems* Home Page (htpp://www.prenhall.com/geosystm) [Prentice Hall]; where the results are posted in the section on this chapter.)

No matter how good storm forecasts become, coastal and lowland property damage will increase until better hazard zoning and development restrictions are in place. The property insurance industry appears to be taking action to promote these improvements. They are requiring tougher building standards to obtain coverage—or, in some cases, they are refusing to insure property.

**FIGURE 3**
***Endeavour* and Luis.**
Space Shuttle *Endeavour* astronauts photographed Hurricane Luis on September 10, 1995. A portion of the robot arm is visible. [Shuttle astronaut photo courtesy of NASA.]

(a)

**FIGURE 8-33**
**Profile of a hurricane.**
Hurricane Gilbert, September 13, 1988: (a) *GOES-7* satellite image;
(b) a stylized portrait of a mature hurricane, drawn from an oblique
perspective (cutaway view shows the eye, rain bands, and windflow
patterns); (c) *SLAR* (side-looking airborne radar) image from an
aircraft flying through the center of the storm. Rain bands of greater
cloud density are false-colored in yellows and reds. The clear sky in
the central eye is dramatically portrayed. [(a) (b) (c) NOAA and the
National Hurricane Center, Miami.]

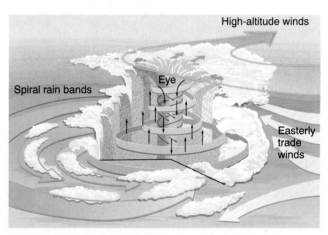

(b)

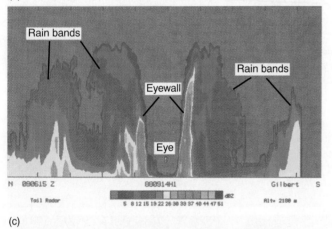

(c)

***Hurricane Andrew–1992.*** The period between August
24 and September 11, 1992, displayed nature's power and
capacity to cause personal, societal, and economic
tragedy. Those 19 days saw Hurricane Andrew (Florida
and Louisiana), Hurricane Iniki (Kauai, Hawaii), and
Typhoon Omar (Guam) strike with record-breaking fury.
Figure 8-34 presents three AVHRR images from August
24, 25, and 26, showing Andrew as it moved across
Florida, westward across the Gulf of Mexico, and onshore
in Louisiana. The central eye is clearly visible, as are the
tight rain bands marking the rapid inward-spiraling, counterclockwise wind flow.

Andrew caused the third-greatest dollar loss from any
natural disaster in U.S. history as it swept across Florida
and Louisiana—property damage exceeded $20 billion.
(Andrew is topped in dollar loss only by the 1993 Midwest
floods and the 1994 Northridge earthquake in Los Angeles, which exceeded $30 billion each.) Sustained winds
were 225 kmph (140 mph), with gusts to 282+ kmph
(175+ mph). Studies recently completed by meteorologist
Theodore Fujita estimated that winds in the eyewall
reached 320 kmph (200 mph) in small vortexes. (See
the Chapter 6 opening photograph for a sense of this
wind devastation.)

The tragedy from Andrew is that the storm
destroyed or seriously damaged 70,000 homes and left
200,000 people homeless between Miami and the
Florida Keys. A year later, 60,000 people still were
homeless and reconstruction was progressing slowly.
Approximately 8% of the agricultural industry in
Florida's Dade County (Miami region) was destroyed
outright—exceeding $1 billion in lost sales.

The remnants of Hurricane Andrew continued northward, reaching Québec and eastern Canada by August
28. Locally, heavy rains were produced as the remnants
of Andrew interacted with weather fronts moving through
the area. In Québec, many local 24-hour precipitation
records were broken.

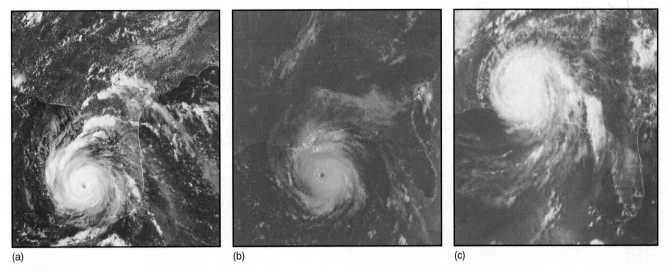

(a)                                    (b)                                    (c)

**FIGURE 8-34**
**Hurricane Andrew.**
*NOAA-10* and *11* AVHRR images of Hurricane Andrew, 1992: (a) August 24 (4:41 P.M. EDT), (b) August 25 (9:28 A.M. EDT), (c) August 26 (4:17 P.M. EDT). Hurricane Andrew moved westward from Florida, recharged itself with plentiful water vapor and latent heat energy as it crossed the warm Gulf of Mexico, and went ashore in Louisiana. Note the time of day and angle of sunlight as it highlights the features of the hurricane. [Images from NOAA as supplied by the EROS Data Center.]

The Everglades, the coral reefs north of Key Largo, some 10,000 acres of mangrove wetlands, and coastal southern pine forests all took significant hits from Andrew. Natural recovery will take years, which offers an opportunity for study. Initial scientific assessments judge the Everglades to be quite resilient because the region's ecosystems have evolved naturally with periodic hurricanes over the millennia. An important fact is that storms damage human structures more than they damage natural systems. Remember this perspective: *Urbanization, agriculture, and water diversion pose a greater ongoing threat to the Everglades than do hurricanes!*

New buildings, apartments, and government offices are opening right next to still-visible rubble and bare foundation pads. Unfortunately, careful hazard planning to guide the settlement of these high-risk areas has never been policy. Thoughtful hazard planning is rarely practiced by public or private decision makers, whether along coastal lowlands, river floodplains, or earthquake fault zones.

The consequence is that our entire society bears the financial cost of planning failure, whether in Florida, California, or the Midwest, not to mention those victims who directly shoulder the physical, emotional, and economic hardship of the event. This recurrent, yet avoidable cycle—construction, devastation, reconstruction, devastation—was reinforced ironically in *Fortune* magazine over 25 years ago after the destruction of Hurricane Camille:

> Before long the beachfront is expected to bristle with new motels, apartments, houses, condominiums, and office buildings. Gulf Coast businessmen, incurably optimistic, doubt there will ever be another hurricane like Camille, and even if there is, they vow, the Gulf Coast will rebuild bigger and better after that.*

---

*Fortune*, October 1969, p. 62.

# Summary and Review—Weather

✔ *Identify* **the requirements for cloud formation and** *explain* **the major cloud classes and types.**

**Weather** is the short-term condition of the atmosphere; **meteorology** is the scientific study of the atmosphere. The spatial implications of atmospheric phenomena and their relationship to human activities strongly link meteorology to physical geography. Analyzing and understanding patterns of wind, air pressure, temperature, and moisture conditions portrayed on daily weather maps is key to numerical (computer) weather forecasting.

A **cloud** is an aggregation of tiny moisture droplets and ice crystals suspended in the air. **Fog** is a cloud that occurs at ground level. Clouds are a constant reminder of the powerful heat-exchange system in the environment. **Moisture droplets** in a cloud form when saturated air and the presence of **cloud-condensation nuclei** lead to *condensation*. Raindrops are formed from moisture droplets through either the **collision-coalescence process** or the **Bergeron ice-crystal process**.

Low clouds, ranging from the surface up to 2000 m (6500 ft) in the middle latitudes, are called **stratus** (flat clouds, in layers) or **cumulus** (puffy clouds, in heaps). When stratus clouds yield precipitation, they are called **nimbostratus**. Sometimes near the end of the day, lumpy, grayish, low-level clouds called **stratocumulus** may fill the sky in patches. Middle-level clouds are denoted by the prefix *alto-*. **Altocumulus** clouds, in particular, represent a broad category that occurs in many different styles. Clouds at high altitude, principally composed of ice crystals, are called **cirrus**. A cumulus cloud can develop into a towering giant called **cumulonimbus** (*-nimbus* denotes precipitation). Such clouds are called *thunderheads* because of their shape and their associated lightning, thunder, surface wind gusts, updrafts and downdrafts, heavy rain, and hail.

> weather (p. 200)
> meteorology (p. 200)
> cloud (p. 200)
> fog (p. 200)
> moisture droplet (p. 201)
> cloud-condensation nuclei (p. 201)
> collision-coalescence process (p. 201)
> Bergeron ice-crystal process (p. 201)
> stratus (p. 202)
> cumulus (p. 202)
> nimbostratus. (p. 202)
> stratocumulus (p. 202)
> altocumulus (p. 202)
> cirrus (p. 202)
> cumulonimbus (p. 202)

1. Specifically, what is a cloud? Describe the droplets that form a cloud.

2. Explain the condensation process: What are the requirements? What two principal processes are discussed in this chapter?

3. What are the basic forms of clouds? Using Table 8-1, describe how the basic cloud forms vary with altitude.

4. Explain how clouds might be used as indicators of the conditions of the atmosphere. Of expected weather.

---

✔ *Identify* **the basic types of fog and** *explain* **the conditions that lead to their formation.**

**Advection fog** forms when air in one place migrates to another place where conditions exist that can cause saturation—for example, when warm, moist air moves over cooler ocean currents. Another type of advection fog forms when cold air flows over the warm water of a lake, ocean surface, or swimming pool. An **evaporation fog**, or steam fog, may form as the water molecules evaporate from the water surface into the cold overlying air. **Upslope fog** is produced when moist air is forced to higher elevations along a hill or mountain. Another fog caused by topography is **valley fog**, formed because cool, denser air settles in low-lying areas, producing fog in the chilled, saturated layer near the ground. **Radiation fog** is caused by radiative cooling of a surface that chills the air layer directly above the surface to the dew-point temperature, creating saturated conditions and fog.

> advection fog (p. 203)
> evaporation fog (p. 203)
> upslope fog (p. 203)
> valley fog (p. 203)
> radiation fog (p. 207)

5. What type of cloud is fog? List and define the principal types of fog.

6. Describe the occurrence of fog in the United States and Canada. Where are the regions of highest incidence?

---

✔ *Describe* **air masses that affect North America and** *relate* **their qualities to source regions.**

Specific conditions of humidity, stability, and cloud coverage occur in regional, homogenous **air masses**. The longer an air mass remains stationary over a region, the more definite its physical attributes become. The homogeneity of temperature and humidity in an air mass sometimes extends through the lower half of the troposphere. Air masses are categorized by their moisture content—**m** for maritime (wetter) and **c** for continental (drier)—and their temperature (a function of latitude)—designated **A** (arctic), **P** (polar), **T** (tropical), **E** (equatorial), and **AA** (antarctic).

air mass (p. 208)

7. How does a source region influence the type of air mass that forms over it? Give specific examples of each basic classification.

8. Of all the air masses, which are of greatest significance to the United States and Canada? What happens to them as they migrate to locations different from their source regions? Give an example of air-mass modification.

---

✔ *Identify* types of atmospheric lifting mechanisms and *describe* four principal examples.

Air masses can be lifted by **convergent lifting** (air flows conflict, forcing some of the air to lift), **convectional lifting** (air passing over warm surfaces gains buoyancy), **orographic lifting** (passage over a topographic barrier), and *frontal lifting*. Conflicting air masses at a front produce a **cold front** (and sometimes a zone of strong wind and rain called a **squall line**) or a **warm front**. Orographic lifting creates wetter windward slopes and drier leeward slopes that fall in the **rain shadow** of the mountain. In North America, **chinook winds** (called *föhn* or *foehn* winds in Europe) are the warm, downslope air flows characteristic of the leeward side of mountains.

A **midlatitude cyclone**, or **wave cyclone**, is a vast low-pressure system that migrates across the continent, pulling air masses into conflict along fronts. These systems are guided by the jet streams of the upper troposphere along seasonally shifting **storm tracks. Cyclogenesis**, the birth of the low-pressure circulation, can occur off the west coast of North America, along the polar front, along the lee slopes of the Rockies, in the Gulf of Mexico, and along the East Coast. A midlatitude cyclone can be thought of as having a life cycle of birth, maturity, old age, and dissolution. An **occluded front** is produced when a cold front overtakes a warm front in the maturing cyclone. Sometimes a **stationary front** develops between conflicting air masses, where air flow is parallel to the front on both sides.

convergent lifting (p. 210)
convectional lifting (p. 211)

orographic lifting (p. 212)
cold front (p. 213)
squall line (p. 215)
warm front (p. 213)
rain shadow (p. 213)
chinook winds (p. 213)
midlatitude cyclone (p. 216)
wave cyclone (p. 216)
storm track (p. 218)
cyclogenesis (p. 217)
occluded front (p. 218)
stationary front (p. 218)

9. Explain why it is necessary for an air mass to rise if there is to be precipitation.

10. When an air mass passes across a mountain range, many things happen to it. Describe each aspect of a mountain crossing by a moist air mass. What is the pattern of precipitation that results?

11. What are the four principal lifting mechanisms that cause air masses to ascend, cool, condense, form clouds, and perhaps produce precipitation? Briefly describe each.

12. Explain how the distribution of precipitation in the state of Washington is influenced by the principles of orographic lifting.

13. Differentiate between a cold front and a warm front as types of frontal lifting.

14. How does a midlatitude cyclone act as a catalyst for conflict between air masses?

15. What is meant by cyclogenesis? In what areas does it occur and why? What is the role of upper-tropospheric circulation in the formation of a surface low?

16. Diagram a midlatitude cyclonic storm during its open stage. Label each of the components in your illustration, and add arrows to indicate wind patterns in the system.

---

✔ *List* the measurable elements that contribute to weather.

*Synoptic analysis* involves the collection of weather data at a specific time, as shown on the three synoptic weather maps. Building a data base of wind, pressure, temperature, and moisture conditions is key to *numerical* (computer-based) *weather prediction* and the development of weather-forecasting models. Preparing a weather report and forecast requires analysis of a daily weather map and satellite images that depict atmospheric conditions. Weather data from these sources include the following:

- Barometric pressure
- Pressure tendency
- Surface air temperature
- Dew-point temperature
- Wind speed and direction
- Type and movement of clouds
- Current weather
- State of the sky
- Precipitation since last observation

17. What is your principal source of weather data, information, and forecasts? Where does your source obtain its data? Have you used the Internet and World Wide Web to obtain weather information?

18. Take a moment to assess your "Weather I.Q." Compare what you knew about the weather and weather forecasts before reading this chapter with what you know now. Listen to some weather forecasts and list terms and concepts that you now understand that you did not understand before. List some questions or topics about which you want to know more. (Such questions about your learning process represent "critical thinking.")

---

✔ **Analyze various types of violent weather and the characteristics of each.**

The violent power of some weather phenomena is displayed in thunderstorms, which produce **lightning** (electrical discharges in the atmosphere), **thunder** (sonic bangs produced by the rapid expansion of air after intense heating by lightning), and **hail** (ice pellets formed within cumulonimbus clouds). A spinning, cyclonic column rising to midtroposphere level—a **mesocyclone**—is sometimes visible as the swirling mass of a cumulonimbus cloud. Dark gray **funnel clouds** pulse from the bottom side of the parent cloud. A **tornado** is formed when the funnel connects with Earth's surface. When tornado circulation occurs over water, a **waterspout** forms.

Within tropical air masses, large low-pressure centers can form along easterly wave troughs. Under the right conditions, a **tropical cyclone** is produced. Depending on wind speeds and central pressure, a tropical cyclone can become a **hurricane**, or **typhoon**, when winds exceed 65 knots (74 mph). As forecasting and the public's perception of weather-related hazards have improved, loss of life has decreased, although property damage continues to increase. Great damage occurs to occupied coastal lands when hurricanes make **landfall** and when winds drive ocean water inland in **storm surges**.

lightning (p. 222)
thunder (p. 222)
hail (p. 223)
mesocyclone (p. 224)
funnel cloud (p. 224)

tornado (p. 224)
waterspout (p. 224)
tropical cyclone (p. 226)
hurricane (p. 228)
typhoon (p. 228)
landfall (p. 229)
storm surge (p. 229)

19. What constitutes a thunderstorm? What type of cloud is involved? What type of air mass would you expect in an area of thunderstorms in North America?

20. Lightning and thunder are powerful phenomena in nature. Briefly describe how they develop.

21. Describe the formation process of a mesocyclone. How is this development associated with that of a tornado?

22. Evaluate the pattern of tornado activity in the United States. Where is Tornado Alley? What generalizations can you make about the distribution and timing of tornadoes?

23. What are the different classifications for tropical cyclones? List the various names used worldwide for hurricanes.

24. What factors contributed to the incredible damage cost of Hurricane Andrew? Why have such damage figures increased, whereas loss of life has decreased over the past 30 years?

25. What forecast factors did scientists use to accurately predict the 1995 Atlantic hurricane season?

 **NetWork**

The *Geosystems Home Page* provides on-line resources for this chapter on the World Wide Web. You will find review exercises, specific updates for items in the chapter, suggested readings, and links to interesting related pathways on the Internet (click on the Table of Contents link and select this chapter). *Geosystems* is at: **http://www.prenhall.com/geosystm**

# 9
# Water Resources

**The Hydrologic Cycle**

**Soil-Water Budget Concept**

**Groundwater Resources**

**Surface Water Resources: Streams**

**Our Water Supply**

**Summary and Review**

### Key Learning Concepts

After reading the chapter, you should be able to:

- *Illustrate* the hydrologic cycle with a simple sketch and *label* it with definitions for each water pathway.

- *Relate* the importance of the water-budget concept to your understanding of the hydrologic cycle, water resources, and soil moisture for a specific location.

- *Construct* the water-balance equation as a way of accounting for the expenditures of water supply and *define* each of the components in the equation and their specific operation.

- *Describe* the nature of groundwater and *define* the elements of the groundwater environment.

- *Define* stream discharge using specific rivers as examples and *explain* the concept of an exotic stream using the Nile and Colorado Rivers.

- *Identify* critical aspects of freshwater supplies for the future and *cite* specific issues related to sectors of use, regions and countries, and potential remedies for any shortfalls.

*Spillway from Oroville Dam, part of the California Water Project, at full discharge.* [Photo by author.]

Our lives are bathed and infused with water. Our own bodies are about 70% water, as are plants and animals. We use water to cook, bathe, wash clothing, and dilute our wastes. We water small gardens and vast agricultural tracts. Most industrial processes would be impossible without water. The physical reality of life is defined by water. It is the essence of our existence and is therefore the most critical resource supplied by Earth systems. Planet Earth is unique in the Solar System as the only planet with water in such quantities. Fortunately, water is a renewable resource, constantly cycling through the environment, endlessly renewed. Even so some 80 countries face impending water shortages; 26 of them face inadequate supplies by the late-1990s.

In the last two chapters, we saw how the exchanges of energy between water and the atmosphere drive Earth's weather systems. The flow of water links the atmosphere, ocean, land, and living things through exchanges of energy and matter. In particular, the energy and moisture exchange between plants and the atmosphere is important to the status of the climate system and the range of responses from plant communities to climate change.

The hydrologic cycle and global water balance give us a model for understanding the global plumbing system, so this important cycle begins the chapter. Water spends time in the ocean, in the air, on the surface, and underground as groundwater. Water availability to plants from precipitation and from the soil is critical to water-resource issues.

We look at the water resource using a water-budget approach—similar in many ways to a money budget—in which we examine water "receipts" and "expenses" at specific locations. Precipitation is the principal receipt of moisture, whereas evaporation and plant transpiration are the principal expenditures. Important to surface-water budgets is water stored as soil moisture and groundwater. Water that soaks into the ground forms a "reservoir" in soil; plants draw on this soil moisture to satisfy their water needs. As you will see, this budget approach can be applied at any scale, from a small garden, to a farm, to a regional landscape.

Streamflow is an important surface-water resource. Water is not always naturally available where and when we want it. Consequently, we rearrange surface-water resources to suit our needs. We drill wells, build cisterns and reservoirs, and dam and divert streams to redirect water either spatially (geographically) or temporally (over time). All of this activity constitutes water-resource management.

This chapter concludes by considering the quantity and quality of the water we withdraw and consume for irrigation, industrial, and municipal uses—our specific water supply. Adequate water supplies in terms of quantity and quality loom as the resource issue for many parts of the world in the next century.

# The Hydrologic Cycle

Vast currents of water, water vapor, ice, and energy are flowing about us continuously in an elaborate, open global plumbing system. Together they form the **hydrologic cycle**, which has operated for billions of years, from the lower atmosphere to several kilometers beneath Earth's surface. The cycle involves the circulation and transformation of water throughout Earth's atmosphere, hydrosphere, lithosphere, and biosphere.

Modern study of the hydrologic cycle involves computer modeling, direct observation, and remote sensing. A better understanding of the hydrologic cycle is central to understanding water resources and global climate change. Such systems as the hydrologic cycle operate in a chaotic manner, making model-building difficult.

## A Hydrologic Cycle Model

Figure 9-1 is a simplified model of this complex system. Let's use the ocean as a starting point for our discussion, although we could jump into the model at any point. More than 97% of Earth's water is in the ocean, and here most evaporation and precipitation occur. We can trace 86% of all evaporation to the ocean. The other 14% comes from the land, including water moving from the soil into plant roots and passing through their leaves (a process called *transpiration*, described later in this chapter).

In the figure, you can see that, of the ocean's evaporated 86%, 66% combines with 12% advected from the land to produce the 78% of all precipitation that falls back into the ocean. The remaining 20% of moisture evaporated from the ocean, plus 2% of land-derived moisture, produces the 22% of all precipitation that falls over land. Clearly, the bulk of continental precipitation comes from the oceanic portion of the cycle.

Figure 9-2 presents a global water balance. The percentages in Figure 9-1 are indicated by the volume (in $1000 \text{ km}^3$) of water flowing along these pathways as shown in Figure 9-2.

***Surface Water.*** Precipitation that reaches Earth's surface follows two basic pathways: It either flows overland or soaks into the soil. What is called **interception** occurs when precipitation strikes vegetation or other ground cover. Intercepted water that drains across plant leaves and down their stems to the ground is *stem flow* and can be an important moisture route to the ground surface. Precipitation that falls directly to the ground, coupled with that which drips or flows onto the ground from vegetation, constitutes *throughfall*. Water soaks into the subsurface through **infiltration**, or penetration of the soil surface. It further permeates soil or rock through downward movement called **percolation**. These concepts are shown in Figure 9-3.

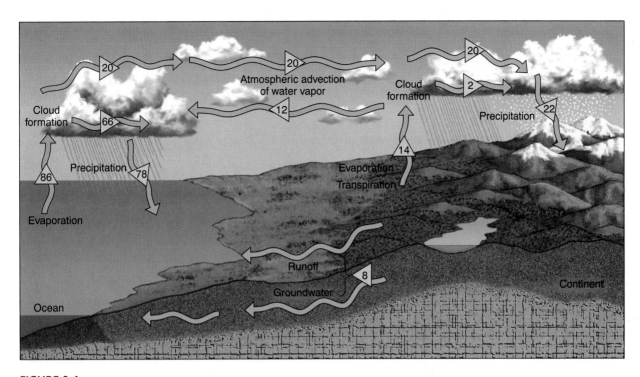

**FIGURE 9-1**

**The hydrologic cycle model.**

The model shows how water travels endlessly through the hydrosphere, atmosphere, lithosphere, and biosphere. The triangles show global average values as percentages. Note that all evaporation (86% + 14% = 100%) equals all precipitation (78% + 22% = 100%), when all of Earth is considered. Locally, various parts of the cycle will vary, creating imbalances.

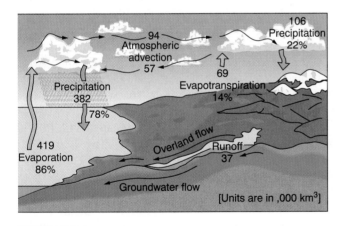

**FIGURE 9-2**

**The global water balance.**

The annual volume of water in all parts of the hydrologic cycle as measured in thousands of cubic kilometers. A balance exists between total evaporation and transpiration and precipitation and between advection in the atmosphere and surface runoff of water (1 km³ × 0.24 = 1 mi³).

The atmospheric advection of water vapor from sea to land and land to sea at the top of Figures 9-1 and 9-2 appears to be unbalanced: 20% (94,000 km³) moving inland but only 12% (57,000 km³) moving out to sea. However, this exchange is balanced by the 8% (37,000 km³) runoff that flows from land to sea. Most of this

runoff—about 95%—comes from surface waters that wash across land as overland flow and streamflow. Only 5% of runoff is slow-moving subsurface groundwater. These percentages indicate that the small amount of water in rivers and streams is very dynamic, whereas the large quantity of subsurface water is sluggish in comparison and represents only a small portion of total runoff.

The residence time for a water molecule in any part of the hydrologic cycle determines its relative importance in affecting Earth's climates. The short time spent by water in transit through the atmosphere (an average of 10 days) is reflected in temporary fluctuations in regional weather patterns. Long residence times, such as the 3000–10,000 years in deep-ocean circulation, groundwater aquifers, and glacial ice, act to moderate temperatures and climates. These slower parts of the cycle work as a "system memory"; the long periods over which heat energy is stored and released buffer the effects of change.

To observe and describe the hydrologic cycle and its related energy budgets, scientists established in 1988 the Global Energy and Water Cycle Experiment (GEWEX). This is part of the World Climate Research Program and represents an important focus in climate-change studies. Now that you are acquainted with the hydrologic cycle, let's examine the concept of the soil-water budget as a method of assessing water resources.

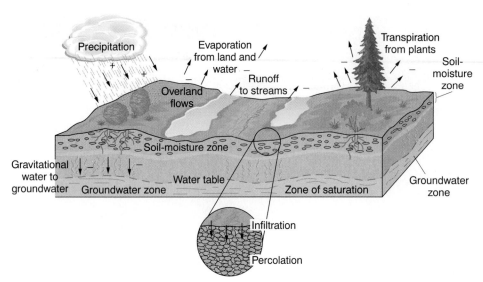

**FIGURE 9-3**
**The soil-moisture environment.**
Precipitation supplies the soil-moisture environment. The principal pathways for water include interception by plants; throughfall to the ground; collection on the surface, forming overland flow to streams; transpiration and evaporation from plants; evaporation from land and water; and gravitational water moving to subsurface groundwater. Water moves from the surface to the soil by infiltration and percolation.

# Soil-Water Budget Concept

A **soil-water budget** can be established for any area of Earth's surface—a continent, country, region, field, or front yard—by measuring the precipitation input and its distribution to satisfy the "demands" of plants, evaporation, and soil moisture storage in the area considered. Such a budget can be constructed for any time frame, from minutes to years.

Think of a soil-water budget as a money budget: precipitation income must be balanced against expenditures of evaporation, transpiration, and runoff. Soil-moisture storage acts as a savings account, accepting deposits and withdrawals of water. Sometimes all expenditure demands are met, and any extra results in a surplus. At other times, precipitation and soil moisture income are inadequate to meet demands, and a deficit, or shortage, results.

Geographer Charles Warren Thornthwaite (1899–1963) pioneered in applied water-resource analysis and worked with others to develop a water-balance methodology. They applied water-balance concepts to geographic problems, especially to irrigation, which requires accurate quantity and timing of water application.

Thornthwaite also developed methods for estimating evaporation and transpiration. He recognized the important relation between water supply and local water demand as an essential climatic element. In fact, his initial use of these techniques was to develop a new climatic classification system.

## *The Soil-Water Balance Equation*

To understand Thornthwaite's water-balance methodology and "accounting" or "bookkeeping" procedures, we must first understand some terms and concepts. Figure 9-3 illustrates the essential aspects of a soil-water

budget. *Precipitation* (mostly rain and snow) provides the moisture input. The object, as with a money budget, is to account for the ways in which this supply is distributed: actual water taken by *evaporation* and plant *transpiration*, extra water that exits in streams and subsurface groundwater, and recharge or utilization of *soil-moisture storage*.

Figure 9-4 organizes the water-balance components into an equation. As in all equations, the two sides must *balance*; that is, the precipitation receipt (left side) must be fully accounted for by expenditures (right side). Follow this water-balance equation as you read the following paragraphs. To help you learn these concepts, the text uses five-letter abbreviations (such as PRECIP) for the components.

***Precipitation (PRECIP) Input.*** The moisture supply to Earth's surface is **precipitation** (PRECIP). It arrives as rain, sleet, snow, and hail. In some climates, PRECIP also is deposited on Earth's surface as dew, frost, and fog. (Recall from the hydrologic cycle that 78% of Earth's PRECIP falls on the ocean, and 22% on land. It is the land portion with which we are concerned here.)

Precipitation is measured with the **rain gauge**. A rain gauge is essentially a large measuring cup, collecting rainfall and snowfall so the water can be measured by depth, weight, or volume (Figure 9-5). Certain conditions affect rain-gauge performance. Wind can cause an undercatch because the drops or snowflakes are not falling vertically. For example, a wind of 37 kmph (23 mph) produces an undercatch as great as 40%, meaning that an actual 1 in. rainfall might gauge at only 0.6 in. Installing windshields above the gauge's opening reduces this error by catching raindrops that arrive at an angle. In addition, when extremely heavy rains hit an area, ordinary buckets or pans may be used to catch the overflow, so it can be measured. According to the World Meteorological Organization, more than 40,000 weather-monitoring stations

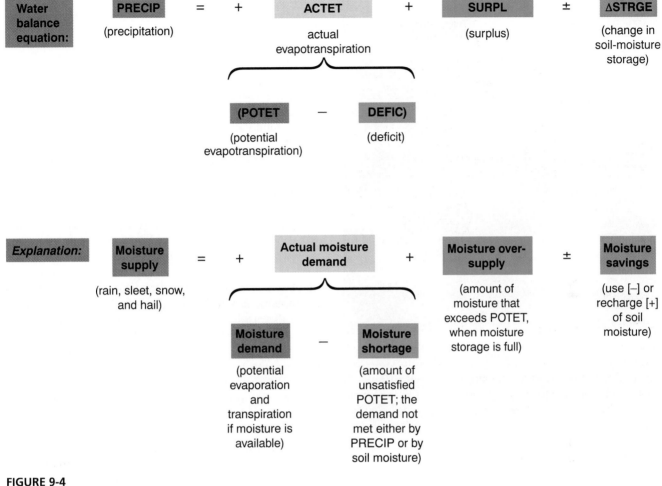

**FIGURE 9-4**
**The water-balance equation explained.**
The outputs (components to the right of the equal sign) are an accounting of expenditures from the precipitation input (to the left).

are operating worldwide, with over 100,000 places measuring precipitation.

Precipitation patterns for the United States and Canada are shown in Figure 9-6. Note the generally wet East and Northwest, and the drier western interiors and extreme north. (Chapter 10 presents a world precipitation map, Figure 10-3.) PRECIP is the principal input to the water-balance equation (Figure 9-4). All the components discussed below are outputs or expenditures of this water.

***Actual Evapotranspiration (ACTET).*** **Evaporation** is the net movement of free water molecules away from a wet surface into air that is less than saturated. **Transpiration** is a cooling mechanism in plants. When a plant transpires, it moves water through small openings (stomata) in the underside of its leaves. The water evaporates, cooling the plant; much as perspiration cools humans. Transpiration is partially controlled by the plants themselves. Control cells around the stomata conserve or release water. Transpired quantities can be significant: On

a hot day, a single tree can transpire hundreds of liters of water; a forest, millions of liters.

Evaporation and transpiration are important water-budget expenditures, and both respond directly to air temperature and humidity. They are reduced when air is cold (can hold less moisture) or when there is high humidity (at or near saturation); both increase when air is hot (can hold more moisture) and dry (not near saturation). Evaporation and transpiration are combined into one term—**evapotranspiration**. (In the hydrologic cycle, Figure 9-1, 14% of evaporation and transpiration occur from land and plants). Now, let's examine ways to estimate evapotranspiration rates.

***Potential Evapotranspiration (POTET).*** Evapotranspiration is an actual expenditure of water. In contrast, **potential evapotranspiration** (POTET) is the amount of water that *would* evaporate and transpire under optimum moisture conditions (adequate precipitation and adequate soil-moisture supply).

**FIGURE 9-5**
**A rain gauge.**
A standard rain gauge is cylindrical, 20.3 cm (8.0 in.) in diameter and 58 cm (23 in.) deep. Water is guided by a funnel into a measuring tube. The device is designed to minimize evaporation, which would cause low readings. [Photo by author.]

Filling a bowl with water and letting it evaporate illustrates this concept: When the bowl becomes dry, is there still an evaporation demand? The demand of course remains, regardless of whether the bowl is dry. Similarly, the amount of water that would evaporate or transpire if water always were available is the POTET, or the amount that would evaporate from the bowl if it constantly were supplied with water. The amount of ultimate POTET demand that went unmet in the dry bowl is the shortage, or deficit (DEFIC). Note that in the water-balance equation, when we subtract the deficit from the potential evapotranspiration we derive what actually happened—ACTET.

***Determining POTET.*** Although exact measurement is difficult, one method of measuring POTET employs an **evaporation pan**, or *evaporimeter* (Figure 9-7). As evaporation occurs, water in measured amounts is automatically replaced in the pan, equaling the amount that evaporated. Mesh screens over the pan protect against mismeasurement due to wind, which accelerates evaporation.

A more elaborate measurement device is a **lysimeter**. A tank, approximately a cubic meter in size or larger,

is buried in a field with its upper surface left open (Figure 9-8). The lysimeter isolates a representative volume of soil, subsoil, and plant cover so that the moisture moving through the sampled area can be measured. A *weighing lysimeter* has this embedded tank resting on a weighing scale. A rain gauge next to the lysimeter measures the precipitation input. The various pathways of this water are then traced: Some water remains as soil moisture, some is incorporated into plant tissues, some drains from the bottom of the lysimeter, and the remainder is credited to evapotranspiration. Thus, given natural conditions, the lysimeter measures actual evapotranspiration. If irrigation supplies adequate water, and optimal soil moisture is available to the plants, water loss will be equivalent to POTET.

Lysimeters and evaporation pans are limited somewhat in availability across North America, but they still provide a data base from which to develop ways of *estimating* POTET. Several methods of estimating POTET on the basis of meteorological data are widely used and easily implemented for regional applications. One of these methods was developed by Thornthwaite. He discovered that monthly mean air temperature and daylength could be used to approximate POTET. These data are readily available, so calculating POTET is easy and fairly accurate for most midlatitude locations. (Recall that daylength is a function of a station's latitude.) His method can work with data from hourly, daily, or annual time frames or with historical data, to recreate water conditions in past environments.

Thornthwaite's method works better in some climates than in others. It ignores warm or cool air movements (wind) across a surface, sublimation and surface-water retention at subfreezing temperatures, poor drainage, and frozen soil (*permafrost*, Chapter 17), although there is allowance for snow accumulations. Nevertheless, we use his water-balance method here because of its great utility as a teaching tool and its overall ease of application in geographic studies to analyze water budgets over large areas with acceptable results.

Figure 9-9 presents POTET values derived by the Thornthwaite method for the United States and Canada. Note that higher values occur in the South, with highest readings in the Southwest (higher average air temperatures and lower relative humidities). Lower POTET values are found at higher latitudes and elevations (lower average temperatures).

Compare this POTET (demand) map with the PRECIP (supply) map in Figure 9-6. The relation between the two determines the remaining components of the water-balance equation. From the two maps, can you identify regions where PRECIP is greater than POTET (for example, the eastern United States)? Or where POTET is greater than PRECIP (for example, the southwestern United States)? Where you live, is the water demand usually met by the precipitation supply? Or does your area experience a natural shortage? One way to determine the answer is to note whether people use sprinklers for lawns and

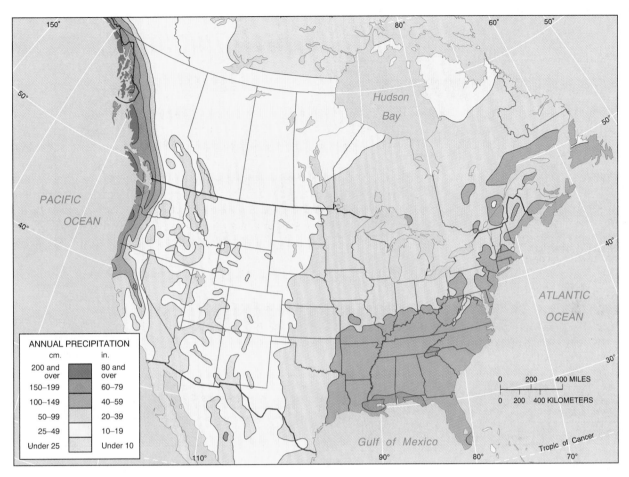

**FIGURE 9-6**
**Precipitation in North America.**
Annual precipitation (water supply, PRECIP) in the United States and Canada. [Adapted from U.S. National Weather Service, U.S. Department of Agriculture, and Environment Canada.]

**FIGURE 9-7**
**Installation of a standard evaporation pan.**
[Photo by author.]

gardens during summer months, and, if they do, whethe the sprinklers are temporary or permanent.

***Deficit (DEFIC).*** The POTET demand can be satisfied in three ways: by PRECIP, by moisture stored in the soil, or through artificial irrigation. If these three sources are

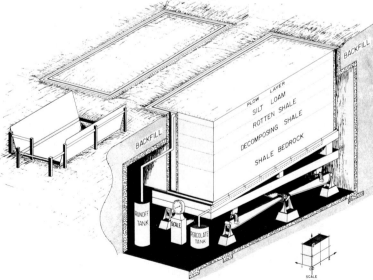

**Lysimeter.**
A weighing lysimeter for measuring evaporation and transpiration. A portion of the soil and its plant cover is actually weighed. [Illustration courtesy of Lloyd Owens, Agricultural Research Service, Coshocton, Ohio.]

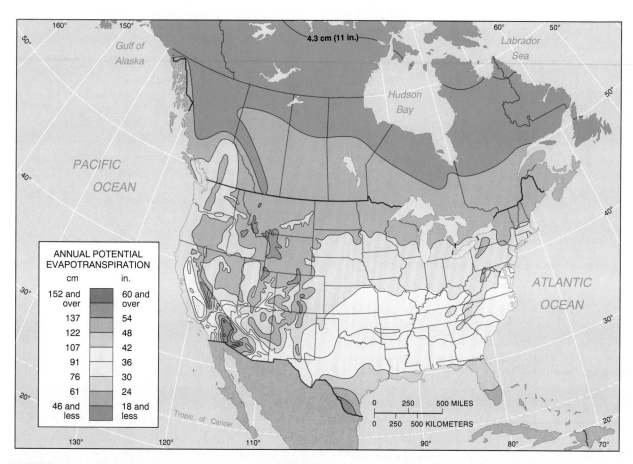

**FIGURE 9-9**
**Potential evapotranspiration (water demand, POTET) for the United States and Canada.**
[From C. W. Thornthwaite, "An approach toward a rational classification of climate," *Geographical Review* 38 (1948): 64.
Adapted by permission from the American Geographical Society. Canadian data adapted from M. Sanderson, "The climates
of Canada according to the new Thornthwaite classification," *Scientific Agriculture* 28 (1948): 501–17.]

inadequate to meet ultimate demand, the location experiences a moisture shortage. This unsatisfied POTET is **deficit** (DEFIC). By subtracting DEFIC from POTET, we determine the **actual evapotranspiration**, or ACTET, that takes place. Under ideal conditions, POTET and ACTET are about the same, so plants do not experience a water shortage; droughts result from deficit conditions.

***Surplus (SURPL).*** If POTET is satisfied and the soil is full of moisture, then additional water input becomes **surplus** (SURPL). This excess water might sit on the surface in puddles, ponds, and lakes, or it might flow across the surface toward stream channels or percolate through the soil underground. The **overland flow** combines with precipitation and groundwater flows into river channels to make up the **total runoff**. Because streamflow, or runoff, is generated mostly from surplus water, the water-balance approach is useful for indirectly estimating streamflow.

***Soil-Moisture Storage (ΔSTRGE).*** A "savings account" of water that receives recharge "deposits" and provides for "withdrawals" is **soil-moisture storage** (ΔSTRGE). This is the volume of water stored in the soil that is accessible to plant roots. The delta symbol Δ means that this

component includes both recharge and utilization (use) of soil moisture; the Δ in math means "change." Soil moisture comprises two categories of water—hygroscopic and capillary—but only capillary water is accessible to plants (Figure 9-10).

**Hygroscopic water** is inaccessible to plants because it is a molecule-thin layer that is tightly bound to each soil particle by the hydrogen bonding of water molecules (Figure 9-10, left). Hygroscopic water exists even in the desert, but it is unavailable to meet POTET demands. Soil is at the **wilting point** when all that remains is this unextractable water; plants wilt and eventually die after a prolonged period of such moisture stress.

**Capillary water** is generally accessible to plant roots because it is held against the pull of gravity in the soil by hydrogen bonds between water molecules (surface tension) and by hydrogen bonding between water molecules and the soil. Almost all capillary water that remains in the soil is **available water** in soil-moisture storage. After some water drains from the larger pore spaces, the amount of available water remaining for plants is termed **field capacity**, or storage capacity. It is removable to meet POTET demands through the action of plant roots and surface evaporation

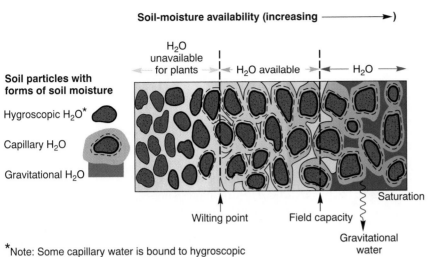

Soil-moisture availability (increasing ⟶)

H$_2$O unavailable for plants ⟷ H$_2$O available ⟷ H$_2$O ⟶

**Soil particles with forms of soil moisture**

Hygroscopic H$_2$O*

Capillary H$_2$O

Gravitational H$_2$O

Saturation

Wilting point     Field capacity     Gravitational water

*Note: Some capillary water is bound to hygroscopic water on soil particle and is also unavailable.

**FIGURE 9-10**
**Types of soil moisture.**

Hygroscopic and gravitational water are unavailable to plants; only capillary water is available. [After D. Steila, *The Geography of Soils*, © 1976, p. 45. Reprinted by permission of Prentice Hall, Inc., Englewood Cliffs, N.J.]

(Figure 9-10, center). Field capacity is specific to each soil type, and the amount can be determined by soil surveys.

When soil becomes saturated after a precipitation event, any water surplus becomes **gravitational water**. It percolates from the shallower capillary zone to the deeper groundwater zone (Figure 9-10, right).

Figure 9-11 shows the relation of soil texture to soil-moisture content. Different plant species send roots to different depths and therefore are exposed to different amounts of soil moisture. For example, shallow-rooted spinach, beans, and carrots send roots down about 65 cm (25 in.) in a silt-loam soil. Deep-rooted alfalfa and shrubs exceed 125 cm (50 in.) depth in such a soil. A soil blend that maximizes available water is best for plants. On the basis of Figure 9-11, can you determine the soil texture with the greatest quantity of available water?

Plant roots exert a tensional force on soil moisture to absorb it. As water transpires through the leaves, a negative pressure gradient is established throughout the plant, thus air pressure pushes more water into the plant roots and through the plant from the soil; the plant acts

like a soda straw. Plants obtain water with the least effort and transpiration is at optimal rates when soil moisture is at field capacity.

As **soil-moisture utilization** removes soil water, the plants must exert greater effort to extract the same amount of moisture. As a result, even though a small amount of water may remain in the soil, plants may be unable to exert enough pressure to use it. The *unsatisfied demand* that results is a *deficit*. Avoiding a deficit and reducing plant growth inefficiencies are the goals of irrigation, for the harder plants must work to get water, the less their yield and growth will be.

Whether by natural precipitation or artificial irrigation, water infiltrates the soil and replenishes available water, a process called **soil-moisture recharge**. The texture and the structure of the soil dictate available pore spaces, or **porosity**. The property of the soil that determines the rate of soil-moisture recharge is its **permeability**. Permeability depends on particle sizes and the shape and packing of soil grains. It is expressed as a rate of the flow of water into the soil. It is generally slower for clays and faster for sands and

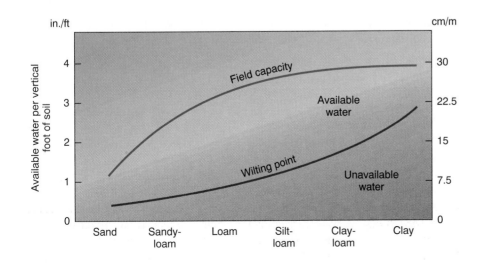

**FIGURE 9-11**
**Soil-moisture availability.**

The relation between soil-moisture availability and soil texture determines the distance between the two curves that show field capacity and wilting point. [After U.S. Department of Agriculture, *1955 Yearbook of Agriculture—Water*, p. 120.]

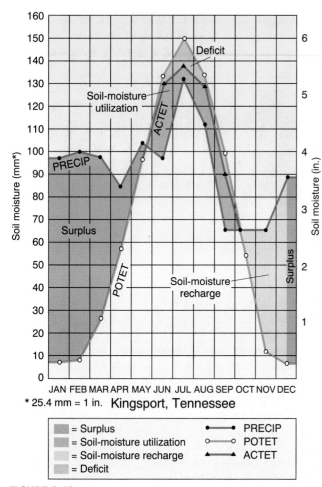

**FIGURE 9-12**

**Sample water budget.**
Annual average water-balance components graphed for Kingsport, Tennessee. The comparison of precipitation inputs and potential evapotranspiration outputs determines the condition of the soil-moisture environment. A typical pattern of spring surplus, summer soil-moisture utilization, a small summer deficit, autumn soil-moisture recharge, and ending surplus highlight the year.

gravels. (Chapter 18 presents more on soil texture, structure, and moisture characteristics.)

Water infiltration is rapid in the first minutes of precipitation and slows as the upper soil layers become saturated, even though the deeper soil is still dry. Agricultural

practices, such as plowing and adding sand or manure to loosen soil structure, can improve both soil permeability and the depth to which moisture can efficiently penetrate to recharge soil-moisture storage.

## Sample Water Budgets

Using all of these concepts, we can graph the water-balance components for several representative cities. Let's begin by looking at Kingsport, in the extreme northeastern corner of Tennessee at 36.6° N 82.5° W, elevation 390 m (1280 ft). (Complete Kingsport data are available on the Geosystems Home Page on Internet.)

Figure 9-12 graphs the water-balance components for the long-term supply and demand, using monthly averages. The monthly values for PRECIP and POTET smooth the actual daily and hourly variability. On the graph, a comparison of PRECIP and POTET by month determines whether there is a *net supply* or a *net demand* for water. There is a net supply (blue line) from October to May, but the warm days from June to September create a net water demand. If we assume a soil-moisture storage capacity of 100 mm (4.0 in.), typical of shallow-rooted plants, the net water-demand months are satisfied through soil-moisture utilization (green area).

Table 9-1 shows the actual data for Kingsport's water-balance equations for the year and for the months of March and September. Here, you can see the interaction of these various components. Check these three equations and perform the functions indicated to see whether we have accounted for all the PRECIP received. Compare March and September in the equations with the same months in the graph in Figure 9-12.

Obviously, not all stations experience the surplus moisture patterns of this humid-continental region. Different climatic regimes experience different relations among water-balance components. Also, medium- and deep-rooted plants produce different results. Figure 9-13 presents water-balance graphs for several other cities in North America and the Caribbean Sea region. Among these examples, compare the summer minimum precipitation of Berkeley, California, with the summer maximum of

## TABLE 9-1

| Annual, March, and September Water Balances for Kingsport, Tennessee. | | | | | | | | | | |
|---|---|---|---|---|---|---|---|---|---|---|
| | *PRECIP* | = | *POTET* | – | *DEFIC* | + | *SURPL* | ± | *Δ STRGE* |
| Annual | 1119 | = | (781 | – | 16 ) | + | 354 | ± | 0 |
| | (44.1) | = | ( 30.7 | – | 0.6) | + | (14.0) | ± | 0 |
| March | 97 | = | ( 24 | – | 0 ) | + | 73 | ± | 0 |
| | (3.8) | = | ( 0.9 | – | 0 ) | + | (2.9) | ± | 0 |
| September | 66 | = | ( 99 | – | 8 ) | + | 0 | – | 25 |
| | (2.6) | = | ( 3.9 | – | 0.3) | + | 0 | – | (1.0) |
| Note: All quantities in millimeters (inches). | | | | | | | | | | |

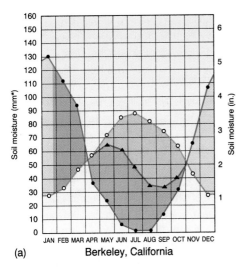

(a) Berkeley, California

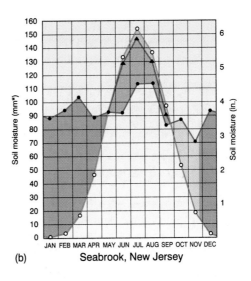

(b) Seabrook, New Jersey

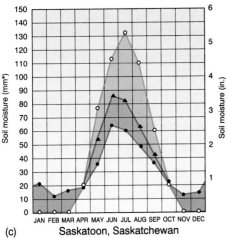

(c) Saskatoon, Saskatchewan

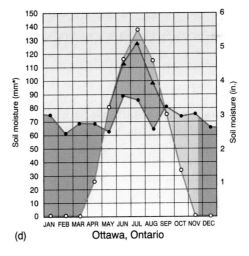

(d) Ottawa, Ontario

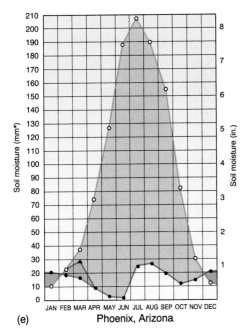

(e) Phoenix, Arizona

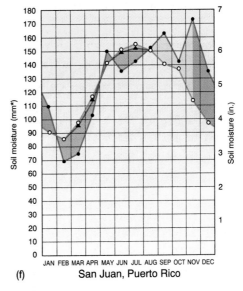

(f) San Juan, Puerto Rico

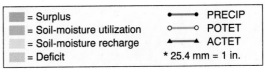

**FIGURE 9-13**

**Sample water budgets for selected stations.**

Sample water-balance regimes for selected stations in North America and the Caribbean region.

Seabrook, New Jersey; the drier prairies of Saskatchewan with the more-humid conditions in Ottawa; and the precipitation receipts of San Juan, Puerto Rico, with the arid desert of Phoenix, Arizona. (The Geosystems Home Page presents PRECIP, POTET, and temperature values, along with other information, for a wide variety of stations worldwide, grouped by climatic classification.)

## Water Budget and Water Resources

Water is distributed unevenly over space and time. Because we require a steady supply, we build large-scale management projects intended to redistribute the water resource either geographically, by moving water from one place to another, or over time, by storing water from time of receipt until it is needed. In this way deficits are reduced, surpluses are held for later release, and water availability is improved to satisfy evapotranspiration and society's demands.

An interesting application of the water-balance approach to water resources is event analysis. Focus Study 9-1 presents Hurricane Camille (1969) and its generally positive effects on regional water resources.

In North America, several large water-management projects already operate: the Bonneville Power Authority in Washington State, the Tennessee Valley Authority in the Southeast, the California Water Project, the Central Arizona Project, and the Churchill Falls and Nelson River Projects of Manitoba. The best sites for multipurpose hydroelectric projects already are taken, so battles among conflicting interests invariably accompany new project proposals.

A particularly ambitious regional water project is the Snowy Mountains Hydroelectric Authority Scheme of Australia, where water-balance variations created the need to relocate water. In the Snowy Mountains, part of the Great Dividing Range in extreme southeastern Australia, precipitation ranges from 100 to 200 cm (40 to 80 in.) a year, whereas interior Australia receives under 50 cm (20 in.) and less than 25 cm (10 in.) farther inland (see the world precipitation map in Figure 10-3). POTET values are greater and PRECIP lower throughout the Australian interior, creating water deficits, compared with conditions in higher elevations of the Snowy Mountains that produce water surpluses.

The Snowy Mountains Scheme was begun in the 1950s and completed more than 20 years later. The plan was to take surplus water that flowed down the Snowy River eastward to the Tasman Sea and reverse the flow to support newly irrigated farmland in the interior of New South Wales and Victoria. Engineering know-how literally rearranged regional water budgets.

Using some of the longest tunnels ever built, vast pumping systems, numerous reservoirs, and power plants, this plan was completed and even expanded to include many tributary streams and drainage systems high in the mountains. The westward-flowing rivers—the Murray, Tumut, and Murrumbidgee—now have increased flow from the diverted rivers, and, as a result, new acreage is now in production in what once was dry outback. These interior lands were formerly served only by wells that drew upon meager groundwater resources. Today, over 2 million acre-feet of irrigation water is pumping into these farmlands from the Snowy Mountains (Figure 9-14).

**FIGURE 9-14**
**The Snowy Mountains Hydroelectric Authority Scheme, Australia.**
Snowy Mountains irrigation water makes possible the opening of new lands in Australia's eastern interior. [Photo by Snowy Mountains Hydroelectric Authority.]

## Focus Study 9-1

## Hurricane Camille, 1969: Water-Balance Analysis Points to Moisture Benefits

Hurricane Camille was one of the most devastating hurricanes of this century. Ironically, it was significant not only for the disaster it brought (256 dead, $1.5 billion damage) but also for the drought it abated.

### Gift from the Storm

Figure 1 shows Camille making landfall on the Gulf Coast west of Biloxi, Mississippi, and then moving north. Hurricane-force winds sharply diminished after landfall, leaving a vast rainstorm that traveled from the Gulf Coast through Mississippi, western Tennessee, Kentucky, and into central Virginia (Figure 2). Severe flooding drowned the Gulf Coast near landfall and the James River basin of Virginia, where torrential rains produced record floods. But, Camille actually had beneficial aspects; it ended a year-long drought along major portions of its track.

According to the weather records of the past 40 years, about one-third of all hurricanes making landfall in the United States have provided beneficial precipitation to local water budgets. Figure 2a maps the precipitation from Camille

(moisture supply) and portrays the storm's track from landfall in Mississippi to the coast of Virginia and Delaware. By comparing actual water budgets along this track with hypothetical water budgets where Camille's rainfall was artificially removed, I was able to analyze the moisture impact of the storm on the other parts of the budget. Figure 2b maps the moisture shortages that were avoided because of Camille's rains—termed "deficit abatement." Over vast portions of the affected area, Camille reduced dry-soil conditions, restored pastures, and filled low reservoirs.

Thus, Camille's monetary benefits inland outweighed its damage by an estimated 2 to 1 ratio. (Of course, the tragic loss of life does not fit into a financial equation.) Hurricanes should be viewed as normal and natural meteorological events that have terrible destructive potential but also contribute to the precipitation regimes of the southern and eastern United States.

### Common Sense Questions

Despite the damage from hurricanes, people continue construction and

development on vulnerable coastal lowlands. Barrier islands from Texas to New York are loaded with homes and businesses. This helps explain the astonishing $20 billion in damage caused by Hurricane Andrew in 1992. As pointed out in Chapter 8, deaths from tropical storms have declined in the past 50 years even as property damage has continued to increase. Insurance companies appear to be taking notice; they are refusing to underwrite coverage for coastal areas that are repeatedly destroyed. These regions will become more and more vulnerable as global sea level continues to rise through the next century.

A geographic information system (GIS) analysis of combined property damage for the past 30 years discloses areas that require zoning restrictions to reduce future damage. Such planning toward *damage avoidance* seems to be a matter of common sense. Yet we remain far from implementation despite the fact that the entire society bears the cost of failing to act. Experience with the politically difficult task of implementing damage-avoidance strategies is described in Focus Study 16-1.

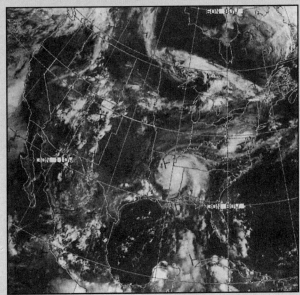

**FIGURE 1**

**Satellite images of Hurricane Camille.**

Hurricane Camille (a) made landfall on the night of August 17, 1969, and (b) continued inland on August 18. These two images were made exactly 24 hours apart. [Photos from NOAA.]

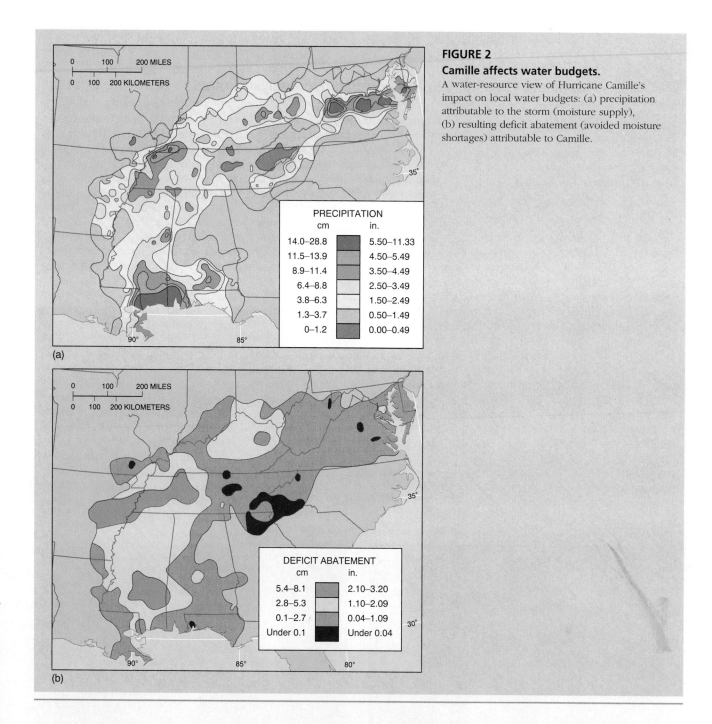

**FIGURE 2**
**Camille affects water budgets.**
A water-resource view of Hurricane Camille's impact on local water budgets: (a) precipitation attributable to the storm (moisture supply), (b) resulting deficit abatement (avoided moisture shortages) attributable to Camille.

# Groundwater Resources

**Groundwater** is an important part of the hydrologic cycle, but it lies beneath the surface, beyond the soil-moisture root zone. It is tied to surface supplies through soil and rock. Groundwater is the largest potential freshwater source in the hydrologic cycle—larger than all surface lakes and streams combined. Between Earth's land surface and a depth of 4 km (13,000 ft) worldwide, some 8,340,000 km³ (2,000,000 mi³) of water resides, a volume comparable to 70 times all the freshwater lakes in the world. Despite this volume and its obvious importance, groundwater is widely abused by pollution and overconsumption in quantities beyond natural replenishment rates.

About 50% of the U.S. population derives a portion of its freshwater from groundwater sources. Between 1950 and 1990, annual groundwater withdrawal increased 160%. In some states, such as Nebraska, groundwater supplies 85% of water needs. In Canada, about 6 million people (two-thirds of them live in rural areas) rely on groundwater for domestic needs—4.2 million m³/day,

1.5 billion m³/year (148.3 million ft³/day, 53 billion ft³/year). Figure 9-15 shows potential groundwater resources in the United States and Canada.

In many ways, groundwater is better than surface water. It is available in many parts of the world that lack dependable runoff. Whereas surface supplies are affected by short-term drought, groundwater is not (although long-term drought affects both). Except in severely polluted areas, groundwater is generally free of sediment, color, and pathogenic (disease) organisms. And chemical composition and temperature remain nearly constant from groundwater sources, for those properties are controlled by subsurface conditions in rock and soil.

## Groundwater Profile and Movement

Figure 9-16 brings together many groundwater phenomena on a single illustration, and it is the basis for the following discussion. Groundwater begins as surplus water, which percolates downward as gravitational water from the zone of capillary water. This excess surface water moves through the **zone of aeration**, where soil and rock are less than saturated (some pore spaces contain air). Eventually, the water reaches an area where subsurface water has accumulated, called the **zone of saturation**. Here, the pores are completely filled with water. Like a hard sponge made of sand, gravel, and rock, the saturation zone stores water in its

**FIGURE 9-15**

**Groundwater resource potential for the United States and Canada.**

Highlighted areas of the United States are underlain by productive aquifers capable of yielding freshwater to wells at 0.2 m³/min or more (for Canada, 0.4 l/s). [Courtesy of Water Resources Council for the United States and the *Inquiry on Federal Water Policy* for Canada.]

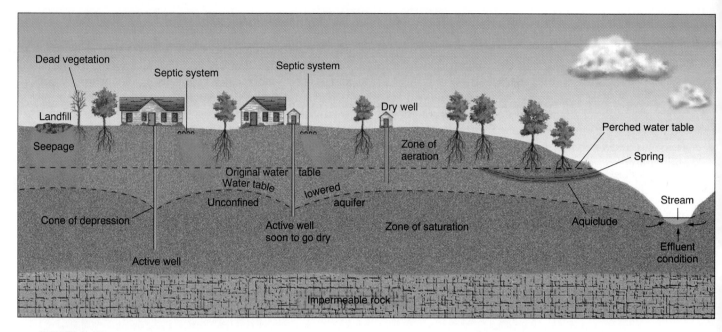

**FIGURE 9-16**
**Groundwater characteristics.**
Subsurface groundwater characteristics, processes, water flows, and human interactions. Begin at the far left and work your way across this integrative illustration.

countless pores and voids. The saturated zone may include the saturated portion of the aquifer and a part of the underlying *aquiclude*, even though the later has such low permeability that its water is inaccessible. (This is the so-called underground reservoir of water that some speak of—not some mystical underground lake, but simply porous rock saturated with water.)

The *porosity* of any rock layer depends on the arrangement, size, and shape of its individual particles, the nature of any cement between them, and their degree of compaction. Subsurface rocks are either *permeable* or *impermeable*. Their permeability depends on whether they conduct water readily (higher permeability) or tend to obstruct its flow (lower permeability). An **aquifer** is a rock layer that is permeable to groundwater flow in *usable* amounts. An **aquiclude** is a body of rock that does not conduct water in usable amounts (also called an aquitard).

The upper limit of the water that collects in the zone of saturation is called the **water table**. It is the contact surface between the zone of saturation and the zone of aeration (all across Figure 9-16). Groundwater movement is controlled by the slope of the water table, which generally follows the contours of the land surface.

## Aquifers, Wells, and Springs

For many people, adequate water supply depends on a good aquifer beneath them that is accessible with a well or exposed on a hillside where water emerges as a spring.

***Confined and Unconfined Aquifers.*** Important to the behavior of an aquifer is whether it is confined or unconfined. A **confined aquifer** is bounded above and below by impermeable layers of rock or sediment (imagine a soaked sponge confined between two wood boards). An **unconfined aquifer** has a permeable layer on top and an impermeable one beneath (imagine a soaked sponge with a newspaper on top and a wood board underneath; see Figure 9-16).

Confined and unconfined aquifers also differ in the size of their recharge area, which is the ground surface where water enters an aquifer to recharge it. For an unconfined aquifer, the **aquifer recharge area** generally extends above the entire aquifer; the water simply percolates down to it. But in a confined aquifer, the recharge area is far more restricted, as you can see in the figure.

Once recharge areas of both confined and unconfined aquifers are identified on the landscape, the prevailing government body should zone them to prohibit pollution discharges, septic and sewage-system installations, or hazardous-material dumping. Of course, an informed population would insist on such action because protection is cheaper than the nearly impossible cleanup or replacement of contaminated water sources. In the illustration, note the improperly located disposal pond on the aquifer recharge area, contaminating wells to the right.

Confined and unconfined aquifers also differ in their water pressure. A well drilled into an unconfined aquifer (Figure 9-16) must be pumped to make the water level rise above the water table. In contrast, the water in a confined aquifer is under the pressure of its own weight, creating a

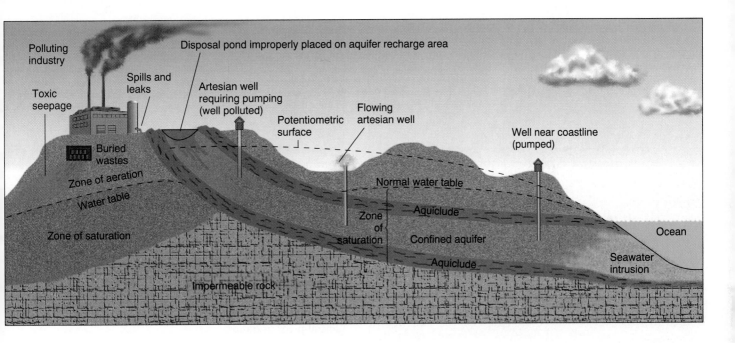

Polluting industry

Toxic seepage

Disposal pond improperly placed on aquifer recharge area

Spills and leaks

Buried wastes

Zone of aeration

Water table

Zone of saturation

Artesian well requiring pumping (well polluted)

Potentiometric surface

Flowing artesian well

Well near coastline (pumped)

Normal water table

Zone of saturation

Aquiclude

Confined aquifer

Aquiclude

Impermeable rock

Ocean

Seawater intrusion

pressure level to which the water can rise on its own, called the **potentiometric surface**.

The potentiometric surface actually can be above ground level (right in figure). Under this condition, **artesian water**, or groundwater confined under pressure, may rise in wells and even flow at the surface without pumping, if the top of the well is lower than the potentiometric surface. (These wells are called artesian, for the Artois area in France where they are common.) In other wells, however, pressure may be inadequate, and the artesian water must be pumped the remaining distance to the surface.

***Wells, Springs, and Streamflows.*** Groundwater movement is controlled by the slope of the water table, which broadly follows the contour of the land surface (Figure 9-16). Groundwater tends to move toward areas of lower pressure and elevation.

Water wells can work only if they penetrate the water table. Too shallow a well will be a "dry well"; too deep a well will punch through the aquifer and into the impermeable layer below, also yielding little water. This is why water wells should be drilled in consultation with a hydrogeologist.

Where the water table intersects the surface, it creates springs (Figure 9-16 left and Figure 9-17). Such an intersection also occurs at lakes and riverbeds. Ultimately, groundwater may enter stream channels to flow as surface water (stream near center in Figure 9-16). In fact during dry periods, the water table may sustain river flows.

Figure 9-18 illustrates the relation between the water table and surface streams in two different climatic settings. In humid climates, where the water table generally supplies a continuous base flow to a stream and is higher than the stream channel, the stream is called *effluent*

because it receives the water flowing out (effluent) from the surrounding ground. The Mississippi River and countless other streams are examples. In drier climates, with lower water tables, water from *influent* streams flows into the adjacent ground, sustaining vegetation along the stream. The Colorado River and the Rio Grande of the American West are examples of influent streams.

## Overuse of Groundwater

As water is pumped from a well, the surrounding water table within an unconfined aquifer may experience **drawdown**, or become lowered. This happens if the pumping rate exceeds the replenishment flow of water into the aquifer, or the horizontal flow around the well. The resultant lowering of the water table around the well is called a **cone of depression** (Figure 9-16, left).

***Overpumping.*** Aquifers frequently are pumped beyond their flow and recharge capacities, a condition known as **groundwater mining**. Today, large tracts experience chronic groundwater overdrafts in the Midwest, West, lower Mississippi Valley, Florida, and the intensely farmed Palouse region of eastern Washington State (Figure 9-19). In many places, the water table or artesian water level has declined more than 12 m (40 ft). In India, approximately 20% of the agricultural districts are mining groundwater beyond recharge rates. In the United States, groundwater mining is of special concern in the great Ogallala aquifer, which is the topic of Focus Study 9-2.

In the Middle East, conditions are even more severe, as detailed in News Report 1. The groundwater resource

**FIGURE 9-17**
**An active spring.**
Springs provide evidence of groundwater and a probable exposed
aquifer near Burney Creek in Northern California. [Photo by author.]

**FIGURE 9-19**
**Central-pivot irrigation.**
Central-pivot irrigation near Las Cruces, New Mexico is used for
growing alfalfa. Each irrigated circle is a quarter section (actual
irrigated area is 51 hectares, 126 acres). [Photo by Comstock.]

beneath Saudi Arabia accumulated over thousands of
years, but, it is not being recharged to any appreciable
degree at present due to the desert climate. Some
researchers have suggested that groundwater in the region
will be depleted by the year 2007, although worsening
water-quality problems will no doubt arise before this
date. Desalinization of seawater to augment diminishing
groundwater supplies is becoming increasingly important
as a freshwater source (Figure 9-20).

***Collapsing Aquifers.*** A possible effect of water
removal from an aquifer is that the aquifer, which is a
layer of rock or sediment, will lose its internal support.

Water in the pore spaces between rock grains is not
compressible, so it adds structural strength to the rock.
If the water is removed through overpumping, air infiltrates the pores. Air is readily compressible, and the
tremendous weight of overlying rock may crush the
aquifer. On the surface, the visible result may be land
subsidence, cracked house foundations, and changes
in drainage. Unfortunately, collapsed aquifers may not
be rechargeable, even if surplus gravitational water
becomes available, because pore spaces may be permanently collapsed.

Houston, Texas, provides an example. Because
groundwater and crude oil were removed, land within

**FIGURE 9-18**
**Groundwater interaction with
streamflow.**
(a) Effluent stream base flow is
partially supplied by a high water
table, characteristic of humid regions.
(b) Influent stream supplies a lower
water table, characteristic of drier
regions.

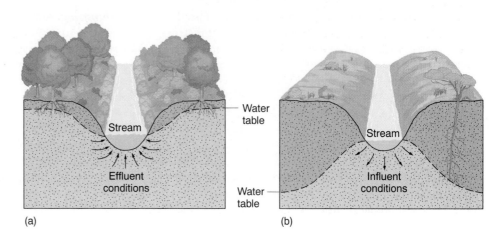

## Focus Study 9-2

### Ogallala Aquifer Overdraft

E arth's largest known aquifer is the Ogallala aquifer (or High Plains aquifer). It lies beneath the American High Plains, a six-state area from southern South Dakota to Texas (Figure 1). For several hundred thousand years, the aquifer's sand and gravel were charged with meltwaters from retreating glaciers. For the past 100 years, however, Ogallala groundwater has been heavily mined, and mining intensified after World War II, when center-pivot irrigation devices were introduced (Figure 2). These large circular devices provide vital water to wheat, sorghums, cotton, corn, and about 40% of the grain fed to cattle in the United States.

The Ogallala aquifer irrigates about one-fifth of all U.S. cropland: 150,000 wells provide water for 5.7 million hectares (14 million acres). Water is being pumped from the aquifer at the rate of 26 billion cubic meters (21 million acre-feet) a year, an increase of more than 300% since 1950.

In the past four decades, the water table in the aquifer has dropped more than 30 m (100 ft), and throughout the 1980s it has averaged a 2 m (6 ft) drop *each year*. The USGS estimates that recovery of the Ogallala aquifer (those portions that have not been crushed or subsided) would take at least 1000 years if groundwater mining stopped today!

The aquifer is most imperiled in Texas, where the saturated layer is shallowest and some 75,000 wells are in operation. The declining water table in Floyd County, Texas, illustrates the overdraft problem (Figure 3). Between 1952 and 1982, pumping costs in Floyd County have increased 221% as water tables dropped (adjusted for increases in the price index for crops), even though irrigated acreage dropped during that same time by almost 15%. These trends continued into the 1990s.

Obviously, billions of dollars of agricultural activity cannot be abruptly halted, but neither can profligate water mining continue. This issue raises tough questions: How best to manage cropland? Can extensive irrigation continue? Can the region continue to meet the demand to produce exports? Can we continue high-volume farming of certain crops that are in chronic oversupply? Should we rethink federal policy on crop subsidies and price supports? What would be the impact on farmers and rural communities from any changes to the system?

Present irrigation practices, if continued, will destroy about half of the Ogallala aquifer resource (and two-thirds of the Texas portion) by the year 2020. Add to this the approximate 10% loss of soil moisture due to climatic warming, as forecast by computer models for this region by 2050, and we have a portrait of a major regional water problem in the new millennium.

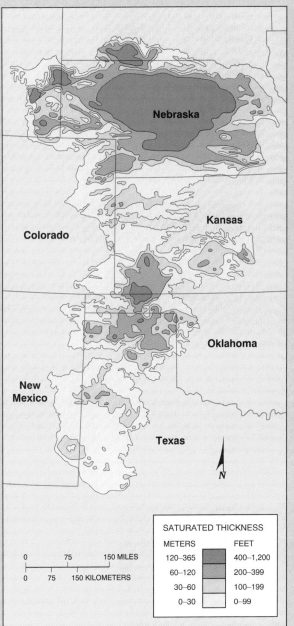

**FIGURE 1**
**Ogallala aquifer.**
The Ogallala is the largest known aquifer in North America. [After D. E. Kromm and S. E. White, "Interstate groundwater management preference differences: The Ogallala region," *Journal of Geography* 86, no. 1 (January–February 1987): 5.]

Nebraska

Kansas

Colorado

Oklahoma

New Mexico

Texas

N

| SATURATED THICKNESS | |
|---|---|
| METERS | FEET |
| 120–365 | 400–1,200 |
| 60–120 | 200–399 |
| 30–60 | 100–199 |
| 0–30 | 0–99 |

0   75   150 MILES
0   75   150 KILOMETERS

**FIGURE 2**
**Central-pivot irrigation.**
Crops are watered by myriad central-pivot irrigation systems in north-central Nebraska. A growing season for corn requires from 10 to 20 revolutions of the sprinkler arm, depending on the weather, applying about 3 cm of water per revolution. The Ogallala aquifer is at depths of over 76 m (250 ft) in this part of Nebraska. [Photo by Comstock.]

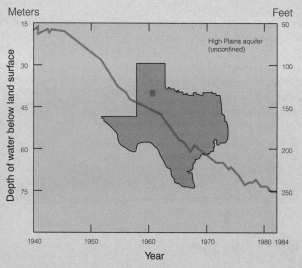

**FIGURE 3**
**Water levels in Floyd County, Texas, wells.**
[After J. E. Schefter, "Declining Ground-Water Levels and Pumping Costs: Floyd County, Texas," USGS Water Supply Paper No. 2275 (Washington, D.C.: Government Printing Office, 1985), p. 114.]

**FIGURE 9-20**
**Water desalination.**
Freshwater is supplied to Saudi Arabia from the Jubail water desalination plant along the Red Sea. Saudi Arabia operates about 20% of the desalination capacity in the world. [Photo by Gamma Liaison Network.]

an 80 km (50 mi) radius of Houston has subsided more than 3 m (10 ft) over the years. Another example is the Fresno area of California's San Joaquin Valley. After years of intensive pumping of groundwater for irrigation, land levels have dropped almost 10 m (33 ft) because of a combination of water removal and soil compaction from agricultural activity.

Aquifers also are destroyed by surface mining ("strip mining") for coal in eastern Texas, northeastern Louisiana, Wyoming, Arizona, and other locales. Aquifer collapse is an increasing problem in West Virginia, Kentucky, and Virginia, where low-sulfur coal is being mined to meet current air-pollution requirements for coal-burning electric power plants.

***Saltwater Encroachment.*** When aquifers are overpumped near the ocean, another problem arises. Along a coastline, fresh groundwater and salty seawater establish a

## News Report 1

### Middle East Water Crisis: Running on Empty

The Persian Gulf states soon may run out of freshwater. Their vast groundwater resource is being over-pumped to such an extent that salty sea-water is encroaching into aquifers *tens of kilometers* inland! By the year 2007, groundwater on the Arabian Peninsula may become undrinkable. Imagine having the hottest issue in the Middle East become water, not oil!

Remedies for groundwater overuse are neither easy nor cheap. In the Persian Gulf area, additional freshwater is being obtained by desalination of seawater, using desalination plants along the coasts. These processing plants remove salt from seawater by distillation and evaporation processes. In fact, approximately 60% of the world's 4000 desalination plants are presently operating in Saudi Arabia and other Persian Gulf states.

A proposed water pipeline may carry water overland from Turkey in the Middle East, through two branches—the Gulf water line would run southeast through Jordan and Saudi Arabia, with extensions into Kuwait, Abu Dhabi, and Oman; the other branch would run south through Syria, Jordan, and Saudi Arabia to the cities of Mecca and Jedda. This pipeline would import 6 million cubic meters (1.68 billion gallons) of water a day some 1500 km (930 mi), the distance from New York City to St. Louis! The $21 billion project cost for this "Peace Pipeline" is estimated to approximate a water system yielding a comparable amount from desalinated water.

Other remedies are possible. Traditional agricultural practices could be modernized to use less water. Urban water use could be made more efficient to reduce groundwater demand. Aquifers and rivers could be shared, although at present no negotiated accords exist for this purpose in the Middle East. The 1991 Persian Gulf War posed a grave threat to desalination facilities in Saudi Arabia, and many in Iraq and Kuwait were damaged or destroyed.

There are severe water shortages in the Middle East....The resources of the Nile, the Tigris-Euphrates, and the Jordan are overextended owing both to natural causes and to those deriving from human behavior....In addition to a severe shortage in the quantity of water, there is a growing concern over water quality.*

---

*N. Kliot, *Water Resources and Conflict in the Middle East* (London: Routledge, 1994), p. 1.

---

natural interface (contact surface). But excessive withdrawal of freshwater can cause this interface to migrate inland. As a result, wells near the shore may become contaminated with saltwater, and the aquifer may become useless as a freshwater source. Figure 9-16 illustrates this *seawater intrusion* (far-right side). Seawater contamination may be halted by pumping freshwater back into the aquifer, but, once contaminated, the aquifer is difficult to reclaim.

### Pollution of Groundwater

When surface water is polluted, groundwater inevitably becomes contaminated because it is recharged from surface-water supplies. Surface water flows rapidly and flushes pollution downstream, but slow-moving groundwater, once contaminated, remains polluted virtually forever.

Pollution can enter groundwater from industrial injection wells (wastes pumped into the ground), septic tank outflows, seepage from hazardous-waste disposal sites, industrial toxic waste, agricultural residues (pesticides, herbicides, fertilizers), and urban solid-waste landfills. As an example, at U.S. gasoline stations, an estimated 10,000 underground gasoline storage tanks are suspected of leaking. You have probably seen the now-commonplace excavations at gas stations to remove old tanks and contaminated soil.

For management purposes, about 35% of pollution is categorized as *point source* (such as a gasoline tank or septic tank); 65% is classed as *nonpoint source* (from a broad area, such as an agricultural field, or urban runoff). Regardless of the spatial nature of the source, pollution can spread over a great distance, as illustrated in Figure 9-16.

***Pollution and the Law.*** Despite widespread publicity about groundwater pollution in the United States, Canada, and Europe, few new protective laws have been established. Major U.S. laws either neglect groundwater protection or do not explicitly prescribe action. These laws include the U.S. Safe Drinking Water Act (1974) and the U.S. Resource Conservation and Recovery Act (1976, 1984).

Some groundwater contamination is reported in all 50 states. In 1988, the U.S. government's General Accounting Office (GAO) surveyed the nation's groundwater protection standards. It found that 38 states have generally worded standards that do not identify pollution concentrations with specific numbers. Such laws are unscientific and difficult to implement because they are so nonspecific and nonuniform. Only 26 of the 38 states have any numerical or empirical groundwater standards. In the face of government inaction, and even attempts to reverse some protection laws, serious groundwater contamination continues nationwide.

One characteristic is the practical irreversibility of groundwater pollution, causing the cost of clean-up to be prohibitively high….It is a questionable ethical practice to impose the potential risks associated with groundwater contamination on future generations when steps can be taken today to prevent further contamination.[*]

# Surface Water Resources: Streams

Earth's streams and rivers contain 1250 km³ (300 mi³) of water, only 0.0001% of all water, and the smallest percentage (0.003%) of any of the freshwater categories we have discussed (see Table 7-1). Yet streams are the portion of the hydrologic cycle on which we most depend, representing four-fifths of all the water we use. Because of their rapid renewal, as compared with lake water or groundwater, streams are the best single measure of available water supply.

Streams may be perennial (constantly flowing; *per ennial* = "through the year") or intermittent. In either case, the total runoff that moves through them comes from surplus surface-water runoff, subsurface throughflow, and groundwater. Figure 9-21 maps annual global river runoff for the world. Highest runoff amounts are along the equator within the tropics, reflecting the continual rainfall along

[*]J. Tripp, "Groundwater Protection Strategies," in *Groundwater Pollution, Environmental and Legal Problems* (Washington, D.C.: American Association for the Advancement of Science, 1984), p. 137. Reprinted by permission.

the ITCZ. Southeast Asia also experiences high runoff, as do northwest coastal mountains in the Northern Hemisphere. In countries having great seasonal fluctuations in runoff, groundwater becomes an important reserve. Regions of lower runoff coincide with Earth's subtropical deserts, rain-shadow areas, and continental interiors, particularly in Asia.

**Discharge** is a stream's flow rate. Discharge usually is expressed as the volume of water that passes a given point per unit of time. For example, the discharge of the Mississippi River at its mouth in Louisiana is 17,300 *cubic meters per second*, or 611,000 *cubic feet per second*. News Report 2 explains some terms used in water measurements.

In volume of runoff, the Western Hemisphere exceeds the Eastern Hemisphere, primarily because the Amazon River (Figure 9-22) discharges far more water than any other stream, and because four of the world's other high-discharge river systems are in the Western Hemisphere (Mississippi-Missouri, Orinoco, in Venezuela; and the St. Lawrence and Mackenzie Rivers in Canada). The Atlantic Ocean receives about 1.5 times more runoff than the Pacific Ocean. Table 9-2 lists the outflow locations, lengths, and discharge values for the world's largest rivers. (Chapter 14 examines more closely the hydrologic aspects of streams and their interaction with the landscape.)

## *Exotic Streams*

Most streamflows increase downstream because the area being drained increases. The Mississippi River is typical. It starts as a small brook in Minnesota and grows to a mighty river pouring into the Gulf of Mexico. But, a

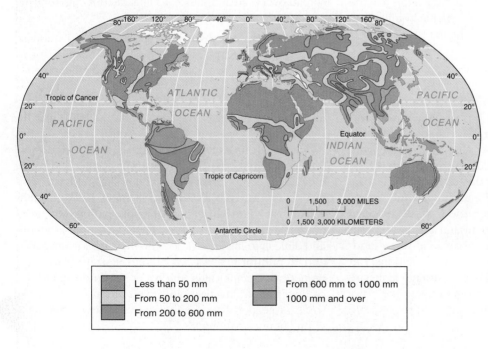

**FIGURE 9-21**
**Annual global river runoff.**
Distribution of runoff is closely correlated with climatic region, as expected. But it is poorly correlated with human population distribution and density. [Data from the Institute of Geography, Russian Academy of Sciences, Moscow, as presented by the World Resources Institute.]

Less than 50 mm
From 50 to 200 mm
From 200 to 600 mm
From 600 mm to 1000 mm
1000 mm and over

# News Report 2

## Measuring Up the Water Resource

In most of the United States, hydrologists measure streamflow in cubic feet per second (ft³/s); Canadians use cubic meters per second (m³/s). In the eastern United States, or for large-scale assessments, water managers use millions of gallons a day (MGD), billions of gallons a day (BGD), or billions of liters per day (BLD). Eventually the metric system will prevail as the United States converts to the international system of measurement units.

In the western United States, where irrigated agriculture is so important, the measure frequently used is acre-feet per year. One acre-foot is an acre of water, 1 foot deep, equivalent to 325,872 gallons (43,560 ft³, or 1234 m³, or 1,233,429 liters). An acre is an area that is about 208 feet on a side and is 0.4047 hectares.

For global measurements, 1 km³ = 1 billion cubic meters = 810 million acre-feet; 1000 m³ = 264,200 gallons = 0.81 acre-feet. For more-local measurements, 1 m³ = 1000 liters = 264.2 gallons.

---

stream can originate in a humid region and subsequently flow through an arid region. In that case, the discharge usually decreases with distance, because of high potential evapotranspiration (POTET) rates in the arid area. Such a stream is called an **exotic stream** (exotic means "of foreign origin").

Exotic streams are exemplified by the Nile River. This great river, Earth's longest as determined by surveys, drains much of northeastern Africa. But as it courses through the deserts of Sudan and Egypt, it loses water instead of gaining it, because of evaporation and withdrawal for agriculture. By the time it empties into the Mediterranean Sea, the Nile's flow has dwindled so much that it ranks only 33rd in discharge.

The United States has exotic streams too, notably the Colorado River. Its flow decreases with distance from its source; in fact, it no longer produces enough discharge to reach its mouth in the Gulf of California! The exotic Colorado River is depleted not only by passage across dry, desert lands but also by upstream removal of water for agriculture and municipal uses—the subject of a Focus Study in Chapter 15.

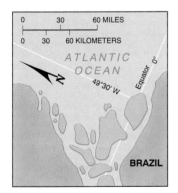

## FIGURE 9-22
**Mouth of the Amazon River.**
The mouth of the Amazon River discharges a fifth of all the freshwater that enters the world's oceans. The mouth of the Amazon is 160 km (100 mi) wide. Millions of tons of sediments are derived from the Amazon's drainage basin, which is as large as the Australian continent. This astronaut's view is to the northeast, over large islands of sediment deposited where the river's discharge leaves the mouth and flows into the Atlantic Ocean. [Space Shuttle photograph from NASA.]

**TABLE 9-2**

| Largest Rivers on Earth Ranked by Discharge Volume | | | | | |
|---|---|---|---|---|---|
| Rank by Discharge Volume | Average Discharge at Mouth (thousands of m³/s) (ft³/s) | River Name [also known as] (Major Tributaries) | Outflow/Location | Length of River (km [mi]) | Rank by Length |
| 1 | 212.5 (7500) | Amazonas [Amazon] (Xingu, Tapajós, Negro, Madeira) | Atlantic Ocean/ Amapá-Pará, Brazil | 6570 (4080) | 2 |
| 2 | 79.3 (2800) | Paraná (Paraguay, Salado) | Atlantic Ocean/Argentina | 3945 (2450) | 16 |
| 3 | 39.7 (1400) | Congo [Zaire] (Kwa, Ubangi, Lualaba) | Atlantic Ocean/ Angola, Zaire | 4630 (2880) | 10 |
| 4 | 38.5 (1360) | Ganges [Padma] (Jamuna Brahmaputra) | Bay of Bengal/ Bangladesh, India | 2898 (1800) | 23 |
| 5 | 21.8 (770) | Chang Jiang [Yangtze] (San, Gan, Han, Xiang) | East China Sea/ Jiangsu, China | 5980 (3720) | 4 |
| 6 | 17.4 (614) | Yenisey (Nizhnyaya Tunguska, Angara) | Gulf of Kara Sea/Siberia | 5870 (3650) | 5 |
| 7 | 17.3 (611) | Mississippi (Missouri, Ohio, Arkansas, Tennessee, Little) | Gulf of Mexico/Louisiana | 6020 (3740) | 3 |
| 8 | 17.0 (600) | Orinoco (Caroní) (distributaries Mánamo, Macareo, Río Grande) | Atlantic Ocean/Venezuela | 2737 (1700) | 27 |
| 9 | 15.5 (547) | Lena (Aldan, Vilyuy) | Laptiv Sea/Siberia | 4400 (2730) | 11 |
| 10 | 14.2 (500) | St. Lawrence (Ottawa) | Gulf of St. Lawrence/ Québec | 3060 (1900) | 21 |
| 17 | 7.9 (280) | Mackenzie (Slave, Peace, Liard, Athabasca) | Beaufort Sea, Northwest Territories | 4240 (2630) | 12 |
| 33 | 2.83 (100) | Bahr el Nîl [Nile] (Kagera, Bahr el Ghazal, El Bahr el Azraq [Blue Nile]) | Mediterranean Sea/Egypt | 6690 (4160) | 1 |

## Internal Drainage

Most streamwater eventually finds its way to progressively larger rivers and into the ocean. In some regions, however, stream drainage does not reach the ocean. Instead, the water leaves the region by means of evaporation or subsurface gravitational flow. Such streams terminate in areas of **internal drainage**.

Portions of Asia, Africa, Australia, Mexico, and the western United States have regions with such internal drainage patterns. Internal drainage dominates the region of the Great Basin of the western United States, as shown on maps in Figures 14-4 and 15-22a. For example, the Humboldt River flows across Nevada to the west, where evaporation and seepage losses make it disappear into the Humboldt "sink." Many streams and creeks flow into the Great Salt Lake, but its only outlet is evaporation, so it exemplifies a region of internal drainage. In the Middle East, the Dead Sea region and, in Asia, the region around the Aral Sea and Caspian Sea have no outlet to the ocean.

# Our Water Supply

Since we are so dependent on water, it would seem that humans should cluster where good water is plentiful. But accessible water supplies are not well correlated with population distribution or the regions where population growth is greatest. Table 9-3 shows estimated world water supplies, estimated population figures for 2010, population doubling times at present growth rates, and a comparison of population and global runoff percentages for each continent. This table gives you an idea of the unevenness of Earth's water supply (tied to climatic variability) and water demand (tied to level of development and per capita consumption). For example, North America's mean annual discharge is 5960 km³ and Asia's is 13,200 km³. However, North America has only 4.8% of the world's population, whereas Asia has 60.4%, with a population doubling time less than half North America's.

Water is critical for survival. Just to produce our food requires enormous quantities of water, for growing, cleaning,

**TABLE 9-3**

| Estimate of Available Global Water Supply Compared by Region and Population | | | | | | | |
|---|---|---|---|---|---|---|---|
| **Region** | **Land Area in Thousands of km²** | **(mi²)** | **Mean Annual Discharge in km³/yr (BGD)** | **Share of Global Stream Runoff (%)** | **Global Population, 2010 (%)** | **Population, 2010 (millions)** | **Population Doubling Time (years)** |
| Africa | 30,600 | (11,800) | 4220 (3060) | 11 | 15.2 | 1069 | 24 |
| Asia | 44,600 | (17,200) | 13,200 (9540) | 36 | 60.4 | 4242 | 42 |
| Australia-Oceania | 8420 | (3250) | 1960 (1420) | 5 | 0.4 | 34 | 60 |
| Europe | 9770 | (3770) | 3150 (2280) | 8.8 | 10.6 | 743 | — |
| North America (Canada, Mexico, US) | 22,100 | (8510) | 5960 (4310) | 15 | 4.8 | 334 | 105 |
| Central and South America | 17,800 | (6880) | 10,400 (7510) | 26 | 8.6 | 601 | 36 |
| Global (excluding Antarctica) | 134,000 | (51,600) | 38,900 (28,100) | — | 100.00 | 7024 | 45 |

processing, and waste disposal (see News Report 3). Yet, we take for granted the quality and quantity of our water supply, because it always seems to be there with little effort on our part. At present, the world's people are withdrawing 30% of the runoff that is accessible. This idea of "accessibility" is important, because about 20% of global runoff is remote and not readily available to meet water demand. Timing of flows is also important, for approximately 50% of total runoff remains beyond use as uncaptured floodwater. For example the greatest river in the world in terms of runoff discharge on Earth—the Amazon—is about 95% remote from population centers, or not readily available.

Water resources are different from other resources in that there is no alternative substance to water. Water-supply shortages loom in at least 80 countries as of 1995. Water shortages increase the probability for international conflict, endanger public health, reduce agricultural productivity,

and damage life-supporting ecological systems. New dams and river-management schemes might increase runoff accessibility by 10% over the next 30 years, but population is projected to grow by 32% in the same time period.

## Water Supply in the United States

The U.S. water supply is derived from surface and groundwater sources that are fed by an average daily precipitation of 4200 BGD (billion gallons a day). That sum is based on an average annual precipitation value of 76.2 cm (30 in.) divided evenly among the 48 contiguous states (excluding Alaska and Hawaii).

The 4200 billion gallons is unevenly distributed across the country and unevenly distributed throughout the year. For example, New England's water supply is so abundant

## News Report 3
### Personal Water Use

On an individual level, statistics indicate that urban dwellers directly use an average 680 liters (180 gallons) of water per person per day, whereas rural populations average only 265 liters (70 gallons) or less per person per day.

However, each of us indirectly accounts for a much greater water demand, because we all consume food and other products produced with water.

For the United States, dividing the total water withdrawal (408 BGD in 1990) by a population of 255 million people (1990) yields a per capita direct and indirect use of 6056 liters (1600 gallons) of water per day! As population and per capita affluence increase, greater demands are placed on the water resource base, which is essentially fixed.

Simply providing the variety of food we enjoy requires voluminous water. For

example, 77 g (2.7 oz) of broccoli requires 42 l (11 gal) of water to grow and process; producing 250 ml (8 oz) of milk requires 182 l (48 gal) of water; producing 28 g (1 oz) of cheese requires 212 l (56 gal); producing 1 egg requires 238 l (63 gal).

*Clearly, the average American lifestyle depends on enormous quantities of water and thus is vulnerable to shortfalls or quality problems.*

**FIGURE 9-23**
**U.S. water budget.**
Daily water budget for the contiguous 48 states in billions of gallons a day (BGD). As of 1990, approximately 30% of available surplus was withdrawn for agricultural, municipal, and industrial use.

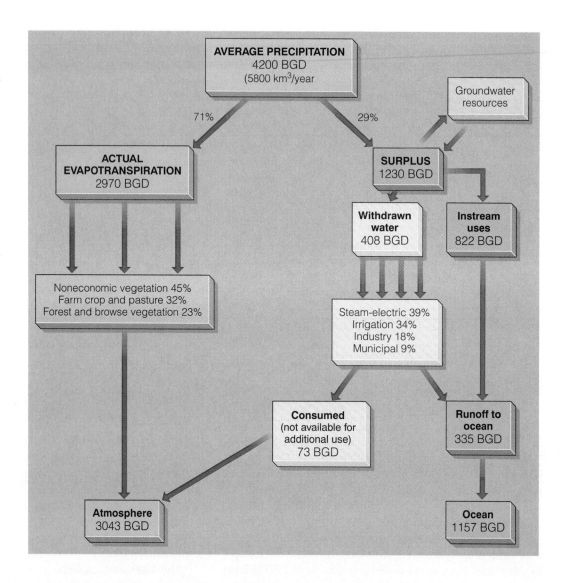

that only about 1% of available water is consumed each year. (The same is true in Canada, where the resource greatly exceeds that in the United States). But in the dry Colorado River Basin mentioned earlier, the discharge is completely consumed. In fact, by treaty and compact agreements, the Colorado actually is budgeted beyond its average discharge! This paradox results from political misunderstanding of the water resource.

Figure 9-23 shows how the U.S. supply of 4200 BGD is distributed. Viewed daily, the national water budget has two general outputs: 71% actual evapotranspiration (ACTET) and 29% surplus (SURPL). The 71% actual evapotranspiration involves 2970 BGD of the daily supply. It passes through nonirrigated land, including farm crops and pasture, forest and browse (young leaves, twigs, and shoots), and nonagricultural vegetation. Eventually, it returns to the atmosphere to continue its journey through the hydrologic cycle. The remaining 29% surplus is what we directly use.

***Instream, Nonconsumptive, and Consumptive Uses.***
The surplus 1230 BGD is runoff, available for withdrawal, consumption, and various instream uses.

- *Instream uses* are those that use streamwater in place: navigation, wildlife and ecosystem preservation, waste dilution and removal, hydroelectric power production, fishing resources, and recreation.

- *Nonconsumptive uses*, or **withdrawal**, remove water from the supply, use it, and then return it to the same supply. Nonconsumptive water is used by industry, agriculture, municipalities, and in steam-electric power generation. A portion of water withdrawn is consumed.

- **Consumptive uses** remove water from a stream but do not return it, so it is not available for a second or third use. Some consumptive examples include: water that evaporates or is vaporized in steam-electric plants.

When water is returned to the system, water quality usually is altered—water is contaminated chemically with pollutants or waste or thermally with heat energy. In Figure 9-23, this portion of the budget is returned to runoff and eventually the ocean. This wastewater represents an opportunity to extend the resource through reuse.

Contaminated or not, returned water becomes a part of all water systems downstream. A good example is New Orleans, the last city to withdraw municipal water from the Mississippi River. New Orleans receives diluted and mixed contaminants added throughout the entire Missouri-Ohio-Mississippi River drainage! This includes the effluent of chemical plants, runoff from millions of acres of farm fields treated with fertilizer and pesticides, treated and untreated sewage, oil spills, gasoline leaks, wastewater from thousands of industries, and turbidity from countless construction sites and from mining, farming, and logging activities. Abnormally high cancer rates among citizens living along the Mississippi River between Baton Rouge and New Orleans have led to the ominous label Cancer Alley for the region.

The estimated U.S. withdrawal of water for 1990 was 408 BGD, an increase of 290% since 1940. The three main uses of this withdrawn water in the United States are steam-electric power (39%), irrigation (34%), industry (18%), and municipalities (9%). In contrast, Canada uses only 7% of its withdrawn water for irrigation and 84% for industry. Figure 9-24 compares regions by their use of withdrawn water. It graphically illustrates the differences between more developed and less developed parts of the world.

## Future Considerations

When precipitation is budgeted, the limits of the water resource become apparent. How can we satisfy the growing demand for water? Water available per person declines as population increases, and individual demand increases with economic development. Thus, world population growth since 1970 has reduced per capita water supplies by a third. Also, pollution limits the water-resource base, so that even before *quantity* constraints are felt, *quality* problems may limit the health and growth of a region (Figure 9-25).

The world's water future is very complex. About 30% of the annual renewable water worldwide is presently utilized. One-half billion people depend on the polluted Ganges River alone. The World Bank estimates that $600 billion in capital investment is needed by 2010 to augment current water resources. In addition, aspects of global climate change will worsen water supply conditions for many regions and therefore increase this cost estimate.

The international aspect of the problem is illustrated by the 200 major river basins in the world (basins where rivers drain into an ocean, lake, or inland sea): 148 rivers are shared by two nations, and 52 are shared by three to ten nations. The Helsinki Rules (1966) guide the analysis of international rivers. They call for a holistic approach involving economic, political, geomorphologic, hydrologic, engineering, demographic, and cultural considerations.

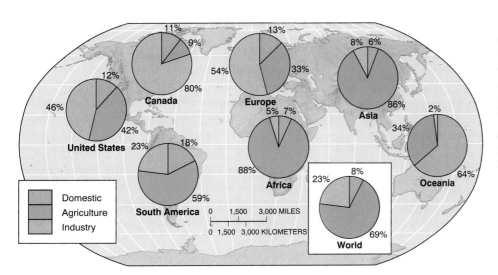

**FIGURE 9-24**
**Water withdrawal by sector.**
In particular, compare industrial water use among the geographic areas, as well as agricultural and municipal uses. [After World Resources Institute, *World Resources, 1992–1993*, from Bureau of Geological and Mining Research, France, and Institute of Geography, Russia, 1988.]

**FIGURE 9-25**
**Fouled waterways.**
Water pollution limits the ability of the functioning environment to renew the resource. The water resource in many regions is limited by poor water quality. [Photo by author.]

Unlike other important commodities such as oil, copper, or wheat, freshwater has no substitutes for most of its uses. It is also impractical to transport the large quantities of water needed in agriculture and industry more than several hundred kilometers. Freshwater is now scarce in many regions of the world, resulting in severe ecological degradation, limits on agriculture and industrial production, threats to human health, and increased potential for international conflict.*

Clearly, cooperation is needed, yet we continue toward a water crisis without a concept of a *world water economy* as a frame of reference.

*S. L. Postal, and others, "Human appropriation of renewable fresh water," *Science* 271, no. 5250 (February 9, 1996), p. 785.

# Summary and Review — Water Resources

✔ *Illustrate* **the hydrologic cycle with a simple sketch and** *label* **it with definitions for each water pathway.**

The flow of water links the atmosphere, ocean, and land through energy and matter exchanges. The **hydrologic cycle** is a model of Earth's water system, which has operated for billions of years from the lower atmosphere to several kilometers beneath Earth's surface. **Interception** occurs when precipitation strikes vegetation or other ground cover. Water soaks into the subsurface through **infiltration**, or penetration of the soil surface. It further permeates soil or rock through vertical movement called **percolation**.

> hydrologic cycle (p. 240)
> interception (p. 240)
> infiltration (p. 240)
> percolation (p. 240)

1. Sketch and explain a simplified model of the complex flows of water on Earth—the hydrologic cycle.

2. What are the possible routes that a raindrop may take on its way to and into the soil surface?

3. Compare precipitation and evaporation volumes from the ocean with those over the land. Describe advection flows of moisture and countering surface and subsurface runoff.

✔ *Relate* **the importance of the water-budget concept to your understanding of the hydrologic cycle, water resources, and soil moisture for a specific location.**

A **soil-water budget** can be established for any area of Earth's surface by measuring the precipitation input and the output of various water demands in the area considered.

Groundwater is a subsurface part of the hydrologic cycle, tied to surface supplies through soil and rock. Groundwater is the largest potential freshwater source in the hydrologic cycle. Streams represent only a tiny fraction of all water (1250 km³, or 300 mi³), the smallest volume of any of the freshwater categories. Yet streams are the portion of the hydrologic cycle on which we most depend, representing four-fifths of all the water we use.

**soil-water budget (p. 242)**

4. How might an understanding of the hydrologic cycle in a particular locale, or a soil-moisture budget of a site, assist you in assessing water resources? Give some specific examples.

---

✔ *Construct* **the water-balance equation as a way of accounting for the expenditures of water supply and** *define* **each of the components in the equation and their specific operation.**

The moisture supply to Earth's surface is **precipitation** (PRECIP), arriving as rain, sleet, snow, and hail. Precipitation is measured with the **rain gauge**. **Evaporation** is the net movement of free water molecules away from a wet surface into air. **Transpiration** is the movement of water through plants and back into the atmosphere; it is a cooling mechanism for plants. Evaporation and transpiration are combined into one term—**evapotranspiration**. The ultimate demand for moisture is **potential evapotranspiration** (POTET), the amount of water that *would* evaporate and transpire under optimum moisture conditions (adequate precipitation and adequate soil moisture). Evapotranspiration is measured with an **evaporation pan** (*evaporimeter*) or the more elaborate **lysimeter**.

A "savings account" of water that receives deposits and provides withdrawals as water-balance conditions change is the **soil-moisture storage** (Δ STRGE). This is the volume of water stored in the soil that is accessible to plant roots. In soil, **hygroscopic water** is inaccessible because it is a molecule-thin layer that is tightly bound to each soil particle by hydrogen bonding. **Capillary water** is generally accessible to plant roots because it is held in the soil by surface tension and hydrogen bonding between water and soil. Almost all capillary water that remains in the soil is **available water** in soil-moisture storage. After water drains from the larger pore spaces, the available water remaining for plants is termed **field capacity**, or storage capacity. When soil is saturated after a precipitation event, surplus water in the soil becomes **gravitational water** and percolates to groundwater.

Unsatisfied POTET is **deficit** (DEFIC). By subtracting DEFIC from POTET, we determine **actual evapotranspiration**, or ACTET. Ideally, POTET and ACTET are about the same, so that plants have sufficient water. If POTET is satisfied and the soil is full of moisture, then additional water input becomes **surplus** (SURPL), which may puddle on the surface, flow across the surface toward stream channels, or percolate through the soil underground. The **overland flow** to streams includes precipitation and groundwater flows into river channels to make up the **total runoff** from the area.

As **soil-moisture utilization** removes soil water, the plants work harder to extract the same amount of moisture. As available water is utilized, soil reaches the **wilting point** (all that remains is unextractable water).

The texture and the structure of the soil dictate available pore spaces, or **porosity**. The soil's **permeability**, or degree to which water can flow through it, determines the rate of **soil-moisture recharge**. Permeability depends on particle sizes and the shape and packing of soil grains.

precipitation  (p. 242)
rain gauge  (p. 242)
evaporation  (p. 243)
transpiration  (p. 243)
evapotranspiration (p. 243)
potential evapotranspiration (p. 243)
evaporation pan (p. 244)
lysimeter (p. 244)
soil-moisture storage (p. 246)
hygroscopic water (p. 246)
capillary water (p. 246)
available water (p. 247)
field capacity (p. 247)
gravitational water (p. 247)
deficit (p. 246)
actual evapotranspiration (p. 246)
surplus (p. 246)
overland flow (p. 246)
total runoff (p. 246)
soil-moisture utilization (p. 247)
wilting point (p. 246)
porosity (p. 248)
permeability (p. 248)
soil-moisture recharge (p. 248)

5. What does this statement mean? "The soil-water budget is an assessment of the hydrologic cycle at a specific site."

6. What are the components of the water-balance equation? Construct the equation and place each term's definition below its abbreviation in the equation.

7. Using the annual water-balance data for Kingsport, Tennessee, work the values through the water-balance "bookkeeping" method. Does the equation balance?

8. Explain how to derive actual evapotranspiration (ACTET) in the water-balance equation.

9. What is potential evapotranspiration (POTET)? How do we go about estimating this potential rate? What factors did Thornthwaite use to determine this value?

10. Explain the operation of soil-moisture storage, soil-moisture utilization, and soil-moisture recharge. Include

discussion of field capacity, capillary water, and wilting point concepts.

11. In the case of silt-loam soil from Figure 9-11, roughly what is the available water capacity? How is this value derived?

12. In terms of water balance and water management, explain the logic behind the Snowy Mountains Scheme in southeastern Australia.

---

✔ *Describe* **the nature of groundwater and** *define* **the elements of the groundwater environment.**

**Groundwater** is a part of the hydrologic cycle, but it lies beneath the surface beyond the soil-moisture root zone. Groundwater replenishment is tied to surface surpluses because groundwater does not exist independently. Excess surface water moves through the **zone of aeration**, where soil and rock are less than saturated. Eventually, the water reaches the **zone of saturation**, where the pores are completely filled with water.

The *permeability* of subsurface rocks depends on whether they conduct water readily (higher permeability) or tend to obstruct its flow (lower permeability). They can even be *impermeable*. An **aquifer** is a rock layer that is permeable to groundwater flow in usable amounts. An **aquiclude** (aquitard) is a body of rock that does not conduct water in usable amounts.

A **confined aquifer** is bounded above and below by impermeable layers of rock or sediment. An **unconfined aquifer** has a permeable layer on top and an impermeable one beneath. The upper limit of the water that collects in the zone of saturation is called the **water table**; it is the contact surface between the zones of saturation and aeration. The **aquifer recharge area** extends over an entire unconfined aquifer. Water in a confined aquifer is under the pressure of its own weight, creating a pressure level to which the water can rise on its own, called the **potentiometric surface**, which can be above ground level. Groundwater confined under pressure is **artesian water**; it may rise up in wells and even flow out at the surface without pumping, if the head of the well is below the potentiometric surface.

As water is pumped from a well, the surrounding water table within an unconfined aquifer will experience **drawdown**, or become lower, if the rate of pumping exceeds the horizontal flow of water in the aquifer around the well. This causes a **cone of depression**. Aquifers frequently are pumped beyond their flow and recharge capacities, a condition known as groundwater mining.

groundwater (p. 252)
zone of aeration (p. 253)
zone of saturation (p. 253)
aquifer (p. 254)

aquiclude (p. 254)
confined aquifer (p. 255)
unconfined aquifer (p. 255)
water table (p. 255)
aquifer recharge area (p. 255)
potentiometric surface (p. 255)
artesian water (p. 255)
drawdown (p. 255)
cone of depression (p. 255)
groundwater mining (p. 255)

13. Are groundwater resources independent of surface supplies, or are the two interrelated? Explain your answer.

14. Make a simple sketch of the subsurface environment, labeling zones of aeration and saturation and the water table in an unconfined aquifer. Then add a confined aquifer to the sketch.

15. At what point does groundwater utilization become groundwater mining? Use the Ogallala aquifer example to explain your answer.

16. What is the nature of groundwater pollution? Can contaminated groundwater be cleaned up easily? Explain.

---

✔ *Define* **stream discharge using specific rivers as examples and** *explain* **the concept of an exotic stream using the Nile and Colorado Rivers.**

**Discharge** is a stream's flow rate, usually expressed as the volume of water that passes a given point per unit of time cubic meters per second. Most streams increase discharge downstream. But some originate in a humid region and flow through an arid region, such that discharge decreases with distance because of high potential evapotranspiration in the arid area. Such an **exotic stream** is exemplified by the Nile River or the Colorado River. Some regions, such as the Great Salt Lake Basin, have **internal drainage** that does not reach the ocean.

discharge (p. 260)
exotic stream (p. 261)
internal drainage (p. 262)

17. What differences exist between the freshwater inflow into the Atlantic and Pacific Oceans? Explain.

18. What are the five largest rivers on Earth in terms of discharge? Relate these to the weather patterns in each area and to regional POTET and PRECIP.

19. Contrast a river such as the Amazon River to an exotic stream such as the Nile or Colorado River.

✔ *Identify* **critical aspects of freshwater supplies for the future and** *cite* **specific issues related to sectors of use, regions and countries, and potential remedies of any shortfalls.**

*Nonconsumptive uses*, or water **withdrawal**, remove water from the supply, use it, and then return it to the stream. **Consumptive uses** remove water from a stream but do not return it, so the water is not available for a second or third use. Americans in the 48 contiguous states withdraw approximately one-third of the available surplus runoff for irrigation, industry, and municipal uses. Water-resource planning regionally and globally, using water-budget principles, is essential if Earth's societies are to have enough water of adequate quality.

withdrawal (p. 264)
consumptive uses (p. 264)

20. Describe the principal pathways involved in the water budget of the contiguous 48 states. What is the difference between withdrawal and consumptive use of water resources? Compare these with instream uses.

21. Characterize each of the sectors withdrawing water: irrigation, industry, and municipalities. What are the present usage trends in more developed and less developed nations?

22. Briefly assess the status of world water resources. What challenges are there in meeting future needs of an expanding population and growing economies?

---

 ## NetWork

The *Geosystems Home Page* provides on-line resources for this chapter on the World Wide Web. You will find review exercises, specific updates for items in the chapter, suggested readings, and links to interesting related pathways on the Internet (click on the Table of Contents link and select this chapter). *Geosystems* is at: **http://www.prenhall.com/geosystm**

# 10

# Global Climate Systems

**Earth's Climate System and Its Components**

**Classification of Climatic Regions**

**Tropical Climates (A)**

**Mesothermal Climates (C)**

**Microthermal Climates (D)**

**Polar Climates (E)**

**Dry Arid and Semiarid Climates (B)**

**Global Climate Change**

**Summary and Review**

## Key Learning Concepts

After reading the chapter, you should be able to:

- *Define* climate and climatology and *explain* the difference between climate and weather.

- *Review* the role of temperature, precipitation, air pressure, and air mass patterns used to establish climatic regions.

- *Review* Köppen's development of an empirical climate classification system and *compare* his with other ways of classifying climate.

- *Describe* the A, C, D, and E climate classification categories and generally *locate* these regions on a world map.

- *Explain* the precipitation and moisture efficiency criteria used to determine the B climates and generally *locate* them on a world map.

- *Outline* future climate patterns from forecasts presented and *explain* the causes and potential consequences.

*Tropical rain forest climate, Kauai, Hawaii.* [Photo by A. R. Christopherson.]

Earth experiences an almost infinite variety of weather. But if we consider the weather at any one place over many years, a clear pattern emerges, and this pattern constitutes climate. Climates are so diverse that no two places on Earth's surface experience exactly the same climatic conditions; in fact, Earth is a vast collection of microclimates. However, broad similarities among local climates permit their grouping into climatic regions.

Early climatologists faced the challenge of what to use as a basis for climate classification. One climatologist, Wladimir Köppen (pronounced KUR-pen), developed a system that uses temperature and precipitation data. Though imperfect, his system is still widely used and easily understood, and we use a modified version of it in this chapter.

Today, climatologists know that intriguing climatic linkages exist in the Earth-atmosphere-ocean system:

- Strong monsoonal rains in West Africa are correlated with the development of intense Atlantic hurricanes. In a few days, an "easterly wave" over West Africa may become a furious hurricane crossing Florida.

- An El Niño in the Pacific is tied to drought-breaking rains in the West, floods in Louisiana and Northern Europe, and a drought in Australia. Yet a subsequent El Niño produces opposite effects.

- The eruption of a volcano in the Philippines is correlated with an increase in atmospheric albedo and lowering of global temperatures.

- Human activities, including pollution, affect climatic patterns and the distribution of plants and animals as natural habitats shift.

Climatologists now are using computer models to simulate these complex interactions in the atmosphere, hydrosphere, lithosphere, and biosphere. Many of the physical elements of the environment that you studied in the first nine chapters of this text now link together to explain climates.

Climatologists are concerned about the changes occurring in global climate patterns. This chapter concludes with a discussion of such changes and their vital implications for society.

# Earth's Climate System and Its Components

The condition of the atmosphere at any given place and time is *weather*. But the consistent behavior of weather over a long time, including its variability, is a region's **climate**. Climate includes not just weather averages but also the extremes experienced at a place. A climate may have temperatures that average above freezing every month yet still threaten severe frost problems for agriculture. Think of climate as long-term patterns and conditions that are dynamic rather than static.

**Climatology**, the study of climate, is the analysis of long-term weather patterns over time and space to find areas of similar weather statistics. Such weather observations, gathered simultaneously from different points within a region, are plotted on maps and are compared to identify **climatic regions**. Observed patterns grouped into regions are at the core of climate classification.

Which weather components combine to produce Earth's climates? They include insolation, temperature, humidity, seasonal precipitation, atmospheric pressure and winds, air masses, types of weather disturbances, and cloud coverage.

The climate where you live may be humid with distinct seasons, or dry with consistent warmth, or moist and cool—almost any combination is possible. There are places where it rains more than 20 cm (8 in.) each month, with monthly average temperatures remaining above 27°C (80°F) year-round. Other places may be rainless for a decade at a time. Expected climatic patterns include episodes of variability, sometimes on a global scale. Of significance is the recurring El Niño phenomenon, the subject of Focus Study 10-1.

Climates greatly influence *ecosystems,* the natural, self-regulating communities of plants and animals that thrive in specific environments. On land, basic climatic regions largely determine where each major ecosystem exists. These ecosystem regions, called *biomes,* are of several types: forest (Figure 10-1), grassland, savanna, tundra, and desert. Unique plant, soil, and animal communities are associated with each biome. (Chapters 19 and 20 present global ecosystems and principal biomes.)

Climates generally are stable over several human generations, but over longer periods climates cycle through periodic change. (A section in Chapter 17 describes some of these long-term cycles of climate change.) Consequently, climates, ecosystems, and biomes should be thought of as continually adapting and responding to conditions.

Climatologists speculate that changes in climate and natural vegetation during the next 50 years could exceed the total of all changes since the peak of the last ice-age episode, some 18,000 years ago. One operative factor behind climatic change is human activity. Figure 10-2 presents a holistic view of Earth's climate system, showing both internal and external processes that influence climate.

## Climate Factors: Insolation, Temperature, Pressure, Air Masses, and Precipitation

The principal elements of climate—insolation, temperature, pressure, air masses, and precipitation—were presented in the first nine chapters. We review them briefly

**FIGURE 10-1**

**A northern forest biome.**

This northern coniferous forest biome reflects the cool, moist climate at this elevation in the Canadian Rockies. [Photo by author.]

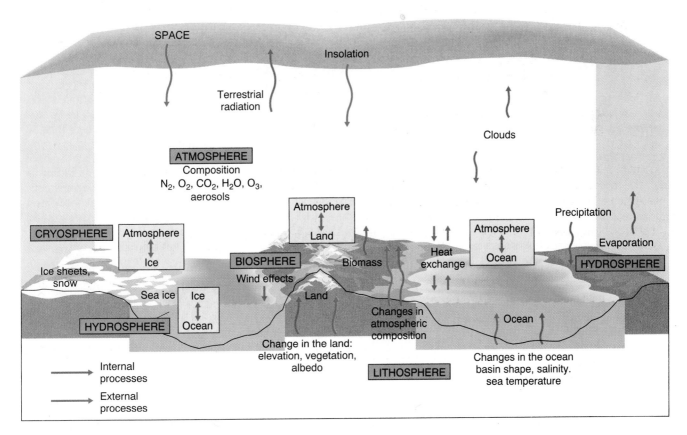

**FIGURE 10-2**

**A schematic of Earth's climate system.**

*Internal processes* that influence climate involve the atmosphere, hydrosphere (streams and ocean), cryosphere (polar ice masses and glaciers), biosphere, and lithosphere (land)—all energized by insolation. All interact to produce climatic patterns. *External processes*, principally from human activity, affect this climatic balance and force climate change. [After J. Houghton, *The Global Climate* (Cambridge: Cambridge University Press, 1984), and the Global Atmospheric Research Program.]

## Focus Study 10-1

### The El Niño Phenomenon

Climate is the consistent behavior of weather over time, but average weather conditions also include extremes that depart from normal. The *El Niño-Southern Oscillation (ENSO)* in the Pacific Ocean forces the greatest interannual variability of temperature and precipitation on a global scale among all natural phenomena.

Revisit Figure 6-24 and you can see that the region off South America's west coast is dominated by the northward-flowing Peru current. These cold waters move toward the equator and join the westward movement of the south equatorial current.

The Peru current is part of the normal counterclockwise circulation of winds and surface ocean currents around the subtropical high-pressure cell dominating the eastern Pacific in the Southern Hemisphere. As a result, an eastern location such as Guayaquil, Ecuador, normally receives 91.4 cm (36 in.) of precipitation each year under dominant high pressure, whereas western islands and maritime landmasses of the Indonesian archipelago receive over 254 cm (100 in.) under dominant low pressure. This normal condition is shown in Figure 1a.

Occasionally, for reasons unexplained so far, surface ocean temperature patterns shift from their usual locations. This realignment affects surface ocean currents and weather on both sides of the Pacific. Sea-surface temperatures of the central and eastern Pacific then rise above normal, sometimes becoming more than 8 C° (14 F°) warmer, replacing the normally cold, upwelling, nutrient-rich water along Peru's coastline. This loss of nutrients affects the food chain and many plankton, fish, and predator birds perish.

Such ocean-surface warming, the "warm pool," may extend to the International Date Line—Figure 1b. See the sea-surface temperature image in Figure 5-8 that shows the Western Pacific Warm Pool.

People fishing the waters off Peru have come to expect a slight warming of surface waters that lasts for a few weeks at irregular intervals. The name El Niño ("the boy child") is used by Peruvians because these episodes usually occur around the traditional December celebration time of Christ's birth (although they have occurred as early as spring and summer).

Higher pressure than normal develops over the western Pacific, and lower pressure over the eastern Pacific. Trade winds normally moving from east to west weaken and can be replaced by an eastward (west-to-east) flow. The shifting of atmospheric pressure and wind patterns across the Pacific, known as the Southern Oscillation, appears to be interrelated to El Niño (Figure 1c). Thus, the designation *ENSO* is derived (El Niño-Southern Oscillation). Scientists at the National Oceanographic and Atmospheric Administration (NOAA) speculate that ENSO events historically occurred in 1877–1878, 1884, 1891, 1899–1900, 1925, 1931, and 1941–1942, among other times. And they are certain that such ENSO events occurred in 1953, 1957–1958, 1965, 1969–1970, 1972–1973, 1976–1977, 1986–1987, 1991–1992, and 1993, with the most intense episode of this century occurring during 1982–1983. The expected interval for recurrence is 3 to 5 years, but it may range from 2 to 12 years.

The anchovy catch off Peru dropped from 12 million tons in 1970 to a low of 0.5 million tons in 1983, following the

warm water blockage of the cold, upwelling current by several ENSO events. Related effects occurred into the midlatitudes: droughts in South Africa, southern India, Australia, and the Philippines; strong hurricanes in the Pacific, including Tahiti and French Polynesia; and flooding in the southwestern United States and mountain states, Bolivia, Cuba, Ecuador, and Peru. The Colorado River flooding shown in the Focus Study in Chapter 15 was in large part attributable to the 1982–1983 El Niño. Yet, as conditions vary, other El Niños have produced drought in the very regions that flooded during a previous episode.

The 1982–1983 El Niño, clearly noted on Figure 1c, strengthened by an extremely strong Southern Oscillation, led scientists to further clarification of complex global interconnections among surface temperatures, pressure patterns in the Pacific, occurrences of drought in some places, excessive rainfall in others, and the disruption of fisheries and wildlife. Estimates place the overall damage from the 1982–1983 ENSO at more than $8 billion worldwide.

The climate of one location is related to climates elsewhere, although it should be no surprise that Earth operates as a vast integrated system. "It is fascinating that what happens in one area can affect the whole world. As to why this happens, that's the question of the century. Scientists are trying to make order out of chaos," said NOAA scientist Alan Strong. Currently, remote-sensing capabilities of satellites and improved surface observations are permitting scientists to monitor sea-surface temperatures and to better predict the timing and intensity of each El Niño.

here. Insolation is the energy input for the climate system, but it varies widely over Earth's surface with latitude (see Chapter 2 and Figures 2-12, 2-13, and 2-14). Daylength and temperature patterns vary diurnally and seasonally. The principal controls of temperature are latitude, altitude, land-water heating differences, and cloud cover. The pattern of world temperatures and their annual ranges is discussed in Chapter 5 (see Figures 5-11, 5-13, and 5-14).

Average temperatures and daylength are the basic factors that help us approximate POTET (potential evap-

otranspiration). Temperature variations are coupled with dynamic forces in the atmosphere to produce Earth's pattern of atmospheric pressure and resulting global wind systems (see Figures 6-13 and 6-15). Important too are the locations and physical characteristics of air masses, those vast bodies of homogeneous air that form over oceanic and continental source regions.

Moisture is the remaining input to climate. The hydrologic cycle transfers moisture, with its tremendous latent heat energy, through Earth's climate system (see

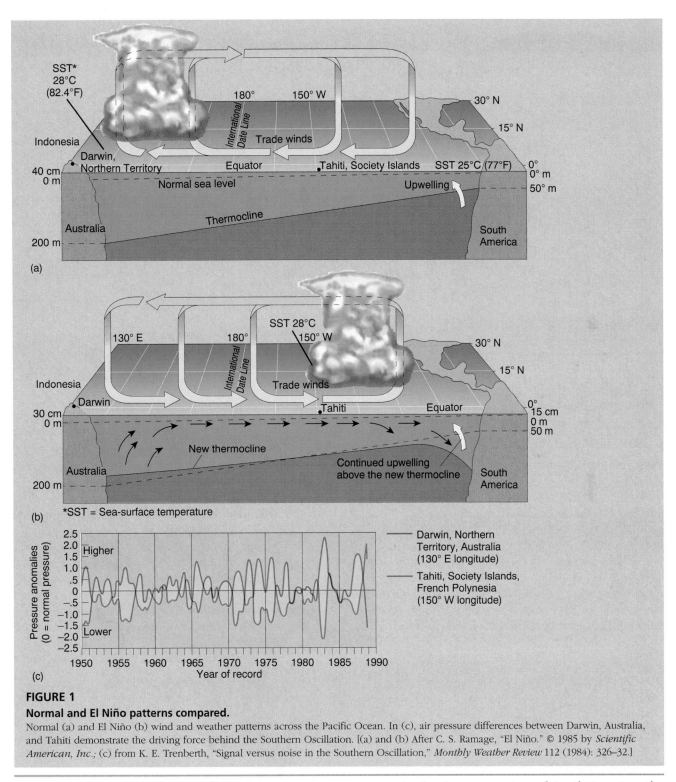

**FIGURE 1**

**Normal and El Niño patterns compared.**

Normal (a) and El Niño (b) wind and weather patterns across the Pacific Ocean. In (c), air pressure differences between Darwin, Australia, and Tahiti demonstrate the driving force behind the Southern Oscillation. [(a) and (b) After C. S. Ramage, "El Niño." © 1985 by *Scientific American, Inc.*; (c) from K. E. Trenberth, "Signal versus noise in the Southern Oscillation," *Monthly Weather Review* 112 (1984): 326–32.]

Figure 9-1). The moisture input to climate is precipitation in all its forms. Figure 10-3 shows the worldwide distribution of precipitation. Its patterns are important, for it is a key climate control factor.

Most of Earth's desert climates and regions of permanent drought are in lands dominated by subtropical high-pressure cells, with bordering lands grading to grasslands and to forests as precipitation increases. The most consistently wet climates on Earth straddle the

equator in the Amazon region of South America, the Zaire region of Africa, and the Indonesian and Southeast Asian area, all of which are influenced by equatorial low pressure and the intertropical convergence zone (ITCZ, see Figure 6-13).

Simply relating the two principal climatic components—temperature and precipitation—reveals climate types (Figure 10-4). Temperature and precipitation patterns, plus other weather factors, provide the key to climate classification.

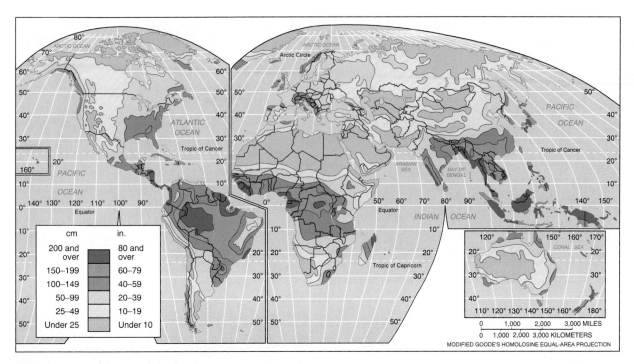

**FIGURE 10-3**
**Worldwide average annual precipitation.**

# Classification of Climatic Regions

The ancient Greeks simplified their view of world climates into three zones: The "torrid zone" referred to warmer areas south of the Mediterranean; the "frigid zone" was to the north; and the area where they lived was labeled the "temperate zone," which they considered the optimum climate. They believed that travel too close to the equator or too far north would surely end in death. But the world is a diverse place, and through exploration and satellite observations, Earth's myriad climatic variations are revealed today.

Note that climate cannot actually be observed and really does not exist at any particular moment. Climate is therefore a conceptual *statistical construction* from measured weather elements according to the assumptions of the classification scheme employed.

**Classification** is the process of ordering or grouping data or phenomena in related categories. Such generalizations are important organizational tools in science and are especially useful for the spatial analysis of climatic regions. A climate classification based on *causative* factors—for example, the interaction of air masses—is called a **genetic classification**. A climate classification based on *statistics* or other data is an **empirical classification**.

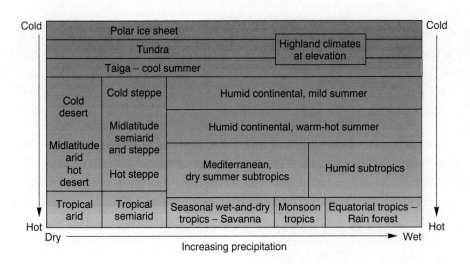

**FIGURE 10-4**
**Climatic relationships.**
Temperature and precipitation schematic reveals climatic relationships.

Climate classifications based on temperature and precipitation data are examples of empirical classifications. This chapter features descriptions of climatic regions that include both genetic factors of causal elements and empirical factors of statistical elements.

Just as there is no single universal climate classification system, neither is there a single set of empirical or genetic elements to which everyone agrees. Any classification system should be viewed as developmental, because it is always open to change and improvement. Such an evolutionary view is healthy scientifically, because a dynamic classification can provide the same stimulus as a theory in generating further exploration and research.

The many climate classifications that have been proposed are based on different assumptions. Genetic classifications explain climates in terms of net radiation, thermal regimes, or air mass dominance over a region. Empirical classifications are descriptive and single out selected weather data. One empirical classification system, published by C. W. Thornthwaite in 1948, identified moisture regions using aspects of the water-budget approach (Chapter 9) and vegetation types. Another empirical classification system, and the one we describe here, is the Köppen classification system.

## The Köppen Classification System

The Köppen classification system, widely used for its ease of comprehension, was designed by Wladimir Köppen (1846–1940), a German climatologist and botanist. His classification began with an article on heat zones in 1884. By 1900, he included plant communities in his selection of some temperature criteria, using a world vegetation map prepared by French plant physiologist Alphonse de Candolle in 1855. He added letter symbols to designate climate types (example: *Af, tropical rain forest*).

Later, Köppen reduced the role played by plants in setting boundaries and moved his system toward strictly climatological empiricism—using weather data. The first wall map showing world climates, co-authored with his student Rudolph Geiger, was introduced in 1928 and soon was widely used. Never completely satisfied with his system, Köppen continued to refine it until his death.

***Classification Criteria.*** The basis of any empirical classification system is the choice of criteria used to draw lines on a map to designate different climates. **Köppen-Geiger climate classification** uses *average monthly temperatures, average monthly precipitation,* and *total annual precipitation* to devise its spatial categories and boundaries. But we must remember that boundaries really are transition zones of gradual change. The trends and overall patterns of boundary lines are more important than their precise placement, especially with the small scales generally used on world maps.

The modified Köppen-Geiger system has its drawbacks. It does not consider winds, temperature extremes, precipitation intensity, amount of sunshine, cloud cover, or net radiation. Yet the system is important because its *correlation with the actual world is reasonable* and the *input data are standardized and readily available.* For our purposes of general understanding, a modified Köppen-Geiger climate classification is useful as an organizational tool for this discussion.

***Köppen's Climatic Designations.*** The Köppen system uses capital letters (A, B, C, D, E, H) to designate climatic categories by latitude, from the equator (A) to the poles (E), plus an H category for the special case of highlands. Of the six categories, all but B are based upon purely temperature criteria:

**A**   Tropical (equatorial regions)
**C**   Mesothermal (humid subtropical, Mediterranean, and marine west coast regions)
**D**   Microthermal (humid continental, subarctic regions)

## News Report 1

### What's in a Boundary?

Originally, Köppen proposed that the isotherm boundary between mesothermal C and microthermal D climates be a coldest month of –3°C (26.6°F) or lower. That might be an accurate criterion for Europe, but for conditions in North America, the 0°C isotherm is considered more appropriate. The difference between the 0°C and –3°C isotherms covers an area about the width of the state of Ohio. Remember, these isotherm lines are really transition zones and do not mean abrupt change from one temperature to another.

A line denoting at least one month below freezing runs from New York City roughly along the Ohio River, trending westward until it meets the dry climates in the southeastern corner of Colorado. The line marking –3°C as the coldest month runs farther north along Lake Erie and the southern tip of Lake Michigan. From year to year, the position of the 0°C isotherm for January can shift several hundred kilometers as weather conditions vary.

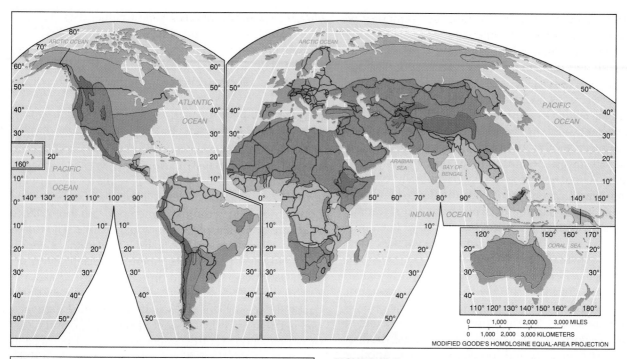

**FIGURE 10-5**
**Climate regions generalized.**
Six general classifications of the Köppen climate classification system.

Tropical climates

Dry arid/semiarid climates

Mesothermal climates

Microthermal climates

Polar climates

Highland climates

**E**  Polar (polar regions)

**H**  Highland (Compared with lowlands at the same latitude, highlands have lower temperatures—recall the normal lapse rate—and more efficient precipitation due to lower POTET demand. This difference justifies the distinction between highlands and surrounding climates.)

Only one climate classification is based on moisture as well:

**B**  Dry (deserts and steppes).

Figure 10-5 shows the distribution of each of these six climate classifications on the land. This generalized map helps you see the spatial pattern of climate. The Geosystems Home Page on the Internet presents weather data for selected stations worldwide, grouped by Köppen climate type.

Within each climate classification, additional lowercase letters are used to signify temperature and moisture conditions. For example, in a *tropical rain forest Af* climate, the *A* tells us that the average coolest month is above 18°C (64.4°F, average for the month), and the *f* indicates that the weather is constantly wet (German *feucht,* for "moist"), with the driest month receiving at least 6 cm (2.4 in.) of precipitation. As you can see in the

map of *A* climates, the *Af tropical rain forest* climate dominates along the equator.

In a *Dfa* climate, the *D* means that the average warmest month is above 10°C (50°F), with at least one month falling below 0°C (32°F); the *f* says that at least 3 cm (1.2 in.) of precipitation falls during every month; and the *a* indicates a warmest summer month averaging above 22°C (71.6°F). Thus, a *Dfa* climate is a humid-continental, hot-summer climate in the microthermal D category. (Using the world climate map in Figure 10-7 and the weather data for your city or town, see if you can determine the Köppen classification for where you live.)

Try not to get lost in the alphabet soup of this system, for it is meant to help you understand complex climates through a simplified set of criteria. For detailed criteria, a box at the end of each climate section presents Köppen's guidelines. To help keep it straight, always say the climate's descriptive name followed by its symbol: for example, *Mediterranean dry-summer Csa, tundra ET,* or *cold midlatitude steppe BSk.* We follow this convention throughout the text.

***A Hypothetical Continent.*** Imagine a large continent that stretches from about 70° north latitude to 60° south, straddling the equator (Figure 10-6). It is relatively low in elevation and affected by surrounding air masses of different characteristics, seasonal patterns of shifting Sun altitude and daylength, and ocean currents of varying temperature and direction. Use this hypothetical continent to get a general idea of the distribution of climatic regions. This may help you understand the world climate map that follows.

The hypothetical continent's climates appear in a logical placement: the warm and wet equatorial region, dry arid and semiarid subtropics in the western area, humid continental climate of the east, marine climates in the northwest and far southern portion, and the arctic, tundra, and polar regions in the extreme north and south. Remembering the description of orographic effects from Chapter 8, what would be the climate response if we put a large mountain range down the western margin of the continent? What effect would increasing ocean temperatures have on the surrounding air masses and therefore climates over land?

## Global Climate Patterns

World climates are presented in Figure 10-7. The following sections describe specific climates, organized around each of the main climate categories. A map showing distribution appears in the introductory section of each climate and includes a few representative cities. **Climographs** are presented for cities that exemplify particular climates. The climographs show monthly temperature and precipitation, location coordinates, average annual temperature, total annual precipitation, elevation, the local population, annual temperature range, annual hours of sunshine (if available, as an indication of cloudiness), and a location map.

Discussions of soils, vegetation, and major terrestrial biomes that fully integrate these global climate patterns are presented in Chapters 18, 19, and 20. Table 20-1 integrates all this information and will enhance your understanding of this chapter, so please refer to it as you read.

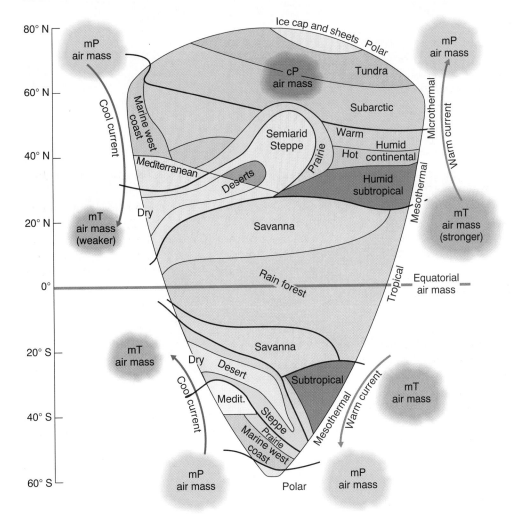

### FIGURE 10-6
### Arrangement of Earth's climates on a hypothetical continent.
Real continents basically follow the pattern shown, but they vary because of their shape, elevation, and latitudinal position.
[Adapted from J. M. Crowley, "Biogeography," *Canadian Geographer* 11 (1967).]

**FIGURE 10-7**
World climates according to the
Köppen-Geiger classification system.

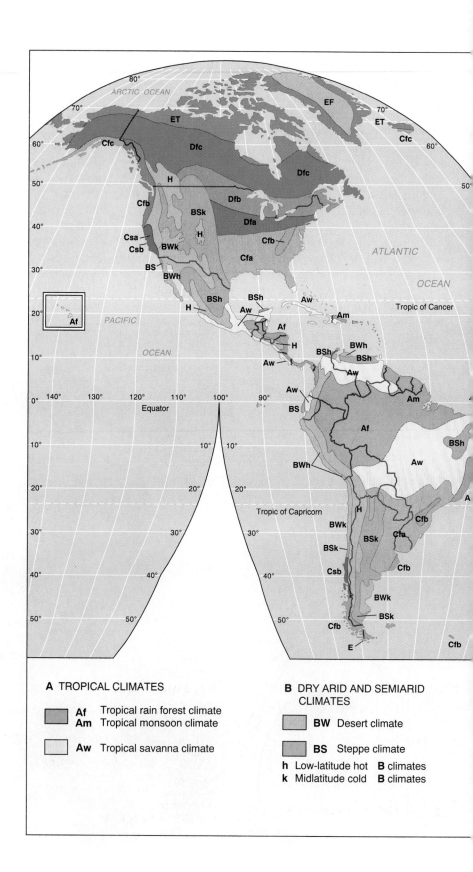

**A** TROPICAL CLIMATES

**Af** Tropical rain forest climate
**Am** Tropical monsoon climate

**Aw** Tropical savanna climate

**B** DRY ARID AND SEMIARID
CLIMATES

**BW** Desert climate

**BS** Steppe climate

**h** Low-latitude hot  **B** climates
**k** Midlatitude cold  **B** climates

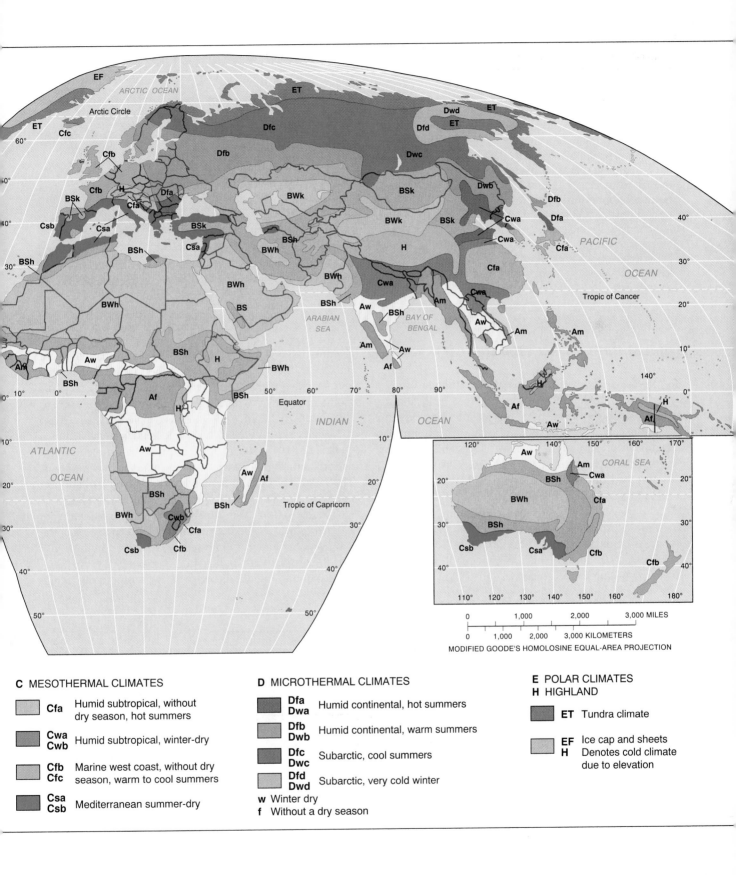

**C** MESOTHERMAL CLIMATES

| | | |
|---|---|---|
| | **Cfa** | Humid subtropical, without dry season, hot summers |
| | **Cwa Cwb** | Humid subtropical, winter-dry |
| | **Cfb Cfc** | Marine west coast, without dry season, warm to cool summers |
| | **Csa Csb** | Mediterranean summer-dry |

**D** MICROTHERMAL CLIMATES

| | | |
|---|---|---|
| | **Dfa Dwa** | Humid continental, hot summers |
| | **Dfb Dwb** | Humid continental, warm summers |
| | **Dfc Dwc** | Subarctic, cool summers |
| | **Dfd Dwd** | Subarctic, very cold winter |

**w** Winter dry
**f** Without a dry season

**E** POLAR CLIMATES
**H** HIGHLAND

| | | |
|---|---|---|
| | **ET** | Tundra climate |
| | **EF H** | Ice cap and sheets Denotes cold climate due to elevation |

MODIFIED GOODE'S HOMOLOSINE EQUAL-AREA PROJECTION

# Tropical Climates (A)

Tropical A climates, the most extensive, occupy about 36% of Earth's surface, including both ocean and land areas. The tropical A climates straddle the equator from about 20° N to 20° S, roughly between the Tropics of Cancer and Capricorn, giving the climate type its name. Tropical A climates stretch northward to the tip of Florida and south-central Mexico, central India, and Southeast Asia. In A climates, the coolest month must be warmer than 18°C (64.4°F), making these climates truly winterless. Consistent daylength and almost perpendicular Sun angle throughout the year generate this warmth.

A climates are subdivided by annual precipitation into *tropical rain forest Af, tropical monsoon Am,* and *tropical savanna Aw.*

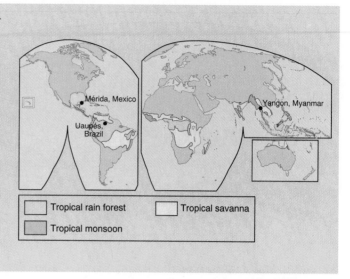

Tropical rain forest ☐  Tropical savanna ☐
Tropical monsoon ☐

## Tropical Rain Forest Climates (Af)

The *tropical rain forest Af* climate is constantly moist and warm. Convectional thunderstorms, triggered by local heating, peak each day from mid-afternoon to late evening inland and earlier in the day where marine influence is strong. This precipitation pattern follows the migrating intertropical convergence zone (ITCZ, Chapter 6). The ITCZ shifts northward and southward with the high summer Sun throughout the year, but it influences *tropical rain forest Af* regions during all 12 months. Not surprisingly, water surpluses are enormous, creating the world's greatest stream discharges in the Amazon and Congo (Zaire) Rivers.

High rainfall sustains lush evergreen broadleaf tree growth, producing Earth's equatorial and tropical rain forests (Figure 10-8). Their leaf canopy is so dense that little light diffuses to the forest floor, leaving the ground surface dim and sparse in plant cover. Dense surface vegetation occurs along riverbanks, where light is abundant. (Widespread deforestation of Earth's rain forest is detailed in Chapter 20.)

High temperature promotes energetic bacterial action in the soil so that organic material is quickly consumed. Heavy precipitation washes away certain minerals and nutrients. The resulting soils are somewhat sterile and can support intensive agriculture only if supplemented by fertilizer.

Uaupés, Brazil (Figure 10-9), is characteristic of *tropical rain forest Af.* On the climograph you can see that the lowest-precipitation month receives nearly 15 cm (6 in.), and the annual temperature range is barely 2 C° (3.6 F°). In all such climates, diurnal (daily) temperature range

**FIGURE 10-8**
**The rain forest.**
The lush equatorial rain forest and Sahgha River, near Quesso Congo, Africa. [Photo by BIOS (M. Gunther)/Peter Arnold, Inc.]

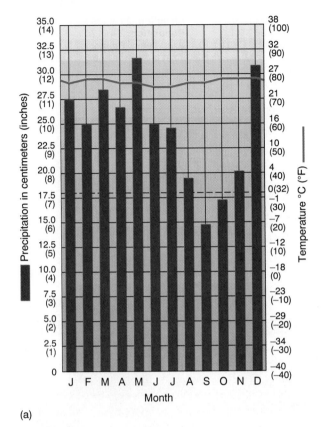

**Station:** Uaupés, Brazil **Af**  **Elevation:** 86 m (282.2 ft)
**Lat/long:** 0°08' S 67°05' W  **Population:** 7500
**Avg. Ann. Temp.:** 25°C (77°F)  **Ann. Temp. Range:**
**Total Ann. Precip.:**  2 C° (3.6 F°)
  291.7 cm (114.8 in.)  **Ann. Hr of Sunshine:** 2018

**FIGURE 10-9**
**Tropical rain forest climate.**
(a) Climograph for Uaupés, Brazil (*tropical rain forest Af*). (b) The rain forest near Uaupés along a tributary of the Rio Negro. [Photo by Will and Deni McIntyre/Photo Researchers, Inc.]

exceeds the annual minimum–maximum range: Day-night temperatures can range more than 11 C° (20 F°), more than five times the annual average range.

The only interruption of *tropical rain forest Af* climates is in the highlands of the South American Andes and in East Africa. There, higher elevations produce lower temperatures; Mount Kilimanjaro is less than 4° south of the equator, but at 5895 m (19,340 ft) it has permanent glacial ice on its summit. Köppen placed such sites in the H (highland) climate designation.

### Tropical Monsoon Climates (Am)

The *tropical monsoon Am* climates feature a dry season that lasts one or more months. Rainfall brought by the ITCZ affects these areas from 6 to 12 months of the year, with the dry season occurring when the convergence zone is not overhead. Yangon, Myanmar (formerly Rangoon, Burma) is an example of this climate type, as illustrated by the climograph and photograph in Figure 10-10.

*Tropical monsoon Am* climates lie principally along coastal areas within the tropical rain forest climatic realm and experience seasonal variation of winds and precipitation. Evergreen trees grade into thorn forests on the drier margins near the adjoining savanna climates.

### Tropical Savanna Climates (Aw)

*Tropical savanna Aw* climates exist poleward of the *tropical rain forest Af* climates. The ITCZ dominates these climates for 6 months or less of the year as it migrates with the high Sun. Summers are wetter than winters because convectional rains accompany the shifting ITCZ. The *w* signifies this winter-dry condition, when rains depart the area and high pressure dominates. Thus, POTET exceeds PRECIP in winter, causing water-budget deficits.

Temperatures fluctuate more in tropical savanna than in tropical rain forest regions. The tropical savanna regime can have two temperature maximums because the Sun's direct rays are overhead twice during the year.

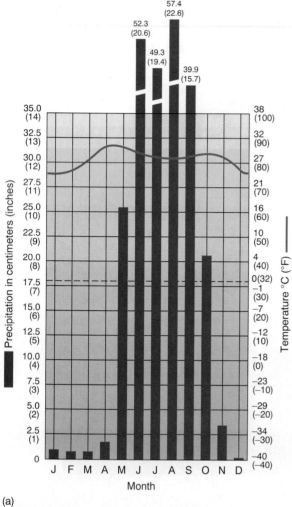

**Station:** Yangon, Myanmar* **Am**
**Lat/long:** 16°47' N 96°10' E
**Avg. Ann. Temp.:** 27.3°C (81.1°F)
**Total Ann. Precip.:**
    252.7 cm (99.5 in.)
*(Formerly Rangoon, Burma)

**Elevation:** 23 m (76 ft)
**Population:** 2,458,700
**Ann. Temp. Range:**
    5.5 C° (9.9 F°)

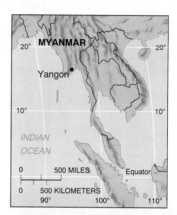

(a)

(b)

**FIGURE 10-10**
**Tropical monsoon climate.**

(a) Climograph for Yangon, Myanmar (formerly Rangoon, Burma)
(*tropical monsoon Am*); (b) The monsoonal forest near Malang, Java
at the Purwodadi Botanical Gardens. [Photo by Tom McHugh/Photo
Researchers, Inc.]

**FIGURE 10-11**
**Tropical savanna climate.**
Characteristic *tropical savanna Aw*
landscape in Kenya, with plants
adapted to seasonally dry water
budgets. [Photo by Stephen J.
Krasemann/DRK Photo.]

*Tropical savanna Aw* vegetation is characterized by dominant grasslands with scattered trees, which are drought-resistant to cope with the highly variable precipitation (Figure 10-11).

Mérida, Mexico, on the northern edge of the Yucatán Peninsula, is a characteristic *tropical savanna Aw* station (Figure 10-12). Temperatures are consistent with tropical A climates, although the comparative winter dryness affords a few months of lower humidity. Summer thunderstorms sometimes are augmented by tropical cyclones. In September 1988, Hurricane Gilbert devastated most of Mérida with rain and strong winds.

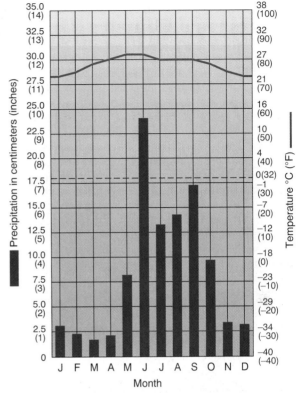

(a)

**FIGURE 10-12**
**Climograph for Mérida, Mexico**
*(tropical savanna Aw).*

**Station:** Mérida, Mexico    **Aw**       **Elevation:** 22 m (72.2 ft)
**Lat/long:** 20°59' N 89°39' W          **Population:** 400,000
**Avg. Ann. Temp.:** 26°C (78.8°F)       **Ann. Temp. Range:** 5 C°(9 F°)
**Total Ann. Precip.:**                  **Ann. Hr of Sunshine:** 2379
            93 cm (36.6 in.)

**Köppen Guidelines**
**Tropical Climates—A**

Consistently warm with all months averaging above 18°C (64.4°F); annual PRECIP exceeds POTET.

**Af —  Tropical rain forest:**
 f  =  All months receive PRECIP in excess of 6 cm (2.4 in.).

**Am —  Tropical monsoon:**
 m  =  A marked short dry season with 1 or more months receiving less than 6 cm (2.4 in.) PRECIP; an otherwise excessively wet rainy season. ITCZ 6–12 months dominant.

**Aw —  Tropical savanna:**
 w  =  Summer wet season, winter dry season; ITCZ dominant 6 months or less, winter water-balance deficits.

# Mesothermal Climates (C)

Mesothermal C climates occupy the second-largest percentage of Earth's land and sea surface, about 27%. When land area alone is considered, they rank only fourth. Together, A and C climates dominate over half of Earth's oceans and about one-third of its land area. More than half the world's population—approximately 55%—resides in C climates. These warm and temperate lands are designated mesothermal, meaning "middle temperature."

The mesothermal C climates, and nearby portions of the microthermal D climates, are regions of great weather variability, for these are the latitudes of greatest air mass conflict. The C climatic region marks the beginning of true seasonality; contrasts in temperature are evidenced by vegetation, soil, and human lifestyle adaptations. Subdivisions of the C classification are based on precipitation variability: *humid subtropical, marine west coast,* and *Mediterranean.*

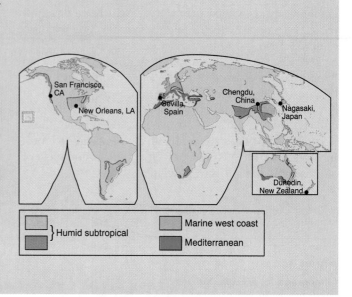

## Humid Subtropical Hot-Summer Climates (Cfa, Cwa)

The *humid subtropical hot-summer* climates *(Cfa, Cwa),* are either moist all year *(f)* or have a pronounced winter-dry period *(w)* and feature a hot summer (indicated by the *a*). During summer, the *humid subtropical hot-summer Cfa* climate is influenced by maritime tropical air masses generated over warm waters off eastern coasts. This warm, moist, unstable air produces convectional showers over land. In fall, winter, and spring, maritime tropical and continental polar air masses interact, generating frontal activity and frequent midlatitude cyclonic storms. Overall, precipitation averages 100–200 cm (40–80 in.) a year.

Nagasaki, Japan (Figure 10-13), is a characteristic *humid subtropical hot-summer Cfa* station. Unlike the precipitation of humid subtropical hot summer cities in the United States (Atlanta, Memphis, Norfolk, New Orleans), however, Nagasaki's summer precipitation is much greater than its winter precipitation because of the effects of the Asian monsoon. The lower precipitation of winter is not quite dry enough to change its classification to a winter-dry *w.*

*Humid subtropical winter-dry Cwa* climates are related to the winter-dry, seasonal pulse of the monsoons. They extend poleward from tropical savanna climates and have a summer month that receives 10 times more precipitation than their driest winter month. A representative station is Chengdu, China. Figure 10-14 demonstrates the strong correlation between precipitation and the high-summer Sun.

The habitability of the *humid subtropical hot-summer Cfa* and *humid subtropical winter-dry Cwa* climates and their ability to sustain populations are borne out by the concentrations in north-central India, the bulk of

China's 1.2 billion people, and the many who live in climatically similar portions of the United States.

## Marine West Coast Climates (Cfb, Cfc)

Marine west coast climates, featuring mild winters and cool summers, dominate Europe and other middle-to-high-latitude west coasts (see Figure 10-7). In contrast to the hot-summer humid climate of the southeastern United States, these regions have a cooler summer. The moderating effects of a marine location and maritime air masses on temperature patterns are discussed in Chapter 5.

Maritime polar air masses—cool, moist, unstable—dominate *marine west coast Cfb* and *Cfc* climates. Weather systems forming along the polar front move into these regions throughout the year, making weather quite unpredictable. Coastal fog, totaling 30 to 60 days, is a part of the moderating marine influence. Frosts are possible and tend to shorten the growing season.

Marine west coast climates are unusually mild for their latitude. They extend along the coastal margins of the Aleutian Islands in the North Pacific, cover the southern third of Iceland in the North Atlantic and coastal Scandinavia, and dominate the British Isles. It is hard to imagine that such high-latitude locations can have average monthly temperatures that are above freezing throughout the year! Unlike marine west coast climates in Europe, these climates in Canada, Alaska, Chile, and Australia are backed by mountains and remain restricted to coastal environs (Figure 10-15). (A graph of temperature at Vancouver, British Columbia, appears in Figure 5-9 with a photo of this *marine west coast Cf* city.)

The climograph for Dunedin, New Zealand, demonstrates the moderate temperature patterns and the annual

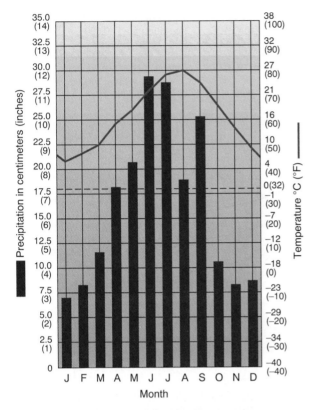

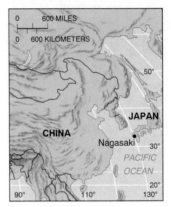

**Station:** Nagasaki, Japan   **Cfa**   **Elevation:** 27 m (88.6 ft)
**Lat/long:** 32°44' N 129°52' E   **Population:** 449,000
**Avg. Ann. Temp.:** 16°C (60.8°F)   **Ann. Temp. Range:**
**Total Ann. Precip.:**                              21 C° (37.8 F°)
                  195.7 cm (77 in.)   **Ann. Hr of Sunshine:** 2131

(a)

**FIGURE 10-13**

**Humid subtropical climate.**

(a) Climograph for Nagasaki, Japan (*humid subtropical Cfa*).
(b) Landscape near Nagasaki. [Photo by Noboru Komine/Photo
Researchers.]

## News Report 2

### *Cwa* Climate Region Sets Precipitation Records

Cherrapunji, India, in the Assam Hills south of the Himalayas, is the all-time precipitation record holder for a single year and for every other time interval from 15 days to 2 years. This region of *humid subtropical Cwa* climate is located just north of Bangladesh, where tropical cyclones, torrential rains, and flooding in 1970, 1988, and 1991 killed hundreds of thousands of people.

Because of the summer monsoons that pour in from the Indian Ocean and the Bay of Bengal, Cherrapunji has received 930 cm (30.5 ft) of rainfall in one month and 2647 cm (86.8 ft) in one year—both world records. In that location the contrast between the dry and wet monsoons is most severe, ranging from dry winds in the winter to torrential rains and floods in the summer.

temperature range for a *marine west coast Cf* station in the Southern Hemisphere (Figure 10-16).

An interesting anomaly occurs in the eastern United States. In portions of the Appalachian highlands, increased elevation moderates summer temperatures in the *humid subtropical hot summer Cfa* classification, producing a *marine west coast Cfb* climate. The climograph for Bluefield, West Virginia (Figure 10-17), reveals marine west coast temperature and precipitation patterns, despite its location in the east. Vegetation similarities between the Appalachians and

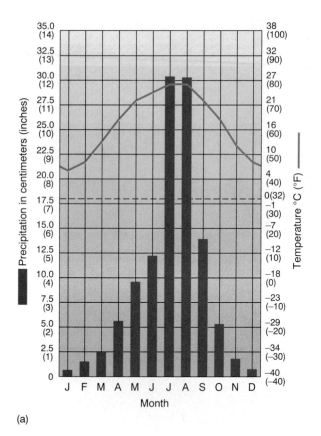

(a)

**Station:** Chengdu, China  **Cwa**   **Elevation:** 498 m (1633.9 ft)
**Lat/long:** 30°40' N 104°04' E    **Population:** 2,260,000
**Avg. Ann. Temp.:** 17°C (62.6°F)   **Ann. Temp. Range:**
**Total Ann. Precip.:**                                    20 C° (36 F°)
      114.6 cm (45.1 in.)   **Ann. Hr of Sunshine:** 1058

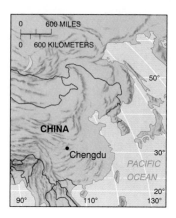

(b)

**FIGURE 10-14**

**Humid subtropical winter-dry climate.**

(a) Climograph for Chengdu, China *(humid subtropical Cwa)*. (b)
Landscape of southern interior China characteristic of this winter-dry
climate. This valley is near Mount Daliang in Sichuan Province.
[Photo by Jin Zuqi, Sovfoto/Eastfoto.]

**FIGURE 10-15**

**A marine west coast climate.**

The natural vegetation of Vancouver Island (*marine west coast Cfb*)
on nearby Cape Scott. [Photo by Michael Collier.]

the Pacific Northwest are quite noticeable and have enticed
many emigrants from the East to settle in these climatically
familiar environments in the West.

## Mediterranean Dry-Summer Climates (Csa, Csb)

Across the planet during summer months, shifting cells
of subtropical high pressure block moisture-bearing winds
from adjacent regions. For example, the continental tropical air mass over the Sahara in Africa shifts northward in
summer over the Mediterranean region and blocks maritime air masses and cyclonic systems. This shifting of stable, warm to hot, dry air over an area in summer and
away from these regions in the winter creates a pronounced dry-summer and wet-winter pattern. The designation specifies that at least 70% of annual precipitation
occurs during the winter months.

**FIGURE 10-16**

**A Southern Hemisphere marine west coast climate.**
(a) Climograph for Dunedin, New Zealand *(marine west coast Cfb).*
(b) Meadow, forest, and mountains on South Island, N.Z.
[Photo by Brian Enting/Photo Researchers, Inc.]

**FIGURE 10-17**

**Climate in the Appalachians.**
(a) Climograph for Bluefield, West Virginia *(marine west coast Cfb).*
(b) Characteristic mixed forest of Dolly Sods Wilderness in the Appalachian highlands. [Photo by David Muench Photography, Inc.]

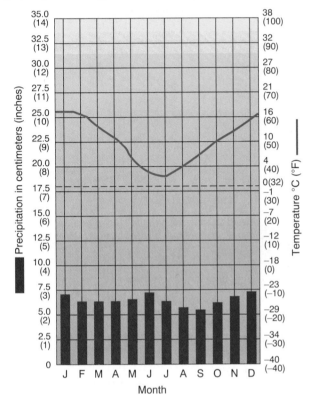

**Station:** Dunedin, New Zealand **Cfb**
**Lat/long:** 45°54' S 170°31' E
**Avg. Ann. Temp.:** 10.2°C (50.3°F)
**Total Ann. Precip.:**
    78.7 cm (31.0 in.)
**Elevation:** 1.5 m (5 ft)
**Population:** 77,500
**Ann. Temp. Range:**
    14.2C° (25.5 F°)

(a)

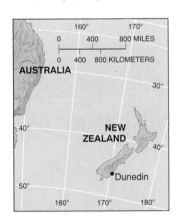

(b)

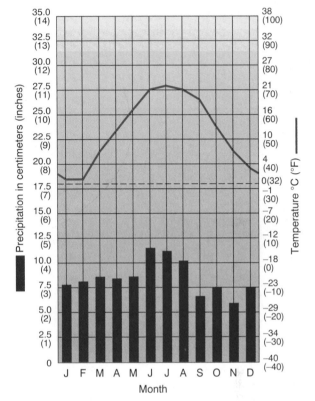

**Station:** Bluefield, West Virginia **Cfb**
**Lat/long:** 37°16' N 81°13' W
**Avg. Ann. Temp.:**
    12°C (53.6°F)
**Total Ann. Precip.:**
    101.9 cm (40.1 in.)
**Elevation:** 780 m (2559 ft)
**Population:** 16,000
**Ann. Temp. Range:**
    21 C° (37.8 F°)

(a)

(b)

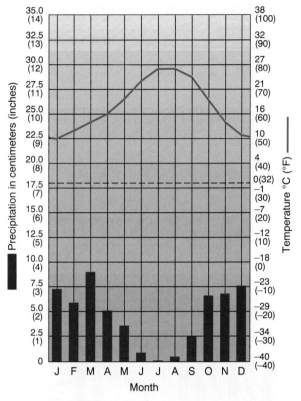

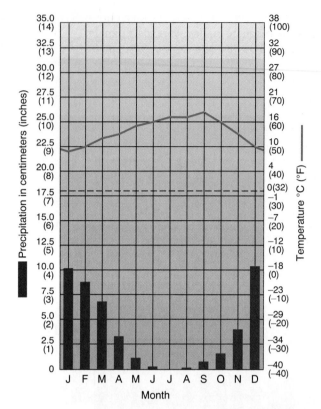

**Station:** Sevilla, Spain **Csa**
**Lat/long:** 37°22' N 6°00' W
**Avg. Ann. Temp.:**
  18°C (64.4°F)
**Total Ann. Precip.:**
  55.9 cm (22 in.)

**Elevation:** 13 m (42.6 ft)
**Population:** 651,000
**Ann. Temp. Range:**
  16 C° (28.8 F°)
**Ann. Hr of Sunshine:**
  2862

(a)

**Station:** San Francisco, California **Csb**
**Lat/long:** 37°37' N 122°23' W
**Avg. Ann. Temp.:**
  14°C (57.2°F)
**Total Ann. Precip.:**
  47.5 cm (18.7 in.)

**Elevation:** 5 m (16.4 ft)
**Population:** 750,000
**Ann. Temp. Range:**
  9 C° (16.2 F°)
**Ann. Hr of Sunshine:**
  2975

(b)

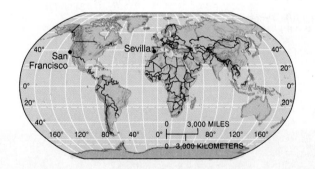

**FIGURE 10-18**

**Mediterranean climates.**

Climographs for (a) Sevilla, Spain (*Mediterranean hot-summer Csa*) and (b) San Francisco (*Mediterranean cool-summer Csb*). (c) The countryside around Olvera, Andalucia, Spain. [Photo by Kaz Chiba/Gamma-Liaison, Inc.]

(c)

Cool offshore currents (the California current, Canary current, Peru current, Benguela current, and West Australian current) produce stability in overlying air masses along west coasts, poleward of subtropical high pressure. The world climate map shows Mediterranean dry-summer climates along the western margins of North America, central Chile, and the southwestern tip of Africa, as well as across southern Australia and the Mediterranean Basin—the climate's namesake region.

Figure 10-18 compares the climographs of Mediterranean dry-summer cities Sevilla (Seville), Spain, and San Francisco. Coastal maritime effects moderate San Francisco's climate, producing a cooler summer. The transition to a hot-summer designation occurs no more than 24–32 km (15–20 mi) inland from San Francisco.

Along most west coasts, the warm, moist air overlying cool ocean water produces frequent summer fog, discussed in Chapter 8. This type of fog is not associated with the Mediterranean region because offshore water temperatures there are higher than water temperatures near other similar climatic regions during the summer months. Instead, the Mediterranean regime features high humidity values.

The *Mediterranean dry-summer* climate brings summer water-balance deficits. Winter precipitation recharges soil moisture, but water use usually exhausts it by late spring. Large-scale agriculture requires irrigation, although some subtropical fruits, nuts, and vegetables are uniquely suited to these conditions. Natural vegetation is a hard-leafed, drought-resistant variety known locally as *chaparral* in the western United States. (Chapter 20 discusses the other local names for this type of vegetation in other parts of the world.)

### Köppen Guidelines
### Mesothermal Climates—C

Warmest month above 10°C (50°F); coldest month above 0°C (32°F) but below 18°C (64.4°F); seasonal climates.

**Cfa, Cwa — Humid subtropical:**
  a = Hot summer; warmest month above 22°C (71.6°F).
  f = Year-round PRECIP.
  w = Winter drought, summer wettest month 10 times more PRECIP than driest winter month.

---

**Cfb, Cfc — Marine west coast:** Mild-to-cool summer
  f = Receives year-round PRECIP.
  b = Warmest month below 22°C (71.6°F) with 4 months above 10°C.
  c = 1—3 months above 10°C.

---

**Csa, Csb — Mediterranean summer dry:**
  s = Pronounced summer drought with 70% of PRECIP in winter.
  a = Hot summer with warmest month above 22°C (71.6°F).
  b = Mild summer; warmest month below 22°C.

# Microthermal Climates (D)

Humid microthermal climates have a long winter with some summer warmth. Here the term *microthermal* means cool temperate to cold. Approximately 21% of Earth's land surface is influenced by these climates, equaling about 7% of Earth's total surface. These climates occur poleward of the mesothermal climates and experience great temperature ranges related to continentality and air mass conflicts.

Because the Southern Hemisphere lacks substantial landmasses, microthermal climates develop there only in highlands. (In Figure 10-7, note the absence of microthermal Southern Hemisphere *D* climates.) Microthermal climates range from a *humid continental hot-summer Dfa* in Chicago to the formidable extremes of a frigid *subarctic Dwd* in Verkhoyansk, Siberia. (The letters *a, b, c*—hot, warm, cool—carry the same meaning that they have for mesothermal *C* climates.)

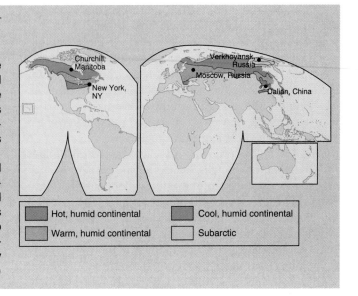

| | | | |
|---|---|---|---|
| ▨ | Hot, humid continental | ▨ | Cool, humid continental |
| ▨ | Warm, humid continental | ☐ | Subarctic |

## Humid Continental Hot-Summer Climates (Dfa, Dwa)

*Humid continental hot-summer Dfa* climates are differentiated by their annual precipitation distribution. Both humid continental moist-all-year (*Dfa*) and winter-dry (*Dwa*) climates are influenced by maritime tropical air masses in the summer. In North America, frequent weather activity is possible between conflicting air masses—maritime tropical and continental polar—especially in winter. The climographs for New York City and Dalian, China, illustrate these two hot-summer mesothermal climates (Figure 10-19).

Originally, forests covered the *humid continental hot-summer Dfa* climatic region of the United States as far west as the Indiana-Illinois border. Beyond that approximate line, tallgrass prairies extended westward to about the 98th meridian (98° W) and the approximate location of the 51 cm (20 in.) isohyet (line of equal precipitation). The shortgrass prairies extend to the west, where precipitation is less.

Deep sod made farming difficult for the first emigrant settlers, as did the climate. However, native grasses soon were replaced with domesticated wheat and barley. Various eastern inventions (barbed wire, the self-scouring steel plow, well-drilling techniques, railroads) helped open the region further. In the United States today, the *humid continental hot-summer Dfa* climatic region is the location of corn, soybean, hog, and cattle production.

The dry winter associated with the vast Asian landmass, specifically Siberia, is exclusively assigned *w* because it is dominated by an extremely dry and frigid winter high-pressure anticyclone. The dry monsoons of southern and eastern Asia are produced in the winter months by this system, as winds blow out of Siberia toward the Pacific and Indian Oceans. The intruding deep cold of continental air is a significant winter feature.

## Humid Continental Mild Summer Climates (Dfb, Dwb)

Soils are thinner and less fertile in the cooler microthermal climates, yet agricultural activity is important and includes dairy cattle, poultry, flax, sunflowers, sugar beets, wheat, and potatoes. Frost-free periods range from fewer than 90 days in the north to as many as 225 days in the south. Overall, precipitation is less than in the hot-summer regions to the south; however, notably heavier snowfall is important to soil moisture recharge when it melts. Various snow-capturing strategies are used, including fences and tall stubble (plant stalks left standing after harvest) left in fields to create snow drifts and thus more moisture retention on the soil.

The dry-winter aspect of the mild-summer climate (*Dwb*) occurs only in Asia, in a far-eastern area poleward of the winter-dry mesothermal climates. A representative *humid continental mild-summer Dwb* climate is Vladivostok, Russia, usually one of only two ice-free ports in that nation. Characteristic Dfb stations are Duluth, Minnesota, and Saint Petersburg, Russia. Figure 10-20 presents a climograph for Moscow, which is at 55° N, or about the same latitude as the southern shore of Hudson Bay in Canada.

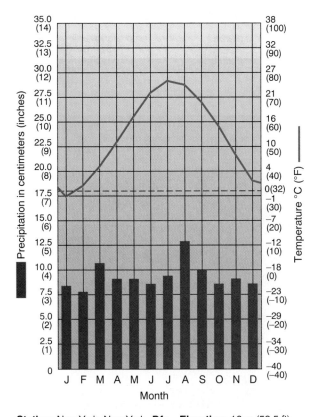

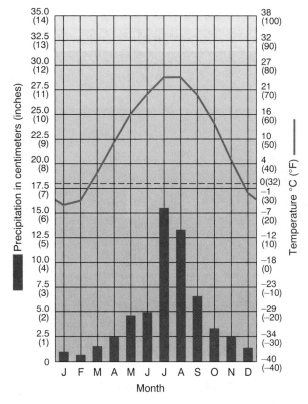

**Station:** New York, New York **Dfa**
**Lat/long:** 40°46' N 74°01' W
**Avg. Ann. Temp.:**
    13°C (55.4°F)
**Total Ann. Precip.:**
    112.3 cm (44.2 in.)

**Elevation:** 16 m (52.5 ft)
**Population:** 7,100,000
**Ann. Temp. Range:**
    24 C° (43.2 F°)
**Ann. Hr of Sunshine:**
    2564

(a)

**Station:** Dalian, China **Dwa**
**Lat/long:** 38°54' N 121°54' E
**Avg. Ann. Temp.:**
    10°C (50°F)
**Total Ann. Precip.:**
    57.8 cm (22.8 in.)

**Elevation:** 96 m (314.9 ft)
**Population:** 4,270,000
**Ann. Temp. Range:**
    29 C° (52.2 F°)
**Ann. Hr of Sunshine:**
    2762

(b)

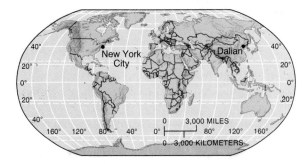

(c)

**FIGURE 10-19**

**Humid continental hot-summer climates.**

Climographs for (a) New York City (*humid continental Dfa*) and
(b) Dalian, China (*humid continental Dwa*). (c) Typical mixed
evergreen and deciduous forest in the Adirondack mountains
of New York. [Photo by Roy Whitehead/Photo Researchers, Inc.]

## *Subarctic Climates (Dfc, Dwc, Dwd)*

Farther poleward, seasonality becomes greater. The short
growing season is more intense during long summer days.
The three cold subarctic climates—*Dfc, Dwc, Dwd*—include

vast stretches of Alaska, Canada, northern Scandinavia, and
Russia. Discoveries of minerals and petroleum reserves have
led to new interest in portions of these regions.

Areas that receive 25 cm (10 in.) or more of precipitation a year on the northern continental margins

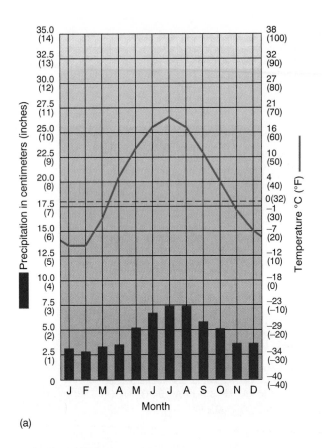

Station: Moscow, Russia    **Dfb**
Lat/long: 55°45' N 37°34' E
Avg. Ann. Temp.: 4°C (39.2°F)
Total Ann. Precip.:
                57.5 cm (22.6 in.)

Elevation: 156 m (511.8 ft)
Population: 9,900,000
Ann. Temp. Range:
                29 C° (52.2 F°)
Ann. Hr of Sunshine:
                1597

(a)

(b)

**FIGURE 10-20**

**Humid continental cool-summer climate.**

(a) Climograph for Moscow, Russia (*humid continental Dfb*). (b) Springtime fields near Saratov, Russia, during the short summer season. [Photo by Wolfgang Kaehler/Gamma-Liaison, Inc.]

and are covered by the so-called snow forest of fir, spruce, larch, and birch are the *boreal forests* of Canada and the *taiga* of Russia. These forests are in transition to the more open northern woodlands and to the tundra region of the far north. Forests thin out to the north when the warmest month drops below an average temperature of 10°C (50°F).

Soils are thin in these lands once scoured by glaciers. Precipitation and POTET both are low, so soils are generally moist and either partially or totally frozen beneath the surface, a phenomenon known as *permafrost*.

The Churchill, Manitoba, climograph (Figure 10-21) shows average monthly temperatures below freezing for 7 months of the year, during which time light snow cover and frozen ground persist. High pressure dominates Churchill during its cold winter—this is the source region for the continental polar air mass. Churchill is representative of the *subarctic Dfc* climate: annual temperature range of 40 C° (72 F°) and low precipitation of 44.3 cm (17.4 in.).

The dry-winter *subarctic Dwc* and *subarctic Dwd* climates occur only within Russia. The intense cold of Siberia and north-central and eastern Asia are difficult to comprehend, for these areas experience a coldest month with an average temperature lower than –38°C (–36.4°F) for 7 months. And as described in Chapter 5, minimum temperatures of below –68°C (–90°F) have been recorded there. Yet summer maximum temperatures in these same areas normally exceed +37°C (+98°F)!

A typical *subarctic Dwd* station is Verkhoyansk, Siberia (Figure 10-22). For 4 months of the year average temperatures fall below –34°C (–29.2°F). Verkhoyansk has probably the world's greatest annual temperature range from winter to summer: a remarkable 63 C° (113.4 F°)! Winters feature brittle metals and plastics, triple-thick window panes, and temperatures that render straight antifreeze a solid.

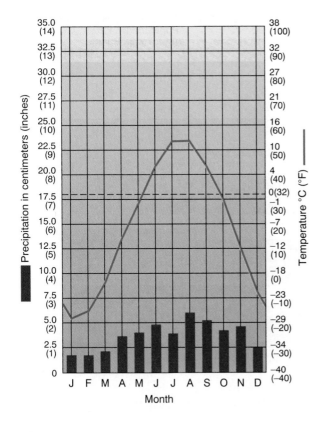

(a)

**FIGURE 10-21**
**Subarctic climate.**
(a) Climograph for Churchill, Manitoba (*subarctic Dfc*). (b) Winter scene in the Churchill area near Hudson Bay. [Photo by Art Wolfe/Tony Stone Images.]

**Station:** Churchill, Manitoba **Dfc**
**Lat/long:** 58°45' N 94°04' W
**Avg. Ann. Temp.:** −7°C (19.4°F)
**Total Ann. Precip.:**
    44.3 cm (17.4 in.)

**Elevation:** 35 m (114.8 ft)
**Population:** 1400
**Ann. Temp. Range:**
    40 C° (72 F°)
**Ann. Hr of Sunshine:**
    1732

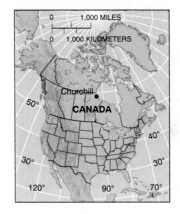

(b)

### Köppen Guidelines
### Microthermal Climates—D

Warmest month above 10°C (50°F); coldest month below 0°C (32°F); cool temperate-to-cold conditions; snow climates. In Southern Hemisphere, only in highland climates.

**Dfa, Dwa — Humid continental:**
  a = Hot summer; warmest month above 22°C (71.6°F).
  f = Year-round PRECIP.
  w = Winter drought.

**Dfb, Dwb — Humid continental:**
  b = Mild summer; warmest month below 22°C (71.6°F).
  f = Year-round PRECIP.
  w = Winter drought.

**Dfc, Dwc, Dwd — Subarctic:**
    Cool summers, cold winters.
  f = Year-round PRECIP.
  w = Winter drought.
  c = 1–4 months above 10°C.
  d = Coldest month below −38°C (−36.4°F), in Siberia only.

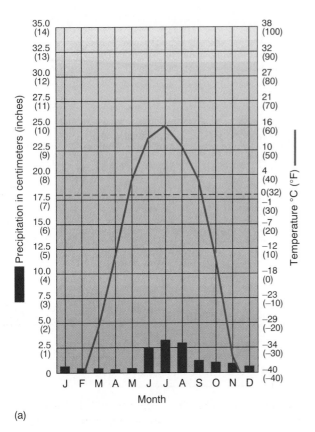

Station: Verkhoyansk, Russia  Dwd  Elevation: 137 m (449.5 ft)
Lat/long: 67°35' N 133°27' E        Population: 1400
Avg. Ann. Temp.: −15°C (5°F)       Ann. Temp. Range:
Total Ann. Precip.:                       63 C° (113.4 F°)
        15.5 cm (6.1 in.)

(a)

(b)

**FIGURE 10-22**
**Extreme subarctic winter-dry climate.**
(a) Climograph for Verkhoyansk, Russia (*subarctic Dwd*). (b) Scene in the town of Verkhoyansk during the short summer.
[Photo by Dean Conger/National Geographic Society.]

## Polar Climates (E)

The polar climates—*tundra ET*, *ice cap EF*, and *polar marine EM*—cover about 19% of Earth's total surface and about 17% of its land area. These climates have no true summer like that in lower latitudes. Poleward of the Arctic and Antarctic Circles, daylength increases in summer until daylight becomes continuous, yet average monthly temperatures never rise above 10°C (50°F). Daylength, which in part determines the amount of insolation received, and low Sun altitude are the principal climatic factors in these frozen and barren regions.

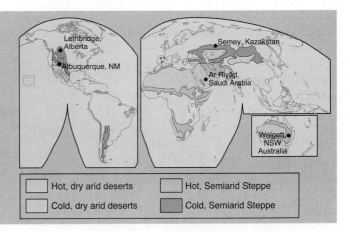

### Tundra ET Climate

In a *tundra ET* climate, land is under continuous snow cover for 8–10 months, but when the snow melts and spring arrives, numerous plants appear—stunted sedges, mosses, flowering plants, and lichens. Much of the area experiences permafrost conditions. The tundra is also the summer home of mosquitoes of legend and black gnats.

Like other microthermal climates, *tundra ET* climates are strictly a Northern Hemisphere occurrence, except

**FIGURE 10-23**
**Antarctica.**
The Antarctic landscape in the daylight at midnight; Earth's frozen freshwater reservoir. [Photo by Wolfgang Kaehler Photography.]

for elevated mountain locations in the Southern Hemisphere and a portion of the Antarctic Peninsula. Because of elevation, the summit of Mount Washington in New Hampshire (1914 m, or 6280 ft) statistically qualifies as a highland *tundra ET* climate of small scale.

## Ice Cap EF Climate

Most of Antarctica falls within the *ice cap EF* climate, as does the North Pole, with all months averaging below freezing. Both regions are dominated by dry, frigid air masses, with vast expanses that never warm above freezing. The area of the North Pole is actually a sea covered by ice, whereas Antarctica is a substantial continental landmass covered by Earth's greatest ice sheet. For comparison, winter minimums at the South Pole (July) frequently drop below the temperature of solid carbon dioxide or "dry ice" (–78°C, or –109°F).

Antarctica is constantly snow-covered but receives less than 8 cm (3 in.) of precipitation each year. Because of constant winds, it is difficult to discern what is new snow and what is old snow being blown around. Antarctic ice is several kilometers thick and is the largest repository of freshwater on Earth (Figure 10-23).

This ice is a vast historical record of Earth's atmosphere. Within it, thousands of volcanic eruptions worldwide have deposited ash layers, and ancient combinations of atmospheric gases lie trapped in frozen bubbles. An analysis of ice cores taken from Greenland and Antarctica is presented in News Report 4, Chapter 17.

***Polar Marine Climate.*** *Polar marine EM* stations are more moderate than other polar climates in winter, with no month below –7°C (20°F), yet they are not as warm as *tundra ET* climates. Because of marine influences, annual temperature ranges are low. This climate exists along the Bering Sea, the tip of Greenland, northern Iceland, Norway, and in the Southern Hemisphere, generally over oceans between 50° S and 60° S. Precipitation, which frequently falls as sleet, is greater in these regions than in continental E climates.

---

**Köppen Guidelines**
**Polar Climates—E**

Warmest month below 10°C (50°F); always cold; ice climates.

**ET — Tundra:**
Warmest month 0–10°C (32–50°F); PRECIP exceeds small POTET demand; snow cover 8–10 months.

**EF — Ice cap:**
Warmest month below 0°C (32°F); PRECIP exceeds very small POTET demand; the polar regions.

**EM — Polar marine:**
All months above –7°C (20°F), warmest month above 0°C; annual temperature range <17°C (30°F).

# Dry Arid and Semiarid Climates (B)

The dry and semiarid climates are the only ones that Köppen classified by precipitation rather than temperature. Dry climates are the world's arid deserts and semiarid regions, with their unique plants, animals, and physical features. The mountains, rock strata, long vistas, and the resilient struggle for life are all magnified by the dryness. Sparse vegetation leaves the landscape bare; POTET exceeds PRECIP throughout *dry arid and semiarid B climates*, creating permanent water deficits, the extent of which distinguishes desert and steppe climatic regions. (Specific annual and daily desert temperature regimes, including the highest record temperatures, are discussed in Chapter 5; surface energy budgets are covered in Chapter 4; desert landscapes in Chapter 15; and desert environments in Chapter 20.)

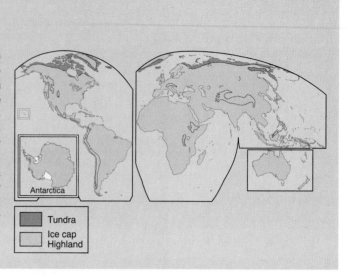

Tundra
Ice cap
Highland

**Desert Characteristics.** Vegetation is typically *xerophytic:* drought-resistant, waxy, hard-leafed, and adapted to aridity and low transpiration loss. Along water courses, plants called *phreatophytes,* or "water-well plants," have roots that penetrate to great depths for the water they need (Figure 10-24).

The *dry arid and semiarid B* climates occupy more than 35% of Earth's land area, clearly the most extensive climate over land. The world climate map in Figure 10-7 reveals the pattern of Earth's dry climates, which cover broad regions between 15° and 30° N and S latitudes. In these areas, subtropical high-pressure cells predominate, with subsiding, stable air and low relative humidity. Under generally cloudless skies, these subtropical deserts extend to western continental margins, where cool, stabilizing ocean currents operate offshore and summer advection fog forms. The Atacama Desert of Chile, the Namib Desert of Namibia, the Western Sahara of Morocco, and the Australian Desert each lie adjacent to a coastline.

Extension of these dry regions into higher latitudes of North and South America is associated with rain shadows, induced by orographic lifting over western mountains. Interior Asia, far distant from any moisture-bearing air masses, is included within the *dry arid and semiarid* climates.

Major subdivisions include: *deserts BW* (PRECIP less than one-half of POTET) and *semiarid steppes BS* (PRECIP more than one-half of POTET). Köppen developed simple formulas to determine the usefulness of rainfall on the basis of the season—whether precipitation falls principally in the winter with a dry summer, in the summer with a dry winter, or is evenly distributed. Winter rains are most effective because they fall at a time of lower moisture demand (Figure 10-25).

### Hot Low-Latitude Desert Climates (BWh)

*Hot low-latitude desert BWh* climates are Earth's true tropical and subtropical deserts and feature annual average temperatures above 18°C (64.4°F). They generally are concentrated on the western sides of continents, although Egypt, Somalia, and Saudi Arabia also fall within this classification. Rainfall is from local summer convectional showers. Some regions receive nearly nothing, whereas others may receive up to 35 cm (14 in.) precipitation a year. A representative *hot low-latitude desert BWh* station is Ar Riyāḍ (Riyadh), Saudi Arabia (Figure 10-26).

**FIGURE 10-24**
**Desert landscape.**
Desert plants are particularly well adapted to the harsh environment of the Mojave desert. [Photo by author.]

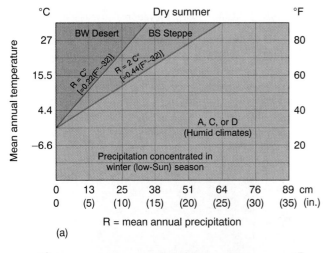

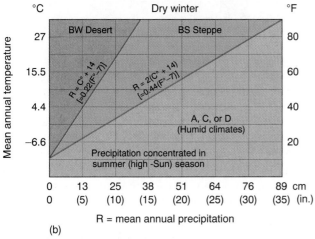

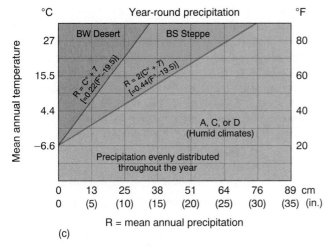

**FIGURE 10-25**

**Determining a dry arid or semiarid climate.**

Graphs demonstrate the moisture criteria used to divide the desert and steppe climates. (a) If the wettest winter month was three times wetter than the driest summer month, Köppen considered the summer dry. (b) The reverse reflects the role of POTET: Winter is considered dry if the wettest summer month receives 10 times the precipitation of the driest winter month. (c) B-climate classification if precipitation is evenly distributed throughout the year.

## Cold Midlatitude Desert Climates (BWk)

*Cold midlatitude desert BWk* climates cover only a small area: the southern countries of the former USSR, the Gobi Desert, and Mongolia in Asia; the central third of Nevada and areas of the American Southwest, particularly at high elevations; and Patagonia in Argentina. Because of lower temperatures and lower POTET values, rainfall must be low for a station to be a *cold midlatitude desert BWk* climate; consequently, total annual average rainfall is only about 15 cm (6 in.).

A representative station is Albuquerque, New Mexico, with 20.7 cm (8.1 in.) of precipitation and an annual average temperature of 14°C (57.2°F) (Figure 10-27). Comparing *cold midlatitude desert BWk* and *hot low-latitude desert BWh* stations reveals an interesting similarity in the annual temperature range and a distinct difference in precipitation patterns.

## Hot Low-Latitude Steppe Climates (BSh)

*Hot low-latitude steppe BSh* climates generally exist around the periphery of hot deserts, where shifting subtropical high-pressure cells create a distinct summer-dry and winter-wet pattern. Average annual precipitation in *hot low-latitude steppe BSh* areas is usually below 60 cm (23.6 in.). Walgett, in interior New South Wales, Australia, provides a Southern Hemisphere example of this climate (Figure 10-28).

This climate is seen around the Sahara's periphery and in Iran, Afghanistan, and the Turkistan, Kazakstan, region. Along the Sahara's southern margin is the Sahel, a drought-tortured region. Human populations in the Sahel have suffered great hardship as desert conditions have gradually expanded over their homelands. The process of desertification is discussed in Chapter 20.

## Cold Midlatitude Steppe Climates (BSk)

The *cold midlatitude steppe BSk* climates occur poleward of about 30° latitude and the *cold midlatitude desert BWk* climates. Such midlatitude steppes are not generally found in the Southern Hemisphere. As with other dry climate classifications, this steppe climate's rainfall is widely variable and undependable, ranging from 20 to 40 cm (7.9 to 15.7 in.). Not all rainfall is convectional, for cyclonic systems penetrate the continents; however, most storms produce little precipitation. For comparison between Asian and North American *cold midlatitude steppe BSk,* consider Semey (Semipalatinsk) in Kazakstan and Lethbridge in Alberta (Figure 10-29).

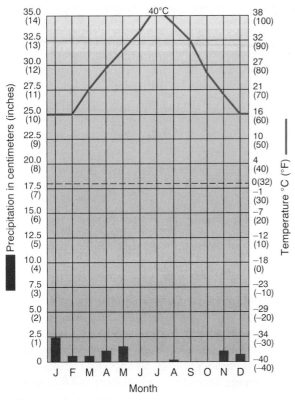

Station: Ar Riyāḍ (Riyadh),    BWh  Elevation: 609 m (1998 ft)
    Saudi Arabia    Population: 1,308,000
Lat/long: 24°42'N 46°43' E    Ann. Temp. Range:
Avg. Ann. Temp.: 26°C (78.8°F)    24 C° (43.2 F°)
Total Ann. Precip.:
    8.2 cm (3.2 in.)

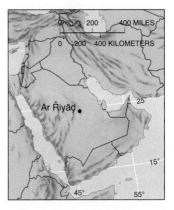

(a)

**FIGURE 10-26**

**Hot low-latitude desert climate.**

(a) Climograph for Ar Riyāḍ (Riyadh), Saudi Arabia (*hot low-latitude desert Bwh*). (b) Photograph of the Arabian desert sand dunes near Ar Riyāḍ. [Photo by Ray Ellis/Photo Researchers, Inc.]

(b)

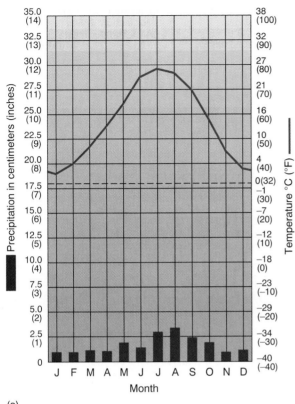

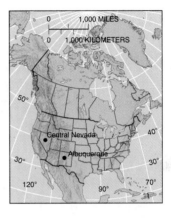

**Station:** Albuquerque, **BWk**
New Mexico
**Lat/long:** 35°03' N 106°37' W
**Avg. Ann. Temp.:** 14°C (57.2°F)
**Total Ann. Precip.:**
　　　　20.7 cm (8.1 in.)

**Elevation:** 1620 m (5315 ft)
**Population:** 370,000
**Ann. Temp. Range:**
　　　　24 C° (43.2 F°)
**Ann. Hr of Sunshine:**
　　　　3420

(a)

## FIGURE 10-27
**Cold midlatitude desert climate.**
(a) Climograph for Albuquerque, New Mexico (*cold midlatitude desert BWk*). (b) Cold desert landscape in north-central Nevada.
[Photo by author.]

(b)

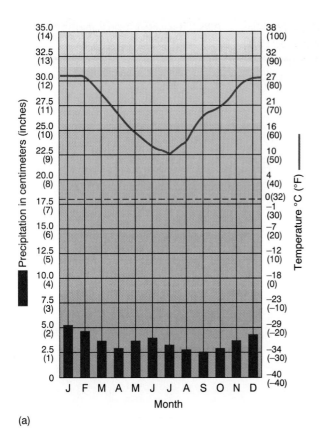

(a)

**Station:** Walgett, New South   **BSh**   **Elevation:** 133 m (436 ft)
Wales, Australia          **Population:** 2160
**Lat/long:** 30°0' S 148°07' E          **Ann. Temp. Range:**
**Avg. Ann. Temp.:** 20°C (68°F)          17 C° (31 F°)
**Total Ann. Precip.:**
45.0 cm (17.7 in.)

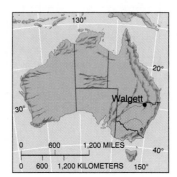

(b)

**FIGURE 10-28**

**Hot low-latitude steppe climate.**

(a) Climograph for Walgett, New South Wales, Australia (*hot low-latitude steppe BSh*). (b) Vast plains characteristic of north-central New South Wales. [Photo by Otto Rogge/Stock Market.]

---

### Köppen Guidelines
### Dry Arid and Semiarid Climates—B

POTET exceeds PRECIP in all B climates. Subdivisions based on PRECIP timing and amount and mean annual temperature.

*Earth's arid climates.*

**BWh** — **Hot low-latitude desert**
**BWk** — **Cold midlatitude desert**
BW  =  PRECIP less than ½ POTET.
  h  =  Mean annual temperature >18°C
      (64.4°F).
  k  =  Mean annual temperature <18°C.

*Earth's semiarid climates.*

**BSh** — **Hot low-latitude steppe**
**BSk** — **Cold midlatitude steppe**
BS  =  PRECIP more than ½ POTET but not
      equal to it.
  h  =  Mean annual temperature >18°C.
  k  =  Mean annual temperature <18°C.

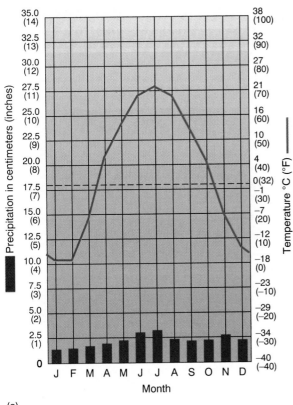

(a)

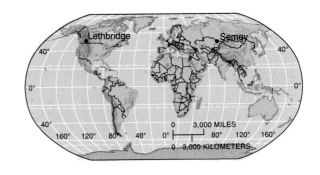

**Station:** Semey (Semipalatinsk), **BSk**
    Kazakstan
**Lat/long:** 50°21′ N 80°15′ E
**Avg. Ann. Temp.:** 3°C (37.4°F)
**Total Ann. Precip.:**
    26.4 cm (10.4 in.)

**Elevation:** 206 m (675.9 ft)
**Population:** 330,000
**Ann. Temp. Range:**
    39 C° (70.2 F°)

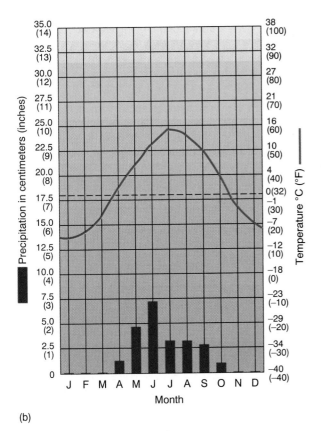

(b)

**Station:** Lethbridge, Alberta    **BSk**
**Lat/long:** 49°42′ N 110°50′ W
**Avg. Ann. Temp.:** 2.9°C (37.3°F)
**Total Ann. Precip.:**
    25.8 cm (10.2 in.)

**Elevation:** 910 m (2985 ft)
**Population:** 58,840
**Ann. Temp. Range:**
    24.3 C° (43.7 F°)

(c)

**FIGURE 10-29**

**Cold midlatitude steppe climate.**

Climographs for (a) Semey (Semipalatinsk) in Kazakstan and (b) Lethbridge in Alberta (*cold midlatitude steppe BSk*).
(c) Canadian prairies and grain elevators of southern Alberta. [(c) Photo by author.]

# News Report 3

## NetWork: Direct to You from the South Pole!

The *South Polar Times* was a newsletter edited by Earnest Shakleton during Robert Scott's Antarctic expeditions that began in 1901. Katy Wallet, a Virginia school teacher, resurrected Shakleton's newsletter as the *New South Polar Times (NSPT)* in 1994. The modern *NSPT* is written by staff members at the Amundsen-Scott Station at the South Pole, 90° S latitude. This station is on the Antarctic Plateau some 2835 m (9300 ft) above sea level (Figure 1). Wallet receives the newsletter by e-mail, does some editing, and e-mails it to George Duckett at Deakin University, Melbourne, Australia. He formats the copy and uploads it onto the Internet.

The following brief excerpts from the *New South Polar Times* cover a 14-month period, including two cherished September equinoxes and their long-awaited sunrises. In these selections, you will get a sense of polar climates and the intrepid human spirit. The staff "reporters" are Tom Jacobs, Dave Whitner, and Katy McNitt.

(a)

(b)

**FIGURE 1**

**Amundsen-Scott Station at the South Pole.**

a) Entrance to Amundsen-Scott station; b) ceremonial South Pole. [Photos by Stuart Klipper, courtesy of the National Science Foundation, Office of Polar Programs.]

**August 1, 1994.** Having the Sun out of the way is great for the four astronomers that are wintering over. This is one of the few places on Earth where an astronomer can watch a star or planet 24 hours a day.…Amundsen-Scott Station is totally supported by airplanes. There are no roads leading to the South Pole. The station is closed 8 months out of the year…no airplanes can land and none of us can leave. We have 27 people wintering over. It is –50°C now. *T. Jacobs*

**August 25, 1994.** There are three types of twilight: astronomical, nautical, and civil, corresponding to when the Sun is 18°, 12°, and 6° below the horizon.…In one year the South Pole goes through one day and one night. When the Sun is up, it stays up for six months and when the Sun is down it stays down for six months. How long is our period of complete darkness? 135 days. One hundred

and thirty-five very dark days and now the Sun is finally lighting up the sky in twilight. If you get up two hours before sunrise you can get an idea of what the sky looks like as I am writing this—now imagine this predawn time lasting for weeks! I am sad to see the stars and aurora slowly fade away. Temperatures: low –59°C, high –38°C. *T. Jacobs*

**September 15, 1994.** The Sun is now six degrees below the horizon. It is really getting bright outside. One week until sunrise!…Everywhere you look the direction is north. We use a coordinate system in which the Greenwich Meridian is called north, longitude 90° east is called east, longitude 180° is called south, and 90° west is called west. *T. Jacobs*

**November 15, 1994.** There aren't any bugs or birds or trees or even fungi that are native to the middle of Antarctica. Which is strange, since the oceans which surround Antarctica house one of the most complex ecosystems in the world. Right now it is a balmy –34.6°C and there is a big halo around the Sun.

A 19 mph summer breeze lifts snow from the Polar Plateau, swirling it around and forming wavelike drifts called *sastrugi. K. McNitt*

**December 12, 1994.** The last few weeks have been very busy here! There are 140 people on-station now, trying to accomplish as much as possible before the station closes in February. Since there isn't much seismic activity in Antarctica, it's a good place to record earthquakes from all over the world. Several groups are studying the Sun and its effects on Earth's atmosphere. Some scientists are measuring snow stakes to see how much snow accumulates near the station per year, and others are drilling ice cores to measure changes in the Earth's atmosphere. *K. McNitt*

**January 5, 1995.** The Navy flies LC-130 transports down here regularly in daylight to bring us the things we need to survive. When the plane lands, we have to work quickly to get everything done. Unlike in warm weather areas, they do not shut off the engines on the airplanes. We bring hoses to the plane and offload extra jet fuel. The fuel goes into one of the three 25,000 gallon, fabric fuel tanks for our generators. *D. Whitner*

**February 13, 1995.** For the next eight months, no planes will land here; we are closing. Weather permitting, one C-141 will fly overhead in June to air-drop medical supplies, fresh produce, mail, and spare parts; otherwise, it's just the 28 of us until next October…28 wondering souls under a tiny aluminum roof in the middle of the coldest, highest, driest, windiest continent on Earth. *K. McNitt*

**February 28, 1995.** Winter didn't waste time arriving either: for the past week we've had winds averaging 20 mph, which fill every open space with drifting snow. Today the air temperature is −46°C, but once you add the wind, the wind-chill temperature feels like −77°C, which is about −106°F! *K. McNitt*

**March 17, 1995.** The Sun is spiraling closer and closer to the horizon, rolling around our field of view like a dog on a leash, slipping slowly toward the Northern Hemisphere. *K. McNitt*

**April 2, 1995.** On March 19 we switched from Daylight Saving Time to New Zealand Standard Time. It takes the Sun two days to set once it touches the horizon. We can see a bright orange sliver above pink clouds, as if the Sun is waving good-bye. By March 30 the sky is still bright, but more stars appear every day. That's my favorite thing about a twilight that lasts for weeks: It gives you plenty of time to identify the stars and planets before the sky gets too crowded! *K. McNitt*

**April 25, 1995.** The temperature dropped below −100°F (−73°C). Brrrr! The National Science Foundation manages all U.S. activity in Antarctica. The buildings are located under an aluminum geodesic dome, 17 m high and 50 meters across. This protects the buildings inside from wind and blowing snow. There are four main buildings under the dome. Temperature under the dome is the same as outside, but the buildings are heated. Most of our food stores are piled along the inside wall of the dome. This is why we have to heat the ice cream in the microwave before we can eat it: It's too cold! *K. McNitt*

**May 16, 1995.** The auroras are going on outside right now: it's not as bright as usual because of the full Moon, but it looks like a swirled green waterspout.…it curls over and twists the whole shape into a bright shimmering fold; at other times, they are swirling rivers of green, white, and pink, and rows of spiky teal needles that bounce and ripple. It is the best entertainment at the South Pole. *K. McNitt*

**June 22, 1995.** Our airdrop: couldn't have asked for better weather, the sky was clear and loaded with stars and the light from a gibbous Moon. The temperature hung at about −71°C (−95°F). Lighted smudge pots marked the drop zone. We could see the strobe lights and parachutes of the dark bundles as they thudded into the drop zone. Each bundle weighs 400 to 1000 pounds and they hit the ice at 65 mph. We received 21 bundles, some marked Do Not Freeze so we hurried and got it all indoors. Soon we were reading letters and newspapers, while munching on apples and bananas. Several science projects were saved by the arrival of spare parts.…Happy Solstice from the South Pole! For many of you in the Northern Hemisphere, this is your longest day of summer, but for us it's Midwinter's Day. The Sun begins its slow spiral back to this side of the world. Twilight is still two months away, but somehow it feels warmer already. *K. McNitt*

**July 12, 1995.** The one problem that has probably taken out more electrical equipment than any other is static electricity. The South Pole is one of the driest places in the world. Between the dryness and the cold, static electricity is always present. *K. McNitt*

**August 6, 1995.** The faint glow of twilight is rolling around the horizon. Next month is the equinox when the Sun will cross the equator again, setting at the North Pole and rising at the South Pole, rolling around and around the horizon, higher and higher in our sky, reaching its highest point during the December solstice…not to set again until March. *K. McNitt*

**September 7, 1995.** We have increased our ozonosonde schedule to two balloon flights a week so we can monitor this year's Antarctic "ozone hole." It's pretty obvious that the total ozone is a lot lower than it was a few weeks ago.…On Sunday there was a cool tangerine glow where the Sun is supposed to be and the air temperature dropped to −78°C (−108°F). There was a little wind, so the wind chill temperature lingered around −106°C (−160°F) all day. Even the snow felt different: It felt slippery and light, like baby powder. *K. McNitt*

**October 3, 1995.** When we go outside we wear several layers of different materials, with a wind-proof "shell" of clothing on the outside. From the South Pole we can measure the depth of the ozone hole, but we have no idea of how wide the ozone thinning is. Last week the Sun crossed the equator and we celebrated sunrise. The best part about sunrise at the South Pole: it takes almost two days, so you can celebrate it over and over. *K. McNitt*

If you want to ask questions of the scientists and staff working at the South Pole, send your queries to: kwallet@pen.k12.va.us. To access the NSPT from the Internet enter: http://205.174.118.254/nspt/home/htm.

# Global Climate Change

Significant climatic change has occurred on Earth in the past and most certainly will occur in the future. There is nothing society can do about long-term influences that cycle Earth through swings from ice ages to warmer periods. However, our global society must address possible short-term changes that are influencing global temperatures within the life span of present generations. Important tools to meet this challenge involve complex computer models of Earth's energy-atmosphere-ocean system.

A cooperative global network of nations participates in the World Weather Watch System and the Global Change Program to gather temperature, weather, and climatic information. The network operates under the supervision of the United Nations Environment Programme (UNEP) and the World Meteorological Organization (WMO). Evidence of this international scientific effort is the ongoing assessment process by the Intergovernmental Panel on Climate Change (IPCC), with reports issued by three Working Groups in 1990, 1992, and 1995.

## Climate Models

Imagine the tremendous task of building a mathematical model of all the climatic components we have discussed over different time frames and at various scales! The challenge is to discern climatic trends in what is essentially a nonlinear (unpredictable), chaotic natural system of weather. The most complex climate model at present, known as a **general circulation model (GCM)**, was developed from mathematical models originally established for forecasting weather. Eight GCMs now operate around the world. At best, they can generalize and be only as accurate as the data input and assumptions built into them.

Thermometers have been in use since the late 1600s, but a significant worldwide density of readings has been available only for about 100 years. Today, approximately 1000 stations worldwide send automated instrument packages into the atmosphere on balloons twice a day. These instruments measure temperature, humidity, and pressure at varying altitudes. New land-based methods employing lasers also obtain these readings and are gradually increasing the data base. You read in Chapter 8 about the exciting development of GPS/MET (News Report 3), which continuously gathers temperature, humidity, and pressure soundings across the globe using the GPS system of satellites. Data availability is rapidly improving.

The first step in describing a climate is to define a manageable portion of Earth's climatic system for study. Climatologists create dimensional "grid boxes" that extend from beneath the ocean to the tropopause, in multiple layers (Figure 10-30). Analysts must deal not only with

the climatic components within each grid layer but also with the interaction among the layers on all sides. Even with today's computers, a program can run for weeks and still give only general indications of climatic patterns.

An important goal of the Global Change Program is to improve computer hardware and software and the data base to create better simulations and predictions of climatic change. A comparative benchmark among the operational GCMs is *climatic sensitivity* to doubling of carbon dioxide levels in the atmosphere. GCMs do not predict specific temperatures, but they do offer various scenarios of global warming.

## Global Warming

Human activities are enhancing the greenhouse effect. There is little scientific doubt that air temperatures are the highest since recordings were begun in earnest more than 100 years ago. In terms of **paleoclimatology**, the science that studies past climates (discussed in Chapter 17), Earth is within 1 C° of equaling the highest average temperature of the past 125,000 years (based on ice-core data). The rate of warming in the past 30 years exceeds any comparable period in the entire measured temperature record, according to NASA scientists, and is thought to be due to a buildup of greenhouse gases. Climatologists Richard

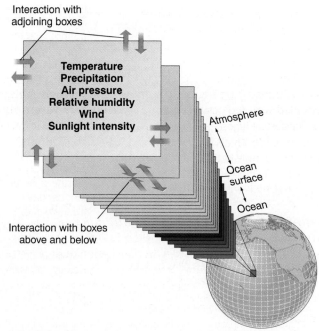

**FIGURE 10-30**

**A general circulation model scheme.**
Temperature, precipitation, air pressure, relative humidity, wind, and sunlight intensity are sampled in myriad grid boxes. In the ocean, sampling is limited, but temperature, salinity, and ocean current data are considered. The interactions within a grid layer, and between layers on all sides, are modeled by a general circulation model program.

Houghton and George Woodwell described the present climatic condition:

> The world is warming. Climatic zones are shifting. Glaciers are melting. Sea level is rising. These are not hypothetical events from a science fiction movie; these changes and others are already taking place, and we expect them to accelerate over the next years as the amounts of carbon dioxide, methane, and other trace gases accumulating in the atmosphere through human activities increase.*

Scientists are attempting to determine the difference between forced fluctuations (human-caused) and unforced fluctuations (natural) as a key to predicting future climate trends. Because the gases that generate temperature changes are human in origin, various management strategies are possible to reduce human-forced changes. Let's begin by examining the problem at its roots.

***Carbon Dioxide and Global Warming.*** *Radiatively active gases* are atmospheric gases, such as carbon dioxide, methane, chlorofluorocarbons (CFCs), and water vapor, that absorb and radiate infrared wavelengths. Carbon dioxide and water vapor are the principal radiatively active gases causing Earth's natural greenhouse effect. They are transparent to light but opaque to the infrared wavelengths radiated by Earth, and thus they transmit light from the Sun to Earth but delay heat-energy loss to space. While detained, this heat energy is absorbed and reradiated over and over, warming the lower atmosphere. As concentrations of these infrared-absorbing gases increase, more heat is maintained in the atmosphere and temperatures increase.

The Industrial Revolution, beginning in the mid-1700s, initiated tremendous burning of fossil fuels. This, coupled with the destruction and inadequate replacement of harvested forests, continues to cause atmospheric carbon dioxide levels to increase by 7.3–9.1 billion metric tons (8–10 billion tons) each year (Figure 10-31). Carbon dioxide alone is thought to be responsible for almost 60% of the global warming trend.

Table 10-1 shows the increasing percentage of carbon dioxide in the lower atmosphere from 1825 to the present and gives estimates for the future. Carbon dioxide in the 1990s is increasing in concentration at the rate of 0.4% per year. The increase appears sufficient to override any natural long-term climatic tendency toward cooling. It also appears sufficient to produce global warming in the next century.

Figure 10-32 shows sources of excessive (non-natural) carbon dioxide by country or region in 1980 and 2025. Developing countries are clearly identified as the

*R. Houghton and G. Woodwell, "Global climate change," *Scientific American,* April 1989, p. 36.

**FIGURE 10-31**
**Industrial source of greenhouse gases.**
Industrial landscapes exemplify the increasing output of carbon dioxide into the atmosphere. [Photo by author.]

sector with the greatest probable growth in fossil fuel consumption and new carbon dioxide production. However, national and corporate policies could alter this forecast by actively steering developing countries toward alternative energy sources (renewable, low temperature, labor intensive) and redirecting industrial countries away from past practices.

***Carbon Dioxide and Policy.*** It is important for the United States to reduce carbon dioxide production and to set a positive example. Yet relatively simple solutions are not being promoted, and throughout the 1980s the United States was uncooperative in worldwide efforts to halt the human-induced increase in carbon dioxide. At present the United States produces 27% of the excess carbon dioxide annually. This level is predicted to rise 15% over 1990 levels by the year 2000. The European Union and a majority of nations favor stabilizing emissions at 1990 levels by the year 2000.

The Office of Technology Assessment (OTA) and the National Academies of Science and Engineering (NAS) separately concluded that the United States could hold to 1990 levels by the year 2015 at little or no additional cost.

**TABLE 10-1**

| Lower Atmosphere Concentration of Carbon Dioxide ($CO_2$) | | |
| --- | --- | --- |
| *Year* | *$CO_2$ Concentration (%)* | *Parts per Million* |
| 1825 | 0.021 | 210 |
| 1888 | 0.028 | 280 |
| 1970 | 0.032 | 320 |
| 1985 | 0.035 | 350 |
| 1995 (estimate) | 0.037 | 370 |
| 2020 (estimate) | 0.055 | 550 |
| 2050 (estimate) | 0.060 | 600 |

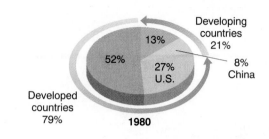

(a)

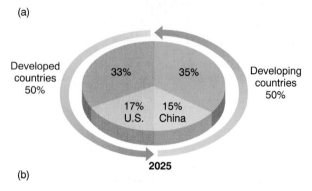

(b)

**FIGURE 10-32**
**Origin of excessive carbon dioxide.**
Countries and regions of origin (a) for excessive carbon dioxide in 1980 and (b) forecast for 2025. [Data from United Nations Environment Programme.]

OTA concluded that "some reductions may even be at a net savings if the proper policies are implemented....The efficiency of practically every end use of energy can be improved relatively inexpensively."[*]

The 1995 report from the Intergovernmental Panel on Climate Change (IPCC)[**] declared that "no regrets" opportunities to slow global change are available in most countries. "No regrets" actions have benefits (reduced energy cost, or reduced emission of regional-scale pollutants) that equal or exceed their cost to society, even when the benefits of slowing climate change is excluded from the calculation. IPCC says these are "measures worth doing anyway." Reducing our contribution to atmospheric carbon dioxide clearly has many benefits.

***Methane and Global Warming.*** Another radiatively active gas contributing to the overall greenhouse effect is *methane* ($CH_4$), which, at more than 1% per year, is increasing in concentration even faster than carbon dioxide is. Air bubbles in ice show that concentrations of methane in the past, between 500 and 27,000 years ago,

[*] See the Office of Technology Assessment, *Changing by Degrees: Steps to Reduce Greenhouse Gases*, February 5, 1991, and *Policy Implications of Greenhouse Warming*, by the Committee on Science, Engineering, and Public Policy of the National Academies of Science, 1991.
[**] Working Group III, *Summary for Policy Makers*. Intergovernmental Panel on Climate Change, 1995)

were approximately 0.7 ppm, whereas current atmospheric concentrations are 1.7 ppm.

Methane is generated by such organic processes as digestion and rotting in the absence of oxygen (anaerobic processes). About 50% of the excess methane comes from bacterial action in the intestinal tracts of livestock and from organic activity in flooded rice fields. Burning of vegetation causes another 20% of the excess, and bacterial action inside the digestive systems of increasing termite populations also is a significant source. Methane is now believed responsible for at least 12% of the total atmospheric warming, complementing the warming caused by the buildup of carbon dioxide.

***Chlorofluorocarbons (CFCs) and Global Warming.***
CFCs are thought to contribute about 25% of the global warming. CFCs absorb infrared in wavelengths missed by carbon dioxide and water vapor in the lower troposphere. As radiatively active gases, CFCs enhance the greenhouse effect and are a cause of stratospheric ozone depletion.

***Warming Indications and the Future.*** Global mean temperatures between 1980 and 1995 included 11 of the warmest years in the history of instrumental measurement (covering at least 140 years). They also included the five warmest years as well: 1995, 1990, 1991, 1988, 1983 (in decreasing order). These occurred despite the 1991 eruption of Mount Pinatubo, which lowered temperatures, leaving 1990 as the record year, with a global average temperature of 15.4°C (59.8°F)—until it was topped in 1995 at 15.5°C. These increases are in line with forecasts from operating GCMs. Comparing annual temperatures and 5-year mean temperatures gives a sense of overall trends. Figure 10-33 shows observed temperatures over the 116 years from 1880 through 1995.

In 1990, IPCC (an initial group of 380 scientists and 63 countries established by the United Nations Environment Programme) found virtual unanimity among greenhouse experts that a warming is on the way, that the consequences will be serious and have social impacts, and that warming rates will accelerate in the near future. However, there is a wide range of uncertainty in estimates of future temperature increase, which is why a minority of scientists persist as critics of the greenhouse warming model.

According to the most recent IPCC assessment, an average warming "low forecast" of 2 C° (3.6 F°) is predicted during 1990 to 2100. This is lower than previous best estimates because lower emission of carbon dioxide and CFCs is predicted, and the cooling effect of sulfate aerosols is considered. The most recent "high forecast" is for a warming of 3.5 C° (6.3 F°).

Regionally, if carbon dioxide concentrations reach 550 ppm by the year 2020 (they are presently at 360 ppm), a temperature increase of 3 C° (5.4 F°) is forecast for the equatorial regions. That equatorial increase translates to a high-latitude increase of about 10 C° (18 F°).

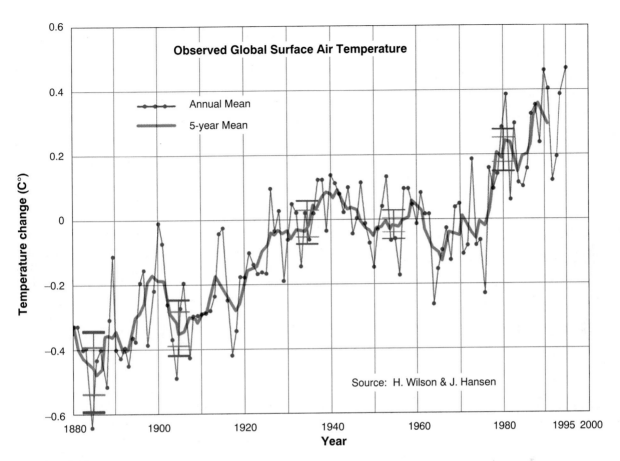

**FIGURE 10-33**

**Global temperature trends.**

Global temperature trends from 1880 to 1995. The 0 baseline represents the 1950–1980 global average.
[Courtesy of J. E. Hansen, Goddard Institute, NASA.]

The Goddard Institute for Space Studies projects a distinct warming for 2000 and 2029, using a moderate-case scenario (Figure 10-34). Consistent with other computer studies, unambiguous warming appears over low-latitude oceans and interior areas of Asia, especially China, with the greatest warming assumed to occur over Arctic and Antarctic regions of sea ice.

By December 1995, all member countries of the United Nations and the World Meteorological Organization were members of IPCC. A joint IPCC-WMO publication stated:

> Human activities, including the burning of fossil fuels..., are increasing the atmospheric concentrations of greenhouse gases (which tend to warm the atmosphere) and, in some regions, aerosols (...which tend to cool the atmosphere). These changes...are projected to change regional and global climate and climate-related parameters such as temperature, precipitation, soil moisture, and sea level.[*]

[*] B. Bolin, J. T. Houghton, and others, *IPCC Second Assessment Synthesis of Scientific-Technical Information Relevant to Interpreting Article 2 of the UN Framework Convention on Climate Change, 1995,* (Geneva, Switzerland: IPCC Secretariat, WMO, October 1995), p. 1.

## Consequences of Climatic Warming

The consequences of uncontrolled atmospheric warming are complex. Regional climate responses are expected as temperature, precipitation, soil moisture, and air mass characteristics change. Although the ability to accurately forecast such regional changes is still evolving, some consequences of warming are forecast.

### Effects on World Food Supply and the Biosphere.

Modern single-crop (monoculture) agriculture is delicate. It is more susceptible to changes in air temperature, water availability, irrigation, and soil chemistry than is traditional multicrop agriculture. The southern and central grain-producing areas of North America are forecast to experience hot, dry weather by the mid-twenty-first century as a result of higher temperatures. Available soil moisture is projected to be at least 10% less throughout the midlatitudes 30 years from now. Scientists are considering changing to late-maturing, heat-resistant crop varieties and adjusting fertilizer applications and irrigation.

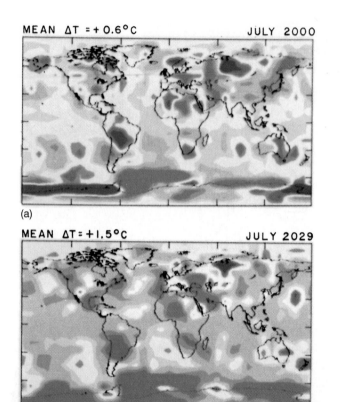

MEAN ΔT = + 0.6°C          JULY 2000

(a)

MEAN ΔT = +1.5°C          JULY 2029

(b)

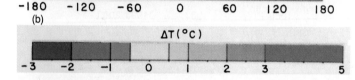

ΔT(°C)

-3   -2   -1   0   1   2   3   5

**FIGURE 10-34**

**July temperature forecasts.**

July temperature projections by the Goddard Institute for Space Studies' general circulation model for the years 2000 and 2029. [From J. E. Hansen, and others, "Global climate changes as forecast by Goddard Institute for Space Studies three-dimensional model," *Journal of Geophysical Research* 43, no. 8 (August 1988): pl. 6.]

Today, the United States sells wheat to China and Russia. If these foreign grain areas become wetter and gain a longer frost-free period than at present, grain sales could shift in the other direction. Canadian harvests would improve according to projections. The possibility exists that billions in agricultural losses in one region could be countered by corresponding gain in another.

Biosphere models predict that a global average of 30% (varying regionally from 15% to 65%) of the present forest cover will undergo major species redistribution—greatest at high latitudes and least in the tropics. Many plant species are already "on the move" to more favorable locations. Land dwellers also must adapt to changing forage. Warming already is stressing some embryos as they reach their thermal limit. Particularly affected are amphibians, whose embryos develop in shallow water.

Studies completed at the universities of Toronto and Wisconsin–Madison suggest that greenhouse warming may warm the Great Lakes, an advantage for various fish species. On the other hand, a warming of small Canadian lakes might

harm aquatic life. (Updated maps that forecast changing biosphere patterns are presented in Chapters 19 and 20.)

Recent studies completed by the U.S. Institute of Public Health, The Netherlands Environmental Protection Agency, The Development Research Centre in India, and others, point to possible health impacts of climate change. Populations previously unaffected by malaria, schistosomiasis, sleeping sickness, dengue fever, and yellow fever would be at greater risk in subtropical and midlatitude areas. A 1987 malaria increase in Rwanda is now linked to a warming of 1 C° in average temperature in that region, where the disease was able to spread to high elevations.

**Melting Glaciers, Melting Ice Sheets, and Sea Level.**
Perhaps the most pervasive climatic effect of global warming is rapid escalation of ice melt. The additional water, especially from continental (on land) ice masses and glaciers, is slowly raising sea level worldwide. Satellite remote-sensing is monitoring global sea-ice and continental-ice. Worldwide measurements confirm that sea level has been gradually rising this century.

Surrounding the margins of Antarctica, and constituting about 11% of its surface area, are numerous *ice shelves*, especially where sheltering inlets or bays exist. Covering many thousands of square kilometers, these ice shelves extend over the sea while still attached to continental ice. The loss of these ice shelves does not significantly raise sea level, for they already displace sea water. The concern is for the possible surge of grounded continental ice that the ice shelves hold back from the sea.

Five of the ice shelves are now in active disintegration at remarkable rates. About 8000 km² (3090 mi²) of ice shelf are gone, changing maps, freeing up islands to circumnavigation, and creating thousands of icebergs. This is a direct result of the 2.5 C° (4.5 F°) temperature increase in the region in the last 40 years.

The 4200 km² (1620 mi²) Larsen Ice Shelf, along the west coast of the Antarctic Peninsula, has been retreating slowly for years, but its retreat accelerated after 1975. During just a few days in 1995, the northernmost area called Larsen-A suddenly disintegrated (Figure 10-35b). One iceberg measured 78 km by 37 km by 200 m thick (48 mi by 23 mi by 655 ft)—about the size of Rhode Island. Larsen-B, to the south, is approaching a critical limit and may go next. The Wordie Ice Shelf, on the west coast of the Antarctic Peninsula, has shown dramatic retreat and disintegration between 1948 and 1992 (Figure 10-35c). As researchers summarized,

> This breakup followed a period of steady retreat that coincided with a regional trend of atmospheric warming. The observations imply that after an ice shelf retreats beyond a critical limit, it may collapse rapidly as a result of disturbed mass balance.*

* H. Rott, P. Skvarca, and T. Nagler, "Rapid collapse of northern Larsen Ice Shelf, Antartica," *Science* 271, (February 9, 1996): 788.

**FIGURE 10-35**

**Disintegrating ice shelves along the Antarctic peninsula.**
(a) Map of the Antarctic Peninsula showing the status of ice shelves. Note the location of the Larsen and Wordie shelves. (b) Retreat of the Larsen Ice Shelf between January and February, 1995, recorded on an *ERS-1* satellite image. (c) Satellite image of the Wordie Ice Shelf retreat as mapped in 1948, 1974, and 1992 (area shown is approximately 90 km by 70 km). [Maps and images provided by the British Antarctic Survey, Cambridge, United Kingdom. All rights reserved.]

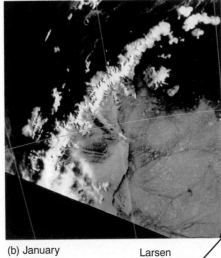

(b) January

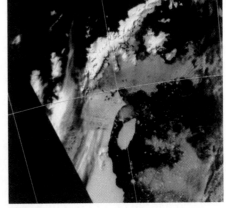

February

Larsen
Ice Shelf

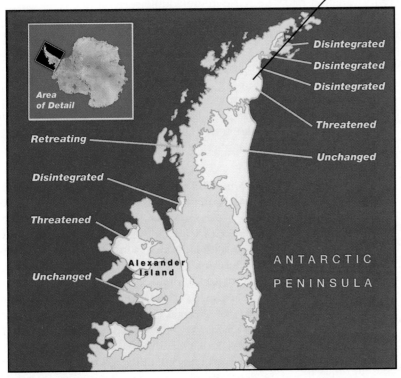

(a)

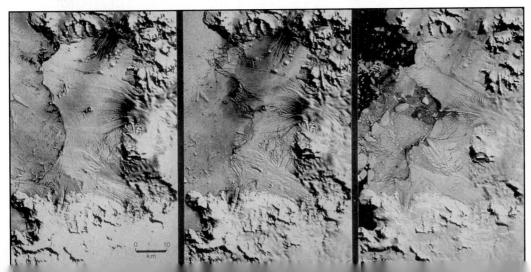

(c)   Wordie Ice Shelf

A loss of polar ice mass, augmented by melting of alpine and mountain glaciers (which have experienced about a 30% decrease in overall ice mass during this century), will affect sea-level rise. The latest IPCC assessment states that "between one-third to one-half of the existing mountain glacier mass could disappear over the next hundred years." In *Climate Change* 1992 (p. 158), IPCC states that "there is conclusive evidence for a worldwide recession of mountain glaciers…This is among the clearest and best evidence for a change in energy balance at the Earth's surface since the end of the last century."

Sea-level rise must be expressed as a range of values that are under constant reassessment. During our century alone, sea-level rise is estimated at 10–20 cm (4–8 in.). Estimates of future sea-level rise range from a low ("best case") of 30–110 cm (1–4 ft) to a high ("worst case") of 6 m (20 ft). The 1995 IPCC forecast for global sea-level rise ranges from 15 to 95 cm (6 to 37 in.), with variation in regional response related to ocean currents and land movements. These increases would continue beyond 2100 even if greenhouse gas concentrations are stabilized.

The Scripps Institution of Oceanography, in La Jolla, California, has kept ocean temperature records since 1916. Significant temperature increases are being recorded to depths of over 300 m (1000 ft) as ocean temperature records are set. Any change in ocean temperature has a profound effect on weather and, indirectly, on agriculture and soil moisture. Even the warming of the ocean itself will contribute about 25% of sea-level rise, simply because of *thermal expansion* of the water.

A quick survey of world coastlines shows that even a moderate rise could bring change of unparalleled proportions. At stake are the river deltas, lowland coastal farming valleys, and low-lying mainland areas, all contending with high water, high tides, and higher storm surges. There will be both internal and international migration of affected human populations, spread over decades, away from coastal flooding if sea level continues to rise.

The position is held by some government leaders that society has adapted to many problems in the past. Capital can be raised and technology harnessed to block the flood and guard the coastlines and deltas. Although such adaptive strategies will be needed, no substantive long-range planning is going forward. Clearly, physical geography is in an important position to synthesize all the spatial variables needed in planning.

## Global Cooling: Volcanoes and Nuclear Winter

The atmospheric cooling effect of volcanic eruptions is well documented historically. In the 1980s and early 1990s, several volcanic eruptions affected global temperatures. Let us look at two examples, El Chichón and Mount Pinatubo.

In 1982, El Chichón erupted in south-central Mexico, sending a cloud of sulfuric acid and ash 25 km (15.5 mi) into the atmosphere. Within 21 days the cloud had circled the globe, increasing atmospheric albedo by about 1% and lowering regional temperatures by about 0.5 C° (0.9 F°). Some of the finer particles remained in the stratosphere through 1985.

In 1991, Mount Pinatubo experienced one of the largest eruptions in this century. The eruption affected temperatures worldwide by injecting aerosols (suspended particles) into the atmosphere. The aerosols formed a stratospheric haze that increased atmospheric albedo by over 1.5%, thereby decreasing insolation reaching the surface. The volcanic aerosols produced a net radiation loss of 2.5 W/m$^2$ at the surface. This increased albedo lowered average temperatures in the Northern Hemisphere by 0.5 C° to 1 C° (0.9 F° to 1.8 F°) and about half that decrease in the Southern Hemisphere.

In addition, worldwide air pollution has increased the turbidity, or murkiness, of the atmosphere. Often overlooked is the contribution of mineral aerosols (fine dust particles) derived from soils disturbed by erosion, overcultivation, and deforestation. Perhaps a combination of increasing air pollution and several volcanic eruptions explains why the warming trend during this century was moderated slightly between 1940 and 1970, only to resume thereafter (see Figure 10-33).

***Nuclear Winter.*** Another potential cooling effect was postulated in "The Atmosphere after a Nuclear War: Twilight at Noon."[*] This landmark study launched an avalanche of scientific analyses, assessments, and general confirmations (discussed further in Chapter 21).

The **nuclear winter hypothesis** now encompasses a whole range of ecological, biological, and climatic impacts associated with the detonation of a relatively small number of nuclear warheads within the biosphere (as few as 100 megatons, equivalent to the yield of about 33 MX missiles). Resulting urban firestorms would produce a great pall of insolation-obscuring smoke. Surface heating would be drastically reduced. Temperatures could drop 25 C° (45 F°) and perhaps more in the midlatitudes. This reality adds an environmental component to negotiations.

[*] P. J. Crutzen and J. W. Birks, "The Atmosphere after a Nuclear War: Twilight at Noon," in a special issue of *Ambio,* Royal Swedish Academy of Sciences, Stockholm, Sweden © 1983.

# Summary and Review—Global Climate Systems

✔ *Define* **climate and climatology and** *explain* **the difference between climate and weather.**

Climate is dynamic, not static. **Climate** is a synthesis of weather phenomena at many scales, from planetary to local, in contrast to weather, which is the condition of the atmosphere at any given time and place. Earth experiences a wide variety of climatic conditions that can be grouped by general similarities into climatic regions. **Climatology** is the study of climate and attempts to discern similar weather statistics and identify **climatic regions**.

> climate (p. 272)
> climatology (p. 272)
> climatic regions (p. 272)

1. Define climate and compare it with weather. What is climatology?
2. Explain how a climatic region synthesizes climate statistics.
3. How does the El Niño phenomenon produce the largest interannual variability in climate? What are some of the changes and effects that occur worldwide?

---

✔ *Review* **the role of temperature, precipitation, air pressure, and air mass patterns used to establish climatic regions.**

Climatic inputs include insolation (pattern of solar energy in the Earth-atmosphere environment), temperature (sensible heat energy content of the air), precipitation (rain, sleet, snow, and hail; the supply of moisture), air pressure (varying patterns of atmospheric density), and air masses (regional-sized homogeneous units of air). Climates are the basic element in ecosystems, the natural, self-regulating communities of plants and animals that thrive in specific environments.

4. How do radiation receipts, temperature, air pressure inputs, and precipitation patterns interact to produce climate types? Give an example from a humid environment and one from an arid environment.
5. Evaluate the relationships among a climatic region, ecosystem, and biome.

---

✔ *Review* **Köppen's development of an empirical climate classification system and** *compare* **his with other ways of classifying climate.**

**Classification** is the process of ordering or grouping data in related categories. A **genetic classification** is based on causative factors, such as the interaction of air masses. An **empirical classification** is one based on statistical data, such as temperature or precipitation.

> classification (p. 276)
> genetic classification (p. 276)
> empirical classification (p. 276)

6. What are the differences between a genetic and an empirical classification system?
7. Describe Köppen's approach to climatic classification. What are the factors used in his system?

---

✔ *Describe* **the A, C, D, and E climate classification categories and generally** *locate* **these regions on a world map.**

The **Köppen-Geiger climate classification** system is a modified empirical classification system and describes distinct environments on Earth: tropical A, mesothermal C, microthermal D, polar E, and arid B climates. Highland climates are assigned an H to denote that they are caused by elevation. The main climatic groups are divided into subgroups based on temperature and the seasonal timing of precipitation. The system uses average monthly temperatures, average monthly precipitation, and total annual precipitation to devise its spatial categories and boundaries. These data are plotted on a **climograph** to display the characteristics of the climate.

> Köppen-Geiger climate classification (p. 277)
> climograph (p. 279)

8. List and discuss each of the principal climate designations. In which one of these general types do you live? Which classification is the only type associated with the annual distribution and amount of precipitation?
9. What is a climograph, and how is it used to display climatic information?
10. Which of the major climate types occupies the most land and ocean area on Earth?
11. Characterize the tropical A climates in terms of temperature, moisture, and location.
12. Using Africa's tropical climates as an example, characterize the climates produced by the seasonal shifting of the ITCZ with the high Sun.

13. Mesothermal C climates occupy the second largest portion of Earth's entire surface. Describe their temperature, moisture, and precipitation characteristics.

14. Explain the distribution of the *humid subtropical Cfa* and *Mediterranean dry-summer Csa* climates at similar latitudes and the difference in precipitation patterns between the two types. Describe the difference in vegetation associated with these two climate types.

15. Which climates are characteristic of the Asian monsoon region?

16. Explain how a *marine west coast Cfb* climate type can occur in the Appalachian region of the eastern United States.

17. What role do offshore currents play in the distribution of the *marine west coast Cfb* climate designation? What type of fog is formed in these regions?

18. Discuss the climatic designation for the coldest places on Earth outside the poles. What do each of the letters in the Köppen classification indicate?

---

✔ ***Explain*** **the precipitation and moisture efficiency criteria used to determine the B climates and generally** ***locate*** **them on a world map.**

The dry and semiarid climates are the only ones that Köppen classified by precipitation rather than temperature. Dry climates are the world's arid deserts and semiarid regions, with their unique plants, animals, and physical features. The *dry arid and semiarid B* climates occupy more than 35% of Earth's land area, clearly the most extensive climate over land.

Major subdivisions are *arid deserts BW* (PRECIP less than one-half of POTET) and *semiarid steppes BS* (PRECIP more than one-half of POTET). Köppen developed simple formulas to determine the usefulness of rainfall on the basis of the season.

19. In general terms, what are the differences among the four desert classifications? How are moisture and temperature distributions used to differentiate these subtypes?

20. Relative to the distribution of dry climates, describe at least three locations where they occur across the globe and the reasons for their presence in these locations.

---

✔ ***Outline*** **future climate patterns from forecasts presented and** ***explain*** **the causes and potential consequences.**

Various activities of present-day society are producing climatic changes, particularly a global warming trend. The 1980s and 1990s were dominated by the highest average annual temperatures experienced since the advent of instrumental measurements. There is a scientific consensus building that global warming is related to the greenhouse effect.

These conditions were further supported by a 1995 update report by the Intergovernmental Panel on Climate Change. IPCC predicted surface-temperature response to a doubling of carbon dioxide with a range of increase between 2 C° and 3.5 C° between the present and 2100. A **general circulation model (GCM)** is used to forecast climate patterns and is evolving to greater capability and accuracy than in the past. People and their political institutions can use GCM forecasts to form policies aimed at reducing unwanted climate change. Natural climatic variability over the span of Earth's history is the subject of **paleoclimatology**.

The possible ecological, biological, and climatic impact from the detonation of a small number of nuclear warheads is described by the **nuclear winter hypothesis**.

> **general circulation model (GCM)** (p. 306)
> **paleoclimatology** (p. 306)
> **nuclear winter hypothesis** (p. 312)

21. Explain climate forecasts. How do general circulation models produce such forecasts?

22. Describe the potential climatic effects of global warming on polar and high-latitude regions. What are the implications of these climatic changes for persons living at lower latitudes?

23. How is climatic change affecting agricultural and food production? Natural environments? Forests? The possible spread of disease?

24. What are the possible global consequences of a nuclear war?

---

 ## NetWork

The *Geosystems Home Page* provides on-line resources for this chapter on the World Wide Web. You will find review exercises, specific updates for items in the chapter, suggested readings, and links to interesting related pathways on the Internet (click on the Table of Contents link and select this chapter). *Geosystems* is at: **http://www.prenhall.com/geosystm**

# Part 3

# The Earth-Atmosphere Interface

*The Endicott Mountains in the Brooks Range, Alaska.* [Photo by Tom Bean.]

Earth is a dynamic planet whose surface is actively shaped by physical agents of change. Part 3 is organized around two broad systems of these agents—endogenic and exogenic. The **endogenic system** (Chapters 11 and 12) encompasses internal processes that produce flows of heat and material from deep below Earth's crust. These processes are powered by radioactive decay, and the materials involved constitute the *solid* realm of Earth. Earth's surface responds by moving, warping, and breaking, sometimes in dramatic episodes of earthquakes and volcanic eruptions.

At the same time, the **exogenic system** (Chapters 13 through 17) involves external processes that set into motion air, water, and ice, all powered by solar energy—this is the *fluid* realm of Earth's environment. These media carve, shape, and reduce the landscape. One such process, called weathering, breaks up and dissolves the crust, eroding materials, transporting them on rivers, winds, coastal waves and flowing glaciers, and depositing them along the way. Thus, Earth's surface is the interface between two vast open systems: one that builds the landscape and one that tears it down.

# 11

# The Dynamic Planet

**The Pace of Change**

**Earth's Structure and Internal Energy**

**Geologic Cycle**

**Plate Tectonics**

**Summary and Review**

## Key Learning Concepts

After reading the chapter, you should be able to:

- *Distinguish* between the endogenic and exogenic systems, *determine* the driving force for each, and *explain* the pace at which these systems operate.

- *Diagram* Earth's interior in cross section and *describe* each distinct layer.

- *Illustrate* the geologic cycle and *relate* the rock cycle and rock types to endogenic and exogenic processes.

- *Describe* Pangaea and its breakup and *relate* several physical proofs that crustal drifting is continuing today.

- *Portray* the pattern of Earth's major plates and *relate* this pattern to the occurrence of earthquakes, volcanic activity, and hot spots.

*The dramatic ramparts of the Matterhorn, near Zermatt, Switzerland. The lake in the foreground is the Riffelsee.* [Photo by Blaine Harrington.]

The twentieth century has been a time of great discovery about Earth's internal structure and dynamic crust, yet much remains undiscovered. Discoveries made during this century have revolutionized our understanding of how continents and oceans came to be arranged as they are. A new era of Earth systems science is emerging, combining various sciences within the study of physical geography. One task of physical geography is to explain the spatial implications of all this new information and its effect on the landscape.

Earth's interior is highly structured, with a core surrounded by concentric shells of material. It is unevenly heated by the radioactive decay of unstable elements. The result is a varied surface, featuring irregular fractures, frequent earthquakes and volcanic activity, and mountain ranges both on continental and oceanic surfaces. Another result is actual change in the amount of Earth's crust.

The U.S. Geological Survey reports that, in an average year, material from Earth's interior increases continental margins and sea floors by 2 km³, or half a cubic mile. At the same time, another 1 km³, or a quarter cubic mile, is "subducted" beneath the surface. The result is a net addition of 1 km³ to Earth's crust. All of this movement of materials results from endogenic forces within Earth—the subject of this chapter.

## The Pace of Change

The **geologic time scale** is a summary timeline of all Earth history, shown in Figure 11-1. It reflects currently accepted names of time intervals for each segment of Earth's history, from vast *eons* through briefer *eras, periods*, and *epochs*. The time scale depicts two important kinds of time: *relative* (what happened in what order) and *absolute* (actual number of years before the present).

*Relative time* is the *sequence* of events, based on the relative positions of rock strata above or below each other. Relative time is based on the important general principle of *superposition*, which states that *rock and sediment always are arranged with the youngest beds "superposed" toward the top of a rock formation and the oldest at the base, if they have not been disturbed.* The study of these sequences is called *stratigraphy.* Thus, relative time places the Precambrian at the bottom (beginning) of the time scale and the Holocene (today) at the top. Important time clues—namely *fossils*, the remains of ancient plants and animals—lie embedded within these strata. Since approximately 3.6 billion years ago, life has left its evolving imprint in the rocks.

*Absolute time*, the actual "millions of years ago" shown on the time scale, is determined by scientific methods such as radiometric dating. These absolute ages refine the time-scale sequence and lend greater accuracy to relative dating sequences. See News Report 1 for more information on radioactivity and Earth's time clock.

The most fundamental principle of Earth science is **uniformitarianism**. Uniformitarianism assumes that *the same physical processes active in the environment today have been operating throughout geologic time.* For example, if streams carve valleys now, they must have done so 500 million years ago. The phrase "the present is the key to the past" describes this principle. Uniformitarianism is supported by evidence emerging from modern exploration and from the landscape record of volcanic eruptions, earthquakes, and exogenic processes. The concept was first proposed by James Hutton in his *Theory of the Earth* (1795) and was later amplified by Charles Lyell in *Principles of Geology* (1830).

In contrast, the principle of **catastrophism** places the vastness of Earth's age and the complexity of its rocks into a shortened time span. Catastrophism holds that Earth is young and that mountains, canyons, and plains formed through catastrophic events that did not require eons of time. But ancient landscapes, such as the Appalachian Mountains, represent a much broader time scale of events (Figure 11-2). Because there is little physical evidence to support catastrophism, it is more appropriately considered a belief than a serious scientific hypothesis.

Geologic time is, however, punctuated by catastrophic events such as massive landslides, earthquakes, and volcanic eruptions. Within the principle of uniformitarianism, these localized events occur as small interruptions in the generally uniform processes that shape the slowly evolving landscape.

To understand the phenomena seen at Earth's surface, we must have knowledge of our planet's internal structure and energy. Let us now journey deep within Earth to see its inner workings.

## Earth's Structure and Internal Energy

Along with the other planets and the Sun, Earth is thought to have condensed and congealed from a nebula of dust, gas, and icy comets about 4.6 billion years ago (Chapter 2). The oldest surface rock yet discovered on Earth (known as the Acasta Gneiss) lies in northwestern Canada and has been radiometrically dated to an age of 3.96 billion years. This age was confirmed in 1989 by scientists from Missouri's Saint Louis University, the Canadian Geological Survey, and Australian National University. Previously, rocks from Greenland held the record age at 3.8 billion years. These discoveries tell us something significant: Earth was forming continental crust nearly 4 billion years ago, during the Archean Eon.

Throughout Earth's formation, heat energy was accumulating. As the protoplanet compacted into a smaller volume, energy from collisions and compression of the debris was transformed into heat energy. In addition,

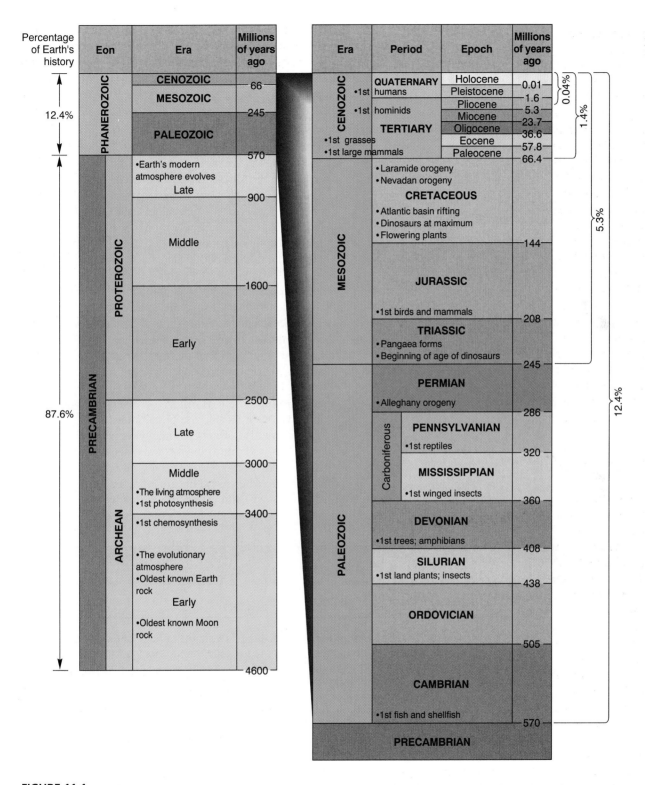

**FIGURE 11-1**

**Geologic time scale.**

The geologic time scale is organized using both relative and absolute dating methods. Relative dating determines the sequence of events and time intervals between them. Absolute dates are determined through technological means, especially radiometric dating. In the column at the left, note that nearly 88% of geologic time occurred during the Precambrian. Highlights of Earth's history also are shown in the figure as bulleted items.

[Data from Geological Society of America.]

**FIGURE 11-2**
**Ancient landscape.**
Ancient rock formations of the Appalachians, which began forming during the Alleghany orogeny over 250 million years ago. The rock outcrop is quartzite, a metamorphic rock. [Photo by Coco McCoy/Rainbow.]

significant energy was trapped in Earth from the decay of unstable forms (isotopes) of uranium, thorium, potassium, and other radioactive elements.

## Earth in Cross Section

As Earth solidified, gravity sorted materials by density. Heavier substances such as iron gravitated slowly to its center, and lighter elements such as silica slowly welled upward to the surface and became concentrated in the crust. Consequently, Earth's interior is sorted into roughly concentric layers, each one distinct in either chemical composition or temperature. Heat energy migrates outward from the center by conduction and by physical convection in the more fluid or plastic layers nearer the surface.

Our knowledge of Earth's *internal differentiation* into these layers has been acquired entirely through indirect evidence, because we are unable to drill more than a few kilometers into Earth's crust. There are several physical properties of Earth materials that enable us to approximate the nature of the interior. For example, when an earthquake or underground nuclear test sends shock waves through the planet, the cooler areas, which generally are more rigid, transmit these **seismic waves** at a higher velocity than do the hotter areas.

Density also affects seismic-wave velocities. Plastic zones simply do not transmit some seismic waves; they absorb them. Some seismic waves are reflected as densities change, whereas others are refracted, or bent, as they travel through Earth. Thus, the distinctive ways in which

---

## News Report 1

### Radioactivity: Earth's Time Clock

The age of Earth and the age of the earliest crustal rock that has been found are astounding, for we think in terms of Earth's trips around the Sun and the pace of our own lives. We need something greater than human time to measure the vastness of geologic time. Nature has provided a way: *radiometric dating*. It is based on the steady decay of certain atoms.

An atom is composed of protons and neutrons in its nucleus. Certain forms of atoms (isotopes) have unstable nuclei; that is, the protons and neutrons do not remain together indefinitely. As particles break away and the nucleus disintegrates, radiation is emitted and the atom decays into a different element—this process is called *radioactivity*.

Radioactivity provides the steady time clock needed to measure the age of ancient rocks. It works because the decay rates for different isotopes have been determined precisely, and they do not vary. These rates are expressed as *half-life*, the time required for one-half of the unstable atoms in a sample to decay. Some examples of unstable elements that become stable elements and their half-lives include uranium-238 to lead-206 (4.5 billion years); thorium-232 to lead-208 (14.1 billion years); potassium-40 to argon-40 (1.3 billion years); and, in organic materials, carbon-14 to nitrogen-14 (5730 years).

The presence of these decaying elements and stable end products in sediment or rock allows scientists to read the radiometric "clock." They compare the amount of original isotope in the sample with the amount of decayed end product in the sample. If the two are in a ratio of 1:1 (equal parts), one half-life has passed. Errors can occur if the sample has been disturbed or subjected to natural weathering processes that might alter its radioactivity. To increase accuracy, investigators may check a sample using more than one radiometric measurement. The date for the Acasta Gneiss, Earth's oldest known rock, was cross-checked using several radiometric methods.

seismic waves pass through Earth and the time they take to travel between two surface points help *seismologists* deduce the structure of Earth's interior (Figure 11-3).

Figure 11-4 illustrates the dimensions of Earth's interior compared with surface distances in North America to give you a sense of size and scale. An airplane flying from Halifax, Nova Scotia, to San Francisco would be traveling the same distance as that from Earth's center to its surface.

***Earth's Core.*** A third of Earth's entire mass, but only a sixth of its volume, lies in its dense core. The **core** is differentiated into two regions—inner core and outer core—divided by a transition zone of several hundred kilometers wide, at an average depth of 5150 km (3200 mi) (see Figure 11-3b). Estimated core temperatures range from 3000°C (5400°F) to as high as 6650°C (12,000°F). The inner core is thought to be solid iron, with a density of 13.5 g/cm³. (For comparison, the density of water is 1.0 g/cm³, and of mercury, a liquid metal, 13.0 g/cm³.)

The inner core is well above the melting temperature of iron at the surface, but it remains solid because of the tremendous pressure. The iron in the core is impure, probably combined with silicon and possibly oxygen and sulfur. The outer core is molten, metallic iron with a lighter density than the inner core, averaging 10.7 g/cm³.

***Earth's Magnetism.*** The fluid outer core generates at least 90% of Earth's magnetic field and the magnetosphere that surrounds and protects Earth from the solar wind. One hypothesis describes circulation patterns in the outer core that are influenced by Earth's rotation. This circulation generates electrical currents, which in turn induce the magnetic field. Think of the outer core flowing thickly at several kilometers per year, over a million times faster than movement in the overlying mantle.

An intriguing feature of Earth's magnetic field is that its polarity sometimes fades to zero and then returns to full strength, with north and south magnetic poles reversed! In the process, the field does not blink on and off but instead diminishes slowly to low intensity then rapidly regains its full strength. This **magnetic reversal** has taken place nine times during the past 4 million years and hundreds of times over Earth's history. The average period of a magnetic reversal is 500,000 years; occurrences possibly vary from as short as several thousand years to as long as tens of millions of years. The obvious question is Why?

Unfortunately, the reasons for these magnetic reversals are unknown at present. However, the spatial patterns they create at Earth's surface are a key tool in understanding the evolution of landmasses and the movements of the continents. When new iron-bearing rocks solidify from molten material (lava) at Earth's surface, the small magnetic particles in the rocks align according to the orientation of the magnetic poles at that time. This alignment is then locked in place as the rocks cool and solidify.

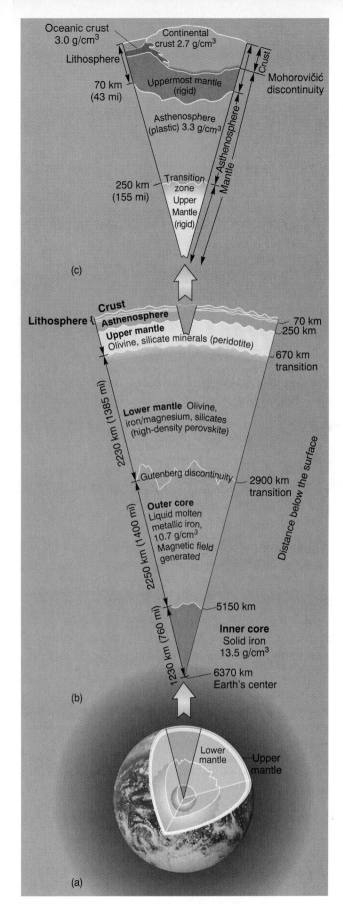

**FIGURE 11-3**
**Earth in cross section.**
(a) Cutaway showing Earth's interior. (b) Earth's interior in cross section, from the inner core to the crust. (c) Detail of the structure of the lithosphere and its relation to the asthenosphere. [Photo from NASA.]

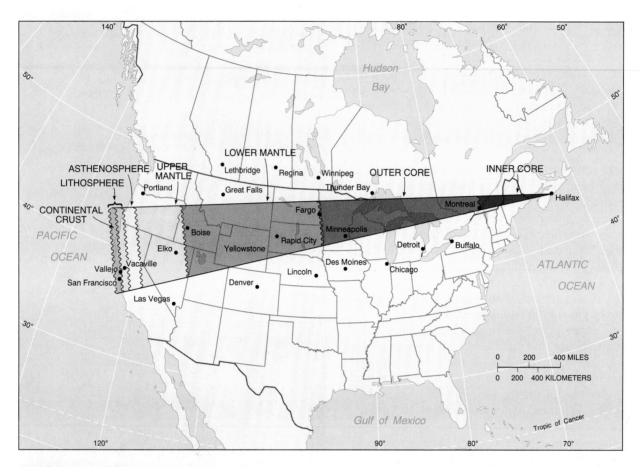

**FIGURE 11-4**
**Distances from core to crust.**
Surface map, using the distance from Halifax, Nova Scotia, to San Francisco, to compare the distance from Earth's center to the surface (6370 km, or 3963 mi). The thickness of the continental crust is the same as the distance between Vallejo (in the eastern portion of San Francisco Bay) and the city of San Francisco.

All across Earth, rocks bear an identical record of magnetic reversals in the form of measurable magnetic "stripes." These illustrate global patterns of changing magnetism. Matching segments allow scientists to reassemble past continental arrangements. Later in this chapter we will see the importance of these magnetic reversals.

An apparent trend in recent geologic time is toward more-frequent reversals. When Earth is without polarity in its magnetic field, a random pattern of magnetism in crustal rocks results. Transition periods last from 4000 to 8000 years. The effects of these low-intensity episodes on life are still speculative, but without a magnetosphere, the surface environment is unprotected from cosmic radiation and solar wind. Given present rates of magnetic field decay, we are perhaps 2000 years away from the next reversal.

***Earth's Mantle.*** Figure 11-3b shows a transition zone of several hundred kilometers wide, at an average depth of about 2900 km (1800 mi) dividing Earth's outer core from its mantle. Scientists at the California Institute of Technology

analyzed more than 25,000 earthquakes and determined that this transition area is uneven, with ragged peak-and-valley–like formations. Some of the motions in the mantle may be created by this rough texture at what is called the *Gutenberg discontinuity.* A *discontinuity* is a place where physical differences occur between adjoining regions in Earth's interior, such as between the outer core and lower mantle.

The **mantle** (lower and upper together) represents about 80% of Earth's total volume. The mantle is rich in oxides of iron and magnesium and silicates (FeO, MgO, and $SiO_2$). They are dense and tightly packed at depth, grading to lesser densities toward the surface (densities average 4.5 g/cm³). The upper mantle is separated from the lower mantle by a broad transition zone of several hundred kilometers, centered about 670 km (415 mi) below the surface. The entire mantle experiences a gradual temperature increase with depth and a stiffening due to increased pressures. The denser lower mantle is thought to contain a mixture of the minerals iron, magnesium, and silicates, with some calcium and aluminum.

The upper mantle is divided into three fairly distinct layers: uppermost mantle, asthenosphere, and upper mantle, shown in Figure 11-3c. Next to the crust is the *uppermost mantle*, a high-velocity zone just below the crust, where seismic waves are transmitted through a rigid, cooler layer. This uppermost mantle, along with the crust, makes up the *lithosphere*, approximately 45–70 km thick.

Below the lithosphere, from about 70 km down to 250 km, is the **asthenosphere**, or plastic layer (from the Greek *asthenos*, meaning "weak"). It contains pockets of increased heat from radioactive decay and is susceptible to slow convective currents in these hotter (and therefore less dense) materials.

Because of its dynamic condition, the asthenosphere is the least rigid region of the mantle, with densities averaging 3.3 g/cm³. About 10% of the asthenosphere is molten in asymmetrical patterns and hot spots. The resulting slow movement in this zone disturbs the overlying crust and creates tectonic activity—the folding, faulting, and general deformation of surface rocks. In return, the movement of the crust apparently influences currents throughout the mantle.

The depth affected by convection currents is the subject of much scientific speculation. One hypothesis states that mixing occurs throughout the entire mantle, upwelling from great depths. Another view states that mixing in the mantle is layered, segregated above and below the 670 km boundary. Presently, evidence indicates some truth in both positions. As an example, there are hot spots on Earth, such as those under Hawaii and Iceland, that appear to be at the top of tall plumes of rising mantle rock that must be anchored in the stiff lower mantle. In other surface regions, slabs of crust descend and penetrate to the lower mantle.

***Lithosphere and Crust.*** The lithosphere includes the **crust** and uppermost mantle to about 70 km (43 mi) depth (Figure 11-3c). An important internal boundary between the crust and the high-velocity portion of the uppermost mantle is another discontinuity, called the **Mohorovičić discontinuity**, or **Moho** for short. It is named for the Yugoslavian seismologist who determined that seismic waves change at this depth owing to sharp contrasts of materials and densities.

Figure 11-3c illustrates the relation of the crust to the rest of the lithosphere and the asthenosphere below. Crustal areas beneath mountain masses extend deep, perhaps to 50–60 km (31–37 mi), whereas the crust beneath continental interiors averages about 30 km (19 mi) in thickness. Oceanic crust averages only 5 km (3 mi). The crust is only 0.01% of Earth's overall mass. Drilling through the crust into the uppermost mantle remains an elusive scientific goal (see News Report 2).

The composition and texture of continental and oceanic crusts are quite different, and this difference is a key to the concept of drifting continents:

- *Continental crust* is basically **granite**; it is crystalline and high in silica, aluminum, potassium, calcium, and sodium. (Sometimes continental crust is called *sial*, shorthand for *si*lica and *al*uminum.) Most important, continental crust is relatively low in density, averaging 2.7 g/cm³.

- Oceanic crust is **basalt**; it is granular and high in silica, magnesium, and iron. (Sometimes oceanic crust is called *sima*, shorthand for *si*lica and *ma*gnesium.) It is denser than continental crust, averaging 3.0 g/cm³.

***A Floating Crust.*** Buoyancy is the principle that something less dense, such as wood, floats in something denser, such as water. The principles of buoyancy and balance were combined in the 1800s into the important principle of **isostasy**. Isostasy explains certain vertical movements of Earth's crust.

Think of Earth's outer crust as floating on the denser layers beneath, much as a boat floats on water. Where the load is greater, owing to glaciers, sediment, or mountains, the crust tends to sink, or ride lower in the asthenosphere. Without that load (for example, when a glacier melts), the crust rides higher, in a recovery uplift known as *isostatic rebound*. Thus, the entire crust is in a constant state of compensating adjustment, or isostasy, slowly rising and sinking in response to its own burdens, as it is pushed and dragged about by currents in the asthenosphere (Figure 11-5).

Because the oceanic crust is denser (3.0 g/cm³) than continental crust (2.7 g/cm³), in a collision of the two, the oceanic crust will be subducted, or will plunge downward, beneath the lighter and more buoyant continental crust.

Earth's crust is the outermost shell: an irregular, brittle layer that resides restlessly on a dynamic and diverse interior. Let us examine the processes at work on this crust and the variety of rock types that compose the landscape.

# Geologic Cycle

Earth's crust is in an ongoing state of change, being formed, deformed, moved, and broken down by physical, chemical, and biological processes. While the endogenic (internal) system is at work building landforms, the exogenic (external) system is busily wearing them down. This vast give-and-take at the Earth-atmosphere-ocean interface is called the **geologic cycle**. It is fueled from two sources—Earth's internal heat and solar energy from space—influenced by the ever-present leveling force of Earth's gravity.

Figure 11-6 illustrates the geologic cycle, combining many of the elements presented in this text. As you can see, the geologic cycle is composed of three subsystems:

- The *hydrologic cycle* is the vast system that circulates water, water vapor, ice, and energy throughout the

## News Report 2
### Drilling the Crust to Record Depths

Scientists wanting to sample mantle material directly have unsuccessfully tried for decades to penetrate Earth's crust to the Moho discontinuity. The longest-lasting deep-drilling attempt is on the northern Kola Peninsula in Russia, 250 km north of the Arctic Circle. Twenty years of high-technology drilling has produced a hole 12 km deep (7.5 mi, or 39,400 ft). Presently, crystalline rock 1.4 billion years old at 180°C (356°F) is being penetrated by diamond-tipped drills.

Oceanic crust is thinner than continental crust and is the object of several drilling attempts. The International Ocean Drilling Program (ODP), a cooperative effort directed by Texas A & M University, drilled a 2.5-km-deep hole in an oceanic rift near the Galápagos Islands from 1975 to 1993, with further drilling planned (Figure 1). An ocean floor of distinct layers of lava and deeper magma chambers was found. But the Moho and Earth's mantle remain untapped.

**FIGURE 1**
**Ocean drilling ship.**
A modern ocean-floor drilling ship, the *JOIDES Resolution*. The ship is operated by the International Ocean Drilling Program. [Photo by ODP/Texas A & M University.]

Earth-atmosphere-ocean environment. This cycle rearranges Earth materials through erosion, transportation, and deposition, and it circulates water as the critical medium that sustains life. (The hydrologic cycle was detailed in Chapter 9.)

- The *rock cycle*, through processes in the atmosphere, crust, and mantle, produces three basic rock types—igneous, sedimentary, and metamorphic. We examine these next.

- The *tectonic cycle* brings heat energy and new material to the surface and recycles old materials to mantle depths, creating movement and deformation of the crust. We discuss the tectonic cycle later in this chapter.

### Rock Cycle

Of Earth's crust, 99% is composed of only eight natural elements. Just two of these—oxygen and silicon—account for 74.3% of the crust (Table 11-1). Oxygen, the most reactive gas in the lower atmosphere, readily combines with other elements. For this reason, the percentage of oxygen is greater in the crust (about 47%) than in the atmosphere (about 21%). The relatively large percentages of lightweight elements such as silicon and aluminum in the crust are explained by the internal differentiation process in which less-dense elements migrate toward the surface, as discussed earlier.

***Minerals and Rocks.*** Earth's elements combine to form minerals. A **mineral** is an inorganic natural compound having a specific chemical formula and possessing a crystalline structure. The combination of elements and the crystal structure give each mineral its characteristic hardness, color, density, and other properties. For example, the common mineral *quartz* is silicon dioxide, formula $SiO_2$, and has distinctive six-sided crystals.

One of the most widespread mineral families on Earth is the *silicates*, because silicon and oxygen are so common and because they readily combine with each other and with other elements. This mineral family includes quartz, feldspar, clay minerals, and numerous gemstones. *Oxides* are a group of minerals in which oxygen combines with metallic elements, such as iron (e.g., hematite, $Fe_2O_3$).

Another important mineral family is the *carbonate* group, which features carbon in combination with oxygen and other elements such as calcium, magnesium, and potassium. An example is the mineral calcite ($CaCO_3$), a form of calcium carbonate.

Of the nearly 3000 minerals, only 20 are common rock-forming minerals. *Mineralogy* is the study of the composition, properties, and classification of minerals. A

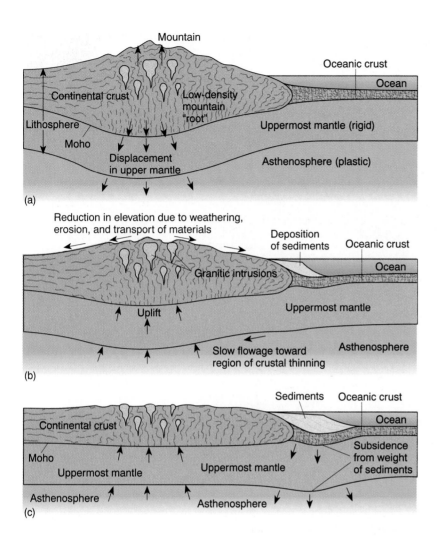

(a)

(b)

(c)

**FIGURE 11-5**
**Isostatic adjustment of the crust.**
Earth's entire crust is in a constant state of compensating adjustment as suggested by these three sequential stages. In (a), the mountain mass slowly sinks, displacing mantle material. In (b), because of the loss of mass through erosion and transportation, the crust isostatically adjusts upward and sediments accumulate in the ocean. As the continental crust thins (c), the heavy sediment load offshore begins to deform the lithosphere beneath the ocean.

**rock** is an assemblage of minerals bound together (such as granite, a rock containing three minerals) or, sometimes, a mass of a single mineral (such as rock salt). Thousands of different rocks have been identified. All can be identified as one of three kinds, depending on the processes that formed them: *igneous* (melted), *sedimentary* (from settling out), and *metamorphic* (altered). Figure 11-7 illustrates these three processes and the interrelations among them that constitute the **rock cycle**. Let's examine each rock-forming process.

## Igneous Processes

A rock that solidifies and crystallizes from a molten state is called an **igneous rock**. Familiar examples are granite, basalt, and lava. Igneous rocks form from **magma**, which is molten rock beneath the surface (hence the name *igneous*, which means "fire-formed" in Latin). Magma is fluid, highly gaseous, and under tremendous pressure. It either *intrudes* into crustal rocks, cools, and hardens or *extrudes* onto the surface as **lava.**

The cooling history of an igneous rock—how fast it cooled and how steadily its temperature dropped—determines its crystalline physical characteristics, or crystallization. Igneous rocks range from coarse-grained (slower cooling, with more time for larger crystals to form) to fine-grained or glassy (faster cooling).

The bulk of crustal rocks are igneous, although they are frequently covered by sedimentary rocks (sandstone, shale, limestone), soil, or oceans. Figure 11-8 illustrates the variety of occurrences of igneous rocks, both on and beneath Earth's surface.

***Intrusive and Extrusive Igneous Rocks.*** Intrusive igneous rock that cools slowly in the crust forms a **pluton**, a general term for any intrusive igneous rock body, regardless of size or shape. It is named after the Roman god of the underworld, Pluto. The largest pluton form is a **batholith**, defined as an irregular-shaped mass with a surface greater than 100 km$^2$ (40 mi$^2$) that has invaded layers of crustal rocks (Figure 11-9a). Batholiths form the mass of many large mountain ranges—for example, the Sierra Nevada batholith in California the Idaho batholith, and the Coast Range batholith of British Columbia and Washington State.

Smaller plutons include the magma conduits of ancient volcanoes that have cooled and hardened. Those that form

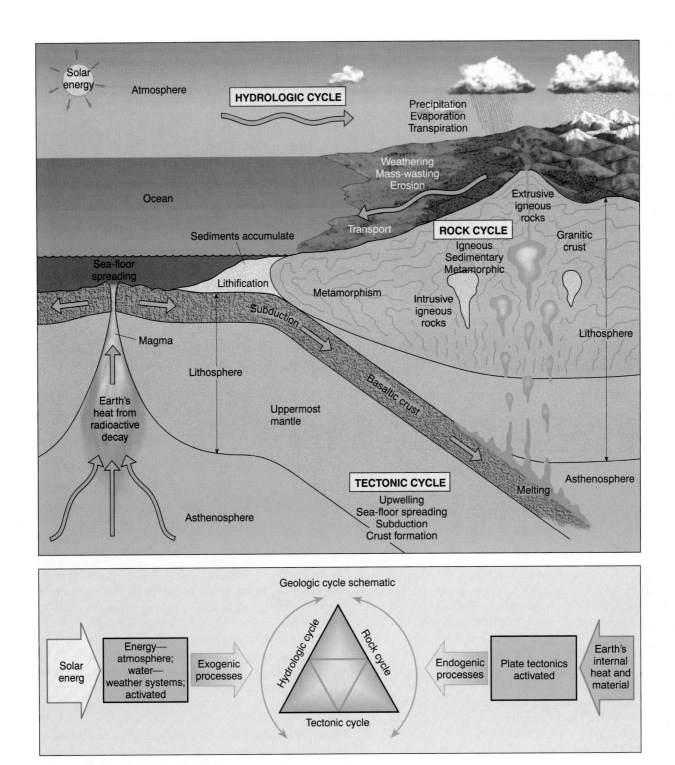

**FIGURE 11-6**

**The geologic cycle.**

The geologic cycle is a model showing the interactive relation among the rock cycle, tectonic cycle, and hydrologic cycle.
Earth's surface is where two dynamic systems interact—the endogenic (internal) and exogenic (external).

parallel to layers of sedimentary rock are termed *sills;* those that cross layers of the rock they invade are called *dikes* (see Figure 11-8). Magma also can bulge between rock strata and produce a lens-shaped body called a *laccolith,* a form of sill. In addition, magma conduits themselves may

solidify in roughly cylindrical forms that stand starkly above the landscape when finally exposed by weathering and erosion. Shiprock *volcanic neck* in New Mexico is such a feature, as suggested by the art in Figure 11-8 and shown in the inset photo. All of these intrusive forms can be

**TABLE 11-1**

| Common Elements in Earth's Crust | |
| --- | --- |
| **Element** | **Percent of Earth's Crust by Weight** |
| Oxygen (O) | 46.6 ⎫ |
| Silicon (SI) | 27.7 ⎭ |
| Aluminum (AL) | 8.1 |
| Iron (Fe) | 5.0 |
| Calcium (Ca) | 3.6 |
| Sodium (Na) | 2.8 |
| Potassium (K) | 2.6 |
| Magnesium (Mg) | 2.1 |
| All others | 1.7 |
| Total | 100.00 |

A quartz crystal ($SiO^2$) consists of Earth's two most abundant elements, silicon (Si) and oxygen (O). Inset photo from *Laboratory Manual in Physical Geology*, 3rd ed., R. M. Busch, ed. © 1993 by Macmillan Publishing Co.

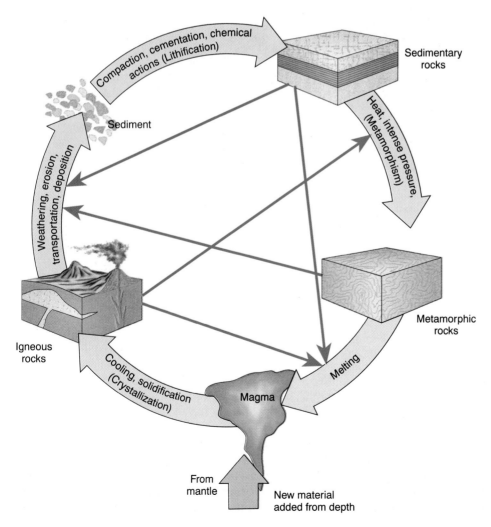

**FIGURE 11-7**

**The rock cycle.**

A rock-cycle schematic demonstrating the relation among igneous, sedimentary, and metamorphic processes. The arrows indicate that each rock type can enter the cycle at various points and be transformed into other rock types. [Adapted by permission from R. M. Busch, ed., *Laboratory Manual in Physical Geology*, 3rd ed.,© 1993 by Macmillan Publishing Co.]

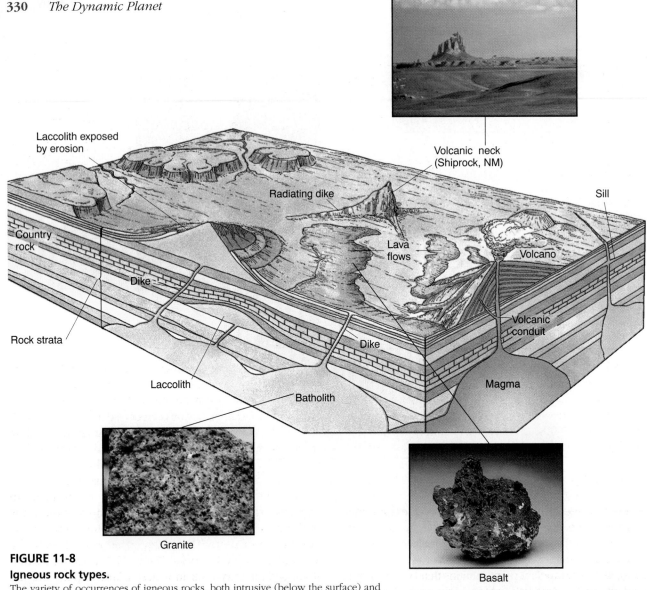

**FIGURE 11-8**

**Igneous rock types.**

The variety of occurrences of igneous rocks, both intrusive (below the surface) and extrusive (on the surface). Inset photographs show samples of granite (intrusive) and basalt (extrusive). [Photos from R. M. Busch, ed., *Laboratory Manual in Physical Geology*, 3rd ed., © 1993 by Macmillan Publishing Co. Photo of Shiprock volcanic neck by author.]

exposed by the weathering action of air, water, and ice.

Extrusive igneous rock, such as lava that cools and forms basalt, is produced by volcanic eruptions and flows (Figure 11-9b). Volcanism is discussed in Chapter 12.

***Classifying Igneous Rocks.*** Igneous rocks usually are categorized by their mineral composition and texture (Table 11-2). The two broad categories are:

1. *Felsic* igneous rocks—derived both in composition and name from *fel*dspar and *silica*. Felsic minerals are generally high in silica, aluminum, potassium, and sodium, and have low melting points. Rocks formed from felsic minerals generally are lighter in color and are less dense than mafic mineral rocks.
2. *Mafic* igneous rocks—derived both in composition and name from *ma*gnesium and *ferric* (Latin for iron).

Mafic minerals are low in silica, high in magnesium and iron, and have high melting points. Rocks formed from mafic minerals are darker in color and of greater density than felsic mineral rocks.

Table 11-2 shows that the same magma that produces coarse-grained granite when it is slowly cooled beneath the surface can form fine-grained *rhyolite* when it is rapidly cooled above the surface (see inset photo). If it cools rapidly, magma having a silica content comparable to granite and rhyolite may form the dark, smoky, glassy-textured rock called *obsidian*, or volcanic glass (see inset photo). Another glassy rock, called *pumice*, forms when escaping gases bubble a frothy texture into the lava. Pumice is full of small holes, is light in weight, and is low enough in density to float in water (see inset photo).

(a)

(b)

**FIGURE 11-9**
**Intrusive and extrusive rocks.**
(a) Exposed granites of the Schoodic Peninsula, Acadia National Park, Maine. (b) Basaltic lava flows on James Island of the Galápagos Islands, Ecuador. [(a) Photo by Adam Jones/Photo Researchers, Inc. (b) photo by Galen Rowell.]

On the mafic side, basalt is the most common fine-grained extrusive igneous rock. It makes up the bulk of the ocean floor, accounting for 71% of Earth's surface. It appears in lava flows such as those on the Galápagos Islands (Figure 11-9b). An intrusive counterpart to basalt, formed by slow cooling of the parent magma, is *gabbro*.

## Sedimentary Processes

The common sedimentary rocks are *sandstone* (sand that became cemented together), *shale* (mud that became compacted into rock), *limestone* (bones and shells that became cemented or calcium carbonate that precipitated in ocean waters), and *coal* (ancient plant remains that became compacted into rock). Most sedimentary rocks are derived from fragments of existing rock or organic materials.

The exogenic processes of weathering, erosion, and transportation generate the sediments needed to form rock. Bits and pieces of former rocks—principally quartz, feldspar, and clay minerals—are eroded and then mechanically transported by water (lake, stream, ocean, overland flow), ice (glacial action), wind, and gravity. They are transported from "higher-energy" sites, where the carrying medium has the energy to pick up and move them, to "lower-energy" sites, where they are deposited.

In addition, some minerals, such as calcium carbonate, are dissolved into solution and form sedimentary deposits by precipitating from those solutions to form rock. This is an important process in the oceanic environment.

The system of *sedimentation* is driven by solar energy and gravity, with water as the principal medium. Existing rock is digested by weathering, picked up and moved by erosion and transportation, and deposited along river, beach, and ocean sites, where burial initiates the rock-forming process. The cementation, compaction, and hardening of sediments into **sedimentary rocks** is called **lithification**.

Characteristically, sedimentary rocks are laid down by wind, water, or ice in horizontally layered beds. A variety of sedimentation forms are created under different environmental conditions. Various cements, depending on availability, fuse rock particles together. Lime, or calcium carbonate ($CaCO_3$) is the most common, followed by iron oxides ($Fe_2O_3$) and silica ($SiO_2$). Particles also can be united by drying (dehydration), heating, or chemical reactions.

The layered strata of sedimentary rocks form an important record of past ages. **Stratigraphy** is the study of the sequence (superposition), thickness, and spatial distribution of strata. These sequences yield clues to the age and origin of the rocks. Figure 11-10a shows a sandstone sedimentary rock in a desert landscape. Note the multiple layers in the formation and how differently they resist weathering processes.

The two primary sources of sedimentary rocks are the mechanically transported bits and pieces of former rock, called *clastic sediments*, and the dissolved minerals in solution, called *chemical sediments*.

**Clastic Sedimentary Rocks.** Clastic sedimentary rocks are derived from weathered and fragmented rocks that are further worn in transport. Table 11-3 lists the range of clast sizes—everything from boulders to microscopic clay particles—and the form they take as lithified rock. Common sedimentary rocks from clastic sediments include siltstone (silt) or mudstone (mud), shale (up to 0.06 mm diameter particles), and sandstone (sand particles that range from 0.06 to 2 mm in diameter).

**Chemical Sedimentary Rocks.** Chemical sedimentary rocks are not formed from physical pieces of broken rock but instead are dissolved minerals, transported in solution and chemically precipitated from solution (they are essentially nonclastic). The most common chemical sedimentary rock is **limestone**, which is lithified calcium

**TABLE 11-2**

## Igneous Rock Minerals

| | Felsic Minerals (feldspars and silica) | | | Mafic minerals (magnesium and iron) | | Ultramafic Minerals (low silica) | |
|---|---|---|---|---|---|---|---|
| General characteristics | Higher ◄――――― Silica content ―――――► Lower | | | | | | |
| | Higher ◄――――― Resistance to weathering ―――――► Lower | | | | | | |
| | ◄― Increased potassium and sodium　Increased calcium, iron and magnesium ―► | | | | | | |
| | Lower ◄――――― Melting temperatures ―――――► Higher | | | | | | |
| | Lighter ◄――――― Coloration ―――――► Darker | | | | | | |
| Mineral families | Quartz | Feldspars | | Mica | Amphibole | Pyroxene | Olivine |
| | | Potassium feldspars | Sodium feldspars　　Calcium feldspars | | | | |
| | SiO$_2$ | K, Al, Si (Orthoclase) | Na, Al, Si, Ca　Ca, Al, Si (plagioclase) (aluminosilicates) | K, Fe, Mg, Al, Si (biotite; black; muscovite; white) | Fe, Mg, Al, Si (complex) (hornblende: black) | Fe, Mg, Si (dark) | Mg, Fe (dark green) (no quarts, no feldspars) |
| Coarse-grained texture (intrusive) slower cooling rate | Granite | Diorite | | | | Gabbro | Peridotite |
| Fine-grained texture (extrusive) faster cooling rate | Rhyolite | Andesite Dacite (sodic feldspar) (Mount St. Helens) | | | | Basalt | |
| Other textures | Obsidian (glassy) | Pumice (vesicular) | | | | Scoria (vesicular) | |

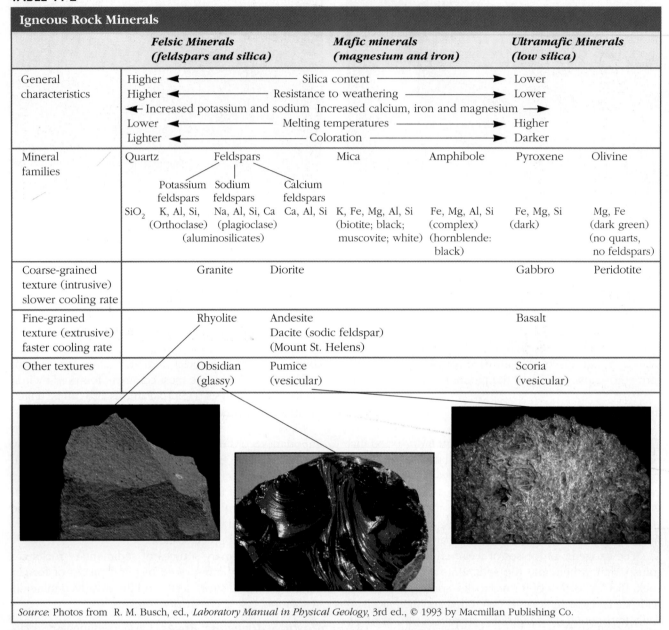

*Source*: Photos from R. M. Busch, ed., *Laboratory Manual in Physical Geology*, 3rd ed., © 1993 by Macmillan Publishing Co.

carbonate, derived from inorganic and organic sources. A similar form is *dolomite*, which is lithified calcium-magnesium carbonate, CaMg(CO$_3$)$^2$. Limestone from marine organic origins is most common; it is biochemical—derived from shell and bone produced by biological activity. Once formed, these rocks are vulnerable to chemical weathering, which produces unique landforms, as discussed in the weathering section of Chapter 13 and as shown in Figure 11-10b.

Chemical sediments from inorganic sources are deposited when water evaporates and leaves behind a residue of salts. These *evaporites* may exist as common salts, such as gypsum or sodium chloride (table salt). They often appear as flat, layered deposits across a dry landscape. This

process is dramatically demonstrated in the pair of photographs in Figure 11-11; one was taken in Death Valley National Park one day after a record 2.57 cm (1.01 in.) rainfall, and the other was taken one month later.

Chemical deposition also occurs in the water of natural hot springs from chemical reactions between minerals and oxygen. This is a sedimentary process related to *hydrothermal activity*. An example is the massive deposit of travertine, a form of calcium carbonate, at Mammoth Hot Springs in Yellowstone National Park (Figure 11-12). Another hydrothermal area and unique landscape occurs near the Nevada-California border at Mono Lake. It is an alkaline lake, high in carbonates and sulfates and the subject of News Report 3.

**FIGURE 11-10**

**Sedimentary rock types.**

Two types of sedimentary rock. (a) Sandstone formation with sedimentary strata subjected to differential weathering; note weaker underlying siltstone. (b) A limestone landscape, formed by chemical sedimentary processes, in County Clare, Ireland. [(a) Photo by author; (b) photo by Tom Bean.]

**TABLE 11-3**

| Clastic Sediment Sizes and Related Rock Form | | |
| --- | --- | --- |
| **Unconsolidated Sediment** | **Grain Size** | **Rock Form** |
| Boulders, cobbles | >80 mm | Conglomerate (breccia, if pieces are angular) |
| Pebbles, gravel | >2 mm | |
| Coarse sand | 0.5–2.0 mm | Sandstone |
| Medium-to-fine sand | 0.062–0.5 mm | Sandstone |
| Silt | 0.002–0.062 mm | siltstone (mudstone) |
| Clay | <0.002 mm | Shale |

***Coal.*** A sedimentary rock of organic origin is coal. **Coal** is really a biochemical rock, very rich in carbon. Plants incorporate carbon into their tissues by taking in carbon dioxide ($CO_2$) during photosynthesis, and it is this process that brings organic carbon into the rock cycle. During the Carboniferous Period (360–286 million years ago) lush vegetation flourished and died in tropical swamps. Coal formed in these bogs through progressive stages.

Pieces of former plants can be easily recognized in *peat*, a brownish, fibrous, compacted organic sediment—the first step in forming coal. Peat has been burned as a low-grade fuel for centuries, producing a low-heat energy output and a great deal of air pollution, both particulates and gases.

Following more compaction and hardening, a soft, coal-like substance known as *lignite* forms. Further lithification leads to *bituminous coal*, the first true sedimentary rock in this sequence and the coal most used in steelmaking and coal-fired power plants. Burial of coal deposits under sedimentary overburden (material covering useful geologic material or bedrock) results in the formation of coal seams, usually sandwiched between beds of shale or sandstone. Where conditions are right, metamorphism of bituminous coal results in *anthracite coal*, a metamorphic rock that is extremely hard and therefore difficult to mine. Some problems of coal use are presented in News Report 4.

(a)

(b)

**FIGURE 11-11**

**Death Valley, wet and dry.**

A Death Valley landscape (a) one day after a record rainfall when the valley was covered by several square kilometers of water only a few centimeters deep. (b) One month later the water had evaporated, and the same valley is coated with evaporites (borated salts). [Photos by author.]

# News Report 3

## The Rescue of Mono Lake

Mono Lake is an alkaline lake near the Nevada-California border. It is high in carbonates and sulfates and presently has an overall salinity of 95‰ (the ocean average is 35‰, or 35 parts per thousand). Carbonates in Mono Lake interact with calcium-rich hot springs to produce a rock called tufa. The tufa deposits grew tall in Mono Lake as long as the springs were beneath the surface of the lake.

However, since the 1940s, tributary streams have been diverted out of the region to Los Angeles, lowering the lake's surface more than 20 m (65 ft), reducing its surface area by more than half, and almost doubling salinity levels from the original 55‰. The lowered lake level exposed chemical sedimentary deposits and alkaline shorelines (Figure 1). Some tufa towers now rise over 10 m (30 ft) above the shoreline and lake surface.

The lower lake level and increased salinity damaged the lake's natural

**FIGURE 1**

**Mono Lake landscape.**

Tufa towers at Mono Lake, California, formed by chemical reactions between spring water and alkali lake water when the lake level was higher. [Photo by author.]

ecology and habitats for wildlife, particularly those of birds, which perished in large numbers in the early 1980s. Strong winds in the area blow across the newly formed alkali flats of evaporites, producing days of severe air pollution from the alkali particles.

After much litigation, and a case that went to the California Supreme Court, the state's water law was significantly changed. Century-old appropriative water rights (first users at water's

edge withdraws whatever they want) were overturned in favor of placing this body of water in the *public trust*—the benefits to the many overshadowed the rights of a few. A master plan was completed and agreed to by all parties, and the future looks bright for this water-troubled region. Tributary flows were restored by court order in April 1991, and the lake level is rising.

## *Metamorphic Processes*

Any rock, either igneous or sedimentary, may be transformed into a **metamorphic rock** by going through profound physical or chemical changes under pressure and

increased temperature. (The name *metamorphic* comes from a Greek word meaning "to change form.") Metamorphic rocks generally are more compact than the original rock and therefore are harder and more resistant to weathering and erosion (Figure 11-13).

**FIGURE 11-12**
**Hydrothermal deposits.**
Mammoth Hot Springs, Yellowstone National Park, is an example of a hydrothermal deposit principally composed of travertine ($CaCO_3$), deposited as a chemical precipitate from heated spring water as it evaporated. [Photo by author.]

Several conditions can cause metamorphism. Most common is when subsurface rock is subjected to high temperatures and high compressional stresses occurring over millions of years. Igneous rocks become compressed during collisions between slabs of Earth's crust (see Plate Tectonics later in this chapter). Sometimes rocks simply are crushed under great weight when a crustal area is thrust beneath other crust. In another setting, igneous rocks may be sheared and stressed along earthquake fault zones, causing metamorphism.

The ancient roots of mountains are composed predominantly of metamorphic rocks. Exposed at the bottom of the inner gorge of the Grand Canyon in Arizona, a Precambrian metamorphic rock called the *Vishnu Schist* is a remnant of such an ancient mountain root (Figure 11-13b). (See News Report 5.)

Another metamorphic condition occurs when sediments collect in broad depressions in Earth's crust and, because of their own weight, create enough pressure in the bottommost layers to transform the sediments into metamorphic rock. Also, molten magma rising within the crust may "cook" adjacent rock, a process called *contact metamorphism.*

Metamorphic rocks may be changed both physically and chemically from the original rocks. If the mineral structure demonstrates a particular alignment after metamorphism, the rock is called *foliated,* and some minerals may appear as wavy striations (streaks or lines) in the new rock (Table 11-4, left photo). On the other hand, parent rock with a more homogeneous (evenly mixed) makeup may produce a *nonfoliated* rock (Table 11-4, right photo). Table 11-4 lists some metamorphic rock types, their parent rocks, and resultant textures.

In this section, you have seen how the rock-forming processes yield the igneous, sedimentary, and metamorphic materials of Earth's crust. Next we look at how huge slabs of the crust itself are moved by **tectonic processes** within Earth, causing the continents to *drift*. The fact that the continents have migrated thousands of kilometers, and continue to migrate, is a revolutionary discovery in the *tectonic cycle* that came to light only in this century.

## News Report 4

### Concerns about Coal Reserves

About 30% to 35% of the world's coal reserves are in the United States and Canada, 25% in Russia and the former Soviet states, 15% in China, and lesser amounts in Australia, Great Britain, India, Poland, and other countries. The largest consumer of coal in the United States is the electrical power industry, which uses 70% of mined coal for the production of steam in power plant boilers. (The steam pressure rotates turbines, which generate the electricity.)

How long will known coal reserves last? Estimates vary and are based on assumptions regarding the annual rate of increase in the mining and consumption of coal. There are at least 500 years of coal remaining in the United States and Canada if the rate of consumption stays about where it was in the 1980s (about 1 billion tons a year). However, if mining of these reserves increases 3%–5% annually, their lifetime will shrink to 100 years or less.

Another concern related to burning coal is an impurity commonly found within it, sulfur. Sulfur is a major air pollutant when released to the atmosphere. The combination of sulfur oxides and water vapor forms acidic precipitation, which causes environmentally damaging acid deposition.

A further problem with coal use is that it produces more carbon dioxide per unit consumed than the other fossil fuels, oil and natural gas. Increased carbon dioxide is enhancing Earth's natural greenhouse effect. Strategies to reduce atmospheric carbon dioxide must include energy conservation and reduced coal use, which would in turn extend the lifetime of this important resource. Coal is a valuable source of chemicals for many important uses, including medicines, so its depletion would be a tragedy for future generations.

(a)

(b)

**FIGURE 11-13**

**Metamorphic rocks.**

(a) A metamorphic rock outcrop in Greenland, the Amitsoq Gneiss, is the second oldest rock found on Earth. (b) Precambrian Vishnu Schist composes these rock cliffs in the inner gorge of the Grand Canyon in northern Arizona. These metamorphic rocks lie beneath many layers of sedimentary rocks deep in the continent. [(a) Photo by Kevin Schafer/Peter Arnold, Inc.; (b) photo by author.]

# Plate Tectonics

Have you ever looked at a world map and noticed that a few of the continental landmasses—particularly South America and Africa—appear to have matching shapes, like pieces of a jigsaw puzzle? The incredible reality is that the continental pieces indeed once were fitted together! Continental landmasses not only migrated to their present locations but continue to move at speeds up to 6 cm (2.4 in.) per year.

We say that the continents are *adrift* because convection currents in the asthenosphere and the rest of the mantle are dragging them around. The key point is that the arrangement of continents and oceans we see today is not permanent but is in a continuing state of change.

A continent such as North America is actually a collage of pieces and fragments of crust that have migrated to form the landscape we know. Through a historical chronology, let us trace the discoveries that produced the *plate tectonics theory*, a major revolution in science.

## A Brief History

As early mapping gained accuracy, some observers noticed the symmetry among continents, particularly the fit between South America and Africa just described. Abraham Ortelius (1527–1598), a geographer, noted the apparent fit of some continental coastlines in his *Thesaurus Geographicus* (1596). In 1620, the English philosopher Sir Francis Bacon noted gross similarities between the edges of Africa and South America (although he did not suggest that they had drifted apart). Benjamin Franklin wrote in 1780 that the crust must be a shell that can break and shift by movements of fluid below! Others wrote—unscientifically—about such apparent relationships, but it was not until much later that a valid explanation was proposed.

In 1912, German geophysicist and meteorologist Alfred Wegener publicly presented in a lecture his idea that Earth's landmasses migrate. His book, *Origin of the Continents and Oceans*, appeared in 1915. Wegener today is regarded

# News Report 5

## Canyon Rocks Harder Than Steel

Tourists visiting the Grand Canyon in Arizona can see some ancient rocks first hand. At the bottom of the inner gorge (the lower 455 m, or 1500 ft), a Precambrian metamorphic rock called the *Vishnu Schist* is over 2 billion years old.

A sample of the rock and a metal file allow visitors to test the rock's hardness against steel. Despite the fact that the schist is harder than the steel, the Colorado River has cut down into the uplifted Colorado Plateau, exposing and

eroding these ancient rocks. Given enough time, the processes of nature will wear away the canyon, only to be replaced by new rock formations produced by Earth's internal processes.

**TABLE 11-4**

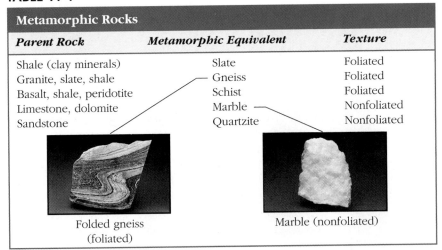

| Metamorphic Rocks | | |
|---|---|---|
| *Parent Rock* | *Metamorphic Equivalent* | *Texture* |
| Shale (clay minerals) | Slate | Foliated |
| Granite, slate, shale | Gneiss | Foliated |
| Basalt, shale, peridotite | Schist | Foliated |
| Limestone, dolomite | Marble | Nonfoliated |
| Sandstone | Quartzite | Nonfoliated |

Folded gneiss
(foliated)

Marble (nonfoliated)

as the father of this concept, which he first called **continental drift**. But scientists at the time, knowing little of Earth's interior structure and bound by previous notions as to how continents and mountains were formed, were unreceptive to Wegener's revolutionary proposal. Thus, a great debate was launched, lasting almost 50 years, until a scientific consensus developed that he was right after all.

Wegener thought that all landmasses were united in one supercontinent approximately 225 million years ago, during the Triassic Period. This one landmass he called **Pangaea**, meaning "all Earth" (see Figure 11-17b). Although his initial model kept the landmasses together far too long and he proposed an incorrect driving mechanism for moving the continents, Wegener's arrangement of Pangaea was right.

To come up with his Pangaea fit, he began studying the research of others, specifically the geologic record (rock strata), the fossil record, and the climatic record for the continents. He concluded that South America and Africa were related in many complex ways. He also concluded that the large midlatitude coal deposits, which today stretch from North America to Europe to China and date to the Permian and Carboniferous Periods (245–360 million years ago), existed because these regions once were more equatorial in location and therefore covered by lush vegetation that became coal.

As modern scientific capabilities built the case for continental drift, the 1950s and 1960s saw a revival of interest in Wegener's concepts and, finally, confirmation. Aided by an avalanche of discoveries, the **plate tectonics** theory today is nearly universally accepted as an accurate model of the way Earth's surface evolves. *Tectonic*, from the Greek *tektonikùs*, meaning "building" or "construction," refers to the deformation of Earth's crust as a result of internal forces. Tectonic processes include upwelling of magma, plate movements, earthquakes, volcanic activity, subduction, warping, folding, and faulting of the crust.

## Sea-Floor Spreading and Production of New Crust

The key to establishing the theory of continental drift was a better understanding of the sea floor. The sea floor has a remarkable feature: an interconnected worldwide mountain chain (ridge) some 64,000 km (40,000 mi) in extent and averaging more than 1000 km (600 mi) in width. A striking view of this great undersea mountain chain opens Chapter 12. How did this global mountain chain get there?

In the early 1960s, geophysicists Harry H. Hess and Robert S. Dietz proposed **sea-floor spreading** as the mechanism that builds this mountain chain and drives continental movement. Hess said that these submarine mountain ranges, called the **mid-ocean ridges**, were the direct result of upwelling flows of magma from hot areas in the upper mantle and asthenosphere and perhaps from the deeper lower mantle. When mantle convection brings magma up to the crust, the crust is fractured, the magma extrudes onto the sea floor, and it cools to form new sea floor. It builds the ridges and spreads laterally. The concept is illustrated in Figure 11-14, which shows how the ocean floor is rifted and scarred along a mid-ocean ridge.

As new crust is generated and the sea floor spreads, magnetic particles in the lava orient with the magnetic field in force at the time. They then become locked in this alignment as the lava chills to form new sea floor. This alignment creates a kind of magnetic tape recording in the sea floor. Each magnetic reversal and reorientation of Earth's polarity is recorded in the continually forming oceanic crust.

Figure 11-15 illustrates just such a recording from the Mid-Atlantic Ridge near Iceland. The colors denote the alternating magnetic polarization. Note the magnetic mirror images that develop on either side of the sea-floor rift as a result of the nearly symmetrical spreading of the sea floor. These periodic reversals of Earth's magnetic

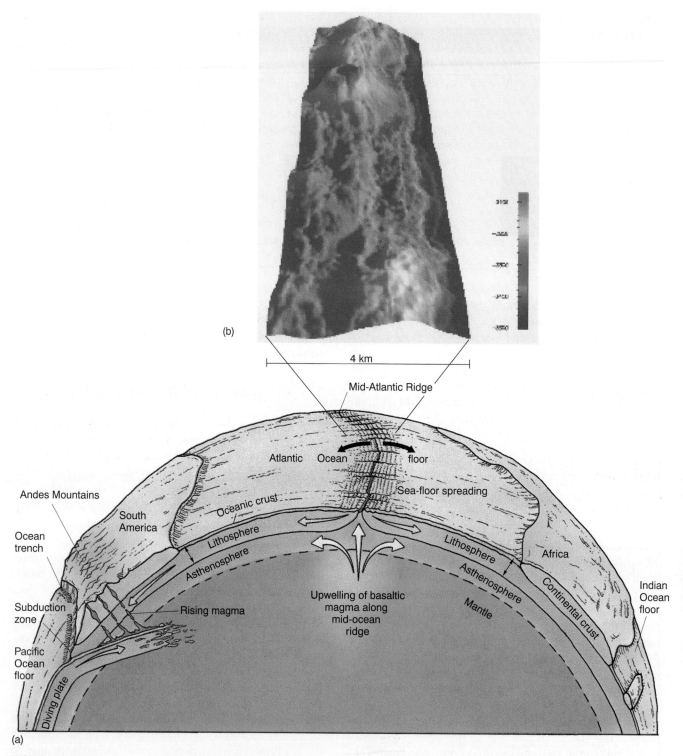

**FIGURE 11-14**

**Crustal movements.**

(a) Sea-floor spreading, upwelling currents, subduction, and plate movements, shown in cross section. Arrows indicate the direction of the spreading. (b) A 4-km wide image of the Mid-Atlantic Ridge showing linear faults, a volcanic crater, a rift valley, and ridges. The image was taken by the TOBI (towed ocean bottom instrument) at approximately 29° N latitude. [(a) After P. J. Wyllie, *The Way the Earth Works*, © 1976, by John Wiley & Sons. Adapted by permission; (b) image courtesy of D. K. Smith, Woods Hole Oceanographic Institute, Woods Hole, Massachusetts. All rights reserved.]

field have proved to be a valuable clue to understanding sea-floor spreading, helping scientists fit together pieces of Earth's crust.

As the age of the sea floor was determined from these sea-floor recordings of Earth's magnetic-field reversals and other measurements, the complex harmony of the two concepts of plate tectonics and sea-floor spreading became clearer. The youngest crust anywhere on Earth is at the spreading centers of the mid-ocean ridges, and, with increasing distance from these centers, the crust gets increasingly older (Figure 11-16). The oldest sea floor is in the western Pacific near Japan, in an area that dates to nearly 200 million years ago (during the Jurassic Period).

Overall, the sea floor is relatively young—nowhere does it exceed 208 million years in age, remarkable when you remember that Earth's age is 4.6 billion years. The reason is that oceanic crust is short-lived—the oldest sections, farthest from the mid-ocean ridges, are slowly plunging beneath continental crust along Earth's deep oceanic trenches. The discovery that the sea floor is so young demolished earlier thinking that the oldest rocks would be found there.

### Subduction of the Crust

In contrast to the upwelling zones along the mid-ocean ridges are the areas of descending crust elsewhere. On the left side of Figure 11-14a note how one plate of the crust is diving beneath another, into the mantle. Recall that the basaltic ocean crust has a density of 3.0 g/cm$^3$, whereas continental crust averages a lighter 2.7 g/cm$^3$. As a result, when continental crust and oceanic crust slowly collide, the denser ocean floor will grind beneath the lighter continental crust, thus forming a **subduction zone**, as shown in the figure.

The world's deep ocean trenches coincide with these subduction zones and are the deepest features on Earth's surface. Deepest is the Mariana Trench near Guam, which descends below sea level to −11,033 m (−36,198 ft); next in depth are the Puerto Rico Trench at −8605 m (−28,224 ft) and, in the Indian Ocean, the Java Trench at −7125 m (−23,376 ft).

The subducted portion of crust is dragged down into the mantle, where it remelts and eventually is recycled as magma, rising again toward the surface through deep fissures and cracks in crustal rock. Volcanic mountains such as the Andes in South America and the Cascade Range from northern California to the Canadian border form inland of these subduction zones. The fact that spreading ridges and subduction zones are areas of earthquake and volcanic activity provides important proof of plate tectonics. Now, using current scientific findings, let us go back and reconstruct the past, and Pangaea.

### The Formation and Breakup of Pangaea

The supercontinent of Pangaea and its subsequent breakup into today's continents represent only the last 225 million years of Earth's 4.6 billion years, or only the most recent 1/23 of Earth's existence. During the other 22/23 of geologic time, other things were happening. The landmasses as we know them were barely recognizable.

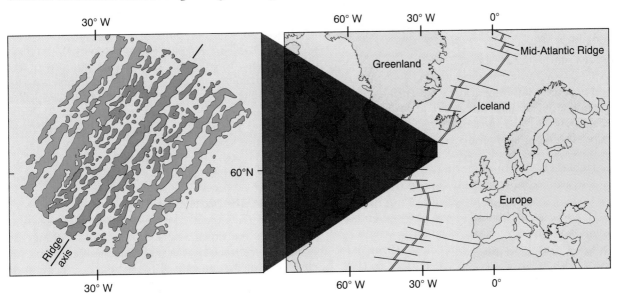

**FIGURE 11-15**

**Magnetic reversals recorded in the sea floor.**

Magnetic reversals recorded in the sea floor south of Iceland along the Mid-Atlantic Ridge. [Magnetic reversals reprinted from J. R. Heirtzler, S. Le Pichon, and J. G. Baron, *Deep-Sea Research* 13, © 1966, Pergamon Press, p. 247.]

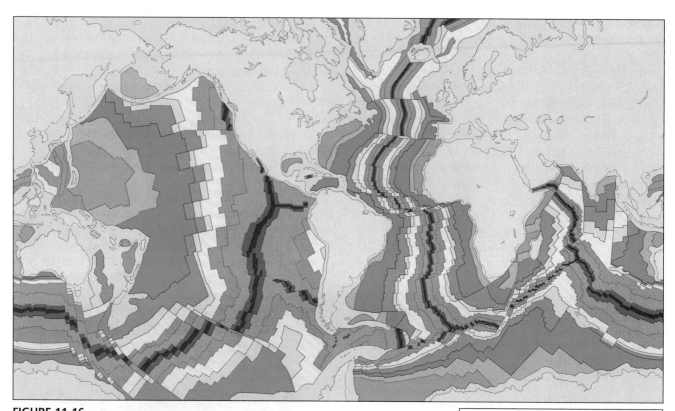

**FIGURE 11-16**
**Relative age of the oceanic crust.**
[Adapted with the permission of W. H. Freeman and Company from *The Bedrock Geology of the World* by R. L. Larson and others. © 1985.]

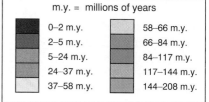

m.y. = millions of years

| | | | |
|---|---|---|---|
| | 0–2 m.y. | | 58–66 m.y. |
| | 2–5 m.y. | | 66–84 m.y. |
| | 5–24 m.y. | | 84–117 m.y. |
| | 24–37 m.y. | | 117–144 m.y. |
| | 37–58 m.y. | | 144–208 m.y. |

***Pre-Pangaea.*** Figure 11-17a shows the pre-Pangaea arrangement of 465 million years ago (during the middle Ordovician Period). The outlines of South America, Africa, India, Australia, and southern Europe appear within the continent called Gondwana in the Southern Hemisphere. What would become North America and Greenland is the inverted continent of Laurentia, which straddles the equator near the location of the present-day Fiji and Samoa island groups. The other land groups are identified.

***Pangaea.*** Figure 11-17b shows an updated version of Wegener's Pangaea, 225–200 million years ago (Triassic-Jurassic Periods). Areas of North America, northern Africa, the Middle East, and Eurasia were near the equator and therefore were covered by plentiful vegetation. The landmasses focused near the South Pole and attached to the sides of Antarctica were covered by ice. Today, those portions of South America, southern Africa, India, Australia, and Antarctica bear unmistakable scars of this shared glacial blanket.

The India plate (the slab of Earth's crust that included modern India) appears larger on the map than the same area today because the Tibetan Plateau formed the northern portion of the plate. We also can see that Africa shared a common connection with both North and South America. Today, the Appalachian Mountains in the eastern United States and the Atlas Mountains of northwestern Africa reflect this common ancestry; they are, in fact, portions of the same mountain range, torn thousands of kilometers apart!

At the time of Figure 11-17b, the Atlantic Ocean did not exist. The only oceans were Panthalassa ("all seas"), which would become the Pacific, and the Tethys Sea (partly enclosed by the African and the Eurasian plates), which would become the Mediterranean Sea, with trapped portions becoming the present-day Caspian Sea.

***Pangaea Breaks Up.*** Figure 11-17c shows the movement of plates that had occurred by 135 million years ago (the beginning of the Cretaceous Period). New sea floor that had formed since the last map is highlighted (gray tint). An active spreading center rifted North America away from landmasses to the east, shaping the coast of Labrador. India was farther along in its journey, with a spreading center to the south and a subduction zone to the north, where the leading edge of the India plate was diving beneath Eurasia. At that time a rift also began in what was to be the South Atlantic, and Antarctica and Australia were pulled into isolation from the other masses.

***Modern Continents Take Shape.*** Figure 11-17d shows the arrangement 65 million years ago (beginning of the Tertiary Period). Sea-floor spreading along the Mid-Atlantic Ridge had grown some 3000 km (almost 1900 mi) in 70 million years. Africa had moved northward about 10° in latitude, leaving Madagascar split from the mainland and opening up the Gulf of Aden. The rifting along what would be the Red Sea had begun. The India plate had moved three-fourths of the way to Asia, as Asia continued to rotate clockwise. Of all the major plates, India traveled the farthest—almost 10,000 km. From 65 million years ago until the present, according to Dietz and Holden, more than half of the ocean floor was renewed.

***The Continents Today.*** Figure 11-17e shows modern geologic time (the late Cenozoic Era). The northern reaches of the India plate have underthrust the southern mass of Asia through subduction, forming the Himalayas in the upheaval created by the collision. Plate motions continue to this day, as noted by the arrows, and will proceed into the future.

Earth's present crust is divided into at least 14 plates, of which about half are major and half are minor, in terms of area (Figure 11-18). These broad plates are composed of literally hundreds of smaller pieces and perhaps dozens of microplates that have migrated together. The arrows in the figure indicate the direction in which each plate is presently moving, and the length of the arrows suggests the rate of movement during the past 20 million years.

Compare Figure 11-18 with the image of the ocean floor in Figure 11-19. In the image, satellite radar-altimeter measurements have recorded the sea-surface height to a remarkable accuracy of 0.03 m, or 1 in. The sea-surface elevation is far from uniform; it is higher or lower in direct response to the mountains, plains, and trenches of the ocean floor beneath it. The small bumps and dips in the sea surface are due to slight differences in Earth's gravity caused by sea-floor topography. For example, a massive mountain on the ocean floor exerts a high gravitational field and attracts water to it, producing a bump; over a trench there is less gravity, resulting in a dip. A mountain 2000 m (6500 ft) high that is 20 km (12 mi) across its base, creates a 2 m rise in the sea surface. Until this use of satellite technology was developed at Scripps Institution of Oceanography and NOAA, these tiny changes in height were not visible. Now we have a comprehensive map of the ocean *floor*, derived from remote sensing of the ocean *surface!*

The illustration of the sea floor that begins Chapter 12 also may be helpful in identifying the plate boundaries in the following discussion. It is interesting to correlate the features illustrated in Figures 11-18, 11-19, and Chapter 12's opening image.

## Plate Boundaries

The boundaries where plates meet clearly are dynamic places, albeit slow-moving. The block diagram inserts in Figure 11-17e show the three general types of motion and interaction that occur along the boundary areas:

- *Divergent boundaries* (lower left in figure) are characteristic of sea-floor spreading centers, where upwelling material from the mantle forms new sea floor ("constructional") and crustal plates are spread apart. The spreading makes these zones of tension. An example noted in the figure is the divergent boundary along the East Pacific rise, which gives birth to the Nazca plate (moving eastward) and the Pacific plate (moving northwestward). Whereas most divergent boundaries occur at mid-ocean ridges, there are a few within continents themselves. An example is the Great Rift Valley of East Africa, where crust is being rifted apart.

- *Convergent boundaries* (upper left) are characteristic of collision zones, where areas of continental and oceanic crust collide. These are zones of compression and crustal loss ("destructional"). Examples include the subduction zone off the west coast of South and Central America and the area along the Japan and Aleutian Trenches. Along the western edge of South America, the Nazca plate collides with and is subducted beneath the South American plate, creating the Andes Mountains chain and related volcanoes in the process. The collision of India and Asia mentioned earlier is another example of a convergent boundary.

- *Transform boundaries* (lower right) occur where plates slide laterally past one another at right angles to a sea-floor spreading center, neither diverging nor converging, and usually with no volcanic eruptions. These are the right-angle fractures stretching across the mid-ocean ridge system worldwide (visible in Figures 11-18, 11-19, and Chapter 12 opening illustration).

***Transform Boundaries.*** In 1965, another piece of the tectonic puzzle fell into place when University of Toronto geophysicist Tuzo Wilson first described the nature of transform boundaries and their relation to earthquake activity. All the spreading centers on Earth's crust feature these perpendicular scars, which stretch out on both sides (Figure 11-20). Some transform faults are a few hundred kilometers long; others, such as those along the East Pacific rise, stretch out 1000 km or more (over 600 mi). Across the entire ocean floor, spreading-center boundaries are the location of **transform faults**. The faults generally are parallel to the direction in which the plate is moving. How do they occur?

You can see in Figure 11-20 that mid-ocean ridges are not simple straight lines. When a mid-ocean rift begins, it

**FIGURE 11-17**

**Continents adrift, from 465 million years ago to the present.**

Observe the formation and breakup of Pangaea and the types of motions occurring at plate boundaries. [(a) from R. K. Bambach, "Before Pangaea: The Geography of the Paleozoic World," *American Scientist* 68 (1980): 26–38, reprinted by permission; (b) through (e) from R. S. Dietz and J. C. Holden, *Journal of Geophysical Research* 75, no. 26 (September 10, 1970): 4939–56, © The American Geophysical Union.]

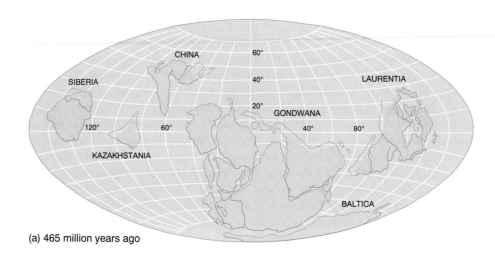

(a) 465 million years ago

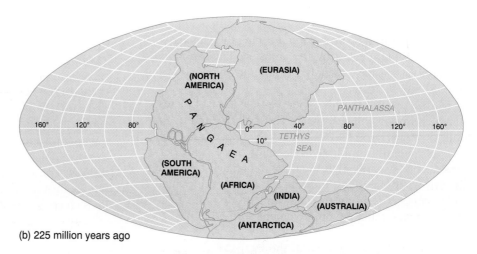

(b) 225 million years ago

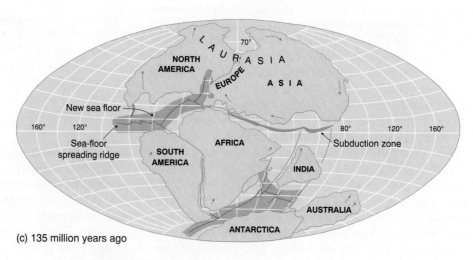

(c) 135 million years ago

opens at points of weakness in the crust. The fractures you see in the figure did not begin as a continuous fracture but as a series of offset breaks in the crust in each portion of the spreading center. As new material rises to the surface, building the mid-ocean ridges and spreading the plates, these offset areas slide past each other, in horizontal faulting motions.

The resulting fracture zone, which follows these breaks in Earth's crust, is active only along the fault section between ridges of spreading centers, as shown in Figure 11-20.

Along transform faults (the fault section between C and D in the figure) the motion is purely one of horizontal displacement—no new crust is formed or old crust

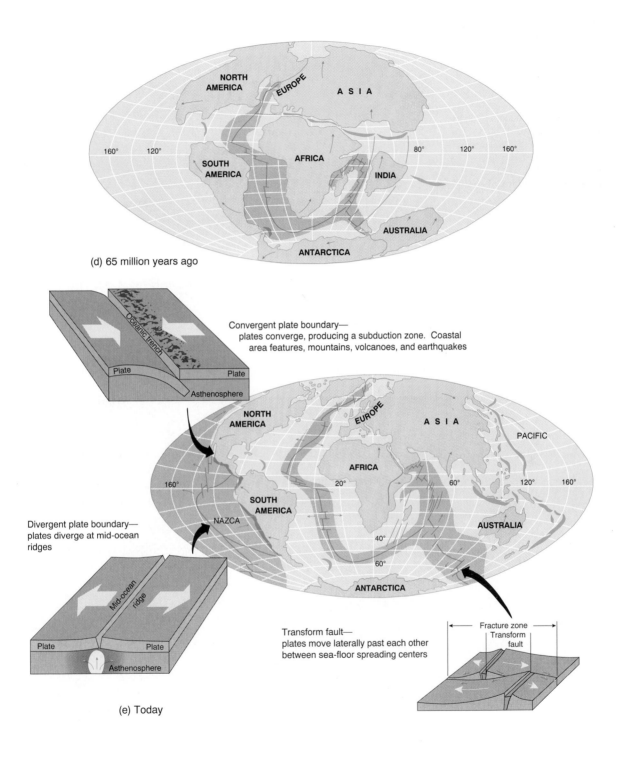

(d) 65 million years ago

Convergent plate boundary—
plates converge, producing a subduction zone. Coastal
area features, mountains, volcanoes, and earthquakes

Divergent plate boundary—
plates diverge at mid-ocean
ridges

Transform fault—
plates move laterally past each other
between sea-floor spreading centers

Fracture zone
Transform
fault

(e) Today

subducted. Beyond the spreading centers, the two sides of the fracture zones are joined and inactive. In fact, the plate pieces on either side of the fracture zone are moving in the same direction, away from the spreading center (the fracture zones between A and B and between E and F in the figure).

The name *transform* was assigned because of this apparent transformation in the direction of the fault movement. The famous San Andreas fault system in California, where continental crust has overridden a transform system, is related to this type of motion. The fault that triggered the 1995 Kobe earthquake in Japan has this type of horizontal motion.

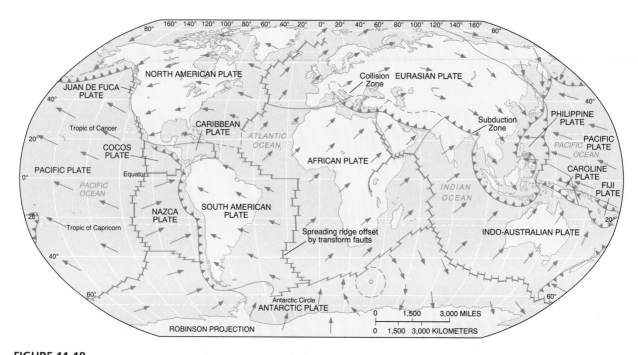

**FIGURE 11-18**

**Earth's 14 lithospheric plates and their movements.**

Each arrow represents 20 million years of movement. The longer arrows indicate that the Pacific and Nazca plates are moving more rapidly than the Atlantic plates. Symbol legend is with Figure 11-21. [Adapted from U.S. Geodynamics Committee.]

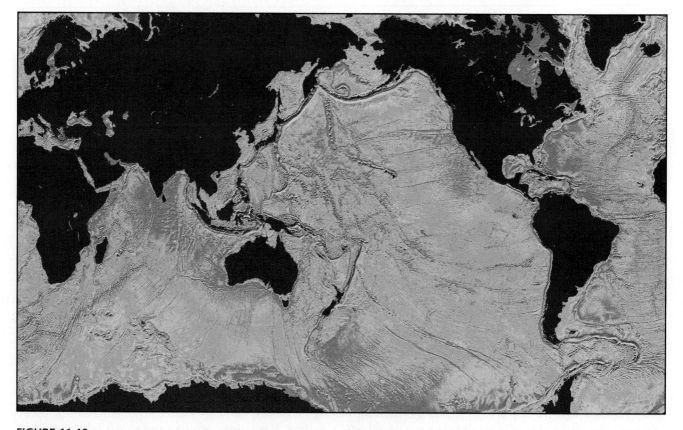

**FIGURE 11-19**

**The ocean floor revealed.**

A global gravity anomaly map derived from *Geosat* and *ERS-1* altimeter data. The radar altimeters measured sea-surface heights. Variation in sea-surface elevation is a direct indication of the topography of the ocean floor. [Global gravity anomaly map image courtesy of D. T. Sandwell, Scripps Institution of Oceanography. All rights reserved, 1995.]

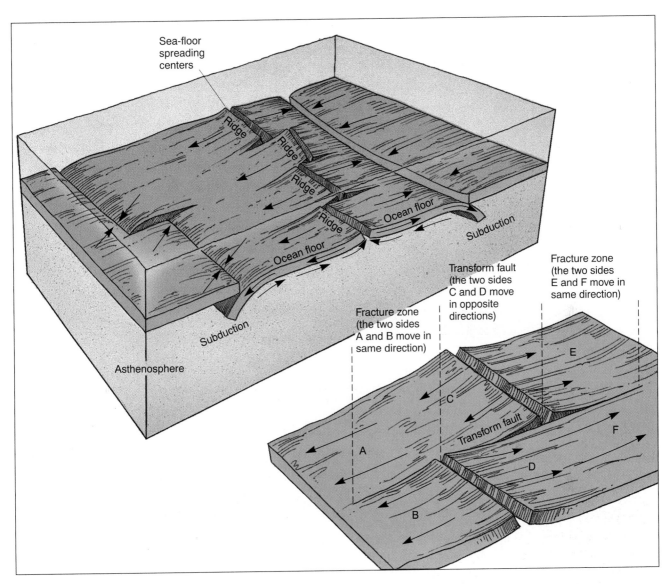

**FIGURE 11-20**
**Transform faults.**
Appearing along fracture zones, a transform fault is only that section between spreading centers where adjacent plates move in opposite directions. [Adapted from B. Isacks, J. Oliver, and L. R. Sykes, *Journal of Geophysical Research* 73, (1968): 5855–99, © The American Geophysical Union.]

## *Earthquake and Volcanic Activity*

Plate boundaries are the primary location of Earth's earthquake and volcanic activity, and the correlation of these phenomena is an important aspect of plate tectonics. Earthquakes and volcanic activity are discussed in more detail in the next chapter, but their general relationship to the tectonic plates is important to point out here.

Earthquake zones and volcanic sites are identified in Figure 11-21. The "ring of fire" surrounding the Pacific Basin, named for the frequent incidence of volcanoes, is evident. The subducting edge of the Pacific plate thrusts deep into the crust and mantle, producing molten material that makes its way back toward the surface, causing active volcanoes along the Pacific Rim. Such processes occur similarly at plate boundaries throughout the world.

## *Hot Spots*

A dramatic aspect of plate tectonics is the estimated 50–100 **hot spots** across Earth's surface (also shown in Figure 11-21). These are individual sites of upwelling material arriving at the surface in tall plumes from the mantle. Hot spots occur beneath both oceanic and continental crust and are probably anchored deep in the stiff lower mantle, tending to remain fixed relative to migrating plates. Thus, the area of a plate that is above a hot spot is locally heated for the brief geologic time it is there (a few hundred thousand or million years).

An isolated hot spot has formed, and continues to form, the Hawaiian-Emperor islands chain (Figure 11-22). The Pacific plate has moved across this hot, upward-erupting plume for almost 80 million years, creating a

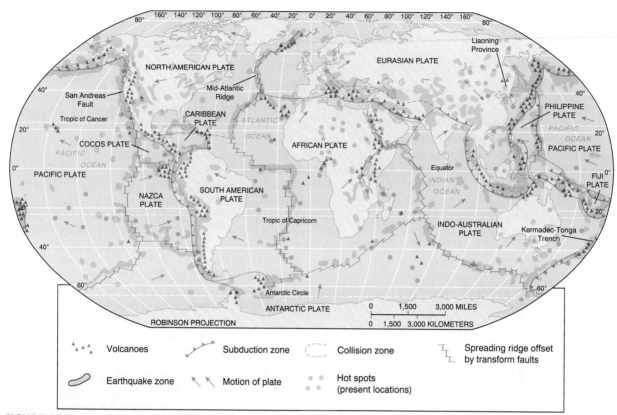

**FIGURE 11-21**

**Earthquake and volcanic activity locations.**

Earthquake and volcanic activity in relation to major tectonic plate boundaries, and principal hot spots. [Earthquake and volcano data from *Earthquakes*, B. A. Bolt, © 1988 W. H. Freeman and Company; reprinted with permission. Hot spots from K. C. Burke and J. T. Wilson, "Hot Spots on the Earth's Surface," *Scientific American*, August 1976, p. 52, © 1976 by Scientific American, Inc. All rights reserved.]

string of volcanic islands that is moving northwestward away from the hot spot. Thus, the ages of the islands or seamounts in the chain increase northwestward from the island of Hawaii, as you can see from the ages marked in the figure. The oldest island in the Hawaiian part of the chain is Kauai, approximately 5 million years old; it is weathered, eroded, and deeply etched with canyons and valleys.

The big island of Hawaii actually took less than 1 million years to build to its present stature. The island is a huge mound of lava, melded from several sea-floor fissures into five volcanoes, rising from the sea floor 5800 m (19,000 ft) to the ocean surface, and then continuing to rise to its highest peak, Mauna Kea, at 4169 m (13,677 ft) elevation above the sea—a total height of almost 10,000 m; it is the highest mountain on Earth, if measured from the sea floor. In all, the island of Hawaii contains about 40,000 km³ of basalt, enough to cover the states of Massachusetts, Connecticut, and Rhode Island to a depth of 1 km!

The youngest island in the chain is still a seamount, a submarine mountain that does not reach the surface. It rises 3350 m (11,000 ft) from the ocean floor but is still 975 m (3200 ft) beneath the ocean surface. Even though this new island will not experience the tropical Sun for about 10,000 years, it is already named Loihi.

To the northwest of this active hot spot in Hawaii, the island of Midway rises as a part of the same system. From there, the Emperor Seamounts stretch northwestward until they reach about 40 million years of age. At that point, this linear island chain shifts direction northward, evidently reflecting a change in the movement of the Pacific plate at that earlier time. At the northernmost extreme, the seamounts that formed about 80 million years ago are now approaching the Aleutian Trench, where they eventually will be subducted beneath the Eurasian plate. Perhaps some 80 million years hence, the Hawaiian Islands too will slowly disappear into the trench. Aloha!

Iceland is an example of an active hot spot sitting astride a mid-ocean ridge—visible on the different maps and images of the sea floor. It also is the best example of a segment of mid-ocean ridge rising above sea level. This hot spot has generated enough material to form Iceland and continues to cause eruptions from deep in the mantle. As a result, Iceland is still growing in area and volume. The youngest rocks are near the center of the island, with rock age increasing toward the eastern and western coasts.

News Report 6 reports on a continental hot spot beneath Yellowstone National Park. Iceland, Hawaii, and Yellowstone are active examples of our changing world on a most dynamic planet.

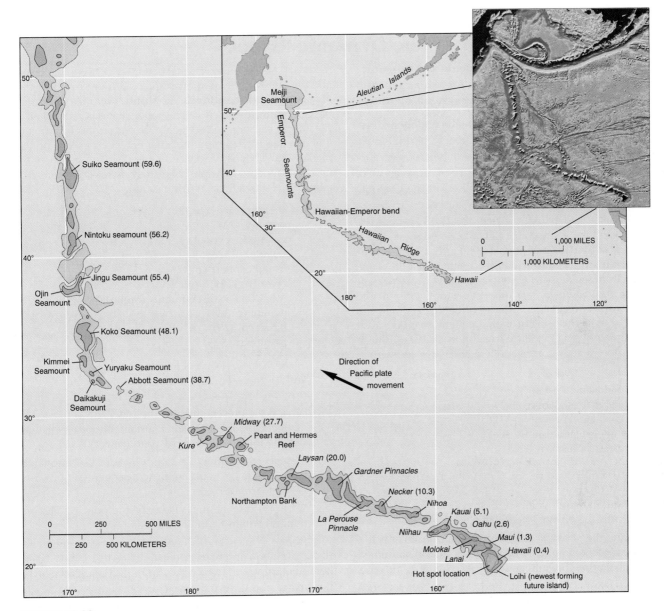

**FIGURE 11-22**
**Hot spot tracks across the North Pacific.**
Hawaii and the linear volcanic chain of islands known as the Emperor Seamounts. The islands and seamounts in the chain are progressively younger toward the southeast. Ages, in millions of years, are shown in parentheses. [After D. A. Clague, "Petrology and K–Ar (Potassium–Argon) Ages of Dredged Volcanic Rocks from the Western Hawaiian Ridge and the Southern Emperor Seamount Chain," *Geological Society of America Bulletin* 86 (1975): 991; inset from global gravity anomaly map image, Scripps Institution of Oceanography. All right reserved.]

## News Report 6

### Yellowstone on the Move

Yellowstone National Park, in northwestern Wyoming, is above a hot spot, as evidenced by Old Faithful Geyser and lava flows. Because the North American plate is moving westward, the hot spot is leaving a track across the western United States, including basaltic flows visible through Idaho, Oregon, and Washington.

Yellowstone and most linear volcanic chains, such as the Hawaiian Islands, have a hot spot at one end, where mantle material works its way up through a passing plate. The track left by these eruptions matches the direction of the plate movement. If the North American plate continues drifting westward, imagine the phenomenon of Yellowstone, including thermal springs, lava flows, and earthquake and volcanic activity, eventually migrating through the Dakotas!

# Summary and Review—The Dynamic Planet

✔ *Distinguish* **between the endogenic and exogenic systems,** *determine* **the driving force for each, and** *explain* **the pace at which these systems operate.**

The Earth-atmosphere interface is where the **endogenic system** (internal), powered by heat energy from within the planet, interacts with the **exogenic system** (external), powered by insolation and influenced by gravity. These systems work together to produce Earth's diverse landscape. The **geologic time scale** is an effective device for organizing the vast span of geologic time. It depicts the *sequence* of Earth's events (relative time) and the approximate *actual dates* (absolute time).

The most fundamental principle of Earth science is **uniformitarianism**. Uniformitarianism assumes that the *same physical processes active in the environment today have been operating throughout geologic time*. In contrast, the principle of **catastrophism** fits the vastness of Earth's age and the complexity of its rocks into a shortened time span. Because there is little physical evidence to support catastrophism, it is more appropriately considered a belief than a serious scientific hypothesis.

> endogenic system (p. 317)
> exogenic system (p. 317)
> geologic time scale (p. 320)
> uniformitarianism (p. 320)
> catastrophism (p. 320)

1. To what extent is Earth's crust active at this time in its history?

2. Define the endogenic and the exogenic systems. Describe the driving forces that energize these systems.

3. How is the geologic time scale organized? What is the basis for the time scale in relative and absolute terms? What era, period, and epoch are we living in today?

4. Contrast uniformitarianism and catastrophism as models for Earth's development.

---

✔ *Diagram* **Earth's interior in cross section and** *describe* **each distinct layer.**

We have learned about Earth's interior from indirect evidence—the way its various layers transmit **seismic waves**. The **core** is differentiated into an inner core and an outer core—divided by a transition zone. Earth's magnetic field is generated almost entirely within the outer core. Polarity reversals in Earth's magnetism are recorded in cooling magma that contains iron minerals. The pattern of **magnetic reversals** frozen in rock helps scientists piece together the story of Earth's mobile crust.

The entire **mantle** represents about 80% of Earth's total volume. It experiences a gradual temperature increase with depth and a stiffening due to increased pressures. The upper mantle is divided into three fairly distinct layers. The uppermost mantle, along with the crust, makes up the *lithosphere*. Below the lithosphere is the **asthenosphere**, or plastic layer. It contains pockets of increased heat from radioactive decay and is susceptible to slow convective currents in these hotter

materials. An important internal boundary between the **crust** and the high-velocity portion of the uppermost mantle is the **Mohorovičić discontinuity**, or **Moho**. *Continental crust* is basically **granite**; it is crystalline and high in silica, aluminum, potassium, calcium, and sodium. *Oceanic crust* is **basalt**; it is granular and high in silica, magnesium, and iron. The principles of buoyancy and balance were combined in the 1800s into the important principle of **isostasy**. Isostasy explains certain vertical movements of Earth's crust.

> seismic waves (p. 322)
> core (p. 323)
> magnetic reversal (p. 323)
> mantle (p. 324)
> asthenosphere (p. 325)
> crust (p. 325)
> Mohorovičić discontinuity (Moho) (p. 325)
> granite (p. 325)
> basalt (p. 325)
> isostasy (p. 325)

5. Make a simple sketch of Earth's interior, label each layer, and list the physical characteristics, temperature, composition, and range of size of each on your drawing.

6. What is the present thinking on how Earth generates its magnetic field? Is this field constant, or does it change? Explain the implications of your answer.

7. Describe the asthenosphere. Why is it also known as the plastic layer? What are the consequences of its convection currents?

8. What is a discontinuity? Describe the principal discontinuities within Earth.

9. Define isostasy and isostatic rebound, and explain the crustal equilibrium concept.

10. Diagram the uppermost mantle and crust. Label the density of the layers in grams per cubic centimeter. What two types of crust were described in the text in terms of rock composition?

---

✔ *Illustrate* **the geologic cycle and** *relate* **the rock cycle and rock types to endogenic and exogenic processes.**

The **geologic cycle** is a model of the internal and external interactions that shape the crust. A **mineral** is an inorganic natural compound having a specific chemical formula and possessing a crystalline structure. A **rock** is an assemblage of minerals bound together (such as granite, a rock containing three minerals), or it may be a mass of a single mineral (such as rock salt). Thousands of different rocks have been identified.

The geologic cycle comprises three cycles: the hydrologic cycle, the tectonic cycle, and the **rock cycle**. The rock cycle describes the three principal rock-forming processes and the rocks they produce. **Igneous rocks** form from **magma**, which is molten rock beneath the surface. Magma is fluid, highly gaseous, and under tremendous pressure. It either *intrudes* into crustal rocks, cools, and hardens or *extrudes* onto the surface as **lava**. Intrusive igneous rock that cools slowly in the crust forms a **pluton**. The largest pluton form is a **batholith**.

The cementation, compaction, and hardening of sediments into **sedimentary rocks** is called **lithification**. These layered strata form important records of past ages. **Stratigraphy** is the study of the sequence (superposition), thickness, and spatial distribution of strata that yield clues to the age and origin of the rocks.

Clastic sedimentary rocks are derived from the bits and pieces of weathered rocks. Chemical sedimentary rocks are not formed from physical pieces of broken rock, but instead are dissolved minerals, transported in solution, and chemically precipitated out of solution (they are essentially nonclastic). The most common chemical sedimentary rock is **limestone**, which is lithified calcium carbonate, $CaCO_3$. Coal is a biochemical sedimentary rock, rich in carbon.

Any rock, either igneous or sedimentary, may be transformed into a **metamorphic rock**, by going through profound physical or chemical changes under pressure and increased temperature.

geologic cycle (p. 325)
mineral (p. 326)
rock (p. 327)
rock cycle (p. 327)
igneous rock (p. 327)
magma (p. 327)
lava (p. 327)
pluton (p. 327)
batholith (p. 327)
sedimentary rock (p. 331)
lithification (p. 331)
stratigraphy (p. 331)
limestone (p. 331)
coal (p. 333)
metamorphic rock (p. 334)

11. Illustrate the geologic cycle and define each component: rock cycle, tectonic cycle, and hydrologic cycle.

12. What is a mineral? A mineral family? Name the most common minerals on Earth. What is a rock?

13. Describe igneous processes. What is the difference between intrusive and extrusive types of igneous rocks?

14. Characterize felsic and mafic minerals. Give examples of both coarse- and fine-grained textures.

15. Briefly describe sedimentary processes and lithification. Describe the sources and particle sizes of sedimentary rocks.

16. What is metamorphism, and how are metamorphic rocks produced? Name some original parent rocks and their metamorphic equivalents.

✔ *Describe* Pangaea and its breakup and *relate* several physical proofs that crustal drifting is continuing today.

The present configuration of the ocean basins and continents are the result of **tectonic processes** involving Earth's interior dynamics and crust. Alred Wegener coined the phrase **continental drift** to describe his idea that the crust is moved by vast forces within the planet. **Pangaea** was the name he gave to a single assemblage of continental crust some 225 million years ago that subsequently broke apart. Earth's crust is fractured into huge slabs or plates, each moving in response to flowing currents in the mantle. The all-encompassing theory of **plate tectonics** includes **sea-floor spreading** along **mid-ocean ridges** and denser oceanic crust diving beneath lighter continental crust along **subduction zones**.

tectonic processes (p. 335)
continental drift (p. 337)
Pangaea (p. 337)
plate tectonics (p. 337)
sea-floor spreading (p. 337)
mid-ocean ridge (p. 337)
subduction zone (p. 339)

17. Briefly review the history of the theory of continental drift, sea-floor spreading, and the all-inclusive plate tectonics theory. What was Alfred Wegener's role?

18. Define upwelling, and describe related features on the ocean floor. Define subduction and explain the process.

19. What was Pangaea? What happened to it during the past 225 million years?

20. Characterize the three types of plate boundaries and the actions associated with each type.

✔ *Portray* the pattern of Earth's major plates and *relate* this pattern to the occurrence of earthquakes, volcanic activity, and hot spots.

Occurrences of often damaging earthquakes and volcanoes are correlated with plate boundaries. Three types of plate boundaries form: divergent, convergent, and transform. Along the offset portions of mid-ocean ridges, horizontal motions produce **transform faults**.

As many as 50 to 100 **hot spots** exist across Earth's surface, where tall plumes of magma, anchored in the lower mantle, remain fixed as drifting plates are penetrated by eruptions.

transform faults (p. 341)
hot spots (p. 345)

21. What is the relation between plate boundaries and volcanic and earthquake activity?

22. What is the nature of motion along a transform fault? Name a famous example of such a fault.

## NetWork

The *Geosystems Home Page* provides on-line resources for this chapter on the World Wide Web. You will find review exercises, specific updates for items in the chapter, suggested readings, and links to interesting related pathways on the Internet (click on the Table of Contents link and select this chapter). *Geosystems* is at: **http://www.prenhall.com/geosystm**

# 12

# Tectonics, Earthquakes, and Volcanism

**The Ocean Floor**

**Earth's Surface Relief Features**

**Crustal Formation Processes**

**Crustal Deformation Processes**

**Orogenesis (Mountain Building)**

**Earthquakes**

**Volcanism**

**Summary and Review**

### Key Learning Concepts

After reading the chapter, you should be able to:

- *Describe* first, second, and third orders of relief and *relate* examples of each from Earth's major topographic regions.

- *Describe* the several origins of continental crust and *define* displaced terranes.

- *Explain* compressional processes and folding; *describe* four principal types of faults and their characteristic landforms.

- *Relate* the three types of plate collisions associated with orogenesis and *identify* specific examples of each.

- *Explain* the nature of earthquakes, their measurement, and the nature of faulting.

- *Distinguish* between an effusive and an explosive volcanic eruption and *describe* related landforms, using specific examples.

Tectonic processes deform, recycle, and reshape Earth's crust. These processes occur sometimes in dramatic episodes but most often in slow, deliberate motions that build the landscape. Continental crust has been forming throughout most of Earth's 4.6 billion-year existence. The arrangement of continents and oceans, the origin of mountain ranges, the topography of the land and the sea floor, and the locations of earthquake and volcanic activity are all evidence of our dynamic Earth.

In this chapter, we examine processes that construct Earth's surface and create world structural regions, often causing hazards for humans in the form of earthquakes and volcanic activities. Principal seismic and volcanic zones occur along plate boundaries, thus linking plate tectonics to the devastation of major earthquakes (Kobe, Japan, 1995) and to the global climatic impact of major eruptions (Mount Pinatubo, Philippines, 1991). Such catastrophic moments can occur without warning.

It was January 17, 1994, at California State University–Northridge. The bookstore was stocked, fees were paid, and registration was complete for 26,000 students to begin classes in two weeks. But at 4:31 A.M., a magnitude 6.8 earthquake hit the Los Angeles metropolitan area. Damage was approximately $30 billion, with $350 million to the campus alone, and 66 people died in the region. Let us learn more about what caused this moment of destruction.

## The Ocean Floor

We begin our look at tectonics and volcanism on the ocean floor, hidden from direct view. The illustration that opens this chapter is a striking representation of Earth with its blanket of water removed, as revealed to us through decades of direct and indirect observation. Careful examination of this portrait gives us a helpful review of the concepts learned in the previous chapter, laying the foundation for this and subsequent chapters. The scarred ocean floor is clearly visible: sea-floor spreading centers marked by oceanic ridges that stretch over 64,000 km (40,000 mi), subduction zones indicated by deep oceanic trenches, and transform faults slicing across oceanic ridges.

On the floor of the Indian Ocean, you can see the wide track along which the India plate traveled northward to its collision with the Eurasian plate. Vast deposits of sediment cover the Indian Ocean floor, south of the Ganges River to the east of India. Sediments derived from the Himalayan Range blanket the floor of the Bay of Bengal (south of Bangladesh) to a depth of 20 km (12.4 mi). These sediments result from centuries of soil erosion in the land of the monsoons.

Isostasy can be examined in action in central and west-central Greenland, where the weight of the ice sheet

has depressed portions of the land far below sea level—now visible with the ice artificially removed by the cartographer (see the extreme upper-left corner of the map). In contrast, the region around Canada's Hudson Bay became ice-free about 8000 years ago and has isostatically rebounded 300 m (1100 ft).

In the area of the Hawaiian Islands, the hot-spot track is marked by a chain of islands and seamounts that you can follow along the Pacific plate from Hawaii to the Aleutians. Subduction zones south and east of Alaska and Japan, as well as along the western coast of South and Central America, are quite visible as dark trenches. Along the left margin of the map you can find Iceland's position on the Mid-Atlantic Ridge.

Follow the East Pacific rise (an ocean ridge and spreading center) northward as it trends beneath the west coast of the North American plate, disappearing under earthquake-prone California. The continents, offshore submerged continental shelves, and the expanse of the sediment-covered abyssal plain are all identifiable on this illustration.

Try to correlate this sea-floor illustration with the maps of crustal plates and plate boundaries shown in Figures 11-18 and the gravity anomaly image in Figure 11-19.

## Earth's Surface Relief Features

**Relief** refers to vertical elevation differences in the landscape. Examples include the low relief of Nebraska and Saskatchewan, medium relief in foothills along mountain ranges, and high relief in the Rockies and Himalayas. The undulating form of Earth's surface, its overall relief, is called **topography**, portrayed so effectively on topographic maps.

The relief and topography of Earth's landforms have played a vital role in human history: High mountain passes have both protected and isolated societies; ridges and valleys have dictated transportation routes; and vast plains have necessitated developing faster methods of communication and travel. Earth's topography has stimulated human invention and adaptation.

### Crustal Orders of Relief

Today's computer capabilities and tools such as the global positioning system (GPS) for determining location and elevation have enhanced our understanding of Earth's relief and topography. Scientists put elevation data in digital form; the data then are available for computer manipulation and display. The resulting *digital elevation models (DEMs)* aid in the scientific analysis of topography, area-altitude distributions, slopes, and local stream-drainage characteristics. The map in Figure 12-14 is from a digitized shaded-relief map of the United States prepared by the U.S. Geological Survey. The entire map is made up of 12 million spot elevations, with less than a kilometer between any two.

For convenience of description, geographers group the landscape's topography into three **orders of relief**. These orders classify landscapes by scale, from vast ocean basins and continents down to local hills and valleys.

***First Order of Relief.*** The first order of relief is the coarsest level of landforms, including huge **continental platforms** and **ocean basins**. Continental platforms are the masses of crust that reside above or near sea level, including the undersea continental shelves along the coastlines. The ocean basins are entirely below sea level and are portrayed in the chapter-opening illustration. Approximately 71% of Earth is covered by water. The remaining dry land appears as continents and islands. The distribution of land and water in evidence today demonstrates a distinct water hemisphere and continental hemisphere, as shown in Chapter 7.

***Second Order of Relief.*** The second order of relief is the intermediate level of landforms, for both continental and ocean-basin features. Continental features in the second order of relief include continental masses, mountain masses, plains, and lowlands. A few examples are the Alps, Canadian and American Rockies, west Siberian lowland, and Tibetan Plateau. The great rock cores ("shields") that form the heart of each continental mass are of this second order. In the ocean basins, second order of relief includes continental rises, slopes, abyssal plains, mid-ocean ridges, submarine canyons, and subduction trenches—all visible in the sea-floor illustration that opens this chapter.

***Third Order of Relief.*** The third and most-detailed order of relief includes individual mountains, cliffs, valleys, hills, and other landforms of smaller scale. These features are identifiable as local landscapes.

***Hypsometry.*** Figure 12-1 is a *hypsographic curve* (from the Greek *hypsos* meaning "height") that shows the distribution of Earth's surface by area and elevation in relation to sea level. Relative to Earth's diameter of 12,756 km, the surface is of low relief, only about 20 km (12 mi) from highest peak to lowest oceanic trench. For perspective, Mount Everest is 8.8 km (5.5 mi) above sea level and the Mariana Trench is 11 km (6.8 mi) below sea level.

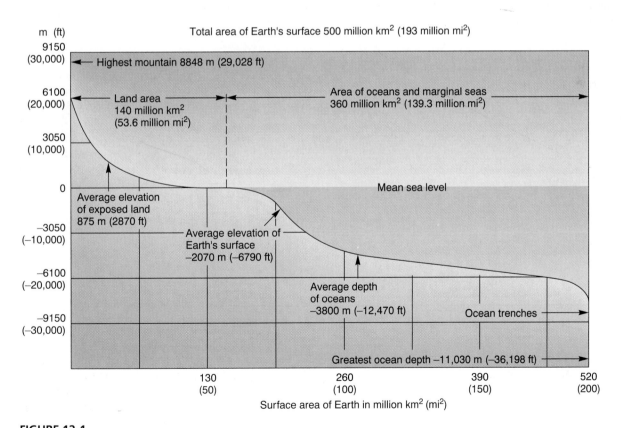

**FIGURE 12-1**

**Earth's hypsometry.**

Hypsographic curve of Earth's surface, charting area and elevation as related to mean sea level. From the highest point above sea level (Mount Everest) to the deepest oceanic trench (Mariana Trench), Earth's overall relief is almost 20 km (12.5 mi).

The average elevation of Earth's solid surface is actually under water: –2070 m (–6790 ft) below mean sea level. The average elevation just for exposed land is +875 m (+2870 ft). For the ocean depths, average elevation is –3800 m (–12,470 ft). From this description you can see that, on the average, the oceans are much deeper than continental regions are high. Overall, the underwater ocean basins, ocean floor, and submarine mountain ranges form Earth's largest area.

### *Earth's Topographic Regions*

The three orders of relief can be further generalized into six topographic regions: plains, high tablelands, hills and low tablelands, mountains, widely spaced mountains, and depressions (Figure 12-2). Each type of topography is defined by an arbitrary elevation or descriptive limit that is in common use (see legend in the figure).

Four of the continents possess extensive *plains*, areas with local relief of less than 100 m (325 ft) and slope angles of 5° or less. Some plains have high elevations of over 600 m (2000 ft); in the United States, the high plains attain elevations above 1220 m (4000 ft). The Colorado Plateau, Greenland, and Antarctica are notable *high tablelands*, with elevations exceeding 1520 m (5000 ft). Africa is dominated by *hills and low tablelands*.

*Mountain* ranges, characterized by local relief exceeding 600 m (2000 ft), occur on each continent. Earth's relief and topography are undergoing constant change as a result of processes that form crust.

## Crustal Formation Processes

How did Earth's continental crust form? What gave rise to the three orders of relief just discussed? Ultimately, the answers to both questions are the combined effects of tectonic activity, which is driven by our planet's internal energy, and the exogenic processes of weathering and erosion, powered by the Sun.

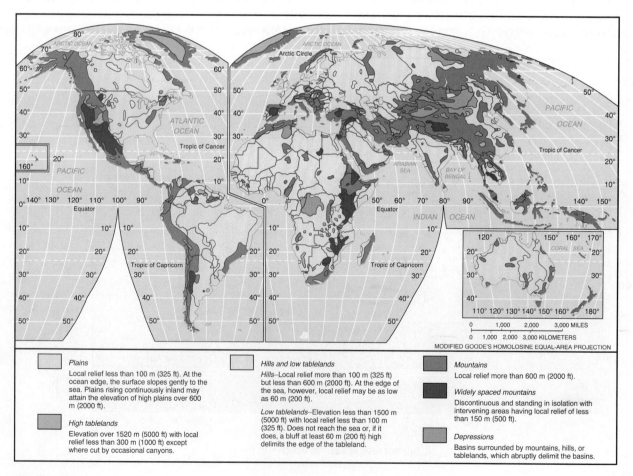

**Plains**
Local relief less than 100 m (325 ft). At the ocean edge, the surface slopes gently to the sea. Plains rising continuously inland may attain the elevation of high plains over 600 m (2000 ft).

**High tablelands**
Elevation over 1520 m (5000 ft) with local relief less than 300 m (1000 ft) except where cut by occasional canyons.

**Hills and low tablelands**
Hills–Local relief more than 100 m (325 ft) but less than 600 m (2000 ft). At the edge of the sea, however, local relief may be as low as 60 m (200 ft).

Low tablelands–Elevation less than 1500 m (5000 ft) with local relief less than 100 m (325 ft). Does not reach the sea or, if it does, a bluff at least 60 m (200 ft) high delimits the edge of the tableland.

**Mountains**
Local relief more than 600 m (2000 ft).

**Widely spaced mountains**
Discontinuous and standing in isolation with intervening areas having local relief of less than 150 m (500 ft).

**Depressions**
Basins surrounded by mountains, hills, or tablelands, which abruptly delimit the basins.

**FIGURE 12-2**
**Earth's topographic regions.**
Earth's topography is characterized as plains, tablelands, hills, mountains, and depressions. [After R. E. Murphy, "Landforms of the World," *Annals of the Association of American Geographers* 58, no. 1 (March 1968). Adapted by permission.]

Tectonic activity generally is slow, requiring millions of years. Endogenic (internal) processes result in gradual uplift and new landforms, with major mountain-building occurring along plate boundaries. These uplifted crustal regions are quite varied, but we think of them in three general categories, all discussed in this chapter: (1) residual mountains and continental cores, formed from inactive remnants of ancient tectonic activity; (2) tectonic mountains and landforms, produced by active folding, faulting, and crustal movements; and (3) volcanic features, formed by the surface accumulation of molten rock from eruptions of subsurface materials. Thus, several distinct processes operate in concert to produce the continental crust we see around us.

## Continental Shields

All continents have a nucleus of ancient crystalline rock on which the continent "grows" with the addition of crustal fragments and sediments. This nucleus is the *craton*, or heartland region, of the continental crust. Cratons generally have been eroded to a low elevation and relief. All are Precambrian, or older than 570 million years; most exceed 2 billion years of age.

A region where a craton is exposed at the surface is called a **continental shield**. Figure 12-3 shows the principal areas of exposed shields. The regions surrounding these shields are covered with layers of sedimentary rock that are quite stable over time. An example of such a stable *platform* is the region that stretches from the east of the Rockies to the Appalachians and northward into central Canada.

## Building Continental Crust

The formation of continental crust is very complex and takes hundreds of millions of years. It involves the entire sequence of sea-floor spreading and formation of oceanic crust, its later subduction and remelting, and its subsequent rise as new magma, all summarized in Figure 12-4.

To understand this process, study Figure 12-4 (and Figure 11-14). Begin with the magma that originates in the asthenosphere and wells up along the mid-ocean ridges. This magma has less than 50% silica, is rich in iron and magnesium, and has a low-viscosity (thin) texture—it tends to flow. It rises to erupt at spreading centers and cools to form new basaltic sea floor, which spreads outward to collide with continental crust along its far edges. This denser oceanic crust plunges beneath the lighter continental crust, into the mantle, where it remelts. The new magma then rises and cools, forming more continental crust, in the form of intrusive granitic igneous rock.

As the subducting oceanic plate works its way under a continental plate, it takes with it trapped seawater and sediment from eroded continental crust. The remelting incorporates the seawater, sediments, and surrounding

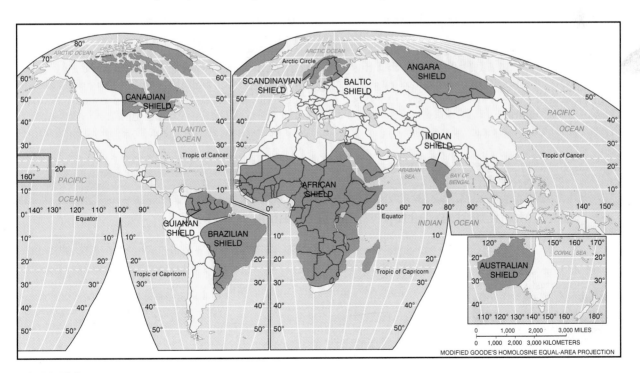

**FIGURE 12-3**
**Continental shields.**
Portions of major continental shields that have been exposed by erosion. [After R. E. Murphy, "Landforms of the World," *Annals of the Association of American Geographers* 58, no. 1 (March 1968). Adapted by permission.]

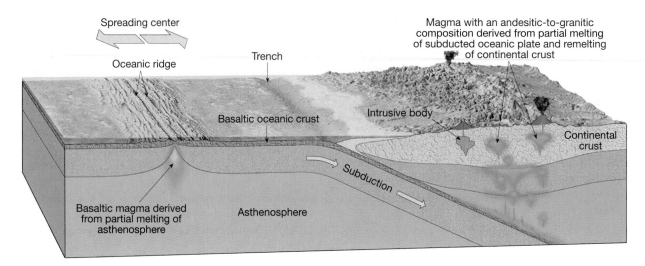

**FIGURE 12-4**
**Crustal formation.**
Material from the asthenosphere upwells along sea-floor spreading centers. Basaltic ocean floor is subducted beneath lighter continental crust, where it melts, along with its cargo of sediments, water, and minerals. This melting generates magma, which makes its way up through the crust to form igneous intrusions and extrusive eruptions. [From E. J. Tarbuck and F. K. Lutgens, *Earth, An Introduction to Physical Geography*, 5th ed., © Prentice Hall, 1996, Fig. 20.20, p. 502.]

crust into the mixture. As a result, the magma (generally called a "melt") that migrates upward from a subducted plate contains 50%–75% silica and aluminum and has a high-viscosity (thick) texture—it tends to block and plug conduits to the surface.

Bodies of such magma may reach the surface in explosive volcanic eruptions, or they may stop short and become subsurface intrusive bodies in the crust, cooling slowly to form granitic crystalline plutons such as batholiths (see Figures 11-8 and 11-9a). Note that this composition is quite different from the magma that rises directly from the asthenosphere at spreading centers. In these processes of crustal formation, you can literally see the cycling of materials in the tectonic cycle.

**Terranes.** Each of Earth's major lithospheric plates actually is a collage of many crustal pieces acquired from a variety of sources. Crustal fragments of ocean floor, curving chains (or arcs) of volcanic islands, and other pieces of continental crust all have been forced against the edges of continental shields and platforms. These slowly migrating crustal pieces, which have become attached or accreted to the plates, are called **terranes** (not to be confused with "terrain," which refers to the topography of a tract of land). News Report 1 discusses the role of terranes in the formation of Alaska through the words of a famous author.

These displaced terranes, sometimes called *microplate* or *foreign terranes*, have histories different from those of the continents that capture them. They are usually framed

## News Report 1

### James Michener on Terranes and Tectonics in *Alaska*

Author James Michener is known for weaving physical geography and geology into his historical novels. In his 1988 book, *Alaska*, Michener tells how successive additions of displaced terranes were pasted onto the shield heartland of North America:

> But the immediate task is to understand how this trivial ancestral [continental] nucleus could aggregate to itself the many additional segments of rocky land which would ultimately

unite to comprise the Alaska we know. Like a spider waiting to grab any passing fly, the nucleus remained passive but did accept any passing terranes—those unified agglomerations of rock considerable in size and adventurous in motion—that wandered within reach....And the great plates of Earth's crust never rest.*

In the region surrounding the Pacific, such accreted terranes are significant. At least 25% of the growth of western North America can be attributed

to the collection of terranes since the early Jurassic Period (190 million years ago). Without these accreted terranes, much of Alaska, British Columbia, Washington, Oregon, and California would not exist today.

* J. A. Michener, *Alaska* (New York: Random House, 1988), pp. 5 and 9. Reprinted by permission.

by fractures and differ in rock composition and structure from their new continental homes.

In the region surrounding the Pacific, accreted terranes are particularly prevalent. At least 25% of the growth of western North America can be attributed to the accretion of terranes since the early Jurassic Period (190 million years ago). A good example is the Wrangell Mountains, which lie just east of Prince William Sound and the city of Valdez, Alaska. The **Wrangellia terranes**—a former volcanic island arc and associated marine sediments from near the equator—migrated approximately 10,000 km (6200 mi) to form the Wrangell Mountains and three other distinct formations along the western margin of the continent (Figure 12-5).

The Appalachian Mountains, extending from Alabama to the Maritime Provinces of Canada, possess bits of land once attached to ancient Europe, Africa, South America, Antarctica, and various oceanic islands. The discovery of terranes, made only in the 1980s, demonstrates one of the ways continents are assembled.

# Crustal Deformation Processes

Rocks, whether igneous, sedimentary, or metamorphic, are subjected to powerful *stress* by tectonic forces, gravity, and the weight of overlying rocks. There are three types of

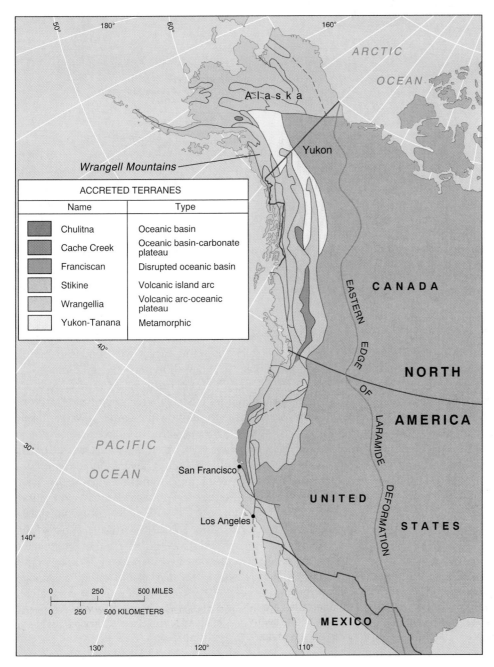

**FIGURE 12-5**
**North American terranes.**
Wrangellia terranes, highlighted among the other terranes along the western margin of North America, occur in four segments. [After "The Growth of Western North America" by D. L. Jones, A. Cox, P. Coney, and M. Beck, illustration by Andrew Tomko, *Scientific American* (November 1982): 71. © 1982 by Scientific American, Inc. All rights reserved.]

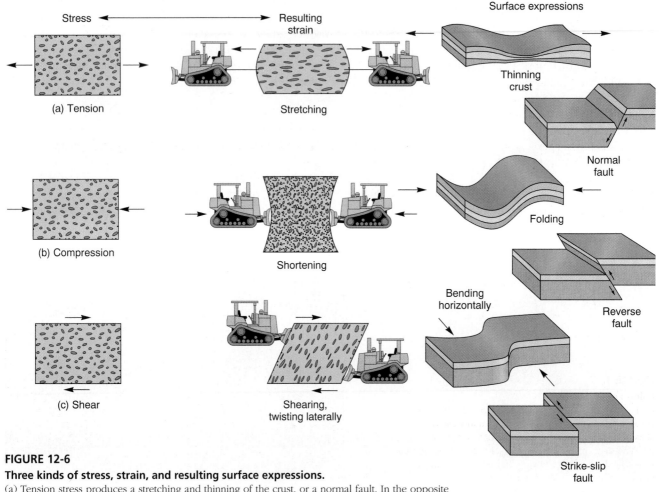

**FIGURE 12-6**
**Three kinds of stress, strain, and resulting surface expressions.**
(a) Tension stress produces a stretching and thinning of the crust, or a normal fault. In the opposite
direction, (b) compression produces a shortening and folding, or reverse faulting, of the crust. In a back-and-
forth horizontal motion, (c) shear stress produces a bending of the crust, and on breaking, a strike-slip fault.

stress: *tension* (stretching), *compression* (shortening), and *shear* (twisting or tearing), as shown in Figure 12-6.

*Strain* is how rocks respond to stress. Strain is expressed in rocks by *folding* (bending) or *faulting* (breaking). Whether a rock bends or breaks depends on several factors including composition and how much pressure is on the rock. An important quality is whether the rock is *brittle* or *ductile*. The patterns created by these processes are clearly visible in the landforms we see today, especially in mountain areas. Figure 12-6 illustrates each type of stress and its resulting strain and surface expressions that develop.

## *Folding*

When rock strata that are layered and flat are subjected to compressional forces, they become deformed (Figure 12-7). Convergent plate boundaries worldwide intensely compress rocks, deforming them in a process known as **folding**. As an analogy, if we take sections of thick fabric, stack

them flat on a table, and then slowly push on opposite ends of the stack, the cloth layers will bend and rumple into folds similar to those shown in Figure 12-7a.

If we then draw a line down the center axis of a resulting ridge, and a line down the center of a resulting trough, we are able to see how the names of the folds are assigned. Along the *ridge* of a fold, layers *slope downward away from the axis*, resulting in an **anticline**. In the *trough* of a fold, however, layers *slope downward toward the axis*, producing a **syncline** (Figure 12-7b).

If the axis of either type of fold is not "level" (horizontal, or parallel to Earth's surface), the layers then *plunge* (dip down) at an angle. Knowledge of how folds are angled to Earth's surface and where they are located is important for the petroleum industry. For example, petroleum geologists know that oil and natural gas collect in the upper portions of folds in permeable rock layers such as sandstone. This is the "anticlinal theory of petroleum accumulation" attributed to I. C. White, founder of the West Virginia Geological Survey.

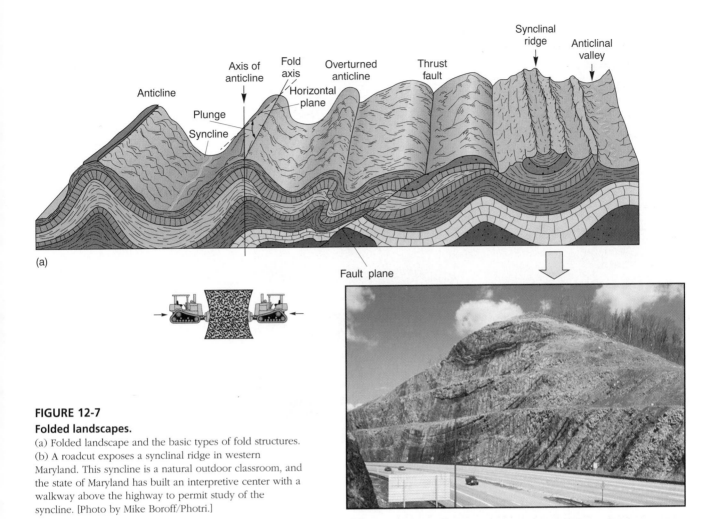

**FIGURE 12-7**
**Folded landscapes.**
(a) Folded landscape and the basic types of fold structures.
(b) A roadcut exposes a synclinal ridge in western
Maryland. This syncline is a natural outdoor classroom, and
the state of Maryland has built an interpretive center with a
walkway above the highway to permit study of the
syncline. [Photo by Mike Boroff/Photri.]

Figure 12-7a further illustrates various folds that have
been weathered and reduced by exogenic processes:

- A residual "synclinal ridge" may form within a syncline, because different rock strata offer greater resistance to weathering processes (Figure 12-7a). An
actual synclinal ridge has been exposed dramatically
in an interstate highway roadcut in Figure 12-7b.

- Compressional forces often push folds far enough
that they actually overturn upon their own strata
("overturned anticline").

- Further stress eventually fractures the rock strata
along distinct lines, and some overturned folds are
thrust upward, causing a considerable shortening of
the original strata ("thrust fault").

The Canadian Rocky Mountains, the Appalachian
Mountains, and areas of the Middle East illustrate well
the complexity of the resulting folded landscape. Satellites let us view many of these structures from an orbital
perspective, as in Figure 12-8. Just north of the Persian
Gulf are the Zagros Mountains of Iran. This area was a
dispersed terrane that separated from the Eurasian plate.
However, the collision produced by the northward push
of the Arabian block is now shoving this terrane back
into Eurasia and forming an active margin known as the
Zagros crush zone, a zone of continuing collision more
than 400 km (250 mi) wide. In the satellite image, anti-
clines form the parallel ridges; active weathering and erosion processes are exposing the underlying strata.

***Broad Warping of the Crust.*** In addition to the rum-
pling of rock strata just discussed, Earth's continental crust
also is subjected to broader warping actions. These actions
produce similar up-and-down bending of strata, but the
bends are far broader than those produced by folding.
Warping forces include mantle convection, isostatic adjustment, and swelling from an underlying hot spot. Warping
features can be small, individual, foldlike structures called
basins and domes (Figure 12-9a, b). Or they can range up
to regional features the size of the Ozark Mountain complex in Arkansas and Missouri, the Colorado Plateau in the
West, the Richat dome in Mauritania, or the Black Hills of
South Dakota (Figure 12-9 c,d).

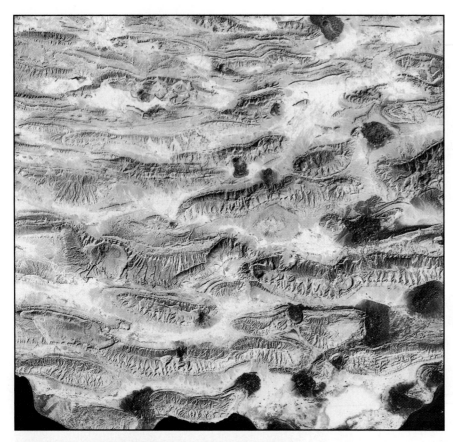

**FIGURE 12-8**

**Folding in the Zagros crush zone, Iran.**
The Zagros Mountains are a product of the Zagros crush zone between the Arabian and Eurasian plates. This area was a dispersed terrane (migrating crustal piece) that separated from the Eurasian plate. However, the collision produced by the northward push of the Arabian block is now shoving this terrane back into Eurasia and forming the folded mountains shown. [NASA image.]

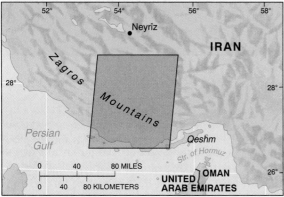

The Hudson Bay region of northern Canada is a striking example of large-scale downwarping, followed by its current upwarping. During the last ice age, the tremendous weight of kilometer-thick glacial ice downwarped this portion of the Canadian Shield. (See the chapter-opening map: Hudson Bay is the large water area in the middle of the Canadian Shield.) Much of this huge bay obviously remains below sea level today, but, with the melting of glacial ice and removal of the depressing weight, the crust is undergoing *isostatic rebound* (principle shown in Figure 11-5). The isostatic rebound of the Hudson Bay region is a broad upwarping that will eventually lead to draining of the bay. The region has uplifted 330 m (1100 ft) over the past 8000 years!

## Faulting

A freshly poured concrete sidewalk is smooth and strong. But stress the sidewalk by driving heavy equipment over it, and the resulting strain might cause a fracture. Pieces on either side of the fracture may move up, down, or horizontally, depending on the direction of stress. Similarly, when rock strata are stressed beyond their ability to remain a solid unit, they express the strain as a fracture. Rocks on either side of the fracture are displaced relative to the other side in a process known as **faulting**. Thus, *fault zones* are areas where fractures in the rock demonstrate crustal movement. At the moment of fracture, a sharp release of energy occurs, called an **earthquake** or *quake*.

The fracture surface along which the two sides of a fault move is called the *fault plane*. The names of the three basic types of faults, illustrated in Figure 12-10, are based on the tilt and orientation of the fault plane: normal (when rocks are pulled apart), thrust or reverse (when rocks are compressed), and strike-slip (when rocks are sheared).

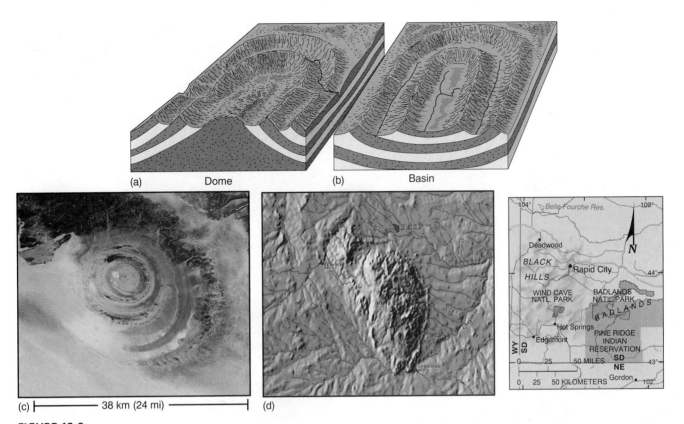

(a) Dome    (b) Basin

(c) |———— 38 km (24 mi) ————|    (d)

**FIGURE 12-9**
**Domes and basins.**
(a) An upwarped dome. (b) A structural basin. (c) The Richat dome structure of Mauritania. (d) The Black Hills of South
Dakota is a dome structure, shown here on a digitized relief map. [(a) and (b) After J. E. Van Riper, *Man's Physical World*,
p. 436, © 1971 by McGraw-Hill; (c) NASA /Mark Marten/Science Source/Photo Researchers, Inc.; (d) from USGS digital
terrain map I-2206, 1992.]

***Normal Fault.*** When forces pull rocks apart, the tension
causes a **normal fault**, or tension fault. When the break
occurs, rock on one side moves vertically along an
inclined fault plane (Figure 12-10a). The downward-shift-
ing side is called the *hanging wall*; it drops relative to the
*footwall block*. The exposed fault plane sometimes is visible along the base of faulted mountains, where individual ridges are truncated by the movements of the fault
and appear as triangular facets at the ends of the ridges.
A cliff formed by faulting is commonly called a *fault scarp*,
or *escarpment*.

***Reverse (Thrust) Fault.*** Compressional forces associated with converging plates force rocks to move *upward*
along the fault plane. This is called a **reverse fault**, or
compression fault (Figure 12-10b). On the surface, it
appears similar to a normal fault, although more collapse
and landslides may occur from the hanging wall component. In England, when miners worked along a reverse
fault, they would stand on the lower side (footwall) and
hang their lanterns on the upper side (hanging wall), giving rise to these terms.

If the fault plane forms a low angle relative to the
horizontal, the fault is termed a **thrust fault**, or over-

thrust fault, indicating that the overlying block has shifted
far over the underlying block (see Figure 12-7, "thrust
fault"). Place your hands palms-down on your desk, with
fingertips together, and slide one hand up over the
other—this is the motion of a low-angle thrust fault, with
one side pushing over the other.

In the Alps, several such overthrusts result from compressional forces of the ongoing collision between the
African and Eurasian plates. Beneath the Los Angeles
Basin, such overthrust faults produce a high risk of earthquakes and have caused many quakes this century,
including the 1994 Northridge earthquake.

***Strike-Slip Fault.*** If movement along a fault plane is horizontal, such as produced along a transform fault, it is
called a **strike-slip fault** (Figure 12-10c). The movement
is described as right-lateral or left-lateral, depending on the
motion perceived when you observe movement on one
side of the fault relative to the other side.

Although strike-slip faults do not produce cliffs
(scarps) as do the other types of faults, they can create
linear rift valleys. This is the case with the San Andreas
fault system of California. The rift valley is clearly visible
in Figure 12-10c, where the edges of the North American

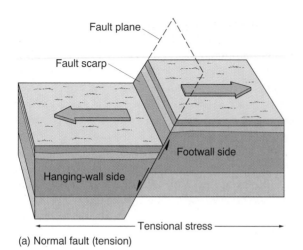

Fault plane

Fault scarp

Hanging-wall side

Footwall side

Tensional stress

(a) Normal fault (tension)

(a)

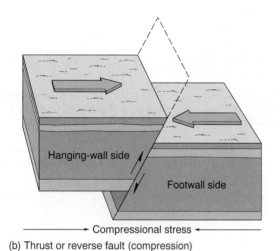

Hanging-wall side

Footwall side

Compressional stress

(b) Thrust or reverse fault (compression)

(b)

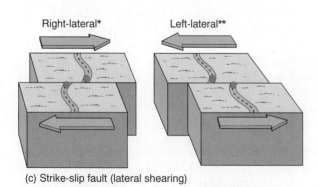

Right-lateral*

Left-lateral**

(c) Strike-slip fault (lateral shearing)

\* Viewed from either dot on each road,
movement to opposite side is *to the right*.
\*\* Viewed from either dot on each road,
movement to opposite side is *to the left*.

(c)

**FIGURE 12-10**

**Types of faults.**

(a) A normal fault produced by tension in the crust; visible along the edge in Tashkent, Uzbekistan. (b) A thrust, or reverse fault, produced by compression in the crust; visible in these offset strata in coal seams and volcanic ash in British Columbia. (c) A strike-slip fault produced by lateral shearing; clearly seen looking north along the San Andreas fault rift zone on the eastern edge of the Coast Ranges in California—Pacific plate to the left, North American plate to the right. [(a) Photo by Fred McConnaughey/Photo Researchers, Inc.; (b) photo by Fletcher and Baylis/Photo Researchers, Inc.; (c) Kevin Schafer/Peter Arnold, Inc.]

and Pacific plates are grinding past one another as a result of transform-fault movement. In its westward drift, the North American plate rode over portions of this sea-floor-spreading center, as shown in Figure 12-17. Consequently, the San Andreas system is a series of faults that are *transform* (associated with a former spreading center), *strike-slip* (horizontal in motion), and *right-lateral* (one side is moving to the right of the other).

***Faults in Concert.*** Combinations of faults can produce distinctive terrains. In the U.S. interior west, the Basin and Range Province experienced tensional forces caused by uplifting and thinning of the crust. This movement cracked the surface to form aligned pairs of normal faults and a distinctive landscape (Figure 12-11a). The term **horst** is applied to upward-faulted blocks; **graben** refers to downward-faulted blocks. One example of a horst and graben landscape is the Great Rift Valley of East Africa (associated with crustal spreading); it extends northward to the Red Sea, which fills the rift formed by parallel normal faults (12-11b). Another example is the Rhine graben, through which the Rhine River flows in Europe.

# Orogenesis (Mountain Building)

We have explained the factors that work to produce crust and have discussed the tectonic forces that bend, warp, and break it. Now let us look at specific mountain-building processes.

**Orogenesis** literally means the birth of mountains (*oros* comes from the Greek for "mountain"). An *orogeny* is a mountain-building episode, occurring over millions of years. It can occur through large-scale deformation and uplift of the crust. It also may include the capture of migrating terranes and cementation of them to the continental margins, and the intrusion of granitic magmas to form plutons. The net result of this accumulating material is a thickening of the crust. The granite plutons often become exposed by erosion following uplift. Uplift is the final act of an orogenic cycle of mountain building. Earth's major chains of folded and faulted mountains, called *orogens*, are remarkably well correlated with the plate tectonics model.

No orogeny is a simple event; many involve previous developmental stages dating back millions of years, and the processes are ongoing today. Major mountain ranges, and the orogens that caused them, include:

- Rocky Mountains of North America (*Laramide orogeny*, 40–80 million years ago)

- Sierra Nevada of California (*Sierra Nevadan orogeny*, 35 million years ago, with older batholithic intrusions dating back 130–160 million years)

- Appalachian Mountains and the Ridge and Valley Province of the eastern United States. (*Alleghany orogeny*, 250–300 million years ago, preceded by at least two earlier orogenies; in Europe the *Hercynian orogeny*)

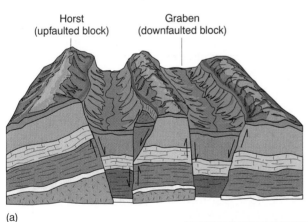

(a)

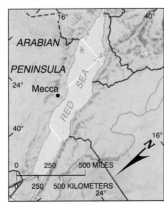

**FIGURE 12-11**
**Faulted landscapes.**
(a) Pairs of faults produce a horst and graben landscape characteristic of the Basin and Range Province in the western United States. (b) The Red Sea occupies a down-dropped block that is part of the rift system that runs through East Africa.
[(b) NASA photo from Gemini.]

(b)

- Alps of Europe (*Alpine orogeny*, 20–120 million years ago and continuing to the present, with many earlier episodes)
- Himalayas of Asia (*Himalayan orogeny*, 45 million years ago, beginning with the collision of the India plate and Eurasia plate and continuing to the present)

## Types of Orogenies

Figure 12-12 illustrates three types of convergent plate collisions that cause orogenesis:

(a) *Oceanic plate–continental plate collision orogenesis.* This is now occurring along the Pacific coast of the Americas and has formed the Andes, the Sierra of Central America, the Rockies, and other western mountains. We see folded sedimentary formations, with intrusions of magma forming granitic plutons at the heart of these mountains. Their buildup has been augmented by capturing of displaced terranes, cemented during their collision with the continental mass. Also, note the associated volcanic activity inland from the subduction zone.

(b) *Oceanic plate–oceanic plate collision orogenesis.* Such collisions can produce either simple volcanic island arcs or more-complex arcs, such as Japan, that include deformation and metamorphism of rocks, and granitic intrusions. These processes have formed the chains of island arcs and volcanoes that continue from the southwestern Pacific to the western Pacific, the Philippines, the Kurils, and on through portions of the Aleutians.

Both collision types, (a) oceanic-continental and (b) oceanic-oceanic, are active around the Pacific Rim. Both are thermal in nature, because the diving plate melts and migrates back toward the surface as molten rock. The region of active volcanoes and earthquakes around the Pacific is known as the **circum-Pacific belt** or, more popularly, the **ring of fire**.

(c) *Continental plate–continental plate collision orogenesis.* Here the orogenesis is quite mechanical; large masses of continental crust are subjected to intense folding, overthrusting, faulting, and uplifting. The converging plates crush and deform both marine sediments and basaltic oceanic crust.

As mentioned earlier, the collision of India with the Eurasian landmass produced the Himalayan Mountains. That collision is estimated to have shortened the overall continental crust by as much as 1000 km (about 600 mi) and to have produced sequences of thrust faults at depths of 40 km (25 mi). The disruption created by that collision has reached far under China, and frequent earthquakes there signal the continuation of this rapid-paced collision. The Himalayas feature the tallest above-sea-level mountains on Earth, including Mount Everest at 8848 m (29,028 ft) elevation.

## The Grand Tetons and the Sierra Nevada

The Sierra Nevada of California and the Grand Tetons of Wyoming are examples of later stages of mountain building. Each is a tilted-fault block mountain range, in which a normal fault on one side of the range has produced a tilted landscape of dramatic relief (Figure 12-13). Magma intruded into those blocks, slowly cooling to form granitic cores of coarsely crystalline rock. After tremendous tectonic uplift and the removal of overlying material through weathering, erosion, and transport, those granitic masses are now exposed in each mountain range. In some areas, overlying material previously covered these batholiths by more than 7500 m (25,000 ft).

Recent research in the Sierra Nevada disclosed that some of the uplift was isostatic, caused by the erosion of overburden and loss of melting ice mass following the last ice age some 18,000 years ago. The accumulation of sediments in the adjoining valley depressed the crust, thus enhancing relief in the landscape.

## The Appalachian Mountains

The old, eroded, fold-and-thrust belt of the eastern United States and southeastern Canada (250–300 million years old) contrasts with the younger, higher mountains of western North America (35–80 million years old). As noted, the *Alleghany orogeny* followed at least two earlier orogenic cycles of uplift and the accretion of several captured terranes.

The original material for the Appalachian Mountains resulted from the collisions that produced Pangaea. In fact, the Atlas Mountains of northwestern Africa were connected to the Appalachians at some time in the past, but the Atlas Mountains, embedded in the African plate, rafted apart from the Applachians. Folded and faulted rock structures in the British Isles and Greenland also demonstrate a past relationship with the Appalachians. The Appalachian Mountain region comprises several physical landscape subregions (refer to locator map in Figure 12-14): the Valley and Ridge Province (elongated sequences of folded sedimentary rock), the Blue Ridge Province (principally of crystalline rock, highest where North Carolina, Virginia, and Tennessee converge), the Piedmont (hilly to gentle terrain along most of the eastern and southern margins of the mountains), and the east coastal plain (from gentle hills to flat plains that extend to the coast).

Figure 12-14 displays the linear folds of the Appalachian system. Each component in the figure

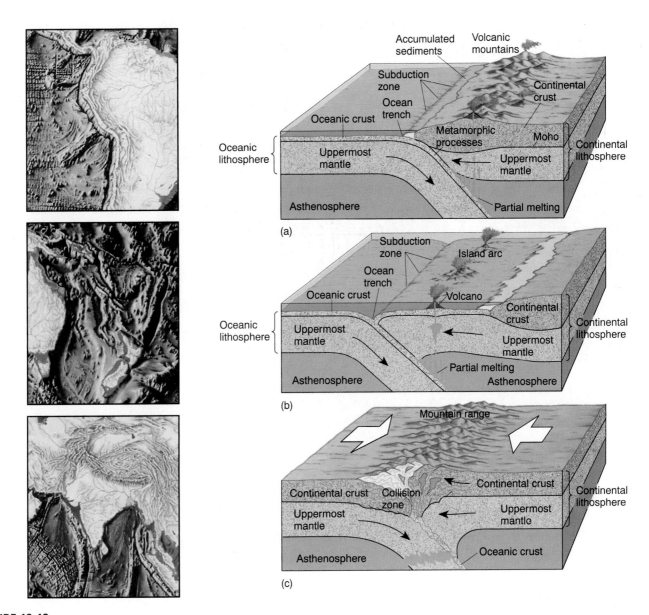

**FIGURE 12-12**
**Three types of plate convergence.**
Real-world examples illustrate three types of crustal collisions. (a) Oceanic–continental (example: Nazca plate–South American plate collision and subduction). (b) Oceanic–oceanic (example: New Hebrides Trench near Vanuatu, 16° S 168° W). (c) Continental–continental (example: India plate and Eurasian landmass collision and resulting Himalayan Mountains). [Illustrations on left from *Floor of the Oceans*, 1975, by Bruce C. Heezan and Marie Tharp © 1980. Reproduced by permission of Marie Tharp.]

demonstrates this folded landscape: the digitized relief map constructed from satellite data, the *Landsat* satellite image, and the topographic map. Note how these dissected ridges are cut through by rivers, forming *water gaps*. These important breaks in the rugged ridges greatly influenced migration, settlement patterns, and the diffusion of cultural traits in the 1700s. The initial flow of people, goods, and ideas was guided by this topography.

## *World Structural Regions*

Examine the first two maps in this chapter (the chapter opener and Figure 12-2) and you will note two vast mountain chains on the continents. In the Western Hemisphere, the *Cordilleran system* stretches from Tierra del Fuego at the southern tip of South America to the massive peaks of Alaska, including the relatively

**FIGURE 12-13**
**Tilted-fault block.**
The Teton Range in Wyoming is an example of a tilted-fault block, a range of scenic beauty and rugged relief. [Photo by Galen Rowell.]

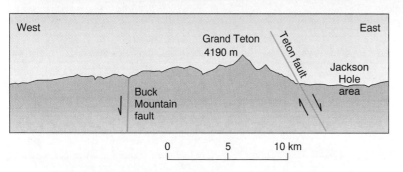

(Locator map for Figure 12-14.)

young Rocky and Andes Mountains along the western margins of the North and South American plates. In the Eastern Hemisphere, the *Eurasian-Himalayan system* stretches from the European Alps across Asia to the Pacific Ocean.

These mountain systems also are shown on the structural regions map as the *Alpine system* (Figure 12-15). The map defines seven fundamental structural regions that possess distinctive types of landscapes, grouped because of their shared physical characteristics. Looking at the distribution of these regions helps summarize the three rock-forming processes (igneous, sedimentary, metamorphic), plate tectonics, landform origins and construction, and orogenesis.

As you examine the map, identify the continental shields at the heart of each landmass. These areas are surrounded by continental platforms composed of sedimentary deposits. Various mountain chains, rifted regions, and isolated volcanic areas are portrayed. On the continent of Australia, you see older mountain sequences to the east, sedimentary layers covering basement rocks west of these ranges, and portions of the original Gondwana, an ancient landscape in the central and western region. (Remember that Gondwana was a landmass that included

Antarctica, Australia, South America, Africa, and the southern portion of India; it broke away from Pangaea some 200 million years ago.)

# Earthquakes

Crustal plates do not glide smoothly past one another. Instead, tremendous friction exists along plate boundaries. The stress of plate motion builds strain in the rocks until friction is overcome and the sides along plate boundaries suddenly break. The two sides of the fault plane then lurch into new positions, moving from centimeters to several meters, and release enormous amounts of seismic energy into the surrounding crust. This energy radiates throughout the planet, diminishing with distance, but sufficient to register on instruments worldwide.

These tectonic forces were brought home to hundreds of millions of television viewers as they watched the 1989 baseball World Series from Candlestick Park near San Francisco. Less than half an hour before the call to "play ball," a powerful earthquake rocked the region, turning sportscasters into newscasters and sports fans into disaster witnesses. Damage exceeded $8 billion, and 67

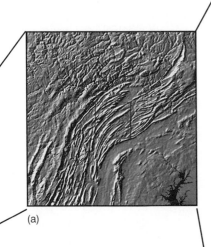

(a)

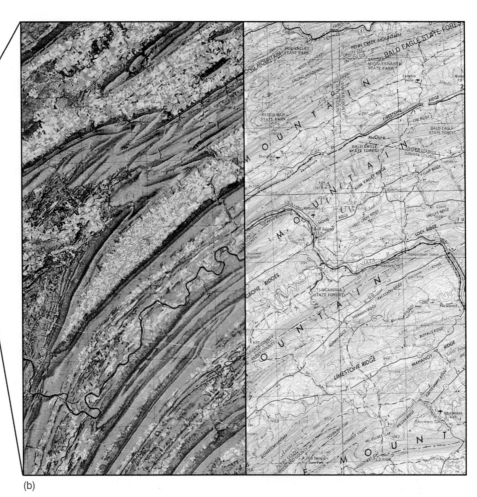

(b)

**FIGURE 12-14**

**The Appalachian Mountains.**

Appalachian Mountain region of central Pennsylvania shown on (a) a digital shaded-relief map originally prepared at a scale of 1:3,500,000 and (b) *Landsat* image (left) and U.S. Geological Survey topographic quadrangle map (right; original at 1:250,000 scale). The highly folded nature of this entire region is clearly visible on the small-scale and intermediate-scale image and maps. [(a) USGS digital terrain map I-2206, 1992; (b) NASA and U.S. Geological Survey.]

people died. Thoughts of the 1906 earthquake that devastated San Francisco probably arose in every spectator's mind (Figure 12-16).

In the Liaoning Province of northeastern China, ominous indications of tectonic activity began in 1970. Foreboding symptoms included land uplift and tilting, increasing minor tremors, and changes in the region's magnetic field—after almost 120 years of quiet.

These precursors of tectonic events continued for almost 5 years before Chinese scientists took the bold step of forecasting an earthquake. In February 1975, some 3 million people were evacuated in what turned out to be a timely manner; the quake struck 6 hours later, within the predicted time frame. Ninety percent of the buildings in the city of Haicheng were destroyed, but thousands of lives were saved, and success was proclaimed—an earthquake had been forecast and preparatory action taken for the first time in history.

However, only 17 months later, at Tangshan in the northeastern province of Hebei (Hopei), 145 km (90 mi) southeast of Beijing, China's capital city, a severe earthquake occurred without warning. No precursors of which scientists were aware permitted a forecast. Consequently,

this 7.4 magnitude quake killed about 250,000 people! (This is the official death toll; other estimates range as high as 650,000.) The earthquake also destroyed 95% of the buildings and 80% of the industrial structures, and it severely damaged more than half the bridges and highways in the area. The jolt threw people against the ceilings of their homes. An old, previously undetected fault had ruptured, and the rocks moved 1.5 m (5 ft) along a stretch 8 km (5 mi) long through Tangshan, devastating large areas. What are the mechanisms that produce such different tectonic events? Why are some expected, whereas most strike in total surprise?

## *Epicenter, Focus, Foreshock, Aftershock*

The subsurface area along a fault plane, where the motion of seismic waves is initiated, is called the *focus*, or hypocenter of an earthquake (see Figure 12-18). The area at the surface directly above the focus is the *epicenter*. Shock waves produced by an earthquake radiate outward through the crust from the focus and epicenter. Some of the seismic waves are conducted throughout the planet to

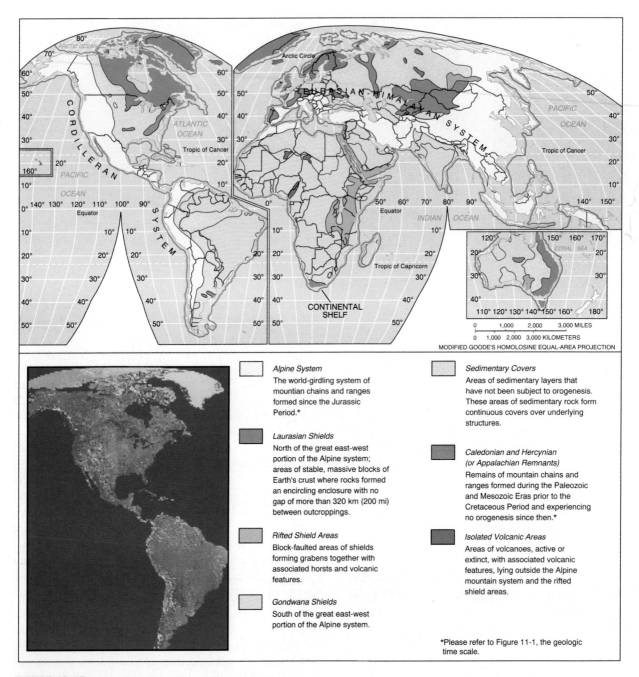

**FIGURE 12-15**

**World structural regions and major mountain systems.**

Some of the regions appear larger than the structures themselves because each region includes related landforms adjacent to the central feature. (Correlate aspects of this map with Figures 12-2 and 12-3.) Structural regions in the Western Hemisphere are visible on the composite *Landsat* image inset. [After R. E. Murphy, "Landforms of the World," *Annals of the Association of American Geographers* 58, no. 1 (March 1968). Adapted by permission. Inset image courtesy of EROS Data Center and the National Geographic Society.]

distant instruments. From these seismic wave patterns, Earth's interior has been defined.

An *aftershock* may occur after the main shock, sharing the same general area of the epicenter; some aftershocks rival the main tremor in magnitude. A *foreshock* also is possible,

preceding the main shock. The pattern of foreshocks is now regarded as an important consideration in the forecast effort. (Before the 1992 Landers earthquake in southern California, at least two dozen foreshocks occurred along the soon-to-erupt portion of the fault.)

**FIGURE 12-16**
**San Francisco, 1906.**
A view of San Francisco, devastated by the 1906 earthquake and subsequent fire. The view is to the east toward Nob Hill and the still-standing Fairmont Hotel. [Photo courtesy of Corbis-Bettmann.]

## *Earthquake Intensity and Magnitude*

Earthquakes associated with faulting are referred to as *tectonic earthquakes*. The number of earthquakes in an area, and their sizes, allow scientists to assess tectonic activity. Earthquakes also can be triggered by underground nuclear explosions, pumping of fluids into faulted rock through injection wells, or fluid pressure introduced in rock fractures created by the presence of a reservoir.

Earthquake vibrations are transmitted as waves of energy throughout Earth's interior and are detected with a **seismograph**, an instrument that records vibrations in the crust. Using this instrument and actual observations, scientists rate earthquakes on two kinds of scales, a qualitative damage intensity scale and a quantitative magnitude-of-energy-released scale.

A damage *intensity scale* is useful in classifying and describing damage to terrain and structures after an earthquake. Earthquake intensity is rated on the arbitrary *Mercalli scale*, a Roman numeral scale from I to XII, representing "barely felt" to "catastrophic total destruction." It was designed in 1902 and was modified in 1931. Table 12-1 shows this scale and the number of quakes in each category expected each year.

***The Richter Scale and Its Revision.*** Earthquake magnitude is estimated according to a system originally designed by Charles Richter in 1935. In this method, the amplitude of seismic waves is recorded on a seismograph located at least 100 km (62 mi) from the epicenter of the quake. That measurement is then charted on the **Richter scale**. The relation of magnitude to energy released is still a useful feature of his scale.

The Richter scale is logarithmic: Each whole number on it represents a 10-fold increase in the measured wave amplitude. Translated into energy, each whole number signifies a 31.5-fold increase in energy released. Thus, a magnitude of 3.0 on the Richter scale represents 31.5 times more energy than a 2.0, and 992 times more energy than a 1.0. It is difficult to imagine the power released by the Tangshan quake, which was rated a 7.6 on the Richter scale.

Today, the Richter scale has been improved and made more quantitative. Revision was needed because, at higher magnitudes, the scale did not properly measure or differentiate between quakes of high intensity. The Northridge quake mentioned earlier is now rated at 6.7 and featured extreme ground acceleration (movement upward), which was underestimated by the old method. Seismologists wanted to know more about what they call

**TABLE 12-1**

| **Earthquake Characteristics and Frequency Expected Each Year** | | | |
|---|---|---|---|
| *Effects in Populated Areas* | *Approximate Intensity (modified Mercalli scale)* | *Approximate Magnitude (Moment Magnitude scale)* | *Number per Year* |
| Nearly total damage | XII | >8.0 | 1 every few years |
| Great damage | X–XI | 7–7.9 | 18 |
| Considerable-to-serious damage to buildings; railroad tracks bent | VIII–IX | 6–6.9 | 120 |
| Felt-by-all, with slight damage to buildings | V–VII | 5–5.9 | 800 |
| Felt-by-some to felt-by-many | III–IV | 4–4.9 | 6200 |
| Not felt, but recorded | I–II | 2–3.9 | 500,000 |
| *Source*: USGS, Earthquake Information Center. | | | |

the *seismic moment* to understand a broader range of possible motions during an earthquake.

The **moment magnitude scale**, in use since 1993, is considered more accurate than Richter's *amplitude magnitude* scale for large earthquakes. Moment magnitude considers the amount of fault slippage produced by the earthquake, the size of the surface (or subsurface) area that ruptured, and the nature of the materials that faulted, including how resistant they were to failure.

A reassessment of past quakes has increased the rating of some and decreased that of others. The 1964 earthquake at Prince William Sound in Alaska has an amplitude magnitude of 8.6, but on the moment magnitude scale it increases to a 9.2. The 1906 San Francisco quake goes from an 8.25 down to a 7.7; the Tangshan quake from a 7.6 to a 7.4. A massive earthquake in Chile in May 1960 had a moment magnitude of 9.5, triggering seismic sea waves, volcanic activity, and floods. It probably is the current record holder for the new scale. Both moment magnitude and Richter's amplitude magnitude ratings for major quakes are shown in Table 12-2. The table includes more earthquakes for the period following 1960 than for the years before 1960. This is not because earthquake frequency has increased; rather, it reflects an effort to include recent events that have affected increased population densities in vulnerable areas. News Report 2 discusses the frequency of earthquakes.

Given this improvement in scientific method and understanding, it is still acceptable to think of these moment magnitudes as related to the Richter scale. Richter is familiar in everyday usage, although moment magnitude is the correct scale and is the scale we use in this book unless stated otherwise.

## The Nature of Faulting

We earlier described types of faults and faulting motions. The specific mechanics of how a fault breaks remain under study, but **elastic-rebound theory** describes the basic process. Generally, two sides along a fault appear to be locked by friction, resisting any movement despite the powerful forces acting on the adjoining pieces of crust. This stress continues to build strain along the fault surfaces, storing elastic energy like a wound-up spring. When the strain buildup finally exceeds the frictional lock, both sides of the fault abruptly move to a condition of less strain, releasing a burst of mechanical energy. This type of sudden movement rocked the region around Kobe, Japan, in 1995, with a magnitude 6.9 quake. See News Report 3 for an account of this event.

Think of the fault plane as a surface with irregularities that act as sticking points, preventing movement, similar to two pieces of wood held together by drops of glue

**TABLE 12-2**

| A Sampling of Significant Earthquakes | | | | | |
|---|---|---|---|---|---|
| **Year** | **Date** | **Location** | **Number of Deaths** | **Mercalli Intensity** | **Moment Magnitude (Richter)** |
| 1556 | Jan. 23 | Shaanxi Province, China | 830,000 | * | * |
| 1737 | Oct. 11 | Calcutta, India | 300,000 | * | * |
| 1812 | Feb. 7 | New Madrid, Missouri | Several | XI–XII | * |
| 1857 | Jan. 9 | Fort Tejon, California | * | X–XI | * |
| 1870 | Oct. 21 | Montreal to Québec, Canada | * | IX | * |
| 1886 | Aug. 31 | Charleston, South Carolina | * | IX | 6.7 |
| 1906 | Apr. 18 | San Francisco, California | 3000 | XI | 7.7 (8.25) |
| 1923 | Sept. 1 | Kwanto, Japan | 143,000 | XII | 7.9 (8.2) |
| 1939 | Dec. 27 | Erzincan, Turkey | 40,000 | XII | 7.6 (8.0) |
| 1964 | Mar. 28 | Southern Alaska | 131 | X–XII | 9.2 (8.6) |
| 1970 | May 31 | Northern Peru | 66,000 | * | 7.9 (7.8) |
| 1971 | Feb. 9 | San Fernando, California | 65 | VII–IX | 6.7 (6.5) |
| 1972 | Dec. 23 | Managua, Nicaragua | 5000 | X–XII | 6.2 (6.2) |
| 1976 | Jul. 28 | Tangshan, China | 250,000 | XI–XII | 7.4 (7.6) |
| 1978 | Sept. 16 | Iran | 25,000 | X–XII | 7.8 (7.7) |
| 1985 | Sept. 19 | Mexico City, Mexico | 7000 | IX–XII | 8.1 (8.1) |
| 1988 | Dec. 7 | Armenia–Turkey border | 30,000 | XII | 6.8 (6.9) |
| 1989 | Oct. 17 | Loma Prieta (near Santa Cruz, California) | 67 | VII–IX | 7.0 (7.1) |
| 1991 | Oct. 20 | Uttar Pradesh, India | 1700 | IX–XI | 6.2 (6.1) |
| 1994 | Jan. 17 | Northridge (Reseda), California | 66 | VII–IX | 6.8 (6.4) |
| 1995 | Jan. 17 | Kobe, Japan | 5500 | XII | 6.9 (6.9) |

*Note:* * Data not available

## News Report 2

### Are Earthquakes on the Increase?

Earthquakes seem to be increasing, and the media are quick to say so in hard-hitting, video-at-eleven stories! It is time for a reality check: The number of magnitude 7.0 quakes has been fairly constant throughout this entire century. In fact, it has decreased slightly over the past few decades worldwide!

Because there are more seismographs, increasing from 350 in 1931 to over 4000 today, and because of instant global communication, the impression of an increase is created. The National Earthquake Information Center in Golden, Colorado, locates the epicenters of about 14,000 earthquakes a year, averaging 38 a day. Since 1970, only the year 1992 exceeded the average of 18 (magnitude 7.0 to 7.9). However, even with a consistent number of quakes, the hazard increases as population centers grow. For instance, on the volcanic, quake-prone islands of Indonesia, the population of 200 million is expected to double in next 43 years!

## News Report 3

### A Tragedy in Kobe, Japan—the Hyogo-ken Nanbu Earthquake

Early on the morning of January 17, 1995, a year to the day after the Northridge quake, a section of the Nojima fault zone, approximately 16 km (10 mi) below the surface, became the focus of a devastating earthquake. On the moment magnitude scale the quake registered 6.9; it killed 5500 people, injured 26,000, and caused over $100 billion in damage. About 200,000 homes, or 10% of all the housing in the metropolitan area, were damaged; some 80,000 buildings completely collapsed; major transportation arteries were severed; and the busy port was brought to a standstill (Figure 1).

For Japan, this was the worst quake loss since the 1923 temblor that hit Tokyo. The Kobe region had large earthquakes in A.D. 868, 1596, and 1916, but lulled by amost 80 years of relative calm, the people there were unsuspecting of this surprise. However, the mapped system of faults and the mechanism of this earthquake were within present seismic understanding. Strain was liberated in a right-lateral, strike-slip motion, similar to movement by the San Andreas fault in California. This event displaced the land along three segments, where displacement of 1.0 to 1.5 m occurred in a rupture zone that stretched over 30 km (18 mi) in length.

The lessons are many, but two stand out: liquefaction and the failure of filled land. Areas of Osaka Bay that had been reclaimed from the sea with landfill were liquefied quickly in the shaking. They failed, collapsing the structures they supported (Figure 1b). These human-made fills and islands also are characteristic of the San Francisco Bay region, where, since the turn of the century, about half the original surface area of the bay has been reclaimed with fill and developed.

On a positive note was the good performance of landfill sites in Kobe and Osaka where drains had been installed and where pilings were in place to support structures. However, the cost of such retrofits in the United States could easily exceed several hundred billion dollars. In the meantime, the resemblance between the geology of the Kobe region and the San Francisco Bay region is troubling. The Kobe lesson raises great concern and a call to action.

(a)

(b)

**FIGURE 1**
**1995 earthquake in Kobe, Japan.**
(a) Catastrophic failure to an elevated freeway; note the failure of the supporting pillars. (b) Ground failure, mainly liquefaction, tilts building to near collapse; note the space between the building and the ground on the right side. [Photos by Haruyoshi Yamaguchi/Sygma.]

rather than an even coating of glue. Research scientists at the USGS and the University of California have identified these small areas of high strain as *asperities*. They are the points that break and release the sides of the fault.

If the fracture along the fault line is isolated to a small asperity break, the quake will be small in magnitude. Clearly, as some asperities break (perhaps recorded as small foreshocks), the strain is increased on surrounding asperities. Thus, small earthquakes in an area may be precursors to a major quake. However, if the break involves the release of strain along several asperities, the quake will be greater in extent and will involve the shifting of massive amounts of crust. The latest evidence points to a wave-like pattern, as rupturing spreads along the fault plane, rather than the entire fault surface giving way at once.

## Earthquakes and the San Andreas Fault

In 1906, San Francisco was a city of 400,000 people (there are several million in the metropolitan region today). In that year, a magnitude 7.7 (8.25 Richter) tremor rocked the city, felling buildings and initiating a firestorm fed by broken gas pipes. After the quake, movement along a fault was evident over a stretch of 435 km (270 mi), prompting intensive research to discover the nature of faulting. (Realize that this occurred 6 years before Wegener's continental drift hypothesis was even proposed!) The elastic-rebound theory developed as a result of research along the rupture, and today quakes, such as the one near San Francisco in 1989, can be interpreted.

The fault system that devastated San Francisco in 1906, and coastal and inland California throughout recorded history, is the San Andreas. The San Andreas fault system provides a good example of a spreading center overridden by an advancing continental plate (Figure 12-17). As shown in the figure, the East Pacific rise developed as a spreading center with associated transform faults (1), while the North American plate was progressing westward after the breakup of Pangaea. Forces then shifted the transform faults toward a northwest-southeast alignment along a weaving axis (2). Finally, the western margin of North America overrode those shifting transform faults (3). In *relative* terms, the motion along this series of faults is right-lateral, whereas in *absolute* terms the North American plate is still moving westward.

The earthquake that disrupted the 1989 World Series, mentioned previously, involved a portion of the San Andreas fault approximately 16 km (9.9 mi) east of Santa Cruz and 95 km (59 mi) south of San Francisco (Figure 12-18). The magnitude was 7.0 (7.1 Richter). A fault had ruptured at a focus unusually deep for the San Andreas system, more than 18 km (11.5 mi) below the surface.

Unlike previous earthquakes—such as the one in 1906, when the plates shifted a maximum of 6.4 m (21 ft) relative to each other—there was no evidence of a fault plane or rifting at the surface. Instead, the fault plane suggested in Figure 12-18 shows the two plates moving horizontally approximately 2 m (6 ft) past each other deep below the surface, with the Pacific plate thrusting 1.3 m (4.3 ft) upward. This is an unusual motion for the San Andreas fault and perhaps is a clue that this fault is more complex than previously thought.

In only 15 seconds, more than 2 km (1.2 mi) of freeway overpass and a section of the San Francisco–Oakland Bay Bridge collapsed, $8 billion in damage was generated, 14,000 people were displaced from their homes, 4000 were injured, and 67 were killed (Figure 12–18b). The damage from this quake was extreme in the region of the Loma Prieta epicenter and in the San Francisco Bay area, where houses and freeways are built on fill and bay mud. As in Mexico City, unstable soils magnified the effects of a distant quake (see News Report 4).

***The Southern California Earthquakes—And the Future.*** Since the mid-1980s, an area east of Los Angeles has experienced seven earthquakes greater than 6.0 magnitude. After a 6.1-magnitude quake in 1992, a 7.4-magnitude tremor rocked the lightly populated area near Landers, California (see the enlargement map of southern California in Figure 12-17). Although it was the single largest quake in California in 30 years, the remote location kept injuries and damage slight. Involved in this series were four different faults and some unknown segments. The main faulting caused a displacement of 6.1 m (20 ft).

For yet unexplained reasons, related earthquakes were recorded over the next few weeks in Mammoth Lakes, about 645 km (400 mi) to the north; at Mount Shasta in northern California; in southern Nevada and Utah; and 1810 km (1125 mi) distant in Yellowstone National Park, Wyoming!

Students at California State University–Northridge, in the San Fernando Valley of southern California, need no reminder of the power of earthquakes. Their campus was near the epicenter of the most expensive earthquake in U.S. history. The Northridge (Reseda) earthquake, a magnitude 6.8, with more than 10,000 aftershocks, caused extensive damage to campus buildings before the beginning of spring classes in 1995 (Figure 12-19). Amazingly, the semester began just 3 weeks late in 450 temporary trailers; a graduation was still held in May!

Across the Los Angeles region, tens of thousands of apartment units and homes were destroyed or damaged by this quake and its extreme ground acceleration (objects hurled upward against the pull of gravity). A deeply buried thrust fault caused the quake, focused 18 km (11 mi)

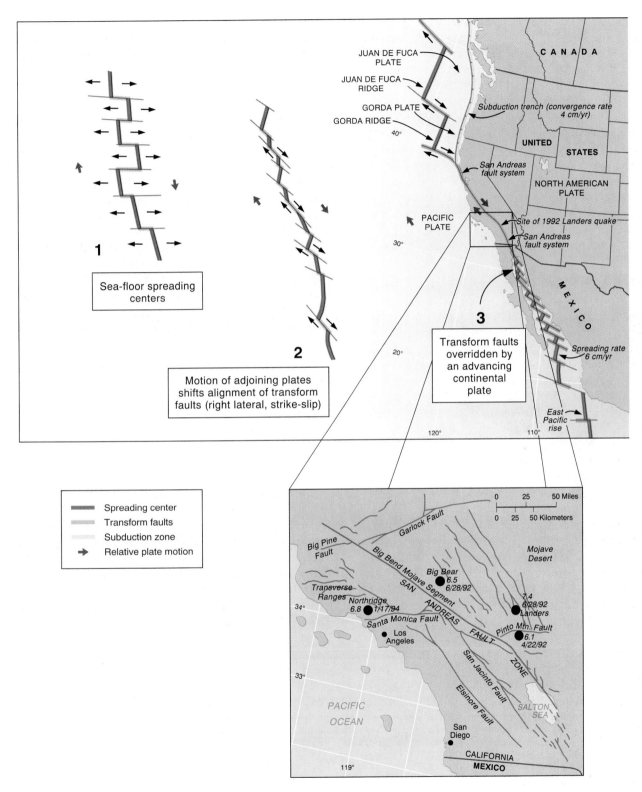

**FIGURE 12-17**

**San Andreas fault formation.**

Formation of the San Andreas fault system as a series of transform faults. Enlargement shows the portion of southern California where the 1992 Landers and the 1994 Northridge (Reseda) earthquakes occurred. Moment magnitudes are shown for four quakes.

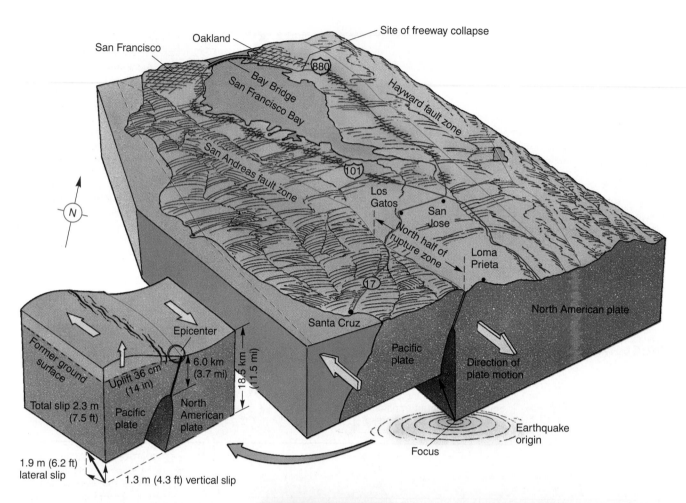

(b)

**FIGURE 12-18**

**Anatomy of an earthquake.**

(a) The fault-plane solution for the 1989 Loma Prieta, California, earthquake shows the lateral and vertical (thrust) movements occurring at depth. There was no surface expression of this fault plane. (b) Route 880 Cypress Freeway collapse. [(a) After P. J. Ward and R. A. Page, *The Loma Prieta Earthquake of October 17, 1989,* (Washington, DC: U.S. Geological Survey, November 1989), p. 1; (b) courtesy of California Department of Transportation.]

beneath the San Fernando Valley (epicenter is shown on Figure 12-17 inset). Scientists think that southern California will be affected by more of these thrust-fault actions as strain continues to build along the San Andreas system of faults—the restless fault line that joins the Pacific and the North American plates.

## Earthquake Forecasting and Planning

Principal zones of earthquake activity occur near plate boundaries, as the shaded areas in Figure 11-21 demonstrate. The correlation is such that earthquakes have provided key diagnostic evidence in understanding plate

**FIGURE 12-19**

**Earthquake damage on a college campus.**

A multilevel parking facility on the California State University–Northridge campus collapsed dramatically in the instant of the earthquake. In addition, damage to 20,000 apartment units across the metropolitan region left thousands homeless after the Northridge (Reseda) earthquake that occurred at 4:31 A.M. , January 17, 1994. [Photo by author.]

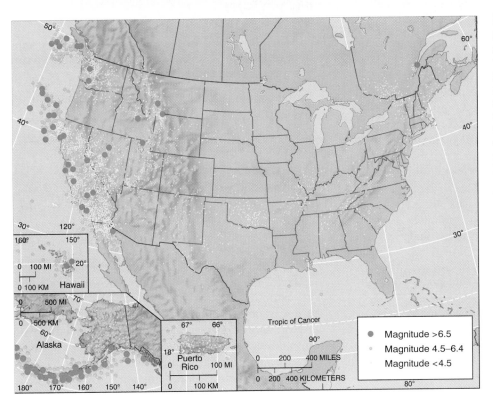

**FIGURE 12-20**

**Seismicity of the United States and southern Canada, 1899–1990.**

Earthquake occurrences of magnitude 4.5 or greater indicate areas of greatest seismic risk. Note the contrast between stable and unstable regions, related to tectonic activity. Active seismic regions include the West Coast, the Wasatch Front of Utah northward into Canada, the central Mississippi Valley, southern Appalachians, and portions of South Carolina, upstate New York, and Ontario. [From the U.S. Geological Survey, National Earthquake Information Center.]

tectonics, and, in turn, that correlation offers a long-term forecast potential for earthquake sites. The map in Figure 12-20 plots the epicenters of earthquakes in the United States and southern Canada between 1899 and 1990. These occurrences give some indication of relative risk by region. The challenge is to discover how to predict the *specific time and place* for a quake in the short term.

Except for isolated successes in China, Japan, and the United States, earthquake prediction has proved to be a much more difficult task than scientists imagined. Prediction is thought to be a matter of assessing earthquake precursors. USGS scientists now think that all earthquakes in a region are related and that the earlier quakes are precursors of later seismic events. Scientists

## News Report 4

## Damage Strikes Far from Epicenters

Normally, the greater the distance from an earthquake's epicenter, the less severe the shock. At a distance of 40 km (25 mi), only 1/10 of the full effect of an earthquake normally is felt. However, two recent earthquakes proved to be an exception to this rule. Scientists think they know why.

Mexico City, currently the world's second-largest city, is built on soft, moist sediments of an ancient lake bed. In the 1985 Mexico City event, two major earthquakes (magnitude 8.1 and 7.6) struck 400 km (250 mi) southwest of Mexico City. Their epicenters were on the sea floor off Mexico and Central America. Despite the distance, seismic waves arrived and set into motion the old lake bed beneath Mexico City, magnifying the shock waves by more than 500%. As a result, 250 buildings collapsed, some 8000 structures experienced damage, and about 7000 people perished.

The damage caused by the Loma Prieta earthquake in 1989 in the San Francisco Bay area, where houses and freeways are built on fill and bay mud, was also at some distance from the epicenter. Like the situation in Mexico City, unstable soils magnified the effects of a distant quake. Scientists now think that, in these events, seismic waves were reflected off the crust-mantle boundary as they traveled from the focal point of the quake. The waves bounced off the Moho discontinuity boundary and back to the surface, where they caused severe damage. Thus, if the ground is unstable, distant effects can be magnified, as they were in Mexico City and in San Francisco.

---

still are trying to assemble an accurate model of the microevents that precede an earthquake. See News Report 5 for more on forecasting.

Actual implementation of an action plan to reduce death, injury, and property damage from earthquakes is very difficult. The political environment adds complexity; sadly, an accurate earthquake prediction would be viewed as a threat to a region's economy (Figure 12-21). If we examine the potential socioeconomic impact of earthquake prediction on an urban community, it is difficult to imagine a chamber of commerce, bank, real estate agent, tax assessor, or politician who would privately welcome such a prediction.

Long-range planning is a complex subject. After the Loma Prieta earthquake in 1989, a committee report by the National Research Council concluded that:

> One of the most jarring lessons from these quakes may be that earthquake professionals have long known many of the things that could have been done to reduce devastation....The cost-effectiveness of mitigation and the importance of closing the knowledge gap among researchers, building professionals, government officials, and the public, are only two of the many lessons from the Loma Prieta that need immediate action.

A valid and applicable generalization is that *humans are unable or unwilling to perceive hazards in a familiar environment.* In other words, we tend to feel secure in our homes and communities, even if they are sitting on a quiet fault zone. Such an axiom of human behavior certainly helps explain why large populations continue to live and work in earthquake-prone settings. Similar questions also can be raised about populations in areas vulnerable to floods, droughts, coastal storm surge, or hurricanes.

# Volcanism

We are reminded of Earth's internal energy by the recent violent eruptions of several volcanoes (years given are when recent action started):

- 1991: Mount Pinatubo (Philippines), Mount Unzen (Japan)

- 1992: Mount Hudson (Chile), Mount Spurr and Mount Redoubt (Alaska)

- 1993: Galeras Volcano (Colombia)

- 1994: Rabaul Caldera (Papua, New Guinea); Merapi, the "Mountain of Fire" (Indonesia); Kliuchevskoi (Russia)

- 1995: Soufriere Hills on Montserrat (West Indies), Mount Etna (Italy), Metis Shoal (Tonga Islands)

Over 1300 identifiable volcanic cones and mountains exist on Earth, although fewer than 600 are active (have had at least one eruption in recorded history). In an average year, about 50 volcanoes erupt worldwide, varying from modest activity to major explosions. Eruptions in remote locations and at depths on the sea floor go largely unnoticed, but the occasional eruption of great magnitude near a population center makes headlines. North America has about 70 volcanoes (mostly inactive) along the western margin of the continent. Mount St. Helens in Washington State is a famous active example, and over 1 million visitors a year travel to Mount St. Helens Volcanic National Monument to see volcanism for themselves.

Volcanoes produce some benefits. These include fertile soils, which develop from lava, as in Hawaii; geothermal energy; and even new real estate in Iceland, Japan, Hawaii, and elsewhere, as lava extends shorelines seaward.

# News Report 5

## Seismic Gaps, Nervous Animals, Dilitancy, and Radon Gas

How do you forecast an earthquake? One approach to is to examine the history of each plate boundary and determine the frequency of past earthquakes, a study called *paleoseismology*. Paleoseismologists construct maps that provide an estimate of expected earthquake activity. An area that is quiet and overdue for an earthquake is termed a *seismic gap*; such an area forms a gap in the earthquake occurrence record and is therefore a place that possesses accumulated strain. The area along the Aleutian Trench subduction zone had three such gaps until the great 1964 Alaskan earthquake filled one of them.

The areas around San Francisco and northeast of Los Angeles represent other such gaps where the fault system appears to be locked by friction and is accumulating strain. The 1989 Loma Prieta earthquake was predicted in 1988 by the U.S. Geological Survey as having a 30% chance of occurring with a 6.5 magnitude within 30 years. The actual quake dramatically filled a portion of the seismic gap in that region. The Nojima fault in the area of Kobe was in another gap.

One potentially positive discovery from the Loma Prieta disaster was that Stanford University scientists found unusually great changes in Earth's magnetic field—about 30 times more than normal—3 hours before the main shock. These measurements were made 7 km (4.3 mi) from the epicenter, raising hopes that short-term warnings might be possible in the future. A valid question being researched is whether animals have the ability to detect these minute changes in magnetic fields, an ability that was perhaps lost in humans. If so, strange pre-quake animal behavior, often reported, might provide forecasting clues.

*Dilitancy* refers to the slight increase in volume of rock produced by small cracks that form under stress and accumulated strain. The affected region may tilt and swell in response to strain. Tiltmeters are installed in suspect areas to measure these changes. Another indicator of dilitancy is an increase in radon, a naturally occurring, slightly radioactive gas, dissolved in groundwater. Presently radon levels are being monitored in thousands of earthquake hazard zones.

Some day accurate earthquake forecasting may be a reality, but several questions remain. How will humans respond to a forecast? Can a major metropolitan region be evacuated for short periods of time? Can cities relocate after a disaster to areas of lesser risk?

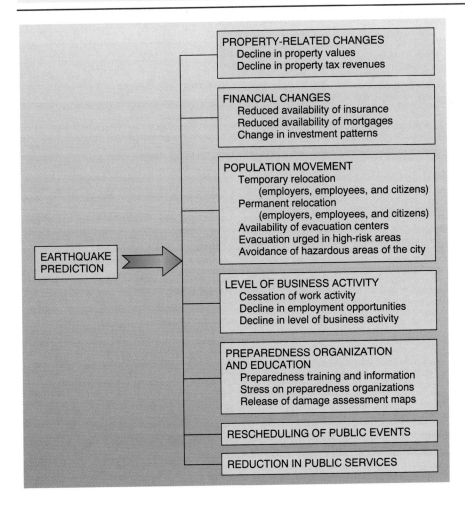

**FIGURE 12-21**

**Socioeconomic impacts of an earthquake prediction.**

Business, political, and monetary interests are at stake as scientists get closer to being able to predict earthquakes. [After J. E. Haas and D. S. Mileti, *Socioeconomic Impact of Earthquake Prediction on Government, Business, and Community* (Boulder: University of Colorado, Institute of Behavioral Sciences, 1976). Used by permission.]

## Volcanic Features

A **volcano** forms at the end of a central vent or pipe that rises from the asthenosphere through the crust into a volcanic mountain. A **crater**, or circular surface depression, usually forms at or near the summit.

Magma rises and collects in a magma chamber deep below the volcano until conditions are right for an eruption. This subsurface magma emits tremendous heat and in some areas it boils groundwater, as seen in the thermal springs and geysers of Yellowstone National Park. The steam given off is a potential energy source if it is accessible. Such **geothermal energy** has provided heating and electricity for more than 90 years in Iceland, New Zealand, and Italy. North of San Francisco, some 1300 megawatts of electrical capacity are generated by turbines that are operated by steam from geothermal wells.

**Lava** (molten rock), gases, and **pyroclastics** (pulverized rock and clastic materials of various sizes ejected violently during an eruption; sometimes described by the more general term *tephra*) pass through the vent to the surface and build a volcanic landform. The fact that lava can occur in many different textures and forms accounts for the varied behavior of volcanoes and the different landforms they build. In this section we will look at five volcanic landforms and their origins: cinder cones, calderas, shield volcanoes, plateau basalts, and composite volcanoes.

A **cinder cone** is a small cone-shaped hill usually less than 450 m (1500 ft) high, with a truncated top formed from cinders that accumulate during moderately explosive eruptions. Cinder cones are made of pyroclastic material and *scoria* (cindery rock, full of air bubbles).

**Calderas.** Another distinctive landform is a large basin-shaped depression called a **caldera** (Spanish for "kettle"). It forms when summit material on a volcanic mountain collapses inward after an eruption or other loss of magma. In the Cape Verde Islands (15° N 24.5° W), Fogo Island has such a caldera in the midst of a volcanic cone, opening to the sea on the eastern side. The best farmland is in the caldera, so the people choose to live and work there. Unfortunately, eruptions began in April 1995, destroying farmland and forcing the evacuation of 5000 people. Previously, eruptions occurred between A.D. 1500 and 1750, followed by six more in the next century, each destroying houses and crops. An example of a caldera in North America is beautiful Crater Lake in southern Oregon. News Report 6 discusses the Long Valley Caldera in California.

## Locations of Volcanic Activity

The location of volcanic mountains on Earth is a function of plate tectonics and hot-spot activity. Volcanic activity occurs in three areas:

1. Along subduction boundaries at continental plate–oceanic plate convergence (such as Mount St. Helens and Kliuchevskoi) or oceanic plate–oceanic plate convergence (Philippines and Japan).
2. Along sea-floor spreading centers on the ocean floor (Iceland) and areas of rifting on continental plates (the rift zone in east Africa).
3. At hot spots, where individual plumes of magma rise to the crust (such as Hawaii and Yellowstone).

# News Report 6

## Is the Long Valley Caldera Next?

Trees are dying in the forests on Mammoth Mountain, in a portion of the Long Valley Caldera, near the California–Nevada border. The oval caldera is about 15 by 30 km (9 by 19 mi), longer in a north-south direction, and sits at 2000 m (6500 ft) elevation, rising to 2600 m (8500 ft) along the western side.

About 1200 tons of carbon dioxide is coming up through the soil in the old caldera each day. Carbon dioxide levels reach 30% to 96% in some soil samples. The source: active and moving magma at some 3 km depth. This and other gas emissions are used elsewhere in the world to forecast a rebirth of volcanic activity. At Mammoth Mountain, these gases signify magmatic activity and may be a portent of things to come.

The Long Valley Caldera was formed by a powerful volcanic eruption 730,000 years ago. The eruption exceeded the volume output of the 1980 Mount St. Helens event by more than 500 times! Volcanic activity of a lesser extent has occurred in the area over the past several hundred thousand years, most recently between A.D. 1720 and 1850 in the Mono Lake area north of Long Valley.

In the late 1970s and again in 1996, swarms of earthquakes shook Long Valley—thousands of quakes overall, some three dozen greater than magnitude 3.0. The quakes are focused at about 11 km beneath the region. As of this writing, the surface has not been lifted or deformed by this activity. The combination of volcanic gas production and earthquakes is drawing scientific and public attention to this area because of the potential for a massive eruption sometime in the next century. The region is a popular tourist mecca and recreational center, with second-home residential development and a growing year-round population.

Figure 12-22 illustrates the three types of volcanic activity, which you can compare with the active volcano sites and plate boundaries shown in Figure 11-21. Figures 12-23 to 12-26 illustrate the various aspects of volcanism in Figure 12-22.

## Types of Volcanic Activity

The variety of forms among volcanoes makes them hard to classify; most fall in transition between one type and another. Even during a single eruption, a volcano may behave in several different ways. The primary factors in determining an eruption type are (1) the magma's chemistry, which is related to its source, and (2) the magma's viscosity. Viscosity is the magma's resistance to flow ("thickness"); it ranges from low viscosity (very fluid) to high viscosity (thick and flowing slowly). We consider two types of eruptions—effusive and explosive—and the characteristic landforms they build.

## Effusive Eruptions

Effusive eruptions are the relatively gentle ones that produce enormous volumes of lava annually on the sea floor and in places such as Hawaii and Iceland. These direct eruptions from the asthenosphere produce a low-viscosity magma that is very fluid and cools to form a dark, basaltic rock (less than 50% silica and rich in iron and magnesium). Gases readily escape from this magma because of its low viscosity, causing a relatively gentle **effusive eruption** that pours out on the surface, with relatively small explosions and little pyroclastics. However, dramatic fountains of basaltic lava sometimes shoot upward, powered by jets of rapidly expanding gases (Figure 12-25).

An effusive eruption may come from a single vent or from the flank of a volcano, through a side vent. If such vents form a linear opening, they are called *fissures*; these sometimes create a dramatic curtain of fire (sheets of molten rock spraying into the air) during eruptions. In Iceland, active fissures are spread throughout the plateau landscape. Rift zones capable of erupting tend to converge on the central crater, or vent, as they do in Hawaii. The interior of such a crater is often a sunken caldera; it may fill with low-viscosity magma during an eruption, forming a molten lake, which then may overflow lava downslope in dramatic rivers and falls of molten rock.

On the island of Hawaii, the continuing Kilauea eruption is the longest in recorded history, active since 1823! During 1989–1990, lava flows from Kilauea actually consumed several visitor buildings in the Hawaii Volcanoes National Park and homes in Kalapana, a nearby town. Kilauea has produced more lava than any other vent on Earth in recorded history.

A typical mountain landform built from effusive eruptions is gently sloped, gradually rising from the surrounding landscape to a summit crater. The shape is similar in outline to a shield of armor lying face up on the ground and therefore is called a **shield volcano**. The shield shape and size of Mauna Loa in Hawaii is distinctive when compared with Mount Rainier in Washington, which is a different type of volcano (explained shortly) and the largest in the Cascade Range (Figure 12-27). The height of the shield is the result of successive eruptions, flowing one on top of another. Mauna Loa is one of five shield volcanoes that make up the island of Hawaii, and it has taken at least 1 million years to accumulate its mass. Mauna Kea is slightly taller, but Mauna Loa is the most massive single mountain on Earth.

***Extensive Igneous Provinces.*** Effusive eruptions send material out through hot spots or elongated fissures, forming extensive sheets of basaltic lava on the surface. The Columbia Plateau of the northwestern United States, some 2 to 3 km thick, results from the eruption of such **plateau basalts**, or *flood basalts* (Figure 12-23). More than double the size of the Columbian Plateau is the Deccan Traps, which dominate west-central India. The Siberian Traps is more than twice the area of that in India and is exceeded only by the Ontong Java Plateau, which covers an area of the sea floor in the Pacific (Figure 12-28). These regions are sometimes referred to as *plateau basalt provinces.* (*Trap* is Dutch for "staircase," referring to the steplike eroded lavas.)

Of these extensive igneous provinces, no presently active sites are in the same league with the largest of the extinct igneous provinces, some of which formed over 200 million years ago. Worldwide, recent scientific findings have determined the volume in these plateau basalt masses to be far greater than previously thought. The hotspot sources are now thought to be linked to the mantle at depths near the core-mantle boundary.

## Explosive Eruptions

Volcanic activity inland from subduction zones produces the well-known explosive volcanoes. Magma produced by the melting of subducted oceanic plate and other materials is thicker (more viscous) than magma from effusive volcanoes; it is 50%–75% silica and high in aluminum. Consequently, it tends to block the magma conduit inside the volcano; the blockage traps and compresses gases, causing pressure to build and forming a possible **explosive eruption**. From this magma, a lighter dacitic rock forms at the surface, as illustrated in the comparison in Figure 12-24 (see *dacite* in Table 11-2).

The term **composite volcano** is used to describe explosively formed mountains. (They are sometimes

**FIGURE 12-22**
**Tectonic settings of volcanic activity.**
Magma rises and lava erupts from rifts, from above
subduction zones, and from thermal plumes at hot spots.
[After U.S. Geological Survey, *The Dynamic Planet*
(Washington, DC: Government Printing Office, 1989).]

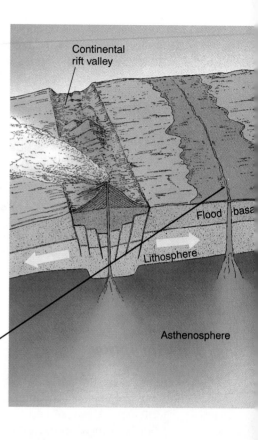

**FIGURE 12-23**
**Two expressions of volcanic activity.**
In the foreground are plateau (flood) basalts characteristic of the
Columbia Plateau in Oregon. Mount Hood, in the background to the
west of the plateau, is a composite volcano, part of the Cascade
Range of active and dormant volcanoes that result from subduction of
the Pacific plate beneath the North American plate. [Photo by author.]

**FIGURE 12-24**
**Extrusive igneous rocks compared.**
Basalt from Hawaii (left, darker rock) and dacite (right, lighter rock)
from Mount St. Helens. What does this coloration tell you about the
rocks' history and composition? Find these rock types in Table 11-2.
[Photo by J. D. Griggs/U.S. Geological Survey.]

called *stratovolcanoes* because they are built up in alternating layers of ash, rock, and lava, but shield volcanoes
also can exhibit a stratified structure, so *composite* is the
preferred term.) Composite volcanoes tend to have steep
sides, are more conical than shield volcanoes, and therefore are also known as *composite cones*. If a single summit vent erupts repeatedly, a remarkable symmetry may
develop as the mountain grows in size, as demonstrated

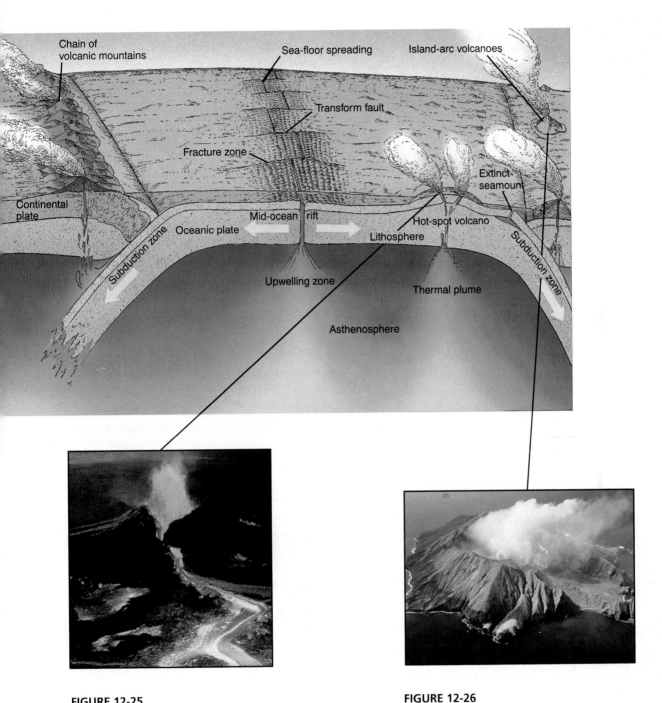

**FIGURE 12-25**
**Volcanic fountaining in Hawaii.**
Kilauea erupts a fountain of lava from its East Rift spatter cone. [Photo by Kepa Maly/U.S. Geological Survey.]

**FIGURE 12-26**
**Volcanic island off the coast of New Zealand.**
White's Isle, New Zealand (37.5° S 177° E). [Photo by Harvey Lloyd/Stock Market.]

by Mount Orizaba in Mexico, Mount Shishaldin in Alaska (Figure 12-29), Mount Fuji in Japan, the pre-1980-eruption shape of Mount St. Helens in Washington, and Mount Mayon in the Philippines.

As the magma in a composite volcano forms a plug near the surface, the blockage in the passage causes tremendous pressure to build, keeping the trapped gases compressed and liquefied. When the blockage can no

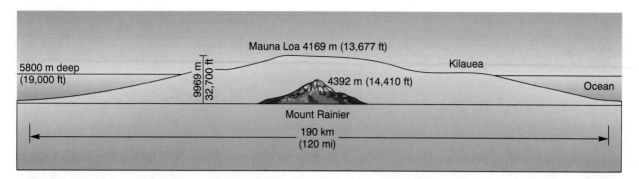

**FIGURE 12-27**

**Shield and composite volcanoes compared.**

Comparison of Mauna Loa in Hawaii, a shield volcano, and Mount Rainier in Washington State, a composite volcano. Their strikingly different profiles signify their different tectonic origins. [After U.S. Geological Survey, *Eruption of Hawaiian Volcanoes* (Washington, DC: Government Printing Office, 1986).]

longer hold back this pressurized inferno, explosions equivalent to megatons of TNT blast the tops and sides off these mountains. This type of eruption produces much less lava than effusive eruptions, but larger amounts of pyroclastics, which include volcanic *ash* (<2 mm in diameter), dust, cinders, *lapilli* (up to 32 mm in diameter), *scoria* (volcanic slag), pumice, and *aerial bombs* (explosively ejected blobs of incandescent lava). Table 12-3 presents some notable composite volcano eruptions. Focus Study 12-1 details the much-publicized 1980 eruption of Mount St. Helens.

***Volcanic Milestones.*** Unlike the volcanoes in Hawaii Volcanoes National Park, where tourists gather at observation platforms to watch the relatively calm effusive eruptions, composite volcanoes do not invite close inspection and can explode with little warning:

• In A.D. 79, Mount Vesuvius buried the city of Pompeii, Italy, in 3 days of pyroclastic eruptions, even though the volcano had been dormant for centuries.

• In 1883 in Indonesia, an entire group of islands centered on the volcano Krakatau (between the big

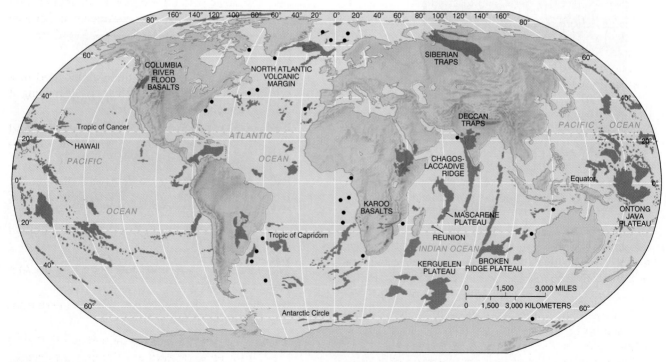

**FIGURE 12-28**

**Earth's extensive igneous provinces.**

Plateau basalts form large igneous provinces across the globe. Many are ancient and inactive. The largest discovered is the Ontong Java Plateau which covers nearly 2 million square kilometers (0.8 million square miles) on the Pacific Ocean floor. [After M. F. Coffin and O. Eldholm, "Large Igneous Provinces," *Scientific American* (October 1993): 42–3. © Scientific American, Inc. All rights reserved.]

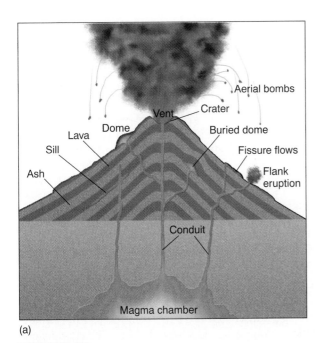

(a)

(b)

**FIGURE 12-29**

**A composite volcano.**

(a) A typical composite volcano with its cone-shaped form. (b) An eruption of Mount Shishaldin, Alaska, a composite volcano.
[(b) Photo courtesy of AeroMap U.S., Inc., Anchorage, Alaska.]

islands of Java and Sumatra) was obliterated in 2 days of explosions. The ash cloud reached into the mesosphere, and the sounds of the blast were heard in central Australia, the Philippines, and even 4800 km (3000 mi) away in the Indian Ocean! (The sound took 4 hours to reach that listener, who noted the noise in a ship's log.) What was not blown into the sky fell back into the collapsed remains of a large caldera 275 m (900 ft) under the sea, creating a great seismic sea wave (tsunami) that killed 36,000 people on neighboring islands.

• In 1902 on the Caribbean island of Martinique, the beautiful port city of Saint-Pierre was devastated in a few minutes by an eruption of Mont Pelée. That eruption featured a lateral blast of incandescent ash and super-hot gases, known as a "glowing cloud," or *nuée ardente*. Despite months of precursory rumbling, ash, hot mudflows, and minor bursts from the mountain, politicians discouraged citizens from fleeing, and all but two of 29,000 people in the town perished.

• In June 1991, after 600 years of dormancy, Mount Pinatubo in the Philippines erupted. The summit of the 1460 m (4795 ft) volcano exploded, devastating many surrounding villages and permanently closing Clark Air Force Base, operated by the United States. Fortunately, scientists from the U.S. Geological Survey and local scientists accurately predicted the eruption, and timely evacuation of the surrounding countryside saved thousands of lives. Although volcanoes are

regional events, their spatial implications can be worldwide. The single volcanic eruption of Mount Pinatubo was significant to the global environment and energy budget, as discussed in Chapters 3, 4, 5, and 10:

— 15–20 million tons of ash and sulfuric acid mist were blasted into the atmosphere, concentrating at 16–25 km (10–15.5 mi) altitude

— 12 km³ (3.0 mi³) of material was ejected and extruded by the eruption (12 times the volume from Mount St. Helens)

— 60 days after the eruption about 42% of the globe was affected (from 20° S to 30° N) by the thin, spreading aerosol cloud in the atmosphere

— colorful twilight and dawn skies were observed worldwide

— an increase in atmospheric albedo of 1.5% (4.3 W/m²) occurred

— an increase in the atmospheric absorption of insolation followed (2.5 W/m²)

— a decrease in net radiation at the surface (2.5 W/m²) and a lowering of Northern Hemisphere average temperatures of 0.5 C° (0.9 F°) were measured

— atmospheric scientists and volcanologists were able to study the eruption aftermath using orbiting sensors aboard satellites and general circulation model simulations on computers

## Focus Study 12-1

## The 1980 Eruption of Mount St. Helens

Probably the most studied and photographed composite volcano on Earth is Mount St. Helens, located 70 km (45 mi) northeast of Portland, Oregon, and 130 km (80 mi) south of the Tacoma-Seattle area of Washington. Mount St. Helens is the youngest and most active of the Cascade Range of volcanoes, which form a line from Mount Lassen in California to Mount Baker in Washington. The Cascade Range is the product of the Juan de Fuca sea-floor spreading center off the coast of northern California, Oregon, Washington, and British Columbia and the plate subduction that occurs offshore, as identified in Figure 12-17 (Figure 1).

The mountain had been quiet since 1857. New activity began in March 1980, with a sharp earthquake registering a magnitude 4.1. The first eruptive outburst occurred one week later, beginning with a 4.5 quake and continuing with a thick black plume of ash and the development of a small summit crater. Ten days later, the first volcanic earthquake, called a harmonic tremor, registered on the many instruments that had been hurriedly placed around the volcano. *Harmonic tremors* are slow, steady vibrations, unlike the sharp releases of energy associated with earthquakes and faulting. Harmonic tremors told scientists that magma was on the move within the mountain.

Also developing was a massive bulge on the north side of the mountain. This bulge indicated the direction of the magma flow within the volcano. A bulge represents the greatest risk from a composite volcano, for it could signal a potential lateral burst through the bulge and across the landscape.

Early on Sunday, May 18, the area north of the mountain was rocked by a magnitude 5.0 quake, the strongest to date. The mountain, with its distended 245 m (800 ft) bulge, was shaken, but nothing happened. Then a second quake (5.1) hit at 8:32 A.M., loosening the bulge and launching the eruption. David Johnson, a volcanologist with the U.S. Geological Survey, was only 8 km (5 mi) from the mountain, servicing instruments, when he saw the eruption begin. He radioed headquarters in Vancouver, Washington, saying, "Vancouver, Vancouver, this is it!" He perished in the eruption.

As the contents of the mountain exploded, a surge of hot gas (about 300°C, or 570°F), steam-filled ash, pyroclastics, and a nuée ardente moved northward, hugging the ground and traveling at speeds up to 400 kmph (250 mph) for a distance of 28 km (17 mi).

The slumping north face of the mountain produced the greatest landslide witnessed in recorded history; about 2.75 km³ (0.67 mi³) of rock, ice, and trapped air, all fluidized with steam, surged at speeds approaching 250 kmph (155 mph). Landslide materials traveled for 21 km (13 mi) into the valley, blanketing the forest, covering a lake, and filling the rivers below. A series of photographs, taken at 10-second intervals from the east looking west, records this sequence (Figure 2). The eruption continued with intensity for 9 hours, first clearing out old rock from the throat of the volcano and then blasting new material.

As destructive as such eruptions are, they also are constructive, for this is the way in which a volcano eventually builds its height. Before the eruption, Mount St. Helens was 2950 m (9677 ft) tall; the eruption blew away 418 m (1370 ft). But today, Mount St. Helens is regrowing by building a lava dome within its crater. The thick lava rapidly and repeatedly plugs and breaks in a series of lesser dome eruptions that may continue for several decades. The dome already is over 300 m (1000 ft) high; so a new mountain is being born from the eruption of the old.

An ever-resilient ecology is recovering as plants and animals reclaim the devastated landscape. Strong interest in the area continues; scientists and over 1 million tourists visited the Mount St. Helens Volcanic National Monument in 1995 (Figure 1).

(a)

(b)

**FIGURE 1**

**Mount St. Helens after the eruption—days and years later.**

(a) The devastated and scorched land shortly after the eruption in 1980. The scorched earth and tree blowdown area covered some 38,950 hectares (95,000 acres). (b) A landscape in recovery as life moves back in and takes hold to establish new ecosystems in 1994. [(a) Photo by Krafft-Explorer/Photo Researchers, Inc.; (b) photo by Chuck Pefley/Tony Stone Images.]

**FIGURE 2**
**The Mount St. Helens eruption sequence and corresponding schematics.**

- A volcano watch continues at Soufriere Hills, a composite volcano on the island of Montserrat in the West Indies (17° N 62° W). The Montserrat Volcano Observatory reports a dozen evacuations since volcanic activity began in 1995, strengthening in 1996. Are these precursors to a main event? Or just routine indigestion that will grow quiet again? The U.S. Volcano Disaster Assistance Program is in place to help local scientists with eruption forecasts.

***Volcano Forecasting and Planning.*** The USGS and the Office of Foreign Disaster Assistance operate the Volcano Disaster Assistance Program (VDAP). The need for such a program is evident in that, over the past 15 years, volcanic activity has killed 29,000 people, forced over 800,000 to evacuate their homes, and caused over $3 billion in damage. The program was established after 23,000 people died in the eruption of Nevado del Ruiz, Colombia, in 1985.

The key to the effort is to set up monitors at the most-threatened sites—a mobile volcano-monitoring system. It was an effort such as this that led to the life-saving evacuation of 60,000 people hours before Mount Pinatubo exploded. In addition, satellite remote sensing is helping VDAP to monitor eruption cloud dynamics, atmospheric emissions and climatic effects, and lava and thermal measurements; to make topographic measurements; to estimate volcanic hazard potential; and to enhance geologic mapping—all in an effort to better understand our dynamic planet.

## TABLE 12-3

| **Notable Composite Volcano Eruptions** | | | |
|---|---|---|---|
| *Date* | *Name of Locations* | *Number of Death* | *Amount extruded (mostly pyroclastics) in km³ (mi³)* |
| Prehistoric | Yellowstone, Wyoming | Unknown | 2400 (576) |
| 4600 B.C. | Mount Mazama (Crater lake, Oregon) | Unknown | 50–70 (12–17) |
| 1900 B.C. | Mount St. Helens | Unknown | 4 (0.95) |
| A.D. 79 | Mount Vesuvius, Italy | 20,000 | 3 (0.7) |
| 1500 | Mount St. Helens | Unknown | 1 (0.24) |
| 1815 | Tambora, Indonesia | 66,000 | 80–100 (19–24) |
| 1883 | Krakatau, Indonesia | 36,000 | 18 (4.3) |
| 1902 | Mont Pelée, Martinque | 29,000 | Unknown |
| 1912 | Mount Katmai, Alaska | Unknown | 12 (2.9) |
| 1943–1952 | Paricutín, Mexico | 0 | 1.3 (0.30) |
| 1980 | Mount St. Helens | 54 | 4 (0.95) |
| 1985 | Nevado del Ruiz, Colombia | 23,000 | 1 (0.24) |
| 1991 | Mount Unzen, Japan | 10 | 2 (0.5) |
| 1991 | Mount Pinatubo, Philippines | 800 | 12 (3.0) |
| 1992 | Mount Spurr/Mount Shishaldin, Alaska | 0 | 1 (0.24) |
| 1993 | Galeras Volcano, Colombia | 5 | 1 (0.24) |
| 1993 | Mount Mayon, Philippines | 0 | 1 (0.24) |
| 1994 | Kliuchevskoi, Russia | 0 | 1 (0.24) |

# Summary and Review—Tectonics, Earthquakes, and Volcanism

✔ ***Describe*** **first, second, and third orders of relief and *relate* examples of each from Earth's major topographic regions.**

Earth's surface is dramatically shaped by tectonic forces generated within the planet. **Relief** is the vertical elevation difference in a local landscape. The undulating physical surface of Earth, including relief, is called **topography**. Convenient descriptive categories are termed **orders of relief**. The coarsest level of landforms includes the **continental platforms** and **ocean basins**; the finest comprises local hills and valleys.

relief (p. 352)
topography (p. 352)
orders of relief (p. 353)
continental platforms (p. 353)
ocean basins (p. 353)

1. How does the map of the ocean floor (chapter-opening illustration) exhibit the principles of plate tectonics? Briefly analyze.

2. What is meant by an "order of relief"? Give an example from each order.

3. Explain the difference between relief and topography.

---

✔ *Describe* **the several origins of continental crust** and *define* **displaced terranes.**

The continents are formed as a result of several processes, including upwelling material from below the crust and migrating portions of crust. The source of material generated by plate tectonic processes determines the behavior and the composition of the crust that results.

A continent has a nucleus of ancient crystalline rock called a *craton*. A region where a craton is exposed is termed a **continental shield**. As continental crust forms, it is enlarged through accretion of dispersed **terranes**. An example is the **Wrangellia terrane** of the Pacific Northwest and Alaska.

> continental shield (p. 355)
> terranes (p. 356)
> Wrangellia terrane (p. 357)

4. What is a craton? Relate this structure to continental shields and platforms, and describe these regions in North America.
5. What is a migrating terrane, and how does it add to the formation of continental masses?
6. Briefly describe the journey and destination of the Wrangellia Terrane.

---

✔ *Explain* **compressional processes and folding;** *describe* **four principal types of faults and their characteristic landforms.**

The crust is deformed by folding, broad warping, and faulting, which produce characteristic landforms. Compression causes rocks to deform in a process known as **folding**, during which rock strata bend and may overturn. Along the *ridge* of a fold, layers *slope downward* away *from the axis*, resulting in an **anticline**. In the *trough* of a fold, however, layers *slope downward* toward *the axis*, producing a **syncline**.

When rock strata are stressed beyond their ability to remain a solid unit, they express the strain as a fracture. Rocks on either side of the fracture are displaced relative to the other side in a process known as **faulting**. Thus, *fault zones* are areas where fractures in the rock demonstrate crustal movement. At the moment of fracture, a sharp release of energy, called an **earthquake**, or *quake*, occurs.

When forces pull rocks apart, the tension causes a **normal fault**, sometimes visible on the landscape as a scarp, or escarpment. Compressional forces associated with converging plates force rocks to move upward, producing a **reverse fault**. A low-angle fault plane is referred to as a **thrust fault**. Horizontal movement along a fault plane that produces a linear rift valley is called a **strike-slip fault**. In the U.S. interior west, the Basin and Range Province is an example of aligned pairs of normal faults and a distinctive landscape. The term **horst** is applied to upward-faulted blocks; **graben** refers to downward-faulted blocks.

> folding (p. 358)
> anticline (p. 358)
> syncline (p. 358)
> faulting (p. 359)
> earthquake (p. 359)
> normal fault (p. 361)
> reverse fault (p. 361)
> thrust fault (p. 361)
> strike-slip fault (p. 361)
> horst (p. 363)
> graben (p. 363)

7. Diagram a simple folded landscape in cross section, and identify the features created by the folded strata.
8. Define the four basic types of faults. How are faults related to earthquakes and seismic activity?
9. How did the Basin and Range Province evolve in the western United States? What other examples exist of this type of landscape?

---

✔ *Relate* **the three types of plate collisions associated with orogenesis and** *identify* **specific examples of each.**

**Orogenesis** is the birth of mountains. An orogeny is a mountain-building episode, occurring over millions of years, that thickens continental crust. It can occur through large-scale deformation and uplift of the crust. It also may include the capture of migrating terranes and cementation of them to the continental margins, and the intrusion of granitic magmas to form plutons.

*Oceanic plate–continental plate* collision orogenesis is now occurring along the Pacific coast of the Americas and has formed the Andes, the Sierra of Central America, the Rockies, and other western mountains. *Oceanic plate–oceanic plate* collision orogenesis produces either simple volcanic island arcs or more-complex arcs such as Japan, the Philippines, the Kurils, and portions of the Aleutians. The region around the Pacific contains expressions of each type of collision in the **circum-Pacific belt**, or the **ring of fire**.

*Continental plate–continental plate* collision orogenesis is quite mechanical; large masses of continental crust, such as the Himalayan Range, are subjected to intense folding, overthrusting, faulting, and uplifting.

> orogenesis (p. 363)
> circum-Pacific belt (p. 364)
> ring of fire (p. 364)

10. Define orogenesis. What is meant by the birth of mountain chains?

11. Name some significant orogenies.

12. Identify on a map Earth's two large mountain chains. What processes contributed to their development?

13. How are plate boundaries related to episodes of mountain building? Do different types of plate boundaries produce differing orogenic episodes and different landscapes?

14. Relate tectonic processes to the formation of the Appalachians and the Alleghany orogeny.

---

✔ ***Explain*** the nature of earthquakes, their measurement, and the nature of faulting.

Earthquakes generally occur along plate boundaries; major ones can be disastrous. Earthquakes result from faults, which are under continuing study to learn the nature of faulting, stress and the buildup of strain, irregularities along fault plane surfaces, the way faults rupture, and the relationship among active faults. Earthquake prediction and improved planning are active concerns of *seismology*, the study of earthquake waves and Earth's interior. Seismic motions are measured with a **seismograph**.

Charles Richter developed the **Richter scale**, a measure of earthquake magnitude. A more precise and quantitative scale that assesses the *seismic moment* is now used, especially for larger quakes; this is the **moment magnitude scale**. The specific mechanics of how a fault breaks are under study, but the **elastic-rebound theory** describes the basic process. In general, two sides along a fault appear to be locked by friction, resisting any movement. This stress continues to build strain along the fault surfaces, storing elastic energy like a wound-up spring. When energy is released abruptly as the rock breaks, both sides of the fault return to a condition of less strain.

> seismograph (p. 369)
> Richter scale (p. 369)
> moment magnitude scale (p. 370)
> elastic-rebound theory (p. 370)

15. Describe the differences in human response between the two earthquakes that occurred in China in 1975 and 1976.

16. Differentiate between the Mercalli and moment magnitude and amplitude scales. How are these used to describe an earthquake? Why has the Richter scale been updated and modified?

17. What is the relationship between an epicenter and the focus of an earthquake? Give examples from the Loma Prieta, California, and Kobe, Japan, earthquakes.

18. What local conditions in Mexico City and San Francisco severely magnified the energy felt in their respective earthquakes?

19. How do the elastic-rebound theory and asperities help explain the nature of faulting?

20. Describe the San Andreas fault and its relationship to ancient sea-floor spreading movements along transform faults.

21. How is the seismic gap concept related to expected earthquake occurrences? Are any gaps correlated with earthquake events in the recent past? Explain.

22. What do you see as the biggest barrier to effective earthquake prediction?

---

✔ ***Distinguish*** between an effusive and an explosive volcanic eruption and *describe* related landforms, using specific examples.

Volcanoes offer direct evidence of the makeup of the asthenosphere and uppermost mantle. A **volcano** forms at the end of a central vent or pipe that rises from the asthenosphere through the crust into a volcanic mountain. A **crater**, or circular surface depression, usually forms at the summit. Areas where magma is near the surface may heat groundwater, producing **geothermal energy**.

Eruptions produce **lava** (molten rock), gases, and **pyroclastics** (pulverized rock and clastic materials ejected violently during an eruption) that pass through the vent to openings and fissures at the surface and build volcanic landforms, such as a **cinder cone**, a small hill, or a large basin-shaped depression called a **caldera**.

Volcanoes are of two general types, based on the chemistry and the viscosity of the magma involved. **Effusive eruption** produces a **shield volcano** (such as Kilauea in Hawaii) and extensive deposits of **plateau basalts**, or flood basalts. Magma of higher viscosity leads to an **explosive eruption** (such as Mount Pinatubo in the Philippines), producing a **composite volcano**. Volcanic activity has produced some destructive moments in history but constantly creates new sea floor, land, and soils.

> volcano (p. 378)
> crater (p. 378)
> geothermal energy (p. 378)
> lava (p. 378)
> pyroclastics (p. 378)
> cinder cone (p. 378)
> caldera (p. 378)
> effusive eruption (p. 379)
> shield volcano (p. 379)
> plateau basalts (p. 379)
> explosive eruption (p. 379)
> composite volcano (p. 379)

23. What is a volcano? In general terms, describe some related features.

24. Where do you expect to find volcanic activity in the world? Why?

25. Compare effusive and explosive eruptions. Why are they different? What distinct landforms are produced by each type? Give examples of each.

26. Describe several recent volcanic eruptions.

---

 **NetWork**

The *Geosystems Home Page* provides on-line resources for this chapter on the World Wide Web. You will find review exercises, specific updates for items in the chapter, suggested readings, and links to interesting related pathways on the Internet (click on the Table of Contents link and select this chapter). *Geosystems* is at: **http://www.prenhall.com/geosystm**

# 13

# Weathering, Karst Landscapes, and Mass Movement

**Landmass Denudation**

**Weathering Processes**

**Karst Topography and Landscapes**

**Mass Movement Processes**

**Summary and Review**

### Key Learning Concepts

After reading the chapter, you should be able to:

- *Define* the science of geomorphology.
- *Illustrate* the forces at work on materials residing on a slope.
- *Define* weathering and *explain* the importance of the parent rock and joints and fractures in rock.
- *Describe* frost action, crystallization, hydration, pressure-release jointing, and the role of freezing water as physical weathering processes.
- *Describe* the susceptibility of different minerals to the chemical weathering processes called hydrolysis, oxidation, carbonation, and solution.
- *Review* the processes and features associated with karst topography.
- *Portray* the various types of mass movements and *identify* examples of each in relation to moisture content and speed of movement.

*Intricately weathered limestone forms the Stone Forest, near Kunming, China.*
[Photo by ChromoSohm/Sohm/Photo Researchers, Inc.]

A benefit you receive from a physical geography course is a new appreciation of the scenery. Whether you go by foot, bicycle, car, train, or plane, travel is an opportunity to experience Earth's varied landscapes and to witness the active processes that produce them.

In the last two chapters, we discussed the endogenic (internal) processes of our planet and how they produce landforms. However, as the landscape is formed, several exogenic (external) processes simultaneously wear and waste it. Perhaps you have noticed rough and broken highways in areas that experience freezing weather. The paving breaks into chunks each winter and develops potholes. Or maybe you have seen older marble structures such as tombstones that rainwater has etched and dissolved. Physical (mechanical) and chemical weathering processes such as these reduce the landscape and release essential minerals from bedrock for soil formation.

Mass movement processes continually operate in and upon the landscape. A news broadcast may describe an avalanche in Colombia, mudflows in Los Angeles, a landslide in Turkey, or lava flows on the slopes of a volcano in Italy. Weathering and mass movements together provide the raw material for erosion.

Here we begin a five-chapter examination of exogenic processes at work on the landscape. This chapter examines weathering and mass movement of the lithosphere. The next four chapters look at specific exogenic agents and their handiwork—river systems, wind-influenced landscapes, coastal processes and landforms, and glaciated regions. Whether you enjoy time along a river, love the desert or coastline, or live in a place where glaciers once carved the land, you will find something of interest in these chapters.

# Landmass Denudation

**Geomorphology** is the science of landforms—their origin, evolution, form, and spatial distribution. It is an important aspect of physical geography. **Denudation** is any process that wears away or rearranges landforms. The principal denudation processes affecting surface materials include *weathering, mass movement, erosion, transportation,* and *deposition,* as produced by the agents of moving water, air, waves, and ice, all influenced by the pull of gravity.

Interactions between the structural elements of the land and denudation processes are complex. They represent a continuing struggle between Earth's internal and external processes, between the resistance of materials and weathering and erosional processes. The 15-story Delicate Arch in Utah is dramatic evidence of this struggle (Figure 13-1). Differing resistances of the rocks, coupled with variations in the processes at work on the rock are carving this delicate sculpture.

Ideally, endogenic processes build *initial landscapes,* whereas exogenic processes develop *sequential landscapes* of low relief, the slowed pace of little change and stability. Various hypotheses have been proposed to model denudation processes and to account for the appearance of the landscape in its developmental stages.

## *Geomorphic Models of Landform Development*

William Morris Davis (1850–1934), famed American geographer and geomorphologist, proposed the erosion cycle, or *geomorphic cycle model.* Davis theorized that a

**FIGURE 13-1**
**Weathered landform.**
Delicate Arch—a dramatic example of differential weathering in Arches National Park, Utah. Resistant rock strata at the top of the structure have helped preserve the arch beneath them as surrounding rock was eroded away. Note the person for scale. [Photo by author.]

landscape undergoes initial uplift that is accompanied by erosion or removal of materials. From the uplifted landscape, streams begin flowing more rapidly, cutting more energetically, both headward (upstream) and deeper. Davis believed that slope angle is gradually reduced over time and that ridges and divides become rounded and lowered over time. According to Davis's cyclic model, the landscape eventually evolves into an old erosional surface (discussed further in Chapter 14).

Davis's theory, which helped launch the science of geomorphology and was innovative at the time, did not account for the processes being observed in modern geomorphology studies. Academic support for his cyclic and evolutionary landform development model declined by 1960. Although not generally accepted today, his thinking about the evolution of landscapes is influential and some of his terminology is still used.

### Dynamic Equilibrium View of Landforms.
Most geomorphologists prefer the **dynamic equilibrium model**, which emphasizes a balance among force, form, and process. This balancing act between tectonic uplift and reduction by weathering and erosion, between the resistance of rocks and the ceaseless attack of weathering and erosion, is summarized in the dynamic equilibrium model.

A landscape is an open system, with highly variable inputs of energy and materials: Uplift creates the *potential energy of position* above sea level and therefore a disequilibrium, an imbalance, between relief and energy. The Sun provides radiant energy that is converted into *heat energy*. The hydrologic cycle imparts *kinetic energy* through mechanical motion. *Chemical energy* is made available from the atmosphere and various reactions within the crust. In response to this input of energy and materials, landforms constantly adjust toward *equilibrium*.

A dynamic equilibrium demonstrates a trend over time. According to current thinking, landscapes in a dynamic equilibrium show ongoing adaptations to the ever-changing conditions of rock structure, climate, local relief, and elevation. Endogenic events (such as earthquakes and volcanic eruptions), or exogenic events (such as heavy rainfall or forest fire), may provide new sets of relationships for the landscape.

Following such destabilizing events, a landform system arrives at a **geomorphic threshold**—the point at which there is enough energy to overcome resistance against movement. At this threshold, the system breaks through to a new equilibrium as the landform adjusts. The pattern over time follows a sequence: (1) equilibrium stability, (2) a destabilizing event, (3) a period of adjustment, and (4) development of a new and different condition of equilibrium stability.

The disturbed hillslope in Figure 13-2 is in the midst of compensating adjustment. Disequilibrium was created by the failure of saturated slopes, causing a landslide into the

**FIGURE 13-2**
**A slope in disequilibrium.**
Unstable, saturated soils gave way, leaving a debris dam partially blocking the river. [Photo by author.]

river. In turn, the new dam of material in the stream threw it into a disequilibrium between its flow and sediment load.

Slow, continuous-change events, such as soil development and erosion, tend to maintain an approximate equilibrium condition. Although less frequent, dramatic events such as a major landslide or dam collapse require longer recovery times before an equilibrium is reestablished. (Figure 1-5 graphically illustrates the distinction between a steady-state equilibrium and a dynamic equilibrium, and the occurrence of a geomorphic threshold).

### Slopes.
Material loosened by weathering is susceptible to erosion and transportation. However, if gravity is to move loosened material downslope, the agents of erosion must overcome the forces of friction, inertia (the tendency of objects at rest to remain at rest), and the cohesion of particles to each other (Figure 13-3a). If the slope angle is steep enough for gravity to overcome frictional forces, or if material is dislodged by the impact of raindrops, hail, falling branches, moving animals, wind, trail bikes, off-road vehicles, or construction equipment, then particle erosion and transport downslope occur.

**Slopes** or hillslopes are curved, inclined surfaces that form the boundaries of landforms. Figure 13-3b illustrates basic slope components that vary with conditions of rock structure and climate. Slopes generally feature an upper *waxing slope* near the top (*waxing* means "increasing"). This convex surface curves downward and grades into the *free face* below. The presence of a free face indicates an outcrop of resistant rock that forms a steep scarp or cliff.

Downslope from the free face is a *debris slope*, which receives rock fragments and materials from above. The condition of a debris slope reflects the local climate. In humid climates, continually moving water carries away material as it arrives, lowering the debris slope. But in arid climates, debris slopes persist and accumulate. A debris slope grades into a *waning slope*,

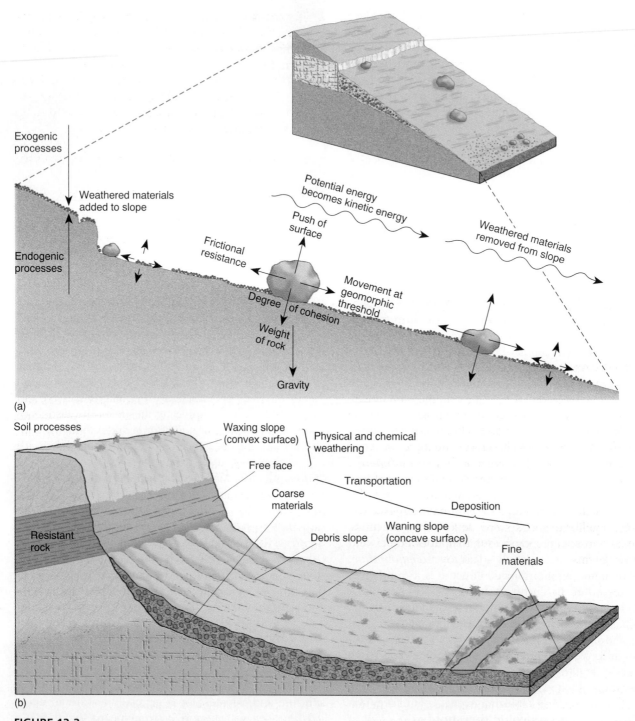

**FIGURE 13-3**
**Slope mechanics and form.**
(a) Directional forces (noted by arrows) act on materials along an inclined slope. (b) The principal elements of a slope.

a concave surface along the base of the slope. This surface of erosional materials gently slopes at a continuously decreasing angle to the valley floor.

A slope is an open system seeking an *angle of equilibrium.* Conflicting forces work simultaneously on slopes to establish an optimum compromise incline that balances these forces. You can identify these slope components

and conditions on the actual hillslope shown in Figure 13-4. When any condition in the balance is altered, all forces on the slope compensate by adjusting to a new dynamic equilibrium.

Slopes, then, are shaped by the relation between rates of weathering and breakup of slope materials, on the one hand, and the rates of mass movement and

**FIGURE 13-4**
**A Wyoming hillslope.**
Compare this slope with the components mentioned in Figure 13-3.
Note the role of the rock outcrop in interrupting the slope and the
rock fragments that are loosened from the outcrop by frost action.
[Photo by author.]

material erosion on the other. A slope is considered *stable* if its strength exceeds these denudation processes, and *unstable* if materials are weaker than these processes.

Why are hillslopes shaped in certain ways? How do slope elements evolve? How do hillslopes behave during rapid, moderate, or slow uplift? These are topics of active study and research. Modern geomorphology emphasizes individual processes and the role of material resistance in soil and rock in an attempt to answer these questions.

Now, with the concepts of landmass denudation, dynamic equilibrium, and slope development in mind, let's examine specific processes that operate to wear away landforms.

# Weathering Processes

Rocks at Earth's surface, and to some depth below it, are exposed to weathering processes. **Weathering** processes either disintegrate both surface and subsurface rock into mineral particles or dissolve them in water. Weathering processes are both physical (mechanical) and chemical.

Weathering does not transport the weathered materials; it simply generates them for erosion and transport by the agents of water, wind, waves, and ice—all influenced by gravity. In most areas, the upper surface of bedrock undergoes continual weathering, creating broken-up rock called **regolith.** Loose surface material comes from further weathering of regolith and from transported and deposited regolith (Figure 13-5a). In some areas, regolith may be missing or undeveloped, thus exposing an outcrop of unweathered bedrock.

**Bedrock** is the *parent rock* from which weathered regolith and soils develop. While a soil is relatively youthful, its parent rock is traceable through similarities in composition. For example, the sand in Figure 13-5c derives its color and character from the parent rock in the background. This unconsolidated fragmental material, known as **sediment**, combines with weathered rock to form the **parent material** from which soil evolves.

## Factors Influencing Weathering Processes

Weathering is greatly influenced by the character of the bedrock: hard or soft, soluble or insoluble, broken or unbroken. *Jointing* in rock is important for weathering processes. **Joints** are fractures or separations in rock that occur without displacement of the sides (as would be the case in faulting). The presence of these usually plane (flat) surfaces greatly increases the surface area of rock exposed to both physical and chemical weathering.

Important controls on weathering rates are climatic elements—precipitation, temperature, and freeze-thaw cycles. Also significant are the position of the water table and water movement (hydraulics).

Another control over weathering rates is the geographic orientation of a slope—whether it faces north, south, east, or west. Orientation controls the slope's exposure to Sun, wind, and precipitation. Slopes facing away from the Sun's rays tend to be cooler, moister, and more vegetated than slopes in direct sunlight. This effect of orientation is especially noticeable in the middle and higher latitudes.

Vegetation is also a factor in weathering. Although vegetative cover can protect rock by shielding it from raindrop impact and because roots stabilize soil, it also produces organic acids from the partial decay of organic matter; these acids contribute to chemical weathering. Plant roots can enter crevices and break up a rock, exerting enough pressure to drive rock segments apart, thereby exposing greater surface area to other weathering processes (Figure 13-6). You may have observed how tree roots can heave the sections of a sidewalk or driveway sufficiently to raise and crack the concrete.

***Climate and Weathering.*** There is a relation among climate (annual precipitation and temperature), physical weathering, and chemical weathering processes. Physical weathering dominates in drier, cooler climates, whereas chemical weathering dominates in wetter, warmer climates, although even in the driest climate some chemical weathering occurs.

You can see this general pattern in Figure 13-7. Extreme dryness reduces weathering rates, as is experienced in desert climates (*low-latitude arid desert BW*). In the hot, wet tropical and equatorial rain forest climates (*tropical rain forest Af*), most rocks weather rapidly, and the weathering extends deep below the surface.

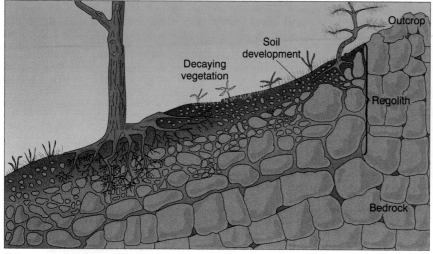

**FIGURE 13-5**

**Regolith, soil, and parent materials.**

(a) A cross section of a typical hillside.
(b) A cliff exposes these hillside components.
(c) These coral-pink dunes
in southern Utah derive their color from
the parent materials in the background.
[Photos by author.]

(a)

(b)

(c)

In climates where freezing is common, another form of weathering is the freeze-thaw action that changes the state of water in rocks, alternately expanding (freezing) it and contracting (thawing) it. The resulting change of volume (ice has about 9% greater volume than liquid water) creates a force great enough to mechanically split the rocks. Freezing actions are important in humid microthermal climates (*humid continental Df*) and in subarctic and polar regimes (*subarctic Dfc* and *Dw* climates and *polar E* climates).

***A Matter of Scale.*** The scale at which we analyze weathering processes is important. Recent research at the microscale has revealed greater complexity in the relation of climate and weathering than was previously known. At the small scale of actual reaction sites on the rock surface, both physical and chemical weathering processes can occur across varied climate types. The presence of hygroscopic water (a molecule-thin water layer on particles) and capillary water (soil water) activates chemical weathering processes, even in the driest landscape.

Imagine all the factors that influence weathering rates as operating in concert—combining actions to produce varied results: climatic influence (precipitation and temperature), soil water and groundwater, rock composition and structure (jointing), slope orientation, vegetation, and microscopic boundary-layer conditions at reaction sites. Of course in all this, *time* is the crucial factor, for these processes require long periods of time to operate.

(a)

(b)

**FIGURE 13-6**
**Organic physical weathering.**
(a) Can you find the tree root growing in and on the jointing fracture in the rock? (b) Roots exert a force on the sides of this joint in the rock. [Photos by author.]

In the complexity of nature, physical and chemical weathering usually operate together with varying emphasis, in a combination of processes. We separate them here for convenience of study.

## Physical Weathering Processes

When rock is broken and disintegrated without any chemical alteration, the process is called **physical weathering** or mechanical weathering. By breaking up rock, physical weathering greatly increases the surface area on which chemical weathering may operate. A single rock that becomes broken into eight pieces has doubled its surface area susceptible to weathering processes. We look briefly at four physical weathering processes: frost action, crystallization, hydration, and pressure-release jointing.

***Frost Action.*** When water freezes, its volume expands as much as 9% (see Chapter 7, Ice, the Solid Phase). This expansion creates a powerful mechanical force called **frost action**, or freeze-thaw action, which can exceed the tensional strength of rock. Repeated freezing and thawing of water break rocks apart. Freezing actions are important in the humid microthermal, subarctic, and polar

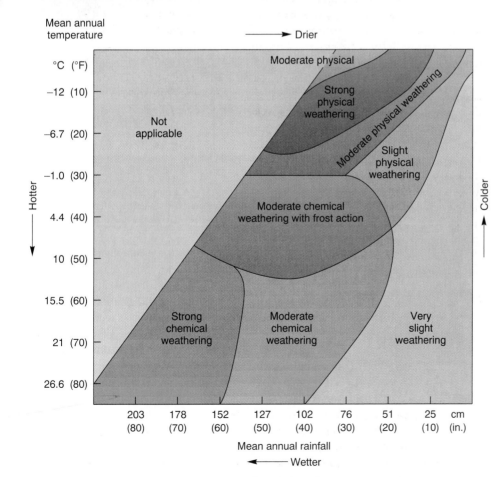

**FIGURE 13-7**
**Climate and weathering.**
Relation between temperature and rainfall and the various types of weathering. Note the opposite extremes in the diagram: Upper right is cold and dry; lower left is hot and wet. [(a) After L. C. Peltier, *Annals of the Association of American Geographers* 40 (1950): 214–36. Adapted by permission.]

climates, and they occur at higher elevations in mountains worldwide. In arctic and subarctic climates, frost action dominates soil conditions and is discussed in further detail in Chapter 17.

The work of ice begins in small openings, gradually expanding until rocks are cleaved (split). Figure 13-8 shows blocks of rock on which this *joint-block separation* occurs along existing joints and fractures. This weathering action, called *frost-wedging*, pushes portions of the rock apart. Cracking and breaking create varied shapes in the rocks, depending on the rock structure. In the photo, the softer supporting rock underneath the slabs is weathering at a faster pace—an example of **differential weathering**.

In Figure 13-8b, frost wedging has loosened rock from these cliffs. The angular pieces of rock fragments cascade down and form a **talus slope**, which is a poorly

(a)

(b)

**FIGURE 13-8**

**Physical weathering examples.**

(a) Physical weathering along joints in rock produces discrete blocks in the back country of Canyonlands National Park in Utah. (b) Frost action loosens and frees rock that falls to the base of the cliff, accumulating as a talus slope, near Wheeler Peak in Great Basin National Park, Nevada. [Photos by author.]

**FIGURE 13-9**

**Rockfall.**

Shattered rock debris from a large rockfall in Yosemite National Park. Another rockfall involving 162,000 tons of granite shocked Yosemite in July, 1996. [Photo by author.]

sorted, cone-shaped deposit of debris at the base of a steep slope. Several talus cones constitute such a slope in the photograph.

Various cultures have used the principle of frost action to quarry rock. Pioneers in the early American West drilled holes in rock, poured water in the holes, and then plugged them. During the cold winter months, expanding ice broke off large blocks along lines determined by the hole patterns. In spring, the pioneers hauled the blocks to town for construction. Frost action also produces unwanted fractures that damage pavement and split water pipes.

Spring can be a risky time to venture into mountainous terrain. As rising temperatures melt the winter's ice, newly fractured rock pieces fall without warning and may even start rock slides. The falling rock pieces may physically shatter on impact—another form of physical weathering (Figure 13-9).

***Crystallization.*** Especially in arid climates, dry weather draws moisture to the surface of rocks. As the water evaporates, dissolved minerals in the water grow crystals. Over time, as the crystals grow and enlarge, they exert a force great enough to spread apart individual mineral grains and begin breaking up the rock. Such *crystallization,* or *salt-crystal growth,* is a form of physical weathering.

In the Colorado Plateau of the Southwest, salty water slowly flows from rock strata. As this salty water evaporates, crystallization loosens the sand grains. Subsequent erosion and transportation by water and wind complete the sculpturing process. The result is that deep indentations develop in sandstone cliffs, especially above impervious layers. Over 1000 years ago, Native Americans built entire villages in these weathered niches at several locations, including Mesa Verde in Colorado and Arizona's Canyon de Chelly (pronounced "canyon duh shay," Figure 13-10).

***Hydration.*** Another process involving water, but little chemical change, is **hydration** (meaning "combination with water"). We list this as a physical weathering process because, when water is absorbed by some minerals, they expand, creating a strong mechanical effect that stresses the rock, forcing grains apart. A cycle of hydration and dehydration can lead to granular disintegration and further susceptibility of the rock to chemical weathering. Hydration works together with carbonation and oxidation to convert feldspar, a common mineral in many rocks, to clay minerals and silica. The hydration process is also at work on the sandstone niches shown in the cliff-dwelling photo.

***Pressure-Release Jointing.*** Recall from Chapter 11 that rising magma that is deeply buried and subjected to high pressure forms intrusive igneous rocks called plutons. They cool slowly and produce coarse-grained, crystalline granitic rocks (a pluton is illustrated in Figure 11-8). As the landscape is subjected to uplift, the regolith overburden is weathered, eroded, and transported away, eventually exposing the pluton as a mountainous batholith.

As the tremendous weight of overburden is removed from the granite, the pressure of deep burial is relieved. Over millions of years, the granite slowly responds with

(a)

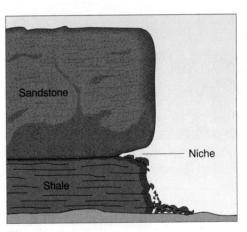

(b)

**FIGURE 13-10**
**Physical weathering in sandstone.**
(a) Cliff dwelling site in Canyon de Chelly, Arizona, occupied by the Anasazi people until about 900 years ago. The niche in the rock was partially formed by crystallization, which forced apart mineral grains and broke up the rock. The dark streaks on the rock are thin coatings of desert varnish, composed of iron oxides with traces of manganese and silica. (b) Water and an impervious sandstone layer helped concentrate weathering processes in the niche. [Photo by author.]

(a)

(b)

**FIGURE 13-11**

**Exfoliation in granite.**

(a) Exfoliated plates of rock are visible in layers, characteristic of
dome formations in granites. The loosened slabs of rock are
susceptible to further weathering and downslope movement. View is
from the east side of Half Dome in Yosemite National Park, California.
(b) Half Dome perspective from the west: Relief is approximately
1500 m (5000 ft) from the top of the dome to the glaciated valley
below. [Photos by author.]

an enormous heave. In a process known as *pressure-
release jointing,* layer after layer of rock peels off in
curved slabs or plates, thinner at the top of the rock structure and thicker at the sides. As these slabs weather, they
slip off in a process called **sheeting**. This *exfoliation
process* creates arch-shaped and dome-shaped features
on the exposed landscape, forming an **exfoliation dome**
(Figure 13-11). Such domes are probably the largest
weathering features on Earth (in areal extent).

## Chemical Weathering Processes

**Chemical weathering** is the actual decomposition
(chemical change) of minerals in rock. Chemical weathering involves reactions between air and water and minerals in rocks. Minerals may combine with oxygen or
carbon dioxide from the atmosphere, or they may dissolve and combine with water in chemical reactions.
Water is important because of its general availability and
its tremendous ability to dissolve. As shown in Figure 13-
7, chemical breakdown is hastened as both temperature
and precipitation increase.

Although individual minerals vary in susceptibility,
all rock-forming minerals are at least slightly responsive
to chemical weathering. A familiar example is the eating
away of cathedral facades and the etching of tombstones
by acidic precipitation. Figure 13-12a illustrates chemical
weathering along joints in granite as chemical processes
dissolve its component minerals. This dissolution in turn
opens spaces for frost action.

As an example of the way chemical weathering
attacks rock, consider **spheroidal weathering**. The
sharp edges and corners of rocks are rounded as the
alteration of minerals progresses through the rock. Joints
in the rock offer more surfaces of opportunity for more
weathering. Water penetrates joints and fractures and
dissolves the rock's weaker minerals or cementing materials. The resulting rounded edges are the basis for the
name *spheroidal.* Spheroidal weathering of rock resembles exfoliation, but it does not result from pressure-
release jointing.

A boulder can be attacked from all sides, shedding
spherical shells of decayed rock like the layers of an
onion. As the rock breaks down, more and more surface
area is exposed for further weathering. This type of
weathering is visible in California's photogenic Alabama
Hills (see News Report 1); note the spheroidal rock formations in Figure 13-12b.

We now look at three chemical weathering processes:
hydrolysis, oxidation, and carbonation and solution.

***Hydrolysis.*** When minerals chemically combine with
water, the process is called **hydrolysis**. Hydrolysis is a
decomposition process that breaks down silicate minerals in rocks. Compared with hydration, a physical process

(a)

**FIGURE 13-12**

**Chemical weathering and spheroidal weathering.**

(a) Chemical weathering processes act on the joints in granite to dissolve weaker minerals, leading to a rounding of the edges of the cracks. The two boulders are erratics left by retreating ice; they too are being attacked by these processes. (b) Rounded granite joint-block outcrops show evidence of spheroidal weathering in the Alabama Hills; the Inyo Mountains are in the distance near the Nevada-California border. [Photos by author.]

(b)

---

 **News Report 1**

## Alabama Hills—A Popular Movie Location

The spheroidal weathering and rock formations in the Alabama Hills have appeared in more movies and commercials than any star. The rugged landscape, just east of Mount Whitney and west of Death Valley, has provided scenic backdrops for films as diverse as *How the West Was Won*, *Gunga Din*, *Bad Day at Black Rock*, and *Tremors*.

Throughout the Alabama Hills, rounded granite joint blocks of chemically and physically weathered granite outcrops appear in a variety of rock forms. In Figure 1 you can imagine a chase scene raising clouds of dust! The colorful, weathered rocks have been used in many television commercials and as alien landscapes in *Star Trek* and other science fiction films and videos. If you ever visit the Alabama Hills, they will seem strangely familiar—you have seen this place before!

**FIGURE 1**

**Alabama Hills, a natural movie set.**

[Photo by author.]

---

in which water is simply attached to minerals in the rock with no chemical reaction, the hydrolysis process involves active participation of water in chemical reactions to produce different compounds and minerals. For example, the weathering of feldspar minerals in granite can be caused by a reaction to the normal mild acids dissolved in precipitation:

feldspar (K, Al, Si, O) + carbonic acid and water → residual clays + dissolved minerals + silica

So the by-products of chemical weathering of feldspar in granite include clay (such as kaolinite) and silica. As clay is formed from some minerals in the granite, quartz ($SiO_2$) particles are left behind. The resistant quartz may wash downstream, eventually becoming sand on some distant beach. Clay minerals become a major component in soil and shale, a common sedimentary rock.

In Table 11-2, page 332, the second line shows resistance to chemical weathering in igneous rocks. On the ultramafic end of the table (right), the low-silica minerals olivine and peridotite are most susceptible to chemical weathering. Stability gradually increases toward the high-silica minerals such as feldspar. On the far left side of the table, quartz is very resistant to chemical weathering. You can see that, because of the nature of the constituent minerals, basalt weathers faster chemically than does granite.

When weaker minerals in rock are changed by hydrolysis, the interlocking crystal network breaks down, so the rock fails, and *granular disintegration* takes place. Such disintegration in granite may make the rock appear etched, corroded, and softened.

*Oxidation.* An example of chemical weathering occurs when oxygen oxidizes (combines with) certain metallic elements to form oxides. This is a chemical weathering process known as **oxidation**. Perhaps the most familiar oxidation form is the "rusting" of iron in rocks or soil that produces a reddish brown stain of iron oxide ($Fe_2O_3$). We have all left a tool or nails outside only to find them weeks later, coated with iron oxide. The rusty color is visible on the surfaces of rock and in heavily oxidized soils such as those in the southeastern United States or the tropics. Here is a simple oxidation reaction in iron:

iron (Fe) + oxygen ($O_2$) → iron oxide (hematite; Fe + O)

As iron is removed from the minerals in a rock, the disruption of the crystal structures in the rock's minerals makes the rock more susceptible to further chemical weathering.

*Carbonation and Solution.* The third form of chemical weathering occurs when a mineral dissolves into **solution**—for example, when sodium chloride (common table salt) dissolves in water. Water is called the universal solvent because it is capable of dissolving at least 57 of the natural elements and many of their compounds. Water vapor readily dissolves carbon dioxide, thereby yielding precipitation containing carbonic acid ($H_2CO_3$). This acid is strong enough to react with many minerals, especially limestone, in a process called **carbonation**.

Carbonation simply means reactions whereby *carbon* combines with minerals. Such chemical weathering transforms minerals that contain calcium, magnesium, potassium, and sodium.

When rainwater attacks formations of limestone (which is calcium carbonate, $CaCO_3$), the constituent minerals are dissolved and wash away with the mildly acidic rainwater:

calcium carbonate + carbonic acid and water →

calcium bicarbonate ($Ca^{2+} + CO_2 + H_2O$)

Walk through an old cemetery and you can observe the carbonation of marble, a metamorphic form of limestone. Weathered limestone and marble, in tombstones or in rock formations, appear pitted and weathered wherever adequate water is available for carbonation. In this era of human-induced increases of acid precipitation, carbonation processes are greatly enhanced (see Focus Study 3-2, Acid Deposition: A Blight on the Landscape).

Entire landscapes composed of limestone are dominated by the chemical weathering process of carbonation. These are the regions of karst topography, which we examine next.

# Karst Topography and Landscapes

Limestone is so abundant on Earth that many landscapes are composed of it (Figure 13-13). These areas are quite susceptible to chemical weathering. Such weathering creates a specific landscape of pitted, bumpy surface topography, poor surface drainage, and well-developed solution channels underground. Remarkable labyrinths of underworld caverns also may develop owing to weathering and erosion caused by groundwater.

These are the hallmarks of **karst topography**, named for the Krš Plateau in Yugoslavia, where these processes were first studied. Approximately 15% of Earth's land area has some karst features, with outstanding examples in southern China, Japan, Puerto Rico, Cuba, the Yucatán of Mexico, Kentucky, Indiana, New Mexico, and Florida.

## Formation of Karst

For a limestone landscape to develop into karst topography, there are several necessary conditions:

- The limestone formation must contain 80% or more calcium carbonate for solution processes to proceed effectively.

- Complex patterns of joints in the otherwise impermeable limestone are needed for water to form routes to subsurface drainage channels.

- There must be an aerated zone between the ground surface and the water table.

- Vegetation cover can determine the amount of organically derived acids available to enhance the solution process.

The role of climate in providing optimum conditions for karst processes remains in debate, although the amount and distribution of rainfall appears important. Karst occurs in arid regions, but it is primarily due to former

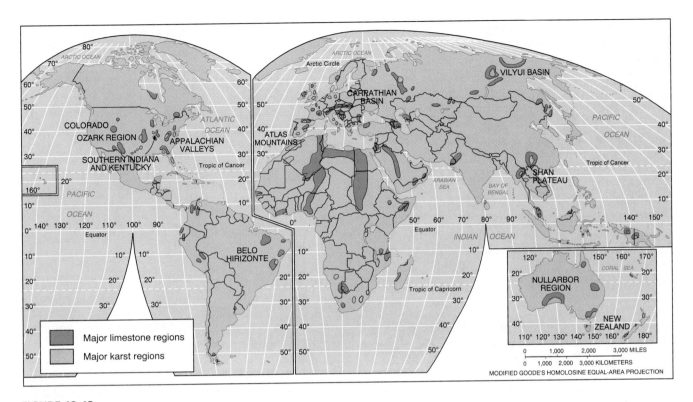

**FIGURE 13-13**
**Karst landscapes.**
Distribution of karst landscapes worldwide. [After R. E. Snead, *Atlas of the World Physical Features*, p. 76.
© 1972 by John Wiley & Sons. Adapted by permission.)

climatic conditions of greater humidity. Karst is rare in the Arctic and Antarctic because the water, although present, is generally frozen.

As with all weathering processes, time is a factor. Earlier in this century, karst landscapes were thought to progress through evolutionary stages of development, as if they were aging. Today, these landscapes are thought to be locally unique, a result of specific conditions, and there is little evidence that different regions evolve sequentially along similar lines. Nonetheless, mature karst landscapes do display certain characteristic forms.

### *Lands Covered with Sinkholes*

The weathering of limestone landscapes creates many **sinkholes**, which form in circular depressions. (Traditional studies may call a sinkhole a *doline.*) If a solution sinkhole collapses through the roof of an underground cavern, a *collapse sinkhole* is formed. A gently rolling limestone plain might be pockmarked by sinkholes with depths of 2 to 100 m (7 to 330 ft) and diameters of 10 to 1000 m (33 to 3300 ft), as shown in Figure 13-14a. Through continuing solution and collapse, sinkholes may coalesce to form a *karst valley*, which is an elongated depression up to several kilometers long.

The area southwest of Orleans, Indiana, has over 1000 sinkholes in just 2.6 km² (1 mi²). In this area, the Lost River, a "disappearing stream," flows more than 12.9 km (8 mi) underground before it resurfaces; its flow is diverted from the surface through sinkholes and solution channels. Its dry bed can be seen on the lower left of the topographic map in Figure 13-14b.

In Florida, several sinkholes have made news because lowered water tables (lowered by pumping) caused their collapse into underground solution caves, taking with them homes, businesses, and even new cars from an auto dealership. One such sinkhole collapsed in a suburban area in 1993 (Figure 13-15).

A complex landscape in which sinkholes intersect is called a *cockpit karst.* The sinkholes can be symmetrically shaped in certain circumstances; one at Arecibo, Puerto Rico, is perfectly shaped for a radio telescope installation (Figure 13-16).

Another type of karst topography forms in the wet tropics, where deeply jointed, thick limestone beds are weathered into gorges, leaving isolated resistant blocks standing. These resistant cones and towers are most remarkable in several areas of China, where an otherwise lower-level plain is interrupted by tower karst up to 200 m (660 ft) high (Figure 13-17).

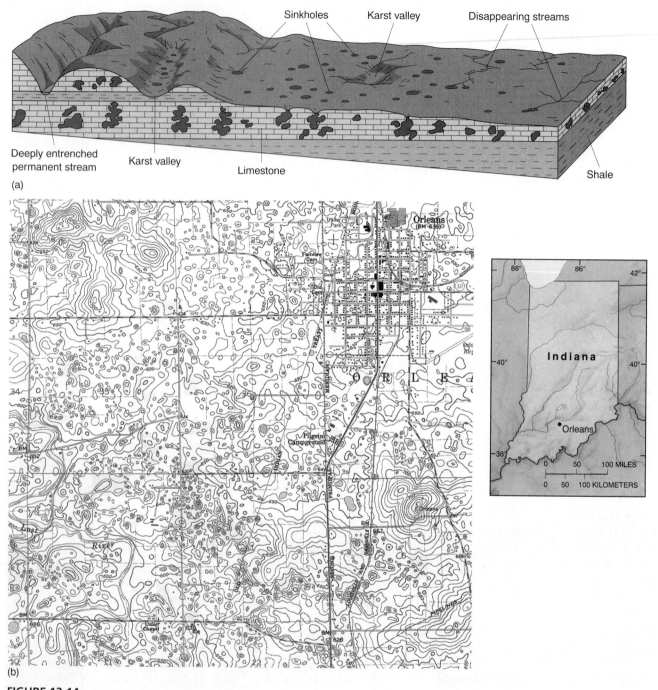

**FIGURE 13-14**

**Features of karst topography.**

(a) Idealized features of karst topography in southern Indiana. (b) Karst topography southwest of the town of Orleans. On average, 1022 sinkholes occur per 1 mi² in this area. On the map, note the contour lines: Depressions are indicated with small *hachures* (tick marks) on the downslope side of contour lines. [(a) adapted from W. D. Thornbury, *Principles of Geomorphology*, illustration by W. J. Wayne, p. 326; © 1954 by John Wiley & Sons. (b) Mitchell, Indiana quadrangle, USGS.]

## *Caves and Caverns*

Most caves are formed in limestone rock, because it is so easily dissolved by carbonation. The largest limestone caverns in the United States are Mammoth Cave in Kentucky, Carlsbad Caverns in New Mexico, and Lehman Cave in Nevada. Carlsbad Caverns are in limestone formations that were deposited 200 million years ago, when the region was covered by shallow seas. Regional uplifts associated with building of the Rockies (the Laramide orogeny, 40–80 million years ago) elevated the region above sea level, leading to active cave formation (Figure 13-18).

(a)

**FIGURE 13-15**
**Sinkholes.**
(a) Florida sinkhole, formed in 1981 in Winter Park, a suburb of Orlando. (b) Karst area 25 km (15.5 mi) north of Winter Park depicted on a topographic map. [(a) Photo by Jim Tuten/Black Star; (b) Orange City quadrangle, USGS.]

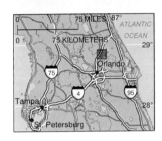

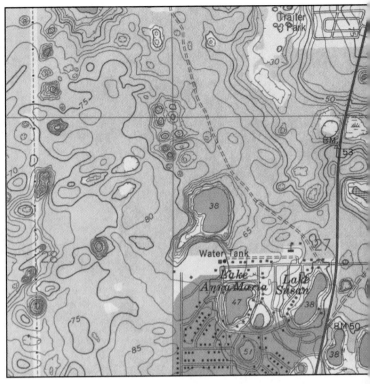

(b)

Caves generally form just beneath the water table, where later lowering of the water level exposes them to further development. *Dripstones* form as water containing dissolved minerals slowly drips from the cave ceiling. Calcium carbonate precipitates out of the evaporating solution, literally one molecular layer at a time, and accumulates at a point below on the cave floor. Depositional features called *stalactites* grow from the ceiling and *stalagmites* build from the floor; sometimes the two grow until they connect and form a continuous *column*. A dramatic subterranean world is thus created. Amateur cavers have made important discoveries about these unique habitats (see News Report 2).

# Mass Movement Processes

On November 13, 1985, at 11 P.M., after a year of earthquakes, 10 months of harmonic tremors, a growing bulge on its northeast flank, and 2 months of small summit eruptions, Nevado del Ruiz in central Colombia, South America, violently erupted in a lateral explosion. The volcano, northernmost of two dozen dormant peaks in the Cordilleran Central of Colombia, had erupted six times during the past 3000 years, killing 1000 people during its last eruption in 1845.

**FIGURE 13-16**
**Deep-space research from a sinkhole.**
Cockpit karst topography near Arecibo, Puerto Rico, provides a natural depression for the dish antenna of a giant radio telescope. The Arecibo Observatory is part of the National Astronomy and Ionosphere Center, which is operated by Cornell University under contract with the National Science Foundation. [Photo courtesy of Cornell University.]

**FIGURE 13-17**
**Tower karst of the Guangxi (Kwangsi) Province, Yangzhou, China.**
Resistant strata protect each tower as weathering removes the surrounding limestone. [Photo by Wolfgang Kaehler/Wolfgang Kaehler Photography.]

But on this night, the familiar pyroclastics, lava, and the blast itself were not the problem. The hot eruption quickly melted about 10% of the ice on the mountain's snowy peak, liquefying mud and volcanic ash, sending a hot mudflow downslope. Such a flow is called a *lahar*, an Indonesian word referring to mudflows of volcanic origin. This lahar moved rapidly down the Lagunilla River toward the villages below. The wall of mud was at least 40 m (130 ft) high as it approached Armero, a regional center with a population of 25,000. The city slept as the lahar buried its homes: 23,000 people were killed there and in other afflicted river valleys; thousands were injured; 60,000 were left homeless. The debris flow from the volcano is now a permanent grave for its victims.

Although less devastating to human life, the mudflow generated by the eruption of Mount St. Helens also was a lahar. Not all mass movements are this destructive, but such processes are important in the denudation of the landscape. And not all mass movements are surface processes; News Report 3 discusses submarine landslides.

## Mass Movement Mechanics

Physical and chemical weathering processes create an overall weakening of surface rock, which makes it more susceptible to the pull of gravity. The term **mass movement** applies to any movement of a body of material,

## News Report 2

### Amateurs Make Cave Discoveries

The exploration and scientific study of caves is called *speleology*. Although professional investigations are carried out by physical and biological scientists, amateur cavers, or "spelunkers," have made many important discoveries.

Cave habitats are unique. They are nearly closed, self-contained ecosystems with simple food chains and great stability. In total darkness, bacteria synthesize inorganic elements and produce organic compounds that sustain many types of cave life, including algae, small invertebrates, amphibians, and fish.

In a cave discovered in 1986, near Movile in southeastern Romania, cave-adapted invertebrates were discovered after millions of years of sunless isolation. Thirty-one of these organisms were previously unknown. Without sunlight, the ecosystem in Movile is based on sulfur-metabolizing bacteria that synthesize organic matter using energy from oxidation processes. These *chemosynthetic* bacteria feed other bacteria and fungi that in turn support cave animals. The sulfur bacteria produce sulfuric acid compounds that may

prove to be important in the chemical weathering of some caves.

The mystery, intrigue, and excitement of cave exploration lie in the variety of dark passageways, enormous chambers that narrow to tiny crawl spaces, strange formations, and underwater worlds that can be accessed only by cave-diving. Many of the major caves were first discovered by private property owners and amateur adventurers, a fact that keeps this popular science/sport very much alive.

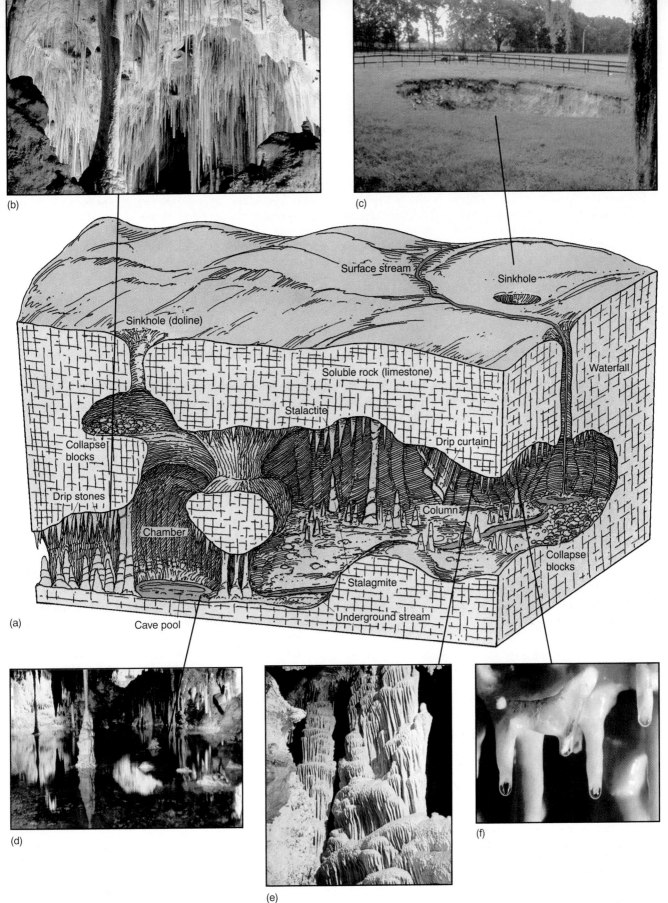

**FIGURE 13-18**

**Cavern features.**

(a) An underground cavern and related forms in limestone. (b) Carlsbad Caverns in New Mexico, where a series of underground caverns includes rooms over 1200 m (4000 ft) long and 190 m (625 ft) wide. Although a national park since 1930, unexplored portions remain. (c) A sinkhole in a Florida pasture. (d) A pool of water in Lehman Caves, Great Basin National Park, Nevada. (e) Dripstone formations in Lehman Caves. (f) Precipitate forms deposits literally one molecular layer at a time. [(b) Photo by Scott T. Smith; (c) photo by Thomas M. Scott, Florida Geological Survey; (d), (e), and (f) courtesy of Great Basin National Park.]

## News Report 3

### Landslides Beneath the Sea

In contrast to surface processes, submarine mass movement is unorganized. Turbulent flows of mud accumulate as deep-sea depositional fans along the continental slope or around islands. The latest discoveries of *submarine landslides* identify them as some of the largest movements of material on Earth.

Especially significant are the 17 giant submarine landslides around the Hawaiian Islands. As an example, the prehistoric Nuuanu slide was formed from a portion of Oahu that collapsed and flowed out across the ocean floor for 235 km (145 mi). Large waves called tsunami were generated by such slides

and must have been devastating to near and distant shores. For instance, there is evidence that a wave with a height of 365 m (1200 ft) hit the island of Lanai after a portion of Mauna Loa collapsed some 300,000 years ago. (More about tsunami in Chapter 16.)

---

propelled and controlled by gravity, such as the lahar just described. Mass movements can range from dry to wet, slow to fast, small to large, and from free-falling to gradual or intermittent (see Figure 13-20). The term *mass movement* is sometimes used interchangeably with **mass wasting**, which is the process involved in mass movements and erosion of the landscape. To combine the concepts, we can say that the mass movement of material works to waste slopes and provide raw material for erosion, transportation, and deposition.

***The Role of Slopes.*** All mass movements occur on slopes under the influence of gravitational stress. If we pile dry sand on a beach, the grains will flow downslope until an equilibrium is achieved. The steepness of the resulting slope depends on the size and texture of the grains; this steepness is called the **angle of repose**. This angle represents a balance of the driving force (gravity) and resisting force (friction and shearing). The angle of repose for various materials commonly ranges between 33° and 37° (from horizontal).

*The driving force* in mass movement is gravity. It works in conjunction with the weight, size, and shape of the surface material; the degree to which the slope is oversteepened (how far it exceeds the angle of repose); and the amount and form of moisture available (frozen or fluid). The greater the slope angle, the more susceptible the surface material is to mass wasting processes.

*The resisting force* is the shearing strength of slope material, that is, its cohesiveness and internal friction, which work against gravity and mass wasting. To reduce shearing strength is to increase shearing stress, which eventually reaches the point at which gravity overcomes friction.

Clays, shales, and mudstones are highly susceptible to hydration (physical swelling in response to the presence of water). If rock strata in a slope are underlain by such substances, the strata will tend toward movement because less driving force energy is required to move material over these unstable surfaces. When clay surfaces are

wet, they deform slowly in the direction of movement, and when saturated they form a viscous fluid with little shearing strength (resistance to movement) to hold back the slope. However, if the rock strata are such that material is held back from slipping, then more driving force energy may be required, such as that generated by an earthquake.

***Madison River Canyon Landslide.*** In the Madison River Canyon near West Yellowstone, Montana, a blockade of dolomite (a magnesium-rich carbonate rock) had held back a deeply weathered and *oversteepened slope* (40° to 60° slope angle) for untold centuries (white area in Figure 13-19). Then, shortly after midnight on August 17, 1959, a 7.5 magnitude earthquake broke the dolomite structure along the foot of the slope. The break released 32 million cubic meters (1.13 billion cubic feet) of mountainside, which moved downslope at 95 kmph (60 mph), causing gale force winds through the canyon. Momentum carried the material more than 120 m (about 400 ft) up the opposite canyon slope, entombing several hundred campers beneath about 80 m (260 ft) of rock and debris.

The mass of material also effectively dammed the Madison River and thus created a new lake, dubbed Quake Lake. The landslide debris dam established a new temporary equilibrium for the river and canyon. A channel was quickly excavated by the U.S. Army Corps of Engineers to prevent a disaster below the dam, for if Quake Lake overflowed the landslide dam, the water would quickly erode a channel and thereby release the entire contents of the new lake onto farmland downstream. These events convey a dramatic example of the role of slopes and tectonic forces in creating massive land movements.

## Classes of Mass Movements

In any mass movement, gravity pulls on a mass until the critical shear-failure point is reached—a geomorphic threshold. The material then can fall, slide, flow, or creep—the four classes of mass movement. These classes

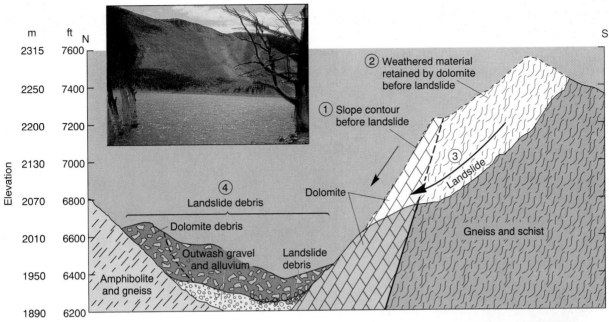

**FIGURE 13-19**

**Madison River Landslide.**

Cross section showing geologic structure of the Madison River Canyon, in Montana, where a landslide was triggered by an earthquake in 1959. (1) Prequake slope contour. (2) Weathered dolomite that failed. (3) Direction of landslide. (4) Landslide debris blocking the canyon and damming the Madison River. [After J. B. Hadley, *Landslides and Related Phenomena Accompanying the Hebgen Lake Earthquake of 17 August 1959,* U.S. Geological Survey Professional Paper 435-K, p. 115. Inset photo by Randall Christopherson.]

are summarized in Figure 13-20. Note the temperature and moisture gradients in the margins of the illustration that show the relation between water content and movement rate. The Madison River Canyon event was a type of *slide,* whereas the Nevado del Ruiz lahar mentioned earlier was a *flow.* We now look at specific mass movement classes.

***Falls and Avalanches.*** This class of mass movement includes rockfalls and debris avalanches. A **rockfall** is simply a volume of rock that falls through the air and hits a surface. During a rockfall, individual pieces fall independently and characteristically form a cone-shaped pile of irregular broken rocks called a *talus slope* at the base of a steep incline (Figure 13-21).

A **debris avalanche** is a mass of falling and tumbling rock, debris, and soil. It is differentiated from a slower debris slide or landslide by the tremendous velocity of onrushing material. This speed often results from ice and water that fluidize the debris. The extreme danger of a debris avalanche results from its tremendous speed and consequent lack of warning (see News Report 4).

In 1962 and again in 1970, debris avalanches roared down the west face of Nevado Huascarán, the highest peak in the Peruvian Andes. The 1962 debris avalanche contained an estimated 13 million cubic meters (460 million cubic feet) of material, burying the city of Ranrahirca and eight other towns, killing 4000 people. The 1970 event was initiated by an earthquake. Upward of 100 million cubic meters (3.53 billion cubic feet) of debris buried the city of Yungay, where 18,000 people perished (Figure 13-22). This avalanche attained velocities of 300 kmph (185 mph), which is especially incredible when you consider the quantity of material involved and the fact that some boulders weighed thousands of tons. The avalanche covered a vertical drop of 4144 m (13,600 ft) and a horizontal distance of 16 km (10 mi) in just a few minutes.

Another dramatic example of a debris avalanche, but one that cost no lives, occurred west of the Saint Elias Range, north of Yakutat Bay in Alaska. Several dozen rockfalls, large debris avalanches, and snow avalanches were triggered by a 7.1 magnitude earthquake in 1979. Approximately 4.7 km² (1.8 mi²) of the Cascade Glacier was covered by the largest single avalanche caused by the earthquake (Figure 13-23). The photo reveals characteristic grooves, lobes, and large rocks associated with these fluid avalanches.

***Landslides.*** A sudden rapid movement of a cohesive mass of regolith or bedrock that is not saturated with moisture is a **landslide**—a large amount of material failing

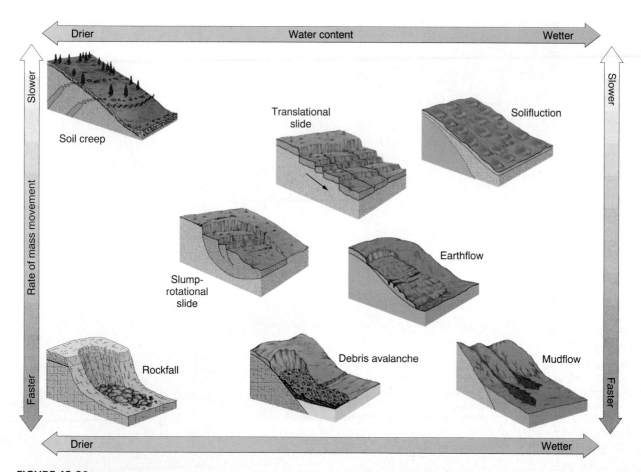

**FIGURE 13-20**

**Mass movement classes.**

Principal types of mass movement and mass wasting events. Variations in water content and rates of movement produce a variety of forms.

simultaneously. Surprise creates the danger, for the downward pull of gravity wins the struggle for equilibrium in an instant. One such surprise event struck near Longarone, Italy, in 1963 and is described in Focus Study 13-1.

To eliminate the surprise element, scientists are using the global positioning system (GPS) to monitor landslide movement. With GPS, scientists measure slight land shifts in suspect areas for clues to possible mass wasting. GPS was applied in two cases in Japan and effectively identified prelandslide movements of 2 to 5 cm per year, providing information to expand the area of hazard concern.

Slides occur in one of two basic forms: translational or rotational (see Figure 13-20 for an idealized view of each). *Translational slides* involve movement along a planar (flat) surface roughly parallel to the angle of the slope, with no rotation. The Madison Canyon landslide described earlier was a translational slide. Flow and creep patterns also are considered translational in nature.

*Rotational slides* occur when surface material moves along a concave surface. Frequently, underlying clay presents an impervious surface to percolating water. As a result, water flows along the clay surface, undermining the overlying block and lubricating the contact, thereby reducing the friction force. The simplest form of rotational slide is a *rotational slump*, in which a small block of land shifts downward. The upper surface of the slide appears to rotate backward and often remains intact. The surface may rotate as a single unit, or it may present a stepped appearance.

On Good Friday in 1964, the Pacific plate plunged a bit further beneath the North American plate near Anchorage, Alaska. More than 80 mass wasting events were triggered by the resulting earthquake. One particular housing subdivision in Anchorage experienced a sequence of translational movements accompanied by rotational slumping. Figure 13-24 shows one view of the segmented ground failure.

***Flows.*** Flows include *earthflows* and more fluid **mudflows**. When the moisture content of moving material is high, the suffix -*flow* is used (see Figure 13-20). Heavy rains can saturate barren mountain slopes and set them

**FIGURE 13-21**
**Talus slope.**
Rockfall and talus deposits at the base of a steep slope.
[Photo by author.]

**FIGURE 13-22**
**Debris avalanche, Peru.**
A 1970 debris avalanche falls more than 4100 m (2.5 mi) down the
west face of Nevado Huascarán, burying the city of Yungay, Peru.
The same area was devastated by a similar avalanche in 1962 and by
others in pre-Columbian times. A great danger remains for the cities
and towns in the valley from possible future mass movements.
[Photo by George Plafker.]

## News Report 4

### Debris Avalanche Destroys Colombian Town

Debris avalanche dangers are related to the tremendous velocity of the rock and material moving downslope. Toez, Colombia, was buried by an avalanche of ice, rock, and mud on an afternoon in June 1994. The avalanche pressed air aside as it rushed downhill, producing hurricane-force winds.

The triggering mechanism of this killer event was a magnitude 6.4 earthquake. The material traveled several kilometers down the slopes of Nevado del Huila, a 5785 m (18,975 ft) volcano, the highest in South America. Hundreds of unsuspecting residents perished, and many more were missing in yet another Andean debris avalanche.

## Focus Study 13-1

### Vaiont Reservoir Landslide Disaster

**Place:** Northeastern Italy, Vaiont Canyon in the Italian Alps, rugged scenery, 680 m (2200 ft) above sea level.

**Location:** 46.3°N 12.3°E, near the border of the Veneto and Friuli-Venezia Giulia regions.

**Situation:** Centrally located in the region, an ideal situation for hydroelectric power production.

**Site:** Steep-sided, narrow glaciated canyon, opening to populated lowlands to the west; ideal site for a narrow-crested, high dam, and deep reservoir.

**The Plan:** Build the second highest dam in the world, using a new thin-arch design, 262 m (860 ft) high, 190 m (623 ft) crest length, impounding a reservoir capacity of 150 million cubic meters (5.3 billion cubic feet) of water (third largest reservoir in the world), and generate hydroelectric power for distribution.

**Geologic Analysis:**

1. Steep canyon walls composed of interbedded limestone and shale; badly cracked and deformed structures; open fractures in the shale are inclined toward the future reservoir body. The steepness of the canyon walls enhance the strong driving forces (gravitational) at work on rock structures.

2. High potential for bank storage (water absorption by canyon walls) into the groundwater system that will increase water pressure on all rocks in contact with the reservoir.

3. The nature of the shale beds is such that cohesion will be reduced as its clay minerals become saturated.

4. Evidence of ancient rockfalls and landslides on the north side of the canyon.

5. Evidence of creep activity along south side of the canyon.

**Political/Engineering Decision:** Begin design and construction of a thin-arch dam at this site.

**Events during Construction and Filling:** Large volumes of concrete were injected into the bedrock as "dental work" in an attempt to strengthen the fractured rock. During reservoir filling in 1960, 700,000 cubic meters (2.5 million cubic feet) of rock and soil slid into the reservoir from the south side. A slow creep of slope materials along the entire south side began shortly thereafter, increasing to about 1 cm per week by January 1963. Creep rate increased as the level of the reservoir rose. By mid-September 1963, the creep rate exceeded 40 cm (16 in.) a day.

This sequence of events set the stage for one of the worst dam disasters in history. Heavy rains began on September 28, 1963. Alarmed at the runoff from the rain into the reservoir and the increase in creep rate along the south wall, engineers opened the outlet tunnels on October 8 in an attempt to bring down the reservoir level—but it was too late.

The next evening, it took only 30 seconds for a landslide of 240 million cubic meters (8.5 billion cubic feet) to crash into the reservoir, shaking seismographs across Europe. A 150 m thick (500 ft) slab of mountainside gave way (2 km by 1.6 km in area; 1.2 mi by 1 mi), sending shock waves of wind and water up the canyon 2 km (1.2 mi) and splashing a 100 m (330 ft) wave over the dam. The former reservoir was effectively filled with bedrock, regolith, and soil that almost entirely displaced its water content (Figure 1). Amazingly, the experimental dam design held.

Downstream, in the unsuspecting town of Longarone, near the mouth of Vaiont Canyon on the Piave River, people heard the distant rumble and were quickly drowned by the 69 m (226 ft) wave of water that came out of the canyon. That night 3000 people perished. As for a lesson or recommendation, we need only to reread the preconstruction geologic analysis that began this Focus Study. The Italian courts eventually prosecuted the persons responsible.

---

moving, as was the case east of Jackson Hole, Wyoming, in the spring of 1925. Material on a slope above the Gros Ventre River (pronounced "grow vaunt") broke loose and slid downslope as a unit. The slide occurred because sandstone formations rested on weak shale and siltstone, which became moistened and soft, offering little resistance to the overlying strata. The slide is still visible after 70 years, as you can see in Figure 13-25.

Because of melted snow and rain, the water content of the Gros Ventre landslide was great enough to classify it as an earthflow. About 37 million cubic meters (1.3 billion cubic feet) of wet soil and rock moved down one side of the canyon and surged 30 m (100 ft) up the other side. The earthflow dammed the river and formed a lake,

as did the landslide across the Madison River in the Hebgen Lake area in 1959. However, in 1925 equipment was not available to excavate a channel, so the new lake filled. Two years later, the lake water broke through the temporary earthflow dam, transporting a tremendous quantity of debris over the region downstream. Malibu, California, a city prone to catastrophes of all kinds, is discussed in News Report 5.

***Creep.*** A persistent, gradual mass movement of surface soil is called **soil creep**. In creep, individual soil particles are lifted and disturbed by the expansion of soil moisture as it freezes, cycles of moistness and dryness, diurnal temperature variations, or grazing livestock or digging animals.

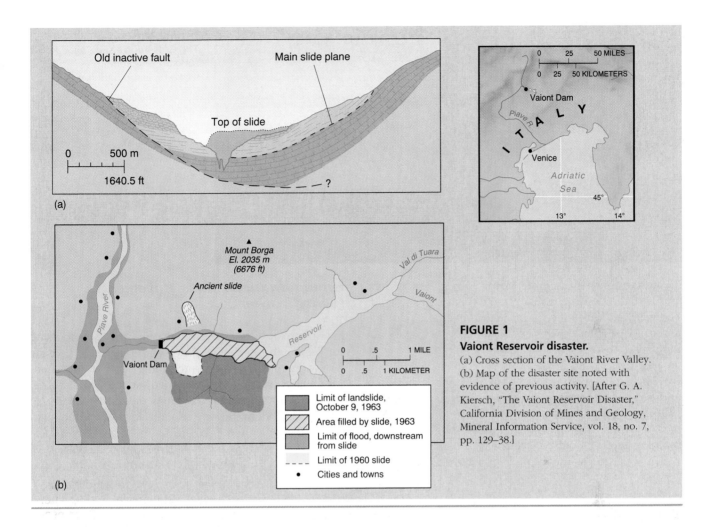

**FIGURE 1**
**Vaiont Reservoir disaster.**

(a) Cross section of the Vaiont River Valley.
(b) Map of the disaster site noted with
evidence of previous activity. [After G. A.
Kiersch, "The Vaiont Reservoir Disaster,"
California Division of Mines and Geology,
Mineral Information Service, vol. 18, no. 7,
pp. 129–38.]

**FIGURE 13-23**
**Debris avalanche, Alaska.**

A debris avalanche covers portions of the Cascade Glacier, west of the
Saint Elias Range in Alaska. The one pictured was the largest of several
dozen triggered by a 1979 earthquake. [Photo by George Plafker.]

**FIGURE 13-24**
**Suburban destruction in Alaska.**

Anchorage, Alaska, landslide, illustrates mass movement patterns
throughout a housing subdivision. [Courtesy of U.S. Geological Survey.]

**FIGURE 13-25**
**The Gros Ventre slide near Jackson, Wyoming.**
[Photo by Steven K. Huhtala.]

In the freeze-thaw cycle, particles are lifted at right angles to the slope by freezing soil moisture, as shown in Figure 13-26. When the ice melts, however, the particles are pulled straight downward by the force of gravity. As the process is repeated, the surface soil gradually creeps its way downslope.

The overall wasting of a slope may cover a wide area and may cause fence posts, utility poles, and even trees to lean downslope. Various strategies are used to arrest the mass movement of slope material—grading the terrain, building terraces and retaining walls, planting ground cover—but the persistence of creep nearly always wins.

## Human-Induced Mass Movements (Scarification)

Every human disturbance of a slope—highway roadcut, surface-mining, or building of a shopping mall, housing development, home, or outbuilding—can hasten mass wasting because the oversteepened surface is destabilized and thrust into a search for a new equilibrium. Imagine the disequilibrium in slope relations created by the highway roadcut pictured in Figure 12-7b. Large open-pit surface mines—such as the Bingham Copper Mine west of Salt Lake City, the Berkeley Pit in Butte, Montana, and

## News Report 5

### Malibu 90263

Malibu, a coastal suburb of Los Angeles, became a city in 1991 and today is fighting for its existence against landslides, mudflows, earthquakes, storm surges, and wildfires (which bare the slopes for more mass wasting). Six federally declared disasters were proclaimed for Malibu in a 4-year period! Mudslides onto the Pacific Coast Highway, the main north-south traffic arterial, are blamed for a $300,000 decrease in sales-tax revenues (Figure 1).

The rugged mountains and canyons that provide the dramatic scenery of the area are disaster-prone. Perhaps an obvious point of hindsight advice is to assess an area for mass movement potential before settling there. City officials now fear that the worst is ahead, for there will certainly be more catastrophes, headlines, and possibly fiscal disaster!

**FIGURE 1**
Malibu, California, another landslide and homes on the move. [Photo by Bob Riha/Gamma-Liaison, Inc.]

**FIGURE 13-26**
**Soil creep and its effects.**

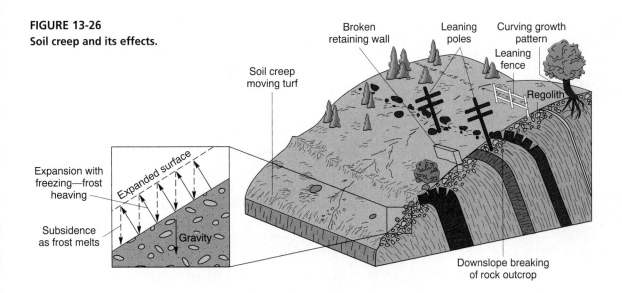

Broken retaining wall

Leaning poles

Curving growth pattern

Leaning fence

Soil creep moving turf

Regolith

Expansion with freezing—frost heaving

Expanded surface

Subsidence as frost melts

Gravity

Downslope breaking of rock outcrop

numerous large coal surface mines in the East and West—are examples of human-induced mass movements, generally called **scarification** (Figure 13-27a).

At the Bingham Copper Mine, a mountain literally was removed (Figure 13-27b). The disposal of tailings (mined ore of little value) and waste material is a significant problem at any surface mine. Such large excavations produce tailing piles that are unstable and susceptible to further weathering, mass wasting, or wind dispersal.

At the abandoned Berkeley Pit, toxic drainage water is threatening the regional aquifers, the Clark Fork River, and local freshwater supply (Figure 13-27c). Residues of copper, zinc, lead, and arsenic lace one creek devoid of life and are in the soil on Butte Hill, where children play. The leaching of toxic materials from tailings and waste piles poses an ever-increasing problem to streams, aquifers, and public health (Figure 13-27d). And wind dispersal is a particular problem with uranium tailings in the West, because of their radioactivity.

Where underground mining is common, particularly for coal in the Appalachians, land subsidence and collapse produce further mass movements. Homes, highways, streams, wells, and property values are severely affected.

***The Scale of Human Scarification.*** Scientists can informally quantify the scale of human-induced scarification for comparison with natural denudation processes. R. L. Hooke, Department of Geology and Geophysics, University of Minnesota, used estimates of U.S. excavations for new housing, mineral production (including the three largest—stone, sand and gravel, and coal), and highway construction. He then prorated these quantities of moved earth for all countries, using their gross national product (GNP), energy consumption, and agriculture's effect on river sediment loads. From these he calculated a global estimate for human earth moving.

Hooke estimated that humans, as a geomorphic agent, annually move 40 to 45 billion tons (40–45 Gt/yr) of the planet's surface. Compare this quantity with natural river sediment transfer (14 Gt/yr), movement through stream meandering (39 Gt/yr), haulage by glaciers (4.3 Gt/yr), movement due to wave action and erosion (1.25 Gt/yr), wind transport (1 Gt/yr), sediment movement by continental and oceanic mountain building (34 Gt/yr), or deep-ocean sedimentation (7 Gt/yr). As Hooke stated,

> *Home sapiens* has become an impressive geomorphic agent. Coupling our earth-moving prowess with our inadvertent adding of sediment load to rivers and the visual impact of our activities on the landscape, one is compelled to acknowledge that, for better or for worse, this biogeomorphic agent may be the premier geomorphic agent of our time.[*]

[*]R. L. Hooke, "On the efficacy of humans as geomorphic agents," *GSA Today*, The Geological Society of America, 4, no. 9 (September 1994): 217–26.

(a)

(b)

(c)

(d)

**FIGURE 13-27**
**Scarification.**
(a) Black Mesa, Arizona, strip-mining for coal. (b) Bingham Canyon, Utah, west of Salt Lake City, strip-mining for copper and other minerals. (c) Abandoned Berkeley Pit copper mine, Butte, Montana. (d) Spoil banks in a West Virginia coal-mining area. [Photos by author.]

# Summary and Review — Weathering, Karst Landscapes, and Mass Movement

✔ *Define* the science of geomorphology.

**Geomorphology** is the science that analyzes and describes the origin, evolution, form, and spatial distribution of landforms. The exogenic system, powered by solar energy and gravity, tears down the landscape through processes of landmass **denudation** involving weathering, mass movement, erosion, transportation, and deposition. Agents of change include moving air, water, waves, and ice. Davis's *geomorphic cycle model* characterized landscapes as evolving through stages from youth to old age. Since the 1960s, research and understanding of the processes of denudation have moved toward the **dynamic equilibrium model**, which considers slope and landform stability to be conse-

quences of the resistance of rock materials to the attack of denudation processes.

> geomorphology (p. 392)
> denudation (p. 392)
> dynamic equilibrium model (p. 393)

1.  Define geomorphology, and describe its relationship to physical geography.
2.  Define landmass denudation. What processes are included in the concept?
3.  Give a brief overview of the geomorphic cycle model.

What was Davis's principal contribution to the models of landmass denudation?

4.  What are the principal considerations in the dynamic equilibrium model?

---

✔ *Illustrate* **the forces at work on materials residing on a slope.**

Slopes are shaped by the relation between rate of weathering and breakup of slope materials and the rate of mass movement and erosion of those materials. A slope is considered stable if it is stronger than these denudation processes; it is unstable if it is weaker. In this struggle against gravity, a slope may reach a **geomorphic threshold**—the point at which there is enough energy to overcome resistance against movement. **Slopes** that form the boundaries of landforms have several general components: *waxing slope, free face, debris slope,* and *waning slope.* Slopes seek an *angle of equilibrium* among the operating forces.

geomorphic threshold (p. 393)
slopes (p. 393)

5.  Describe conditions on a hillslope that is right at the geomorphic threshold. What factors might push the slope beyond this point?

6.  Given all the interacting variables, do you think a landscape ever reaches a stable, old-age condition? Explain.

7.  What are the general components of an ideal slope?

8.  Relative to slopes, what is meant by an "angle of equilibrium"? Can you apply this concept to the photograph in Figure 13-2?

---

✔ *Define* **weathering and** *explain* **the importance of parent rock and joints and fractures in rock.**

**Weathering** processes disintegrate both surface and subsurface rock into mineral particles or dissolve them in water. The upper layers of surface material undergo continual weathering and create broken-up rock called **regolith**. Weathered **bedrock** is the *parent rock* from which regolith forms. The unconsolidated, fragmented material that develops after weathering is called **sediment**, which along with weathered rock forms the **parent material** from which soil evolves.

Important in weathering processes are **joints**, the fractures and separations in the rock. Jointing opens up rock surfaces on which weathering processes operate. Factors that influence weathering include: character of the bedrock (hard or soft, soluble or insoluble, broken or unbroken), climatic elements (temperature, precipitation, freeze-thaw cycles), position of the water table, slope orientation, surface vegetation and its subsurface roots, and time.

weathering (p. 395)
regolith (p. 395)
bedrock (p. 395)
sediment (p. 395)
parent material (p. 395)
joints (p. 395)

9.  Describe weathering processes operating on an open expanse of bedrock. How does regolith develop? How is sediment derived?

10. Describe the relationship between mesoscale climatic conditions and rates of weathering activities.

11. What is the relation among parent rock, parent material, regolith, and soil?

12. What role do joints play in the weathering process? Give an example from one of the illustrations in this chapter.

---

✔ *Describe* **frost action, crystallization, hydration, pressure-release jointing, and the role of freezing water as physical weathering processes.**

**Physical weathering** refers to the breakup of rock into smaller pieces with no alteration of mineral identity. The physical action of water when it freezes (expands) and thaws (contracts) is a powerful agent in shaping the landscape. This **frost action** may break apart any rock. Working in joints, expanded ice can produce *joint-block separation* through the process of *frost-wedging*. Different rocks offer differing resistance to weathering and produce a pattern on the landscape of **differential weathering**. Frost action loosens rock that falls from a steep cliff, producing a **talus slope** of poorly sorted debris at the base of the slope.

Another process of physical weathering is *crystallization*; as crystals in rock grow and enlarge over time they force apart mineral grains and break up rock. **Hydration** occurs when a mineral absorbs water and expands, thus creating a strong mechanical force that stresses rocks.

As overburden is removed from a granitic batholith, the pressure of deep burial is relieved. The granite slowly responds with *pressure-release jointing*, with layer after layer of rock peeling off in curved slabs or plates. As these slabs weather, they slip off in a process called **sheeting**. This *exfoliation process* creates an arch-shaped or dome-shaped feature on the exposed landscape, forming an **exfoliation dome**.

physical weathering (p. 397)
frost action (p. 397)
differential weathering (p. 398)
talus slope (p. 398)
hydration (p. 399)
sheeting (p. 400)
exfoliation dome (p. 400)

13. What is physical weathering? Give an example.

14. What is the interplay between the resistance of rock structures and weathering variabilities?
15. Why is freezing water such an effective physical weathering agent?
16. What weathering processes produce a granite dome? Describe the sequence of events.

---

✔ *Describe* **the susceptibility of different minerals to the chemical weathering processes called hydrolysis, oxidation, carbonation, and solution.**

**Chemical weathering** is the chemical decomposition of minerals in rock. It can cause **spheroidal weathering**, in which chemical weathering occurs in cracks in the rock. As cementing and binding materials are removed, the rock begins to disintegrate, and sharp edges and corners become rounded.

**Hydrolysis** breaks down silicate minerals in rock, as in the chemical weathering of feldspar into clays and silica. Water is not just absorbed, as in hydration, but actively participates in chemical reactions. **Oxidation** is the reaction of oxygen with certain metallic elements, the most familiar example being the rusting of iron, producing iron oxide. **Solution** is chemical weathering. For instance, a mild acid such as carbonic acid in rainwater will cause **carbonation**, wherein carbon combines with certain minerals, such as calcium, magnesium, potassium, and sodium.

> chemical weathering (p. 400)
> spheroidal weathering (p. 400)
> hydrolysis (p. 400)
> oxidation (p. 402)
> solution (p. 402)
> carbonation (p. 402)

17. What is chemical weathering? Contrast this set of processes to physical weathering.
18. What is meant by the term spheroidal weathering? How is this formed?
19. What is hydrolysis? How does it affect rocks?
20. Iron minerals in rock are susceptible to which form of chemical weathering? What characteristic color is associated with this?
21. With what kind of minerals do carbon compounds react, and under what circumstances does this reaction occur? What is this weathering process called?

---

✔ *Review* **the processes and features associated with karst topography.**

**Karst topography** refers to distinctively pitted and weathered limestone landscapes. Surface circular **sinkholes** form and may extend to form a *karst valley*. A sinkhole may collapse through the roof of an underground cavern, forming a *collapse sinkhole*. The formation of caverns is part of karst processes and groundwater erosion. Limestone caves feature many unique erosional and depositional features, producing a dramatic subterranean world.

> karst topography (p. 402)
> sinkholes (p. 403)

22. Describe the development of limestone topography. What is the name applied to such landscapes? From what area was this name derived?
23. Differentiate among sinkholes, karst valleys, and cockpit karst. Within which form is the radio telescope at Arecibo, Puerto Rico?
24. In general, how would you characterize the region southwest of Orleans, Indiana?
25. What are some of the unique erosional and depositional features you find in a limestone cavern?

---

✔ *Portray* **the various types of mass movements and** *identify* **examples of each in relation to moisture content and speed of movement.**

Any movement of a body of material, propelled and controlled by gravity, is called **mass movement**, also called **mass wasting**. The **angle of repose** of loose sediment grains represents a balance of driving and resisting forces on a slope. Mass movement of Earth's surface produces some dramatic incidents, including **rockfalls** (a volume of rock that falls), **debris avalanches** (mass of tumbling, falling rock, debris, and soil at high speed), **landslides** (a large amount of material failing simultaneously), **mudflows** (material in motion with a high moisture content), and **soil creep** (a persistent movement of individual soil particles that are lifted by the expansion of soil moisture as it freezes, by cycles of wetness and dryness, and temperature variations, or the impact of grazing animals). In addition, human mining and construction activities have created massive **scarification** of landscapes.

> mass movement (p. 406)
> mass wasting (p. 408)
> angle of repose (p. 408)
> rockfalls (p. 409)
> debris avalanche (p. 409)
> landslides (p. 409)
> mudflows (p. 410)
> soil creep (p. 412)
> scarification (p. 415)

26. Define the role of slopes in mass movements—angle of repose, driving force, resisting force, and geomorphic threshold.

27. What events occurred in the Madison River Canyon in 1959?

28. What are the classes of mass movement? Describe each briefly and differentiate among these classes.

29. Name and describe the type of mudflow associated with a volcanic eruption.

30. Describe the difference between a landslide and what happened on the slopes of Nevado Huascarán.

31. What is scarification, and why is it considered a type of mass movement? Give several examples of scarification. Why are humans a significant geomorphic agent?

---

 **NetWork**

The *Geosystems Home Page* provides on-line resources for this chapter on the World Wide Web. You will find review exercises, specific updates for items in the chapter, suggested read-

ings, and links to interesting related pathways on the Internet (click on the Table of Contents link and select this chapter). *Geosystems* is at: **http://www.prenhall.com/geosystm**

# 14

# River Systems and Landforms

**Fluvial Processes and Landscapes**

**Streamflow Characteristics**

**Floods and River Management**

**Summary and Review**

## Key Learning Concepts

After reading the chapter, you should be able to:

- *Define* the term *fluvial* and *outline* the fluvial processes: erosion, transportation, and deposition.

- *Construct* a basic drainage basin model and *identify* different types of drainage patterns, with examples.

- *Describe* the relation among velocity, depth, width, and discharge and *explain* the various ways that a stream erodes and transports its load.

- *Develop* a model of a meandering stream, including point bar, undercut bank, and cutoff, and *explain* the role of stream gradient in these flow characteristics.

- *Define* a floodplain and *analyze* the behavior of a stream channel during a flood.

- *Differentiate* the several types of river deltas and *detail* each.

- *Explain* flood probability estimates and *review* strategies for mitigating flood hazards.

*Drai Sap Falls, near Ban Me Thout, Central Highlands, Vietnam.*
[Photo by Wolfgang Kaehler Photography.]

Earth's rivers and waterways form vast arterial networks that drain the continents. They also shape the continents by removing the products of weathering, mass movement, and erosion and transporting them downstream. To call rivers "Earth's lifeblood" is no exaggeration, inasmuch as rivers redistribute mineral nutrients important for soil formation and plant growth and serve society in many ways.

Rivers not only provide essential water supplies; they also process waste (diluting and transporting it), provide critical cooling water for manufacturing and power generation, and form one of the world's most important transportation networks. Rivers have been important in the geography of human history, influencing where settlements were built, where livelihoods were made, and where borders were drawn. This chapter discusses the dynamics of river systems and their landforms.

At any one moment, approximately 1250 km³ (300 mi³) of water is flowing through Earth's waterways. Even though this volume is only 0.003% of all freshwater, the work performed by this energetic flow makes it a dominant agent of landmass denudation. Of the world's rivers, those with the greatest flow volume (*discharge*) are the Amazon and the Paraná of South America, the Zaire (Congo) of Africa, the Ganges of India, and the Chang Jiang (Yangtze) of Asia. In North America, the greatest discharges are from the Missouri-Ohio-Mississippi, Saint Lawrence, and Mackenzie river systems. (Table 9-2, page 262, shows the largest rivers, ranked by discharge; Figure 9-21 is a map portraying North America's surface runoff, our basic water supply.)

# Fluvial Processes and Landscapes

Stream-related processes are termed **fluvial** (from the Latin *fluvius*, meaning "river"). Geographers seek to describe stream patterns and the fluvial processes that created them. Fluvial systems, like all natural systems, have characteristic processes and produce predictable landforms. Yet a stream system can behave with randomness and disorder. The term *river* is applied to a trunk stream or an entire river system. *Stream* is a more general term not necessarily related to size: There is some overlap in usage between the two terms.

Insolation and gravity power the hydrologic cycle, so they are the driving forces of fluvial systems. Individual streams vary greatly, depending on the climate in which they operate, the composition of the surface and topography over which they flow, the nature of vegetation and plant cover, and the length of time they have been operating in a specific setting.

Water dislodges, dissolves, or removes surface material in the process called **erosion**. Streams produce *fluvial erosion*, in which weathered sediment is picked up

for **transport** to new locations. Thus, a stream is a mixture of water and solids; the solids are carried in solution, suspension, and by mechanical transport. Materials are laid down by another process, **deposition**. **Alluvium** is the general term for the clay, silt, and sand deposited by running water.

## Base Level of Streams

American geologist and ethnologist John Wesley Powell (1834–1902) was a director of the U.S. Geological Survey, first director of the U.S. Bureau of Ethnology, explorer of the Colorado River, and a pioneer in understanding the landscape (Figure 14-1). In 1875 he put forward the idea of **base level**, or a level below which a stream cannot erode its valley. In general, the *ultimate base level* is sea level (the average level between high and low tides). As shown in Figure 14-2, you can imagine the base level as a surface extending inland from sea level, inclined gently upward under the continents. Ideally, this is the lowest practical level for all denudation processes.

Of course, Powell recognized that not every landscape has degraded all the way to sea level; clearly, other intermediate base levels are in operation. A *local base level*, or temporary one, may control the lower limit of local streams. The local base level may be a river, a lake, hard and resistant rock, or a human-made dam (Figure 14-2). In arid landscapes, with their intermittent precipitation, local control is provided by valleys, plains, or other low points.

Over time, the work of streams modifies the landscape dramatically. Landforms are produced by the two basic processes, erosive action of flowing water and deposition of stream-transported materials. Let's begin our study by examining a basic fluvial unit—the drainage basin.

## Drainage Basins

Figure 14-3 illustrates drainage basin concepts. Every stream has a **drainage basin**, ranging in size from tiny to vast. Every drainage basin is defined by ridges that form *drainage divides*; that is, the ridges are the dividing lines that control into which basin precipitation drains. Drainage divides define a **watershed**, the catchment area of the drainage basin. In any drainage basin, water initially moves downslope in a thin film called **sheetflow**, or *overland flow*. High ground that separates one valley from another, and directs sheet flow, is termed an *interfluve*. Surface runoff concentrates in *rills*, or small-scale downhill grooves, which may develop into deeper *gullies* and a stream course in a valley (Figure 14-3).

### Drainage Divides and Basins. The United States and Canada are divided by several high drainage divides,

**FIGURE 14-1**
**Powell's journey to base level.**
John Wesley Powell developed his base level concept during extensive exploration of the West. He first ventured down the Colorado River in 1869 in heavy oak boats, drifting by Dead Horse Point and the bend in the river visible in the canyon. [Photo by author.]

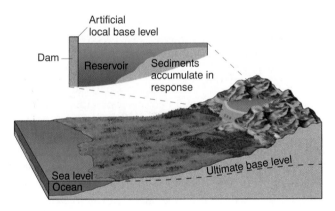

**FIGURE 14-2**
**Ultimate and local base levels.**
The concepts of ultimate base level (sea level) and local base level (natural, such as a lake, or artificial, as shown). Note how base level curves gently upward from the sea as it is traced inland; this is the theoretical limit for stream erosion.

called **continental divides**. These are extensive mountain and highland regions that separate drainage basins, sending flows to the Pacific, the Gulf of Mexico, the Atlantic, Hudson Bay, or the Arctic Ocean. The principal drainage divides and drainage basins in the United States and Canada are mapped in Figure 14-4. These divides form water-resource regions and provide a spatial framework for water planning.

A major drainage basin system is made up of many smaller drainage basins, which in turn comprise even smaller basins, each divided into specific watersheds. Each drainage basin gathers and delivers its precipitation and sediment to a larger basin, concentrating the volume into the main stream. A good example is the great Mississippi-Missouri-Ohio river system (Figure 14-4).

Consider the travels of rainfall in north-central Pennsylvania. This water feeds hundreds of small streams that flow into the Allegheny River. At the same time, rainfall in southern Pennsylvania feeds hundreds of streams that flow into the Monongahela River. The two rivers then join at Pittsburgh to form the Ohio River. The Ohio flows southwestward and at Cairo, Illinois, connects with the Mississippi River, which eventually flows on past New Orleans into the Gulf of Mexico. Each contributing tributary, large or small, adds its discharge and sediment load to the larger river.

In our example, sediment weathered and eroded in north-central and southern Pennsylvania is transported thousands of kilometers and accumulates on the floor of the Gulf of Mexico, where it forms the Mississippi River delta.

***Drainage Basins Are Open Systems.*** Drainage basins are open systems in which inputs include precipitation and the minerals and rocks of the regional geology. Energy and materials are redistributed as the stream constantly adjusts with its landscape. System outputs of water and sediment exit through the mouth of the river, into a lake, another river, or the ocean.

Change that occurs in any portion of a drainage basin can affect the entire system, because the stream adjusts to carry the appropriate load of sediment relative to its discharge and velocity. If a region is brought to a threshold where it can no longer maintain its present form, the relations within the drainage basin system are destabilized, initiating a transition period to a more stable condition. A

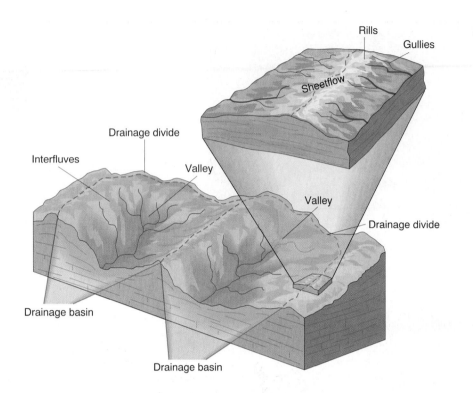

**FIGURE 14-3**
**A drainage basin.**
The drainage basin and its watershed are separated from other basins by a drainage divide.

stream drainage system constantly struggles toward equilibrium among the interacting variables of discharge, transported load, channel shape, and channel steepness.

***Delaware River Basin.*** Let us look at the Delaware River basin, within the Atlantic Ocean drainage region (Figure 14-5). The Delaware River headwaters are in the Catskill Mountains of New York. This basin encompasses 33,060 km² (12,890 mi²) and includes parts of five states in the river's length, 595 km (370 mi) from headwaters to the mouth.

The entire basin lies within a humid, temperate climate and receives an average annual precipitation of 120 cm (47.2 in.). Topography varies from low-relief coastal plains to the Appalachian Mountains in the north. The river provides water for an estimated 20 million people, not only within the basin but to cities outside the basin as well. Several major conduits export water from the Delaware River. Note on the map the Delaware Aqueduct to New York City (in the north) and the Delaware & Raritan Canal (near Trenton). Several reservoirs in the drainage basin allow some control over water flow and storage for dry periods. The river system ends at Delaware Bay, which eventually enters the Atlantic Ocean.

The USGS is studying the Delaware River basin to determine the potential impact of future climate change on water resources. The specific concern is global warming. The long-term study includes projected changes in streamflow, irrigation demand, reduction in soil-moisture storage, possible saltwater intrusion into the basin near the ocean, and problems associated with sea-level rise. A decrease in precipitation caused by changing climate patterns would cause significant disruption of the water supply to the surrounding metropolitan region.

## Drainage Density and Patterns

A primary feature of any drainage basin is its density. **Drainage density** is determined by dividing the total length of all stream channels in the basin by the area of the basin. The number and length of channels in a given area express the landscape's regional topography and surface appearance. For example, Figure 14-6 reveals a very high drainage density in a humid climate. In contrast, the typical desert has a very low drainage density.

The **drainage pattern** is the arrangement of channels in an area. Patterns are quite distinctive, for they are determined by the combination of regional steepness, variable rock resistance, variable climate, variable hydrology, relief of the land, and structural controls imposed by the underlying rocks. Consequently, the drainage pattern of any land area on Earth is a remarkable visual summary of every characteristic—geologic and climatic—of that region. The *Landsat* image and topographic map in Figure 14-6 are of the Ohio River drainage near the junction of West Virginia, Ohio, and Kentucky. The high-density drainage pattern and intricate dissection of the land occur because the region has generally level, easily eroded sandstone, siltstone, and shale strata and a humid mesothermal climate. Fluvial action and other denudation processes are responsible for this dissected topography.

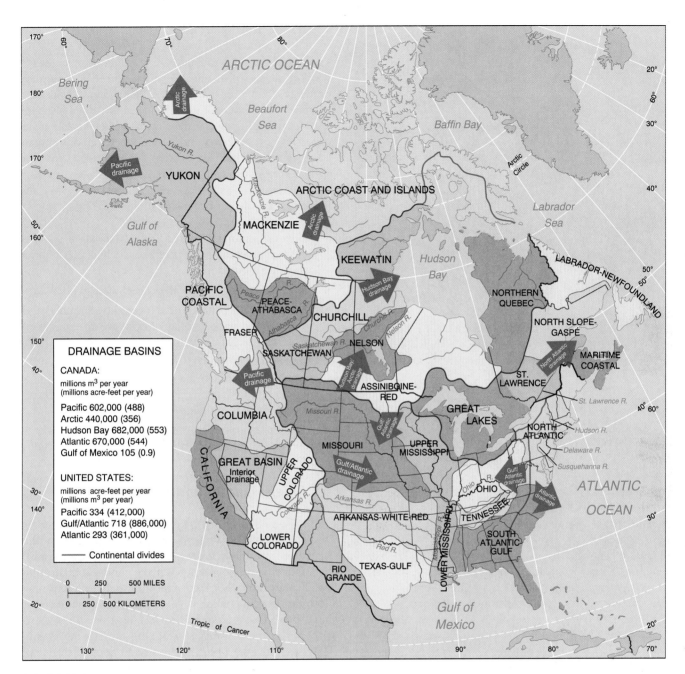

**FIGURE 14-4**

**Drainage basins and continental divides.**

Continental divides (blue lines) separate the major drainage basins that empty into the Pacific, Atlantic, Gulf of Mexico, and to the north through Canada into Hudson Bay and the Arctic Ocean. Subdividing these large-scale basins are major river basins. [After U.S. Geological Survey; *The National Atlas of Canada*, 1985, Energy, Mines, and Resources Canada; and Environment Canada, *Currents of Change–Inquiry on Federal Water Policy*–Final Report 1986.]

***Common Drainage Patterns.*** The seven most common drainage patterns are shown in Figure 14-7. A most familiar pattern is *dendritic drainage* (Figure 14-7a). This tree-like pattern (Greek *dendron* = tree) is similar to that of many natural systems, such as capillaries in the human circulatory system, the vein patterns in leaves, and tree roots. Energy expended by this drainage system is efficient

because the overall length of the branches is minimized. Figure 14-6 shows dendritic drainage; on the satellite image and topographic map, you can trace the branching pattern of streams.

The *trellis drainage* pattern (Figure 14-7b) is characteristic of dipping or folded topography. Such drainage exists in the nearly parallel mountain folds of the Ridge

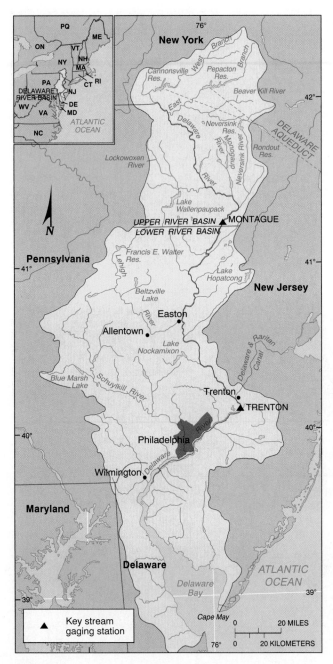

**FIGURE 14-5**

**The Delaware River drainage basin.**

The basin is under study by the USGS to assess the potential impact of global warming on a river system. [After U.S. Geological Survey, 1986, "Hydrologic events and surface water resources," *National Water Summary 1985*, Water Supply Paper 2300 (Washington, DC: Government Printing Office), p. 30.]

and Valley Province in the East. Refer to Figure 12-14, a satellite image and map of this region that shows its distinctive drainage pattern. Here drainage patterns are influenced by rock structures of variable resistance and folded strata. The principal streams are directed by the parallel folded structures, whereas smaller dendritic tributary

streams are at work on nearby slopes, joining the main streams at right angles, like a plant trellis.

The inset sketch to Figure 14-7b suggests that a headward-eroding part of one stream could break through a drainage divide and *capture* the headwaters of another stream in the next valley, and indeed this does happen. The sharp bends in two of the streams in the illustration are called *elbows of capture* and are evidence that one stream has breached a drainage divide. This type of capture, or *stream piracy*, also can occur in other drainage patterns.

The remaining drainage patterns in Figure 14-7 are responses to other specific structural conditions:

- A *radial* drainage pattern (c) results when streams flow off a central peak or dome, such as occurs on a volcanic mountain.

- *Parallel* drainage (d) is associated with steep slopes.

- A *rectangular* pattern (e) is formed by a faulted and jointed landscape, which directs stream courses in patterns of right-angle turns.

- *Annular* patterns (f) are produced by structural domes, with concentric patterns of rock strata guiding stream courses. Figure 12-9c provides an example of annular drainage on a dome structure.

- In areas having disrupted surface patterns, such as the glaciated shield regions of Canada, northern Europe, and some parts of Michigan and other states, a *deranged* pattern (g) is in evidence, with no clear geometry in the drainage and no true stream valley pattern.

These seven drainage patterns are directed by the structure and relief of the land. Drainage patterns also occur that are discordant with the landscape through which they flow. For example, a drainage system may flow in apparent conflict with older, buried structures that have been uncovered by erosion, so that the streams appear to be superimposed. Where an existing stream flows as rocks are uplifted, the stream keeps its original course, cutting into the rock in a pattern contrary to its structure. Such a stream is called a *superposed stream* (the stream cuts across weak and resistant rocks alike). A few examples include: Wills Creek, cutting a water gap through Haystack Mountain at Cumberland, Maryland; the Columbia River through the Cascade Mountains of Washington; and the River Arun that cuts across the Himalayas.

# Streamflow Characteristics

A mass of water positioned above base level in a stream has potential energy. As the water flows downslope (downstream) under the influence of gravity, this energy becomes kinetic energy. The rate of this conversion from

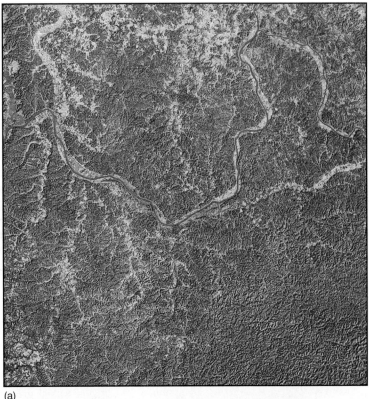

(a)

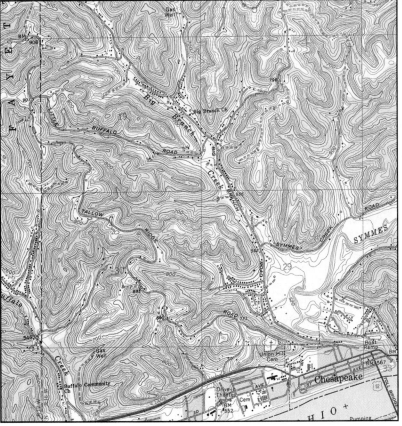

(b)

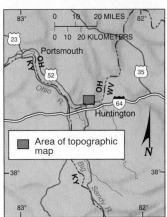

**FIGURE 14-6**

**A stream-dissected landscape.**

(a) Dendritic drainage pattern of a highly
dissected topography around the junction of the
West Virginia, Ohio, and Kentucky. All of these
borders are formed by rivers. (b) A portion of
the USGS topographic map covering the area
north of Huntington, West Virginia, reveals the
intricate complexity of dissected landscapes.
(The image and map are not at the same scale.)
[(a) *Landsat* image from NASA; (b) Huntington
Quadrangle, Eros Data Center/USGS.]

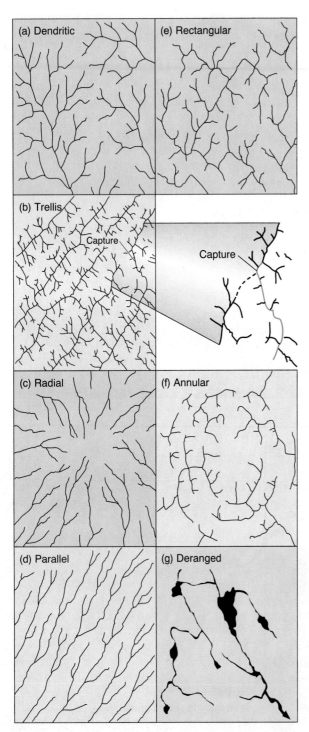

**FIGURE 14-7**

**The seven most common drainage patterns.**

Each pattern is a visual summary of all the geologic and climatic conditions of its region. [After A. D. Howard, "Drainage analysis in geological interpretation: A summation," *Bulletin of American Association of Petroleum Geologists* 51 (1967), p. 2248. Adapted by permission.]

potential to kinetic energy depends on the steepness of the stream channel.

Stream channels vary in *width* and *depth*. The streams that flow in them vary in *velocity* and in the *sediment load*

they carry. All of these factors increase with increasing discharge. Discharge is calculated by multiplying the velocity of the stream by its width and depth for a specific cross section of the channel, as stated in the simple expression:

$$Q = wdv$$

where $Q$ = discharge; $w$ = channel width; $d$ = channel depth; and $v$ = stream velocity. As $Q$ increases, some combination of channel width, depth, and stream velocity increases. Discharge is expressed either in cubic meters per second (m³/s) or cubic feet per second (cfs).

Figure 14-8 illustrates the relation of discharge to width, depth, and velocity. The graphs show that mean velocity increases with greater discharge, despite the common misperception that downstream flow becomes more sluggish. (The increased velocity downstream often is masked by the apparent smooth, quiet flow of the water.) Not surprisingly, discharge also increases with width and depth.

Given the interplay of channel width and depth and stream velocity, the cross section of a stream varies over time, especially during heavy floods. Figure 14-9 shows changes in the San Juan River channel in Utah that occurred during a flood. Greater discharge increases the velocity and therefore the carrying capacity of the river as the flood progresses. As a result, the river's ability to scour materials from its bed is enhanced. Such scouring represents a powerful clearing action, especially in the excavation of alluvium. Scouring also can provide new recreational areas and wildlife habitats (see News Report 1).

You can see in Figure 14-9 that the San Juan River's channel was deepest on October 14, when floodwaters were highest (blue line). Then, as the discharge returned to normal, the kinetic energy of the river was reduced, and the bed again filled as sediments were redeposited. You can see this process in the progression of the red, green, and purple lines. The process graphed in Figure 14-9 moved a depth of about 3 m (10 ft) of sediment from this cross section of the stream channel. Such adjustments in a stream channel occur as the stream system continuously works toward an equilibrium, to balance discharge, velocity, and sediment load.

### Stream Erosion

A stream's erosional turbulence and abrasion carve and shape the landscape through which it flows. **Hydraulic action** is the work of *turbulence* in the water. Running water causes hydraulic squeeze-and-release action that loosens and lifts rocks. As this debris moves along, it mechanically erodes the streambed further, through the process of **abrasion**, with rock particles grinding and carving the streambed like liquid sandpaper.

The upstream tributaries in a drainage basin usually have small and irregular discharges, and most of the stream's energy is expended in turbulent eddies. As a result, hydraulic action in these upstream sections is at maximum,

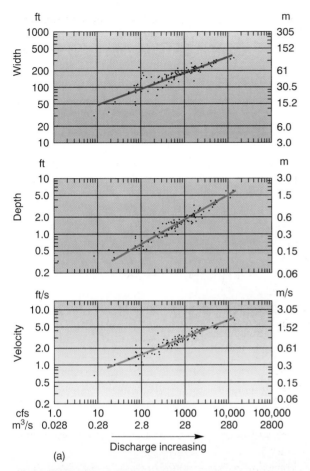

(b)

**FIGURE 14-8**
**Effects of stream discharge.**
(a) Relation of stream width, depth, and velocity to stream discharge of the Powder River at Locate, Montana. Discharge is shown in cubic meters per second (m³/s) and cubic feet per second (cfs). (b) The Powder River area near the Wyoming border in southeastern Montana. [(a) After L. Leopold and T. Maddock Jr., *The Hydraulic Geometry of Stream Channels and Some Physiographic Implications*, U.S. Geological Survey Professional Paper 252 (Washington, DC: Government Printing Office, 1953), p. 7; (b) photo by Joyce Wilson.]

whereas the coarse-textured load of such a stream is small. The downstream portions of a river, however, move much larger volumes of water past a given point and carry larger suspended loads of sediment (Figure 14-10).

Stream velocity determines rates of erosion and deposition. Sediment particles are deposited onto the streambed at slower velocities, whereas they are eroded at higher velocities. Interestingly, finer clay particles have such a cohesiveness among themselves that a stream can actually erode coarser sands, as well as noncohesive silts and clays, more easily than the fine clays.

## Stream Transport

You may have watched a river or creek after a rainfall, the water colored brown by the heavy sediment load being transported. The amount of material available to a stream depends on topographic relief, the nature of rock and soil through which the stream flows, climate, vegetation, and human activity in a drainage basin. *Competence*, which is a stream's ability to move particles of a specific size, is a function of stream velocity. *Capacity* is the total possible load that a stream can transport. Eroded materials are transported by four processes: solution, suspension, saltation, and traction; each is shown in action in Figure 14-11.

*Solution* refers to the **dissolved load** of a stream, especially the chemical solution derived from minerals such as limestone or dolomite or from soluble salts. The main contributor of material in solution is chemical weathering. Sometimes the undesirable salt content that hinders human use of some rivers comes from dissolved rock formations and from springs in the stream channel; as an example, the San Juan and Little Colorado Rivers that flow into the Colorado River near the Utah-Arizona border add dissolved salts to the system.

The **suspended load** consists of fine-grained, clastic particles (bits and pieces of rock). They are held aloft in the stream, with the finest particles not deposited until the stream velocity slows nearly to zero. Turbulence in the water, with random upward motion, is an important mechanical factor in holding a load of sediment in suspension.

**Bed load** refers to coarser materials that are dragged along the streambed by **traction** or are rolled and bounced along by **saltation** (from the Latin *saltim*, which means "by leaps or jumps"). At times, it is difficult to distinguish traction from saltation of bed load. Particles transported by saltation are too large to remain in suspension, a distinction directly related to a stream's velocity and its ability to retain particles in suspension. With increased kinetic energy, parts of the bed load are rafted upward and become suspended load. This type of action is demonstrated in the flood-induced channel deepening of the San Juan River shown in Figure 14-9. Saltation is also a process in wind transport of materials (see Chapter 15).

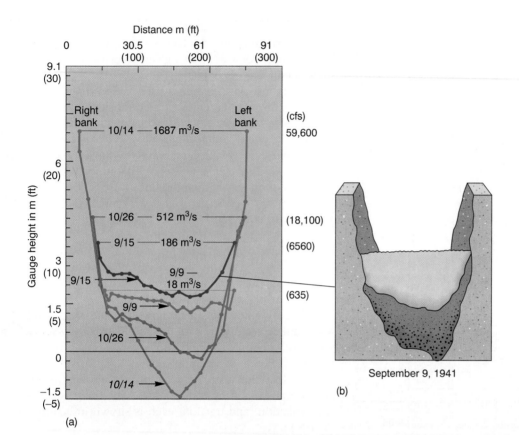

(a)

September 9, 1941

(b)

**FIGURE 14-9**
**A flood affects a stream channel.**
(a) Stream channel cross sections showing the progress of a 1941 flood on the San Juan River near Bluff, Utah. (b) Detail of stream-channel profile on September 9, 1941. [After L. Leopold and T. Maddock Jr., *The Hydraulic Geometry of Stream Channels and Some Physiographic Implications*, U.S. Geological Survey Professional Paper 252, (Washington, DC: Government Printing Office, 1953), p. 32.]

# News Report 1

## Scouring the Grand Canyon for New Beaches and Habitats

Glen Canyon Dam sealed the Colorado River gorge north of the Grand Canyon, near Lees Ferry and the Utah-Arizona border (see the map in Focus Study 15-1, Figure 1). Impoundment of water began in 1963, as did the inundation of many canyons upstream from the dam by the new Lake Powell. The lake began collecting the tremendous sediment load of the Colorado River, reducing the volume of water in the river below the dam and eliminating the natural seasonal fluctuations in river discharge. The production of hydroelectricity at Glen Canyon further affected the Grand Canyon with highly variable water releases keyed to electrical turbine operations. Water in the canyon rose and fell as much as 4.3 m (14 ft) as the distant lights and air conditioning of Las Vegas and Phoenix went on and off.

All of these changes affected the canyon. Over the years, the beaches were starved for sand, and channels filled with sediment. Fisheries were disrupted, and backwater channels were depleted of nutrients. In 1996, an unprecedented experiment began: The Grand Canyon was artificially flooded with a 45,000 cfs (1260 m³/s) release from Glen Canyon. The flow lasted for 7 days and then was reduced to 8000 cfs (224 m³/s). Flows were decreased gradually to allow the newly formed beaches to drain, in contrast to the rapid opening of the pipes that initiated the flood. Lake Powell dropped 1 m (3.3 ft), and Lake Mead, downstream behind Hoover Dam, rose 0.8 m (2.6 ft).

The results were positive: 35% more beach area was created by the scouring of sand and sediments from the channel. Approximately 80% of the aggradation took place in the first 40 hours and was completed by the 100-hour mark. Numerous backwater channels were created, flush with fresh nutrients for the humpback chub and other endangered fish species. In contrast to these benefits were a few negatives in the form of existing ecosystem disruption, although a full assessment is incomplete. This is a new approach to recreating beaches and habitats through dam operations on the troubled Colorado River.

The first Spanish explorers to visit the Grand Canyon reported in their journals that they were kept awake at night by the thundering sound of boulders tumbling along the Colorado River bed (a combination of traction and saltation). Such sounds today are substantially lessened because of the reduced velocity and discharge of the Colorado resulting from the many dams and control facilities that now trap sediments and reduce bed load capacity.

(a)

(b)

**FIGURE 14-10**

**Stream velocity and discharge increase together.**

(a) A low-volume, low-velocity (but turbulent) mountain stream, near Alanje Rio, Costa Rica. (b) The high-velocity, high-discharge downstream portion of the Mississippi River near Natchez, Mississippi. [(a) Photo by Stephen J. Krasemann/DRK Photos; (b) photo by author.]

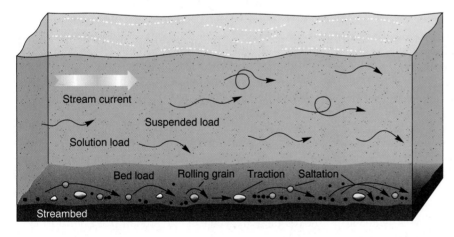

**FIGURE 14-11**

**Fluvial transport.**

Fluvial transportation of eroded materials through saltation, traction, suspension, and solution.

Figure 14-12 shows the annual water discharge and suspended sediment load for the Colorado River at Yuma, Arizona, from 1905 to 1964. The completion of Hoover Dam in the 1930s dramatically reduced suspended sediment. Addition of Glen Canyon Dam upstream from Hoover Dam in 1963 further reduced stream competence and capacity. In fact, Lake Powell, which formed upstream behind the artificial base level of Glen Canyon Dam, is forecast to fill with sediment over the next 100 years. River flows through the Grand Canyon today frequently are blue-green rather than the sediment-laden muddy red of the past and can fluctuate up to 4.3 m (14 ft) as hydroelectric production at Glen Canyon adjusts to daily electrical demands in distant cities.

If the load (bed and suspended) exceeds a stream's capacity, sediments accumulate as **aggradation** (the opposite of degradation) and the stream channel builds up through deposition. With excess sediment, a stream becomes a maze of interconnected channels that form a **braided stream** pattern (Figure 14-13). Braiding often

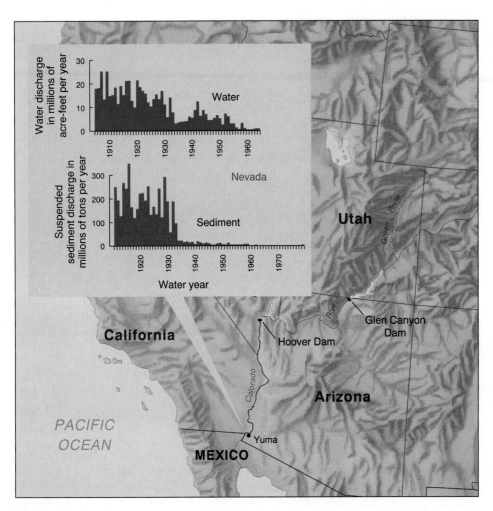

**FIGURE 14-12**
**Water and sediment discharge at Yuma, Arizona, 1905–1964.** Note that both river discharge and sediment load declined sharply in the 1930s. [After U.S. Geological Survey, 1985, "Hydrologic events, selected water-quality trends, and ground water resources," *National Water Summary 1984*, Water Supply Paper 2275 (Washington, DC: Government Printing Office), p. 55.]

occurs when reduced discharge lowers a stream's transporting ability, such as after flooding, or when a landslide occurs upstream, or from increased load where weak banks of sand or gravel exist. Locally, braiding also may result from a new sediment load from glacial meltwaters, as in Alaska's Chitina River in the photo.

## Flow and Channel Characteristics

Streams have two general types of flow, laminar and turbulent. *Laminar flow* is a streamlined flow of water in which individual clay and other fine particles move along evenly in generally parallel flows. Natural streams have stretches of laminar flow only when the water is moving slowly and the channel surfaces are smooth.

Flow becomes *turbulent* in streams of greater velocity—shallow streams, or where the channel is rough, as in a section of rapids. Small eddies are caused by friction between streamflows and the channel sides and bed. Complex turbulent flows propel sand, pebbles, and even boulders, increasing the suspension, traction, and saltation. As

bed load material is rolled, lifted, and dragged along, the channel is further abraded. The *laminar* velocity of a stream is the greatest flow rate at which the flow remains laminar; above this rate, flows are called turbulent.

Flow characteristics of a stream are best seen in a cross-sectional view. The greatest velocities in a stream are near the surface at the center (Figure 14-14), corresponding to the deepest part of the stream channel. Velocities decrease closer to the sides and bottom of the channel because of the frictional drag on the water flow. The portion of the stream flowing at maximum velocity moves diagonally across the stream from bend to bend; this diagonal movement is known as the flow *crossing*.

Laminar and turbulent flow form three types of stream channels: *braided, straight,* or *meandering.* To form these channel patterns, where slope is gradual, stream channels develop a sinuous (snakelike) form, weaving across the landscape. This action produces a **meandering stream**. The term *meander* comes from the ancient Greek Maiandros River in Asia Minor (the present-day Menderes River in Turkey), which had a meandering channel pattern. The tendency to meander is evidence of a struggle

**FIGURE 14-13**
**A braided stream.**
Braided stream pattern in Chitina River, Wrangell–Saint Elias National Park, Alaska. This stream reflects excessive sediment load associated with glacial meltwaters. [Photo by Tom Bean.]

in the system between self-organizing order (equilibrium) and chaotic disorder in nature.

The outer portion of each meandering curve is subject to the fastest water velocity and therefore the greatest scouring erosive action; it can be the site of a steep bank called an **undercut bank**, or *cutbank* (Figure 14-14). On the other hand, the inner portion of a meander experiences the slowest water velocity and thus receives sediment fill, forming a deposit called a **point bar**. As meanders develop, these scour-and-fill features gradually work their way downstream. As a result, the landscape near a meandering river bears meander scars of residual deposits from previous river channels. The photograph of the Itkillik River in Alaska, Figure 14-15a, shows both meanders and meander scars.

Meandering streams create a remarkable looping pattern on the landscape. Meanders gradually form loops, as shown in the four-picture sequence in Figure 14-15b. In (1), the stream erodes its outside bank as the curve migrates downstream, forming a neck. In (2), the narrowing neck of land created by the looping meander eventually erodes through and forms a *cutoff* in (3). A cutoff marks an abrupt change in the stream's lateral movements—the stream becomes straighter.

When the former meander becomes isolated from the rest of the river, the resulting **oxbow lake** (4) may gradually fill with silt or may again become part of the river when it floods. The Mississippi River is many miles shorter today than it was in the 1830s because of artificial cutoffs that were dredged across meander necks to improve navigation and safety. News Report 2 looks at the difficulty of using a meandering stream as a political boundary.

## Stream Gradient

Every stream runs downhill under the pull of gravity. In the course of this process, each stream develops its own **gradient**, which is the rate of elevation decline from its headwaters to its mouth. This decline is far from linear; characteristically, the *longitudinal profile* (side view) of a stream features a steeper slope upstream and a gentler slope downstream, forming an uneven concave shape (Figure 14-16). This gradient is concave for complex reasons related to the stream's having just enough energy to transport the sediment load it receives. The longitudinal profile of streams can be expressed mathematically, enabling scientists to categorize, study, and predict stream behavior.

An important concept is that of the **graded stream**, nicely defined by J. H. Mackin, a geomorphologist:

> A graded stream is one in which, over a period of years, slope is delicately adjusted to provide, with available *discharge* and with prevailing *channel characteristics*, just the velocity required for transportation of the load supplied from the drainage basin.[*]

Attainment of a graded condition does not mean that the stream is at its lowest gradient, but rather that it represents a present balance (a dynamic equilibrium) among erosion, transportation, and deposition over time along a specific portion of the stream. Both high-gradient and low-gradient streams can achieve a graded condition. The longitudinal profile of a graded stream is called a *profile of equilibrium*, a parabolic curve, gently flattening toward the mouth.

One problem with applying the graded stream concept is that an individual stream can have both graded and ungraded portions and may have graded sections without having an overall graded slope. Variations and interruptions are the rule rather than the exception. A profile of equilibrium may not be smooth throughout its course and cannot exist for long, for it represents a theoretical perfect balance. With streams, as in all of nature, change is the only constant.

Stream gradient may be affected by tectonic uplift of the landscape, which changes the base level. If tectonic

[*] J. H. Mackin, "Concept of the graded river," *Geological Society of America Bulletin*, 59 (1948): 463.

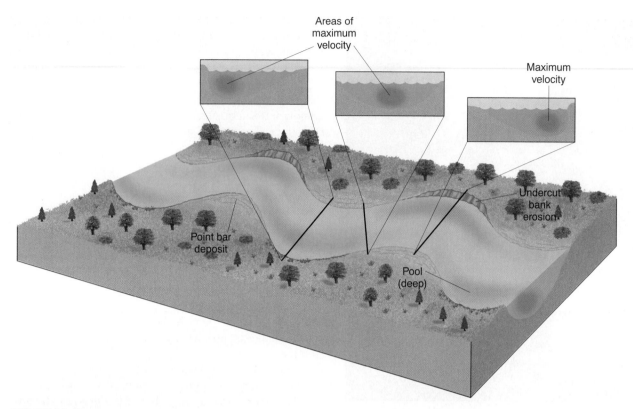

**FIGURE 14-14**

**Meandering stream profile.**

Aerial view and cross sections of a meandering stream, showing the location of maximum flow velocity, point-bar deposits, and areas of undercut bank erosion.

forces slowly lift the landscape, the stream gradient will increase, stimulating renewed erosional activity. Imagine this occurring to the landscape in Figure 14-1. The stream flowing through the landscape would become *rejuvenated*. With rejuvenation, river meanders would actively return to downcutting and deepening themselves and eventually would become *entrenched meanders* in the landscape. Figure 14-17 depicts an actual rejuvenated landscape.

***Nickpoints.*** When the longitudinal profile of a stream shows an abrupt change in gradient, such as at a waterfall or an area of rapids, the point of interruption is termed a **nickpoint** (also spelled *knickpoint*). At a nickpoint, the conversion of potential energy to concentrated kinetic energy works to eliminate the nickpoint. Figure 14-18 shows a stream with two such interruptions.

Nickpoints can result when a stream flows across a zone of hard, resistant rock or from various tectonic uplift episodes, such as might occur along a fault line. Temporary blockage in a channel, caused by a landslide or a logjam, also could be considered a nickpoint; when the logjam breaks, the stream quickly readjusts its channel to its former grade.

Waterfalls are interesting and beautiful gradient breaks. At the edge of a fall, a stream is free-falling, moving at high velocity under the acceleration of gravity, causing increased abrasion on the channel below. The increased abrasion and hydraulic action generally undercut the waterfall. Eventually they will cause the rock ledge at the lip of the fall to collapse, and the waterfall will shift a bit farther upstream (Figure 14-18). Thus, a nickpoint migrates upstream, sometimes for kilometers, until it is gone. The height of the waterfall is gradually reduced as debris accumulates at its base.

At Niagara Falls on the Ontario–New York border, glaciers advanced over the region and then receded. In doing so, they exposed resistant rock strata that are underlain by less-resistant shales. As this less-resistant material continues to weather away, the overlying rock strata collapse, allowing the falls to erode farther upstream toward Lake Erie (Figure 14-19). As this example demonstrates, a nickpoint is a relatively temporary and mobile feature on the landscape. News Report 3 presents an interesting attempt to delay the natural loss of Niagara Falls.

***Do Stream Gradients and Landscapes Evolve?*** In Chapter 13, we introduced William Morris Davis, a geomorphologist who founded the Association of American Geographers in 1904. His evolutionary concepts of erosion also included fluvial processes. Drawing from G. K. Gilbert, an important geomorphologist and his contemporary, Davis incorporated the graded-stream concept into his model, identifying erosion stages in a *cyclic model* he called *youth*, *maturity*, and *old age*.

(a)

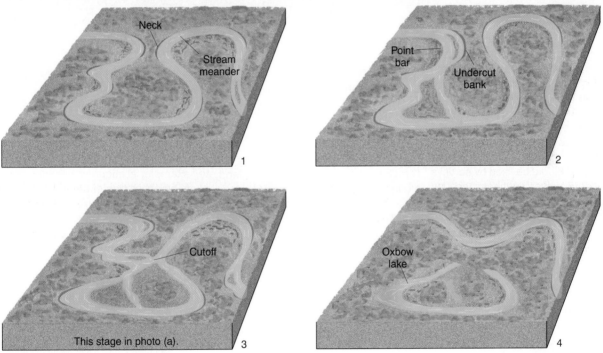

(b)

**FIGURE 14-15**

**Meandering stream development.**

(a) Itkillik River in Alaska. (b) Development of a river meander and oxbow lake simplified in four stages.

[(a) U.S. Geological Survey photo.]

Davis thought that, following uplift and the consequent drop in base level, a stream would quickly down-cut narrow V-shaped canyons in its youth, leaving broad interfluvial uplands. Features such as waterfalls, rapids, and lakes would be eliminated early as the stream evolved to a more graded profile. He believed that, in a stream's mature stage, it would achieve a nearly graded profile and begin to widen its valley by cutting laterally into

banks and the surrounding upland areas. Late maturity would be achieved when the floodplain broadened and the low stream gradient produced a wide, meandering flow pattern and floodplain.

Today, the *functional model* of *dynamic equilibrium* is supported by geomorphologists. The dynamic equilibrium model emphasizes the effects of individual processes interacting on streams and hillslope systems. Stream form

# News Report 2

## Rivers Make Poor Political Boundaries

Streams often are used as natural political boundaries. Commonly, where a stream forms a boundary, the boundary line is drawn down the middle of the stream. It is easy to see how boundary disputes might arise when they are based on river channels that shift around. For example, the Ohio, Missouri, and Mississippi Rivers can shift their positions quite rapidly during times of flood, so the boundaries based upon them should shift too. However, the boundaries are not changed and existing land ownership remains set.

Carter Lake, Iowa, provides a fascinating example (Figure 1). The Nebraska-Iowa border originally was placed midchannel in the Missouri River. But in 1877, the meander loop that curved around the town of Carter Lake in Iowa was cut off by the river, leaving the town "captured" by Nebraska! The old boundary marked along the former meander bend still is used as the state line. The oxbow lake created by the cutoff was named Carter Lake.

This event illustrates why boundaries always should be fixed by surveys independent of river locations, because border disputes inevitably result when river channels change course. Such surveys have been completed along the Rio Grande near El Paso, Texas, and along the Colorado River between Arizona and California, permanently establishing political boundaries separate from shifting river channels. In the latter example, a midpoint between the bluffs on either side of the floodplain is used as the state line. Meanwhile, Carter Lake is still in Iowa.

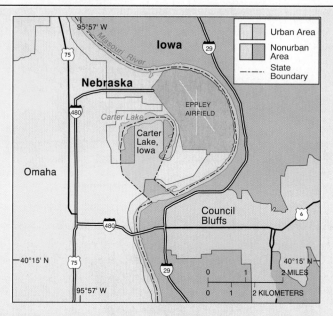

**FIGURE 1**

**A stranded town.**

Carter Lake, Iowa, sits within the curve of a former meander that was cut off by the Missouri River. The city and oxbow lake remain part of Iowa even though they are stranded within Nebraska.

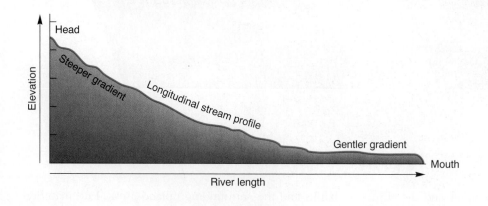

**FIGURE 14-16**

**An ideal longitudinal profile.**

Idealized cross section of the longitudinal profile of a stream, showing its gradient. Upstream segments have a steeper gradient; downstream, the gradient is gentler. The middle and lower portions in the illustration appear graded, or in dynamic equilibrium.

and behavior result from complex interactions of slope, discharge, and load, all of which are variable within different climates and with different rock types. Landscapes simply do not provide enough clear evidence to support a cyclic model of evolution. Regardless, Davis's work was a breakthrough in understanding landscapes, and many of his terms persist.

As suggested by S. A. Schumm and R. W. Lichty, two modern geomorphologists, the validity of cyclic or functional landscape models may depend on the *time frame*.

(a)  (b)

**FIGURE 14-17**

**Entrenched meanders.**

Rejuvenated (uplifted) Colorado Plateau landscape incised by entrenched meanders of the San Juan River near Mexican Hat, Utah. [(a) Photo by Betty Crowell; (b) photo by Randall M. Christopherson.]

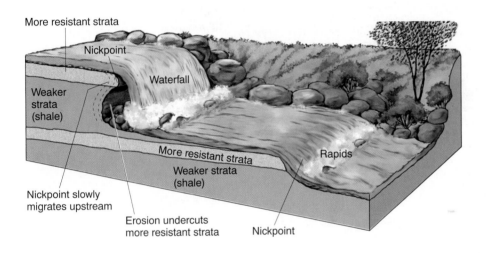

**FIGURE 14-18**

**Nickpoints interrupt a stream profile.**

Longitudinal stream profile showing nickpoints produced by resistant rock strata. Potential energy is converted into kinetic energy and concentrated at the nickpoint, accelerating erosion, which will eventually eliminate the feature.

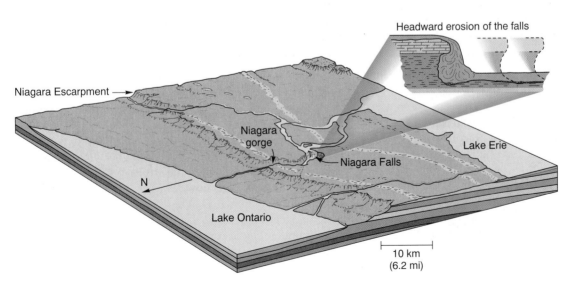

**FIGURE 14-19**

**Retreat of Niagara Falls.**

Headward retreat of Niagara Falls from the Niagara escarpment. It has taken the falls about 12,000 years to reach this position at a pace of about 1.3 m (4.3 ft) per year. [After W. K. Hamblin, *Earth's Dynamic Systems*, 6th ed., (Upper Saddle River, N.J.: Macmillan Publishing, an imprint of Prentice Hall, Inc. © 1992), fig. 12.15, p. 246. Used by permission.]

## News Report 3

## Niagara Falls Closed for Inspection

A waterfall is a break in stream gradient—a nickpoint. The ledge over which the waterfall cascades is undercut by the energy and turbulence concentrated in the free-falling water. Niagara Falls is an example of natural processes laboring to eliminate this beautiful tourist attraction, reducing this portion of the river to a series of mere rapids. In fact, the falls have retreated more than 11 km (6.8 mi) from the steep face of the Niagara escarpment (cliff) during the past 12,000 years (see Figure 14-19).

Millions of tourists visit these dramatic falls along the U.S.–Canadian border. At times, engineers have used control facilities upstream to reduce flows over the American Falls at Niagara. Once the flows have been reduced to a trickle, the area beneath the falls can be examined for possible reinforcement to save the waterfall (Figure 1). Although such inspections are infrequent, it is quite a shock to visitors to arrive and see the falls turned off as engineers scramble around the site!

**FIGURE 1**

**The American Falls turned off for inspection.**

Niagara Falls, with the American Falls portion almost completely shut off by upstream controls for inspection. Horseshoe Falls in the background is still flowing. Such inspections allow engineers to assess the progress of natural processes that are working to eliminate the Niagara Falls nickpoint. [Photo courtesy of the New York Power Authority.]

---

Let us consider three time frames: geologic time, graded time, and steady time. Over the long span of *geologic time*, cyclic models of evolutionary development might explain the disappearance of entire mountain ranges through denudation, for example. *Graded time* is between the two time frames, and within it lies the realm of dynamic equilibrium conditions. Whereas, *steady time* applies to short-term adjustments ongoing in a drainage basin.

### Stream Deposition

After weathering, mass movement, erosion, and transportation, deposition is the next logical event in a sequence. In *deposition*, a stream deposits alluvium, or unconsolidated sediments, thereby creating depositional landforms, such as floodplains, terraces, or deltas.

As discussed earlier, stream meanders tend to migrate downstream through the landscape. Over time, the landscape near a meandering river comes to bear meander scars of residual deposits from former, abandoned channels. Former point-bar deposits leave low-lying ridges, creating a *bar-and-swale relief* (a swale is a gentle low

area). The *Landsat* image in Figure 14-20 exhibits characteristic meandering scars: meander bends, oxbow lakes, natural levees, point bars, and undercut banks. The image of northwestern Mississippi shows the portion of the Mississippi River that forms the Mississippi-Arkansas border.

**Floodplains.** The flat low-lying area flanking a stream channel that is subjected to recurrent flooding is a **floodplain**. It is formed when the river overflows its channel during times of high flow. Thus, when floods occur, the floodplain is inundated. When the water recedes, it leaves behind alluvial deposits that generally mask the underlying rock under their accumulating thickness. The present river channel is embedded in these alluvial deposits. Figure 14-21 illustrates a characteristic floodplain and a representative topographic map of an area near Philipp, Mississippi.

On either bank of most streams, **natural levees** develop as by-products of flooding. When flood waters arrive, the river overflows its banks, loses velocity as it spreads out, and drops a portion of its sediment load to form the levees. Larger sand-sized particles drop out first, forming the principal component of the levees, with finer

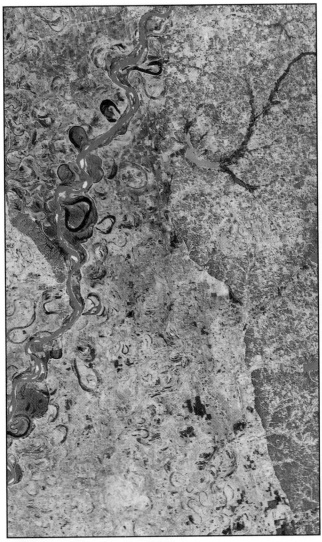

(a)

(b)

**FIGURE 14-20**

**Meander scars.**

The Mississippi River forms a portion of the Mississippi-Arkansas border near Senatobia, Mississippi. Characteristic meander patterns and scars of former channels are visible in the *Landsat* image. [Image by GEOPIC, Earth Satellite Corporation.]

silts and clays deposited farther from the river. Successive floods increase the height of the levees (levée is French for "raising"). The levees may grow in height until the river channel becomes elevated, or *perched* above the surrounding floodplain.

On the topographic map (Figure 14-21b), you can see the natural levees represented by several contour lines that run immediately adjacent to the Tallahatchie River. These contour lines (5-ft interval) denote a height of 10 to 15 feet (3 to 4.5 m) above the river and the adjoining floodplain. Next time you have an opportunity to see a river and its floodplain, look for levees (they may be low and subtle).

Notice on Figure 14-21 an area labeled **backswamp** and a stream called a **yazoo tributary**. The natural levees and elevated channel of the river prevent this tributary from joining the main channel, so it flows parallel to the river and through the backswamp area. (The name comes from the Yazoo River in the southern part of the Mississippi floodplain.)

People build cities on floodplains despite the threat of flooding because floodplains are nearly level and they are next to water. People often are encouraged by government assurances of artificial protection from floods and disaster assistance if floods occur. This assistance represents an indirect subsidy to those who chose to develop the floodplain. Government assistance may be provided by building artificial levees on top of natural levees. Artificial levees do increase the capacity in the channel, but they also lead to even greater floods when they are overtopped by flood waters or when they fail. The catastrophic floods along the Mississippi River and its tributaries in 1993 illustrate the risk of building settlements on floodplains. Damage estimates from those floods, detailed in News Report 4, exceeded $30 billion (Figure 14-22).

Perhaps the best use of some floodplains is for agriculture, because inundation generally delivers nutrients to the land with each new alluvial deposit. But floodplains that are covered with coarse sediment—sand and gravel—are not suitable for agriculture. Are there river floodplains where you live? If so, what is your impression of present land-use patterns, people's hazard perception, and local planning and zoning?

***Stream Terraces.*** As explained earlier, several factors may rejuvenate stream energy and stream-landscape relations so that a stream can scour downward with renewed vigor and increased erosion. The resulting entrenchment of the river deeper into its own floodplain produces **alluvial terraces** on either side of the valley, which look like topographic steps above the river. Alluvial terraces generally appear paired at similar elevations on each side of the valley (Figure 14-23). If more than one set of paired terraces is present, the valley probably has undergone more than one episode of rejuvenation. The flat terrace areas, above the lowest section of floodplain along the river, have always been a location for settlement.

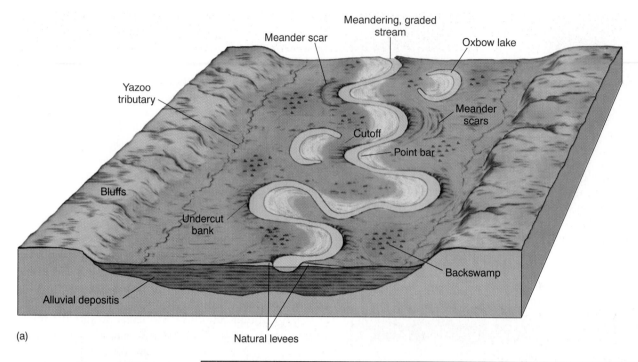

(a)

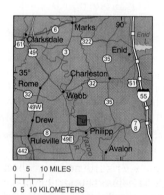

**FIGURE 14-21**
**A floodplain.**
(a) Typical floodplain landscape
and related landscape features.
(b) A portion of the Philipp,
Mississippi, topographic map
quadrangle. [(b) Topographic
map from Eros Data Center/USGS.]

(b)

## News Report 4

## The 1993 Midwest Floods

The Mississippi and Missouri Rivers are no strangers to flooding. The damage a flood might cause increases as more people settle on the vulnerable floodplains. The widespread floods of 1993 in the upper Mississippi and lower Missouri River basins exceeded peak discharge records at nearly 100 gaging stations, making it one of the greatest floods in U.S. history.

In spring 1993, a series of low-pressure systems (with their counterclockwise winds) stalled in the West, and high pressure (with its clockwise winds) dominated the Eastern Seaboard. These systems combined to produce a region of sustained convergence, instability, and thundershowers over the Midwest. Precipitation was 150% to 200% above normal for most cities in the flooded area (Figure 1). By late June, many reservoirs were filled, and soils were saturated by record rainfall over the region. Then came July, when some stations received 75 cm (30 in.) of precipitation in the first 3 weeks!

As 10,000 km (6200 mi) of levees were overtopped, more than 1000 levees were breached (broken through). Water reached almost 10 m (32 ft) above flood stage in some areas, flooding many cities and towns, and 6 million hectares (14 million acres) of forest and farmland, in the Missouri and Mississippi River drainage basins. The result was a Presidential Disaster Declaration for 487 counties in Illinois, Iowa, Kansas, Minnesota, Missouri, the Dakotas,

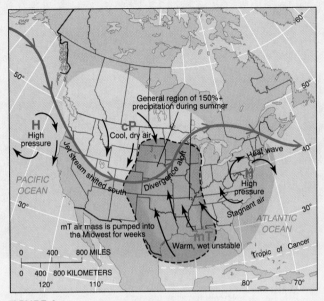

**FIGURE 1**

**Summer 1993 weather map summary.**

Weather pattern summarized for the summer of 1993. Atmospheric conditions produced a steady flow of moisture-rich, unstable, maritime tropical air from the Gulf of Mexico for a prolonged period of time.

Nebraska, and Wisconsin. Overall, damage estimates may top $30 billion.

Some cities had prepared for the disaster through improved levee construction and hazard zoning of susceptible floodplains. Many others had postponed raising local taxes for such action and had done little to prevent flood damage. This unevenness between areas created political friction. Three

remarkable aspects of "The Great Flood of 1993" were that the flood stage along some rivers was sustained for over a month, flood crests set new historical records, and many locations experienced multiple crests. This flood event was a painful reminder of the power of nature in our lives and the need for improved hazard perception, preparation, and avoidance strategies.

---

If the terraces on either side of the valley do not match in elevation, then entrenchment actions must have been continuous as the river meandered from side to side, with each meander cutting a terrace slightly lower in elevation—a condition of *unpaired terraces*. Thus, alluvial terraces represent what originally was a depositional feature (a floodplain) that subsequently has been eroded by its own stream because the stream experienced changes in stream load and capacity.

**River Deltas.** The mouth of a river is where it reaches a base level. The river's forward velocity rapidly decelerates as it enters a larger body of water. The reduced velocity causes the transported load to quickly exceed the river's

carrying capacity. Coarse sediments such as sand and gravel drop out first and are deposited closest to the river's mouth. Finer clays are carried farther and form the extreme end of the deposit. This depositional plain that forms at the mouth of a river is called a **delta** for its characteristic triangular shape, named after the Greek letter delta (Δ). The significance of the Nile River delta was perceived by Herodotus in ancient times (see News Report 5).

Each flood stage deposits a new layer of alluvium over the surface of the delta so that it grows outward. At the same time, river channels divide into smaller courses known as *distributaries*, which appear as a reverse of the dendritic drainage pattern of tributary streams discussed earlier. Here are a few examples:

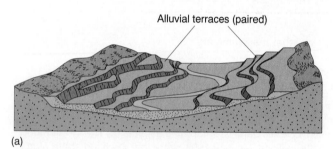

Alluvial terraces (paired)

(a)

(b)

**FIGURE 14-23**
**Alluvial stream terraces.**
(a) Alluvial terraces are formed as a stream cuts into a valley.
(b) Alluvial terraces along the Rakaia River, in New Zealand.
[(a) After W. M. Davis, *Geographical Essays* (New York: Dover, 1964 [1909]), p. 515; (b) photo by Bill Bachman/Photo Researchers, Inc.]

- The Ganges River delta features an intricate pattern of distributaries in a *braided delta*. Bountiful alluvium carried from deforested slopes upstream provides excess sediment that is deposited to form many deltaic islands (Figure 14-24).

- The Nile River delta is an *arcuate* (arc-shaped) *delta* (Figure 14-25). Also arcuate are the Danube River delta in Romania where it enters the Black Sea and the Indus River delta. (See News Report 5 for an update on the condition of the disappearing Nile Delta.)

- The Tiber River in Italy has an *estuarine delta*, one that is in the process of filling an **estuary**, which is the seaward mouth of a river where the river's fresh-water encounters seawater.

- The Mississippi River has produced a *bird-foot delta*, a long channel with many distributaries and sediments carried beyond the tip of the delta into the Gulf of Mexico.

***Mississippi River Delta.*** The Mississippi River delta has an interesting history. Over the past 120 million years, the Mississippi has collected sediments throughout its vast basin and deposited them into the Gulf of Mexico. During the past 5000 years, the river has formed a succession of seven distinct deltaic complexes along the Louisiana coast (Figure 14-26).

Each generalized lobe in the illustration reflects distinct course changes in the Mississippi River, probably where the river broke through its natural levees, probably during episodes of severe flooding. In 1966, C. R. Kolb and J. R. Van Lopik, two engineering geologists, prepared a map of this deltaic history. It shows the progressively smaller

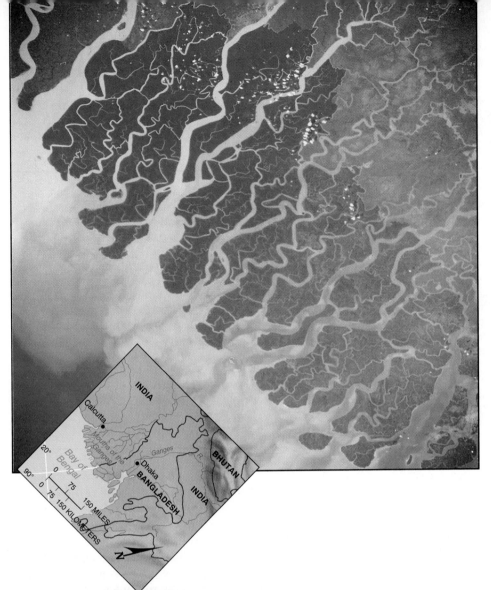

**FIGURE 14-24**
**The Ganges River enters the Bay of Bengal.**
The complex distributary pattern in the "many mouths" of the Ganges River delta in Bangladesh and extreme eastern India is visible from orbit. [Space Shuttle photo courtesy of NASA.]

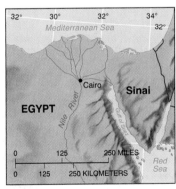

**FIGURE 14-25**
**The Nile River delta.**
The arcuate Nile River delta. Intensive agricultural activity is noted in false color (red) in the delta and along the Nile River floodplain. [Image courtesy of Earth Satellite Corporation.]

## News Report 5

### The Nile Delta Is Disappearing

People along the Nile River have depended on its regular flow and annual floods for millennia. Herodotus noted in the fifth century B.C. how the people farmed the fields in the floodplain and delta regions. After harvesting their crops, they retreated from the area to their homes. They would await the annual floods that brought fresh silt and nutrients for next year's planting. This cycle of fertility continued until the completion of the Aswan High Dam in 1964 and a partial interruption in the supply of sediment to the delta. Herodotus stated in *The History, Book Two*, "...in the part called the Delta, it seems to me that if the Nile no longer floods...for all time to come, the Egyptians will suffer."

The Aswan High Dam is blamed for holding back a lot of sediment that formerly replenished the fields in the delta. After the dam was built, the delta was no longer expanding but instead had become dominated by wave erosion along the coast. A significant amount of sediment was still in the river as it entered the delta system, but the delta coastline continues to actively recede.

D. J. Stanley, an oceanographer at the Smithsonian Institution, has proposed an intriguing answer to what is happening in addition to the impact of the dam. Over the centuries, more than 10,000 km (6200 mi) of canals have been built in the delta to augment the natural distributary system. As the river discharge enters the network of canals, flow velocity is reduced, stream competence and capacity are lost, and sediment load is deposited far short of where the delta touches the Mediterranean Sea. River flows no longer effectively reach the sea. The Nile Delta is receding from the coast at an alarming 50 to 100 m (165 ft to 330 ft) per year. Seawater is intruding farther inland in both surface water and groundwater. Human action and reaction to this evolving situation will no doubt determine the delta's future. If Herodotus could only see us now!

---

size and changing configuration of the delta (Figure 14-26a). The seventh and current delta has been building for at least 500 years.

The Mississippi River delta clearly is dynamic over time. *Landsat* images recorded in 1973 and 1989 demonstrate the changes that occurred over just a 16-year span (Figure 14-26c, d). The main channel persists because of much effort and expense directed at maintaining the artificial levee system.

To further complicate this situation, compaction and the tremendous weight of the sediments in the Mississippi River create isostatic adjustments in Earth's crust. These adjustments are causing the entire region of the delta to subside, thereby placing ever-increasing stress on natural and artificial levees and other structures along the lower Mississippi.

The city of New Orleans is now almost entirely below river level, with some sections of the city below sea level. Severe flooding is a certainty for existing and planned settlements unless further intervention or urban relocation occurs. The building of multiple flood control structures and extensive reclamation efforts by the U.S. Army Corps of Engineers apparently have only delayed the peril, as demonstrated by recent flooding.

An additional problem for the lower Mississippi Valley is the possibility, in a worst-case flood, that the river could break from its existing channel and seek a new route to the Gulf of Mexico. If you examine the map in Figure 14-26b, you can see that the Mississippi's obvious alternative is the Atchafalaya River. The Atchafalaya would

provide a much shorter route to the Gulf of Mexico, less than one-half the present distance, and it has a steeper gradient than the Mississippi. For the Mississippi to bypass New Orleans entirely would be a blessing, for it would remove the flood threat, but it would be a financial disaster, as a major U.S. port would silt in and seawater would intrude into freshwater resources.

At present, artificial barriers block the Atchafalaya from reaching the Mississippi at the point shown; without the floodgates they do connect. The Old River Control Project (1963) maintains three structures and a lock about 320 km (200 mi) from the Mississippi's mouth to keep these rivers in their channels. The potential causes of a breach could include both high water and sediment deposition. A major flood is only a matter of time, and residents should prepare for the river-channel change.

***Why Some Rivers Lack Deltas.*** The Amazon River, Earth's highest-discharge stream, exceeds 175,000 m³/s (6.2 million cfs) discharge and carries sediments far into the deep Atlantic offshore. Yet the Amazon lacks a true delta. Its mouth, 160 km (100 mi) wide, has formed an underwater deltaic plain deposited on a sloping continental shelf. As a result, the Amazon's mouth is braided into a broad maze of islands and channels (see Space Shuttle photo in Figure 9-22).

Other rivers also lack deltaic formations if they do not produce significant sediment or if they discharge into strong erosive currents. The Columbia River of the U.S. Northwest lacks a delta because offshore currents remove sediment before it can accumulate into a delta.

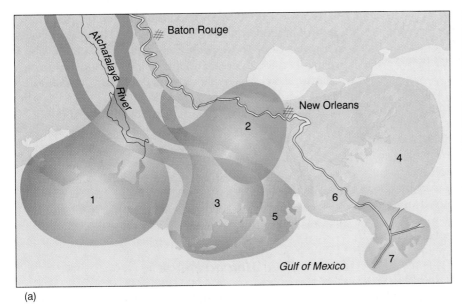

(a)

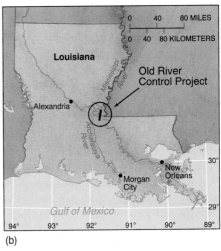

(b)

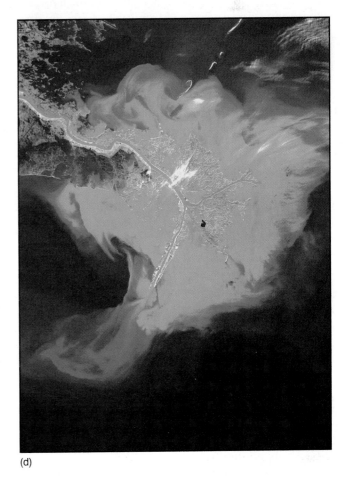

(c)                                                     (d)

**FIGURE 14-26**

**The Mississippi River delta.**

(a) Evolution of the present delta, from 5000 years ago (1) to present (7). (b) Location map of the Old Control Structure and potential "capture point" (arrow) of Atchafalaya River. The bird-foot delta of the Mississippi River exhibits change over time, as shown in these two *Landsat* images from 1973 (c) and 1989 (d). The continuous supply of sediments, focused by controlling levees, extends ever farther into the Gulf of Mexico, although subsidence of the delta and rising sea level have diminished the overall surface area. [(a) Adapted from C. R. Kolb and J. R. Van Lopik, "Depositional environments of the Mississippi River deltaic plain," in M. L. Shirley, ed., *Deltas in Their Geologic Framework* (Houston: Houston Geological Society, 1966), p. 22. Adapted by permission. (b) *Landsat* images courtesy of EROS Data Center.]

# Floods and River Management

Throughout history, civilizations have settled floodplains and deltas, especially since the agricultural revolution of 10,000 years ago, when the fertility of floodplain soils was discovered. Early villages generally were built away from the area of flooding, or on stream terraces, because the floodplain was dedicated exclusively to farming. However, as commerce grew, competition for sites near rivers grew, because these locations were important for transportation. Port and dock facilities were built, as were river bridges. Because water is a basic industrial raw material used for cooling and for diluting and removing wastes, waterside industrial sites became desirable. These competing human activities on vulnerable flood-prone lands require planning to avoid losing lives and property during floods.

Catastrophic floods continue to be a threat, especially in poor nations. Bangladesh is perhaps the most persistent example. A combination of human abuse of the land and intense monsoonal rains and tropical cyclones in 1988 and 1991 created devastating floods over the country's vast alluvial plain, which sprawls over an area the size of Alabama (130,000 km², or 50,000 mi²). Bangladesh is one of the most densely populated countries on Earth, and more than *three-fourths* of it is on floodplains.

The flooding severity is magnified as a consequence of human economic activities. Excessive forest harvesting in the upstream portions of the Ganges-Brahmaputra River watersheds increased runoff. Over time, the increased load carried by the river was deposited in the Bay of Bengal, creating new islands (see Figure 14-24). These islands, barely above sea level, had become sites of new farming villages, and as a result, about 150,000 people perished in the 1988 and 1991 floods. When the floodwaters finally did recede, the lack of freshwater—coupled with crop failures, epidemics, and pestilence—led to famine and the death of more people, in the tens of thousands. About 30 million people, the population of California, were left homeless, and many of the alluvial islands had disappeared.

## Rating Floodplain Risk

A **flood** is a high water level that overflows the natural (or artificial) levees along any portion of a stream. Both floods and the floodplains they might occupy are rated statistically for the expected time intervals between floods. Thus, you hear about "10-year floods," "50-year floods," and so on. A *10-year flood* is the greatest level of flooding that is likely to occur once every 10 years. This also means that such flooding has only a 10% likelihood of occurring in any one year and is likely to occur about 10 times each century. For any given floodplain, such a frequency indicates a moderate threat.

A 50-year or 100-year flood is of greater and perhaps catastrophic consequence, but it is also less likely to occur in a given year. These probability ratings of flood levels are mapped for an area, and the defined floodplains that result are then labeled as a "50-year floodplain," or a "100-year floodplain."

These statistical estimates are probabilities that events will occur randomly during any single year of the specified period. Of course, two decades might pass without a 50-year flood, or a 50-year level of flooding could occur three years in a row. The record-breaking Mississippi River Valley floods in 1993 easily exceeded a 1000-year flood probability. See Focus Study 14-1 for more about floodplain hazards and management strategies.

## Streamflow Measurement

Understanding flood patterns in a drainage basin is as complex as understanding the weather, for floods and weather are equally variable, and both include a level of unpredictability. Measuring and analyzing the behavior of each large watershed and stream enable engineers and concerned parties to develop the best possible flood management strategy. Unfortunately, reliable data often are not available for small basins or for the changing landscapes of urban areas.

The key to flood avoidance or management is to possess extensive measurements of *streamflow*, a stream's discharge and its pattern (Figure 14-27). Once the cross section of a stream is fully measured, only the stream level is needed to determine discharge (using the calculation: discharge = width × depth × velocity. A *staff guage* (a pole marked with water levels) is placed in a stream, as shown in the figure, to measure stream level. Another method involves a *stilling well* on the stream bank with a guage mounted in it to measure stream level. A movable current meter can be used to sample stream velocity at various locations.

Approximately 11,000 stream gaging stations (hydrologists use this spelling) are used in the United States (an average of over 200 per state). Of these, 7000 are operated by the U.S. Geological Survey and have continuous recorders for stage (level) and discharge. Many of these stations automatically send telemetry data to satellites, from which information is retransmitted to regional centers (Figure 14-27). Environment Canada's Water Survey of Canada maintains more than 3000 gaging stations.

**Hydrographs.** A graph of stream discharge over time for a specific place is called a **hydrograph**. The hydrograph in Figure 14-28 shows the relation between precipitation input (the bar graph) and stream discharge (the curves). During dry periods, at low-water stages, the flow is described as base flow and is largely maintained by input from local groundwater (purple line).

## Focus Study 14-1

## Floodplain Strategies

Management and mitigation of flood hazards is a complex process. In 1969, Hurricane Camille's remnants combined with an existing low-pressure area and moved from west to east over the entire James River Basin in Virginia. According to unofficial measurements, rainfall amounts reached 78.7 cm (31 in.) within the 3-day storm period. The basin's narrow, steep valleys quickly concentrated runoff into a flood surge that peaked at Richmond, Virginia, as a 1000-year flood occurrence. This James River flood was the greatest since recordkeeping began in 1771.

Most of the residents of the mountain hollows, hamlets, and towns were asleep when the storm began. Little warning was possibleRapidly rising streams and landslides caused by the unprecedented rainfall destroyed homes as the occupants slept....The raging floods passed out of the small headwater streams and consolidated in the normally placid James River.*

Ironically, the remnants of Hurricane Agnes in 1972—only 3 years later—produced severe floods in that same region. Those floods also had a magnitude reaching the 1000-year probability of occurrence. So much for feeling safe with probability estimates for potential disasters!

Detailed measurements of streamflows and flood records have been kept rigorously in the United States only for about 100 years, in particular since the 1940s. At any selected location along any given stream, the *probable maximum flood* (PMF) is a hypothetical flood of such a magnitude that there is virtually no possibility it will be exceeded. Because floods are produced by the collection and concentration of rainfall, hydrologists speak of a corollary, the *probable maximum precipitation* (PMP) for a given drainage basin, which is an

*H. J. Thompson, "The James River flood of August 1969," *Weatherwise* 22, no. 5 (October 1969): 183.

amount of rainfall so great that it will never be exceeded.

These parameters are used by hydrologic engineers to establish a *design flood* against which to take protective measures. For urban areas near creeks, planning maps often include survey lines for a 50-year or a 100-year floodplain; such maps have been completed for most U.S. urban areas. The design flood usually is used to enforce planning restrictions and special insurance requirements. Restrictive zoning using these floodplain designations is an effective way of avoiding potential damage.

Unfortunately, such political action is not generally implemented, and the scenario all too often goes like this: (1) Minimal zoning precautions are not carefully supervised; (2) a flooding disaster occurs; (3) the public is outraged at being caught off guard; (4) businesses and homeowners are surprisingly resistant to stricter laws and enforcement; (5) eventually another flood refreshes the memory and promotes more knee-jerk planning. As strange as it seems, *there is little indication that our risk perception improves as the risk increases*.

### A Few Strategies

A planning strategy in some large river systems is to develop artificial floodplains. This is done by constructing bypass channels to accept seasonal or occasional floods. When not flooded, the bypass channel can serve as farmland, often benefiting from the occasional soil-replenishing inundations. When the river reaches flood stage, large gates called weirs are opened, allowing the water to enter the bypass channel. This alternate route relieves the main channel of the burden of carrying the entire discharge.

Dams (an artificial structure placed in a river channel) and reservoirs (an impoundment of water behind a dam; a human-made lake) are common streamflow control methods within a watershed. For conservation purposes, a dam holds back seasonal peak flows for distribution

during low-water periods. In this way, streamflows are regulated to assure year-round water supplies. Dams also are constructed for flood control, to hold back excess flows for later release at more moderate discharge levels. Adding hydroelectric power production to these functions of conservation and flood control can define a modern multipurpose reclamation project.

The function of reservoir impoundment is to provide flexible storage capacity within a watershed to regulate river flows, especially in a region with variable precipitation. Figure 1 shows one reservoir during drought conditions and during a time of wetter weather 6 years later.

### Reservoir Considerations

Unfortunately, the multipurpose benefits of reservoir construction are countered by some negative consequences. The area upstream from a dam becomes permanently drowned. In mountainous regions, this may mean loss of white water rapids and recreational sections of a river. In agricultural areas, the ironic end result may be that a hectare of farmland is inundated upstream to preserve a hectare of farmland downstream.

Furthermore, dams built in warm and arid climates lose substantial water to evaporation, compared with the free-flowing streams they replace. Reservoirs in the southwestern United States can lose 3–4 meters (10–13 ft) of water a year. Also, sedimentation can reduce the effective capacity of a reservoir and can shorten a dam's life span, as mentioned earlier regarding Glen Canyon Dam.

Large multipurpose projects invariably produce political conflict over territorial rights and questions of public trust versus private right to the environment. Vast environmental disruption for the sake of economic gain no longer appears popular with the public. The James Bay Project in central and northern Québec is a case in point.

(a)

(b)

**FIGURE 1**
**Reservoir extremes.**
Comparative photographs of the New Hogan reservoir, central California, during (a) dry and (b) wet weather conditions. [Photos by author.]

Launched in 1970 by Hydro-Québec and only one-third complete at this time, the project might eventually include 215 dams, 25 power stations, and 20 river diversions. Many unexpected environmental problems have arisen because no environmental impact studies were completed at the outset. The early stages remain a huge experiment with fragile ecosystems. Well into the planning phase, the public learned that corporate interests sought inexpensive, publicly generated electric power, and that much of it was for export to the United States—all at public expense.

The second major phase of the James Bay Project, the Great Whale project (named for a local river), may never be completed because of the success of conservation programs begun by utilities in the northeastern United States and court challenges to assess impacts before further construction. In addition, the state of New York in 1992 withdrew its offer to buy 1000 megawatts of power from the Hydro-Québec project.

**Geologic Assessment**

One final concern is the need for great care in geologic assessment of any dam site, for dam failures occur more often than many people realize. In Chapter 13, the Vaiont Reservoir disaster was detailed. In 1972, two U.S. dams failed, one near Rapid City, South Dakota (237 dead), and another at Buffalo Creek, West Virginia (118 dead). The U.S. General Accounting Office estimates that about 1900 unsafe dams exist near urban areas in all parts of the country, in a variety of designs and sizes, and under different circumstances.

As a case in point, Teton Dam collapsed in 1976, near Rexburg, Idaho, releasing a large lake—more than 303 billion liters (80 billion gallons) of water—destroying 41,000 hectares (100,000 acres) of farmland, killing 16,000 head of livestock, and causing more than $1 billion in property damage (Figure 2). Congressional testimony at the time disclosed: "The principal human cause of failure of the Teton Dam was very poor site selection." After engineering surveys disclosed specific geologic problems at the chosen site, construction of Teton Dam continued anyway. The dam survived less than one month after filling began!

When rainfall occurs in some portion of the watershed, the runoff collects and is concentrated in streams and tributaries. The amount, location, and duration of the rainfall episode determine the *peak flow*. Also important is the nature of the surface in a watershed; for example, a hydrograph for a specific portion of a stream changes after a forest fire or urbanization of the watershed.

Human activities have enormous impact on water flow in a basin. The effects of urbanization are quite dramatic, both increasing and hastening peak flow, as you can see by comparing preurban stream flow (red curve) and urbanized stream flow (blue) in Figure 14-28. In fact, urban areas produce runoff patterns quite similar to those of deserts. The sealed surfaces of the city drastically

**FIGURE 2**
**Disaster remnants.**
The failed remains of Teton Dam in Idaho. [Photo by author.]

reduce infiltration and soil-moisture recharge; their effect is similar to that of the hard, nearly barren surfaces of the desert. A significant part of urban flooding occurs because of the alteration in surfaces.

A useful engineering tool is the *unit hydrograph*, which depicts a unit depth, such as a centimeter or inch, of effective rainfall spread uniformly over a drainage basin and received during a specific period (1 to 24 hours, depending on basin size). The unit hydrograph demonstrates how a basin behaves under various rainfall patterns and is used to design water-management facilities and to assess potential impact on cities downstream.

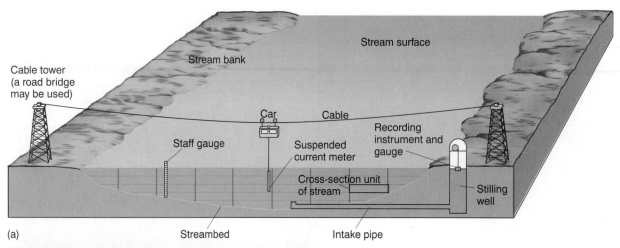

(a)

(b)

**FIGURE 14-27**
**Streamflow measurement.**
(a) A typical streamflow measurement installation may use a variety of devices: staff gauge, stilling well with recording instrument, and suspended current meter. (b) An automated hydrographic station sends telemetry to a satellite for collection by the USGS. [Photo courtesy of California Department of Water Resources.]

## A Final Thought about Floods

The benefit of any levee, bypass, or other project intended to prevent flood destruction is measured in avoided damage and is used to justify the cost of the protection facility. Thus, ever-increasing damage leads to the justification of ever-increasing flood control structures. All such strategies are subjected to cost-benefit analysis, but bias is a serious drawback because such an analysis usually is prepared by an agency or bureau with a vested interest in building more flood control projects.

As suggested in an article titled "Settlement Control Beats Flood Control,"[*] there are other ways to protect populations than with enormous, expensive, sometimes environmentally disruptive projects. Strictly zoning the floodplain is one approach. However, the flat, easily developed floodplains near pleasant rivers are desirable for housing, and thus weaken political resolve. A reasoned zoning strategy would set aside the floodplain for farming or passive recreation, such as a riverine park, golf course, or plant and wildlife sanctuary, or for other uses that are not hurt by natural floods. This study concludes that "urban and industrial losses would be largely obviated by set-back levees and zoning and thus cancel the biggest share of the assessed benefits which justify big dams."

[*]Walter Kollmorgen, *Economic Geography* 29, no. 3 (July 1953): 215.

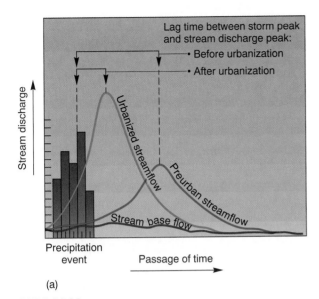

(a)

(b)

**FIGURE 14-28**

**Urban flooding.**

(a) Effect of urbanization on a typical stream hydrograph. Normal base flow is indicated with a dark blue line. The purple line indicates discharge after a storm, before urbanization. Following urbanization, stream discharge dramatically increases, as shown by the light blue line. (b) Severe flooding of an urban area in Linda, California, after a levee break on the Sacramento River in 1986. [(b) Photo from California Department of Water Resources.]

# Summary and Review — River Systems and Landforms

✔ *Define* **the term fluvial and** *outline* **the fluvial processes: erosion, transportation, and deposition.**

River systems, fluvial processes and landscapes, floodplains, and river control strategies are important to populations in affected areas and as demands for limited water resources increase. Stream-related processes are called **fluvial**. Water dislodges, dissolves, or removes surface material in the process called **erosion**. Streams produce *fluvial erosion*, in which weathered sediment is picked up for **transport** to new locations. Sediments are laid down by another process, **deposition**. **Alluvium** is the general term for the clay, silt, and sand deposited by running water.

**Base level** is the lowest elevation limit of stream erosion. A *local* base level occurs when something interrupts the stream's ability to achieve base level, such as is created by a dam or a landslide that blocks a stream channel.

> fluvial (p. 422)
> erosion (p. 422)
> transport (p. 422)
> deposition (p. 422)
> alluvium (p. 422)
> base level (p. 422)

1. What role is played by rivers in the hydrologic cycle?

2. Define the term *fluvial*. What is a fluvial process?

3. What is the sequence of events that takes place as a stream dislodges material?

4. Explain the base level concept. What happens to a local base level when a reservoir is constructed?

---

✔ *Construct* **a basic drainage basin model and** *identify* **different types of drainage patterns, with examples.**

The basic fluvial system is a **drainage basin**, which is an open system. *Drainage divides* define the **watershed** catchment area of the drainage basin. In any drainage basin, water initially moves downslope in a thin film called **sheetflow**, or *overland flow*. This surface runoff concentrates in *rills*, or small-scale downhill grooves, which may develop into deeper *gullies* and a stream course in a valley. A high ground that separates one valley from another and directs sheet flow is termed an *interfluve*. Extensive mountain and highland regions act as **continental divides** that separate major drainage basins.

**Drainage density** is determined by the number and length of channels in a given area and is an expression of a landscape's topographic surface appearance. **Drainage**

**pattern** refers to the arrangement of channels in an area as determined by the steepness, variable rock resistance, variable climate, hydrology, relief of the land, and structural controls imposed by the landscape. There are seven basic drainage patterns generally found in nature: dendritic, trellis, radial, parallel, rectangular, annular, and deranged.

> drainage basin (p. 422)
> watershed (p. 422)
> sheetflow (p. 422)
> continental divides (p. 423)
> drainage density (p. 424)
> drainage pattern (p. 424)

5. What is the spatial geomorphic unit of an individual river system? How is it determined on the landscape? Define the several relevant key terms used.

6. On Figure 14-4, follow the Allegheny River system to the Gulf of Mexico, analyze the pattern of tributaries, and describe the channel. What role do continental divides play in this drainage?

7. Describe drainage patterns. Define the various patterns that commonly appear in nature. What drainage patterns exist in your home town? Where you attend school?

---

✔ ***Describe* the relation among velocity, depth, width, and discharge and *explain* the various ways that a stream erodes and transports its load.**

Stream channels vary in *width* and *depth*. The streams that flow in them vary in *velocity* and in the *sediment load* they carry. All of these factors may increase with increasing discharge. *Discharge* is calculated by multiplying the velocity of the stream by its width and depth for a specific cross section of the channel.

**Hydraulic action** is the work of *turbulence* in the water. Running water causes hydraulic squeeze-and-release action to loosen and lift rocks and sediment. As this debris moves along, it mechanically erodes the streambed further, through a process of **abrasion**.

*Solution* refers to the **dissolved load** of a stream, especially the chemical solution derived from minerals such as limestone or dolomite or from soluble salts. The **suspended load** consists of fine-grained, clastic particles held aloft in the stream, with the finest particles not deposited until the stream velocity slows nearly to zero. **Bed load** refers to coarser materials that are dragged along the stream bed by **traction** or are rolled and bounced along by **saltation**. If the load in a stream exceeds its capacity, sediments accumulate as **aggradation** as the stream channel builds through deposition. With excess sediment, a stream becomes a maze of interconnected channels that form a **braided stream** pattern.

> hydraulic action (p. 428)
> abrasion (p. 428)
> dissolved load (p. 429)
> suspended load (p. 429)
> bed load (p. 429)

> traction (p. 429)
> saltation (p. 429)
> aggradation (p. 431)
> braided stream (p. 431)

8. What was the impact of flood discharge on the channel of the San Juan River near Bluff, Utah? Why did these changes take place?

9. How does stream discharge do its erosive work? What are the processes at work in the channel?

10. Differentiate between stream competence and stream capacity.

11. How does a stream transport its sediment load? What processes are at work?

---

✔ ***Develop* a model of a meandering stream, including point bar, undercut bank, and cutoff, and *explain* the role of stream gradient in these flow characteristics.**

Where the slope is gradual, stream channels develop a sinuous form called a **meandering stream**. The outer portion of each meandering curve is subject to the fastest water velocity and can be the site of a steep **undercut bank**. On the other hand, the inner portion of a meander experiences the slowest water velocity and forms a **point bar** deposit. When a meander neck is cut off as two undercut banks merge, the meander becomes isolated and forms an **oxbow lake**.

Every stream develops its own **gradient** and establishes a longitudinal profile. A portion of the stream is designated a **graded stream** when the stream is adjusted among available discharge, channel characteristics, its velocity, and the load supplied from the drainage basin. An interruption in a stream's longitudinal profile is called a **nickpoint**. This can occur as the stream flows across hard resistant rock or after tectonic uplift episodes.

> meandering stream (p. 432)
> undercut bank (p. 433)
> point bar (p. 433)
> oxbow lake (p. 433)
> gradient (p. 433)
> graded stream (p. 433)
> nickpoint (p. 434)

12. Describe the flow characteristics of a meandering stream. What is the pattern of flow in the channel? What are the erosional and depositional features and the typical landforms created?

13. Explain these statements: (a) All streams have a gradient, but not all streams are graded. (b) Graded streams may have ungraded segments.

14. Why is Niagara Falls an example of a nickpoint? Without human intervention, what do you think would eventually take place at Niagara Falls?

15. What is meant by "the validity of cyclic or equilibrium models depends on which of three time frames is being considered?" Explain and discuss.

---

✔ *Define* **a floodplain and** *analyze* **the behavior of a stream channel during a flood.**

Floodplains have been an important site of human activity throughout history. Rich soils, bathed in fresh nutrients by floodwaters, attract agricultural activity and urbanization. Despite our knowledge of historical devastation by floods, floodplains are settled, raising issues of human hazard perception. The flat low-lying area along a stream channel that is subjected to recurrent flooding is a **floodplain**. It is formed when the river overflows its channel during times of high flow. On either bank of most streams, **natural levees** develop as by-products of flooding. On the floodplain, **backswamps** and **yazoo tributaries** may develop. The natural levees and elevated channel of the river prevent a yazoo tributary from joining the main channel, so it flows parallel to the river and through the backswamp area. **Alluvial terraces** are formed by the entrenchment of a river into its own floodplain.

> floodplain (p. 438)
> natural levees (p. 438)
> backswamp (p. 439)
> yazoo tributary (p. 439)
> alluvial terraces (p. 439)

16. Describe the formation of a floodplain. How are natural levees, oxbow lakes, backswamps, and yazoo tributaries produced?

17. Can you identify any of the features listed in question 16 on the Philipp, Mississippi, topographic quadrangle in Figure 14-21b?

18. Are there any floodplains near where you live or where you go to college? Have you seen any of these features?

---

✔ *Differentiate* **the several types of river deltas and** *detail* **each.**

A depositional plain formed at the mouth of a river is called a **delta**. When the mouth of a river enters the sea and is inundated by the sea in a mix with freshwater it is called an **estuary**.

> delta (p. 441)
> estuary (p. 442)

19. What is a river delta? What are the various deltaic forms? Give some examples.

20. How might life in New Orleans change in the next century? Explain.

21. Describe the Ganges River delta. What factors upstream explain its form and pattern? Assess the consequences of settlement on this delta.

22. What is meant by the statement "the Nile River delta is disappearing."

---

✔ *Explain* **flood probability estimates and** *review* **strategies for mitigating flood hazards.**

A **flood** occurs when high water overflows the natural or artificial levees of a stream. Both floods and the floodplains they might occupy are rated statistically for the expected time interval between floods. A 10-year flood is the greatest level of flooding that is likely once every 10 years. A graph of stream discharge over time for a specific place is called a **hydrograph**. A *unit hydrograph* is a useful engineering tool that depicts a unit depth, such as a centimeter or inch, of effective rainfall spread uniformly over a drainage basin and received during a specific period.

Collective efforts by government agencies undertake to reduce flood probability. Such management attempts include the construction of artificial levees, bypasses, straightened channels, diversions, dams, and reservoirs. Society is still learning how to live in a sustainable way with Earth's dynamic river systems.

> flood (p. 446)
> hydrograph (p. 446)

23. Specifically, what is a flood? How are such flows measured and tracked?

24. Differentiate between a hydrograph from a natural terrain and one from an urbanized area.

25. What do you see as the major consideration regarding floodplain management? How would you describe the general attitude of society toward natural hazards and disasters?

26. What do you think the author of the article "Settlement Control Beats Flood Control" meant by the title? Explain your answer, using information presented in the chapter.

---

# NetWork

The *Geosystems Home Page* provides on-line resources for this chapter on the World Wide Web. You will find review exercises, specific updates for items in the chapter, suggested readings, and links to interesting related pathways on the Internet (click on the Table of Contents link and select this chapter). *Geosystems* is at: **http://www.prenhall.com/geosystm**

# 15

# Eolian Processes and Arid Landscapes

**The Work of Wind**

**Overview of Desert Landscapes**

**Summary and Review**

## Key Learning Concepts

After reading the chapter, you should be able to:

- *Characterize* the unique work accomplished by wind and eolian processes.
- *Describe* eolian erosion, including deflation, abrasion, and the resultant landforms.
- *Describe* eolian transportation; and *explain* saltation and surface creep.
- *Identify* the major classes of sand dunes and *present* examples within each class.
- *Define* loess deposits, their origins, locations, and landforms.
- *Portray* desert landscapes and *locate* these regions on a world map.

*Totem Pole and the Yei-Bi-Chei Dancers formation in Monument Valley Tribal Park, Navajo Nation, along the Utah-Arizona border.* [Photo by author.]

ind is an agent of geomorphic change. Like
moving water, moving air (wind) causes erosion, transportation, and deposition of materials. Like moving water, moving air is a fluid, and it
behaves similarly, although it has a lower viscosity (it is
"thinner") than water. Wind's effectiveness as a geomorphic agent has been the subject of much debate; in fact,
wind at times was thought to produce major landforms.
Scientists presently regard wind as having relatively minor
effects on weathering and erosion, but it is significant
enough to deserve our attention.

Wind processes modify and move sediment in deserts
and along coastlines in a variety of climates. Wind may
contribute to soil formation in places far distant from the
point of origin, bringing fine material from regions where
glaciers deposited it. Elsewhere, fallow fields (those not
planted) give up their soil resource to destructive wind
erosion. Scientists are only now getting an accurate picture of the amount of wind-blown dust that fills the atmosphere. In this chapter we examine the work of wind,
associated processes, and resulting landforms. For convenience of organization, we include the discussion of
arid lands in this chapter.

Earth's dry lands stand out in stark contrast on the
water planet. In desert environments, an overall lack of
moisture and stabilizing vegetation allows wind to create
extensive sand seas and dunes of infinite variety. The polar
regions are deserts as well, so there should be no surprise
that Yuma, Arizona, and stations in Antarctica receive the
same amount of annual precipitation. These polar and
high-latitude deserts possess unique features related to their
cold, dry environment—aspects covered in Chapter 17.

Arid landscapes display unique landforms and life
forms: "Instead of finding chaos and disorder the observer
never fails to be amazed at a simplicity of form, an exac-
titude of repetition and a geometric order" in the desert.[*]

## The Work of Wind

The work of the wind—erosion, transportation, and deposition—is called **eolian** (also spelled aeolian.) The word
comes from Aeolus, ruler of the winds in Greek mythology. Much eolian research was accomplished by a British
Army major, Ralph Bagnold, who was stationed in Egypt
in 1925. An engineering officer who spent much of his
time in the deserts west of the Nile, Bagnold measured,
sketched, and developed hypotheses about the wind and
desert forms. Imagine the scene in the 1920s: A Model-T
Ford chugs across the desert west of Cairo taking Bagnold
to his research area; rolls of chicken wire in the back are

[*]R.A. Bagnold, *The Physics of Blown Sand and Desert Dunes* (London:
Methuen, 1941).

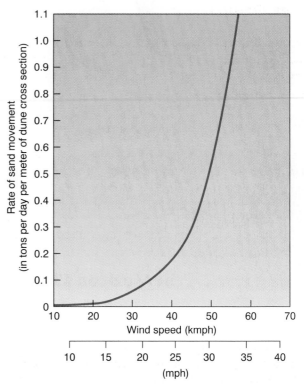

**FIGURE 15-1**
**Sand movement and wind velocity.**
Sand movement relative to wind velocity, as measured over a square
meter cross section of ground surface. [After R. A. Bagnold, *The
Physics of Blown Sand and Desert Dunes* (London: Methuen, 1941.
Adapted by permission.]

laid down over treacherous stretches of sand! Bagnold's
often-cited work, *The Physics of Blown Sand and Desert
Dunes*, was published in 1941 after he had completed
wind-tunnel simulation of windy desert conditions.

The actual ability of wind to move materials is small
compared with that of other transporting agents such as
water and ice, because air is so much less dense than those
other media. Yet, over time, wind accomplishes enormous
work. Bagnold studied the ability of wind to transport sand
over the surface of a dune. Figure 15-1 shows that a steady
wind of 50 kmph (30 mph) can move approximately one-
half ton of sand per day over a square meter section of
dune. The graph also demonstrates how rapidly the
amount of transported sand increases with wind speed.

Grain size is important in wind erosion. Intermedi-
ate-sized grains are moved most easily. It is the largest
and the smallest sand particles that require the strongest
winds to move. The large particles are heavier, and thus
require stronger winds. Small particles are difficult to
move because they exhibit a mutual cohesiveness and
because they usually present a smooth surface (aerodynamic) to the wind.

## Eolian Erosion

Two principal wind-erosion processes are **deflation**, the removal and lifting of individual loose particles, and **abrasion**, the grinding of rock surfaces with a "sand-blasting" action by particles captured in the air. Deflation and abrasion produce a variety of distinctive landforms and landscapes.

*Deflation.* Deflation literally blows away loose or non-cohesive sediment. Fine materials are eroded away by wind deflation and moving water, leaving behind a concentration of pebbles and gravel called **desert pavement**. Resembling a cobblestone street, desert pavement protects underlying sediment from further deflation and water erosion (Figure 15-2). Desert pavements are so common that many provincial names are used for them—for example, *gibber plain* in Australia; *gobi* in China; and in Africa, *lag gravels* or *serir*, or *reg* desert if some fine particles remain. Water is an important factor in the formation of desert pavement, washing away fine materials and concentrating and cementing the remaining rock pieces.

Heavy recreational activity damages fragile desert landscapes, especially in the arid lands of the United States, where over 14 million off-road vehicles (ORVs) are now in use. Such vehicles crush plants and animals; disrupt desert pavement, leading to greater deflation; and create ruts that easily concentrate sheetwash to form gullies. Measures to restrict their use to specific areas, preserving the remaining desert, are controversial. Military activities can also threaten delicate desert landscapes (see News Report 1).

Wherever wind encounters loose sediment, deflation may remove enough material to form basins called **blowout depressions**. These range from small indentations less than a meter wide up to areas hundreds of meters wide and many meters deep. Chemical weathering, although slow in the desert owing to the lack of water, is important in the formation of a blowout, for it removes the cementing materials that give particles their cohesiveness. (Remember, that in arid climates chemical weathering is active on the surfaces of particles at the microscopic level, utilizing hygroscopic and capillary water.)

Large depressions in the Sahara Desert are at least partially formed by deflation. The enormous Munkhafad el Qaṭṭâra (Qaṭṭâra Depression) just inland from the Mediterranean Sea in the Western Desert of Egypt, which covers 18,000 km² (7000 mi²), is now about 130 m (427 ft) below sea level at its lowest point.

*Abrasion.* You may have seen work crews sandblasting surfaces on buildings and bridges to clean them or on streets to remove unwanted markings. Sandblasting uses a stream of compressed air filled with sand grains to quickly and smoothly abrade a surface. Abrasion by wind-blown

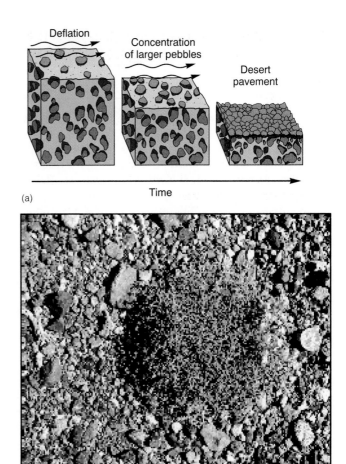

(a)

**FIGURE 15-2**
**Desert pavement.**
(a) Desert pavement is formed from larger rocks and fragments left after deflation and sheetwash. (b) A typical desert pavement. [Photo by author.]

particles is nature's slower version of sandblasting, and it is especially effective at polishing exposed rocks when the abrading particles are hard and angular.

Variables that affect the rate of abrasion include the hardness of surface rocks, wind velocity, and wind constancy. Abrasive action is restricted to the area immediately above the ground, usually no more than a meter or two in height, because sand grains are lifted only a short distance.

Rocks exposed to eolian abrasion appear pitted, grooved, or polished. They usually are aerodynamically shaped in a specific direction, according to the consistent flow of airborne particles carried by prevailing winds. Rocks that bear such evidence of eolian erosion are called **ventifacts** (literally, "artifacts of the wind").

*Yardangs.* On a larger scale, deflation and abrasion are capable of streamlining rock structures that are aligned parallel to the most effective wind direction, leaving behind distinctive, elongated ridges called **yardangs**.

## News Report 1

### War Tears up the Pavement

A serious environmental impact of the 1991 Persian Gulf War was the disruption of desert pavement. Thousands of square kilometers of stable desert pavement were shattered by bombardment with many thousands of tons of explosives and were disrupted by the movement of heavy vehicles and equipment.

The resulting loosened sand and silt is now available for deflation and thus threatens cities and farms with increased dust and sand accumulations. Given the firepower available in modern technological warfare, environmental assessment would be a wise addition to strategic planning.

**FIGURE 15-3**
**A yardang.**
A wind-sculpted landform in the Tassilin Ajjer, Algeria, desert region of Saharan Africa. [Photo by Mario Fantin/Photo Researchers, Inc.]

These wind-sculpted features can range from meters to kilometers in length and up to many meters in height. Some are large enough to be seen from spacecraft (see News Report 2). Abrasion is concentrated on the windward end of each yardang, with deflation operating on the leeward portions (Figure 15-3). The Sphinx in Egypt, perhaps partially formed as a yardang, suggesting a head and body to the ancients. Some scientists think this shape led them to complete the bulk of the sculpture artificially with masonry.

### *Eolian Transportation*

As mentioned in Chapter 6, atmospheric circulation is capable of transporting fine material, such as volcanic debris, worldwide within days. Wind exerts a drag, or frictional pull, on surface particles until they become airborne, just as water in a stream picks up sediment (again, think of air as a fluid). The distance that wind is capable

## News Report 2

### Yardangs from Mars

Wind deflation and abrasion streamline rock structures and leave behind distinctive, elongated yardangs. On Earth, some yardangs are large enough to be detected on satellite imagery. The Ica Valley of southern Peru contains yardangs reaching 100 m (330 ft) in height and several kilometers in length, and yardangs in the Lut Desert of Iran attain 150 m (490 ft) in height.

Spacecraft in orbit around Mars use remote sensors to study the Martian surface. Images have disclosed curious features on Mars that suggest yardangs, or wind-sculpted formations. These images provide clues as to the nature of the Martian surface: a windy place of frequent dust storms and enough wind-blown sediment to sandblast yardangs.

of transporting particles varies greatly with particle size. Only the finest dust particles travel significant distances, so the finer material suspended in a *dust storm* is lifted much higher than the coarser particles of a *sand storm*, which may be lifted only about 2 m (6.5 ft).

People living in areas of frequent dust storms are faced with infiltration of very fine particles into their homes and businesses through even the smallest cracks. (Figure 3-7 illustrates such dust storms in the Nevada desert, the blowing alkali dust in the Andes mountains, and the reddish dust of an Australian storm.) People living in desert regions, where frequent sand storms occur, contend with the sandblasting of painted surfaces and etched window glass.

*Saltation.* The term *saltation* was used in Chapter 14 to describe movement of particles by water. The term also describes the wind transport of grains along the ground, grains usually larger than 0.2 mm (0.008 in.). About 80% of wind transport of particles is accomplished by this skipping and bouncing action (Figure 15-4a). Compared with fluvial transport, in which saltation is accomplished by hydraulic lift, eolian saltation is executed by aerodynamic lift, elastic bounce, and impact (compare Figure 15-4a with Figure 14-11). Grains lift much higher in air than in water

because of air's lower viscosity. On impact, grains hit other grains, with little cushioning, and knock them into the air.

Saltating particles crash into other particles, knocking them both loose and forward (Figure 15-4b). This type of movement is **surface creep**, which slides and rolls particles too large for saltation and affects about 20% of the material being transported. Once in motion, particles continue to be transported by lower wind velocities. In a desert or along a beach, you can hear the myriad saltating grains of sand produce a slight hissing sound, almost like steam escaping, as they bounce along and collide with surface particles. Both human and natural sand erosion and transport from a beach are slowed by conservation measures such as the introduction of stabilizing native plants, the use of fences, and the restriction of pedestrian traffic to walkways (Figure 15-5).

Through processes of weathering, erosion, and transportation, mineral grains are removed from parent rock and redistributed elsewhere. In Figure 13-5c, you can see the relation between the composition and color of the sandstone in the background and the derived sandy surface in the foreground. (Also note the deflation areas in the photo.) Wind action is not significant in the weathering process that frees individual grains of sand from the parent rock, but it is active in relocating the weathered grains.

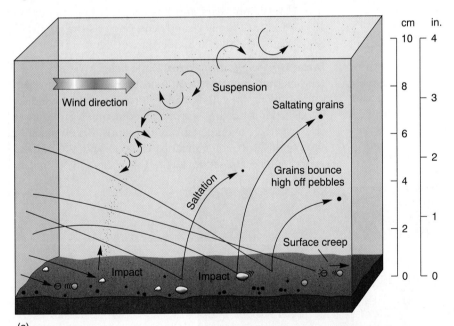

(a)

(b)

**FIGURE 15-4**
**How the wind moves sand.**
(a) Eolian suspension, saltation, and surface creep are mechanisms of sediment transportation. Compare with the saltation and traction that occur in another fluid, water, in Figure 14-11. (b) Sand grains saltating along in the surface in the Stovepipe Wells dune field. [Photo by author.]

**FIGURE 15-5**
**Preventing sand transport.**
Further erosion and transport of coastal dunes can be controlled by stabilizing native plants and fences and by confining pedestrian traffic to walkways. [Photo by author.]

## Eolian Depositional Landforms

The smallest features shaped by individual saltating grains are *ripples* (Figure 15-6). Ripples form in crests and troughs, positioned transversely (at a right angle) to the direction of the wind. Their formation is influenced by the length of time particles are airborne. Eolian ripples are similar to fluvial ripples, although the impact of saltating grains is slight in water.

Many people have never been in a desert; they know this arid landscape only from the movies, which leave the impression that most deserts are covered by sand. Instead, desert pavements predominate across most subtropical arid landscapes; only about 10% of desert areas are covered with sand. Sand grains generally are deposited as transient ridges or hills called **dunes**.

A dune is a wind-sculpted accumulation of sand. An extensive area of dunes, such as that found in North Africa, is characteristic of an **erg desert**, which means **sand sea**. The Grand Erg Oriental in the central Sahara exceeds 1200 m (4000 ft) in depth and covers 192,000 km² (75,000 mi²), the area of Nebraska. This sand sea has been active for over 1.3 million years and has average dune heights of 120 m (400 ft). Similar sand seas, such as the Ar Rub'al Khālī Erg, are active in Saudi Arabia (Figure 15-7).

***Dune Movement and Form.*** Dune fields, whether in arid regions or along coastlines, tend to migrate in the direction of strong prevailing winds. Strong seasonal winds or winds from a passing storm may sometimes prove more effective than average prevailing winds. When saltating sand grains encounter small patches of sand, their kinetic energy (motion) is dissipated and they start to accumulate;

**FIGURE 15-6**
**Sand ripples.**
Myriad sand ripple patterns later may become lithified into fixed patterns in rock. The area in the photo looks vast but is only about 1 m wide. [Photo by author.]

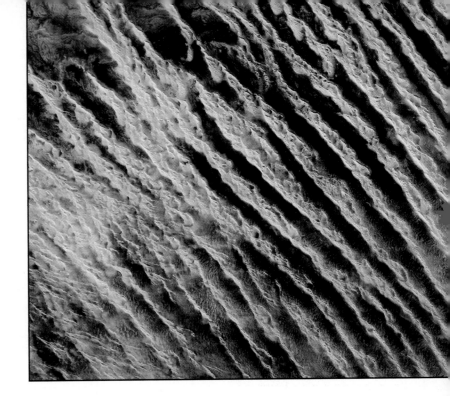

**FIGURE 15-7**

**A sand sea.**

The grand Ar Rub'al Khālī Erg dominates southern Saudi Arabia. Effective southwesterly winds shape the pattern and direction of the transverse and barchanoid dunes. This scene is approximately 80 km (50 mi) across. [SPOT image by CNES, Reston, Virginia. Used by permission.]

a dune is born. As height increases above 30 cm (12 in.) a **slipface** and characteristic dune features form.

Study the dune model in Figure 15-8 and you can see that winds characteristically create a gently sloping *windward side* (stoss side), with a more steeply sloped slipface on the *leeward side*. A dune usually is asymmetrical in one or more directions. The angle of a slipface is the steepest angle at which loose material is stable—its *angle of repose*. Thus, the constant flow of new material makes a slipface a type of *avalanche slope*. Sand builds up as it moves over the crest of the dune to the brink, then it avalanches as the slipface continually adjusts, seeking its angle of repose (usually 30° to 34°). In this way, a dune migrates downwind, as suggested by the successive dune profiles in Figure 15-8.

Dunes that move actively are called *freedunes* and reflect most dynamically the interaction between fluid atmospheric winds and moving sand. However, because the sand is moving close to the ground, it may encounter many obstructions, such as stabilizing vegetation or rock outcrops. The result is a *tied dune*, or one that is fixed in place. The ever-changing form of these eolian deposits is part of their beauty, eloquently described by one author: "I see hills and hollows of sand like rising and falling waves. Now at midmorning, they appear paper white. At dawn they were fog gray. This evening they will be eggshell brown."[*]

Dunes have many wind-shaped styles that make classification difficult. We can simplify dune forms into three classes—*crescentic, linear,* and *star* dunes (summarized in Figure 15-9).

[*]J. E. Bowers, *Seasons of the Wind* (Flagstaff, Ariz.: Northland Press, 1985), p. 1.

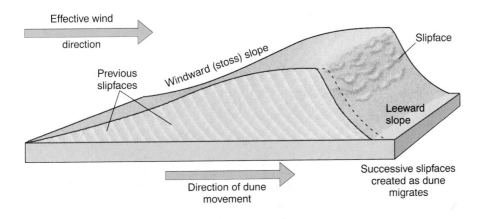

**FIGURE 15-8**

**Dune cross section.**

Successive slipfaces exhibit a distinctive pattern as the dune migrates in the direction of the effective wind.

| Class | Type | Description |
|---|---|---|
| Crescentic | Barchan | Crescent-shaped dune with horns pointed downwind. Winds are constant with little directional variability. Limited sand available. Only one slipface. Can be scattered over bare rock or desert pavement or commonly in dune fields. |
| | Transverse | Asymmetrical ridge, transverse to wind direction (right angle). Only one slipface. Results from relatively ineffective wind and abundant sand supply. |
| | Parabolic | Role of anchoring vegetation important. Open end faces upwind with U-shaped "blow-out" and arms anchored by vegetation. Multiple slipfaces, partially stabilized. |
| | Barchanoid ridge | A wavy, asymmetrical dune ridge aligned transverse to effective winds. Formed from coalesced barchans; look like connected crescents in rows with open areas between them. |

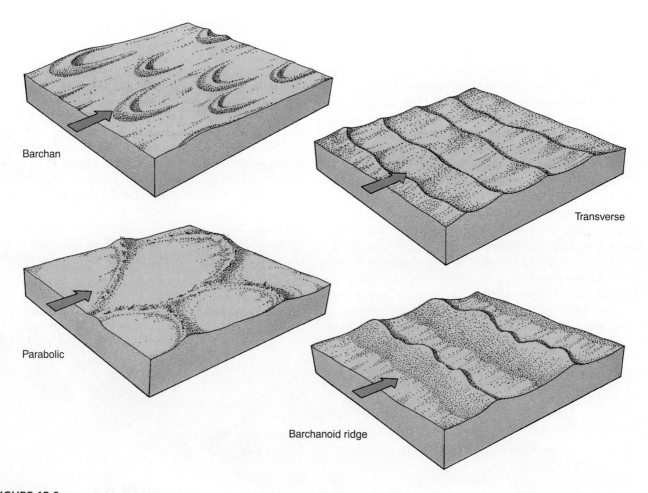

Barchan

Transverse

Parabolic

Barchanoid ridge

**FIGURE 15-9**

**Major dune forms.**

Arrows show wind direction. [Adapted from E. D. McKee, *A Study of Global Sand Seas*, U.S. Geological Survey Professional Paper 1052 (Washington, D.C.: U.S. Government Printing Office, 1979).]

| Class | Type | Description |
|-------|------|-------------|
| Linear | Longitudinal | Long, slightly sinuous, ridge-shaped dune, aligned parallel with the wind direction; two slipfaces. Can be 100 m high and 100 km long. The "draas" is up to 400 m high. Results from strong effective winds varying in one direction. |
|  | Seif | After Arabic word for "sword"; a more sinuous crest and shorter than longitudinal dunes. Rounded toward upwind direction and pointed downwind. |
| Star dune |  | The giant of dunes. Pyramidal or star-shaped with three or more sinuous radiating arms extending outward from a central peak. Slipfaces in multiple directions. Results from effective winds shifting in all directions. Tend to form isolated mounds in high effective winds and connected sinuous arms in low effective winds. |
| Other | Dome | Circular or elliptical mound with no slipface. Can be modified into barchanoid forms. |
|  | Reversing | Asymmetrical ridge form intermediate between star dune and transverse dune. Wind variability can alter shape between forms. |

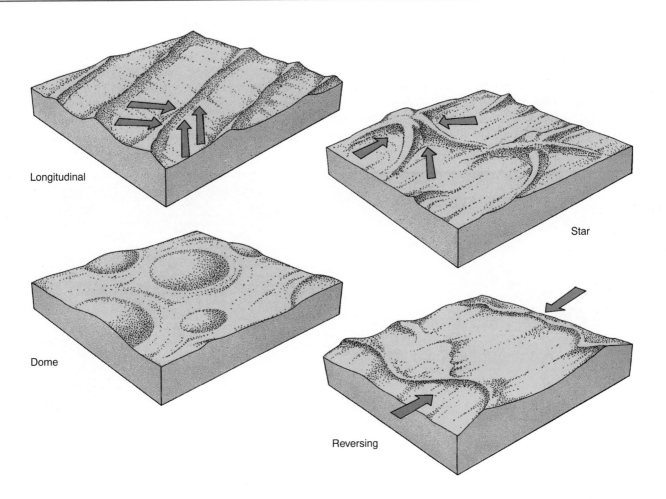

Longitudinal

Star

Dome

Reversing

*Crescentic dunes* are crescent-shaped ridges of sand that form in response to a fairly unidirectional wind pattern. The crescentic group is most common, with related forms including *barchan dunes* (limited sand available), *transverse dunes* (abundant sand), *parabolic dunes* (controlled by vegetation), and *barchanoid ridges* (rows of coalesced barchans).

*Linear dunes* generally form long parallel ridges separated by sheets of sand or bare ground. Linear dunes characteristically are much longer than they are wide; some exceed 100 km (60 mi) in length. Winds producing these dunes are principally bidirectional, so the slipface alternates from side to side. Related forms include *longitudinal dunes* and the *seif* (Arabic for "sword"), which is a sharper, narrower linear dune with a sinuous crest.

*Star dunes* are the mountainous giants of the sandy desert. They form in response to complicated, changing wind patterns and have multiple slipfaces. They are pinwheel-shaped, with several radiating arms rising and joining to form a common central peak. The best examples of star dunes are in the Sahara, where they approach 200 m (650 ft) in height (Figure 15-10).

Figure 15-11 correlates active sand regions with deserts (tropical, continental interior, and coastal). It is interesting to note the limited extent of desert area covered by active sand dunes—only about 10% of all continental land between 30° N and 30° S. Also noted on the map are dune fields in humid climates such as along coastal Oregon, the south shore of Lake Michigan (Figure 15-11b), along the Gulf and Atlantic coastlines, in Europe, and elsewhere.

These same dune-forming principles and terms (e.g., dune, barchan, slipface) apply to snow-covered landscapes. *Snow dunes* are formed as wind deposits snow in drifts. In semiarid farming areas, capturing drifting snow with fences and tall stubble left in fields contributes significantly to soil moisture when the snow melts.

## Loess Deposits

Approximately 15,000 years ago, in several episodes, Pleistocene glaciers retreated in many parts of the world, leaving behind large glacial outwash deposits of fine-grained clays and silts. These materials were blown great distances by the wind and redeposited in unstratified, homogeneous deposits. These deposits were given the name **loess** (pronounced "luss") by peasants working along the Rhine River Valley in Germany. No specific landforms were created; instead, loess covered existing landforms with a thick blanket of material that assumed the general topography of the existing landscape.

Because of its own binding strength, loess weathers and erodes into steep bluffs, or vertical faces. At Xi'an, Shaanxi Province, China, a loess wall has been excavated for dwelling space (Figure 15-12a). When a bank is cut into a loess deposit, it generally will stand vertically, although it can fail if saturated (Figure 15-12b).

**FIGURE 15-10**
**Mountains of the desert.**
**Star dune in the Namib Desert of Namibia in southwestern Africa.**
**[Photo by Comstock.]**

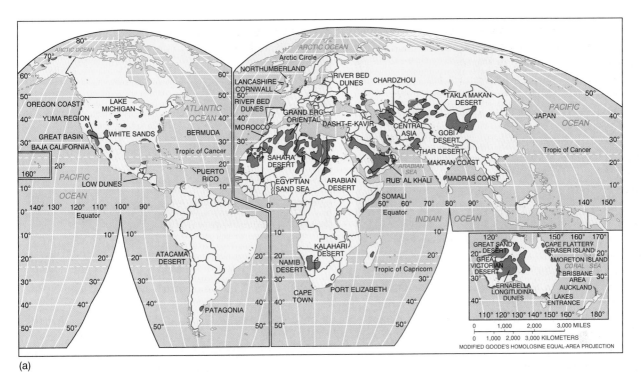

(a)

(b)

**FIGURE 15-11**
**Sandy regions of the world.**
(a) Worldwide distribution of active and stable sand regions. (b) Sand dunes along the shore of Lake Michigan in Dunes State Park in Michigan. [After R. E. Snead, *Atlas of World Physical Features*, p. 134, © 1972 by John Wiley & Sons. Adapted by permission of John Wiley & Sons, Inc. (b) Photo by Carol Niffenegger.]

Figure 15-13 shows the worldwide distribution of loess deposits. Significant accumulations throughout the Mississippi and Missouri valleys form continuous deposits 15–30 m (50–100 ft) thick. Loess deposits also occur in eastern Washington State and Idaho. This silt explains the fertility of the soils in these regions, for loess deposits are well drained, deep, and have excellent moisture retention. Loess deposits also cover much of Ukraine, central Europe, China, the Pampas-Patagonia regions of Argentina, and lowland New Zealand. The soils derived from loess are some of Earth's "breadbasket" farming regions.

In Europe and North America, loess is thought to be derived mainly from glacial and periglacial sources. The vast deposits of loess in China, covering more than 300,000 km² (116,000 mi²), are thought to be derived from wind-blown desert sediment rather than glacial sources. Accumulations in the Loess Plateau of China exceed 300 m (1000 ft) thickness, forming complex weathered badlands and some good agricultural land. These wind-blown deposits are interwoven with much of Chinese history and society. New research also has discovered that plumes of wind-blown dust from African deserts have

moved across the Atlantic Ocean to enrich soils of the Amazon rain forest of South America, the southeastern United States, and the Caribbean islands. Transport of loess soils occurred in the catastrophic Dust Bowl in the 1930s in the U.S.; see News Report 3.

# Overview of Desert Landscapes

Dry climates occupy about 26% of Earth's land surface and, if all semiarid climates are considered, perhaps as much as 35% of all land, constituting the largest single climatic region on Earth (see Figures 10-5 and 10-7 for the location of these *arid deserts BW* and *semiarid steppe BS* climate regions, and Figure 20-4 for the distribution of these desert environments).

Deserts occur worldwide as topographic plains, such as the Great Sandy and Simpson deserts of Australia, the Arabian Desert, the Kalahari Desert, and portions of the extensive Taklimakan Desert, which covers some 270,000 km² (105,000 mi²) in the central Tarim Basin of China. Deserts

(a)

(b)

**FIGURE 15-12**

**Example of loess deposits.**

(a) Loess formation in Xi'an, Shaanxi Province, China, has sufficient structural strength to permit excavation for dwelling rooms. (b) A loess bluff along the Arikaree River in extreme northwestern Cheyenne County, Kansas. [(a) Photo by Betty Crowell; (b) photo by Steve Mulligan Photography.]

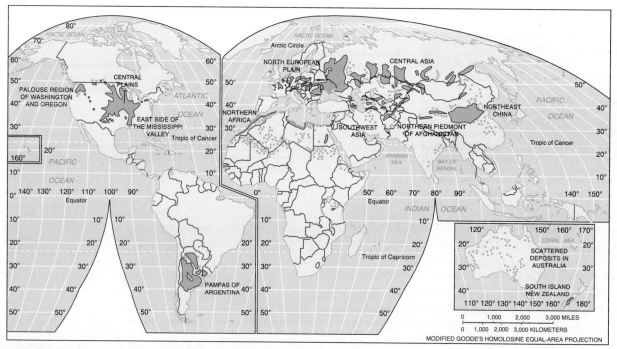

**FIGURE 15-13**

**Worldwide loess deposits.**

Dots represent small scattered loess formations. [After R. E. Snead, *Atlas of World Physical Features*, p. 138, © 1972 by John Wiley & Sons. Adapted by permission of John Wiley & Sons, Inc.]

also are found in mountainous regions: interior Asia, from Iran to Pakistan, and in China and Mongolia. In South America, lying between the ocean and the Andes, is the rugged Atacama Desert. Now, let us look at the link between climate and Earth's deserts.

## Desert Climates

The spatial distribution of dry lands is related to three climatological settings: to subtropical high-pressure cells between 15° and 35°, both N and S latitudes (see Figures

## News Report 3

### The Dust Bowl

Deflation and wind transport of loess soils produced a catastrophe in the American Great Plains in the 1930s—the Dust Bowl. Over a century of overgrazing and intensive agriculture left soil susceptible to drought and eolian processes. The deflation of many centimeters of soil occurred in southern Nebraska, Kansas, Oklahoma, Texas, and eastern Colorado.

Fine sediments were lifted by winds to form severe dust storms. The transported dust darkened the skies of Midwestern cities and drifted over farmland. Streetlights were left on throughout the day in Kansas City, St. Louis, and other midwestern cities and towns. Such episodes can devastate economies, cause tremendous loss of topsoil, and even bury farmsteads. Southeastern Australia experienced severe dust storms in 1993 that included consequences similar to those of the American Dust Bowl.

---

6-13 and 6-15), to the rain shadow on the lee side of mountain ranges (see Figure 8-17), or to areas at great distance from moisture-bearing air masses, such as central Asia. Figure 15-14 portrays this distribution according to the modified Köppen climate classification used in this text and presents photographs of four major desert regions.

Desert areas possess unique landscapes created by the interaction of intermittent precipitation events, weathering processes, and wind. Rugged, hard-edged desert landscapes of cliffs and scarps contrast sharply with the vegetation-covered, rounded and smoothed slopes characteristic of humid regions.

The daily surface energy balance for El Mirage, California, presented in Figure 4-21a, highlights the high sensible heat conditions and intense ground heating in the desert. Such areas receive a high input of insolation through generally clear skies, and they experience high radiative heat losses at night. A typical desert water balance shows high potential evapotranspiration (POTET) demand, low precipitation supply, and prolonged summer deficits (see, for example, Figure 9-13e for Phoenix, Arizona). Fluvial processes in the desert generally are characterized by intermittent running water, with hard, poorly vegetated desert pavement yielding high runoff during rainstorms (Figure 15-15).

### Desert Fluvial Processes

Precipitation events in a desert may be rare indeed, a year or two apart, but when they do occur, a dry streambed can fill with a torrent called a **flash flood**. Such channels may fill in a few minutes and surge briefly during and after a storm. Depending on the region, such a dry streambed is known as a **wash**, an *arroyo* (Spanish), or a *wadi* (Arabic). A desert highway that crosses a wash usually is posted to warn drivers not to proceed if rain is in the vicinity, for a flash flood can suddenly arise and sweep away anything in its path.

When washes fill with surging flash flood waters, a unique set of ecological relationships quickly develops. Crashing rocks and boulders break open seeds that respond to the timely moisture and germinate. Other plants and animals also spring into brief life cycles as the water irrigates their limited habitats.

At times of intense rainfall, remarkable scenes fill the desert. Figure 15-16 shows two photographs taken just one month apart in a sand dune field in Death Valley, California. A rainfall event produced 2.57 cm (1.01 in.) of precipitation in one day, in a place that receives only 4.6 cm (1.83 in.) in an average year. The stream in the photograph continued to run for hours and then collected in low spots on hard, underlying clay surfaces. The water was quickly consumed by the high evaporation demand so that, in just a month, these short-lived watercourses were dry and covered with accumulations of alluvial materials.

As runoff water evaporates, salt crusts may be left behind on the desert floor. This intermittently wet and dry low area in a region of closed drainage is called a **playa**, site of an *ephemeral lake* when water is present. Accompanying our earlier discussion of evaporites, Figure 11-11 shows such a playa in Death Valley, covered with salt precipitate just one month after this record rainfall event.

Permanent lakes and continuously flowing rivers are uncommon features in the desert, although the Nile River and the Colorado River are notable exceptions. Both these rivers are exotic streams having their headwaters in a wetter region and the bulk of their course through arid regions. The Colorado River and its problem of overuse in an arid land is described in detail in Focus Study 15-1.

***Alluvial Fans.*** In arid climates, a prominent landform is the **alluvial fan**, which occurs at the mouth of a canyon where it exits into a valley. The fan is produced by flowing water that abruptly loses velocity as it leaves the constricted channel of the canyon and therefore drops layer upon layer of sediment along the base of the mountain block. Water then flows over the surface of the fan and produces a braided drainage pattern, shifting from channel to channel with each precipitation event (Figure 15-17). A continuous apron, or **bajada**

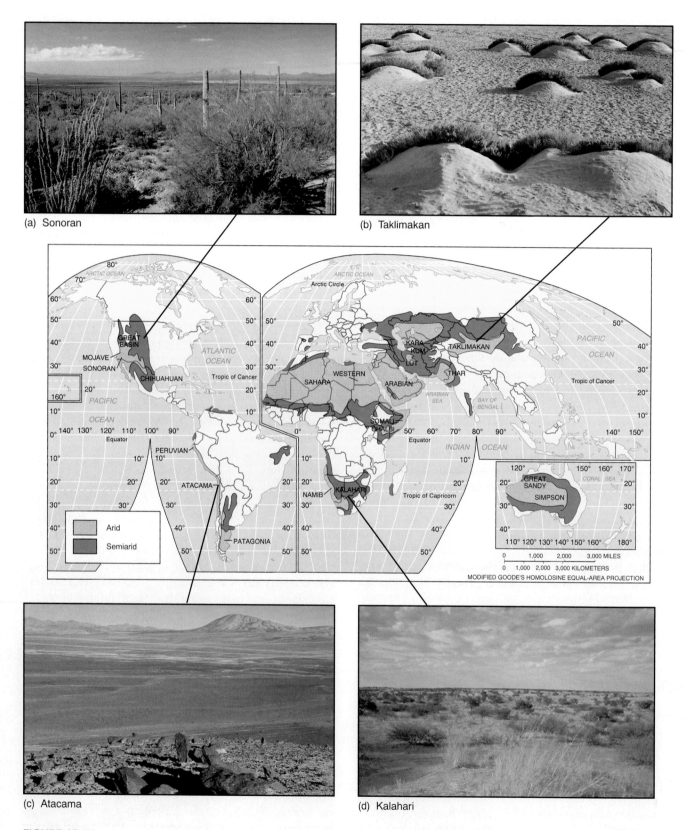

(a) Sonoran

(b) Taklimakan

(c) Atacama

(d) Kalahari

**FIGURE 15-14**

**The world's dry regions.**

Worldwide distribution of arid lands (*arid desert BW* climates) and semiarid lands (*semiarid steppe BS* climates) based on the Köppen climatic classification system. (a) Sonoran Desert in the American southwest. (b) Taklimakan Desert in central Asia. (c) Atacama Desert in subtropical Chile near Baquedano. (d) Kalahari Desert in south-central Africa. [Photos: (a) by author, (b) by Wu Chunzhan/New China Pictures Co./East Photo, (c) by Jacques Jangoux/Photo Researchers, Inc., and (d) by Nigel J. Dennis/Photo Researchers, Inc.]

**FIGURE 15-15**
**Canyonland vista.**
View from Grandview Point, Canyonlands National Park, Utah, eastward to the Manti–La Sal Mountains some 60 km (37 mi) distant. [Photo by author.]

(Spanish for "slope"), may form if individual alluvial fans coalesce into one sloping surface (see Figure 15-22). Fan formation of any sort is reduced in humid climates because perennial streams constantly carry away much of the alluvium.

An interesting aspect of an alluvial fan is the natural sorting of materials by size. Near the mouth of the canyon at the apex of the fan, coarse materials are deposited, grading slowly to pebbles and finer gravels with distance out from the mouth. Then sands and silts are deposited, with the finest clays and salts carried in suspension and solution all the way to the valley floor. Dissolved minerals accumulate as evaporite deposits on the valley floor left after evaporation.

Well-developed alluvial fans also can be a major source of groundwater. Some cities—San Bernardino, California,

for example—are built on alluvial fans and extract their municipal water supplies from them. However, because water resources in an alluvial fan are recharged from surface supplies and these fans are in arid regions, groundwater mining and overdraft beyond recharge rates are common. In other parts of the world, such water-bearing alluvial fans are known as *qanat* (Iran), *karex* (Pakistan), and *foggara* (western Sahara).

## Desert Landscapes

Contrary to popular belief, deserts are not wastelands, for they abound in specially adapted plants and animals. Moreover, the limited vegetation, intermittent rainfall, intense insolation, and distant vistas produce starkly beautiful landscapes. And all deserts are not the same: For example, North American deserts have more vegetation cover than do the generally barren Saharan expanses.

The shimmering heat waves and related mirage effects in the desert are products of light refraction through layers of air that have developed a temperature gradient near the hot ground. The desert's enchantment is captured in the book *Desert Solitaire:*

> Around noon the heat waves begin flowing upward from the expanses of sand and bare rock. They shimmer like transparent, filmy veils between my sanctuary in the shade and all the sun-dazzled world beyond. Objects and forms viewed through this tremulous flow appear somewhat displaced or distorted....The great Balanced Rock floats a few inches above its pedestal, supported by a layer of superheated air. The buttes, pinnacles, and fins in the windows area bend and undulate beyond the middle ground like a painted backdrop stirred by a draft of air.[*]

[*]E. Abbey, *Desert Solitaire* (New York: McGraw-Hill, 1968), p. 154. Copyright © 1968 by Edward Abbey.

(a)

(b)

**FIGURE 15-16**
**An improbable river in Death Valley.**
The Stovepipe Wells dune field of Death Valley, California, shown (a) the day after a 2.57 cm (1.01 in.) rainfall and (b) one month later (identical location). [Photos by author.]

## Focus Study 15-1

## The Colorado River: A System Out of Balance

An *exotic stream* has headwaters in a humid region of water surpluses but then flows mostly through arid lands for the rest of its journey to the sea. Exotic streams have few incoming tributaries. Consequently, an exotic stream has a discharge pattern that is opposite that of a typical stream: Instead of discharge increasing downstream, it decreases. The Nile and Colorado Rivers are outstanding examples.

In the case of the Nile, the East African mountains and plateaus provide a humid source area. The Nile first rises in remote headwaters as the Kagera River in the eastern portion of the Lake Plateau country of East Africa. On its way to Lake Victoria, it forms the partial boundary of Tanzania, Rwanda, and Uganda. The Nile itself then rises out of the lake and continues on its 6650 km (4132 mi) course to its mouth on the Mediterranean Sea near Cairo.

### The Colorado River Basin

The Colorado rises on the high slopes of Mount Richthofen (3962 m, or 13,000 ft) in Rocky Mountain National Park and flows almost 2317 km (1440 mi) to where a trickle of water disappears in the sand, kilometers short of its former mouth in the Gulf of California (Figure 1).

Orographic precipitation totaling 102 cm (40 in.) per year (mostly snow) falls in the Rockies, feeding the Colorado headwaters. But at Yuma, Arizona, near the river's end, annual precipitation is a scant 8.9 cm (3.5 in.), an extremely small amount when compared with the annual potential evapotranspiration demand in the Yuma region of 140 cm (55 in.).

From its source region, the Colorado River quickly leaves the humid Rockies and spills out into the arid desert of western Colorado and eastern Utah. At Grand Junction, Colorado, on the Utah border, annual precipitation is only 20 cm (8 in.). After carving its way through the intricate labyrinth of canyonlands in Utah, the river enters Lake Powell, 945 m (3100 ft) lower in elevation than the river's source area upstream in the Rockies some 982 km (610 mi) away. The Colorado then flows

through the Grand Canyon chasm, formed by its own erosive power. The canyon's mystical beauty is evident to all who visit its rim and to those who venture within its depths.

West of the Grand Canyon, the river turns southward, tracing its final 644 km (400 mi) as the Arizona-California border. Along this stretch sits Hoover Dam, just east of Las Vegas; Davis Dam, built to control the releases from Hoover; Parker Dam for the water needs of Los Angeles; three more dams for irrigation water (Palo Verde, Imperial, and Laguna); and finally, Morelos Dam at the Mexican border. Mexico owns the end of the river and whatever water is left. Overall, the drainage basin encompasses 641,025 km² (247,500 mi²) of mountain, basin-and-range, plateau, canyon, and desert landscapes, in parts of seven states and two countries. A discussion of the Colorado River is included in this chapter because of its crucial part in the history of the Southwest and its role in the future of this region.

### Dividing Up the Colorado's Dammed Water

John Wesley Powell (1834–1902), the first person of record to successfully navigate the Colorado River through the Grand Canyon, was the first director of the U.S. Bureau of Ethnology and later director of the U.S. Geological Survey (1881–1892). Powell perceived that the challenge of the West was too great for individual efforts and believed that solutions to problems such as water availability could be met only through private cooperative efforts. His 1878 study (reprinted 1962), *Report of the Lands of the Arid Region of the United States*, is a conservation landmark.

Today, Powell probably would be skeptical of the intervention by government agencies in building large-scale reclamation projects, although Lake Powell is named after him despite his probable opposition were he alive. An anecdote in Wallace Stegner's *Beyond the Hundredth Meridian* relates that, at an 1893 international irrigation conference held in Los Angeles, Powell

observed development-minded delegates bragging that the entire West could be conquered and reclaimed from nature. Powell spoke against that sentiment: "I tell you, gentlemen, you are piling up a heritage of conflict and litigation over water rights, for there is not sufficient water to supply the land."* He was booed from the hall. But history has shown Powell to be correct.

The Colorado River Compact was signed by six of the seven basin states in 1923. (The seventh, Arizona, signed in 1944, the same year as the Mexican Water Treaty.) With this compact, the Colorado River basin was divided into an upper basin and a lower basin, arbitrarily separated for administrative purposes at Lees Ferry near the Utah-Arizona border (Figure 1). Congress adopted the Boulder Canyon Act in 1928, authorizing Hoover Dam as the first major reclamation project on the river. Also authorized was the All-American Canal into the Imperial Valley, which required an additional dam. Los Angeles then began its project to bring Colorado River water 390 km (240 mi) from still another dam and reservoir on the river to their city.

Shortly after Hoover Dam was finished and downstream enterprises were thus offered flood protection, the other projects were quickly completed. There are now eight major dams on the river and many irrigation works. The latest effort to redistribute Colorado River water is the Central Arizona Project, which carries water to the Phoenix area.

### Highly Variable River Flows

The flaw in all this planning and water distribution is that exotic streamflows are highly variable, and the Colorado is no exception. In 1917, the discharge measured at Lees Ferry totaled 24 million acre-feet (maf), whereas in 1934 it dropped by nearly 80 percent, to only 5.03 maf. In 1977, the discharge dropped again to 5.02 maf, but in 1984 it rose to an all-time high of 24.5 maf. In addition to this variability, approximately

*W. Stegner, *Beyond the Hundredth Meridian* (Boston: Houghton Mifflin Co., 1954), p. 343.

**FIGURE 1**

The Colorado River basin, showing division of the upper and lower basins near Lees Ferry in northern Arizona.
(a) Headwaters of the Colorado River near Mount Richthofen in the Colorado Rockies. (b) Hoover Dam
spillways in rare operation. (c) Davis Dam in full release. (d) The flooded Topock Estates subdivision near
Needles, California, in 1983. (e) Central Arizona Project facility west of Phoenix. [(a) Photo by Randall
Christopherson; (b–d) photos by author; (e) photo by Tom Bean/DRK Photo.]

70% of the year-to-year discharge occurs between April and July; the other 30% is spread over the balance of the year.

The average flows between 1906 and 1930 were almost 18 maf a year, but averages dropped to 13 maf during the past 60 years. As a planning basis for the Colorado River Compact, the government used average river discharges from 1914 up to the treaty signing in 1923, an exceptionally high average of 18.8 maf. That amount was perceived as more than enough for the upper and lower basins each to receive 7.5 maf and, later, for Mexico to receive 1.5 maf in the 1944 Mexican Water Treaty.

We might question whether proper long-range planning should rely on the providence of high variability. Tree-ring analyses of past climates have disclosed that the only other time Colorado discharges were at the 1914–1923 level was between A.D. 1606 and 1625! However, government thinking about Colorado River flows seems never to have accepted this reality, for the dependable flows of the river have been consistently overestimated. This denial is shown in an estimated budget for the river (Table 1): clearly the situation is out of balance, for there is not enough discharge to meet budgeted demands. A few wet years in the late 1980s and early 1990s only delay the inevitable shortfall crisis.

Presently, the seven states ideally want rights to a total of 25 maf. When added to the guarantee for Mexico this comes up to 26.5 maf of wants. And six states share one common opinion: California's right to the water must be limited.

## Water Loss at Glen Canyon Dam and Lake Powell

Glen Canyon Dam was completed and began water impoundment (Lake Powell) in 1963, 27.4 km (17 mi) north of the basin division point at Lees Ferry. The deep, fluted, inner gorges of many canyons, including the Glen, Navajo, Labyrinth, and Cathedral, slowly were flooded by advancing water. Glen Canyon Dam's primary purpose, according to the Bureau of Reclamation, is to regulate flows between the upper and lower basins. An additional benefit is the production of hydroelectric power, which is sold wholesale at a low rate to utilities across the Southwest. Also, there is a growing recreation and tourism industry

## TABLE 1

| Estimated Colorado River Budget | |
| --- | --- |
| **Water Demand** | **Quantity (maf)\*** |
| In-basin consumptive uses (75% agricultural)\*\* | 11.5 |
| Central Arizona Project (rising to 2.8 maf) | 1.0 |
| Mexican allotment (1944 Treaty) | 1.5 |
| Evaporation from reservoirs | 1.5 |
| Bank storage at Lake Powell | 0.5 |
| Phreatophytic losses (water-demanding plants) | 0.5 |
| Budgeted total demand | 16.5 |
| 1930–1980 average flow of the river | 13.0 |

*Source:* Bureau of Reclamation, Colorado River Board, states of Arizona and California.
\*1 million acre-feet = 325,872 gallons; 1.24 million liters.
\*\*7.0 maf in lower basin; includes 5.5 pumped out-of-basin, 5.1 of which goes to California.

around Lake Powell, as previously inaccessible desert scenery now can be reached by boat.

Many thought that Lake Mead, behind Hoover Dam downstream, could have served the primary administrative function of flow regulation, especially considering the serious water-loss problems with the Glen Canyon dam and reservoir:

- *4% loss to the rock.* The porous Navajo sandstone underlies most of the Lake Powell reservoir at Glen Canyon Dam. This sandstone absorbs an estimated 4% of the river's overall annual discharge as *bank storage*, water that is not practical to retrieve. The higher the lake level, the greater the loss into the sandstone. To date, approximately 12 maf have disappeared into this highly permeable rock.

- *4% loss to the air.* Lake Powell is an open body of water in an arid desert, where hot, dry winds accelerate evaporative losses. Another 4% of the Colorado's overall discharge is lost annually from the reservoir in this manner.

- *4% loss to the plants.* Now-permanent sand bars and banks have stabilized along the regulated river, allowing water-demanding plants called phreatophytes to establish and extract an additional 4% of the river flow.

Combined, these annual losses total approximately 1.5 maf, or 12% of the Colorado's long-term average flow—all attributable to Glen Canyon Dam and Lake Powell. Given the chronic deficits in the overall Colorado River budget, we can

now seriously question the rationale behind this single project.

## Flood Control in 1983

Intense precipitation and heavy snowpack in the Rockies, attributable to the 1982–1983 El Niño (see Focus Study 10-1), led to record-high discharge rates on the Colorado, testing the controllability of one of the most regulated rivers in the world. Federal reservoir managers were not prepared for the high discharge, since they had set aside Lake Mead's primary purpose (flood control) in favor of competing water and power interests. What followed was a human-caused flood.

The only time the spillways at Hoover Dam had ever operated was more than 40 years earlier, when the reservoir capacity was artificially raised for a test; now they were opened to release the floodwaters. Davis Dam, which regulates releases from Hoover Dam, was within 30 cm (1 ft) of overflow, a real problem for a structure made partially of earth fill. In addition, Glen Canyon Dam was over capacity and at risk; it actually was damaged by the volume of discharge tearing through its spillways. The decision to increase releases to save these and other upstream facilities doomed towns and homeowners along the river, especially in subdivisions near Needles, California.

We might wonder what John Wesley Powell would think if he were alive today to witness such errant attempts to control the mighty and variable Colorado. He foretold such a "heritage of conflict and litigation."

**FIGURE 15-17**

**An alluvial fan.**

The photo shows an alluvial fan in a desert landscape. The topographic map shows the Cedar Creek alluvial fan. (Topographic map is the Ennis Quadrangle, 15-minute series, scale 1:62,500, contour interval = 40 ft; latitude/longitude coordinates for mouth of canyon are 45°2′ N 111°35′ W.) [Photo by author; Eros Data Center/ USGS map.]

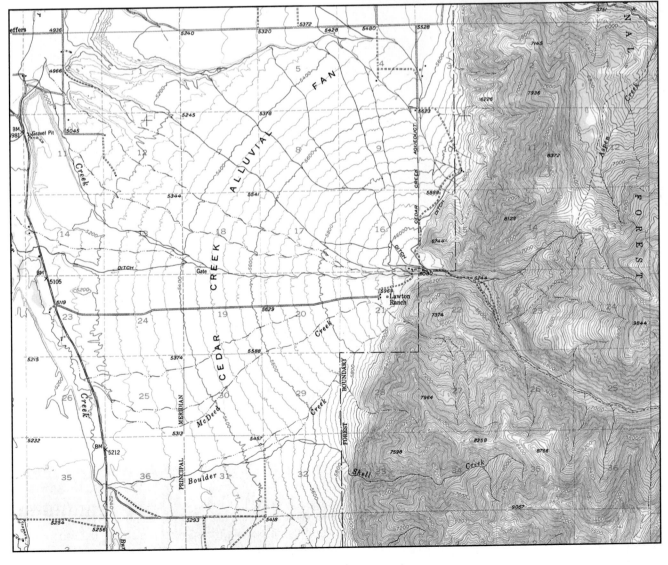

The buttes, pinnacles, and mesas of arid landscapes are resistant horizontal rock strata that have eroded differentially. Removal of the less-resistant sandstone strata produces unusual desert sculptures—arches, pedestals, and delicately balanced rocks (Figure 15-18). Specifically, the upper layers of sandstone along the top of an arch or butte are more resistant to weathering and protect the sandstone rock beneath.

The removal of all surrounding rock through differential weathering leaves enormous buttes as residuals on the landscape. If you imagine a line intersecting the tops of the Mitten Buttes shown in Figure 15-19, you can gain some idea of the quantity of material that has been removed. These buttes exceed 300 m (1000 ft) in height, similar to the Chrysler Building in New York City or First Canadian Place in Toronto.

**FIGURE 15-18**
**A balanced rock—differential weathering.**
Balanced Rock in Arches National Park, Utah, where writer-naturalist Edward Abbey (quoted in text) worked as a ranger years before it became a park. [Photo by author.]

Desert landscapes are places where stark erosional remnants stand above the surrounding terrain as knobs or hills. Such a bare, exposed rock, called an *inselberg* (island mountain), is exemplified by Uluru (Ayers) Rock in Australia (Figure 15-20).

In a desert area, weak surface material may weather to a complex, rugged topography, usually of relatively low and varied relief. Such a landscape is called a *badland*, probably so named because it offered little economic value and was difficult to traverse in nineteenth-century wagons. The Badlands region of the Dakotas and northcentral Arizona (the Painted Desert) are of this form.

Sand dunes that existed in some ancient deserts have lithified, forming sandstone structures that bear the imprint of *cross-stratification*. When such a dune was accumulating, sand cascaded down its slipface, and distinct bedding planes (layers) were established that remained after the dune lithified (Figure 15-21). Ripple marks, animal tracks, and fossils also are found preserved in these sandstones, which originally were eolian-deposited sand dunes.

***Basin and Range Province.*** A *province* is a large region that is characterized by several geologic or physiographic traits. Characterizing the **Basin and Range Province** of the western United States are alternating basins and mountain ranges that lie in the rain shadow of mountains to the west (Figure 15-22). The physiography and geography combine to give the province a dry climate, few permanent streams, and *interior drainage patterns*—drainage basins that lack any outlet to the ocean (see Figure 14-4, "Great Basin" and interior drainage).

The vast Basin and Range Province—almost 800,000 km² (300,000 mi²)—was a major barrier to early settlers in their migration westward. The combination of desert climate and north-south trending mountain ranges presented harsh challenges. Today, when you traverse U.S. Highway 50 across Nevada you cross five passes (horsts) of over 1950 m (6400 ft) and numerous basins (grabens). Throughout the drive you are reminded where you are by prideful signs that plainly state: "The Loneliest Road in America." It is difficult to imagine crossing this topography with wagons and oxen.

How the Basin and Range formed is interesting. As the North American plate lumbered westward, it overrode former oceanic crust and hot spots at such a rapid pace that slabs of subducted material literally were run over. The crust was stretched, creating a landscape fractured by many faults. The present landscape consists of nearly parallel sequences of *horsts* (upward-faulted blocks, which are the "ranges") and *grabens* (downward-faulted blocks, which are the "basins," or valleys). Figure 15-22c shows this pattern of normal faults.

John McPhee captured the feel of this desert province in his book *Basin and Range*:

(a)

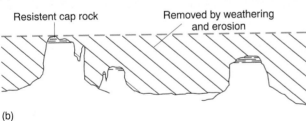

Resistent cap rock     Removed by weathering and erosion

(b)

**FIGURE 15-19**
**Monument Valley landscape.**
(a) Mitten Buttes, Merrick Butte, and rainbow in Monument Valley, Navajo Tribal Park, along the Utah-Arizona border. (b) A schematic of the tremendous removal of material by weathering, erosion, and transport. [Photo by author.]

Supreme over all is silence. Discounting the cry of the occasional bird, the wailing of a pack of coyotes, silence—a great spatial silence—is pure in the Basin and Range. It is a soundless immensity with mountains in it. You stand...and look up at a high mountain front, and turn your head and look fifty miles down the valley, and there is utter silence.[*]

[*]J. McPhee, *Basin and Range* (New York: Farrar, Straus, Giroux, 1981), p. 46.

Basin-and-range relief is abrupt, and rock structures are angular and rugged. As the ranges erode, transported materials accumulate to great depths in the basins, gradually producing extensive desert plains. The basin's elevation averages roughly 1200–1500 m (4000–5000 ft) above sea level, with mountain crests rising higher by some 900–1500 m (3000–5000 ft). Death Valley, California, is the lowest of these basins, with an elevation of −86 m (−282 ft). However, to the west of the valley, the Panamint Range rises to 3368 m (11,050 ft) at Telescope Peak—almost 3.5 vertical kilometers (2.2 mi) of desert relief!

**FIGURE 15-20**
**Australian landmark.**
Uluru (Ayers) Rock in Northern Territory, Australia, is an isolated mass of weathered rock. The formation is 348 m (1145 ft) high and is 2.5 km long by 1.6 km wide. Uluru Rock is sacred to Aboriginal tribes and has been protected in Uluru National Park since 1950. [Photo by Porterfield/Chickering.]

**FIGURE 15-21**
**Cross bedding.**
In sandstone, the bedding pattern, called cross-stratification, in these sandstone rocks tells us about patterns that were established in the dunes before lithification. [Photo by author.]

In Figure 15-22b, note the **bolson**, a slope-and-basin area between the crests of two adjacent ridges in a dry region of interior drainage. Death Valley (Figure 15-23), California, is part of the Basin and Range Province and a dramatic example of these arid-land features.

Figure 15-22 also identifies a *playa* (central salt pan), a *bajada* (coalesced alluvial fans), and a mountain front in retreat from weathering and erosional attack. A *pediment* is an area of bedrock that is layered with a thin veneer, or coating, of alluvium. It is an erosional surface, as opposed to the depositional surface of the bajada.

Vast arid and semiarid lands remain an enigma on the water planet. They challenge our technology, courage, and personal need for water. These lands hold a mysterious fascination, perhaps because they are so lacking in the moisture that infuses our lives.

# Summary and Review—Eolian Processes and Arid Landscapes

✔ *Characterize* the unique aspects of the wind and eolian processes.

Winds are produced by the movement of the atmosphere in response to pressure differences. Wind is a geomorphic agent of erosion, transportation, and deposition. **Eolian** processes modify and move sand accumulations along coastal beaches and deserts. Wind's ability to move materials is small compared with that of water and ice.

eolian p. (456)

1. Who was Ralph Bagnold? What was his contribution to eolian studies?
2. Explain the term *eolian* and its application in this chapter. How would you characterize the ability of the wind to move material?

✔ *Describe* eolian erosion, including deflation, abrasion, and the resultant landforms.

Two principal wind-erosion processes are **deflation**, the removal and lifting of individual loose particles, and **abrasion**, the "sandblasting" of rock surfaces with particles captured in the air. Fine materials are eroded by wind deflation and moving water, leaving behind concentrations of pebbles and gravel called **desert pavement**. Wherever wind encounters loose sediment, deflation may remove enough material to form basins. Called **blowout depressions**, they range from small indentations less than a meter wide up to areas hundreds of meters wide and many meters deep. Rocks that bear evidence of such eolian erosion are called **ventifacts**. On a larger scale, deflation and abrasion are capable of streamlining rock structures, leaving behind distinctive, elongated ridges called **yardangs**.

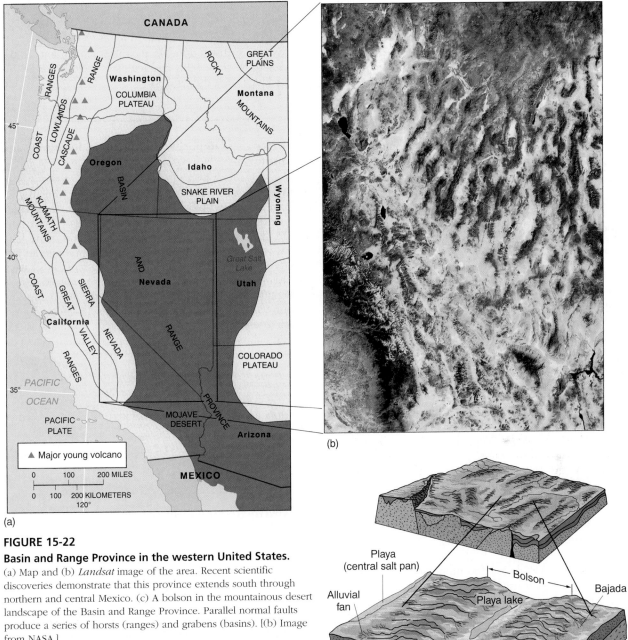

(a)

## FIGURE 15-22

### Basin and Range Province in the western United States.

(a) Map and (b) *Landsat* image of the area. Recent scientific discoveries demonstrate that this province extends south through northern and central Mexico. (c) A bolson in the mountainous desert landscape of the Basin and Range Province. Parallel normal faults produce a series of horsts (ranges) and grabens (basins). [(b) Image from NASA.]

## FIGURE 15-23

### Desert playa.

Death Valley demonstrates a central playa, parallel mountain ranges, alluvial fans, and bajadas along the base of the range. The region is now designated Death Valley National Park. [Photo by author.]

deflation (p. 457)
abrasion (p. 457)
desert pavement (p. 457)
blowout depressions (p.457)
ventifacts (p. 457)
yardangs (p. 457)

3. Describe the erosional processes associated with moving air.

4. Explain deflation and the evolutionary sequence that produces desert pavement.

5. How are ventifacts and yardangs formed by the wind?

---

✔ *Describe* **eolian transportation;** *explain* **saltation and surface creep.**

Wind exerts a drag or frictional pull on surface particles until they become airborne. Only the finest dust particles travel significant distances, so the finer material suspended in a *dust storm* is lifted much higher than the coarser particles of a *sand storm*. Saltating particles crash into other particles, knocking them both loose and forward. The motion called **surface creep** slides and rolls particles too large for saltation.

surface creep (p. 459)

6. Differentiate between a dust storm and a sand storm.

7. What is the difference between eolian saltation and fluvial saltation?

8. Explain the concept of surface creep.

---

✔ *Identify* **the major classes of sand dunes and** *present* **examples within each class.**

In arid and semiarid climates and along some coastlines where sand is available **dunes** accumulate. A dune is a wind-sculpted accumulation of sand. An extensive area of dunes, such as that found in North Africa, is characteristic of an **erg desert**, which means **sand sea**. When saltating sand grains encounter small patches of sand, their kinetic energy (motion) is dissipated and they start to accumulate into a dune. As height increases above 30 cm (12 in.) a **slipface** and characteristic dune features are formed. Dune forms are broadly classified as *crescentic, linear,* and *star*.

dunes (p. 460)
erg desert (p. 460)
sand sea (p. 460)
slipface (p. 461)

9. What is the difference between an erg and a reg desert? Which type is a sand sea? Are all deserts covered by sand? Explain.

10. What are the three classes of dune forms? Describe the basic types of dunes within each class. What do you think is the major shaping force for sand dunes?

11. Which form of dune is the mountain giant of the desert? What are the characteristic wind patterns that produce such dunes?

---

✔ *Define* **loess deposits, their origins, locations, and landforms.**

Eolian transported materials contribute to soil formation in distant places. Wind-blown **loess** deposits occur worldwide and can develop into good agricultural soils. These fine-grained clays and silts are moved by the wind many kilometers, where they are redeposited in unstratified, homogeneous deposits. The binding strength of loess causes it to weather and erode in steep bluffs, or vertical faces.

Significant accumulations throughout the Mississippi and Missouri valleys form continuous deposits 15–30 m (50–100 ft) thick. Loess deposits also occur in eastern Washington State, Idaho, much of Ukraine, central Europe, China, the Pampas-Patagonia regions of Argentina, and lowland New Zealand.

loess (p. 464)

12. How are loess materials generated? What form do they assume when deposited?

13. Name a few examples of significant loess deposits on Earth.

---

✔ *Portray* **desert landscapes and** *locate* **these regions on a world map.**

Dry and semiarid climates occupy about 35% of Earth's land surface. The spatial distribution of these dry lands is related to subtropical high-pressure cells between 15° and 35° N and S, to rain shadows on the lee side of mountain ranges, or to areas at great distance from moisture-bearing air masses, such as central Asia.

Water events are rare, yet running water is still the major erosional agent in deserts. Precipitation events may be rare, but when they do occur, a dry streambed fills with a torrent called a **flash flood**. Depending on the region, such a dry streambed is known as a **wash**, an *arroyo* (Spanish), or a *wadi* (Arabic). As runoff water evaporates, salt crusts may be left behind on the desert floor. This intermittently wet and dry low area in a region of closed drainage is called a **playa**, site of an *ephemeral lake* when water is present.

In arid climates, a prominent landform is the **alluvial fan** at the mouth of a canyon where it exits into a valley. The fan is produced by flowing water that abruptly loses velocity as it leaves the constricted channel of the canyon and deposits a layer of sediment along the mountain block. A continuous apron, or **bajada**, may form if individual alluvial fans coalesce. A *province* is a large region that is characterized by several geologic or physiographic traits. The **Basin and Range Province** of the western United States consists of alternating basins and mountain ranges. A slope-and-basin area between the crests of two adjacent ridges in a dry region of interior drainage is termed a **bolson**.

> flash flood (p. 467)
> wash (p. 467)
> playa (p. 467)
> alluvial fan (p. 467)
> bajada (p. 467)
> Basin and Range Province (p. 474)
> bolson (p. 476)

14. Characterize desert energy and water balance regimes. What are the significant patterns of occurrence for arid landscapes in the world?

15. How would you describe the water budget of the Colorado River? What was the basis for agreements regarding distribution of the river? Why has thinking about the river's discharge been so optimistic?

16. Describe a desert bolson from crest to crest. Draw a simple sketch with the components of the landscape labeled.

17. Where is the Basin and Range Province? Briefly describe its appearance and character.

---

 ## NetWork

The *Geosystems Home Page* provides on-line resources for this chapter on the World Wide Web. You will find review exercises, specific updates for items in the chapter, suggested readings, and links to interesting related pathways on the Internet (click on the Table of Contents link and select this chapter). *Geosystems* is at: **http://www.prenhall.com/geosystm**

# 16

# The Oceans, Coastal Processes and Landforms

**Global Oceans and Seas**

**Coastal System Components**

**Coastal System Actions**

**Coastal System Outputs**

**Human Impact on Coastal Environments**

**Summary and Review**

## Key Learning Concepts

After reading the chapter, you should be able to:

- *Describe* the chemical composition of seawater and the physical structure of the ocean.
- *Identify* the components of the coastal environment and *list* the physical inputs to the coastal system, including tides and mean sea level.
- *Describe* wave motion at sea and near shore and *explain* coastal straightening as a product of wave refraction.
- *Identify* characteristic coastal erosional and depositional landforms.
- *Describe* barrier islands and their hazards as they relate to human settlement.
- *Assess* living coastal environments: corals, wetlands, salt marshes, and mangroves.
- *Construct* an environmentally sensitive model for settlement and land use along the coast.

*Coastal cliffs, Nullarbor National Park, South Australia.*
[Photo by M. P. Kahl/DRK Photo.]

Walk along a shoreline and you witness the dramatic interaction of Earth's vast oceanic, atmospheric, and lithospheric systems. At times, the ocean attacks the coast in a stormy rage of erosive power; at other times, the moist sea breeze, salty mist, and repetitive motion of the water are gentle and calming. Few have captured this confrontation between land and sea as well as biologist Rachel Carson:

> The edge of the sea is a strange and beautiful place. All through the long history of Earth it has been an area of unrest where waves have broken heavily against the land, where the tides have pressed forward over the continents, receded, and then returned. For no two successive days is the shoreline precisely the same. Not only do the tides advance and retreat in their eternal rhythms, but the level of the sea itself is never at rest. It rises or falls as the glaciers melt or grow, as the floors of the deep ocean basins shift under its increasing load of sediments, or as the earth's crust along the continental margins warps up or down in adjustment to strain and tension. Today a little more land may belong to the sea, tomorrow a little less. Always the edge of the sea remains an elusive and indefinable boundary.*

Despite such variability, many people live and work near the ocean because of commerce, shipping, fishing, and tourism. A recent U.N. population survey estimates that about two-thirds of Earth's population lives in or very near coastal regions. In the United States, a bit over 50% of the people live on the 10% of the land that is designated as *coastal* (this includes the Great Lakes.). Therefore, an understanding of coastal processes and landforms is important to humanity. And because these processes along coastlines oftentimes produce dramatic change, they are essential to consider in planning and development.

We begin the chapter with a brief look at our global oceans and seas. The physical and chemical properties of the sea distinguish it from the waters of the continent. The coastlines are areas of dynamic change and beauty, where oceans and seas confront the land. Coastal processes are organized in this chapter as a system of specific inputs (components and driving forces), actions (movements and processes), and outputs (results and consequences). We conclude with a look at the considerable human impact on coastal environments.

## Global Oceans and Seas

The ocean is one of Earth's last great scientific frontiers and is of great interest to geographers. Remote sensing from orbit, aircraft, vessels, and submersibles is providing

a wealth of data and a new capability to understand the oceanic system. The pattern of sea-surface temperatures is presented in Figure 5-8 and ocean currents in Figure 6-24. The world's oceans, their area, volume, and depth, are illustrated in Figure 7-4. The locations of oceans and major seas are given alphabetically in Figure 16-1.

### Chemical Composition of Seawater

Water is called the "universal solvent" dissolving at least 57 of the 92 elements found in nature. In fact, most natural elements and the compounds they form are found in the seas as dissolved solids, or *solutes*. Thus, seawater is a solution, and the concentration of dissolved solids is called **salinity**.

Long ago, during the evolutionary atmosphere and living atmosphere, heavy precipitation produced runoff from the land. This outflow delivered enormous quantities of solids to the ocean. About a billion years ago and after much accumulation, the salinity of the oceans achieved a steady-state equilibrium. Any *influx* of material from physical, chemical, and biological processes was countered by an *efflux*, or removal, in the form of marine sediments, formation of sulfide and silicate compounds, clays, and particulate skeletal remains (biogenic silica).

The oceans remain a remarkably homogeneous mixture. The ratio of individual salts does not change, despite minor fluctuations in overall salinity. In 1874 the British HMS *Challenger* sailed around the world, taking surface and depth measurements and collecting samples of seawater. Analyses of those samples demonstrated the uniform composition of seawater.

***Ocean Chemistry.*** Ocean chemistry is a result of complex exchanges among seawater, the atmosphere, minerals, bottom sediments, and living organisms. In addition, significant flows of mineral-rich water enter the ocean through hydrothermal (hot water) vents in the ocean floor. (These vents are called "black smokers" for the dense, black mineral-laden water that spews from them.) The uniformity of seawater results from complementary chemical reactions and continuous mixing—after all, the ocean basins interconnect, and water circulates among them.

Seven elements account for more than 99% of the dissolved solids in sea water. They are (with their ionic form); chlorine (as chloride $Cl^-$), sodium (as $Na^+$), magnesium (as $Mg^{2+}$), sulfur (as sulfate $SO_4^{2-}$), calcium (as $Ca^{2+}$), potassium (as $K^+$), and bromine (as bromide $Br^-$). Seawater also contains dissolved gases (such as carbon dioxide, nitrogen, and oxygen), suspended and dissolved organic matter, and multitude of trace elements.

Commercially, only sodium chloride (common table salt), magnesium, and bromine are extracted in any significant amount. Future mining of minerals from the sea floor is technically feasible, although it remains uneconomical.

*"The Marginal World," in *The Edge of the Sea* by Rachel Carson. © 1955 by Rachel Carson, © renewed 1983 by R. Christie (Boston: Houghton Mifflin), p. 11.

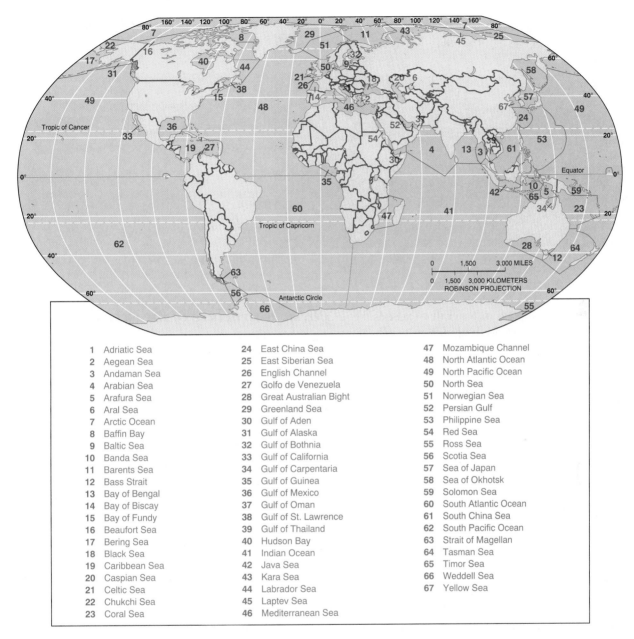

**FIGURE 16-1**

**Principal oceans and seas of the world.**

A sea is generally smaller than an ocean and is near a landmass; sometimes the term refers to a large, inland, salty body of water.

| | | | | | |
|---|---|---|---|---|---|
| 1 | Adriatic Sea | 24 | East China Sea | 47 | Mozambique Channel |
| 2 | Aegean Sea | 25 | East Siberian Sea | 48 | North Atlantic Ocean |
| 3 | Andaman Sea | 26 | English Channel | 49 | North Pacific Ocean |
| 4 | Arabian Sea | 27 | Golfo de Venezuela | 50 | North Sea |
| 5 | Arafura Sea | 28 | Great Australian Bight | 51 | Norwegian Sea |
| 6 | Aral Sea | 29 | Greenland Sea | 52 | Persian Gulf |
| 7 | Arctic Ocean | 30 | Gulf of Aden | 53 | Philippine Sea |
| 8 | Baffin Bay | 31 | Gulf of Alaska | 54 | Red Sea |
| 9 | Baltic Sea | 32 | Gulf of Bothnia | 55 | Ross Sea |
| 10 | Banda Sea | 33 | Gulf of California | 56 | Scotia Sea |
| 11 | Barents Sea | 34 | Gulf of Carpentaria | 57 | Sea of Japan |
| 12 | Bass Strait | 35 | Gulf of Guinea | 58 | Sea of Okhotsk |
| 13 | Bay of Bengal | 36 | Gulf of Mexico | 59 | Solomon Sea |
| 14 | Bay of Biscay | 37 | Gulf of Oman | 60 | South Atlantic Ocean |
| 15 | Bay of Fundy | 38 | Gulf of St. Lawrence | 61 | South China Sea |
| 16 | Beaufort Sea | 39 | Gulf of Thailand | 62 | South Pacific Ocean |
| 17 | Bering Sea | 40 | Hudson Bay | 63 | Strait of Magellan |
| 18 | Black Sea | 41 | Indian Ocean | 64 | Tasman Sea |
| 19 | Caribbean Sea | 42 | Java Sea | 65 | Timor Sea |
| 20 | Caspian Sea | 43 | Kara Sea | 66 | Weddell Sea |
| 21 | Celtic Sea | 44 | Labrador Sea | 67 | Yellow Sea |
| 22 | Chukchi Sea | 45 | Laptev Sea | | |
| 23 | Coral Sea | 46 | Mediterranean Sea | | |

***Average Salinity: 35‰.*** There are several ways to express salinity (dissolved solids by volume) in seawater. Here are examples, using the worldwide average value:

- 3.5% (%, or parts per hundred)
- 35,000 ppm (parts per million)
- 35,000 mg per liter
- 35 g/kg
- 35‰ (‰ = parts per thousand), the most common notation

Salinity worldwide normally varies between 34‰ and 37‰; variations are attributable to atmospheric conditions above the water and to the volume of freshwater inflows. In equatorial water, precipitation is great throughout the year, diluting salinity values to slightly lower than average (34.5‰). In subtropical oceans—where evaporation rates are greatest because of the influence of hot, dry subtropical high-pressure cells—salinity is more concentrated, increasing to 36‰. Figure 16-2 plots the difference between evaporation and precipitation and salinity by latitude to illustrate this slight spatial variability.

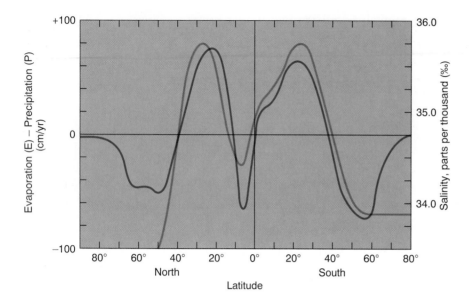

**FIGURE 16-2**
**Variation in ocean salinity by latitude.**
Salinity (green line) is principally a function of climatic conditions. Specifically important is the moisture relation expressed by the difference between evaporation and precipitation (E – P) (purple line). [G. Wüst, 1936.]

The term **brine** is applied to water that exceeds the average of 35‰ salinity. **Brackish** applies to water that is less than 35‰ salts. In general, oceans are lower in salinity near landmasses because of freshwater runoff and river discharges. Extreme examples include the Baltic Sea (north of Poland and Germany) and the Gulf of Bothnia (between Sweden and Finland), which average 10‰ or less salinity because of heavy freshwater runoff and low evaporation rates.

On the other hand, the Sargasso Sea, within the North Atlantic subtropical gyre, averages 38‰. The Persian Gulf has a salinity of 40‰ as a result of high evaporation rates in a nearly enclosed basin. Deep pockets, or "brine lakes," along the floor of the Red Sea and the Mediterranean Sea register up to a salty 225‰.

## Physical Structure of the Ocean

The basic physical structure of the ocean is layered, as shown in Figure 16-3. The figure also graphs four key parameters of the ocean, each of which varies with increasing depth: average temperature, salinity, dissolved carbon dioxide, and dissolved oxygen level.

The ocean's surface layer is warmed by the Sun and is wind-driven. Variations in water temperature and solutes are blended rapidly in a *mixing zone* that represents only 2% of the oceanic mass. Below this is the *thermocline transition zone*, a region of decreasing temperature gradient that lacks the motion of the surface. Friction dampens the effect of surface currents, and colder water temperatures at the lower margin tend to inhibit any convective movements.

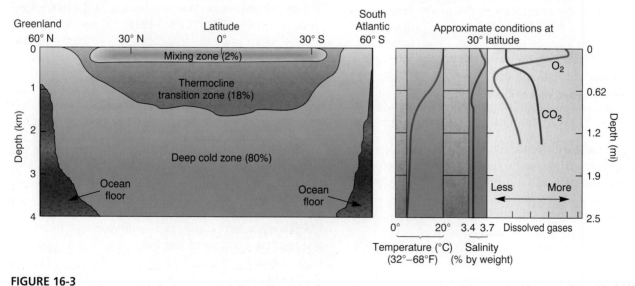

**FIGURE 16-3**

**The ocean's physical structure.**

Schematic of average physical structure observed throughout the ocean's vertical profile as sampled along a line from Greenland to the South Atlantic. Temperature, salinity, and dissolved gases are shown plotted by depth.

From a depth of 1–1.5 km (0.6–0.9 mi) to the bottom, temperature and salinity values are quite uniform. Temperatures in this *deep cold zone* are near 0°C (32°F). Water in the deep cold zone does not freeze, however, because of its salinity and intense pressures at those depths; seawater freezes at about –2°C (28.4°F) at the surface. The coldest water is at the bottom, except near the poles, where the coldest water may be near or at the surface.

***Effect on Pollution.*** Many believe that pollution pumped into the sea is mixed throughout its mass (the "solution-to-pollution-is-dilution" rationale). But the physical structure just described exposes the inaccuracy of that supposition. In reality, only the surface layers actively and rapidly mix. Deep currents are slow-moving masses. It takes thousands of years for deep currents to flow among the North Atlantic, Indian, and North Pacific Oceans, or for the colder, saltier flows that sink near Antarctica to flow northward along the ocean floor past the equator (see Figure 6-25).

The United States enacted the Ocean Dumping Act Ban (1988) to immediately halt industrial waste disposal at sea. The dumping of municipal sewage sludge was to cease completely by the end of 1991. New York City, at the cost of large fines, continued to dump through 1992. There are 110 dumping sites in operation, principally for dredged (excavated) materials. There is constant political pressure to open the oceans and coastal environments for dumping of most municipal, industrial, and radioactive wastes.

# Coastal System Components

We know that the continents were formed over many millions of years. However, most of Earth's coastlines are relatively new, existing in their present state as the setting for continuous change. A dynamic equilibrium exists among the energy of waves, tides, wind, and currents, the supply of materials, the slope of the coastal terrain, and the fluctuation of relative sea level. This interaction produces coastlines of erosional and depositional features of infinite variety and beauty.

An example of a unique coastal environment is the 17,000 km (10,200 mi) of shoreline and connecting waterways of the Great Lakes of North America. This environment includes shoreline systems, dunes and beaches, tributary streams, biological diversity, and dependent human population. It is discussed in Chapter 19, Focus Study 19-1.

## Inputs to the Coastal System

Inputs to the coastal environment include many elements we have already discussed:

- *Solar energy* input drives the atmosphere and the hydrosphere. Prevailing winds, weather systems, and climate are produced by conversion of insolation to kinetic energy.
- *Atmospheric winds*, in turn, generate ocean currents and waves, key inputs to the coastal environment.
- *Climatic regimes*, which result from insolation and moisture, strongly influence coastal geomorphic processes.
- *The nature of coastal rock* is important in determining rates of erosion and sediment production.
- *Human activities* are an increasingly significant input to coastal change.

All of these inputs occur within the ever-present influence of gravity's pull, not only from Earth but from the Moon and Sun. Gravity provides the potential energy of position for materials in motion and generates the tides.

## The Coastal Environment and Sea Level

The coastal environment is called the **littoral zone**, from the Latin word for "shore." Figure 16-4 illustrates the littoral zone and includes specific components discussed later in the chapter. The littoral zone spans some land as well as water. Landward, it extends to the highest water line that occurs on shore during a storm. Seaward, it extends to where water is too deep for storm waves to move sediments on the sea floor (usually around 60 m, or 200 ft). The specific contact line between the sea and the land is the *shoreline*, although this line shifts with tides, storms, and sea-level adjustments. The *coast* continues inland from high tide to the first major landform change and may include areas considered to be part of the coast in local usage.

Because the level of the ocean varies, the littoral zone naturally shifts position from time to time. A rise in sea level causes submergence of land, whereas a drop in sea level exposes new coastal areas. Changes in the littoral zone also are initiated by uplift and subsidence of the land itself.

***Sea Level.*** Sea level is extremely important. Every elevation you see in an atlas or on a map is referenced to *mean sea level*. Yet this average sea level changes daily with the tides and over the long term with changes in climate, tectonic plate movements, and glaciation. Thus, *sea level* is a relative term. At present, there exists no international system to determine exact sea level over time. NOAA is establishing a Global Absolute Sea Level Monitoring System, discussed later in this chapter, to continually observe sea level.

**Mean sea level (MSL)** is a value based on average tidal levels recorded hourly at a given site over many years. MSL varies spatially because of ocean currents and waves, tidal variations, air temperature and pressure differences, ocean temperature variations, slight variations in Earth's gravity, and changes in oceanic volume (see News Report 1).

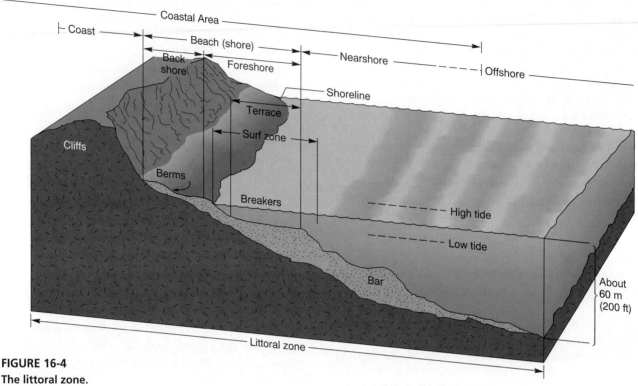

**FIGURE 16-4**
**The littoral zone.**
The littoral zone includes the coast, beach, and nearshore environments.

At present, the overall U.S. MSL is calculated at approximately 40 locations along the coastal margins of the continent. These sites are being upgraded with new equipment in the Next Generation Water Level Measurement System, specifically along the U.S. and Canadian Atlantic coasts, Bermuda, and the Hawaiian Islands. The *NAVSTAR* satellites that make up the new Global Positioning System (GPS) make possible the correlation within a network of ground- and ocean-based measurements. These measurements are augmented by new remote-sensing technology, including the *TOPEX/Poseidon* satellite launched in 1992. This satellite has two radar altimeters that measure sea level at any one location every 10 days between 66° N and 66° S latitudes. These measurements are made to an astonishing precision of 4 mm, or about twice the thickness of the cover of this textbook (Figure 16-5)!

# Coastal System Actions

The coastal system is the scene of complex tidal fluctuation, winds, waves, ocean currents, and the occasional impact of storms. These forces shape landforms ranging from gentle beaches to steep cliffs, and they sustain delicate ecosystems.

## *Tides*

**Tides** are complex daily oscillations in sea level, ranging worldwide from barely noticeable to several meters. They are experienced to varying degrees along every ocean shore around the world and to a lesser extent along the shores of larger lakes, such as the Great Lakes. Tidal action is a relentless energy agent for geomorphic change.

## News Report 1

### Sea Level Varies Along U.S. Coastline

Sea level varies along the full extent of North American shorelines. The mean sea level (MSL) of the U.S. Gulf Coast is about 25 cm (10 in.) higher than that of Florida's east coast, which is the lowest in North America. MSL rises northward along the eastern coast, to 38 cm (15 in.) higher in Maine than in Florida. Along the U.S. western coast, MSL is higher than Florida by about 58 cm (23 in.) in San Diego and by about 86 cm (34 in.) in Oregon.

Overall, North America's Pacific coast MSL averages about 66 cm (26 in.) higher than the Atlantic coast MSL. MSL is affected by differences in ocean currents, air pressure and wind patterns, water density, and water temperature.

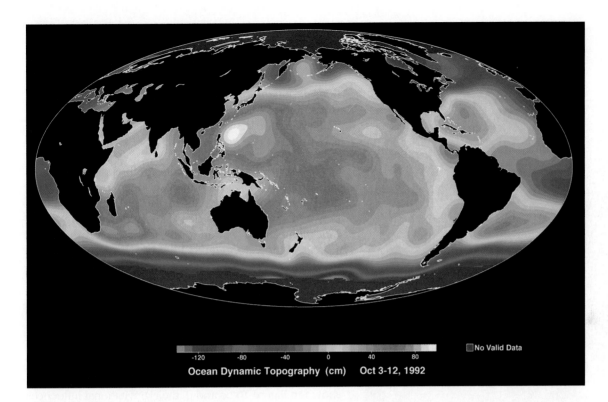

**FIGURE 16-5**
**Ocean topography as revealed by satellite.**
Sea-level data recorded by the radar altimeter aboard the *TOPEX/Poseidon* satellite, October 3–12, 1992. The color scale is given in centimeters above or below Earth's geoid. The overall relief portrayed in the image is about 2 m (6.6 ft). The maximum sea level is in the western Pacific Ocean (white). The minimum is around Antarctica (blue and purple). [Image from JPL and NASA's Goddard Space Flight Center. *TOPEX/Poseidon* is a joint U.S.-French Mission.]

As tides flood (rise) and ebb (fall), the daily migration of the shoreline landward and seaward causes significant changes that affect sediment erosion and transportation.

Tides also are important in human activities, including navigation, fishing, and recreation. Tides are especially important to ships because the entrance to many ports is limited by shallow water, and thus high tide is required for passage. At other times, tall-masted ships need a low tide to clear overhead bridges. Tides also exist in large lakes, but because the tidal range is small tides are difficult to distinguish from changes caused by wind. Lake Superior, for instance, has a tidal variation of only about 5 cm (2 in.).

***Causes of Tides.*** Tides are produced by the gravitational pull of both the Sun and the Moon. (Earth's astronomical relation to the Sun and the Moon is discussed in Chapter 2.) The Sun's influence is only about half that of the Moon's because of the Sun's greater distance from Earth, although it still is a significant force. Figure 16-6 illustrates the relation among the Moon, the Sun, and Earth and the generation of variable tidal bulges on opposite sides of the planet.

The gravitational pull of the Moon tugs on Earth's atmosphere, oceans, and lithosphere. The same is true for the Sun, to a lesser extent. Earth's solid and fluid surfaces all experience some stretching as a result of this gravitational pull. The stretching raises large *tidal bulges* in the atmosphere (which we can't see), smaller tidal bulges in the ocean, and very slight bulges in the rigid lithosphere. Our concern here is the tidal bulges in the ocean.

Gravity and inertia are essential elements in understanding tides. Gravity is the force of attraction between two bodies. Inertia is the tendency of objects to stay still if motionless or to keep moving in the same direction if in motion. The gravitational effect on the side of Earth facing the Moon or Sun is greater than that experienced by the far side, where inertial forces are slightly greater. This difference exists because gravitational influences decrease with distance. It is this difference in the net force of gravitational attraction and inertia that generates the tides.

Figure 16-6a shows the Moon and the Sun in *conjunction* (lined up with Earth), a position in which the sum of their gravitational forces produces a large tidal bulge. The corresponding tidal bulge on Earth's opposite side is primarily the result of the farside water's remaining in position (being left behind) because its inertia exceeds the gravitational pull of the Moon and Sun. In effect, from this inertial point of view, as the nearside water and Earth are drawn toward the Moon and Sun, the farside water is left behind because of the slightly

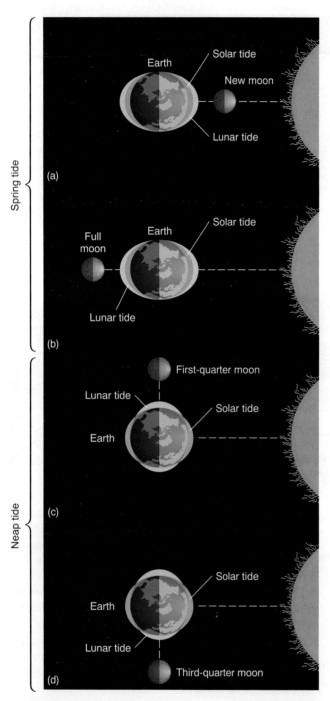

**FIGURE 16-6**
**The cause of tides.**
Gravitational relations of Sun, Moon, and Earth combine to produce spring tides (a, b) and neap tides (c, d). (Tides are greatly exaggerated for illustration.)

weaker gravitational pull. This arrangement produces the two opposing tidal bulges on opposite sides of Earth.

Tides appear to move in and out along the shoreline, but they do not actually do so. Instead, Earth's surface rotates into and out of the relatively "fixed" tidal bulges

as Earth changes its position in relation to the Moon and Sun. Every 24 hours and 50 minutes, any given point on Earth rotates through two bulges as a direct result of this rotational positioning. Thus, every day, most coastal locations experience two high (rising) tides, known as **flood tides**, and two low (falling) tides, known as **ebb tides**. The difference between consecutive high and low tides is considered the *tidal range*.

***Spring and Neap Tides.*** The combined gravitational effect of the Sun and Moon is strongest in the conjunction alignment and results in the greatest tidal range between high and low tides, known as **spring tides**. (*Spring* means to "spring forth"; it has no relation to the season of the year.) Figure 16-6b shows the other alignment that gives rise to spring tides, when the Moon and Sun are at *opposition*. In this arrangement, the Moon and Sun cause separate tidal bulges, affecting the water nearest to each of them. In addition, each bulge is augmented by the left-behind water resulting from the pull of the body on the opposite side.

When the Moon and the Sun are neither in conjunction nor in opposition but are more or less in the positions shown in Figure 16-6c and d, their gravitational influences are offset and counteract each other, producing a lesser tidal range known as **neap tide**. (*Neap* means "without the power of advancing.")

Tides also are influenced by other factors, including ocean basin characteristics (size, depth, and topography), latitude, and shoreline shape. These cause a great variety of tidal ranges. For example, some locations may experience almost no difference between high and low tides. The highest tides occur when open water is forced into partially enclosed gulfs or bays. (For information on tidal bores, see News Report 2.) The Bay of Fundy in Nova Scotia records the greatest tidal range on Earth, a difference of 16 m (52.5 ft) (Figure 16-7a, b).

***Tidal Power.*** The fact that sea level changes daily with the tides suggests an opportunity: Could these predictable flows be harnessed to generate electricity? The answer is yes, given the right conditions. Bays and estuaries tend to focus tidal energy, concentrating it in a smaller area than in the open ocean. This circumstance provides an opportunity to construct a dam with water gates, locks to let ships through, and turbines to generate power. A bay or estuary under consideration must have a narrow entrance suitable for dam construction, and it must experience a tidal range of flood and ebb tides large enough to turn turbines (at least 5 m, or 16 ft).

Only 30 locations in the world are suited for tidal power generation. At present, only three are producing electricity. Two are outside North America—an experimental 1-megawatt station in Russia since 1969 (at Kislaya-Guba Bay on the White Sea) and a facility in France since

# News Report 2

## Tidal Bores Are True Tidal Waves

A tidal current entering a bay or a river is visible as a *tidal bore*. It can cause turbulence as it moves within a constricting estuary or bay, against an opposing tidal current, or against a river's discharge. Most tidal bores range to 1 m in height. Because they are predictable, they pose little danger.

A tidal bore (from *bara*, meaning "wave") is a true *tidal wave*. The media incorrectly describe seismic sea waves (tsunami) as "tidal waves" in their reporting, but tsunami are not tidal. They are triggered by earthquakes, underwater landslides, and sea-floor volcanoes.

A tidal bore entering the mouth of the Amazon River, the Ganges Delta, or the Severn River in England can raise the water level in the river channel a meter or more. Exceptional tidal bores occur in the Chian Tang Kiang River in China, measuring 2.5–3.5 m (8.2–11.5 ft) higher than normal river level. This is a real *tidal wave*.

(a)

(b)

(c)

**FIGURE 16-7**

**Tidal range and tidal power.**

Tidal range is great in some bays and estuaries, such as Halls Harbor near the Bay of Fundy at flood tide (a) and ebb tide (b). The flow of water between high and low tide is ideal for turning turbines and generating electricity, as is done at the Annapolis Tidal Generating Station (c), near the Bay of Fundy in Nova Scotia, in operation since 1984. [(a), (b) Photos by Imagery; (c) photo courtesy of Nova Scotia Power Incorporated.]

1967 (on the Rance River estuary on the Brittany coast). The tides in the Rance estuary fluctuate up to 13 m (43 ft), and power production has been almost continuous there, providing an electrical generating capacity of a moderate 240 megawatts (about 20% of the capacity of Hoover Dam).

The third area is the Bay of Fundy in Nova Scotia, Canada. At one of several favorable sites on the Bay, the Annapolis Tidal Generating Station was built in 1984 as a test. Nova Scotia Power Incorporated operates this

20-megawatt plant (Figure 16-7c). According to the Canadian government, tidal power generation at ideal sites is economically competitive with fossil fuel plants.

Future Canadian possibilities for tidal power development include sites in British Columbia (Observatory Bay near Prince Rupert and Sechelt Inlet near Vancouver) and Labrador (Ungava Bay). China is also looking toward further development. Expansion depends on the growth in electrical demand, the price of fossil fuels, any resolution of problems with nuclear power, and accommodation of environmental concerns over construction along coastlines.

## Waves

Friction between moving air (wind) and the ocean surface generates undulations of water called **waves**. Waves travel in groups, called *wave trains*. Waves vary widely in scale; on a small scale, a moving boat creates a wake of small waves; at a larger scale, storms generate large groups of wave trains.

A stormy area at sea is called a *generating region* for large wave trains, which radiate outward in all directions. The ocean is crisscrossed with intricate patterns of waves traveling in all directions. The waves seen along a coast may be the product of a storm center thousands of kilometers away.

Regular patterns of smooth, rounded waves, called **swells**, are the mature undulations of the open ocean. As waves leave the generating region, wave energy continues to run in these swells, which can range from small ripples to very large flat-crested waves. A wave leaving a deep water generating region tends to extend its wavelength horizontally for many meters. Tremendous energy occasionally accumulates to form unusually large waves. One moonlit night in 1933, the U.S. Navy tanker *Ramapo* reported a wave in the Pacific higher than their mainmast, at about 34 m (112 ft)!

As you watch waves in open water, it appears that water is migrating in the direction of wave travel, but only a slight amount of water is actually advancing. It is the *wave energy* that is moving through the flexible medium of water. Water within a wave in the open ocean is simply transferring energy from molecule to molecule in simple cyclic undulations, called *waves of transition* (Figure 16-8). Individual water particles move forward only slightly, forming a vertically circular pattern.

The diameter of the paths formed by the orbiting water particles decreases with depth. As a deep-ocean wave approaches the shoreline and enters shallower water (10–20 m or 30–65 ft), the orbiting water particles are vertically restricted. This restriction causes more-elliptical, flattened, orbits to form near the bottom. This change from circular to elliptical orbits for the water particles slows the entire wave, although waves keep arriving. The resultant effects are closer-spaced waves, growing in height and steepness, with sharper wave crests. As the crest of each wave rises, a point is reached when its height exceeds its vertical stability and the wave falls into a characteristic **breaker**, crashing onto the beach (Figure 16-8b).

In a breaker, the orbital motion of transition gives way to form *waves of translation*, *both* energy and water move toward shore. The slope of the shore determines wave style. Plunging breakers indicate a steep bottom profile, whereas spilling breakers indicate a gentle, shallow bottom profile. In some areas, unexpected high waves can arise suddenly (see News Report 3).

*Wave Refraction.* In general, wave action tends to straighten a coastline. Where waves approach an irregular coast, they bend around headlands, which are protruding landforms generally composed of resistant rocks (Figure 16-9). The submarine topography refracts (bends) approaching waves. The refracted energy is focused around headlands and dissipates energy in *coves*, *bays*, and the submerged coastal valleys between headlands. Thus, headlands receive the brunt of wave attack along a coastline. This **wave refraction** redistributes wave energy so that different sections of the coastline vary in erosion potential, with the long-term effect of straightening the coast.

Figure 16-10 shows waves approaching the coast at an angle, for they usually arrive at some angle other than parallel. As the waves enter shallow water, they are refracted and generate a current parallel to the coast, zigzagging in the prevalent direction of the incoming waves. This **longshore current**, or *littoral current*, depends on wind direction and wave direction. A longshore current is generated only in the surf zone and works in combination with wave action to transport large amounts of sand, gravel, sediment, and debris along the shore as *longshore drift*.

Particles on the beach also are moved along as **beach drift**, shifting back and forth between water and land with each *swash* and *backwash* of surf. Individual sediment grains trace arched paths along the beach. You have perhaps stood on a beach and heard the sound of myriad sand grains and seawater, in the backwash of surf. These dislodged materials are available for transport and eventual deposition in coves and inlets and can represent a significant volume.

*Tsunami, or Seismic Sea Wave.* An occasional wave that momentarily but powerfully influences coastlines is the *tsunami*. **Tsunami** is Japanese for "harbor wave," named for its devastating effect where its energy is focused in harbors. Tsunami often are reported incorrectly as "tidal waves," but they have no relation to the tides (see News Report 2). They are formed by sudden, sharp motions in the sea floor, caused by earthquakes, or submarine landslides, or eruptions of undersea volcanoes. Thus they properly are called *seismic sea waves*.

A tsunami typically begins when a large undersea disturbance generates a solitary wave of great wavelength, or sometimes a group of two or three long waves. These waves generally exceed 100 km (60 mi) in wavelength, but they are only a meter or so in height. Because of their great wavelength, tsunami are affected by the topography of the deep-ocean floor and are refracted by rises

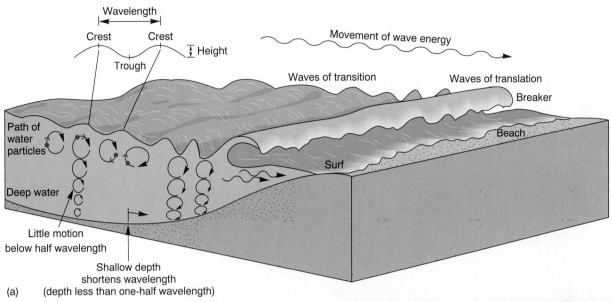

**FIGURE 16-8**

**Wave formation and breakers.**
(a) Wave structure and the orbiting tracks of water particles change from circular motions and swells in deep *waves of transition* to more elliptical orbits near the bottom in shallow *waves of translation*. (b) Cascades of waves attack the shore. [Photo by Bobbé Christopherson.]

(b)

## News Report 3

### Killer Waves Strike Beachcombers

Signs along portions of the California, Oregon, Washington, and British Columbia coastline warn beachcombers to watch for "killer waves." It is a good idea to learn to recognize killer wave conditions before you venture along the shore in these areas.

As various wave trains move along in the open sea, they interact by *interference*. These interfering waves sometimes align so that the wave crests and troughs from one wave train are in phase with those of another. When this *in-phase* condition occurs, the height of the waves is increased, sometimes dramatically. The resulting "killer waves" or "sleeper waves" can sweep in unannounced and overtake unsuspecting victims.

On the other hand, when wave trains are *out of phase* with each other, they cancel each other and their heights are reduced. Maybe you have noticed the changing pattern of the surf (bigger waves and smaller waves) caused by wave interference. When you observe the breakers along a beach, the changing beat of the surf actually is produced by the patterns of wave interference that occurred in far-distant areas of the ocean.

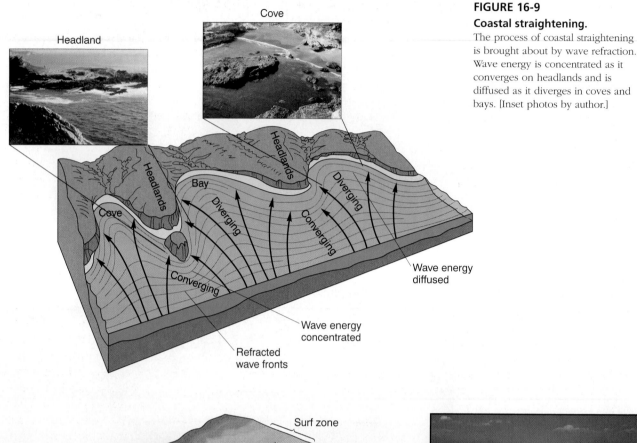

**FIGURE 16-9**
**Coastal straightening.**
The process of coastal straightening
is brought about by wave refraction.
Wave energy is concentrated as it
converges on headlands and is
diffused as it diverges in coves and
bays. [Inset photos by author.]

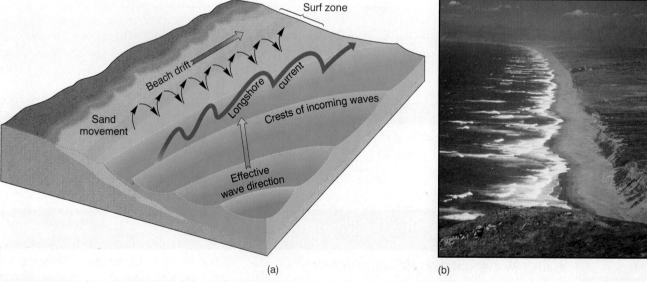

(a)                    (b)

**FIGURE 16-10**

**Longshore current and beachdrift.**

(a) Longshore currents are produced as waves approach the surf zone and shallower water. Longshore and beach drift result
as substantial volumes of material are moved along the shore. (b) Processes at work along Point Reyes Beach, Point Reyes
National Seashore. [(b) Photo by author.]

and ridges. They travel at great speeds in deep-ocean
water—velocities of 600 to 800 kmph (375 to 500 mph)
are not uncommon—but they often pass unnoticed on
the open sea because their great length makes the slow
rise and fall of water hard to observe.

As a tsunami approaches a coast, however, the shallow water forces the wavelength to shorten. As a result, the
wave height may increase up to 15 m (50 ft) or more. Such
a wave has the potential to devastate a coastal area, causing property damage and death. For example, in 1992, the

citizens of Casares, Nicaragua, were surprised by a 12 m (39 ft) tsunami that took 270 lives. An earthquake along an offshore subduction zone triggered this destructive wave. The Nicaraguan tusunami was the first to be observed and analyzed by a new seismic network. Two tsunami in 1994 killed 173 in Java, Indonesia, and 45 people in the Philippines; both waves were triggered by earthquakes.

In the Alaskan earthquake of 1964, a tsunami generated by undersea landslides radiated across the Pacific Ocean. It reached Crescent City, California, in 4 hours 3 minutes; Hilo, Hawaii, in 5 hours 24 minutes; Hokkaido, Japan, in 6 hours 44 minutes; and La Punta, Peru, in 15 hours 34 minutes. Curious onlookers in both California and Oregon were killed by this tsunami generated in distant Alaska.

Hawaii is vulnerable to tsunami because of its position in the open central Pacific, surrounded by the ring of fire that outlines the Pacific Basin. The U.S. Army Corps of Engineers has reported 41 damaging tsunami in Hawaii during the past 142 years—statistically, 1 every 3.5 years. Figure 16-11 illustrates tsunami travel times across the Pacific Ocean to Honolulu; for instance, it would take a tsunami 10 hours to reach Hawaii from the Philippines.

Because tsunami travel such great distances so quickly and are undetectable in the open ocean, accurate forecasts are difficult and occurrences are often unexpected. A warning system now is in operation for nations surrounding the Pacific, where the majority of tsunami occur. A warning is issued whenever seismic stations detect a significant quake under water, where it might generate a tsunami. Warnings always should be heeded, despite the many false alarms, for the causes lie hidden beneath the ocean and are difficult to monitor in any consistent manner.

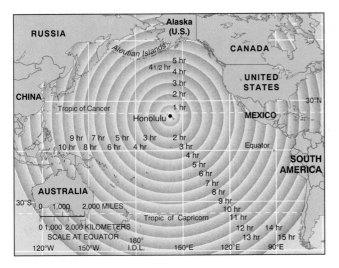

**FIGURE 16-11**
**Tsunami travel times to Honolulu, Hawaii.**
[After NOAA.]

## Sea-Level Changes

Over the long term, sea-level fluctuations expose a great range of coastal landforms to tidal and wave processes. Sea-level changes are initiated either by tectonic forces that raise or lower coastlines or by glacio-eustatic processes that affect the quantity of Earth's water that is stored as ice. Because of complex interactions, it is difficult to isolate individual factors. However, as average global temperatures cycle through cold or warm climatic spells, the quantity of ice locked up in the ice sheets of Antarctica and Greenland and in hundreds of mountain glaciers can increase or decrease accordingly. (Chapter 10 presents the climatic implications of a greenhouse-effect warming pattern on high-latitude and polar environments, which contain enormous deposits of ice.)

If Antarctica and Greenland ever became ice-free (ice sheets fully melted), sea level would rise at least 65 m (215 ft) worldwide. Indeed, sea level has risen throughout the most recent geologic epoch, the Holocene (the past 10,000 years), and generally ever since the end of the Pleistocene Ice Age, which began about 1.65 million years ago.

At the peak of the most recent Pleistocene glaciation about 18,000 B.P. (years before the present), sea level was about 130 m (430 ft) lower than it is today. Just 100 years ago sea level was 38 cm (15 in.) lower along the coast of southern Florida. Venice, Italy, has experienced a rise of 25 cm (10 in.) since 1890. Elsewhere, increases of 10 to 20 cm (4 to 8 in.) this century are common. This rate of rise of 20 to 40 cm (8 to 16 in.) per 100 years is about *six times faster* than any observed from historical records or recent geologic time.

Sea level will rise not only from melted ice, but about 25% will be the result of thermal expansion of warmer water (water is densest at 4°C, or 39°F, and expands at higher temperatures). Measurements by the Scripps Institute of Oceanography demonstrate a 0.8 C° (1.44 F°) warming since 1950 off southern California, other studies indicate a warming along the entire U.S. West Coast. Similar ocean warming trends are being reported worldwide.

Given these trends and the predicted climatic change, sea level could rise and be potentially devastating for many coastal locations. A rise of only 0.3 m (1 ft) would cause shorelines worldwide to move inland an average of 30 m (100 ft)! This elevated sea level would inundate some valuable real estate along coastlines worldwide. Some 20,000 km², 7800 mi², of land along North American shores alone would be drowned, at a staggering loss of $650 billion. A 95 cm (3.1 ft) sea-level rise could inundate 15% of Egypt's arable land, 17% of Bangladesh, and many island nations and communities.

However, great uncertainty exists in these forecasts, and the pace of the rise should be slow. In 1995, the Intergovernmental Panel on Climate Change (IPCC, discussed in Chapter 10) estimated the range of expected sea-level rise at between 15 cm and 95 cm (0.5 ft and 3.1 ft) between now and 2100. These increases would continue beyond 2100 even if greenhouse gas concentrations are stabilized. Despite this uncertainty, planning should start now along coastlines worldwide because preventive strategies are cheaper than recovery costs from possible destruction. Insurance underwriters have begun the process by refusing coverage for shoreline properties vulnerable to rising sea level.

# Coastal System Outputs

As you can see, coastlines are very active portions of continents, with energy and sediment being continuously delivered to a narrow environment. The action of tides, currents, wind, waves, and changing sea level produces a variety of erosional and depositional landforms. We look first at erosional coastlines, then at depositional coastlines.

## Erosional Coastal Processes and Landforms

The active margin of the Pacific Ocean along North and South America is a typical erosional coastline. *Erosional coastlines* tend to be rugged, of high relief, and tectonically active, as expected from their association with the leading edge of drifting lithospheric plates (review the plate tectonics discussion in Chapter 11). Figure 16-12 presents features commonly observed along an erosional coast.

*Sea cliffs* are formed by the undercutting action of the sea. As indentations slowly grow at water level, a sea cliff becomes notched and eventually will collapse and retreat. Other erosional forms evolve along cliff-dominated coastlines, including *sea caves, sea arches*, and *sea stacks*. As erosion continues, arches may collapse, leaving isolated stacks in the water. The coasts of southern England and Oregon are prime examples of such erosional landscapes.

Wave action can cut a horizontal bench in the tidal zone, extending from a sea cliff out into the sea. Such a structure is called a **wave-cut platform**, or *wave-cut terrace*. If the relation between the land and sea level has

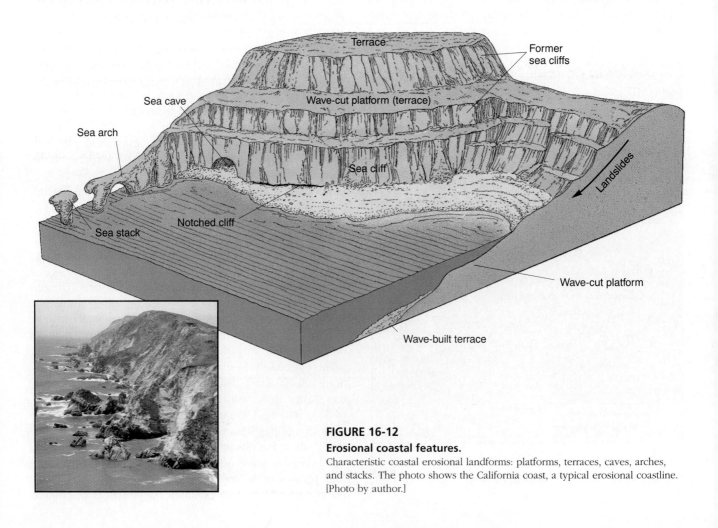

**FIGURE 16-12**
**Erosional coastal features.**
Characteristic coastal erosional landforms: platforms, terraces, caves, arches, and stacks. The photo shows the California coast, a typical erosional coastline. [Photo by author.]

**FIGURE 16-13**
**Marine terraces.**
Marine terraces formed as wave-cut platforms along the Pacific Coast, near Bixby Bridge along the dramatic Cabrillo Highway. [Photo by Nancy Sanford/Stock Market.]

changed over time, multiple platforms or terraces may rise like stairsteps back from the coast. These marine terraces are remarkable indicators of a changing relation between the land and sea, with some terraces more than 365 m (1200 ft) above sea level. A tectonically active region, such as the California coast, has many examples of multiple wave-cut platforms (Figure 16-13).

## Depositional Coastal Processes and Landforms

*Depositional coasts* generally are located along land of gentle relief, where sediments are available from many sources. Such is the case with the Atlantic and Gulf coastal plains of the United States, which lie along the relatively passive, trailing edge of the North American lithospheric plate. These depositional coasts also are influenced by erosional processes and inundation, particularly during storm activity.

Characteristic landforms deposited by waves and currents are illustrated in Figure 16-14. One notable deposition landform is the **barrier spit**, which consists of material deposited in a long ridge extending out from a coast. It partially crosses and blocks the mouth of a bay. Classic examples include Sandy Hook, New Jersey (south of New York City), and Cape Cod, Massachusetts. The photo in Figure 16-14 shows a typical barrier spit forming partway across the mouth of Prion Bay in Southwestern National Park in Tasmania.

If a spit grows to completely cut off the bay from the ocean and form an inland lagoon, it is called a **bay barrier**, or *baymouth bar*. Spits and barriers are made up of materials that have been eroded and transported by *littoral drift* (beach and longshore drift combined). For much sediment to accumulate, offshore currents must be weak; strong currents carry material away before it can be deposited. Tidal flats and salt marshes are characteristic low-relief features wherever tidal influence is greater than wave action. If these deposits completely cut off the bay from the ocean, an inland **lagoon** is formed.

A **tombolo** occurs when sediment deposits connect the shoreline with an offshore island or sea stack (Figures 16-14 and 16-15). The tombolo forms when sediments accumulate on an underwater wave-built terrace.

**Beaches.** Of all the features associated with a depositional coastline, beaches probably are the most familiar. Beaches vary in type and permanence, especially along

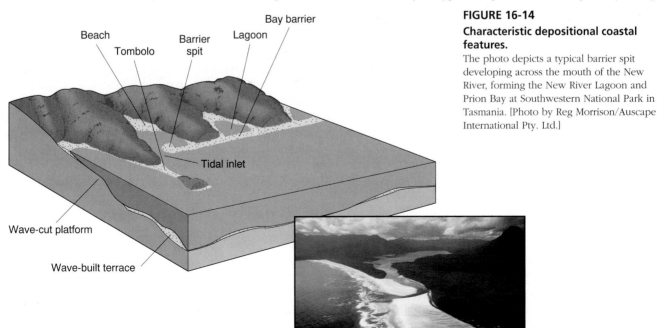

**FIGURE 16-14**
**Characteristic depositional coastal features.**
The photo depicts a typical barrier spit developing across the mouth of the New River, forming the New River Lagoon and Prion Bay at Southwestern National Park in Tasmania. [Photo by Reg Morrison/Auscape International Pty. Ltd.]

Beach
Tombolo
Barrier spit
Lagoon
Bay barrier
Tidal inlet
Wave-cut platform
Wave-built terrace

**FIGURE 16-15**
**A tombolo.**
A tombolo at Point Sur along the central California coast where sediment deposits connect the shore with an island. [Photo by Bobbé Christopherson.]

coastlines dominated by wave action. Technically, a **beach** is that place along a coast where sediment is in motion; deposited by waves and currents. Material from the land temporarily resides on the beach while it is in active transit along the shore. You may have experienced a beach at some time, along a sea coast, a lake shore, or even a stream. Perhaps you have even built your own "landforms" in the sand, only to see them washed away by the waves: a lesson in erosion.

On average, the beach zone spans from about 5 m (16 ft) above high tide to 10 m (33 ft) below low tide (see Figure 16-4). However, the specific definition varies greatly along individual shorelines. Worldwide, beaches are dominated by sands of quartz ($SiO_2$) because it resists weathering and therefore remains after other minerals are removed. In volcanic areas, beaches are derived from wave-processed lava. These black-sand beaches are found in Hawaii and Iceland, for example.

Many beaches, such as those in southern France and western Italy, lack sand and are composed of pebbles and cobbles—a type of "shingle beach." Some shores have no beaches at all; scrambling across boulders and rocks may be the only way to move along the coast. The coasts of Maine and portions of Canada's Atlantic provinces are classic examples. These coasts, composed of resistant granite rock, are scenically rugged but have few beaches.

A beach acts to stabilize a shoreline by absorbing wave energy, as is evident by the amount of material that is in almost constant motion (see "sand movement" in Figure 16-10). Some beaches are stable. Others cycle seasonally: They accumulate during the summer, are moved offshore by winter storm waves, forming a submerged bar, and are redeposited onshore the following summer. Protected areas along a coastline tend to accumulate sediment, which can lead to large coastal sand dunes. Prevailing winds often drag such coastal dunes inland, sometimes burying trees and highways.

Figure 16-4 shows the location of beach elements—the back shore, berm, and foreshore (which includes the surf zone). Beach profiles fall into two broad categories, according to prevailing conditions. During quiet, low-energy periods, both backshore and foreshore areas experience net accumulations of sediment and are well defined. However, a storm brings increased wave action and high-energy conditions to the beach, usually reducing its profile to a sloping foreshore only. Sediments from the backshore area do not disappear but may be moved by wave action to the nearshore (submerged) area, only to be returned by a later low-energy surf. Therefore, beaches go through constant change, especially in this era of rising sea levels.

***Maintaining Beaches.*** Changes in coastal sediment transport can thwart human activities—beaches are lost, harbors are closed, and coastal highways and beach houses can be inundated with sediment. Thus, people use various strategies to interrupt longshore drift and beach drift. The goal is either to halt sand accumulation or to force accumulation in a desired way through construction of engineered structures, called "hard" shoreline protection.

Figure 16-16 illustrates common approaches: a *jetty* to block material from harbor entrances; a *groin* to slow drift action along the coast, and a *breakwater* to create a zone of still water near the coastline. However, interrupting the littoral drift that is the natural replenishment for beaches may lead to unwanted changes in sediment distribution downcurrent. Careful planning and impact assessment should be part of any strategy for preserving or altering a beach.

*Beach nourishment* refers to the artificial replacement of sand along a beach (see News Report 4). Through such efforts, a beach that normally experiences a net loss of sediment will instead show a net gain. In contrast to hard structures, this hauling of sand to replenish a beach is considered "soft" shoreline protection. Enormous energy and material must be committed to counteract the relentless energy that nature expends along the coast. Years of human effort and expense can be erased by a single storm, as occurred when offshore islands along the Gulf Coast were completely eliminated by Hurricane Camille in 1969.

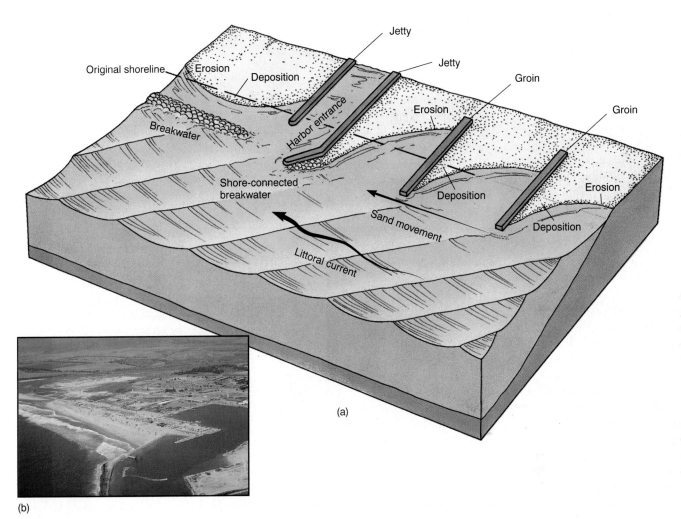

(a)

(b)

**FIGURE 16-16**

**Interfering with the littoral drift of sand.**

(a) Various constructions attempt to control littoral drift (beach drift and longshore drift) along a coast: breakwater, jetty, and groin. (b) Aerial photograph of coastal constructions. Note the disruption of sediment movement along the coast at the entrance to Oceanside, California, harbor. [(b) Photo by Craig Everts.]

## News Report 4

### Engineers Nourish a Beach

The city of Miami, Florida, and surrounding Dade County have spent almost $70 million since the 1970s in a continuing effort to rebuild their beaches. Sand is transported to the replenishment area from a source area. To maintain a 200 m wide beach (660 ft), planners determine net sand loss per year and set a schedule for replenishment. In Miami Beach, an 8-year replenishment cycle is maintained. During Hurricane Andrew,] in 1992, the replenished Miami Beach is thought to have prevented millions of dollars in shoreline structural damage.

Unforeseen environmental impact may accompany the addition of sand to a beach, if the sand is from an unmatched source, in terms of its ecological relationship of the site. If the new sands do not match (physically and chemically) the existing varieties, disruption of coastal marine life is possible. The U.S. Army Corps of Engineers, which operates the Miami replenishment program, is running out of "borrowing areas" for sand that matches the natural sand of the beach. A proposal to haul a different type of sand from the Bahamas is being studied as to possible environmental consequences.

***Barrier Formations.*** Barrier chains are long, narrow, depositional features, generally of sand, that form offshore roughly parallel to the coast. Common forms are **barrier beaches**, or the broader, more extensive landform, **barrier islands**. Tidal variation in the area usually is moderate to low, with adequate sediment supplies coming from nearby coastal plains. Figure 16-17 illustrates the many features of barrier chains, using North Carolina's famed Outer Banks. It includes Cape Hatteras, across Pamlico Sound from the mainland. The area presently is designated as one of three national seashore reserves supervised by the National Park Service.

On the landward side of a barrier formation are tidal flats, marshes, swamps, lagoons, coastal dunes, and beaches, visible in Figure 16-17. Barrier beaches appear to adjust to sea level and may naturally shift position from time to time in response to wave action and longshore currents. A break in a barrier forms an inlet that connects a bay with the ocean. The name *barrier* is appropriate, for these formations take the brunt of storm energy and actually shield the mainland.

Barrier beaches and islands are quite common worldwide, lying offshore of nearly 10% of Earth's coastlines. Examples are found offshore of Africa, India's eastern coast, Sri Lanka, Australia, Alaska's northern slope, and

(a)

**FIGURE 16-17**
**Barrier island chain.**

(a) *Landsat* image of barrier island chain along the North Carolina coast. Hurricane Emily swept past Cape Hatteras in August 1993, causing damage and beach erosion. You can see key depositional forms: spit, island, beach, lagoon, and inlet. A *sound* is a large inlet of the ocean; Pamlico Sound forms an ideal example. (b) View from Cape Hatteras lighthouse illustrates the narrow strand of sand that stands between the ocean and the mainland. [(a) Image from NASA; (b) photo by Mike Booher].

(b)

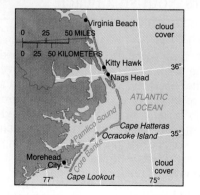

the shores of the Baltic and Mediterranean Seas. Earth's most extensive chain of barrier islands is along the U.S. Atlantic and Gulf Coast, extending some 5000 km (3100 mi) from Long Island to Texas and Mexico.

***Barrier Island Origin and Hazards.*** Various hypotheses have been proposed to explain the formation of barrier islands. They may begin as offshore bars or low ridges of submerged sediment near shore and then gradually migrate toward shore as sea level rises.

Because many barrier islands seem to be migrating landward, they are an unwise choice for homesites or commercial building. Nonetheless, they are a common choice, even though they take the brunt of storm energy. The hazard represented by the settlement of barrier islands was made graphically clear when Hurricane Hugo assaulted South Carolina in 1989.

Hurricane Hugo made a destructive pass over Puerto Rico, the Virgin Islands, and the northeastern Caribbean and then turned northwestward. The storm attacked the Grand Strand barrier islands off the northern half of South Carolina's coastline, most affecting Charleston and the South Strand portion of the islands. The storm made landfall as the worst hurricane to strike there in 35 years.

In the Charleston area, beachfront houses, barrier-island developments, and millions of tons of sand were swept away; up to 95% of the single-family homes in one community were destroyed. The southern portion of one island was torn away, and one of every four homes was destroyed. Some homes that had escaped the last major storm, Hurricane Hazel in 1954, were lost to Hugo's peak winds of 209 kmph (130 mph).

In comparable dollars, Hugo caused nearly $4 billion in damage, compared with Hazel's $1 billion. The increased damage partially resulted from expanded construction during the intervening years. With increased development and continuing real estate appreciation, each future storm can be expected to cause ever-increasing capital losses.

## Biological Processes: Coral Formations

Not all coastlines form by purely physical processes. Some form as the result of biological processes, such as coral growth. A **coral** is a simple marine animal with a small, cylindrical, saclike body (polyp); it is related to other marine invertebrates, such as anemones and jellyfish. Corals secrete calcium carbonate ($CaCO_3$) from the lower half of their bodies, forming a hard calcified external skeleton.

Corals live in a *symbiotic* relationship with algae: They live together in a mutually helpful arrangement, each dependent on the other for survival. Corals cannot photosynthesize, but they do ingest some of their own nourishment. Algae perform photosynthesis and convert solar energy to chemical energy in the system, providing the coral with about 60% of its nutrition and assisting the coral with the calcification process. In return, corals provide the algae with nutrients.

Figure 16-18 shows the distribution of currently living coral formations. Corals thrive in warm tropical oceans, so the difference in ocean temperature between the western coasts and eastern coasts of continents is critical to their distribution. Western coastal waters tend to be cooler, thereby discouraging coral activity, whereas eastern coastal currents are warmer and thus enhance coral growth.

Living colonial corals range in distribution from about 30° N to 30° S and occupy a very specific ecological zone defined by a maximum depth of 55 m (180 ft), salinity of 27‰–40‰, and water temperature of 18°–29°C (64°–85°F). Corals require clear, sediment-free water and consequently do not locate near the mouths of sediment-charged freshwater streams. For example, note the lack of these structures along the U.S. Gulf Coast.

***Coral Reefs.*** Both solitary and colonial corals exist. It is the colonial corals that produce enormous structures. Their skeletons accumulate, forming a biological derived coral rock. Through many generations, live corals near the ocean's surface build on the foundation of older coral skeletons, which in turn may rest upon a volcanic seamount or some other submarine feature built up from

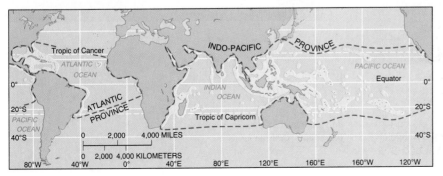

**FIGURE 16-18**
**Worldwide distribution of living coral formations.**
Yellow areas also include prolific reef growth and atoll formation. The red dotted line marks the geographical limits of coral activity. [After J. L. Davies, *Geographical Variation in Coastal Development*, (Essex, England: Longman House, 1973). Adapted by permission.]

the ocean floor. By this process, *coral reefs* are formed. Thus, a coral reef is biologically derived sedimentary rock. It can assume one of several distinctive shapes.

The principal reef shapes are fringing reefs, barrier reefs, and atolls. In 1842, Charles Darwin hypothesized for an evolution of reef formation. He suggested that, as reefs developed around a volcanic island and the island itself gradually subsided, an equilibrium between the subsidence of the island and the upward growth of the corals is maintained. This idea, generally accepted today, is portrayed in Figure 16-19. Note the specific examples of each reef stage: *fringing reefs* (platforms of surrounding coral rock), *barrier reefs* (forming enclosed lagoons), and *atolls* (circular, ring-shaped).

The most extensive fringing reef is the Bahamian platform in the western Atlantic, covering some 96,000 km² (37,000 mi²). The largest barrier reef, the Great Barrier Reef along the shore of Queensland State in Australia, exceeds 2025 km (1260 mi) in length, is 16–145 km (10–90 mi) wide, and includes at least 200 coral-formed islands and keys (coral islets or barrier islands).

**Coral Bleaching.** Normally colorful corals are turning stark white as they expel their own nutrient-supplying algae, a phenomenon known as *bleaching*. Exactly why the corals eject their symbiotic partner is unknown, for without algae the corals die. Scientist are tracking their unprecedented bleaching and dying worldwide. The

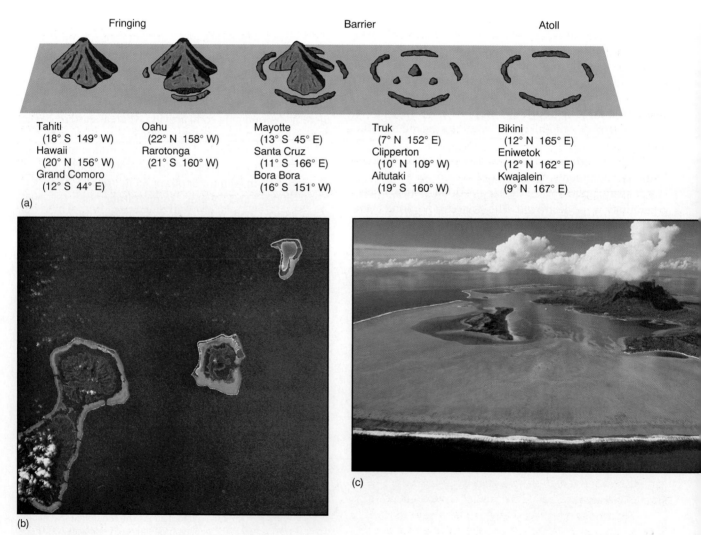

**FIGURE 16-19**
**Coral forms.**
(a) Common coral formations in a sequence of reef growth formed around a subsiding volcanic island: fringing reefs, barrier reefs, and an atoll. (b, c) Space Shuttle and aerial photographs of the atolls in Bora Bora, Society Islands, 16° 30′ S 151° 45′ W. [(a) After D. R. Stoddart, *The Geographical Magazine* 63 (1971): 610. (b) Space Shuttle photo from NASA; (c) photo by Harvey Lloyd/Stockmarket.]

**FIGURE 16-20**
**Coastal salt marsh.**
Salt marshes are productive ecosystems commonly occurring poleward of 30° latitude in both hemispheres. This is Bolinas Lagoon on the Pacific Coast in central California. The tide is high in the photo; at low tide an extensive tidal flat is uncovered. [Photo by author.]

Caribbean, Australia, Japan, Indonesia, Kenya, Florida, Texas, and Hawaii are experiencing this phenomenon. The bleaching is due to a loss of colorful algae (red-brown to green) from within and upon the coral.

Possible causes include local pollution, disease, sedimentation, and changes in salinity. Another possible cause is the 1 C° to 2 C° (1.8 F° to 3.6 F°) warming of sea-surface temperatures, as stimulated by greenhouse warming of the atmosphere. In the western Caribbean, water temperatures held at 30°C (86°F) for 2 months in 1995, triggering a loss of algae. During the 1982–1983 ENSO (Chapter 10), areas of the Pacific Ocean were warmer than normal and widespread coral bleaching occurred. Coral bleaching worldwide is continuing as average ocean temperatures climb higher. Further bleaching through the decade will affect most of Earth's coral-dominated reefs and may be an indicator of serious and enduring environmental trauma. If present trends continue, most living corals on Earth could perish over the next decade.

## Wetlands, Salt Marshes, and Mangrove Swamps

Some coastal areas have great *biological productivity* (plant growth) stemming from trapped organic matter and sediments. Such a rich coastal marsh environment can greatly outproduce a wheat field in raw vegetation per acre. This means that coastal marshes can support rich wildlife habitats. Unfortunately, these wetland ecosystems are quite fragile and are threatened by human development.

**Wetlands** are saturated with water enough of the time to support *hydrophytic vegetation* (plants that grow in water or very wet soil). Wetlands usually occur on poorly drained soils. Geographically, they occur not only along coastlines but also as northern bogs (peatlands with high water tables), as potholes in prairie lands, as cypress swamps (with standing or gently flowing water), as river bottomlands and floodplains, and as arctic and subarctic environments that experience permafrost during the year.

***Coastal Wetlands.*** Coastal wetlands are of two general types—**salt marshes** and **mangrove swamps**. In the Northern Hemisphere, salt marshes tend to form north of the 30th parallel, whereas mangrove swamps form equatorward of that line. This distribution is dictated by the occurrence of freezing conditions, which control the survival of mangrove seedlings. Roughly the same latitudinal limits apply in the Southern Hemisphere.

Salt marshes usually form in estuaries and behind barrier beaches and spits. An accumulation of mud produces a site for the growth of *halophytic* (salt-tolerant) plants. This vegetation then traps additional alluvial sediments and adds to the salt marsh area. Because salt marshes are in the intertidal zone (between the farthest reaches of high and low tides), sinuous, branching channels are produced as tidal waters flood into and ebb from the marsh (Figure 16-20).

Sediment accumulation on tropical coastlines provides the site for mangrove trees, shrubs, and other small trees. The prop roots of the mangrove are constantly finding new anchorages. They are visible above the water line but reach below the water surface, providing a habitat for a multitude of specialized life forms. Mangrove swamps often secure and fix enough material to form islands (Figure 16-21).

The World Resources Institute and the U.N. Environment Programme estimate that, from preagricultural times to today, mangrove losses are running between 40% (e.g., Cameroon and Indonesia) and nearly 80% (e.g., Bangladesh and Philippines). Deliberate removal was a common practice by many governments because of a falsely conceived fear of disease or pestilence in these swamplands. The development of Florida's Marco Island, described later, is an example of mangrove loss due to urbanization.

(a)

(b)

**FIGURE 16-21**

**Mangroves.**

Mangroves tend to grow equatorward of 30° latitude. (a) Mangroves along the East Alligator River (12° S latitude) in Kakadu National Park, Northern Territory, Australia. (b) Mangroves retain sediments and can form anchors for island formations such as Aldabra Island (9° S), Seychelles. [(a) Photo by Belinda Wright/ DRK Photo; (b) photo by Wolfgang Kaehler Photography.]

# Human Impact on Coastal Environments

Development of modern societies is closely linked to estuaries, wetlands, barrier beaches, and coastlines. Estuaries are important sites of human settlement because they provide natural harbors, a food source, and convenient sewage and waste disposal. Society depends upon daily tidal flushing of estuaries to dilute the pollution created by waste disposal. Thus, the estuarine and coastal environment is vulnerable to abuse and destruction if development is not carefully planned.

Before World War II only 28 of 280 barrier islands in the United States were developed to any degree. Today, about 50% of estuarian zones in the United States and Canada have been altered or destroyed, and by the end of this decade most barrier islands will have been developed and occupied.

Barrier islands and coastal beaches do migrate over time. Houses that were more than a mile from the sea in the Hamptons along the southeastern shore of Long Island, New York, are now within only 30 m (100 ft) of it. The time remaining for these homes can be quickly shortened by a single hurricane or by a sequence of storms with the intensity of a series that occurred in 1962 or the severe storms that struck the U.S. and Canadian Atlantic Coast in 1992.

Despite our understanding of beach and barrier-island migration, the effect of storms, and warnings from scientists and government agencies, coastal development proceeds. Society behaves as though beaches and barrier islands are stable, fixed features, or as though they can be engineered to be permanent. Experience has shown that severe erosion generally cannot be prevented, as we have seen with the cliffs of southern California, the shores of the Great Lakes, the New Jersey shore, the Gulf Coast, and the devastation of coastal South Carolina by Hurricane Hugo. Shoreline planning is the topic of Focus Study 16-1.

## Marco Island, Florida: An Example of Impact

An example of intense barrier-island development is the Marco Island area on the southwestern Florida coast about 24 km (15 mi) south of Naples. Figure 16-22a shows Marco Island as it was in 1952: approximately 5300 acres of subtropical mangrove habitat and barrier island terrain, formed from river sediments and old shell and reef fragments.

Development began in 1962, despite tropical storms that occasionally assault the area, as Hurricane Donna had done in 1960. Development included artificial landfill for housing sites and general urbanization of the entire island. Scientists regarded the island and estuarine habitat as highly productive and unique in its mixture of mangrove species, endangered bird species, and a productive aquatic environment. Consequently, numerous challenges and court actions attempted to halt development. But construction activities continued, and by 1984 Marco Island had been completely developed (Figure 16-22b).

In 1984, a geographic information system (GIS, see Chapter 1) was used to study Marco Island, incorporating existing maps, remote-sensing techniques, and a knowledge of the ecology of an estuarine mangrove community. In Figure 16-22, you can compare the 1952 and 1984 digitized inventory maps and a 1984 color-infrared aerial photograph, to see the impact of urban development on this island. Note that the only remaining natural community is restricted to a very limited portion along the extreme perimeter.

The key to protective environmental planning and zoning is to *allocate responsibility and cost in the event of a disaster.* An ideal system places a hazard tax on land, based on assessed risk, and restricts the government's responsibility to fund reconstruction or an individual's right to reconstruct on frequently damaged sites. Comprehensive mapping of erosion-hazard areas would help avoid the ever-increasing costs from recurring disasters. The development of Marco Island might have assumed a different direction had such measures been in place in 1962.

This type of planning should include all coastlines (ocean, sea, or lake), floodplains, earthquake and volcanic hazard areas, toxic contamination sites, and important scenic and scientific areas.

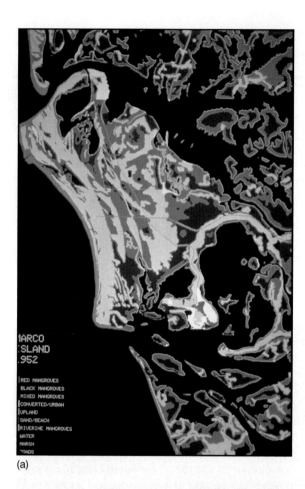

(a)

(b)

## FIGURE 16-22

### Mangrove loss at Marco Island.

Digitized inventories of Marco Island, Florida, showing dramatic development from (a) 1952 to (b) 1984. (c) 1984 color-infrared photograph of the island. [From S. Patterson, *Mangrove Community Boundary Interpretation and Detection of Areal Changes on Marco Island, Florida*, Biological Report 86 (10), for National Wetlands Research Center, U.S. Fish and Wildlife Service, August 1986, pp. 23, 46, 49.]

(c)

## Focus Study 16-1

## An Environmental Approach to Shoreline Planning

Coastlines are places of wonderful opportunity. They also are zones of specific constraints. Poor understanding of this resource and a lack of environmental analysis often go hand in hand. Ecologist and landscape architect Ian McHarg, in *Design with Nature*, discusses the New Jersey shore. He shows how proper understanding of a coastal environment could have avoided problems from major storms and coastal development. Much of what is presented here applies to coastal areas elsewhere.

Figure 1 illustrates the New Jersey shore from ocean to back bay. Let's walk across this landscape and discover how it should be treated under ideal conditions. The principles here can apply to coastal environments elsewhere.

**Beaches and Dunes—Where to Build?**

We begin our walk at the water's edge and proceed across the beach. Sand beaches are the primary natural defense against the ocean; they act as bulwarks against the pounding of a stormy sea. But they are susceptible to pollution and require environmental protection to control

nearshore dumping of dangerous materials. The shoreline tolerates recreation, but not construction, because of its shifting, changing nature during storms, daily tidal fluctuations, and the potential effects of rising sea level. In recent years, high water levels in the Great Lakes and along the southern California coast attest to the vulnerability of shorelines to erosion and inundation. Wave erosion attacks cliff formations and undermines, bit by bit, the foundations of houses and structures built too close to the edge (Figure 2).

Walking inland, we encounter the primary dune. It is even more sensitive than the beach: It is fragile, easily disturbed, and vulnerable to erosion, and it cannot tolerate the passage of people trekking to the beach. Delicate plants struggle to hold the sand in place. Primary dunes are like human-made dikes in the Netherlands: They are the primary defense against the sea, so they should not be disturbed by development or heavy traffic. Carefully controlled access points to the beach should be designated, and restricted access should be enforced, for even foot

traffic can cause destruction. (See the beach stabilization example in Figure 15-5.)

The trough behind the primary dune is relatively tolerant of limited recreation and building. The plants that fix themselves to the surface, sending roots down to fresh groundwater reserves and anchoring the sand in the process. Thus, if construction should inhibit the surface recharge of that water supply, the natural protective ground cover could die and destabilize the environment. Or subsequent saltwater intrusion might contaminate well water for human use. Clearly, groundwater resources and the location of recharge aquifers must be considered in planning.

Behind the trough is the secondary dune, a second line of defense against the sea. It, too, is tolerant of some use yet is vulnerable to destruction. Next is the backdune, more suitable for development than any zone between it and the sea. Further inland are the bayshore and the bay, where no dredging or filling and only limited dumping of treated wastes and toxics should be permitted. This zone is tolerant to intensive recreation. In reality,

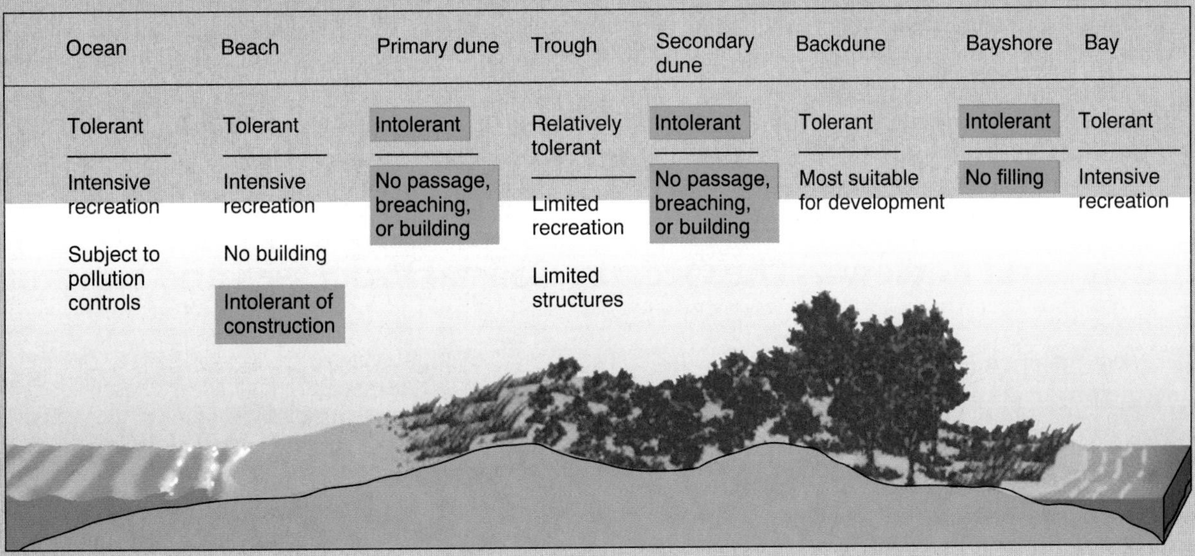

| Ocean | Beach | Primary dune | Trough | Secondary dune | Backdune | Bayshore | Bay |
|---|---|---|---|---|---|---|---|
| Tolerant | Tolerant | Intolerant | Relatively tolerant | Intolerant | Tolerant | Intolerant | Tolerant |
| Intensive recreation | Intensive recreation | No passage, breaching, or building | | No passage, breaching, or building | Most suitable for development | No filling | Intensive recreation |
| | | | Limited recreation | | | | |
| Subject to pollution controls | No building | | Limited structures | | | | |
| | Intolerant of construction | | | | | | |

**FIGURE 1**
Coastal environment—a planning perspective from ocean to bay. [After *Design with Nature* by Ian McHarg. © 1969 by Ian L. McHarg. Adapted by permission of Doubleday, a division of Bantam Doubleday Dell Publishing Group, Inc.]

of course, the opposite of such careful assessment and planning prevails.

## A Scientific View and a Political Reality

Before an intensive government study completed in 1962, no analysis of coastal hazards had been done, outside of academic circles. Thus, what is common knowledge to geographers, botanists, biologists, and ecologists in the classroom and laboratory still has not filtered through to the general planning and political processes. As a result, on the New Jersey shore and along much of the Atlantic and Gulf Coasts, improper development of the fragile coastal zone led to extensive destruction during storms in 1962, 1992, and numerous hurricanes of this century.

When the storms of 1962 hit, scientists were studying the area. They forecast a strong probability that the Cape May portion of the New Jersey shore could be 90 m (300 ft) farther inland by the year 2010, a little more than a decade from now. There is a 15% chance that the shore could be even twice as far inland (180 m, or 600 ft) by that time. Similar estimates persist along the entire East Coast, making it clear that society must reconcile *ecology* and *economics* if these coastal environments are to be sustained.

South Carolina enacted their Beach Management Act of 1988 (modified 1990) to apply some of McHarg's principles. Vulnerable coastal areas

**FIGURE 2**
Coastal erosion and a failed house foundation. [Photo by author.]

are protected from new construction or rebuilding. Now guided by law are structure size, replacement limits for damaged structures, and placement of structures on lots, although liberal interpretation is allowed. The act sets standards for different coastal forms: beach and dunes, eroding shorelines, or a hypothetical baseline along a coast that lacks dunes or has unstabilized inlets. In its first 2 years, more than 70 lawsuits protested the act as an invalid seizure of private property without compensation. Implementation

has been difficult, political pressure intense, and results mixed. Similar measures in other states have faced the same difficult path.

Ian McHarg voices an optimistic hope: "May it be that these simple ecological lessons will become known and incorporated into ordinance [law] so that people can continue to enjoy the special delights of life by the sea."[*]

[*]From *Design with Nature* by Ian McHarg, p. 17. © 1969 by Ian L. McHarg. Published by Bantam Doubleday Dell Publishing Group, Inc.

# Summary and Review—The Oceans, Coastal Processes and Landforms

✔ **Describe** **the chemical composition of seawater and the physical structure of the ocean.**

Water is called the "universal solvent," dissolving at least 57 of the 92 elements found in nature. Most natural elements and the compounds they form are found in the seas as dissolved solids. Seawater is a solution, and the concentration of dissolved solids is called **salinity**. **Brine** exceeds the average 35‰ (parts per thousand) salinity; **brackish** applies to water that is less than 35‰. The ocean is divided by depth into a narrow mixing zone at the surface, a thermocline transation zone, and the deep cold zone.

salinity (p. 482)
brine (p. 484)
brackish (p. 484)

1. Describe the salinity of seawater: its composition, amount, and distribution.

2. Analyze the latitudinal distribution of salinity as shown in Figure 16-2. Why is salinity less along the equator and greater in the subtropics?

3. What are the three general zones relative to physical structure within the ocean? Characterize each by temperature, salinity, dissolved oxygen, and dissolved carbon dioxide.

✔ *Identify* **the components of the coastal environment and** *list* **the physical inputs to the coastal system, including tides and mean sea level.**

The coastal environment is called the **littoral zone** and exists where the tide-driven, wave-driven sea confronts the land. Inputs to the coastal environment include solar energy, wind and weather, ocean currents and waves, climatic variation, and the nature of coastal rock. **Mean sea level (MSL)** is based on average tidal levels recorded hourly at a given site over many years. MSL varies spatially because of ocean currents and waves, tidal variations, air temperature and pressure differences, ocean temperature variations, slight variations in Earth's gravity, and changes in oceanic volume.

**Tides** are complex daily oscillations in sea level, ranging worldwide from barely noticeable to many meters. Tides are produced by the gravitational pull of both the Moon and the Sun. Most coastal locations experience two high (rising) **flood tides**, and two low (falling) **ebb tides** every day. The difference between consecutive high and low tides is the *tidal range*. **Spring tides** exhibit the greatest tidal range, when the Moon and Sun are either in conjunction or opposition. **Neap tides** produce a lesser tidal range.

littoral zone  (p. 485)
mean sea level (MSL) (p. 485)
tide (p. 486)
flood tide (p. 488)
ebb tide (p. 488)
spring tide (p. 488)
neap tide (p. 488)

4. What are the key terms used to describe the coastal environment?

5. Define mean sea level. How is this value determined? Is it constant or variable around the world? Explain.

6. What interacting forces generate the pattern of tides?

7. What characteristic tides are expected during a new Moon or a full Moon? During the first-quarter and third-quarter phases of the Moon? What is meant by a flood tide? An ebb tide?

8. Is tidal power being used anywhere to generate electricity? Explain briefly how such a plant would utilize the tides to produce electricity. Are there any sites in North America? Where are they?

---

✔ *Describe* **wave motion at sea and near shore and** *explain* **coastal straightening as a product of wave refraction.**

Friction between moving air (wind) and the ocean surface generates undulations of water that we call **waves**. Wave energy in the open sea travels through water, but the water itself stays in place. Regular patterns of smooth, rounded waves, called **swells**, are the mature undulations of the open ocean. Near shore, the restricted depth of water slows the wave, forming *waves of translation*, in which *both* energy and water actually move forward toward shore. As the crest of each wave rises, the wave falls into a characteristic **breaker**.

**Wave refraction** redistributes wave energy so that different sections of the coastline vary in erosion potential. Headlands are eroded, whereas coves and bays receive materials, with the long-term effect of straightening the coast. The **longshore current** of water moving parallel to the shore produces the *longshore drift* of sand, sediment, and gravel and particles, and materials moving along the beach transported as **beach drift**—together these make up the overall littoral drift.

A **tsunami** is a seismic sea wave triggered by an undersea landslide or earthquake. It travels at great speeds in the open sea and gains height as it comes ashore, posing a coastal hazard.

wave (p. 490)
swell (p. 490)
breaker (p. 490)
wave refraction (p. 490)
longshore current (p. 490)
beach drift (p. 490)
tsunami (p. 490)

9. What is a wave? How are waves generated, and how do they travel across the ocean? Does the water travel with the wave? Discuss the process of wave formation and transmission.

10. Describe the refraction process that occurs when waves reach an irregular coastline. Why is the coastline straightened?

11. Define the components of beach drift and the longshore current and longshore drift.

12. Explain how a seismic sea wave attains such tremendous velocities. Why is it given a Japanese name?

---

✔ *Identify* **characteristic coastal erosional and depositional landforms.**

An *erosional coast* features wave action that cuts a horizontal bench in the tidal zone, extending from a sea cliff out into the sea. Such a structure is called a **wave-cut platform**, or *wave-cut terrace*. In contrast, *depositional coasts* generally are located along land of gentle relief, where sediments are available from deposition from many sources. Characteristic landforms deposited by waves and currents: a **barrier spit** (material deposited in a long ridge extending out from a coast); a **bay barrier**, or *baymouth bar* (when a spit cuts off the bay from the ocean and forms an inland **lagoon**); a **tombolo** (where sediment deposits connect the shoreline with an offshore island or sea stack); and a **beach** (land along the shore where sediment is in motion, deposited by waves and currents). A beach helps to stabilize the shoreline, although it may be unstable seasonally.

wave-cut platform (p. 494)
barrier spit (p. 495)
bay barrier (p. 495)
lagoon (p. 495)
tombolo (p. 495)
beach (p. 496)

13. What is meant by an erosional coast? What are the expected features of such a coast?

14. What is meant by a depositional coast? What are the expected features of such a coast?

15. How do people attempt to modify littoral drift? What strategies are used? What are the positive and negative impacts of these actions?

16. Describe a beach—its form, composition, function, and evolution.

17. What success has Miami had with beach replenishment? Is it a practical strategy?

---

✔ *Describe* **barrier islands and their hazards as they relate to human settlement.**

Barrier chains are long, narrow, depositional features, generally of sand, that form offshore roughly parallel to the coast. Common forms are **barrier beaches**, and the broader, more extensive **barrier islands**. Barrier formations are transient coastal features, constantly on the move, and they are a poor, but common, choice for development.

barrier beach (p. 498)
barrier island (p. 498)

18. On the basis of the information in the text and any other sources at your disposal, do you think barrier islands and beaches should be used for development? If so, under what conditions? If not, why not?

19. After the Grand Strand off South Carolina was destroyed by Hurricane Hazel in 1954, settlements were rebuilt, only to be hit by Hurricane Hugo 35 years later, in 1989. Why do these recurring events happen to human populations? Compare the impact of the two storms.

---

✔ *Assess* **living coastal environments: corals, wetlands, salt marshes, and mangroves.**

A **coral** is a simple marine invertebrate with a small, cylindrical, saclike body (polyp). Corals live in a *symbiotic* (mutually helpful) relationship with algae; each is dependent on the other for survival. Over generations, corals accumulate in large reef structures.

**Wetlands** are land saturated with water and support specific plants adapted to wet conditions. They occur along coastlands and inland in bogs, swamps, and river bottomlands. Coastal wetlands form as **salt marshes** poleward of the 30th parallel in each hemisphere and **mangrove swamps** equatorward of these parallels.

coral (p. 499)
wetlands (p. 501)
salt marsh (p. 501)
mangrove swamp (p. 501)

20. How are corals able to construct reefs and islands?

21. Describe a trend in corals that is troubling scientists, and discuss some possible causes.

22. Why are the coastal wetlands poleward of 30° N and S latitude different from those that are equatorward? Describe the differences.

---

✔ **Construct an environmentally sensitive model for settlement and land use along the coast.**

23. Describe the condition of Marco Island, Florida. Was a rational model used to assess the environment prior to development? What economic and political forces were involved?

24. What type of environmental analysis is needed for rational development and growth in a region like the New Jersey shore? Evaluate South Carolina's approach to coastal hazards and protection.

---

 **NetWork**

The *Geosystems Home Page* provides on-line resources for this chapter on the World Wide Web. You will find review exercises, specific updates for items in the chapter, suggested readings, and links to interesting related pathways on the Internet (click on the Table of Contents link and select this chapter). *Geosystems* is at: **http://www.prenhall.com/geosystm**

# 17

# Glacial and Periglacial Processes and Landforms

**Rivers of Ice**

**Glacial Processes**

**Glacial Landforms**

**The Pleistocene Ice Age Epoch**

**Deciphering Past Climates: Paleoclimatology**

**Arctic and Antarctic Regions**

**Summary and Review**

## Key Learning Concepts

After reading the chapter, you should be able to:

- *Differentiate* between alpine and continental glaciers and *describe* their principal features.
- *Describe* the process of glacial ice formation and *portray* the mechanics of glacial movement.
- *Describe* characteristic erosional and depositional landforms created by alpine glaciation and continental glaciation.
- *Explain* the Pleistocene ice age epoch and related glacials and interglacials and *describe* some of the methods used to study paleoclimatology.

*Alpine glaciers on Mont Blanc A. du midi, Chamonix, France.*
[Photo by Peter Miller/Photo Researchers, Inc.]

About 77% of Earth's freshwater is frozen, with the bulk of that ice sitting restlessly in just two places—Greenland and Antarctica. The remaining ice covers various mountains and fills some alpine valleys. More than 29 million cubic kilometers (7 million cubic miles) of water is tied up as ice. These deposits provide an extensive frozen record of Earth's climatic history over the past several million years and perhaps some clues to its climatic future.

This chapter focuses on Earth's extensive ice deposits—their formation, movement, and the ways in which they produce various erosional and depositional landforms. Glaciers, transient landforms themselves, leave in their wake a variety of landscape features. The fate of glaciers is intricately tied to change in global temperature, which ultimately concerns us all. We discuss the methods used to decipher past climates—the science of *paleoclimatology*—and the clues they give us to understand the future.

Finally, in Focus Study 17-1, we examine the cold, near-glacial world of permafrost and periglacial processes. Approximately 20% of Earth's land area is subject to freezing conditions and frost action characteristic of periglacial regions. These are areas near former and existing glaciers. Reminders of the last ice age occur in many parts of the globe.

# Rivers of Ice

A **glacier** is a large mass of ice, resting on land or floating as an ice shelf in the sea adjacent to land. Glaciers are not frozen lakes or groundwater ice. Instead, they form by the continual accumulation of snow that recrystallizes under its own weight into an ice mass. Glaciers are not stationary; they move slowly under the pressure of their own great mass and the pull of gravity. In fact, they flow only meters each year in streamlike patterns, as you can see in the lower photo in Figure 17-1.

Today, about 11% of Earth's land area is dominated by these slowly flowing rivers of ice. But during colder episodes in the past, as much as 30% of continental land was covered by glacial ice. Through these "ice ages," below-freezing temperatures prevailed at lower latitudes than they do today, allowing snow to accumulate year after year.

Glaciers form in such areas of permanent snow, both at high latitudes and at high elevations at any latitude. A **snowline** is the lowest elevation where snow can survive year-round; specifically, it is the lowest line where winter snow accumulation persists throughout the summer. Glaciers form on some high mountains along the equator, such as in the Andes Mountains of South America and on Mount Kilimanjaro in Tanzania. On equatorial mountains, the snowline is around 5000 m (16,400 ft); on midlatitude mountains, such as the European Alps, snowlines average 2700 m (8850 ft); and in southern Greenland, snowlines are as low as 600 m (1970 ft).

## *Types of Glaciers*

Glaciers are as varied as the landscape itself. They fall within two general groups, based on their form, size, and flow characteristics: alpine glaciers and continental glaciers.

***Alpine Glaciers.*** With few exceptions, a glacier in a mountain range is called an **alpine glacier**, or *mountain glacier*. The name comes from the Alps mountains of central Europe, where such glaciers abound. Alpine glaciers form in several subtypes. One prominent type is a *valley glacier*, literally a river of ice confined within a valley that originally was formed by stream action. Such glaciers range in length from only 100 m (325 ft) to over 100 km (60 mi).

In Figure 17-1a, at least a dozen valley glaciers are identifiable in the *Landsat* image of the Alaska Range. Several are named on the map. The other image (Figure 17-1b) is a high-altitude photograph of the Eldridge and Ruth Glaciers (also named on the map), which fill valleys as they flow from source areas near Mount McKinley.

As a valley glacier flows slowly downhill, the mountains, canyons, and river valleys beneath its mass are profoundly altered by its erosive passage. Some of the debris created by the glacier's excavation is transported on the ice, visible as dark streaks and bands on the ice being transported for deposition elsewhere; other portions of its debris load are carried within or along its base (see Figures 17-3 and 17-5c).

Most alpine glaciers originate in a mountain *snowfield* that is confined in a bowl-shaped recess. This scooped-out erosional landform at the head of a valley is called a **cirque**. A glacier that forms in a cirque is called a *cirque glacier*. Several cirque glaciers may jointly feed a valley glacier, as shown on the topographic map and photograph in Figure 17-2 (note section 8 on the map). The Kuskulana Glacier, which flows off the map to the west, is supplied by numerous cirque glaciers feeding several tributary valley glaciers.

Wherever several valley glaciers pour out of their confining valleys and coalesce at the base of a mountain range, a *piedmont glacier* is formed and spreads freely over the lowlands. Malaspina Glacier is an excellent example of a piedmont glacier (Figure 17-3). The debris deposits on the surface of the ice form beautiful streaked patterns as the glacier fans out over 5000 km² (1950 mi²) of the coastal plain. This glacier is at the foot of the Mount St. Elias Range in southern Alaska, ending in Yakutat Bay. A *tidal glacier*, such as the Columbia Glacier on Prince William Sound in Alaska, ends in the sea,

**FIGURE 17-1**

**Glaciers in south-central Alaska.**

(a) Valley glaciers in the Alaska Range of Denali National Park. (b) Oblique infrared image of Eldridge and Ruth Glaciers, with Mount McKinley at upper left, taken at 18,300 m (60,000 ft). [(a) *Landsat* image from NASA; (b) Alaska High Altitude Aerial Photography from EROS Data Center.]

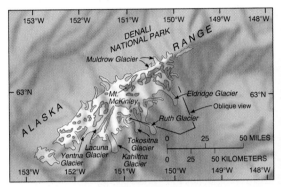

(a)

(b)

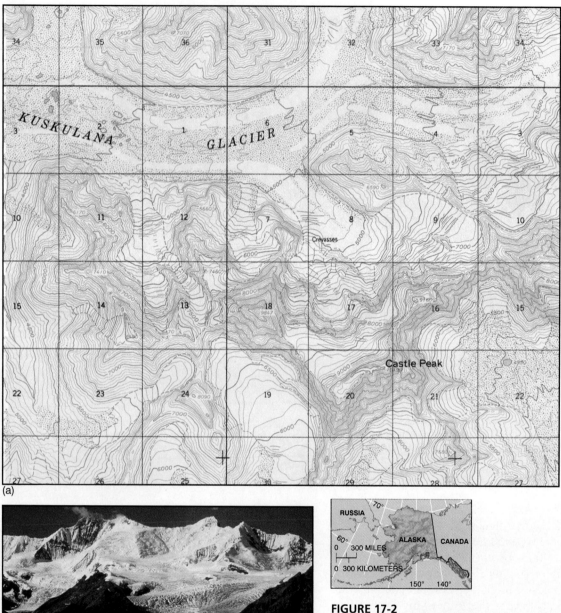

(a)

(b)

**FIGURE 17-2**

**Topographic map from southeastern Alaska.**

(a) Numerous cirque glaciers depicted on a topographic map of southeastern Alaska (McCarthy C-7 Quadrangle, 1:63,360 scale, 100-ft contour interval). The blue areas represent active glaciers, and stripped brown areas on the ice are moraines. (b) A dramatic scene near the area shown on the map depicts the active Kennicott glacier. [(a) Courtesy of U.S. Geological Survey; (b) photo by Pat and Tom Leeson/Photo Researchers, Inc.]

*calving* (breaking off) to form floating ice called **icebergs** (Figure 17-3c). Icebergs usually form wherever glaciers meet the ocean.

***Continental Glaciers.*** On a much larger scale than individual alpine glaciers, a continuous mass of ice is called a **continental glacier**. In its most extensive form, it is an **ice sheet**. Most of Earth's glacial ice exists in the ice sheets that blanket 80% of Greenland (1.8 million cubic kilometers, or 0.43 million cubic miles) and 90% of Antarctica (13.9 million cubic kilometers, or 3.3 million cubic miles). Antarctica alone has 91% of all the glacial ice on the planet.

The Antarctic and Greenland ice sheets have such enormous mass that large portions of each landmass beneath the ice have been isostatically depressed below

(a)

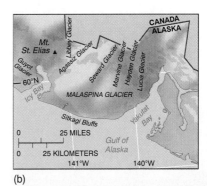

(b)

(c)

**FIGURE 17-3**
**Piedmont glacier and ice calving into the sea.**
(a) Malaspina Glacier in southeastern Alaska. This piedmont glacier is nearly the size of Rhode Island and covers some 5000 km² (1950 mi²); its thickest point exceeds 600 m (1970 ft). About 70% of Malaspina's ice comes from the Seward Glacier (upper center), which is fed by ice fields in the St. Elias Mountain Range (b). (c) When glaciers reach the sea, large blocks begin *calving* off, forming icebergs, such as from Childs glacier near Cordova, Alaska. [(a) High-altitude photo courtesy of AeroMap U.S., Inc., Anchorage, Alaska; (c) photo by Vanessa Vick/Photo Researchers, Inc.]

sea level. Each ice sheet is more than 3000 m (10,000 ft) deep, burying all but the highest peaks.

Two additional types of continuous ice cover associated with mountain locations are *ice caps* and *ice fields*. An **ice cap** is roughly circular and, by definition, covers an area of less than 50,000 km² (19,300 mi²). An ice cap completely buries the underlying landscape. The volcanic island of Iceland features several ice caps (Figure 17-4a).

An **ice field** is not extensive enough to form the characteristic dome of an ice cap; instead, it extends in a characteristic elongated pattern in a mountainous region. A fine example is the Patagonian ice field of Argentina and Chile, one of Earth's largest. It attains only 90 km (56 mi) width, but stretches 360 km (224 mi), from 46° to 51° S latitude (equivalent in latitude from South Dakota to central Manitoba in the Northern Hemisphere). In an ice field, ridges and peaks are visible above the buried terrain; the term *nunatak* refers to these peaks, visible in Figure 17-4b.

Continuous ice sheets or ice caps are drained by more rapidly moving, solid *ice streams* that form around their periphery, moving to the sea or to lowlands. Such frozen ice streams flow from the edges of Greenland and Antarctica and appear to be doing so at an accelerated pace.

An *outlet glacier* flows out from an ice sheet or ice cap but is constrained by a mountain valley or pass.

# Glacial Processes

A glacier is a dynamic body, moving relentlessly downslope at rates that vary within its mass, excavating the landscape through which it flows. A glacier is composed of dense ice that is formed from snow and water through a process of compaction, recrystallization, and growth. A glacier's *budget* consists of net gains or losses of this glacial ice, which determine whether the glacier expands or retreats. Let us now look at glacial ice formation, mass balance, movement, and erosion before we discuss the fascinating landforms produced by these processes.

## *Formation of Glacial Ice*

Consider for a moment the nature of ice. You may be surprised to learn that ice is both a mineral (an inorganic natural compound of specific chemical makeup and crystalline structure) and a rock (a mass of one or more

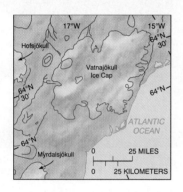

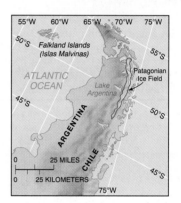

**FIGURE 17-4**

**Ice cap and ice field.**

(a) The Vatnajökull ice cap in southeastern Iceland (*jökull* means "ice cap" in Danish). (b) The southern Patagonian ice field of Argentina. [(a) *Landsat* image from NASA; (b) photo by Cosmonauts G. M. Greshko and Yu V. Romanenko, *Salyut 6*.]

minerals). Ice is a frozen fluid, a trait that it shares with igneous rocks. The accumulation of snow in layered deposits is similar to sedimentary rock formations. And to give birth to a glacier, snow and ice are transformed under pressure, recrystallizing into a type of metamorphic rock. Glacial ice is a remarkable material!

The essential input to a glacier is snow that accumulates in a snowfield, a glacier's accumulation zone (Figure 17-5a). Snowfields usually are at the highest elevation of an ice sheet, ice cap, or head of a valley glacier, usually in a cirque. (Highland snow accumulation sometimes is the product of orographic processes; review Chapter 8.) Avalanches from surrounding mountain slopes can add to the snowfield.

As the snow accumulation deepens in sedimentary-like layers, the increasing thickness results in increased weight and pressure on underlying ice. Rain and summer snowmelt then contribute water, which stimulates further

melting, and that meltwater seeps down into the snow-field and refreezes.

Snow that survives the summer and into the following winter begins a slow transformation into glacial ice. Air spaces among ice crystals are pressed out as snow packs to a greater density. The ice recrystallizes and consolidates under pressure. In a transition step to glacial ice, snow becomes **firn**, which has a compact, granular texture.

As this process continues, many years pass before dense glacial ice is produced. Formation of **glacial ice** is analogous to metamorphic processes: Sediments (snow and firn) are pressured and recrystallized into a dense metamorphic rock (glacial ice). In Antarctica, glacial ice formation may take 1000 years because of the dryness of the climate (minimal snow input), whereas in wet climates the time is reduced to just a few years because of rapid, constant snow input to the system.

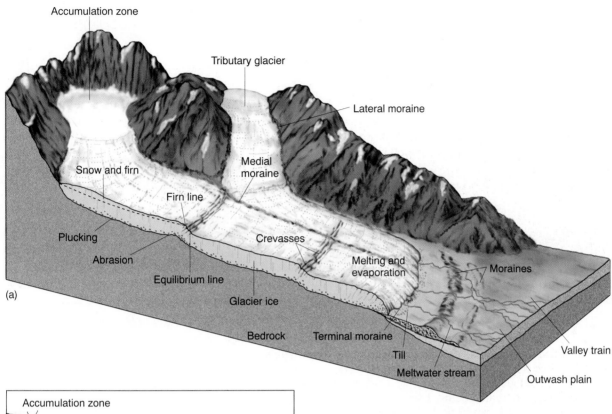

(a)

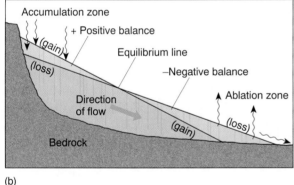

(b)

## FIGURE 17-5

**A retreating alpine glacier and mass balance.**
(a) Cross section of a typical retreating alpine glacier. (b) Annual mass balance of a glacial system, showing how the relation between accumulation and ablation controls the location of the equilibrium line. (c) John Hopkins Glacier, Glacier Bay National Park, Alaska, demonstrates many of the features in the illustration. Note the formation of the medial moraine where the two glaciers merge.
[(c) Photo by Frank S. Balthis.]

(c)

## *Glacial Mass Balance*

A glacier is an open system, with *inputs* of snow and *outputs* of ice, meltwater, and water vapor. At its upper end, a glacier is fed by snowfall and other moisture in the *accumulation zone* (Figure 17-5a). This area ends at the **firn line**, indicating where the winter snow and ice accumulation survived the summer melting season.

Toward a glacier's lower end, it is wasted (reduced) through several processes: melting on the surface, internally,

and at its base; ice removal by deflation; the calving of ice blocks; and sublimation (recall from Chapter 7 that this is the direct evaporation of ice). Collectively, these losses are called **ablation**.

The zone where accumulation gain balances ablation loss is the *equilibrium line* (Figure 17-5b). This area of a glacier generally coincides with the firn line. Glaciers achieve a positive net balance of mass—grow larger—during cold periods with adequate precipitation. In warmer times, the equilibrium line migrates up-glacier, and the glacier retreats—grows smaller—because of its negative net balance. Internally, gravity continues to move a glacier forward even though its lower terminus might be in retreat owing to ablation. The mass-balance loses from the South Cascade glacier provides a case in point (News Report 1).

Highways of ice that flow from accumulation areas high in the mountains are marked by trails of transported debris called *moraines* (Figure 17-5c). A *lateral moraine* accumulates along the sides as rock is loosened by abrasion and plucking from the valley walls. A *medial moraine* forms down the middle when two glaciers merge

and their lateral moraines combine, forming an elongated deposit on and within the combined glacier.

### *Glacial Movement*

Like all minerals, ice has specific properties of hardness, color, melting point (quite low in the case of ice), and brittleness. We know the properties of ice best from those brittle little cubes in the freezer. But glacial ice has different properties, depending on where it is in a glacier. In a glacier's depths, glacial ice behaves in a plastic manner, distorting and flowing in response to weight and

## News Report 1

## South Cascade Glacier Loses Mass

The net mass balance of the South Cascade Glacier in Washington State demonstrated significant losses between 1955 and 1995. In just one year (September 1991 to October 1992), the terminus of the glacier retreated 38 m (125 ft), resulting in major changes to the surface and sides of the glacier. There

was more than a 2% loss of the glacier's mass in just that one year. The net accumulation and net wastage are illustrated in Figure 1.

The reasons for these losses are complex. Possible causes are increasing average air temperature and decreasing precipitation. A comparison of the trend of

this glacier's mass balance with that of others in the world shows that temperature changes apparently are causing widespread reductions in middle- and lower-elevation glacial ice. The wastage (ice loss) from alpine glaciers worldwide is thought by some to contribute over 25% to the rise in sea level.

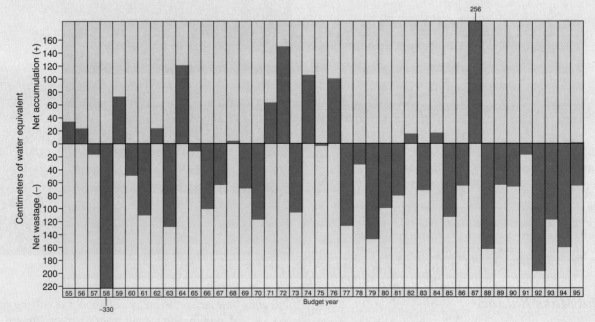

**FIGURE 1**

**South Cascade Glacier net mass balance, 1955–1995.**

Over the past 40 years, a negative net mass balance has dominated this shrinking glacier. Data in centimeters; 2.54 cm per in. (Note: 1958 is a large net wastage year; 1987 is a large net accumulation year.) [Data from R.M. Krimmel, *Mass Balance, Meteorological, and Runoff Measurements at South Cascade Glacier, 1995 Balance Year.* U.S. Geological Survey Open-File Report, Tacoma, Washington.]

pressure from above and the degree of slope below. In contrast, the glacier's upper portion is more like the every-day ice we know, quite brittle.

A glacier's rate of flow ranges from almost nothing to a kilometer or two per year on a steep slope. The rate of snow accumulation in the formation area is critical to the pace of glacial movement.

Glaciers are not rigid blocks that simply slide downhill. The greatest movement within a valley glacier occurs *internally*, below the rigid surface layer, where the underlying zone moves plastically forward (Figure 17-6a). At the same time, the base creeps and slides along, varying its speed with temperature and the presence of any lubricating water or saturated sediment beneath the ice. This *basal slip* usually is much slower than the internal plastic flow of the glacier, so the upper portion of the glacier flows ahead of the lower portion. The difference in speed stretches the glacier's brittle surface ice.

In addition, the pressure may vary in response to unevenness in the landscape beneath the ice. Basal ice may be melted by compression at one moment, only to refreeze later. This process is called *ice regelation*, meaning to refreeze, or re-gel. Regelation is important because it facilitates downslope movement and because the process incorporates rock debris into the glacier. Consequently, a glacier's basal ice layer, which can extend tens of meters above its base, has a much greater debris content than the ice above.

A flowing glacier can develop vertical cracks known as **crevasses** (Figure 17-6a, b). These result from friction with valley walls, or tension from stretching as the glacier passes over convex slopes, or compression as the glacier passes over concave slopes. Traversing a glacier,

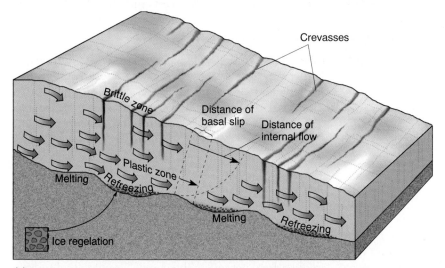

(a)

(b)

**FIGURE 17-6**
**Glacial movement.**
(a) Cross section of a glacier, showing its forward motion and brittle cracking at the surface and flow along its basal layer. (b) Surface crevasses and cracks are evidence of a glacier's forward motion (near the Don Sheldon Amphitheater, Denali National Park, Alaska). [(b) Photo by Michael Collier.]

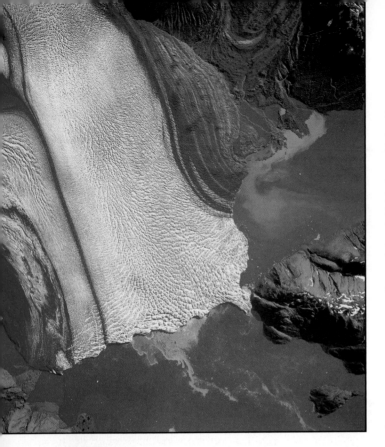

**FIGURE 17-7**
**A glacier surges.**
The Hubbard Glacier surged across Russell Fjord in Alaska, effectively damming the fjord and temporarily creating Russell Lake. [*Landsat* infrared image from EROS Data Center.]

whether an alpine glacier or an ice sheet, is dangerous because a thin veneer of snow sometimes masks the presence of a crevasse.

Tributary valley glaciers merge to form a *compound valley glacier*. The flowing movement of a compound valley glacier is different from that of a river with tributaries. Tributary glaciers flow into a compound glacier and merge alongside one another by extending and thinning rather than by blending, as do rivers. Each tributary maintains its own patterns of transported debris (dark streaks, visible on the John Hopkins Glacier in Figure 17-5c and the photo in Figure 17-2).

**Glacier Surges.** Although glaciers flow plastically and predictably most of the time, some will lurch forward with little or no warning in a **glacier surge**. A surge is not quite as abrupt as it sounds; in glacial terms, a surge can be tens of meters per day. The Jakobshavn Glacier in Greenland, for example, is known to move between 7 and 12 km (4.3 and 7.5 mi) a year.

In the spring of 1986, Hubbard Glacier and its tributary Valerie Glacier surged across the mouth of Russell Fjord in Alaska, cutting it off from contact with Yakutat Bay (Figure 17-7). This area, the St. Elias Mountain Range in southeastern Alaska, is fed by annual snowfall that averages more than 850 cm (335 in.) a year, so the surge event had been predicted. But the rapidity of the surge was surprising. The fjord was dubbed Russell Lake for the time it

remained impounded. The glacier's movement exceeded 34 m (112 ft) per day during the peak surge, an enormous increase over its normal rate of 15 cm (6 in.) per day.

The exact cause of such a glacial surge is being studied. Some surge events result from a buildup of water pressure under the glacier, sometimes enough to actually float the glacier slightly, detaching it from its bed, during the surge. As a surge begins, icequakes are detectable, and ice faults are visible. Surges can occur in dry conditions as well, as the glacier plucks rock from its bed and moves forward. Another cause of glacial surges could be the presence of a water-saturated layer of sediment, a so-called soft bed, beneath the glacier. This is a deformable layer that cannot resist the tremendous sheer stress produced by the moving ice of the glacier. Scientists examining cores taken from several ice streams now accelerating through the West Antarctic Ice Sheet think they have identified this cause—although water pressure is still important.

**Glacial Erosion.** The way in which a glacier erodes the land is similar to a large excavation project, with the glacier hauling debris from one site to another for deposition. The passing glacier mechanically plucks rock material and carries it away. Debris is carried on its surface and is also transported internally, or *englacial*, embedded within the glacier itself. There is evidence that rock pieces actually freeze to the basal layers of the glacier in a *glacial plucking* process and, once embedded, enable the glacier to scour and sandpaper the landscape as it moves—a process called **abrasion**. This abrasion and gouging produce a smooth surface on exposed rock, which shines with *glacial polish* when the glacier retreats. Larger rocks in the glacier act much like chisels, gouging the underlying surface and producing glacial striations parallel to the flow direction (Figure 17-8).

**FIGURE 17-8**
**Glacial sandpapering polishes rock.**
Glacial polish and striations are examples of glacial erosion near a Canadian lake. [Photo by Leland Brun/Photo Researchers, Inc.]

# Glacial Landforms

Glacial erosion and deposition produce distinctive land-forms that differ greatly from the way the land looked before the ice came and went. You might expect all glaciers to create the same landforms, but alpine and continental glaciers each generate their own characteristic landscapes. We look first at erosional landforms created by alpine glaciers, then at their depositional landforms. Finally, we examine the landscape results of continental glaciation.

## Erosional Landforms Created by Alpine Glaciation

Alpine glaciers create spectacular, dramatic landforms that bring to mind the Canadian Rockies, the Swiss Alps, or the vaulted peaks of Torres del Paine pictured on the cover of this book. Geomorphologist William Morris Davis depicted the stages of a valley glacier in drawings published in 1906 and redrawn here in Figure 17-9. Study of these figures reveals the handiwork of ice as sculptor:

- In (a), you see typical stream-cut valleys as they exist before glaciation. Note the prominent **V** shape.

- In (b), you see the same landscape during subsequent glaciation. Glacial erosion and transport are actively removing much of the regolith (weathered bedrock) and the soils that covered the stream valley landscape. As the cirque walls erode away, sharp ridges form, dividing adjacent cirque basins. These **arêtes** ("knife-edge" in French) become the sawtooth, serrated ridges in glaciated mountains. Two eroding cirques may reduce an arête to a saddle-like depression or pass, called a **col**. A **horn** (pyramidal peak) results when several cirque glaciers gouge an individual mountain summit from all sides. Most famous is the Matterhorn in the Swiss Alps, but many others occur worldwide.

- In (c), you see the same landscape at a time of warmer climate when the ice has retreated. The glaciated valleys now are **U**-shaped, greatly changed from their previous stream-cut **V** form. You can see the oversteepened sides and the straightened course of the valleys. Physical weathering from the freeze-thaw cycle has loosened rock along the steep cliffs, where it has fallen to form *talus slopes* along the valley sides.

Note that in the cirques where the valley glaciers originated, small mountain lakes called **tarns** have formed. One cirque contains small, circular, stair-stepped lakes called **paternoster** ("our father") lakes for their resemblance to rosary (religious) beads. Paternoster lakes may have formed from the differing resistance of rock to glacial processes or from damming by glacial deposits.

The valleys carved by tributary glaciers are left stranded high above the valley floor, because the primary glacier eroded the valley floor so deeply. These *hanging valleys* are the sites of spectacular waterfalls.

Where a glacial trough intersects the ocean, the glacier can continue to erode the landscape, even below sea level. As the glacier retreats, the trough floods and forms a deep **fjord** in which the sea extends inland, filling the lower reaches of the steep-sided valley (Figure 17-10). The fjord may be flooded further by rising sea level or by changes in the elevation of the coastal region. All along the glaciated coast of Alaska, glaciers now are in retreat, thus opening many new fjords that previously were blocked by ice. Coastlines with notable fjords include those of Norway (Figure 17-10), Greenland, Chile, the South Island of New Zealand, Alaska, and British Columbia.

## Depositional Landforms Created by Alpine Glaciation

You have just seen how glaciers excavate tremendous amounts of material and create fascinating landforms in the process. Glaciers produce a different set of distinctive landforms when they melt and deposit their debris cargo, at the glacier's *terminus* (end). Figure 17-5a shows that, after the glacier melts, debris accumulates to mark the former margins of the glacier, both its end and sides.

**Glacial drift** is the general term for *all* glacial deposits, both unsorted and sorted. Sediments deposited by glacial meltwater are sorted by size and are termed **stratified drift**. Direct ice deposits leave unstratified and unsorted debris called **till**.

As a glacier flows to a lower elevation, a wide assortment of rock fragments become *entrained* (carried along) on its surface, embedded within its mass, or in its base. As the glacier melts, this unsorted cargo is deposited on the ground surface. Such till is poorly sorted and is difficult to cultivate for farming, but the clays and finer particles can provide a basis for soil development.

***Moraines.*** **Moraine** is the name for specific landforms produced by the deposition of glacial sediments. Several types of moraines are exhibited by Hole-in-the-Wall Glacier, shown in Figure 17-11. A **lateral moraine** forms along each side of a glacier. If two glaciers with lateral moraines join, a **medial moraine** may form (see Figures 17-5 and 17-9). A deposition of till that is generally spread across a surface is called a *ground moraine*, or till plain, and may hide the former landscape. Such plains are found in portions of the U.S. Midwest.

Eroded debris that is dropped at the glacier's farthest extent is called a **terminal moraine**. However, there also may be *end moraines*, formed at other points where a glacier paused after reaching a new equilibrium between growth and ablation. Both terminal and end moraine

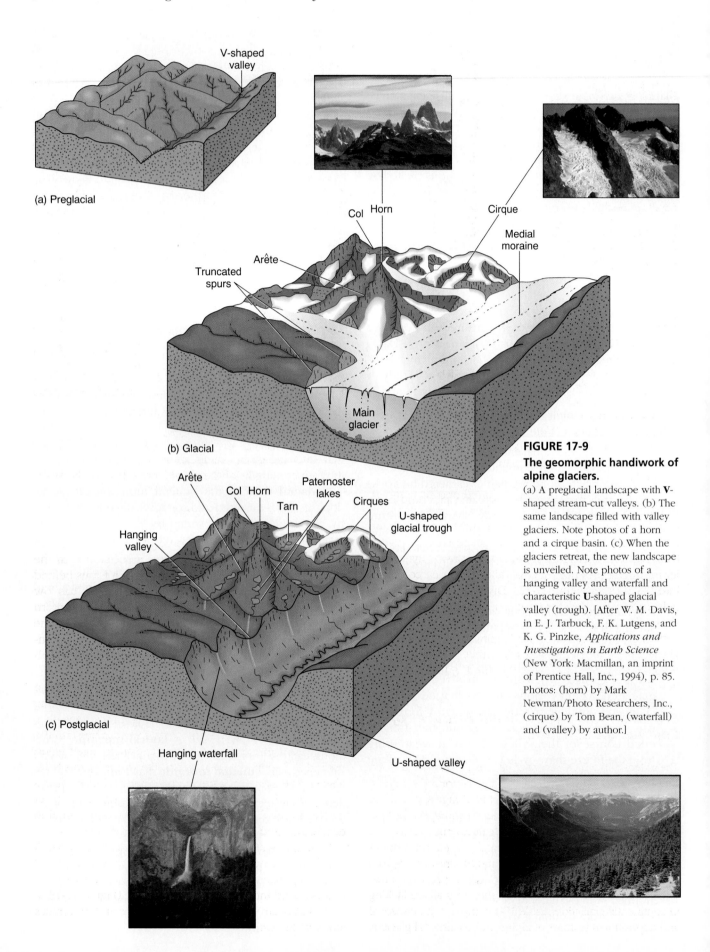

**FIGURE 17-9**

**The geomorphic handiwork of alpine glaciers.**

(a) A preglacial landscape with **V**-shaped stream-cut valleys. (b) The same landscape filled with valley glaciers. Note photos of a horn and a cirque basin. (c) When the glaciers retreat, the new landscape is unveiled. Note photos of a hanging valley and waterfall and characteristic **U**-shaped glacial valley (trough). [After W. M. Davis, in E. J. Tarbuck, F. K. Lutgens, and K. G. Pinzke, *Applications and Investigations in Earth Science* (New York: Macmillan, an imprint of Prentice Hall, Inc., 1994), p. 85. Photos: (horn) by Mark Newman/Photo Researchers, Inc., (cirque) by Tom Bean, (waterfall) and (valley) by author.]

**FIGURE 17-10**
**Norwegian fjord.**
A cruise ship sails along a coastal fjord in Norway. [Photo by Desjardins/Rapho/Photo Researchers, Inc.]

**FIGURE 17-11**
**Alpine glacier depositional features.**
Medial, lateral, terminal, and ground moraine deposits are evident in this photo. These features were produced by the Hole-in-the-Wall Glacier, Wrangell–St. Elias National Park, Alaska. Also, note the kettle pond in the foreground. [Photo by Tom Bean.]

forms are clearly visible in Figure 17-11. Lakes may form behind terminal and end moraines after a glacier's retreat, with the moraine acting as a dam. A type of end moraine that forms during significant pauses in a glacier's retreat is named a *recessional moraine*.

All of these till types are unsorted and unstratified. In contrast, streams of glacial meltwater can carry and deposit sorted and stratified glacial drift beyond a terminal moraine. Meltwater-deposited material downvalley from a glacier is called a *valley train deposit*. Peyto Glacier in Alberta produces such a valley train that continues into Peyto Lake (Figure 17-12). Distributary stream channels appear braided across its surface. The picture also shows the milky meltwater associated with glaciers, laden with finely ground "rock flour." Meltwater is produced by glaciers at all times, not just when they are retreating. A retreating glacier may leave behind other debris as described in News Report 2.

### Erosional and Depositional Features of Continental Glaciation

The extent of the most recent continental glaciation in North America and Europe, 18,000 years ago, is portrayed several pages ahead in Figure 17-17. When these huge sheets of ice advanced and retreated, they produced some, but not all, of the erosional and depositional features characteristic of alpine glaciation. Because continental glaciers form under different circumstances—not in mountains, but across broad, open landscapes—the intricately carved alpine features, lateral moraines, and medial moraines all are lacking in continental glaciation. Table 17-1 compares the erosional and depositional features of alpine and continental glaciers.

Figure 17-13 illustrates some of the most common erosional and depositional features associated with the retreat of a continental glacier. A **till plain** forms behind end moraines; it features *unstratified* coarse till, has low and rolling relief, and has a deranged drainage pattern (see Figure 14-7g). Beyond the moraineal deposits lie the **outwash plains** of *stratified drift* featuring stream channels that are meltwater-fed, braided, and overloaded with sorted and deposited materials.

Figure 17-13a shows a sinuously curving, narrow ridge of coarse sand and gravel called an **esker**. It forms along the channel of a meltwater stream that flows beneath a glacier, in an ice tunnel, or between ice walls beneath the glacier. As a glacier retreats, the steep-sided esker is left behind in a pattern roughly parallel to the path of the glacier. The ridge may not be continuous and in places may even appear to be branched, following the path set by the subglacial watercourse. Commercially valuable deposits of sand and gravel are quarried from some eskers.

Another feature of outwash plains is a **kame**, a small hill, knob, or mound of poorly sorted sand and gravel that is deposited directly by water, by ice in crevasses, or in ice-caused indentations in the surface (Figure 17-13c). Kames also can be found in deltaic forms and in terraces along valley walls.

**FIGURE 17-12**
**A valley train deposit.**
Peyto Glacier in Alberta, Canada.
Note valley train, braided stream, and
glacial meltwater. [Photo by author.]

**TABLE 17-1**

| Features | Alpine (Valley) Glacier | Continental Glaciation |
|---|---|---|
| **Erosional** | | |
| Striations, polish, etc. | Common | Common |
| Cirques | Common | Absent |
| Horns, arêtes, cols | Common | Absent |
| U-shaped valleys, truncated spurs, hanging valleys | Common | Rare |
| Fjords | Common | Absent |
| **Depositional** | | |
| Till | Common | Common |
| Terminal moraines | Common | Common |
| Recessional moraines | Common | Common |
| Ground moraines | Common | Common |
| Lateral moraines | Common | Absent |
| Medial moraines | Common, easily destroyed | Absent |
| Drumlins | Rare or absent | Locally common |
| Erratics | Common | Common |
| Stratified drift | Common | Common |
| Kettles | Common | Common |
| Eskers, crevasse fillings | Rare | Common |
| Kames | Common | Common |
| Kame terraces | Common | Present in hilly country |

Source: Adapted from L. D. Leet, S. Judson, and M. Kauffman,
*Physical Geology*, 5th ed., © 1978, p. 317. Reprinted by permission
of Prentice Hall, Inc., Englewood Cliffs, N.J.

Sometimes an isolated block of ice, perhaps more than a kilometer across, remains in a ground moraine, an outwash plain, or valley floor after a glacier has retreated. As much as 20 to 30 years are required for it to melt. In the interim, material continues to accumulate around the melting ice block. When the block finally melts, it leaves behind a steep-sided hole. Such a feature then frequently fills with water. This is called a **kettle**. Thoreau's Walden Pond, mentioned in the quotation in Chapter 7, is such a glacial kettle and is pictured in Figure 17-13b.

Glacial action also forms two types of streamlined hills. One is erosional, called a roche moutonnée, and the other is depositional, called a drumlin. A **roche moutonnée** ("sheep rock" in French) is an asymmetrical hill of exposed bedrock. Its gently sloping upstream side (stoss side) has been polished smooth by glacial action, whereas its downstream side (lee side) is abrupt and steep where rock pieces were plucked by the glacier (Figure 17-14).

A **drumlin** is deposited till that has been streamlined in the direction of continental ice movement, blunt end upstream and tapered end downstream (the opposite of a roche moutonnée). "Swarms" of drumlins occur across the landscape in portions of New York and Wisconsin, among other areas. Sometimes their shape is that of an elongated teaspoon bowl, lying face down. They attain lengths of 100–5000 m (300 ft to over 3 mi) and heights up to 200 m (650 ft).

Figure 17-15 shows a portion of a topographic map for the area south of Williamson, New York, which experienced continental glaciation during the last ice age. In studying the map, can you identify the numerous drumlins? In what direction do you think the continental glaciers moved across this region? (Look for gentle, tapered slopes in the downstream direction of each drumlin, left behind as the glacier retreated.)

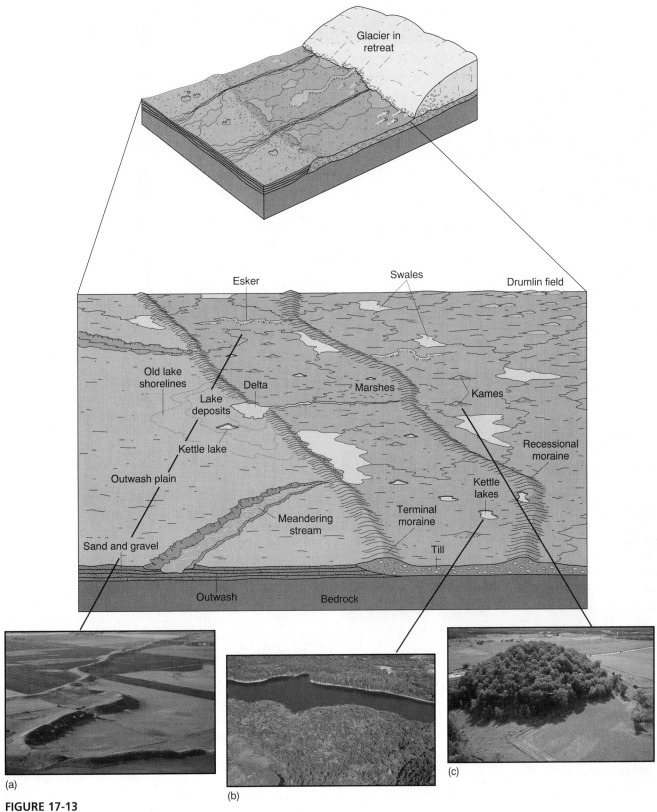

**FIGURE 17-13**

**Continental glacier depositional features.**

Common depositional landforms produced by glaciers. (a) An esker through farmland near Dahlen, North Dakota. (b) Walden Pond is a kettle surrounded by mixed forest in northeastern Massachusetts. (c) A kame covered by a woodlot near Campbellsport, Wisconsin. [Illustration from R. M. Busch, ed., *Laboratory Manual in Physical Geology*, 3rd ed. (New York: Macmillan, an imprint of Prentice Hall, Inc., 1993), p. 188. Photos: (a) and (c) by Tom Bean; (b) by Dan McCoy.]

## News Report 2

## Glacial Erratic Marks a Famous Grave

The grave of Louis Agassiz (1807–1873), an early glacial theorist and the author of two books on glaciers, is marked by a polished and striated boulder. Retreating glaciers leave behind large rocks (sometimes house-sized), boulders, and cobbles that are "foreign" in composition and origin from the ground on which they were deposited.

These *glacial erratics*, lying in strange locations with no obvious means of transport, were an early clue to Agassiz and others that blankets of ice once had covered the land. Agassiz was an early proponent of a continental glaciation hypothesis. It is fitting that compelling evidence of Earth's "ice age" marks his resting place.

# The Pleistocene Ice Age Epoch

Imagine almost a third of Earth's land surface buried beneath ice sheets and glaciers—most of Canada, the northern Midwest, England, and northern Europe, and many mountain ranges, beneath thousands of meters of ice! This is how it was at the height of the Pleistocene Epoch of the late Cenozoic Era. In addition, periglacial regions along the margins of the ice during the last ice age covered about twice their present areal extent.

(a)

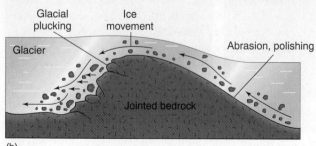

(b)

**FIGURE 17-14**
**Glacial erosion streamlined rock.**
Roche moutonnée, as exemplified by Lambert Dome in the Tuolumne Meadows area of Yosemite National Park, California. [Photo by author.]

The Pleistocene is thought to have begun about 1.65 million years ago and is one of the more prolonged cold periods in Earth's history. It featured not just one glacial advance and retreat, but at least 18 expansions of ice over Europe and North America, each obliterating and confusing the evidence from the one before.

The term **ice age** is applied to any such extended period of cold (not a single brief cold spell). An ice age is a period of generally cold climate that includes one or more *glacials*, interrupted by brief warm spells known as *interglacials*. Each glacial and interglacial is given a name that is usually based on the location where evidence of the episode is prominent—for example, "Wisconsinan glacial." Apparently, glaciation takes about 90,000 years, whereas deglaciation is rapid, requiring less than 10,000 years.

Modern research techniques include the examination of ancient ratios of oxygen isotopes, depths of coral growth in the tropics, analysis of sediments worldwide, and analysis of the latest ice cores from Greenland and Antarctica. These techniques have opened the way for a new chronology and understanding. Glaciologists currently recognize the Illinoian glacial and Wisconsinan glacial periods, with the Sangamon interglacial between them. These events span the 300,000-year period prior to our present Holocene Epoch (Figure 17-16).

The chart shows that the Illinoian glacial actually consisted of two glacials (occurring during oxygen isotope stages 6 and 8), as did the Wisconsinan (stages 2 and 4), which is dated at 10,000 to 35,000 years ago. The oxygen isotope glacial/interglacial stages on the chart are numbered back to stage 23 at approximately 900,000 years ago. To overcome local bias in core records, investigators correlate oxygen isotope data with other indicators worldwide.

## *Continental Glaciers Change the Landscape*

The continental ice sheets covered portions of Canada, the United States, Europe, and Asia about 18,000 years ago, as illustrated on the polar map projection in Figure 17-17. In North America, the Ohio and Missouri River systems mark the southern terminus of continuous ice at its greatest extent during the Pleistocene.

**FIGURE 17-15**
**Glacially deposited streamlined features.**
(a) Topographic map south of Williamson, New York, featuring numerous drumlins. (7.5-minute series quadrangle map, originally produced at a 1:24,000 scale, 10-ft contour interval.)
(b) A drumlin field in Dodge County, Wisconsin.
[(a) USGS map; (b) photo by Tom Bean.]

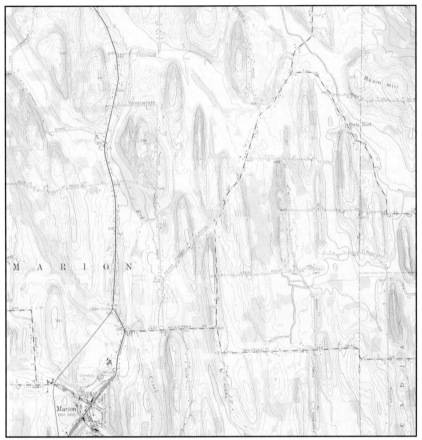

(a)

(b)

The ice sheets ranged in thickness to over 2 km (1.2 mi). As these glaciers retreated, they exposed a drastically altered landscape that includes the rocky soils of New England, the polished and scarred surfaces of Canada's Atlantic Provinces, the sharp crests of the Sawtooth Range and Tetons of Idaho and Wyoming, the scenery of the Canadian Rockies and the Sierra Nevada, the Great Lakes of the United States and Canada, the fjords of New Zealand, Norway, and Chile, the Matterhorn of Switzerland, and much more.

The continental glaciers came and went several times over the region we know as the Great Lakes (Figure 17-18). The ice enlarged and deepened stream valleys to form the basins of the future lakes. This complex history produced five lakes that today cover 244,000 km$^2$ (94,000 mi$^2$) and hold some 18% of all the lake water on Earth. Figure 17-18(a)–(d) shows the final formation of the Great Lakes, which involved two advancing and two retreating stages—between 13,200 and 10,000 years before the present. During the final retreat, tremendous quantities of glacial meltwater flowed into the gouged, isostatically depressed basins. Drainage at first was to the Mississippi River via the Illinois River, to the St. Lawrence River via the Ottawa River, and to the Hudson River in the east. In recent times, drainage has shifted through the St. Lawrence system.

The ice sheet had disappeared by 7000 years ago. Figure 17-18e illustrates major moraines left during various glacial retreats from the upper Midwest. A lobate form to these moraines is easily spotted on this simplified map. The edge of an ice sheet is not even; instead it expands and retreats in *ice lobes* with curved fronts. At each of these terminal moraines the glacial retreat stalled, and various moraines formed as the ice receded from those fronts.

Study of these glaciated landscapes is important, for we can better understand paleoclimatology (past climates) and discover the mechanisms that produce ice ages and climatic change.

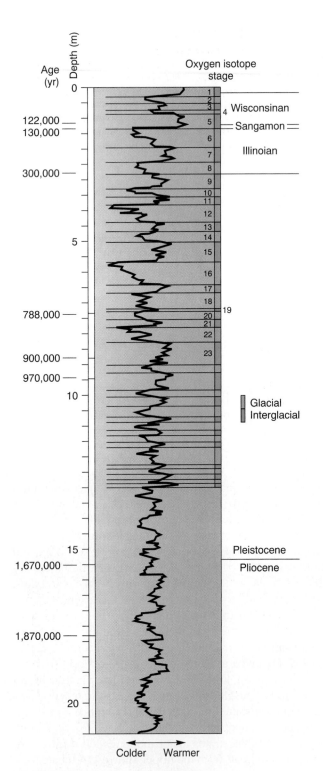

**FIGURE 17-16**

**Temperature record of the past 2 million years.**

Pleistocene temperatures, determined by oxygen isotope fluctuations in fossil planktonic foraminifera (tiny marine organisms having a calcareous shell) from deep-sea cores. Twenty-three stages cover 900,000 years, with names assigned for the past 300,000 years. [After N. J. Shackelton and N. D. Opdyke, *Oxygen-Isotope and Paleomagnetic Stratigraphy of Pacific Core V28-239, Late Pliocene to Latest Pleistocene,* Geological Society of America Memoir 145. © 1976 by the GSA. Adapted by permission.]

### Lowered Sea Levels and Temperatures

Scientists worldwide participated in the *CLIMAP* project (Climate: Long-Range Investigation, Mapping, and Prediction). Using oxygen isotope ratios from sea-floor core samples, they constructed sea-surface temperature maps for 18,000 years ago, when the last major extent of Pleistocene glacial ice prevailed.

Sea levels at that time were approximately 100 m (330 ft) lower than today because so much of Earth's water was frozen and tied up in the glaciers instead of being in the ocean (see tan areas along continental margins in Figure 17-19). Imagine the coastline of New York being 100 km farther east; Alaska and Russia connected by land across the Bering Straits, and England and France joined by a land bridge. In fact, sea ice extended southward into the North Atlantic and Pacific and northward in the Southern Hemisphere about 50% farther than it does today (light brown ocean areas in Figure 17-19).

Sea-surface temperatures during February were 1.4 C° (2.5 F°) lower than today's global average, and in August they were 1.7 C° (3.1 F°) lower. During the coldest portion of the Pleistocene Ice Age, air temperatures were as much as 12 C° (22 F°) colder than today's average, although milder periods ranged to within 5 C° (9 F°) of present air temperatures.

### Paleolakes

Figure 17-20 portrays the West dotted with large lakes 12,000–30,000 years ago. Except for the Great Salt Lake in Utah (a remnant of the former Lake Bonneville noted on the map) and a few smaller lakes, only dry basins, ancient shorelines, and lake sediments remain today. These ancient lakes are called **paleolakes**.

The term *pluvial* (Latin: "rain") describes any period of wet conditions, such as occurred during the Pleistocene Epoch. During pluvial periods, lake levels increased in arid regions. The drier periods between pluvials are called *interpluvials.* Interpluvials are marked with *lacustrine deposits,* which are lake sediments that form terraces along former shorelines.

Are these lakes of glacial origin like the Great Lakes? Earlier researchers attempted to correlate pluvial and glacial ages, given their coincidence during the Pleistocene. However, few sites actually demonstrate such a simple relation. For example, in the western United States, the estimated volume of melted ice from glaciers is only a small portion of the actual water volume that was in the paleolakes. Also, these lakes tend to predate glacial times and are correlated instead with periods of wetter climate, or periods thought to have had lower evaporation rates. The term *paleolake* is increasingly used in the scientific literature to describe these lakes and to separate their occurrence from specific glacial stages.

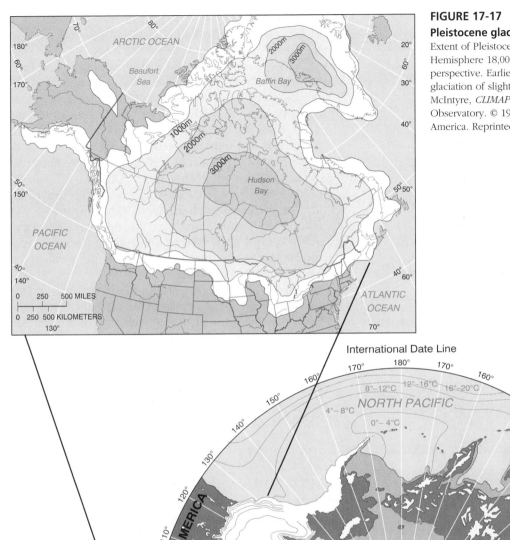

**FIGURE 17-17**
**Pleistocene glaciation.**
Extent of Pleistocene glaciation in the Northern Hemisphere 18,000 years ago, viewed from a polar perspective. Earlier episodes produced continental glaciation of slightly greater extent. [From A. McIntyre, *CLIMAP Project*, Lamont-Dougherty Earth Observatory. © 1981 by the Geological Society of America. Reprinted by permission.]

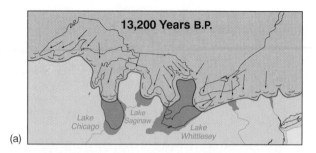

(a)

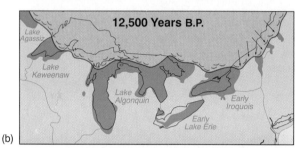

(b)

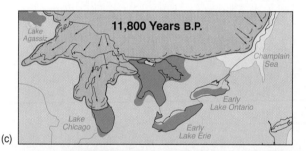

(c)

(d)

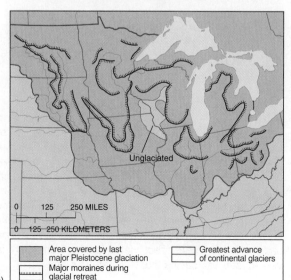

(e)

Paleolakes existed in North and South America, Africa, Asia, and Australia. Today, the Caspian Sea in Kazakstan and southern Russia has a level of 30 m (100 ft) below mean world sea level, but ancient shorelines are visible about 80 m (265 ft) above lake level.

In North America, the two largest late Pleistocene paleolakes were in the Basin and Range Province of the West—Lake Bonneville and Lake Lahontan. Respectively, these two lakes were eight times and six times the size of their present-day remnants. Figure 17-20 shows these and other paleolakes at their highest level and the few remaining modern lakes in dark blue. New evidence explains that the occurrence of these lakes in North America was related to specific changes in the polar jet stream that steered storm tracks across the region, creating pluvial conditions. Changes in jet stream position were influenced by the continental ice sheet.

The Great Salt Lake, near Salt Lake City, Utah, and the Bonneville Salt Flats in western Utah are remnants of Lake Bonneville, which at its greatest extent covered more than 50,000 km² (19,500 mi²) and reached depths of 300 m (1000 ft), spilling over into the Snake River drainage to the north. Today, it is a closed basin with no drainage except an artificial outlet to the west where excess water from the Great Salt Lake is pumped during floods. (see News Report 3).

# Deciphering Past Climates: Paleoclimatology

We observe glacials and interglacials because Earth's climate has fluctuated in and out of warm and cold ages. Evidence for this fluctuation now is being traced in ice cores from Greenland and Antarctica, in layered deposits of silts and clays, in the extensive pollen record from ancient plants, and in the relation of past coral productivity to sea level. This evidence is analyzed with radioactive dating methods and other techniques. One especially interesting fact is emerging from these studies: We humans (*Homo erectus* and *H. sapiens* of the last 1.9 million years) have never experienced Earth's normal (more moderate, less extreme) climate. We are living in a climatic anomaly, not the climate most characteristic of Earth's entire 4.6 billion year span.

**FIGURE 17-18**

**Late stages of Great Lakes formation.**

(a–d) Four "snapshots" of the Great Lakes' evolving development during the retreat of the Wisconsinan glaciation. Note the change in stream drainage between (b) and (d). Time is in years before the present. (e) Maximum extent of glaciation and the pattern of terminal moraines. [a–d After *The Great Lakes–An Environmental Atlas and Resource Book*, Environment Canada, U.S. EPA, Brock University, and Northwestern University (Toronto: Environment Canada, 1987), p. 7. (e) After Charles Denny, USGS.]

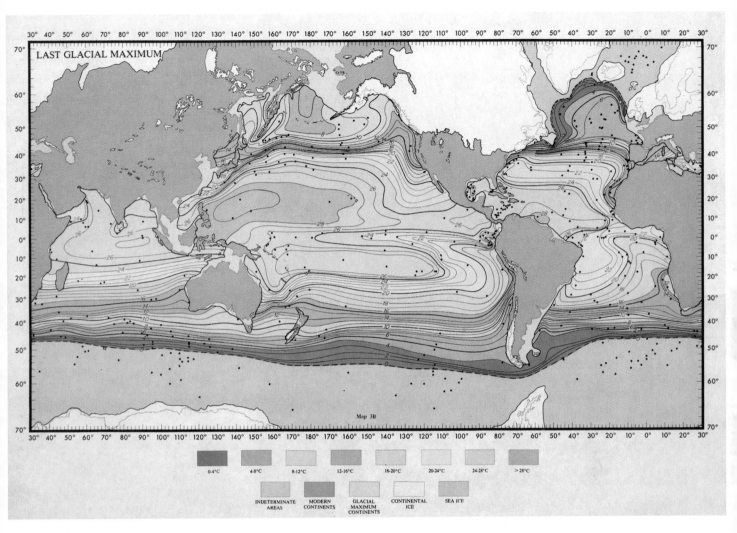

**FIGURE 17-19**

**Ocean temperatures 18,000 years ago.**

Ocean surface temperatures in August and the location of glaciers 18,000 years ago. White areas show the maximum extent of the most recent glaciation. The lowered level of the ocean during this maximum uncovered continental shelf areas (the cream-colored areas). [From A. McIntyre, *CLIMAP Project*, Lamont-Dougherty Earth Observatory. © 1981 by the Geological Society of America. Reprinted by permission.]

## News Report 3

### The Great Salt Lake Floods

Great Salt Lake in Utah has been experiencing dynamic change during the 1980s and 1990s. The lake level began rising in 1981, following increased rainfall and snowpack meltwater from the adjoining Wasatch Mountains. Shifts in storm tracks favored increased precipitation in the area during 1982–1986. Evaporation is the lake's only output from the closed Salt Lake Basin, so the lake level rose.

By the mid-1980s, the lake achieved its highest level in historical times, flooding an interstate highway, transcontinental railroad tracks, resorts, and thousands of acres of grazing land. Fortunately, the summers following the highest lake level also had high evaporation rates, and precipitation dropped to 50% below average.

The beds of the railroad tracks and the interstate highway were raised to prevent their loss if the lake level again rises. Also, massive pumps were installed to draw down the lake level when it gets too high. These pumps take water from the Great Salt Lake a short distance to the Newfoundland Evaporation Basin—an area about one-fourth the size of the lake. This ancient paleolake vividly demonstrates how climate change can effect water levels—especially in basins that lack natural drainage.

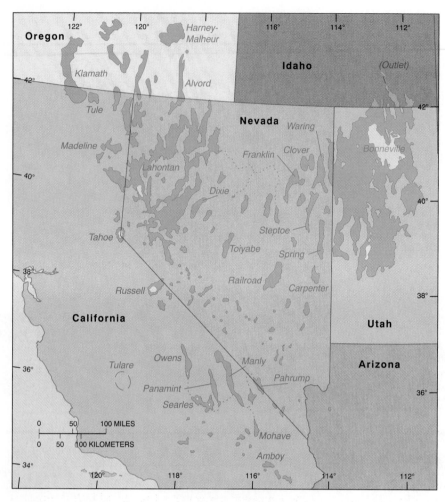

**FIGURE 17-20**

**Paleolakes in the western United States.**

Paleolakes of the western United States at their greatest extent 12,000 to 30,000 years ago, a recent pluvial period. Lake Lahontan and Lake Bonneville were the largest. The Great Salt Lake in Utah is a remnant of Lake Bonneville; Pyramid Lake in western Nevada and Honey Lake in California are remnants of ancient Lake Lahontan; Mono Lake in California is what remains of pluvial Lake Russell. With few exceptions, most paleolakes are dry basins today. [After R. F. Flint, *Glacial and Pleistocene Geology.* © 1957 by John Wiley & Sons. Adapted by permission.]

Apparently, Earth's climates slowly fluctuated until the past 1.2 billion years, when temperature patterns with cycles of 200–300 million years became more pronounced. The most recent cold episode was the Pleistocene Epoch, which began in earnest 1.65 million years ago and through which we may still be progressing. The Holocene epoch began approximately 10,000 years ago, when average temperatures abruptly increased 6 C° (11 F°). This period we live in may represent an end to the Pleistocene, or it may be merely a mild interglacial time. Figure 17-21 details the climatic record of the past 160,000 years.

## Medieval Warmth and Little Ice Age Cold

In A.D. 1001, Leif Eriksson inadvertently ventured onto the North American continent, perhaps the first European to do so. He and his fellow Vikings were favored by a medieval warming episode as they sailed the less-frozen North Atlantic to settle Iceland and Greenland. The mild climatic episode that lasted from about A.D. 800 to 1200 is known as the *Medieval Warm Period.* During the warmth, vineyards were planted far into England some 500 km north of present-day commercial plantings. Oats and barley were planted in Iceland, and wheat was planted as far north as Trondheim, Norway. The shift to warmer, wetter weather influenced migration and settlement northward in North America, Europe, and Asia.

However, from approximately 1200–1350 through 1800–1900, a *Little Ice Age* took place. Parts of the North Atlantic froze, and many key mountain passes in Europe were blocked by expanding glaciers. Snowlines in Europe lowered about 200 m (650 ft) in the coldest years. The Greenland colonies were deserted. Cropping patterns changed, and northern forests declined, along with human population. In the winter of 1779–80, New York's Hudson and East Rivers and the entire Upper Bay froze over. People walked and hauled heavy loads across the ice between Staten and Manhattan Islands!

However, the Little Ice Age was not consistently cold throughout its 700-year reign. Reading the record of the Greenland ice cores, scientists have found many mild years among the harsh. More accurately, this was a time of rapid, short-term climate fluctuations that lasted only decades.

Ice cores drilled in Greenland have revealed a record of annual snow and ice accumulation that, when correlated with other aspects of the core sample, is indicative of air temperature. Figure 17-22 presents this record since A.D. 500. The warmth of the Medieval Warm Period is evident, whereas the Little Ice Age appears mixed with colder conditions around 1200, 1500, and after about 1800.

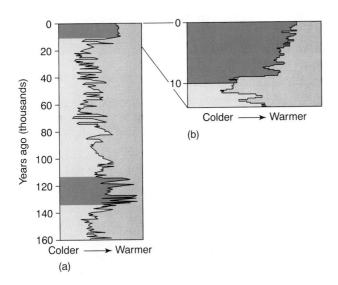

**FIGURE 17-21**

**Recent climates determined from an ice core.**

(a) Temperature patterns during the past 160,000 years. Note the cold spell that interrupted the earlier interglacial (darker colored band, 115,000–135,000 years ago). (b) Higher resolution of the last 12,000 years. The cold period known as the "Younger Dryas" was intensifying at the beginning of this record. Warming was beginning by 11,700 years ago and abruptly began to increase by the Holocene. [Courtesy of the Greenland Ice Core Project (GRIP)].

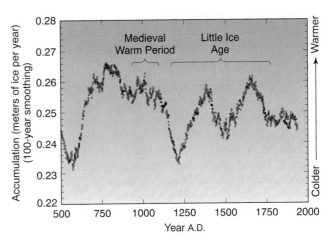

**FIGURE 17-22**

**Snow accumulation and temperature in Greenland over the past 1500 years.**

This accumulation record of snow and ice at the GISP-2 ice-core site is analogous to warmer and colder climatic conditions in Greenland—warmer times yield greater accumulations than colder times because of the moisture capacity of warmer air. Note the Medieval Warm Period and consistent temperatures as compared with the chaotic record of the Little Ice Age. [From D. A. Meese and others, "The accumulation record from the GISP-2 core as an indicator of climate change throughout the Holocene," *Science* 266 (December 9, 1994): 1681].

## What Are the Mechanisms of Climate Fluctuation?

What mechanisms cause short-term fluctuations? And why is Earth pulsing through long-term climatic changes that span several hundred million years? The ice age concept is being researched and debated with an unprecedented intensity for three principal reasons: (1) Continuous ice cores from Greenland and Antarctica are providing a new, detailed record of weather and climate patterns, volcanic eruptions, and trends in the biosphere (discussed in News Report 4). (2) To understand present and future climate change and to refine general circulation models, we must understand the natural variability of the atmosphere and climate. And (3) global warming, and its relation to ice ages, is a major concern.

Because past occurrences of low temperature appear to have followed a pattern, researchers have looked for causes that also are cyclic in nature. They have identified a complicated mix of interacting variables that appear to influence long-term climatic trends. Let's take a look at several of them.

**Climate and Celestial Relations.** As our Solar System revolves around the distant center of the Milky Way, it crosses the plane of the galaxy approximately every 32 million years. At that time, Earth's plane of the ecliptic aligns parallel to the galaxy's plane, and we pass through regions in space of increased interstellar dust and gas, which may have some climatic effect.

Other possible astronomical factors were developed by Milutin Milankovitch (1879–1954), a Yugoslavian astronomer who studied Earth-Sun orbital relations. Milankovitch wondered whether the development of an ice age was related to seasonal astronomical factors—Earth's revolution around the Sun, rotation, and tilt—extended over a longer time span (Figure 17-23). In summary:

- Earth's elliptical orbit about the Sun is not constant. The shape of the ellipse varies by more than 17.7 million kilometers (11 million miles) during a 100,000-year cycle, from nearly circular to an extreme ellipse (Figure 17-23a).

- Earth's axis "wobbles" through a 26,000-year cycle, in a movement much like that of a spinning top winding down. Earth's wobble is called *precession*. As you can see in Figure 17-23b, precession changes the orientation of hemispheres and landmasses to the Sun.

- Earth's present axial tilt of 23.5° varies from 22° to 24° during a 40,000-year period (Figure 17-23c).

Milankovitch calculated, without the aid of today's computers, that the interaction of these Earth-Sun relations creates a 96,000-year climatic cycle. His glaciation model assumes that changes in astronomical relations affect the amounts of insolation received.

Milankovitch died in 1954, his ideas still not accepted by a skeptical scientific community. Now, in the era of modern computers, remote-sensing satellites, and world-wide efforts to decipher past climates, Milankovitch's

# News Report 4

## GRIP and GISP-2: Boring Ice for Exciting History

The Greenland Ice Core Project (GRIP) was launched in 1989. A site was selected near the summit of the Greenland ice sheet at 3200 m (10,500 ft) so that the maximum thickness of ice history would be drilled through (Figure 1). After 3 years, drilling hit bedrock 3030 m (9940 ft) below the site—or, in terms of time, 250,000 years into the past. The core is 10 cm (4 in.) in diameter.

In 1990, about 32 km (20 mi) west of the summit, the Greenland Ice Sheet Project (GISP-2) began to bore back through time (Figure 1). GISP-2 reached bedrock in 1993. This core is slightly larger in diameter, at 13.2 cm (5.2 in.), and collects about twice the data as GRIP. The existence of a second core helps scientists compensate for any folds or disturbed sections they encountered below 2700 m, or about 115,000 years, in the first core. Also, clarifying the record are correlations with Antarctic ice cores, core samples of marine sediments from the oceans and Lake Baykal in Siberia, and coral studies that indicate past sea levels.

What is being discovered from a 3030 m ice core? Locked into the core are air bubbles of past atmospheres, which indicate ancient gas concentrations. Of special interest are the greenhouse gases, carbon dioxide and methane. Chemical and physical properties of the atmosphere and the snow that accumulated each year are frozen in place.

Pollutants are locked into the core record for study. For example, during cold

**FIGURE 1**

**Greenland ice core locations.**

GRIP is at 72.5° N 37.5° W.
GISP-2 is at 72.6° N 38.5° W.
[Map courtesy of GISP-2 Science Management Office.]

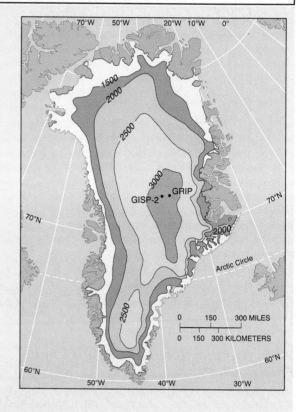

periods, high concentrations of dust were present, brought by winds from distant dry lands. An invaluable record of past volcanic eruptions is included in the layers, as if on a calendar. Even the exact beginning of the Bronze Age is recorded in the ice core—about 3000 B.C. When the Greeks and later the Romans began smelting copper, they produced ash and smoke that the winds carried to this distant place. The presence of ammonia indicates ancient forest fires

at lower latitudes. And, as an important analog of past temperatures on the ice-sheet surface when each snowfall occurred, the ratio between stable forms of oxygen is measured.

GRIP and GISP-2 have greatly refined the paleoclimatology of the late Cenozoic Era, helping us to understand Earth's dynamic climate system. Our chances of predicting future patterns have improved, thanks to these efforts.

---

valuable work has stimulated much research to explain climatic cycles and has experienced some confirmation. A roughly 100,000-year climatic cycle is confirmed in such diverse places as ice cores in Greenland and the accumulation of sediment in Lake Baykal, Siberia.

***Climate and Solar Variability.*** If the Sun significantly varies its output over the years, as some other stars do, that variation would seem a convenient and plausible cause of ice-age timing. However, lack of evidence that the Sun's radiation output varies significantly over long cycles argues against this hypothesis. Nonetheless, inquiry about the Sun's variability actively continues.

***Climate and Tectonics.*** Major glaciations also can be associated with plate tectonics because some landmasses have migrated to higher, cooler latitudes. Chapters 11 and 12 explain that the shape and orientation of landmasses and ocean basins have changed greatly during Earth's history. Continental plates have drifted from equatorial locations to polar regions and vice versa, thus exposing the land to a gradual change in climate. Gondwana (the southern half of Pangaea) experienced extensive glaciation that left its mark on the rocks of parts of present-day Africa, South America, India, Antarctica, and Australia. Landforms in the Sahara, for example, bear the markings of even earlier glacial activity. These markings are partly

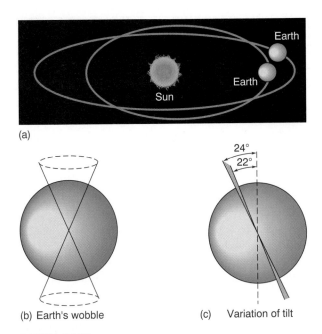

(a)

(b) Earth's wobble

(c) Variation of tilt

**FIGURE 17-23**
**Astronomical factors that may affect broad climatic cycles.**
(a) Earth's elliptical orbit varies widely during a 100,000-year cycle, stretching out to an extreme ellipse. (b) Earth's 26,000-year axial wobble. (c) Variation in Earth's axial tilt every 40,000 years.

explained by the fact that portions of Africa were centered near the South Pole during the Ordovician Period, 465 million years ago (see Figure 11-17a).

Episodes of mountain building over the past billion years have forced mountain summits above the snowline, where snow remains after the summer melt. Mountain chains influence downwind weather patterns and jet stream circulation, which in turn guides weather systems. More dust was present during glacial periods, suggesting drier weather and more extensive deserts beyond the frozen regions.

***Climate and Atmospheric Factors.*** A volcanic eruption might produce lower temperatures for a year or two. The lower temperatures could initiate a buildup of long-term snow cover at high latitudes. These high-albedo snow surfaces then would reflect more insolation away from Earth, to further enhance cooling. Climatic effects from the eruption of Mount Pinatubo in the Philippines in 1991 are being closely studied, for the eruption caused a temporary cooling.

The fluctuation of atmospheric greenhouse gases could trigger higher or lower temperatures. An ice core taken at Vostok, the Russian research station near the geographic center of Antarctica, contained trapped air samples from more than 200,000 years in the past. It showed carbon dioxide levels varying from more than 290 ppm to a low of nearly 180 ppm (Figure 17-24). Higher levels of carbon dioxide generally are thought to be correlated with each interglacial, or warmer period. During the

1990s, atmospheric carbon dioxide reached 370 ppm, higher than at any time in the past 200,000 years, principally owing to anthropogenic (human-created) sources.

***Climate and Oceanic Circulation.*** Finally, oceanic circulation patterns have changed. For example, the Isthmus of Panama formed about 3 million years ago and effectively separated the circulation of the Atlantic and Pacific Oceans. Changes in ocean basin configuration, surface temperatures, and salinity and in upwelling and downwelling rates affect air mass formation and air temperature.

Our understanding of Earth's climate—past, present, and future—is unfolding. We are learning that climate is a multicyclic system controlled by an interacting set of cooling and warming processes, all founded on celestial relations, tectonic factors, atmospheric variables, and changes in oceanic circulation.

# Arctic and Antarctic Regions

Climatologists use environmental criteria to define the Arctic and the Antarctic regions. The *Arctic region* is defined by the purple line in Figure 17-25. This is the 10°C (50°F) isotherm for July (the Northern Hemisphere summer). This line coincides with the visible treeline—the boundary between the northern forests and tundra. The Arctic Ocean is covered by two kinds of ice: floating sea ice (frozen seawater) and glacier ice (frozen freshwater). This ice pack thins in the summer months and sometimes breaks up. Scientists are discovering more about the behavior of the Arctic Ocean ice during the last ice age. (News Report 5).

The *Antarctic region* is defined by the *Antarctic convergence*, a narrow zone that extends around the continent

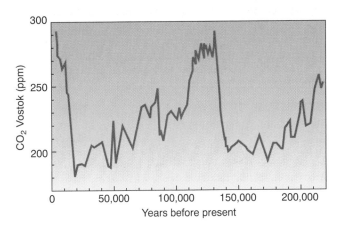

**FIGURE 17-24**
**Vostok ice core and past carbon dioxide concentrations.**
Carbon dioxide, trapped in air bubbles in Antarctic ice, was measured from the Vostok ice core. [Adapted by permission from J. Jouzel and others, "Extending the Vostok ice-core record of paleoclimate to the penultimate glacial period," *Nature* 364 (July 29, 1993): 411. © 1993 Macmillan Magazines Ltd.]

(a)

(b)

**FIGURE 17-25**

**The Arctic and Antarctic regions.**

In (a), note the 10°C (50°F) isotherm in midsummer, which designates the Arctic region. In (b), the Antarctic convergence designates the Antarctic region. Arrows on the ice sheet show the general direction of ice movement on Antarctica.

as a boundary between colder Antarctic water and warmer water at lower latitudes. This boundary follows roughly the 10°C (50°F) isotherm for February (the Southern Hemisphere summer) and is located near 60° S latitude (purple line in the figure). The Antarctic region that is covered just with sea ice represents an area greater than North America, Greenland, and Western Europe combined!

### The Antarctic Ice Sheet

Antarctica is a continent-sized landmass and therefore is much colder overall than the Arctic, which is an ocean. In simplest terms, Antarctica can be thought of as a continent covered by a single enormous glacier, although it contains distinct regions such as the East Antarctic and West Antarctic ice sheets, which respond differently to slight climatic variations. This ice sheet is in constant motion, as indicated in Figure 17-25b.

Ice sheet edges that enter coastal bays form extensive ice shelves, with sharp ice cliffs rising up to 30 m (100 ft) above the sea. Large tabular islands of ice are formed when sections of the shelves break off and move out to sea (Figure 17-26). These ice islands can be very large; several in the late 1980s and again in 1995 exceeded the area of Rhode Island. The disintegration of some of these ice shelves caused by higher temperatures is discussed in Chapter 10 and illustrated in Figure 10-35.

Individual outlet glaciers flow toward the periphery of the ice sheet in Antarctica. The Byrd Glacier moves through the Transantarctic Mountains, draining an area of more than 1,000,000 km² (386,000 mi²) as it flows 180 km (112 mi) and drops 5000 m (3.1 mi) in elevation. This glacier exceeds 22 km (13.7 mi) in width at its narrowest point and, where its thickness is 1000 m (3300 ft), it maintains a flow rate of approximately 840 m (2750 ft) per year, as fast as some surging glaciers.

News Report 3 in Chapter 10 describes the scientific importance of Antarctica in the words of the people who work there. A place so remote from civilization is an excellent laboratory for sampling past and present human and natural variables that are transported to it by atmospheric and oceanic circulation. Within this complex environment, scientists are attempting to decipher past climatic rhythms and future climatic trends. Glacial processes, landforms, and Earth's frozen climatic record in the ice sheets all provide clues for further scientific exploration.

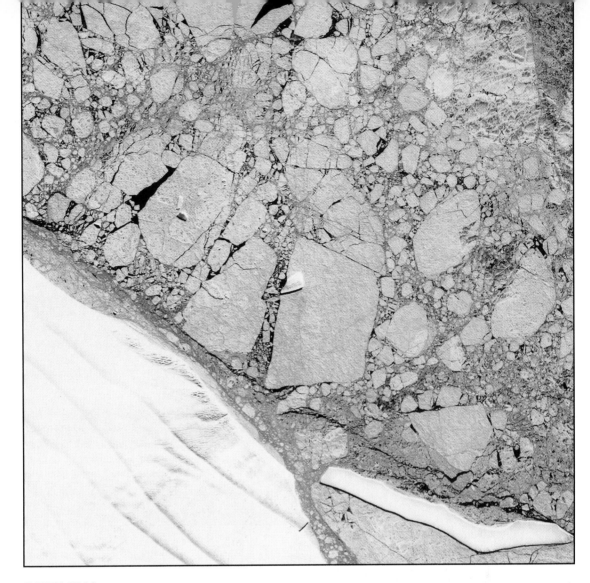

**FIGURE 17-26**
**The Antarctic ice shelf breaks up.**
Icebergs break off from the Filchner Ice Shelf into the Weddell Sea, Antarctica (78° S 40° W). [*SPOT* Image courtesy of CNES. All rights reserved. Used by permission.]

## News Report 5

### An Arctic Ice Sheet?

Questions remain about the nature of the ice in the Arctic Ocean during the Pleistocene Ice Age. Some scientists believe that the ice was similar to the thin sea ice we see today. Others think that the region was blanketed under deep ice cover. The latter were pleased when sonar images of the ocean floor, made in 1990, were published in 1994. These images showed deep gouges and grooves in bottom sediments on the floor of the Fram Strait between Greenland and Norway.

Imagine icebergs twice the size of the largest in Antarctic water, extending up to 300 m (1000 ft) high and as much as 700 m (2300 ft) deep. These enormous ice vessels scarred the ocean floor with their deep keels. In another area, the sonar disclosed submerged ridges that were scraped smooth. These super-icebergs could have originated from a floating ice sheet or from glaciers surrounding the Arctic Basin. Now the question is to what extent the North American and European ice sheets were joined across the Arctic Ocean—the jury is still out!

# Focus Study 17-1

## Periglacial Landscapes

In 1909, geologist W. Lozinski coined the term **periglacial** to describe cold-climate processes, landforms, and topographic features that exist along the margins of glaciers, past and present. Periglacial regions occupy over 20% of Earth's land surface. These areas have either near-permanent ice or are at high elevation, and the ground is seasonally snow-free. Under these conditions, a unique combination of periglacial processes operates, including permafrost (Figure 1), frost action, and ground ice.

Climatologically, these regions are in *subarctic Dfc, Dfd,* and *polar E* climates (especially *tundra ET* climate). Such climates occur either at high latitude (tundra and boreal forest environments) or at high elevation in lower-latitude mountains (alpine environments). These periglacial

regions are dominated by processes that are related to physical weathering, mass movement (Chapter 13), climate (Chapter 10), and soil (Chapter 18).

### Geography of Permafrost

When soil or rock temperatures remain below 0°C (32°F) for at least 2 years, **permafrost** ("permanent frost") develops. An area of permafrost that is not covered by glaciers is considered periglacial. Note that this criterion is based solely on *temperature* and has nothing to do with how much or how little water is present. Two other factors also contribute to permafrost: the presence of fossil permafrost from previous ice-age conditions and the insulating effect of snow cover or vegetation that inhibits heat loss.

Permafrost regions are divided into two general categories, continuous and discontinuous. They merge along a general transition zone. *Continuous permafrost* is the region of severest cold and is perennial, roughly poleward of the –7°C (19°F) mean annual temperature isotherm (purple area in Figure 1). Continuous permafrost affects all surfaces except those beneath deep lakes or rivers. The depth of continuous permafrost may exceed 1000 m (3300 ft), averaging approximately 400 m (1300 ft).

Unconnected patches of *discontinuous permafrost* gradually coalesce poleward toward the continuous zone. Permafrost becomes scattered or sporadic until it gradually disappears equatorward of the –1°C (30.2°F) mean annual temperature isotherm (dark blue area on

### FIGURE 1

**Permafrost distribution.**

Distribution of permafrost in the Northern Hemisphere. Alpine permafrost is noted except for small occurrences in Hawaii, Mexico, Europe, and Japan. Subsea permafrost occurs in the ground beneath the Arctic Ocean along the margins of the continent noted. Note the towns of Resolute and Coppermine in the Northwest Territories and Hotchkiss in Alberta. A cross section of the permafrost beneath these towns is shown in Figure 2. [Adapted from T. L. Péwé, "Alpine permafrost in the contiguous United States: A review," *Arctic and Alpine Research* 15, no. 2 (May 1983): 146. © Regents of the University of Colorado. Used by permission.]

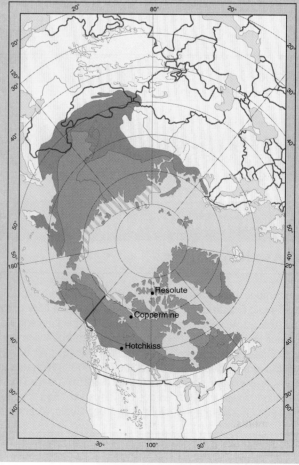

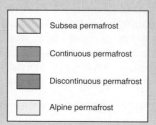

map). In the discontinuous zone, permafrost is absent on sun-exposed south-facing slopes, areas of warm soil, or areas insulated by snow. In the Southern Hemisphere, north-facing slopes experience increased warmth.

As much as 50% of Canada and 80% of Alaska are affected by permafrost of either type. In central Eurasia, the effects of continentality and elevation produce discontinuous permafrost that extends equatorward to the 50th parallel.

Areas of discontinuous permafrost feature a mixture of *cryotic* (frozen) and *noncryotic* ground. In addition to these two types of ground, zones of high-altitude *alpine permafrost* extend to lower latitudes, as shown on the map. Microclimatic factors such as slope orientation and snow cover are important in the alpine environment.

In the Mackenzie Mountains of Canada (62° N), continuous permafrost extends down to an elevation of 1200 m (4000 ft), and discontinuous permafrost occurs throughout. The Colorado Rockies (40° N) experience continuous permafrost down to an elevation of 3400 m (11,150 ft) and discontinuous permafrost to 1700 m (5600 ft).

**Behavior of Permafrost.** We have looked at the spatial distribution of permafrost; let us now examine how permafrost behaves. Figure 2 is a stylized cross section from approximately 75° N to 55° N, using the three sites located on the map in Figure 1. The **active layer** is the zone of seasonally frozen ground that exists between the subsurface permafrost layer and the ground surface.

The active layer is subjected to consistent daily and seasonal freeze-thaw cycles. This cyclic melting of the active layer affects as little as 10 cm (4 in.) depth in the north (Ellesmere Island, 78° N), up to 2 m (6.6 ft) in the southern margins (55° N) of the periglacial region, and 15 m (50 ft) in the alpine permafrost of the Colorado Rockies (40° N).

The depth and thickness of the active layer and permafrost zone change slowly in response to climatic change. Higher temperatures degrade (reduce) permafrost and increase the thickness of the active layer; lower temperatures gradually aggrade (increase) permafrost depth and reduce active layer thickness. Although somewhat sluggish in response, the active layer is a dynamic open system driven by energy gains and losses in the subsurface environment. As you might expect, most permafrost exists in disequilibrium with environmental conditions and therefore actively adjusts to inconstant climatic conditions.

A *talik* is unfrozen ground that may occur above, below, or within a body of discontinuous permafrost or beneath a water body in the continuous region. Taliks occur beneath deep lakes and may extend to bedrock and noncryotic soil beneath large deep lakes (Figure 2). Taliks form connections between the active layer and groundwater, whereas in continuous permafrost groundwater is essentially cut off from surface water. In this way, permafrost disrupts aquifers and taliks, leading to water supply problems.

### Ground Ice and Frozen Ground Phenomena

In regions of permafrost, frozen subsurface water is termed *ground ice*. The moisture content of areas with ground ice varies from nearly none in drier regions to almost 100% in saturated soils. From the area of maximum energy loss, freezing progresses through the ground along a *freezing front*, or boundary between

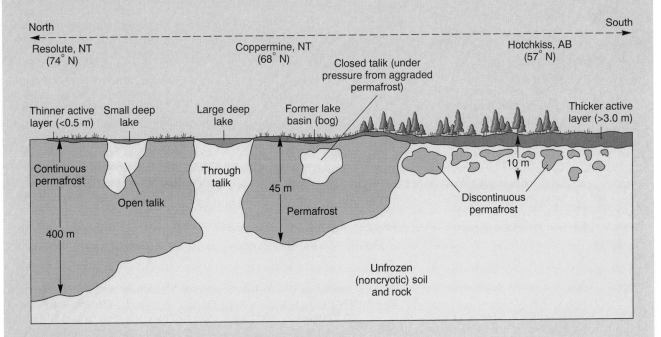

**FIGURE 2**
**Periglacial environments.**
Cross section of a typical periglacial region in northern Canada, showing typical forms of permafrost, active layer, talik, and ground ice. The three sites noted are shown on the map in Figure 1.

frozen and unfrozen soil. The presence of frozen water in the soil initiates geomorphic processes associated with *frost action* and the expansion of water as it freezes (Chapters 7 and 13).

Ground ice may occur as:

- Common *pore ice* (subsurface water frozen in the soil's pore spaces)
- Lenses (horizontal bodies) and veins (channels extending in any direction)
- *Segregated ice* (layers of buried ice that increase in mass by accreting water as the ground freezes, producing layers of relatively pure ice)
- *Intrusive ice* (the freezing of water injected under pressure, as in a pingo, discussed shortly)
- *Wedge ice* (surface water entering a crack and freezing)

Various aspects of frost action as a geomorphic agent are discussed in Chapter 13 ("Physical Weathering Processes" and the section "Classes of Mass Movements," which describes soil creep). Some forms of ground ice may occur at the surface, during episodes of *icing*, in which a river or spring forms freezing layers of surface ice. Such icings can occur on slopes as well as level ground. Spring flooding along rivers can result from the presence of icings; it is a particular problem in rivers draining into the Arctic Ocean.

**Frost Action Processes.** The 9% expansion of water as it freezes produces strong mechanical forces. Such frost action shatters rock, producing angular pieces that form a *block field*, or *felsenmeer*. It accumulates as part of the arctic and alpine periglacial landscape, particularly on mountain summits and slopes.

If sufficient water freezes, the saturated soil and rocks are subjected to *frost-heaving* (vertical movement) and *frost-thrusting* (horizontal movement). Boulders and rock slabs may be thrust to the surface. Soil horizons (layers) may be disrupted by frost action and appear to be stirred or churned, a process termed *cryoturbation*. Frost action also can produce contractions in soil and rock, opening up cracks for ice wedges to form. Also, there is a tremendous increase in pressure in the soil as ice expands, particularly if there are multiple freezing fronts trapping unfrozen soil and water between them.

An *ice wedge* develops when water enters a crack in the permafrost and freezes (Figure 3). Thermal contraction in ice-rich soil forms a tapered crack—wider at the top, narrowing toward the bottom. Repeated seasonal freezing and melting progressively enlarge the wedge, which may widen from a few millimeters to 5–6 m (16–20 ft) and up to 30 m (100 ft) in depth. Widening may be small each year, but after many years the wedge can become significant, like the sample shown.

The annual thickness of ice added is marked by thin layers of sediment that form foliations in the wedge. Thus, the age of a wedge can be determined by counting the foliations, like counting annual accumulations recorded in an ice core. In summer, when the active layer thaws, the wedge itself may not be visible. However, the presence of a wedge beneath the surface is noticeable where the ice-expanded sediments form raised, upturned ridges.

**Frost Action Landforms.** Large areas of frozen ground (soil-covered ice) can develop a heaved-up, circular, ice-cored mound called a *pingo*. It rises above the flat landscape, occasionally exceeding 60 m (200 ft) height. Pingos rise when freezing water expands, sometimes as a result of pressure developed by artesian water injected into permafrost (Figure 4).

A *palsa* (from the Swedish word for "elliptical") is a rounded or elliptical mound of peat that contains thin perennial ice lenses rather than an ice core, as in a pingo. Palsas can be 2–30 m (6–100 ft) wide by 1–10 m (3–30 ft) high and usually are covered by soil or vegetation over a cracked surface.

The expansion and contraction of frost action results in the transport of stones and boulders. As the water-ice volume changes and the ice wedge deepens, coarser particles are moved toward the surface. An area with a system of ground ice and frost action develops sorted and unsorted accumulations of rock at the surface that take the shape of polygons called *patterned ground*.

Patterned ground may include polygons of sorted rocks that coalesce into stone polygon nets (Figure 5). Various terms are in use to describe such ice-wedge and stone polygon forms: nets, circles, hummocks (vegetation-covered), and stripes (elongated polygons formed on a hillslope). Ice wedges sometimes are found beneath the perimeter of each cell;

however, questions still exist as to how such patterned ground actually develops and the degree to which such ground forms are the vestiges of past eras.

**Hillslope Processes: Solifluction and Gelifluction.** Soil drainage is poor in areas of permafrost and ground ice. The active layer of soil and regolith is saturated with soil moisture during the thaw cycle (summer), and the whole layer commences to flow from higher to lower elevation if the landscape is even slightly inclined. Such soil flows are generally called *solifluction*; in the presence of ground ice, the more specific term *gelifluction* is applied.

In this ice-bound type of soil flow, movement up to 5 cm (2 in.) per year can occur on slopes as gentle as a degree or two. Gentle downslope movement of saturated surface material can occur in various climatic regimes outside of periglacial regions.

The cumulative effect of this landflow can be an overall flattening of a rolling landscape, with identifiable sagging surfaces and scalloped and lobed patterns in the downslope soil movements (Figure 6). Other types of periglacial mass movement include failure in the active layer, producing translational and rotational slides and rapid flows associated with melting ground ice. Periglacial mass movement processes are related to slope dynamics and processes discussed in Chapter 13.

**Thermokarst Landscapes.** As ground ice melts, irregular features develop across the landscape, creating *thermokarst* topography. It results from thermal subsidence and erosion caused by ice-wedge melting and poor drainage. (Note that the term refers only to the topographic style and has nothing to do with the solution processes and chemical weathering that cause limestone karst.)

Thermokarst topography is hummocky, marked by cave-ins, bogs, small depressions, pits, standing water, and small lakes. More thermokarst landforms are found in Siberia and Scandinavia than in North America. In Canada and Alaska, rounded thaw lakes, or cave-in lakes, are thermokarst features.

**Humans and Periglacial Landscapes**

In areas of permafrost and frozen ground phenomena, people face several

(a)

**FIGURE 3**

**Evolution of an ice wedge.**

(a) An ice wedge and ground ice in northern Canada. (b) Sequential illustration of ice-wedge formation. [(a) Photo by H. M. French. (b) Adapted from A. H. Lachenbruch. "Mechanics of thermal contraction and ice-wedge polygons in permafrost," *Geological Society of America Bulletin Special Paper 70* (1962).]

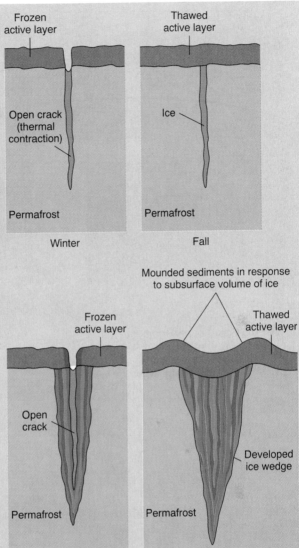

(b)

**FIGURE 4**

**A pingo.**

An ice-cored pingo resulting from hydraulic pressures that have pushed the mound upward above the landscape. Coastal erosion has exposed the ice core. This pingo is near Tuktoyaktuk, Mackenzie Delta, Canada. [Photo by H. M. French.]

related problems. Because thawed ground above the permafrost zone frequently shifts, highways and rail lines become warped, twisted, and fail, and utility lines are disrupted. In addition, any building placed directly on frozen ground will "melt" into the defrosting soil, creating subsidence in structures (Figure 7).

Construction in periglacial regions dictates placing structures above the ground to allow air circulation beneath. This air flow allows the ground to cycle through its normal annual temperature pattern. Utilities such as water and sewer lines must be built above ground in "utilidors" to protect them from freezing and

thawing ground (Figure 8). Likewise, the Alaskan oil pipeline was constructed above ground on racks to avoid melting the frozen ground, causing shifting that could rupture the line.

The effects of global warming are already showing up in Siberia, where the active layer is now more than twice the

(a)

(b)

**FIGURE 5**
**Patterned ground phenomena.**
(a) Aerial view of polygonal nets formed in Alaska, a fairly widespread frozen-ground phenomenon. (b) Surface appearance of patterned ground in tundra. [(a) Photo from the Marbut Collection, Soil Science Society of America, Inc., (b) photo by Len Rue, Jr.]

**FIGURE 6**
**Soil flowage in periglacial environments.**
Gelifluction (solifluction) lobes on a hillside near the Yukon-Northwest Territories border. [Photo by Joyce Lundberg.]

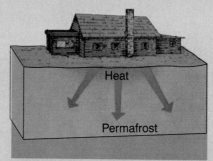

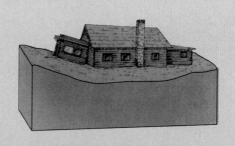

**FIGURE 7**
**Permafrost melting and structure collapse.**
Building failure due to the melting of permafrost south of Fairbanks, Alaska. [Adapted from U.S. Geological Survey. Photo by Steve McCutcheon, illustration based on U.S. Geological Survey pamphlet "Permafrost" by L. L. Ray.]

(a)

(b)

**FIGURE 8**
**Special structures for permafrost.**
(a) Proper construction in periglacial environments requires placement of buildings above ground and running water and sewage lines in elevated "utilidors" (Inuvik, Northwest Territories).
(b) Supporting the Trans-Alaska oil pipeline on racks protects the permafrost from heat.
[(a) Photo by Joyce Lundberg; (b) photo by Peter J. Williams.]

depth it was in the past. Buildings constructed with shallow pilings for support will certainly fail if this trend continues. In an era of global change, the regions of permafrost may provide further climatic indicators.

**Review—Focus Study 17-1**

periglacial (p. 538)
permafrost (p. 538)
active layer (p. 539)

1. In terms of climatic types, describe the areas on Earth where periglacial landscapes occur. Include both higher latitude and higher altitude climate types.

2. Define two types of permafrost, and differentiate their occurrence on Earth. What are the characteristics of each?
3. Describe the active zone in permafrost regions, and relate the degree of development to specific latitudes.
4. What is a talik? Where might you expect to find taliks, and to what depth do they occur?
5. What is the difference between permafrost and ground ice?
6. Describe the role of frost action in the formation of various landform types in the periglacial region.

7. Relate some of the specific problems humans encounter in developing periglacial landscapes.

# Summary and Review — Glacial and Periglacial Processes and Landforms

✔ *Differentiate* between alpine and continental glaciers and *describe* their principal features.

More than 77% of Earth's freshwater is frozen. Ice covers about 11% of Earth's surface, and periglacial features occupy another 20% of ice-free but cold-dominated landscapes. A **glacier** is a mass of ice sitting on land or floating as an ice shelf in the ocean next to land. Glaciers form in areas of permanent snow. A **snowline** is the lowest elevation where snow occurs year-round and its altitude varies by latitude—higher near the equator, lower poleward.

A glacier in a mountain range is an **alpine glacier**. If confined within a valley, it is termed a *valley glacier*. The area of origin is a snowfield, usually in a bowl-shaped erosional landform called a **cirque**. Where alpine glaciers flow down to the sea, they calve and form **icebergs**. A **continental glacier** is a continuous mass of ice on land. Its most extensive form is an **ice sheet**; a smaller, roughly circular form is an **ice cap**; and the least extensive form, usually in mountains, is an **ice field**.

glacier (p. 512)
snowline (p. 512)
alpine glacier (p. 512)
cirque (p. 512)
iceberg (p. 514)
continental glacier (p. 514)
ice sheet (p. 514)
ice cap (p. 514)
ice field (p. 515)

1. Describe the location of most freshwater on Earth today.
2. What is a glacier? What is implied about existing climate patterns in a glacial region?
3. Differentiate between an alpine glacier and a continental glacier.
4. Name the three types of continental glaciers. What is the basis for dividing continental glaciers into types? Which type covers Antarctica?

✔ *Describe* **the process glacial ice formation and** *portray* **the mechanics of glacial movement.**

Snow becomes glacial ice through accumulation, increasing thickness, pressure on underlying layers, and recrystallization. Snow progresses through transitional steps from **firn** (compact, granular) to a denser **glacial ice** after many years.

A glacier is an open system with inputs and outputs that can be analyzed through observation of the growth and wasting of the glacier itself. A glacier is fed by snowfall and is wasted by **ablation** (losses from its upper and lower surfaces and along its margins). Accumulation and ablation achieve a mass balance in each glacier. A **firn line** is the lower extent of a fresh snow-covered area.

As a glacier moves downhill, vertical **crevasses** may develop. Sometimes a glacier will move rapidly in a **glacier surge**. The presence of water along the basal layer appears to be important in glacial movements. As a glacier moves, it plucks rock pieces and debris, incorporating them into the ice, and these scour and sandpaper underlying rock through **abrasion**.

> firn (p. 516)
> glacial ice (p. 516)
> ablation (p. 517)
> firn line (p. 517)
> crevasse (p. 519)
> glacier surge (p. 520)
> abrasion (p. 520)

5. Trace the evolution of glacial ice from fresh fallen snow.
6. What is meant by glacial mass balance? What are the basic inputs and outputs underlying that balance?
7. What is meant by a glacier surge? What do scientists think produces surging episodes?

---

✔ *Describe* **characteristic erosional and deposition-al landforms created by alpine glaciation and continental glaciation.**

Extensive valley glaciers have profoundly reshaped mountains worldwide, carving **V**-shaped stream valleys into **U**-shaped glaciated valleys, producing many distinctive erosional and depositional landforms. As cirque walls erode away, sharp **arêtes** (sawtooth, serrated ridges) form, dividing adjacent cirque basins. Two eroding cirques may reduce an arête to a saddlelike **col**. A **horn** results when several cirque glaciers gouge an individual mountain summit from all sides, forming a pyramidal peak. An ice-carved rock basin left as a glacier retreats may fill with water to form a **tarn**; a string of tarns separated by moraines is called **paternoster lakes**. Where a glacial valley trough joins the ocean, and the glacier retreats, the sea extends inland to form a **fjord**.

All glacial deposits, whether ice-borne or meltwater-borne, constitute **glacial drift**. Direct deposits from ice, called **till**, are unstratified and unsorted. Glacial meltwater deposits are sorted and are called **stratified drift**. Specific landforms produced by the deposition of drift are **moraines**. A **lateral moraine** forms along each side of a glacier; merging glaciers with lateral moraines form a **medial moraine**; and eroded debris dropped at the glacier's terminous is a **terminal moraine**.

Continental glaciation leaves different features than does alpine glaciation. A **till plain** forms behind end moraines, featuring unstratified coarse till, low and rolling relief, and deranged drainage. Beyond the morainal deposits, **outwash plains** of stratified drift feature stream channels that are melt-water-fed, braided, and overloaded with debris that is sorted and deposited across the landscape. An **esker** is a sinuously curving, narrow ridge of coarse sand and gravel that forms along the channel of a meltwater stream beneath a glacier. A **kame** is a small hill, knob, or mound of poorly sorted sand and gravel that is deposited directly by water or by ice in crevasses.

An isolated block of ice left by a retreating glacier becomes surrounded with debris; when the block finally melts, it leaves a steep-sided **kettle**. Glacial action forms two stream-lined hills: the erosional **roche moutonnée** is an asymmetrical hill of exposed bedrock, gently sloping upstream and abruptly sloping downstream; the depositional **drumlin** is deposited till, streamlined in the direction of continental ice movement (blunt end upstream and tapered end downstream).

> aréte (p. 521)
> col (p. 521)
> horn (p. 521)
> tarn (p. 521)
> paternoster lakes (p. 521)
> fjord (p. 521)
> glacial drift (p. 521)
> till (p. 521)
> stratified drift (p. 521)
> moraine (p. 521)
> lateral moraine (p. 521)
> medial moraine (p. 521)
> terminal moraine (p. 521)
> till plain (p. 523)
> outwash plain (p. 523)
> esker (p. 523)
> kame (p. 523)
> kettle (p. 524)
> roche moutonnée (p. 524)
> drumlin (p. 524)

8. How does a glacier accomplish erosion?
9. Describe the evolution of a **V**-shaped stream valley to a **U**-shaped glaciated valley. What features are visible after the glacier retreats?
10. How is an iceberg generated?
11. Differentiate between two forms of glacial drift—till and outwash.
12. What is a morainal deposit? What specific moraines are created by alpine and continental glaciers?
13. What are some common depositional features encountered in a till plain?

14. Contrast a roche moutonnée and a drumlin regarding appearance, orientation, and the way each forms.

---

✔ ***Explain*** **the Pleistocene ice age epoch and related glacials and interglacials and *describe* some of the methods used to study paleoclimatology.**

An **ice age** is any extended period of cold. The late Cenozoic Era has featured pronounced ice-age conditions in an epoch called the Pleistocene. During this time, alpine and continental glaciers covered about 30% of Earth's land area in at least 18 glacials, punctuated by interglacials of milder weather. Beyond the ice, **paleolakes** formed because of wetter conditions. Evidence of ice-age conditions is gathered from ice cores drilled in Greenland and Antarctica, from ocean sediments, from coral growth in relation to past sea levels, and from rock. The study of past climates is *paleoclimatology*.

The apparent pattern followed by these low-temperature episodes indicates cyclic causes. A complicated mix of interacting variables appears to influence long-term climatic trends: celestial relations, solar variability, tectonic factors, atmospheric variables, and oceanic circulation.

ice age (p. 526)
paleolake (p. 528)

15. What is paleoclimatology? Describe Earth's past climatic patterns. Are we experiencing a normal climate pattern in this era, or have scientists noticed any significant trends?

16. Define an ice age. When was the most recent? Explain "glacial" and "interglacial" in your answer.

17. Summarize what science has learned about the causes of ice ages by listing and explaining at least four possible factors in climate change.

18. Describe the role of ice cores in deciphering past climates. What record do they preserve? Where were they drilled?

19. What criteria define the Arctic and Antarctic regions? Is there any coincidence in these criteria and the distribution of Northern Hemisphere forests on the continents?

---

 ## NetWork

The *Geosystems Home Page* provides on-line resources for this chapter on the World Wide Web. You will find review exercises, specific updates for items in the chapter, suggested readings, and links to interesting related pathways on the Internet (click on the Table of Contents link and select this chapter). *Geosystems* is at: **http://www.prenhall.com/geosystm**

# *Part 4*

# Soils, Ecosystems, and Biomes

*Hoh Rain Forest, Olympic National Park.* [Photo by Jack W. Dykinga.]

Earth is the home of the only known biosphere in the Solar System—a unique, complex, and interactive system of abiotic (nonliving) and biotic (living) components working together to sustain a tremendous diversity of life. Energy enters the biosphere through conversion of solar energy by photosynthesis in the leaves of plants. Soil is the essential link among the lithosphere, plants, and the rest of Earth's physical systems. Thus, soil helps sustain life.

Life is organized into a feeding hierarchy from producers to consumers, ending with decomposers. Taken together, the soils, plants, animals, and all abiotic components produce aquatic and terrestrial ecosystems, known as biomes. Today we face crucial issues—preservation of the diversity of life in the biosphere and the survival of the biosphere itself. These important applied topics are considered in Part 4.

# 18

# The Geography of Soils

**Soil Characteristics**

**Soil Properties**

**Soil Formation Factors and Management**

**Soil Classification**

**Summary and Review**

## Key Learning Concepts

After reading the chapter, you should be able to:

- *Define* soil and soil science and *describe* a pedon, polypedon, and typical soil profile.
- *Describe* soil properties of color, texture, structure, consistence, porosity, and soil moisture.
- *Explain* basic soil chemistry, including cation-exchange capacity, and *relate* these concepts to soil fertility.
- *Evaluate* principal soil formation factors, including the human element.
- *Describe* the eleven soil orders of the Soil Taxonomy classification system and *explain* their general occurrence.

*Potato rows in West Cape, Prince Edward Island, Canada.* [Photo by Lionel Stevenson/Photo Researchers, Inc.]

Earth's landscape generally is covered with soil. **Soil** is a dynamic natural material composed of fine particles in which plants grow, and which contains both mineral fragments and organic matter. If you have ever planted a garden, tended a house plant, or been concerned about famine and soil loss, this chapter will interest you. A knowledge of soil is at the heart of agriculture and food production.

**Soil science** is interdisciplinary, involving physics, chemistry, biology, mineralogy, hydrology, taxonomy, climatology, and cartography. Physical geographers are interested in the spatial patterns formed by soil types and the environmental factors that interact to produce them. As an integrative science, physical geography is well suited to the study of soils.

*Pedology* concerns the origin, classification, distribution, and description of soil (*ped* from the Greek *pedon*, meaning "soil" or "earth"). Pedology is at the center of learning about soil as a natural body, but it does not dwell on its practical uses. *Edaphology*, from the Greek *edaphos*, which means "soil" or "ground", focuses on soil as a medium for sustaining higher plants. Edaphology emphasizes plant growth, fertility, and the differences in productivity among soils. Pedology gives us a general understanding of soils and their classification, whereas edaphology reflects society's concern for food and fiber production and the management of soils to increase fertility and reduce soil losses.

Soil science deals with a complex substance whose characteristics vary from kilometer to kilometer, and even centimeter to centimeter. In many locales, an *agricultural extension service* can provide specific information and perform a detailed analysis of local soils. Soil surveys and local soil maps are available for most counties in the United States and for the Canadian provinces. (Your local phone book may list the U.S. Department of Agriculture, Natural Resources Conservation Service, or Agriculture Canada's Canada Soil Survey, or you may contact the appropriate department at a local college or university.)

# Soil Characteristics

Classifying soils is similar to classifying climates, because both involve complex interacting variables. Before we look at soil classification, let us examine the physical properties that distinguish soils as they develop through time and in response to climate, relief, and topography.

## Soil Profiles

Just as a book cannot be judged by its cover, so soils cannot be evaluated at the surface only; doing so produces a misleading picture. Instead, a soil profile should be studied from the surface to the deepest extent of plant roots, or to where regolith or bedrock is encountered. Such a profile, called a **pedon**, is a hexagonal column measuring 1 to 10 m² in top surface area (Figure 18-1). At the sides of the pedon, the various layers of the soil profile are visible in cross section and are labeled with letters.

*A pedon is the basic sampling unit used in soil surveys.* Many pedons together in one area make up a **polypedon**, which has distinctive characteristics differentiating it from surrounding polypedons. A polypedon is an essential soil individual, comprising an identifiable *series* of soils in an area. It can have a minimum dimension of about 1 m² and no specified maximum size. *The polypedon is the soil unit used in preparing local soil maps.*

## Soil Horizons

Each layer exposed in a pedon is a **soil horizon**. A horizon is roughly parallel to the pedon's surface and has characteristics distinctly different from horizons directly above or below. The boundary between horizons usually is distinguishable in the field, on the basis of the properties of color, texture, structure, consistence (meaning soil consistency or cohesiveness), porosity, the presence or absence of certain minerals, moisture, and chemical processes (Figure 18-2). Soil horizons are the building blocks of soil classification. However, because they can vary in almost countless ways, we use the simple model of an ideal pedon and soil profile shown in Figure 18-1.

At the top of the soil profile is the *O* (organic) *horizon*, named for its organic composition, derived from plant and animal litter that was deposited on the surface and transformed into humus. **Humus** is not just a single material; it is a complex mixture of decomposed and synthesized organic materials, usually dark in color. Microorganisms work busily on this organic debris, performing a portion of the *humification* (humus-making) process. The O horizon is 20%–30% or more organic matter, which is important because of its ability to retain water and nutrients and for the way it acts in a complementary manner to clay minerals.

At the bottom of the soil profile is the *R* (rock) *horizon*, consisting of either unconsolidated (loose) material or consolidated bedrock. When bedrock weathers into regolith, it may or may not contribute to overlying soil horizons. The A, E, B, and C horizons mark differing mineral strata between O and R. These middle layers are composed of sand, silt, clay, and other weathered by-products. (Table 11-3 presents a description of grain sizes of these weathered soil particles.)

In the *A horizon*, humus and clay particles are particularly important, for they provide essential chemical links between soil nutrients and plants. This horizon usually is richer in organic content, and hence darker, than lower horizons. The A horizon grades into the *E*

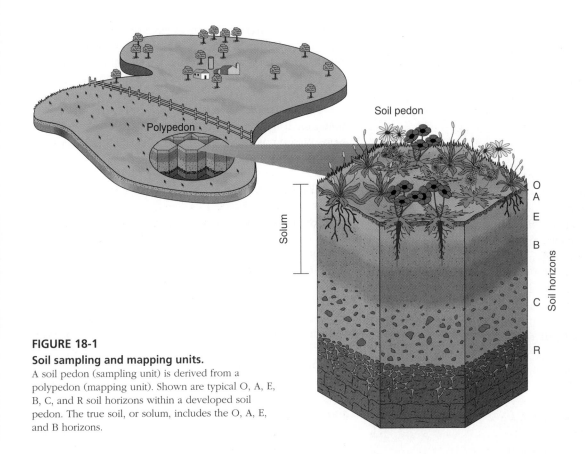

**FIGURE 18-1**
**Soil sampling and mapping units.**
A soil pedon (sampling unit) is derived from a polypedon (mapping unit). Shown are typical O, A, E, B, C, and R soil horizons within a developed soil pedon. The true soil, or solum, includes the O, A, E, and B horizons.

*horizon*, made up of coarse sand, silt, and resistant minerals; generally it is lighter in color.

From the E horizon, clays and oxides of aluminum and iron are leached (removed by water) and are carried to lower horizons with the water as it percolates through the soil. This process of removal of fine particles and minerals by water is termed **eluviation**; thus the E designation for this horizon. The greater the precipitation, the higher the rate of eluviation.

In contrast to the A and E horizons, *B horizons* accumulate clays, aluminum, and iron. B horizons are dominated by **illuviation**, a depositional process. (Eluviation is a removal process; illuviation is depositional.) B horizons may exhibit reddish or yellowish hues because of the illuviated presence of mineral and organic oxides. Some materials occurring in this soil zone also may have formed in place from weathering processes rather than arriving there by *translocation*, or migration. In the humid tropics, these layers often develop to some depth.

The combination of the A and E horizons with their eluviation removals and the B horizon with its illuviation accumulations is designated the **solum**, considered the true definable soil of the pedon. The A, E, and B horizons experience the most active soil processes.

The *C horizon* is weathered bedrock or weathered parent material. It excludes the bedrock itself, which is beyond biological influences. This zone is identified as *regolith* (although the term sometimes is used to include the solum as well). The C horizon is not much affected by soil operations in the solum and lies outside the biological influences experienced in the shallower horizons. Plant roots and soil microorganisms are rare in the C horizon. It lacks clay concentrations and generally is made up of carbonates, gypsum, or soluble salts, or of iron and silica, which form cemented soil structures. In dry climates, calcium carbonate commonly forms the cementing material of these hardened layers.

Soil scientists using the U.S. classification system employ letter suffixes to further designate special conditions within each soil horizon—for example, Ap (A horizon has been plowed) or Bt (B horizon is of illuviated clay)—see Table 18-1. (Soil horizon suffix designations used in the Canadian System of Soil Classification are presented in Appendix C.)

# Soil Properties

Soils are complex and varied, as this section reveals. Observing a real soil profile will help you identify color, texture, structure, and other soil properties. A good opportunity to observe soil profiles is at a construction site or excavation, perhaps on your campus, or at a roadcut along a highway.

**FIGURE 18-2**
**A typical soil profile.**
This is a Mollisol pedon in southeastern South Dakota. The parent material is glacial till, and the soil is well drained. Distinct carbonate nodules are visible in the lower B and upper C horizons. [Photo from American Society of Agronomy.]

## Soil Color

Color is important, for it sometimes suggests composition and chemical makeup. If you look at exposed soil, color may be the most obvious trait. Among the many possible hues are the reds and yellows found in soils of the southeastern United States (high in iron oxides), the blacks of prairie soils in portions of the U.S. grain-growing regions and Ukraine (richly organic), or white-to-pale hues found in soils containing silicates and aluminum oxides. However, color can be deceptive: Soils of high humus content are often dark, yet clays of warm-temperate and tropical regions with less than 3% organic content are some of the world's blackest soils.

To standardize color descriptions, soil scientists describe a soil's color by comparing it with a *Munsell Color*

*Chart* (developed by artist and teacher Albert Munsell in 1913). These charts display 175 colors arranged by *hue* (the dominant spectral color, such as red), *value* (degree of darkness or lightness), and *chroma* (purity and saturation of the color, which increase with decreasing grayness). Each color is identified by a name and a Munsell notation, so soil scientists can make worldwide comparisons of soil color. Soil color is checked against the chart at various depths within a pedon (Figure 18-3).

## Soil Texture

Soil texture, perhaps a soil's most permanent attribute, refers to the mixture of sizes of its particles and the proportion of different sizes. Individual mineral particles are called *soil separates.* All particles smaller in diameter than 2 mm (0.08 in.), such as very coarse sand, are considered part of the soil. Larger particles such as pebbles, gravel, or cobbles, are not part of the soil.

Figure 18-4 is a *soil texture triangle* showing the relation of sand, silt, and clay concentrations in soil. Each corner of the triangle represents a soil consisting solely of the particle size noted (although rarely are true soils composed of a single separate). Every soil on Earth is defined somewhere in this triangle.

Figure 18-4 includes the common designation **loam**, which is a balanced mixture of sand, silt, and clay that is beneficial to plant growth. A sandy loam with clay content below 30% (lower left) usually is considered ideal

**FIGURE 18-3**
**A Munsell Soil Color Chart page.**
A soil sample is viewed through the hole to match it with a color on the chart. Color is judged by hue, value, and chroma. [Photo by Macbeth.]

**TABLE 18-1**

| Soil Horizon Suffix Designations | |
|---|---|
| **Symbol** | **Description** |
| b | A buried soil horizon; soils that resulted from some previous set of soil-forming processes and that now are buried beneath another solum |
| ca | Accumulation of carbonates of alkaline elements, commonly calcium; usually lime ($CaCo_3$) or dolomite ($MgCO_3$) in concentrations greater than that of the parent material; may appear in A, B, or C horizons |
| f | Frozen soil; applied to layers in the soil that are permanently frozen, as in permafrost |
| g | Strong gleying; condition usually found in stagnant, poorly aerated water; usually dull gray or blue-gray because of the removal of oxygen from iron compounds; may appear mottled or spotted |
| h | Accumulations of humus; distinct coating of sand and silt particles with illuviated humus or organic material in the B horizon; appears as a dark coating |
| ir | Accumulation of iron; distinct coating or cementing of silts and sands with illuviated iron; appears reddish |
| m | Strong cementation; applied to hardened horizons that are more than 90% continuous, such as caliche or desert hardpan; not applicable to the firmness associated with fragipans |
| p | Plowing or other disturbance; used with the A horizon to designate disturbance by cultivation or pasturage; Ap used even when plowing may have gone into what was once B horizon |
| sa | Accumulation of soluble salts; similar to ca, in that salts are present in an amount greater than that of parent materials |
| t | Illuviated clay; accumulations of silicate clays as masses, pore fillings, or coatings on sand and silt within B horizons; transported there from horizons above |
| x | Fragipan formation, firmness or brittleness found in A, B, and C horizon |

*Source:* U.S. Department of Agriculture, *Soil Taxonomy,* Agricultural Handbook No. 436, (Washington, DC: U.S. Government Printing Office, 1975).

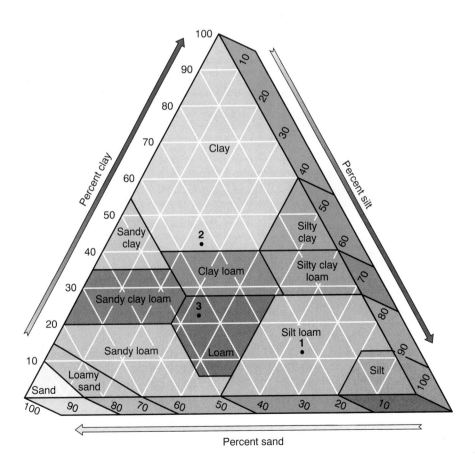

**FIGURE 18-4**

**Soil texture triangle.**

Soil texture is assessed by measuring the ratio of clay, silt, and sand. As an example, points 1, 2, and 3 designate samples taken at three different horizons in the *Miami silt loam* in Indiana. [After U.S. Department of Agriculture Soil Conservation Service, *Soil Taxonomy,* Agricultural Handbook No. 436 (Washington, DC: U.S. Government Printing Office, 1975).]

**TABLE 18-2**

| Textural Analysis of Miami Silt Loam | | | |
|---|---|---|---|
| *Sample Points* | *% Sand* | *% Silt* | *% Clay* |
| 1 = A horizon | 21.5 | 63.4 | 15.0 |
| 2 = B horizon | 31.1 | 25.0 | 43.4 |
| 3 = C horizon | 42.4 | 34.0 | 23.5 |

*Source:* J. E. Van Riper, *Man's Physical World*, p. 570. Copyright © 1971 by McGraw-Hill. Adapted by permission.

by farmers because of its water-holding characteristics and ease of cultivation. Soil texture is important in determining water-retention and water-transmission traits.

Consider a soil type in Indiana called the *Miami silt loam*. Samples from it are plotted on the soil texture triangle as points 1, 2, and 3. A sample taken near the surface in the A horizon is recorded at point 1; in the B horizon at point 2; and in the C horizon at point 3. Textural analyses of these samples are summarized in Table 18-2. Note that silt dominates the surface, clay the B horizon, and sand the C horizon.

The U.S. Department of Agriculture *Soil Survey Manual* presents guidelines for estimating soil texture by feel, a relatively accurate method when used by an experienced person. However, laboratory methods using graduated sieves and separation by mechanical analysis in water allow more-precise measurements. An interesting practice related to soil texture is described in News Report 1.

## Soil Structure

Soil *texture* describes the *size* of soil particles, but soil *structure* refers to the *arrangement* of them. Structure can partially modify the effects of soil texture. The smallest natural lump or cluster of particles is a *ped*. The shape of soil peds determines which of the structural types the soil exhibits: crumb or granular, platy, blocky, prismatic, or columnar (Figure 18-5).

Peds separate from each other along zones of weakness, creating voids (pores) that are important for moisture storage and drainage. Rounded peds have more pore space between them and greater permeability than other shapes. They are therefore better for plant growth than are blocky, prismatic, or platy peds, despite comparable fertility. Terms used to describe soil structure include fine, medium, or coarse. Adhesion among peds ranges from weak to strong.

## Soil Consistence

In soil science, the term *consistence* is used to describe the consistency of a soil or cohesion of its particles. Consistence is a product of texture (particle size) and structure (ped shape). Consistence reflects a soil's resistance to breaking and manipulation under varying moisture conditions:

- A *wet soil* is sticky between the thumb and forefinger, ranging from a little adherence to either finger, to

## News Report 1

### Geophagy: Eating Clay

Fine clays are eaten in some societies—not out of hunger or a food dare, but as a regular part of the diet. This practice is known as *geophagy*. Fine clays are the preferred texture, not gritty silt and sand. Shaped into thin tablets or cookies and baked in ovens or sun-dried, the clay is sold in the marketplace. These fine clays may contain iron, calcium, magnesium, potassium, and zinc, among other minerals. The taste is a bit tart, like citrus.

Before we turn up our noses at the practice, we should realize that most everyone reading this has consumed edible clays in the form of kaolin, the main ingredient in common over-the-counter antidiarrhea medicine. In some countries, eating clay is the only available way to reduce abdominal pain in persons afflicted with hookworms, to reduce heartburn, or to obtain a dietary supplement during pregnancy. The association with pregnancy is particularly strong, and is a related craving. In addition, geophagy may be part of religious practices in Africa and Latin America.

In the United States, the practice is focused in the southern states. Favorite clay-gathering sites become a family tradition and clays may be sent to relatives who have moved from a traditional clay source. In your travels, you might see signs warning: "Do not eat clays collected from road sides that have been sprayed for weed control." Now you know what these strange signs are referring to. In an ultimate sense, geophagy is truly an act of directly participating in soil science!

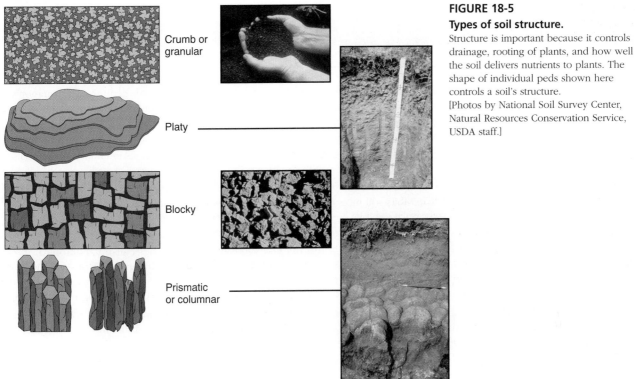

**FIGURE 18-5**
**Types of soil structure.**
Structure is important because it controls drainage, rooting of plants, and how well the soil delivers nutrients to plants. The shape of individual peds shown here controls a soil's structure.
[Photos by National Soil Survey Center, Natural Resources Conservation Service, USDA staff.]

Crumb or granular

Platy

Blocky

Prismatic or columnar

sticking to both fingers, to stretching when the fingers are moved apart. Plasticity, the quality of being moldable, is roughly measured by rolling a piece of soil between your fingers and thumb to see whether it rolls into a thin strand.

- A *moist soil* is filled to about half of field capacity (the usable water capacity of soil), and its consistence grades from loose (noncoherent), to *friable* (easily pulverized), to firm (not crushable between thumb and forefinger).

- A *dry soil* is typically brittle and rigid, with consistence ranging from loose, to soft, to hard, to extremely hard.

Soil particles are sometimes cemented together, to some degree. Soils are described as weakly cemented, strongly cemented, or *indurated* (hardened). Calcium carbonate, silica, and oxides or salts of iron and aluminum all can serve as cementing agents. The cementation of soil particles that occurs in various horizons is a function of consistence and may be continuous or discontinuous.

### Soil Porosity

Soil porosity, permeability, and moisture storage are discussed in Chapter 9. Pores in the soil horizon control the movement of water—its intake, flow, and drainage—and air ventilation. Important porosity factors are pore *size*, pore *continuity* (whether they are interconnected), pore *shape* (whether they are spherical, irregular, or tubular), pore *orientation* (whether pore spaces are vertical, horizontal, or random), and pore *location* (whether they are within or between soil peds).

Porosity is improved by the biotic actions of plant roots, animal activity such as the tunneling of gophers or worms, and human intervention through soil manipulation (plowing, adding humus or sand, or planting soil-building crops). Much of a farmer's soil preparation work before planting is done to improve soil porosity.

### Soil Moisture

Reviewing Figures 9-10 and 9-11 (soil moisture types and availability) will help you understand this section. Plants operate most efficiently when the soil is at *field capacity*, which is the maximum water availability for plant use after large pore spaces have drained. Field capacity is determined by soil type. The depth to which a plant sends its roots determines the amount of soil moisture to which the plant has access. If soil moisture is removed below field capacity, plants must exert increased energy to obtain available water. This moisture removal inefficiency continues until the plant's wilting point is reached. Beyond this point, plants are unable to extract the water they need, and they die.

Soil moisture regimes and their associated climate types shape the biotic and abiotic properties of the soil more than any other factor. The U.S. Natural Resources Conservation Service (formerly the Soil Conservation Service) recognizes five soil moisture regimes based on Thornthwaite's water-balance principles (see "The Water-Budget Concept" in Chapter 9), (Table 18-3).

**TABLE 18-3**

| Principal Soil Moisture Regimes | |
|---|---|
| **Regime** | **Description** |
| Aquic (L. *aqua*, "water") | *The groundwater table lies at or near the surface*, so the soil is almost constantly wet, as in bogs, marshes, and swamps, a reducing environment with virtually no dissolved oxygen present. Commonly, groundwater levels fluctuate seasonally; a small borehole will produce standing, stagnant water. |
| Aridic (torric) (L. *aridis*, "dry," and L. *torridus*, "hot and dry") | *Soils in this regime are dry more than half the time*, with soil temperatures at a depth of 50 cm (19.7 in.) above 5°C (41°F). In some or all areas, soils are never moist for as long as 90 consecutive days. This regime occurs mainly in arid climates, although where surface structure inhibits infiltration and recharge or where soils are thin and therefore dry, this regime occurs in simiarid regions as well. |
| Udic (L. *udus*, "humid") | *Soils have little or no moisture deficiency throughout the year*, specifically during the growing season. The water balance exhibits soil moisture surpluses that flush through the soils during one season of the year. If the water balance exhibits a moisture surplus in all months of the year, the regime is called *perudic*, with adequate soil moisture always available to plants. |
| Ustic (L. *ustus*, "burnt," implying dryness) | *This regime is intermediate between aridic and udic regimes*, it includes the semiarid and tropical wet-dry climates. Moisture is available but is limited, with a prolonged deficit period following the period of soil-moisture utilization during the early portion of the growing season. Temperature is important in determining the moisture efficiency of available water in this borderline moisture regime. |
| Xeric (L. *xeros*, "dry") | *This regime applies to those few areas that experience a Mediterranean climate*, dry and warm summer, rainy and cool winter. There are at least 45 consecutive dry days during the 4 months following the summer solstice. Winter rains effectively leach the soils. |

*Source*: U.S. Department of Agriculture, *Soil Taxonomy*, Agricultural Handbook No. 436, (Washington, DC: U.S. Government Printing Office, 1975).

## Soil Chemistry

Recall that soil pores may be filled with air, or water, or a mixture of the two. Consequently, soil chemistry involves both air and water. The atmosphere within soil pores is mostly nitrogen, oxygen, and carbon dioxide. Nitrogen concentrations are about the same as in the atmosphere, but oxygen is less and carbon dioxide is greater because of ongoing respiration processes.

Water present in soil pores is called the *soil solution*. It is the medium for chemical reactions in soil. This solution is critical to plants as their source of nutrients, and it is the foundation of *soil fertility*. Carbon dioxide combines with the water to produce carbonic acid, and various organic materials combine with the water to produce organic acids. These acids are then active participants in soil processes, as are dissolved alkalies and salts.

To understand how the soil solution behaves, let us go through a quick chemistry review. An *ion* is an atom, or group of atoms, that carries an electrical charge (examples: $Na^+$, $Cl^-$, $HCO_3^-$). An ion has either a positive charge or a negative charge. For example, when NaCl (sodium chloride) is dissolved in solution, it separates into two ions: $Na^+$, a *cation* (positively charged ion), and $Cl^-$, an *anion* (negatively charged ion). Some ions in soil carry single charges, whereas others carry double or even triple charges (e.g., sulfate, $SO_4^{2-}$; and aluminum, $Al^{3+}$).

**Retaining Ions for Plants.** Ions in soil are retained by **soil colloids**. These tiny particles of clay and organic material (humus) carry a negative electrical charge and consequently attract any positively charged ions in the soil (Figure 18-6). The positive ions, many metallic, are critical to plant growth. If it were not for the negatively charged soil colloids, the positive ions would be leached away by the soil solution and thus would be unavailable to plant roots.

Individual clay colloids are thin and platelike, with parallel surfaces that are negatively charged (Figure 18-6). They are more chemically active than silt and sand particles but less active than organic colloids. Metallic cations attach to the surfaces of the colloids by *adsorption* (not *ab*sorption, which means "to enter"). Colloids can exchange cations between their surfaces and the soil solution, an ability called **cation-exchange capacity (CEC)**, which is the measure of soil fertility. A high CEC means that the soil colloids can store or exchange more cations from the soil solution, an indication of good soil fertility (unless there is a complicating factor, such as the soil's being too acid).

Therefore, **soil fertility** is the ability of soil to sustain plants. Soil is fertile when it contains organic substances and clay minerals that absorb water and adsorb certain elements needed by plants. Billions of dollars are expended to create fertile soil conditions, yet the future of Earth's most fertile soils is threatened because soil erosion is increasing worldwide.

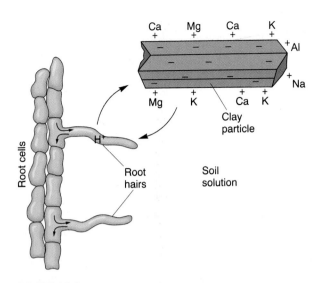

**FIGURE 18-6**
**Soil colloids and cation-exchange capacity (CEC).**
This typical soil colloid retains mineral ions by adsorption to its surface (opposite charges attract). This holds the ions until they are absorbed by root hairs.

### *Soil Acidity and Alkalinity*

A soil solution may contain significant hydrogen ions ($H^+$), the cations that stimulate acid formation. The result is a soil rich in hydrogen ions, or an *acid soil*. On the other hand, a soil high in base cations (calcium, magnesium, potassium, sodium) is a basic or *alkaline soil*. Such acidity and alkalinity is expressed on the pH scale (Figure 18-7).

Pure water is nearly neutral, with a pH of 7.0. Readings below 7.0 represent increasing acidity. Readings above 7.0 represent increasing alkalinity. Acidity usually is regarded as strong at 5.0 or lower on the pH scale, whereas 10.0 or above is considered strongly alkaline.

Today, the major contributor to soil acidity is acid precipitation (rain, snow, fog, or dry deposition). Acid rain actually has been measured below pH 2.0—an incredibly low value for natural precipitation, as acid as lemon juice. Because most crops are sensitive to specific pH levels, acid soils below pH 6.0 require treatment to raise the pH. This soil treatment is accomplished by the addition of bases in the form of minerals that are rich in base cations, usually lime (calcium carbonate, $CaCO_3$). Increased acidity in the soil solution accelerates the chemical weathering of mineral nutrients and increases their depletion rates.

---

## Soil Formation Factors and Management

Soil-forming factors are both *passive* (parent material, topography and relief, and time) and *dynamic* (climate, biology, and human activities). These factors work together as a system to form soils. The roles of these factors are considered here and in the soil order discussions that follow.

### *Natural Factors*

Physical and chemical weathering of rocks in the upper lithosphere provides the raw mineral ingredients for soil formation. These are called *parent materials*, and their composition, texture, and chemical nature help determine the type of soil that forms. Clay minerals are the principal weathered by-products in soil. Also important in soil is the organic content present and all that is living in soil, from bacteria, algae, fungi, worms, and insects.

Climate types correlate closely with soil types worldwide. The moisture, evaporation, and temperature regimes of climates determine the chemical reactions, organic activity, and eluviation rates of soils. Not only is the present climate important, but many soils exhibit the imprint of past climates, sometimes over thousands of years. Most notable is the effect of glaciations. Among other contributions, glaciation produced the *loess* soil materials that have been windblown thousands of kilometers to their present locations (Chapter 15).

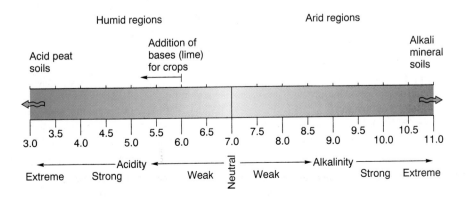

**FIGURE 18-7**
**pH scale.**
The pH scale measures acidity (lower pH) and alkalinity (higher pH). (The complete pH scale ranges between 0 and 14.)

The organic content of soil is determined by vegetation and by animal and bacterial activity. The chemical makeup of the vegetation contributes to the acidity or alkalinity of the soil solution. For example, broadleaf trees tend to increase alkalinity, whereas needleleaf trees tend to produce higher acidity. Thus, when civilization moves into new areas and alters the natural vegetation by logging or plowing, the affected soils are likewise altered, often permanently.

Topography also affects soil formation. Slopes that are too steep cannot have full soil development because gravity and erosional processes remove materials. Lands that are nearly level inhibit soil drainage and can become waterlogged. The compass orientation of slopes is important because it controls exposure to sunlight. In the Northern Hemisphere, a south-facing slope is warmer overall through the year because it receives direct sunlight. Water-balance relations are affected because north-facing slopes are colder, causing slower snowmelt and slower evaporation, providing more moisture for plants than is available on south-facing slopes, which tend to dry faster.

### The Human Factor

Human intervention has a major impact on soils. Millennia ago, farmers in most cultures learned to plant slopes "on the contour"—to make rows or mounds around a slope at the same elevation, not vertically up and down the slope. Planting on the contour prevents water from flowing straight down the slope and thus reduces soil erosion. It was common to plant and harvest a floodplain but to live on higher ground nearby. Floods were celebrated as blessings that brought water, nutrients, and more soil to the land. Society is drifting away from these commonsense strategies.

All of the identified factors in soil development (climate, biological activity, parent material, landforms and topography, and human activity) require *time* to operate. A few centimeters thickness of prime farmland soil may require *500 years* to mature. Yet, this same thickness is being lost annually through soil erosion when the soil-holding vegetation is removed and the land is plowed regardless of topography. Over the same period, exposed soils may be completely leached of needed cations, thereby losing their fertility.

Some 35% of farmlands are losing soil faster than it can form—a loss exceeding 23 billion metric tons (25 billion tons) per year. Soil depletion and loss are at record levels from Iowa to China, Peru to Ethiopia, the Middle East to the Americas. The impact on society is potentially disastrous as population and food demands increase (see News Report 2).

Soil erosion can be compensated for in the short run by using more fertilizer, increasing irrigation, and by planting higher-yielding strains. But the potential yield from prime agricultural land will drop by as much as 20% over the next 20 years if only moderate erosion continues. One 1995 study tabulated the market value of lost nutrients and other variables in the most comprehensive soil-erosion study to date. The sum of direct damage (to agricultural land) and indirect damage (to streams, society's infrastructure, and human health) was estimated at over $25 billion a year in the United States and hundreds of billions of dollars worldwide. (Of course, this is a controversial assessment in the agricultural industry.) The cost to bring erosion under control in the United States is estimated at approximately $8.5 billion, or about 30 cents on every dollar of damage and loss. Figure 18-8 maps regions of soil loss.

Contour plowing reduces *sheet erosion* (which removes topsoil) and simultaneously improves water retention in the soil. However, the demands of today's

## News Report 2

### Soil Is Slipping Through Our Fingers

- The U.S. General Accounting Office estimated that from 3 to 5 million acres of prime farmland are lost *each year* in the United States through mismanagement or conversion to nonagricultural uses. About half of all cropland in the United States and Canada is experiencing excessive rates of soil erosion.
- A 1984 report to the Canadian Environmental Advisory Council estimated that the organic content of cultivated prairie soils has declined by as much as 37–48% compared with noncultivated native soils. In Ontario and Québec, losses of organic content increased to as much as 50%, and losses are even higher in the Atlantic Provinces, which were naturally low in organic content before cultivation.
- A 1995 study completed at Cornell University concluded that soil erosion is a major environmental threat to the sustainability and productive capacity of agriculture. Estimated annual dollar losses to soil erosion exceed $25 billion.
- Since 1950, as much as 30% of the world's farmable land has been lost to soil erosion, and the rate continues at 10 million hectares (about 25 million acres) per year.
- The world's human population is growing at the rate of 7.5 million people a month (net increase), increasing the demand for food and agricultural productivity.

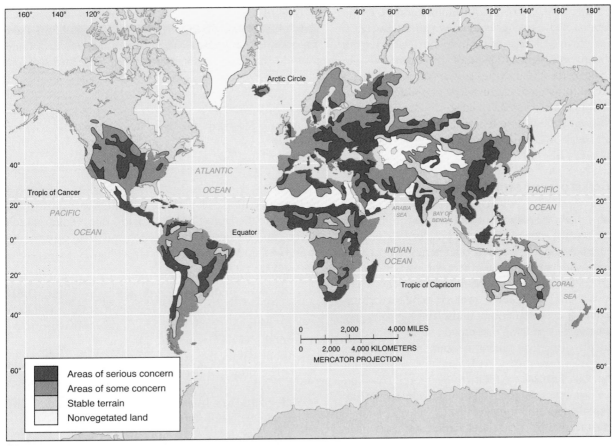

(a)

(b)

**FIGURE 18-8**
**Soil degradation.**
(a) Approximately 1.2 billion hectares (3.0 billion acres) of Earth's soils suffer degradation through erosion caused by human misuse and abuse. (b) Typical loss through soil sheet erosion on a northeastern Wisconsin farm. One millimeter of soil lost from an acre weighs about 5 tons. [(a) *A Global Assessment of Soil Degradation*, adapted from United Nations Environment Programme, International Soil Reference and Information Centre, "Map of Status of Human-Induced Soil Degradation," Sheet 2, Nairobi, Kenya, 1990. (b) Photo by D. P. Burnside/Photo Researchers, Inc.]

large farm equipment have forced farmers to alter their practice of contouring because the equipment cannot make the tight turns that such proper plowing requires. Other effective soil-holding practices also are being eliminated to accommodate the larger equipment, such as terracing (similar to contour plowing), planting tree rows as wind breaks, leaving grass strips at the edges of fields, and planting shelter belts.

Our spatial impact on the precious soil resource must be recognized, and protective measures must be taken. There is a need for cooperative international action. Critical is a long-term assessment of cost (continued behavior) and benefits (mitigated damage and preservation of the market value of prime soils).

## Soil Classification

Classification of soils is complicated by the continuing interaction of properties and processes just discussed. This creates thousands of distinct soils, with well over 15,000 soil series identified in the United States and Canada alone. Not surprisingly, different classification systems are in use worldwide. The United States, Canada, the United Kingdom, Germany, Australia, Russia, and the United Nations Food and Agricultural Organization (FAO) each has its own soil classification system. Each system reflects the environment of its country; for example, the National Soil Survey Committee of Canada developed a system suited to its great expanses of boreal forest, tundra, and cool climatic regimes (detailed in Appendix C). A new revision of the U.S. system will include some of the Canadian material when the revision is published in 1998.

## A Brief History of Soil Classification

Nineteenth-century Russians contributed greatly to modern soil science because they were first to consider soils as independent natural bodies with their own genesis. They saw that genetic soil types could be organized into an orderly global pattern. By contrast, American and European scientists still considered soil to be a geologic product, or simply a mixture of chemical compounds. The Russians saw soil as a system, determining that broad interrelations exist among the physical environment, vegetation, and soils. In 1883, Russian soil scientist V. V. Dukuchaev published a monograph that organized soils into rough groups based on observable properties, most of which resulted from climatic and biological soil-forming processes.

The first formal U.S. soil classification system is credited to Curtis F. Marbut, who derived his approach from Dukuchaev's system. Published in the 1935 *Atlas of American Agriculture* and in the *Yearbook of Agriculture, 1938*, Marbut's classification of *great soil groups* recognized the importance of soils as products of dynamic natural processes. But as the data base of soil information grew, the system became inadequate.

In 1951, the U.S. Soil Conservation Service began researching a new soil classification system. That process went through seven exhaustive reviews, culminating in what is called the Seventh Approximation in the 1960s. The current version was published in 1975: *Soil Taxonomy–A Basic System of Soil Classification for Making and Interpreting Soil Surveys*; we refer to it simply as **Soil Taxonomy**. Much of the information in this chapter is derived from that keystone publication.

***U.S. Soil Taxonomy.*** The U.S. Soil Taxonomy system is based on soil properties and morphology (form and structure) actually seen in the field. Thus, it is open to addition, change, and modification as the sampling data base grows. The system recognizes the importance of interactions between humans and soils and the changes that humans have introduced, both purposely and inadvertently.

The classification system divides soils into six categories, creating a hierarchical sorting system (Table 18-4). The smallest, most-detailed category is the soil series, which ideally includes only one polypedon but may

**TABLE 18-4**

| U.S. Soil Taxonomy | |
| --- | --- |
| *Soil Category* | *Number of Soils Included* |
| Orders | 11 |
| Suborders | 47 |
| Great groups | 230 |
| Subgroups | 1200 |
| Families | 6000 |
| Series | 15,000 |

include adjoining polypedons. In sequence from smallest category to the largest, the Soil Taxonomy recognizes *soil series, soil families, soil subgroups, soil great groups, soil suborders,* and *soil orders.*

***Pedogenic Regimes.*** Before the Soil Taxonomy system was devised, **pedogenic regimes** were used. These regimes keyed specific soil-forming processes to climatic regions. Although each pedogenic process may be active in several soil orders and in different climates, we discuss them within the soil order where they commonly occur. The principal pedogenic regimes are:

- **laterization:** a leaching process in humid and warm climates, discussed with Oxisols

- **salinization:** a process that concentrates salts in soils in climates with excessive POTET (potential evapotranspiration) rates, discussed with Aridisols

- **calcification:** a process that produces an illuviated accumulation of calcium carbonates, in continental climates discussed with Mollisols and Aridisols

- **podzolization:** a process of soil acidification associated with forest soils in cool climates, discussed with Spodosols

- **gleization:** a process that includes an accumulation of humus and a thick, water-saturated gray layer of clay beneath, usually in cold, wet climates and poor drainage conditions

Such climate-based regimes are convenient for relating climate and soil processes. However, *the Soil Taxonomy system recognizes the great uncertainty and inconsistency in basing soil classification on such climatic variables.*

## Diagnostic Soil Horizons

To identify specific soil series within the Soil Taxonomy, the U.S. Natural Resources Conservation Service describes diagnostic horizons in a pedon. A *diagnostic horizon* reflects a distinctive physical property (color, texture, structure, consistence, porosity, moisture) or a dominant soil process (discussed with the soil types).

In the solum (A, E, and B horizons), two diagnostic horizons may be identified, the epipedon and the subsurface:

- The **epipedon** (literally, "over the soil") is the diagnostic horizon at the surface. It may extend downward through the A horizon, even including all or part of an illuviated B horizon. It is visibly darkened by organic matter and sometimes is leached of minerals. Six recognized epipedons are shown in Table 18-5. (Excluded from the epipedon are alluvial deposits, eolian deposits, and cultivated areas, because soil-forming processes have lacked the time to erase these short-lived characteristics.)

- The **subsurface diagnostic horizon** originates below the surface at varying depths. It may include part of the A or B horizon or both. Many subsurface diagnostic horizons have been identified, as shown in Table 18-5.

The presence or absence of either of these diagnostic horizons usually distinguishes a soil for classification.

Table 18-5 summarizes both surface and subsurface diagnostic horizons and their general characteristics, as described in the Soil Taxonomy.

## The 11 Soil Orders of the Soil Taxonomy

At the heart of the Soil Taxonomy are 11 general soil orders, listed in Table 18-6. Their worldwide distribution is shown in Figure 18-9a, and their distribution in the United States in Figure 18-9b. Please consult this table and these maps as you read the following descriptions. Because the Soil Taxonomy evaluates each soil order on its own characteristics, there is no priority to the classification. However, you will find a progression in this discussion, for the 11 orders are arranged loosely by latitude, beginning along the equator as in Chapter 10, "Global Climate Systems," and in Chapter 20, "Terrestrial Biomes."

**TABLE 18-5**

| **Major Features of Diagnostic Horizons** | | |
|---|---|---|
| *Name* | *Derivation* | *Characteristics* |
| **Surface Horizons (Epipedons)** | | |
| Mollic (A) | *mollis*, "soft" | Thick, dark-colored, humus-rich; base cation dominant; strong structure |
| Anthropic (A) | *anthropos*, "having to do with humans" | Mollic-like, modified by long and continued human use and addition of water to soil |
| Umbric (A) | *umbra*, "shade," hence "dark" | Identical to mollic, except the low base cation content; dark coloration not related to humus |
| Histic (O) | *histos*, "tissue" | Thin layer of peat or muck in wet places; saturated at least 30 consecutive days a year |
| Plaggen (A) | *plaggen*, "sod" | Human-made ≥50 cm thick, produced by long, continuing manuring (W. Europe) |
| Ochric (A) | *ochros*, "pale" | Light color, low humus content, hard and massive when dry, too marginal to fit other epipedons |
| **Subsurface Horizons** | | |
| Argillic | argilla, "white clay" | Illuviated clay accumulation |
| Agric | *ager*, "field" | Illuvial layer formed below cultivation |
| Natric | *natrium*, "sodium" | Argillic that is high in sodium |
| Spodic | *spodos*, "wood ashes" | Accumulation of organic matter, Fe, and Al |
| Placic | *plax*, "flatstone" | Thin, dark-reddish, iron-cemented pan |
| Cambic | *cambiare*, "to exchange" | Altered or changed by physical movement or chemistry |
| Oxic | *oxide*, "oxide" | Highly weathered mix of Fe, Al oxides; >30 cm thick |
| Duripan | *durus*, "hard" | Silica-cemented hardpan; not softened by water |
| Fragipan | *fragilis*, "brittle" | Weakly cemented brittle pan; loam mixture |
| Albic | *albus*, "white" | Light, pale layer lacking clay, Fe, and Al |
| Calcic | calcium | Accumulation of $CaCO_3$ or $CaCO_3 \cdot MgCO_3$ |
| Gypsic | gypsum | Accumulation of gypsum (hydrous calcium sulfate) |
| Salic | salts | Accumulation of >2% soluble salts |
| Plinthite | *plinthos*, "brick" | Iron-rich, humus-poor; becomes ironstone under repeated wet-dry cycles |

*Source:* U.S. Department of Agriculture, *Soil Taxonomy*, Agricultural Handbook No. 436, (Washington, DC: U.S. Government Printing Office, 1975).

**TABLE 18-6**

| Soil Taxonomy Soil Orders . | | | | | | |
|---|---|---|---|---|---|---|
| Order | Derivation of Term | Marbut Equivalent (Canadian) | General Location and Climate | World Land Area (%) | U.S. Land Area (%) | Description |
| Oxisols | Fr. *oxide*, "oxide" Gr. *oxide*, "acid or sharp" | Latosols lateritic soils | Tropical soils, hot humid areas | 9.2 | 0.02 | Maximum weathering and eluviation, oxic horizon, continuous plinthite layer |
| Aridisols | L. *aridos*, "dry" | Reddish desert, gray desert, sierozems | Desert soils, hot dry areas | 19.2 | 11.5 | Limited alteration of parent material, low climate activity, ochric epipedon, light color, subsurface illuviation of carbonates |
| Mollisols | L. *mollis*, "soft" | Chestnut, chernozem (Chernozemic) | Grassland soils; subhumid, semiarid lands | 9.0 | 24.6 | Noticeably dark with organic material, base saturation high, mollic epipedon; friable surface with well-structured horizons |
| Alfisols | Invented syllable | Gray-brown pod-zolic, degraded chernozem (Luvisol) | Moderately weath-ered forest soils, humid temperate forests | 14.7 | 13.4 | B horizon high in clays, moderate to high degree of base saturation, argillic or natric horizons, no pronounced color change with depth |
| Ultisols | L. *ultimus*, "last" | Red-yellow podzolic, reddish yellow lateritic | Highly weathered forest soils, subtropical forests | 8.5 | 12.9 | Similar to Alfisols, B horizon high in clays, generally low amount of base saturation, strong weathering in argillic horizons, redder than Alfisols |
| Spodosols | Gr. *spodos* or L. *spodus*, "wood ash" | Podzols, brown podzolic (Pod-zol) | Northern conifer forest soils, cool humid forests | 5.4 | 5.1 | Illuvial B horizon of Fe/Al clays, humus; without structure, partially cemented; spodic horizon; highly leached, strongly acid; coarse texture of low bases |
| Entisols | Invented syllable from *recent* | Azonal soils, tundra | Recent soils, profile undeveloped, all climates | 12.5 | 7.9 | Limited development; inherited properties from parent material; ochric epipedon; few specific properties; young soils lacking horizons |
| Inceptisols | L. *inceptum*, "beginning" | Ando, subarctic brown forest lithosols, some humic gleys (Brunisol, Cryosol with permafrost, Gleysol wet) | Weakly developed soils, humid regions | 15.8 | 18.2 | Intermediate development; embryonic soils, cambic horizon, but few diagnostic features; further weathering possible |
| Vertisols | L. *verto*, "to turn" | Grumusols (1949) tropical black clays | Expandable clay soils; subtropics, tropics; sufficient dry period | 2.1 | 1.0 | Forms large cracks on drying, self-mixing action, contains >30% in swelling clays, ochric epipedon |
| Histosols | Gr. *histos*, "tissue" | Peat, muck, bog (Organic) | Organic soils, wet places | 0.8 | 0.5 | Peat or bog, >20% organic matter, much with clay >40 cm thick, surface organic layers, no diagnostic horizons |
| Andisols | L. *ando*, "volcanic ash" | — | Areas affected by frequent volcanic activity (formerly within Inceptisols and Entisols) | >1.0 | >1.0 | Volcanic parent materials, particularly ash and volcanic glass; weathering and mineral transformation important; high CEC and organic content, generally fertile |

***Oxisols.*** The intense moisture, temperature, and uniform daylength of equatorial latitudes profoundly affect soils. These generally old landscapes, exposed to tropical conditions for millennia or hundreds of millennia, are deeply developed. They have greatly altered minerals (except in certain newer volcanic soils in Indonesia—the Andisols). Oxisols are among the most mature soils on Earth. Distinct horizons usually are lacking where these soils are well drained (Figure 18-10). Related vegetation is the luxuriant and diverse tropical and equatorial rain forest. Oxisols include five suborders.

**Oxisols** are so called because they have a distinctive horizon of iron and aluminum oxides. The concentration of oxides results from heavy precipitation, which leaches soluble minerals and soil constituents from the A horizon. Typical Oxisols are reddish (from the iron oxide) and yellowish (from the aluminum oxides); with a weathered claylike texture, sometimes in a granular soil structure that is easily broken apart. The high degree of eluviation removes basic cations and colloidal material to lower illuviated horizons. Thus, Oxisols are low in CEC (cation-exchange capacity) and fertility, except in regions augmented by alluvial or volcanic materials.

Consequently, Oxisols have a subsurface diagnostic horizon that is highly weathered, contains iron and aluminum oxides, is at least 30 cm (12 in.) thick, and lies within 2 m (6.5 ft) of the surface (Figure 18-10). Figure 18-11 illustrates *laterization*, the leaching process that operates in well-drained soils in warm, humid tropical and subtropical climates. If Oxisols are subjected to repeated wetting and drying, an *ironstone hardpan* develops (this is a hardened soil layer in the lower A or in the B horizon). It is called a *plinthite* (from the Greek *plinthos*, meaning "brick"). This form of soil, also called a *laterite*, can be quarried in blocks and used as a building material (Figure 18-12).

Simple agricultural activities can be conducted with care in these soils. Early cultivation practices, called *slash-and-burn shifting cultivation*, were adapted to these soil conditions and formed a unique style of crop rotation. The scenario went like this: People in the tropics cut down (slashed) and burned the rain forest in small tracts and then cultivated the land with stick and hoe. Mineral nutrients in the organic material and short-lived fertilizer input from the fire ash would quickly be exhausted. After several years, the soil lost fertility through leaching by intense rainfall, so the people shifted cultivation to another tract and repeated the process. After many years of moving from tract to tract, the group returned to the first patch to begin the cycle again. This practice protected the limited fertility of the soils somewhat, allowing periods of recovery.

However, this orderly land rotation was halted by the invasion of foreign plantation interests, development by local governments, vastly increased population pressures, and conversion of vast forest tracts to pasturage. Permanent tracts of cleared land put tremendous pressure on the remaining forest and brought disastrous consequences. When Oxisols are disturbed, soil loss can exceed a thousand tons per square kilometer per year, not to mention the greatly increased extinction rate for plant and animal species that accompanies such destruction. The regions dominated by the Oxisols and rain forests are rightfully the focus of much worldwide environmental attention.

***Aridisols.*** The largest single soil order occurs in the world's dry regions. **Aridisols** (desert soils) occupy approximately 19% of Earth's land surface and some 12% of U.S. land (Figure 18-9). A pale, light soil color near the surface is diagnostic (Figure 18-13a).

Not surprisingly, the water balance in Aridisol regions has marked periods of soil-moisture deficit and generally inadequate soil moisture for plant growth. High potential evapotranspiration and low precipitation produce very shallow soil horizons. Usually there is no period greater than 3 months when the soils have adequate moisture. Lacking water and therefore lacking vegetation, Aridisols also lack organic matter of any consequence. Low precipitation means infrequent leaching, yet Aridisols are leached easily when exposed to excessive water, for they lack a significant colloidal structure.

*Salinization* is common in Aridisols, resulting from excessive potential evapotranspiration rates in deserts and semiarid regions. Salts dissolved in soil water migrate to surface horizons and are deposited there as the water evaporates. These deposits appear as subsurface salic (salty) horizons, which will damage or kill plants when the horizons occur near the root zone. The salt accumulation associated with a desert playa is an example of extreme salinization (see Chapter 15).

Obviously, salinization complicates farming in Aridisols. The introduction of irrigation water may either waterlog poorly drained soils or lead to salinization. Nonetheless, vegetation does grow where soils are well drained and low in salt content. If large capital investments are made in water, drainage, and fertilizers, Aridisols possess much agricultural potential (Figure 18-14). In the Nile and Indus river valleys, for example, Aridisols are intensively farmed with a careful balance of these environmental factors, although thousands of acres of once-productive land, not so carefully treated, now sit idle and salt-encrusted. In California, the Kesterson Wildlife Refuge was reduced to a toxic waste dump in the early 1980s by contaminated agricultural drainage. Focus Study 18-1 elaborates on the Kesterson tragedy. News Report 3 describes efforts to deal with salinization.

***Mollisols.*** **Mollisols** (grassland soils) are some of Earth's most significant agricultural soils. There are seven recog-

**FIGURE 18-9**
**Soil Taxonomy.**
(a) Worldwide distribution of the 11 soil orders. Another category, soils developed from volcanic ash, is not completely represented on this map. (b) Soil Taxonomy orders in the United States. [Adapted from U.S. Department of Agriculture Natural Resources Conservation Service.]

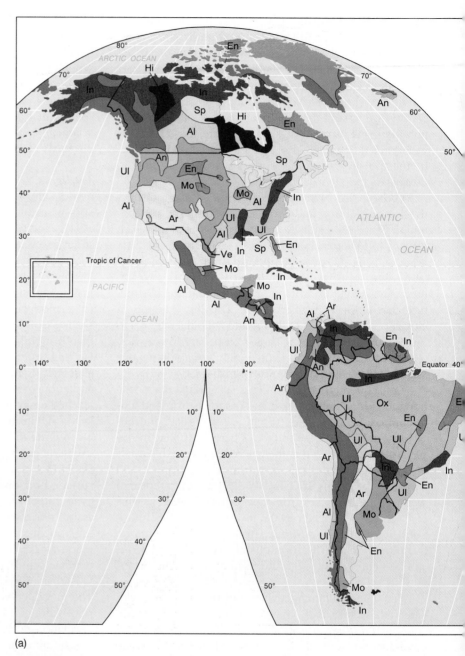

(a)

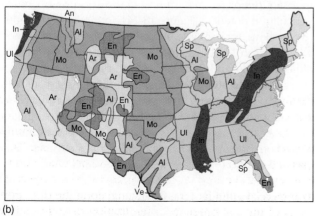

(b)

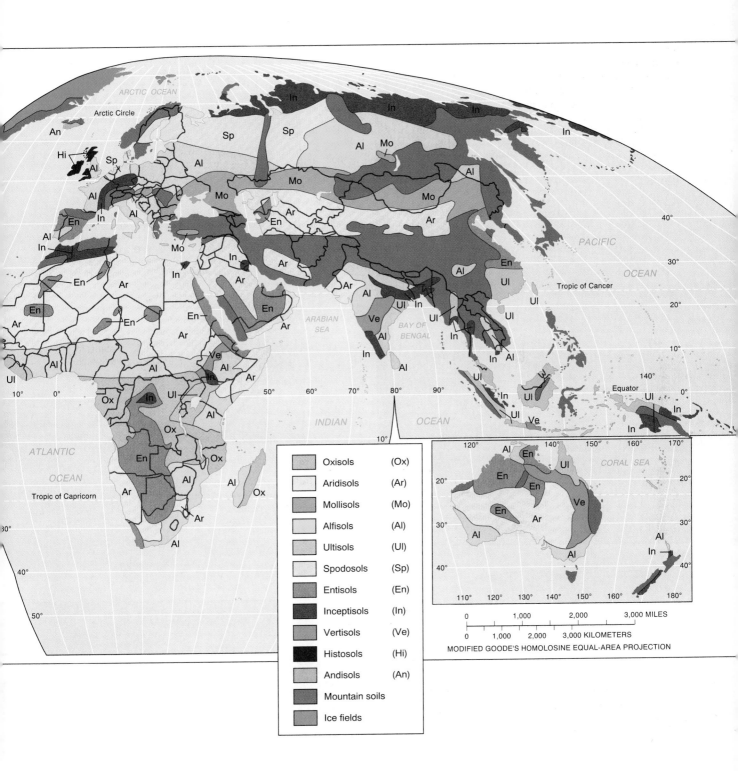

| | Oxisols | (Ox) |
| | Aridisols | (Ar) |
| | Mollisols | (Mo) |
| | Alfisols | (Al) |
| | Ultisols | (Ul) |
| | Spodosols | (Sp) |
| | Entisols | (En) |
| | Inceptisols | (In) |
| | Vertisols | (Ve) |
| | Histosols | (Hi) |
| | Andisols | (An) |
| | Mountain soils | |
| | Ice fields | |

MODIFIED GOODE'S HOMOLOSINE EQUAL-AREA PROJECTION

nized suborders, which vary in fertility. The dominant diagnostic horizon is the *mollic epipedon*, a dark, organic surface layer some 25 cm (10 in.) thick (Figure 18-15). As the Latin name implies (words using the same root, *mollis*, are "mollify," "emollient"), Mollisols are soft, even when dry. They have granular or crumbly peds, loosely arranged when dry. These humus-rich soils are high in basic cations (calcium, magnesium, and potassium) and have a high cation-exchange capacity and therefore, high

fertility. In soil moisture, these soils are intermediate between humid and arid.

Soils of the world's steppes and prairies belong to this soil group: the North American Great Plains, the Pampas of Argentina, and the region from Manchuria in China westward to Europe. Agriculture ranges from large-scale commercial grain farming to grazing along the drier portions of the soil order. With fertilization or soil-building practices, high crop yields are common.

(a)

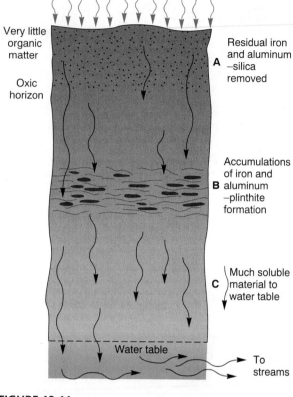

**FIGURE 18-11**
**Laterization.**

The laterization process is characteristic of moist tropical and subtropical climate regimes.

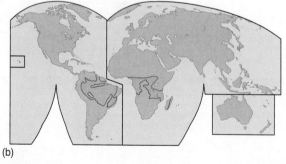

(b)

**FIGURE 18-10**
**Oxisol.**

(a) Deeply weathered Oxisol profile in central Puerto Rico.
(b) General map of worldwide distribution. [Photo from American Society of Agronomy.]

**FIGURE 18-12**
**Oxisols are used for building material.**

Here plinthite is being quarried in India. [Photo by Henry D. Foth.]

(a)

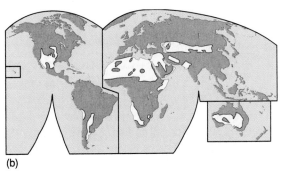

(b)

**FIGURE 18-13**
**Aridisol.**
(a) Soil profile from central Arizona. (b) General map of worldwide distribution. [Photo from American Society of Agronomy.]

**FIGURE 18-14**
**Agriculture in arid lands.**
The Coachella Valley of southeastern California is a desert in agricultural bloom. Providing drainage for excessive water application in such areas often is necessary to prevent salinization of the rooting zone. [Photo by author.]

## Focus Study 18-1

### Selenium Concentration in Western Soils

Irrigated agriculture has increased greatly since 1800, when only 8 million hectares (about 20 million acres) were irrigated worldwide. Today, approximately 200 million hectares (about 500 million acres) are irrigated, and this figure is on the increase. Representing about 15% of Earth's agricultural land, irrigated land accounts for nearly 35% of the harvest.

Two related problems common in irrigated lands are salinization and waterlogging, especially in arid lands that are poorly drained. In many areas, production has decreased and even ended because of salt buildup in the soils. Examples include areas along the Tigris and Euphrates Rivers, the Indus River valley, sections of South America and Africa, and the western United States.

### Irrigation in the West

About 95% of the irrigated acreage in the United States lies west of the 98th meridian. This region is increasingly troubled with salinization and waterlogging problems. But at least nine sites in the West, particularly California's western San Joaquin Valley, are experiencing related contamination of a more serious nature—increasing selenium concentrations. As parent materials weathered, selenium-rich alluvium washed into the semiarid valley, forming the soils that needed only irrigation water to become productive.

After 1960, large-scale irrigation efforts intensified, resulting in subtle increases in selenium concentrations in soil and water. In trace amounts, selenium is a dietary requirement for animals and humans, but in higher amounts it is toxic to both. Toxic effects of selenium were reported during the 1980s in some domestic animals grazing on grasses grown in selenium-rich soils in the Great Plains.

Drainage of agricultural wastewater poses a particular problem in semiarid and arid lands, where river discharge is inadequate to dilute and remove field runoff. One solution to prevent salt accumulations and waterlogging is to place field drains beneath the soil at depths of 3–4 m (10–13 ft). Such drains collect gravitational water from fields that have been purposely overwatered to keep salts away from the effective rooting depth of the crops. But agricultural drain water must go somewhere, and for the San Joaquin Valley of central California this problem triggered a 15-year controversy.

### Death of Kesterson

Central California's potential drain outlets are to the ocean, or San Francisco Bay, or the central valley. But all these suggested destinations failed to pass environmental impact assessments under Environmental Policy Act requirements. Nonetheless, by the late 1970s, about 80 miles of a drain were finished, even though no formal plan or adequate funding had been completed—a drain was built with no outlet! In the absence of any plan, large-scale irrigation continued, supplying the field drains with salty, selenium-laden runoff that made its way to the Kesterson National Wildlife Refuge in the northern portion of the San Joaquin Valley east of San Francisco. The unfinished drain abruptly stopped at the boundary to the refuge.

The selenium-tainted drainage only took 3 years to destroy the wildlife refuge, which was officially declared a contaminated toxic waste site (Figure 1). Aquatic life forms (e.g., marsh plants, plankton, and insects) had taken in the selenium, which then made its way up the food chain and into the diets of higher life forms in the refuge. According to U.S. Fish and Wildlife Service scientists, the toxicity moved through the food chain and genetically damaged and killed wildlife. For example, birth defects and deaths were widely reported in all varieties of birds that nested at Kesterson; approximately 90% of the exposed birds perished or were injured. Because this wildlife refuge was a major migration flyway and stopover point for birds from throughout the Western Hemisphere, this destruction of the refuge also violated several multinational wildlife protection treaties.

Such damage to wildlife presents a real warning to human populations—remember where we are in the food chain. The field drains were sealed and removed in 1986, following a court order that forced the federal government to uphold existing laws. Irrigation water then immediately began backing up in the corporate farmlands, producing both waterlogging and selenium contamination.

Since 1985, more than 0.6 million hectares (1.5 million acres) of irrigated Aridisols and Alfisols have gone out of production in California, marking the end of several decades of irrigated farming in climatically marginal lands. Severe cutbacks in irrigated acreage no doubt will continue, underscoring the need to preserve prime farmlands in wetter regions and to understand essential soil processes.

Frustrated agricultural interests have asked the federal government to finish the drain, either to San Francisco Bay or to the ocean. However, neither option appears capable of passing an environmental impact analysis. Another strategy is to allow irrigated lands to start pumping again into the former wildlife refuge because it is now a declared toxic dump site anyway! Government agencies are still wrestling with these options in the 1990s. There are nine such threatened sites in the West; Kesterson was simply the first to fail.

**FIGURE 1**

Salt-encrusted soil and plants at the contaminated Kesterson National Wildlife Refuge. [Photo by Gary R. Zahm.]

(a)

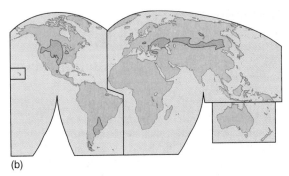

(b)

**FIGURE 18-15**
**Mollisol.**
(a) Soil profile from central Iowa. (b) General map of
worldwide distribution. [Photo from American Society
of Agronomy.]

The "fertile triangle" of Ukraine, Russia, and western portions of the Russian Federation is of this soil type. It stretches from the Caspian Sea westward in a widening triangle toward Central Europe (Figure 18-16). The soils in the fertile triangle are known as the *chernozem* in other classification systems. The remainder of the Russian landscape presents varied soils of lower fertility and productivity.

In North America, the Great Plains straddle the 98th meridian, which is coincident with the 51 cm (20 in.) isohyet of annual precipitation—wetter to the east and drier to the west. The Mollisols here mark the historic division between the short- and tallgrass prairies (Figure 18-17).

*Calcification* is a soil process characteristic of some Mollisols and adjoining marginal areas of Aridisols. Calcification is the accumulation of calcium carbonate or magnesium carbonate in the B and C horizons. Calcification forms a calcic subsurface diagnostic horizon, which is thickest along the boundary between dry and humid climates (Figure 18-18).

When cemented or hardened, these deposits are called *caliche*, or *kunkur*, they occur in widespread soil formations in central and western Australia, the Kalahari region of interior Southern Africa, and the High Plains of the west-central United States, among other places. (Soils dominated by the processes of calcification and salinization were formerly known as pedocals.)

***Alfisols.*** **Alfisols** (moderately weathered forest soils) are spatially the most widespread of the soil orders, extending in five suborders from near the equator to high latitudes. Representative Alfisol areas include Boromo and Burkina Faso (interior western Africa); Fort Nelson, British Columbia; the states near the Great Lakes, and the valleys of central California. Most Alfisols have a grayish brown to reddish *ochric epipedon* and are considered moist versions of the Mollisol soil group. Moderate eluviation is present, as well as a subsurface horizon of illuviated clays because of a pattern of increased precipitation (Figure 18-19).

Alfisols have moderate to high reserves of basic cations and are fertile. However, productivity depends on moisture and temperature. Alfisols usually are supplemented by a moderate application of lime and fertilizer in areas of active agriculture. Some of the best farmland in the United States stretches from Illinois, Wisconsin, and Minnesota through Indiana, Michigan, and Ohio to Pennsylvania and New York. This land produces grains, hay, and dairy products. The soil here is an Alfisol subgroup called Udalfs, which are characteristic of humid continental, hot summer climates.

The Xeralfs, another Alfisol subgroup, are associated with the moist winter, dry summer pattern of the Mediterranean climate. These naturally productive soils are farmed intensively for subtropical fruits, nuts, and special crops that can grow in only a few locales worldwide (e.g., California grapes, olives, citrus, artichokes, almonds, figs).

## News Report 3

### Drainage Tiles, But Where to Go?

In the California valleys of Imperial, Coachella, and San Joaquin, and in the Wellton-Mohawk district of extreme southern Arizona, the fields have been excavated, drainage tiles (perforated ceramic pipe) laid in place, and the fields replaced. The field drains carry away excess irrigation water, which has been purposely applied to leach salts away from plant roots. The problem has been to find a place to dump the salt- and chemical-laden drainage water.

In the Imperial and Coachella Valleys, the drainage water was sent to the Salton Sea (Figure 1)—a large inland sea formed by an irrigation accident in 1906 (a broken canal levee allowed the Colorado River to flow into the desert of southern California for 18 months). For

**FIGURE 1**

Soil drainage canal collects contaminated water from field drains and directs it into the Salton Sea. [Photo by author.]

the Wellton-Mohawk area, the contaminated drainage water was pumped back into the Colorado River just above the Mexican border. By the late 1950s,

salty Colorado River water was killing every plant it touched in Mexico; even tolerant cotton was wiped out by the 2500 ppm salinity in the river. The Mexican government formally protested, and a commission was established to look into the matter.

Fifteen years later, in 1974, the "Colorado River Basin Salinity Control Act" was passed and plans began for a desalinization plant near Yuma, Arizona. The plant went into service in the late 1980s, pumping desalted water back to the river and saline sludge directly to the ocean. The irony is that it would be cheaper for the government to buy the subsidized irrigated orchards and fields and allow the land to revert to desert than it is to operate the desalinization works.

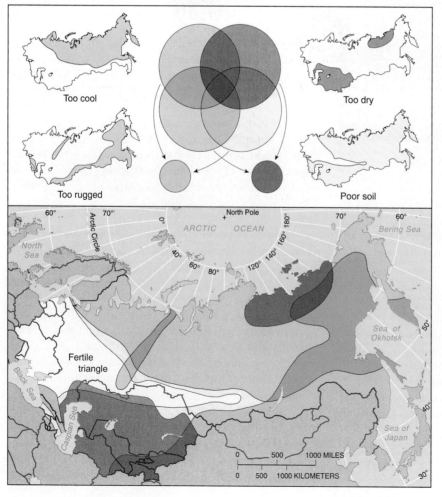

**FIGURE 18-16**
**Soil fertility.**

Top: four main factors influence soil fertility. Bottom: The fertile triangle of western portions of the Russian Federation (former Soviet Union). The white area is highest in soil fertility (no negative factors). Areas shown with any color or combination are limited for the reasons shown. The least-fertile area in Russia is northeastern Siberia. [After P. W. English and J. A. Miller, *World Regional Geography: A Question of Place*, p. 183, © 1989. Adapted by permission of John Wiley & Sons.]

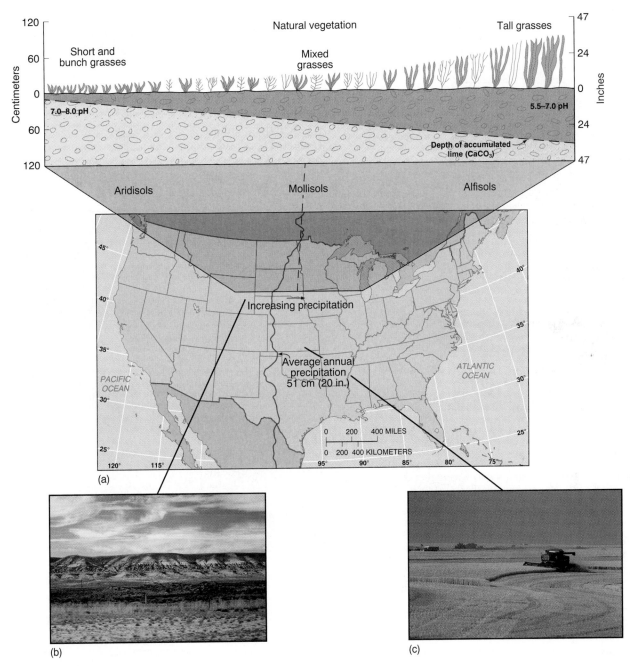

**FIGURE 18-17**
**Soils of the Midwest.**
(a) Aridisols (to the west), Mollisols, and Alfisols (to the east)—a soil continuum in the north-central U.S. and southern Canadian prairies. Graduated changes that occur in soil pH and the depth of accumulated lime are shown. (b) Bunch grasses and shallow soils of Wyoming. (c) Wheat harvest from rich Mollisols near Hardtner, Kansas [(a) Adapted from N.C. Brady, *The Nature and Properties of Soils*, 10th ed., © 1990 by Macmillan Publishing Company. Adapted by permission. (b) Photo by author. (c) Photo by Garry D. McMichael/Photo Researchers, Inc.]

***Ultisols.*** Farther south in the United States are the **Ultisols** (highly weathered forest soils) and their five suborders. An Alfisol might degenerate into an Ultisol, given time and exposure to increased weathering under moist conditions. These soils tend to be reddish because of residual iron and aluminum oxides in the A horizon (Figure 18-20).

The relatively high precipitation in Ultisol regions causes greater mineral alteration and more eluvial leaching than in other soils. Therefore the level of basic cations is lower and the soil fertility is lower. Fertility is further reduced by certain agricultural practices and the effect of soil-damaging crops such as cotton and tobacco, which

POTET equal to or
greater than PRECIP

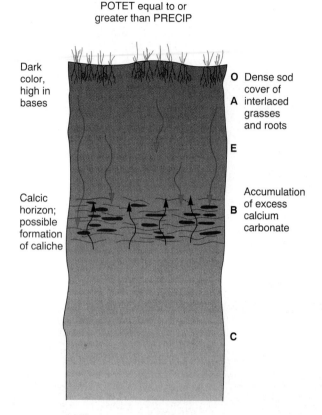

Dark color, high in bases

Calcic horizon; possible formation of caliche

O  Dense sod cover of
A  interlaced grasses and roots

E

B  Accumulation of excess calcium carbonate

C

**FIGURE 18-18**
**Calcification in soil.**

Calcification process in Aridisol/Mollisol soils occurs in climatic regimes that have potential evapotranspiration equal to or greater than precipitation.

deplete nitrogen and expose soil to erosion. However, these soils respond well if subjected to good management—for example, crop rotation restores nitrogen, and certain cultivation practices prevent sheetwash and soil erosion. Much needs to be done to achieve sustainable management of these soils.

***Spodosols.*** The **Spodosols** (northern coniferous forest soils) and their four suborders occur generally to the north and east of the Alfisols. They are in cold and forested moist regimes (*humid continental mild summer Dfb* climates) in northern North America and Eurasia, Denmark, the Netherlands, and southern England. Because there are no comparable climates in the Southern Hemisphere, this soil type is not identified there. Spodosols form from sandy parent materials, shaded under evergreen forests of spruce, fir, and pine. Spodosols with more moderate properties form under mixed or deciduous forests (Figure 18-21).

Spodosols lack humus and clay in the A horizons. An eluviated *albic horizon*, sandy and leached of clays and irons, lies in the A horizon instead, above a horizon of illuviated organic matter and iron and aluminum oxides

(Figure 18-21c). The surface horizon receives organic litter from base-poor, acid-rich trees (evergreens), which contribute to acid accumulations in the soil. The acidic soil solution effectively leaches clays, iron, and aluminum, which are passed to the upper diagnostic horizon. An ashen-gray color is common in these subarctic forest soils and is characteristic of a formation process called *podzolization.*

When agriculture is attempted, the low basic cation content of Spodosols requires the addition of nitrogen, phosphate, and potash (potassium carbonate), and perhaps crop

(a)

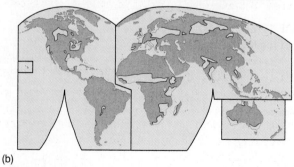

(b)

**FIGURE 18-19**
**Alfisol.**

(a) Soil profile from central California. (b) General map of worldwide distribution. [Photo from American Society of Agronomy.]

(a)

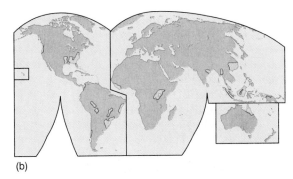

(b)

**FIGURE 18-20**
**Ultisol.**
(a) Soil profile from the upper coastal plain of central North Carolina.
(b) General map of worldwide distribution. (c) A type of Ultisol in
rural Alabama bears its characteristic color. [(a) Photo from American
Society of Agronomy. (c) Photo by author.]

(c)

rotation as well. A soil *amendment* such as limestone can significantly increase crop production by raising the pH of these acidic soils. For example, the yields of several crops (corn, oats, wheat, and hay) grown in specific Spodosols in New York State were increased up to a third with the application of 1.8 metric tons (2 tons) of limestone per 0.4 hectare (1.0 acre) during each 6-year rotation.

***Entisols.*** The **Entisols** (recent, undeveloped soils) lack vertical development of their horizons. The five suborders of this soil group are based on differences in parent materials and climatic conditions, although the presence of Entisols is not climate-dependent, for they occur in many climates worldwide. Entisols are true soils, but they have not had sufficient time to generate the usual horizons.

Entisols generally are poor agricultural soils, although those formed from river silt deposits are quite fertile. The same conditions that inhibit complete development also prevent adequate fertility—too much or too little water, poor structure, and insufficient accumulation of weathered nutrients. Active slopes, alluvium-filled floodplains,

poorly drained tundra, tidal mud flats, dune sands and erg (sandy) deserts, and plains of glacial outwash all are characteristic regions that have these soils. Figure 18-22 shows an Entisol in a desert climate where shales formed the parent material.

***Inceptisols.*** **Inceptisols** (weakly developed soils) and their six suborders are inherently infertile. They are weakly developed young soils, although they are more developed than the Entisols. Inceptisols include a wide variety of different soils, all holding in common a lack of maturity with evidence of weathering just beginning. Inceptisols are associated with moist soil regimes and are regarded as eluvial because they demonstrate a loss of soil constituents throughout their profile but retain some weatherable minerals. This soil group has no distinct illuvial horizons.

Inceptisols include the soils of most of the arctic tundra; glacially derived till and outwash materials from New York down through the Appalachians; and alluvium on the Mekong and Ganges floodplains.

(a)

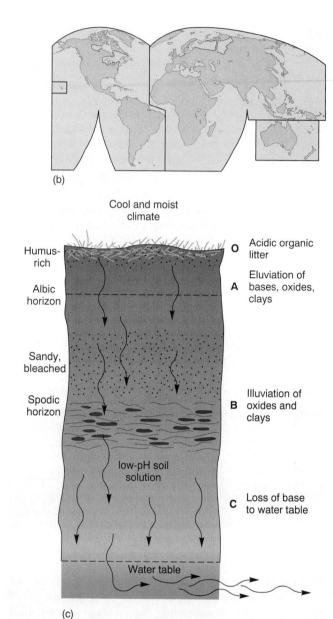

(b)

Cool and moist climate

Humus-rich

Albic horizon

Sandy, bleached

Spodic horizon

low-pH soil solution

Water table

O — Acidic organic litter

A — Eluviation of bases, oxides, clays

B — Illuviation of oxides and clays

C — Loss of base to water table

(c)

**FIGURE 18-21**

**Spodosol.**

(a) Soil profile from northern New York. (b) General map of worldwide distribution. (c) The podzolization process is typical in cool and moist climatic regimes. (d) A type of Spodosol formed under a characteristic pine-forest cover. [(a) Photo from American Society of Agronomy. (d) Photo by Galen Rowell/Peter Arnold, Inc.]

(d)

***Andisols.*** **Andisols** (volcanic parent materials) and seven suborders occur in areas of volcanic activity. These soils formerly were classified as Inceptisols and Entisols, but in 1990 they were placed in this new order. Andisols are derived from volcanic ash and glass. Previous soil horizons frequently are found buried by ejecta from repeated eruptions. Volcanic soils are unique in their mineral content because they are recharged by eruptions. For an example, see Figures 19-27 and 19-28, recovery and succession of pioneer species in the developing soil north of Mount St. Helens.

Weathering and mineral transformations are important in this soil order. Volcanic glass weathers readily into allophane (a noncrystalline aluminum silicate clay mineral that acts as a colloid) and oxides of aluminum and iron. Andisols feature a high CEC and high water-holding ability and develop moderate fertility, although phosphorus availability is an occasional problem. In Hawaii, the fertile Andisol fields produce sugar cane and pineapple as important cash crops (Figure 18-23a). Andisol distribution is small in areal extent; however, such soils are locally important in the volcanic ring of fire surrounding the Pacific Rim. Figure 18-23b shows vegetables and upland rice being grown in these soils in Japan.

***Vertisols.*** **Vertisols** (expandable clay soils) are heavy clay soils. They contain more than 30% swelling clays (clays that swell significantly when they absorb water), such as *montmorillonite*. They are located in regions experiencing highly variable soil moisture balances through the seasons. These soils occur in areas of subhumid to semiarid moisture and moderate to high temperature. Vertisols frequently form under savanna and

**FIGURE 18-22**
**Entisol.**
A characteristic Entisol: young, undeveloped soil forming from a shale parent material in the desert near Zabriskie Point in Death Valley, California. [Photo by author.]

(a)

(b)

**FIGURE 18-23**
**Andisols in agricultural production.**
(a) Fertile Andisols planted with sugar cane on Kauai, Hawaii. The fiftieth state is the largest U.S. producer of sugar cane. (b) Vegetables and upland rice production on newly reclaimed Andisols in Japan. [(a) Photo by Wolfgang Kaehler Photography; (b) photo from American Society of Agronomy.]

(a)

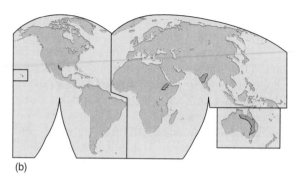

(b)

**FIGURE 18-24**
**Vertisol.**
(a) Soil profile in the Lajas Valley of Puerto Rico.
(b) General map of worldwide distribution.
[Photo from American Society of Agronomy.]

Despite the fact that clay soils are plastic and heavy when wet, with little available soil moisture for plants, Vertisols are high in bases and nutrients and thus are some of the better farming soils where they occur. For example, they occur in a narrow zone along the coastal plain of Texas and in a section along the Deccan region of India. Vertisols often are planted with grain sorghums, corn, and cotton.

*Histosols.* **Histosols** (organic soils), including four suborders, are formed from accumulations of thick organic matter. In the midlatitudes, when conditions are right, beds of former lakes may turn into Histosols, with water gradually replaced by organic material to form a bog and layers of peat. (Lake succession and bog or marsh formation are discussed in Chapter 19.) Histosols also form in small, poorly drained depressions. Here, Histosols have produced conditions ideal for significant deposits of sphagnum peat to form. This material can be cut, baled, and sold as a soil amendment. In other locales, dried peat has served for centuries as a low-grade fuel. The area southwest of Hudson Bay in Canada is typical of Histosol formation. (A complete treatment of the Canadian System of Soil Classification is presented in Appendix C, which includes a map showing the location of these organic soils in Canada.)

grassland vegetation in tropical and subtropical climates and are sometimes associated with a distinct dry season following a wet season. Although widespread, individual Vertisol units are limited in extent.

Vertisol clays are black when wet, but not because of organics: Rather, the blackness is due to specific mineral content. They range from brown to dark gray. These deep clays swell when moistened and shrink when dried. In the drying process, vertical cracks may form, as wide as 2–3 cm (0.8–1.2 in.) and up to 40 cm (16 in.) deep. Loose material falls into these cracks, only to disappear when the soil again expands and the cracks close. After many such cycles, soil contents tend to invert or mix vertically, bringing lower horizons to the surface (Figure 18-24).

# Summary and Review —The Geography of Soils

✔ *Define* **soil and soil science and** *describe* **a pedon, polypedon, and typical soil profile.**

**Soil** is the portion of the land surface in which plants can grow. It is a dynamic natural body composed of fine materials and contains both mineral and organic matter. **Soil science** is the interdisciplinary study of soils involving physics, chemistry, biology, mineralogy, hydrology, taxonomy, climatology, and cartography. *Pedology* deals with the origin, classification, distribution, and description of soil. *Edaphology* specifically focuses on the study of soil as a medium for sustaining the growth of higher plants.

The basic sampling unit used in soil surveys is the **pedon**. The **polypedon** is the soil unit used to prepare local soil maps and may contain many pedons. Each discernible layer in an exposed pedon is a **soil horizon**. The horizons are designated O (it contains **humus**, a complex mixture of decomposed and synthesized organic materials), A (rich in humus and clay, darker), E (zone of **eluviation**, the removal of fine particles and minerals by water), B (zone of **illuviation**, the deposition of clays and minerals translocated from elsewhere), C (regolith, weathered bedrock), and R (bedrock). Soil horizons A, E, and B experience the most active soil processes and together are designated the **solum**.

> soil (p. 550)
> soil science (p. 550)
> pedon (p. 550)
> polypedon (p. 550)
> soil horizon (p. 550)
> humus (p. 550)
> eluviation (p. 551)
> illuviation (p. 551)
> solum (p. 551)

1. Soils provide the foundation for animal and plant life and therefore are critical to Earth's ecosystems. Why is this true?

2. What are the differences among soil science, pedology, and edaphology?

3. Define polypedon and pedon, the basic units of soil.

4. Characterize the principal aspects of each soil horizon. Where does the main accumulation of organic material occur? Where does humus form? Explain the difference between the eluviated layer and the illuviated layer. Which horizons constitute the solum?

✔ *Describe* **soil properties of color, texture, structure, consistence, porosity, and soil moisture.**

We use several physical properties to classify soils. Color suggests composition and chemical makeup. Soil texture refers to the size of individual mineral particles and the proportion of different sizes. For example, **loam** is a balanced mixture of sand, silt, and clay. Soil structure refers to the arrangement of soil peds, which are the smallest natural cluster of particles in a soil. The cohesion of soil particles to each other is called soil consistence. Soil porosity refers to the size, alignment, shape, and location of spaces in the soil. Soil moisture refers to water in the soil pores and its availability to plants.

> loam (p.552)

5. Soil color is identified and compared using what technique?

6. Define a soil separate. What are the various sizes of particles in soil? What is loam? Why is loam regarded so highly by agriculturists?

7. What is a quick, hands-on method for determining soil consistence?

8. Summarize the five soil-moisture regimes common in mature soils.

✔ *Explain* **basic soil chemistry, including cation-exchange capacity, and** *relate* **these concepts to soil fertility.**

Particles of clay and organic material form negatively charged **soil colloids** that attract and retain positively charged mineral ions in the soil. The capacity to exchange ions between colloids and roots is an ability called the **cation-exchange capacity (CEC)**. CEC is a measure of **soil fertility**, the ability of soil to sustain plants. Fertile soil contains organic substances and clay minerals that absorb water and retain certain elements needed by plants.

> soil colloids (p. 556)
> cation-exchange capacity (CEC) (p. 556)
> soil fertility (p. 556)

9.  What are soil colloids? How are they related to cations and anions in the soil? Explain cation-exchange capacity.

10. What is meant by the concept of soil fertility?

---

✔ *Evaluate* **principal soil formation factors, including the human element.**

Environmental factors that affect soil formation include parent materials, climate, vegetation, topography, and time. Human influence is having great impact on Earth's prime soils. Essential soils for agriculture and their fertility are threatened by mismanagement, destruction, and conversion to other uses. Much soil loss is preventable through the application of known technologies, improved agricultural practices, and government policies.

11. Briefly describe the contribution of the following factors and their effect on soil formation: parent material, climate, vegetation, landforms, time, and humans.

12. Explain some of the details that support the concern over loss of our most fertile soils. What cost estimates have been placed on soil erosion?

---

✔ *Describe* **the eleven soil orders of the Soil Taxonomy classification system and** *explain* **their general occurrence.**

The **Soil Taxonomy** classification system is used in the United States and is built around an analysis of various diagnostic horizons and 11 soil orders, as actually seen in the field. The system divides soils into six hierarchical categories: series, families, subgroups, great groups, suborders, and orders.

Specific soil-forming processes keyed to climatic regions (not a basis for classification) are called **pedogenic regimes**: **laterization** (leaching in warm and humid climates), **salinization** (collection of salt residues in surface horizons in hot, dry climates), **calcification** (accumulation of carbonates in the B and C horizons in drier continental climates), **podzolization** (soil acidification in forest soils in cool climates), and **gleization** (humus and clay accumulate in cold, wet climates with poor drainage).

The Soil Taxonomy system uses two diagnostic horizons to identify soil: the **epipedon**, or the surface soil, and the **subsurface diagnostic horizon**, or the soil below the surface at various depths. (For an overview and definition of the 11 soil orders, please refer to Table 18-6, "Soil Taxonomy Soil Orders.") The 11 soil orders are: **Oxisols** (tropical soils), **Aridisols** (desert soils), **Mollisols** (grassland soils), **Alfisols** (moderately weathered, temperate forest soils), **Ultisols** (highly weathered, subtropical forest soils), **Spodosols** (northern conifer forest soils), **Entisols** (recent, undeveloped soils), **Inceptisols** (weakly developed, humid region soils), **Andisols** (soils formed from volcanic materials), **Vertisols** (expandable clay soils), and **Histosols** (organic soils).

Soil Taxonomy (p. 560)
pedogenic regimes (p. 560)
laterization (p. 560)
salinization (p. 560)
calcification (p. 560)
podzolization (p. 560)
gleization (p. 560)
epipedon (p. 561)
subsurface diagnostic horizon (p. 561)
Oxisols (p. 563)
Aridisols (p. 563)
Mollisols (p. 563)
Alfisols (p. 569)
Ultisols (p. 571)
Spodosols (p. 572)
Entisols (p. 573)
Inceptisols (p. 573)
Andisols (p. 574)
Vertisols (p. 575)
Histosols (p. 576)

13. Summarize the brief history of soil classification described in this chapter. What led soil scientists to develop the new Soil Taxonomy classification system?

14. What is the basis of the Soil Taxonomy system? How many orders, suborders, great groups, subgroups, families, and soil series are there?

15. Define an epipedon and a subsurface diagnostic horizon. Give a simple example of each.

16. Locate each soil order on the world map and on the U.S. map as you give a general description of it.

17. How was slash-and-burn shifting cultivation, as practiced in the past, a form of crop and soil rotation and conservation of soil properties?

18. Describe the salinization process in arid and semiarid soils. What associated soil horizons develop?

19. Which of the soil orders are associated with Earth's most productive agricultural areas?

20. What is the significance to plants of the 51 cm (20 in.) isohyet in the Midwest relative to soils, pH, and lime content?

21. Describe the podzolization process associated with northern coniferous forest soils. What characteristics are associated with the surface horizons? What strategies might enhance these soils?

22. What former Inceptisols now form a new soil order? Describe these soils as to location, nature, and formation processes. Why do you think they were separated into their own order?

23. Why has a selenium contamination problem arisen in western U.S. soils? Explain the impact of agricultural practices, and tell why you think this is or is not a serious problem.

 **NetWork**

The *Geosystems Home Page* provides on-line resources for this chapter on the World Wide Web. You will find review exercises, specific updates for items in the chapter, suggested readings, and links to interesting related pathways on the Internet (click on the Table of Contents link and select this chapter). *Geosystems* is at: **http://www.prenhall.com/geosystm**

# 19

# Ecosystem Essentials

**Ecosystem Components and Cycles**

**Stability and Succession**

**Summary and Review**

### Key Learning Concepts

After reading the chapter, you should be able to:

- *Define* ecology, biogeography, and the ecosystem concept.
- *Describe* communities, habitats, and niches.
- *Explain* photosynthesis and respiration and *derive* net photosynthesis and the world pattern of net primary productivity.
- *List* abiotic ecosystem components and *relate* those components to ecosystem operations.
- *Explain* trophic relationships in ecosystems.
- *Define* succession and *outline* the stages of general ecological succession in both terrestrial and aquatic ecosystems.

*Flowering bromeliads in El Yunque Rain Forest, Puerto Rico.* [Photo by Tom Bean.]

Diversity is an impressive feature of the living Earth. The diversity of organisms, both plant and animal, is a response to the interaction of the atmosphere, hydrosphere, and lithosphere, all powered by solar energy. This interaction produces a variety of conditions within which the biosphere exists. The diversity of life also results from the complex interplay of living organisms themselves.

The biosphere, the sphere of life and organic activity, extends from the ocean floor to about 8 km (5 mi) altitude in the atmosphere. The biosphere includes myriad ecosystems from simple to complex, each operating within general spatial boundaries. An **ecosystem** is a self-sustaining association of living plants and animals and their nonliving physical environment. In an ecosystem, a change in one component causes changes in others, as systems adjust. (Once again, this is dynamic equilibrium at work.)

Earth's biosphere itself is a collection of ecosystems within the natural boundary of the atmosphere. Natural ecosystems are open systems for both energy and matter, with almost all ecosystem boundaries functioning as transition zones rather than as sharp demarcations. Small ecosystems—for example, forests, seas, mountaintops, deserts, beaches, islands, lakes, ponds—make up the larger whole.

**Ecology** is the study of the relationships between organisms and their environment and among the various ecosystems in the biosphere. The word *ecology*, developed by German naturalist Ernst Haeckel in 1869, is derived from the Greek *oikos* ("household", or "place to live") and *logos* ("study of"). **Biogeography** is the study of the distribution of plants and animals, the diverse spatial patterns they create, and the physical and biological processes, past and present, that produce this distribution across Earth. To better understand an ecosystem, biogeographers look across the ages and reassemble Earth's tectonic plates to recreate previous environmental relationships.

The degree to which modern society understands ecosystems will determine the extent of our success as a species and the long-term survival of a habitable Earth:

> The time is ripe to step up and expand current efforts to understand the great interlocking systems of air, water, and minerals nourishing the Earth. . . . Moreover, without vigorous action toward that goal, nations will be seriously handicapped in trying to cope with proven and suspected threats to ecosystems and to human health and welfare resulting from alterations in the cycles of carbon, nitrogen, phosphorus, sulfur, and related materials. . . . Society depends upon this life-support system of planet Earth.[*]

Humans are Earth's most influential biotic (living) agent. This is not arrogance; it is fact, because we powerfully influence every ecosystem on Earth. In this chapter, we explore ecosystems, their stability, their resilience,

and how one ecosystem succeeds another. The next chapter applies these principles, examining Earth's major terrestrial ecosystems.

# Ecosystem Components and Cycles

An ecosystem is a complex of many variables, all functioning independently yet in concert, with complicated flows of energy and matter (Figure 19-1). An ecosystem includes both biotic (living) and abiotic (nonliving) components. Nearly all ecosystems depend upon a direct input of solar energy; the few limited ones that exist in dark caves or on the ocean floor depend upon chemical reactions (chemosynthesis).

Ecosystems are divided into subsystems, with the biotic portion composed of producers (plants), consumers (animals), and decomposers (bacteria and fungi). The abiotic flows in an ecosystem include gaseous, hydrologic, and mineral cycles. Figure 19-2 illustrates the essential elements of an ecosystem.

## Communities

A convenient biotic subdivision within an ecosystem is a community. A **community** is formed by interactions among populations of living animals and plants at a particular time. An *ecosystem* is the interaction of many communities with the abiotic physical components of its environment.

Some examples help to clarify these concepts. In a *forest ecosystem*, a specific community may exist on the forest floor, whereas another community functions in the canopy of leaves high above. A diverse community exists in the habitat along the edge of the forest—songbirds, deer, and plants that require open shade, such as woody shrubs and small trees (Figure 19-3). Similarly, within a *lake ecosystem*, the plants and animals that flourish in the bottom sediments form one community, whereas those near the surface form another. A community is identified in several ways—by its physical appearance, the species present and the abundance of each, the complex patterns of their interdependence, and the trophic (feeding) structure of the community.

Within a community, two concepts are important: habitat and niche. **Habitat** is the type of environment in which an organism resides or is biologically adapted to live. In terms of physical and natural factors, most species have specific habitat requirements with definite limits and a specific regimen of sustaining nutrients.

**Niche** (French *nicher*, "to nest") refers to the function, or occupation, of a life form within a given community. It is the way an organism obtains and sustains the physical, chemical, and biological factors it needs to survive. A niche has several facets. Among these are a habitat niche,

---

[*]G.F. White and M.K. Tolba, *Global Life Support Systems*, United Nations Environment Programme Information, No. 47 (Nairobi, Kenya: United Nations, 1979), p. 1.

## FIGURE 19-1
### The web of life.

"Life devours itself: everything that eats is itself eaten; everything that can be eaten is eaten; every chemical that is made by life can be broken down by life; all the sunlight that can be used is used. . .The web of life has so many threads that a few can be broken without making it all unravel, and if this were not so, life could not have survived the normal accidents of weather and time, but still the snapping of each thread makes the whole web shudder, and weakens it. . .You can never do just one thing: the effects of what you do in the world will always spread out like ripples in a pond." [Quotation from Friends of the Earth and Amory Lovins, The United Nations Stockholm Conference, *Only One Earth* (London: Earth Island Limited, 1972), p. 20. Photo by author.]

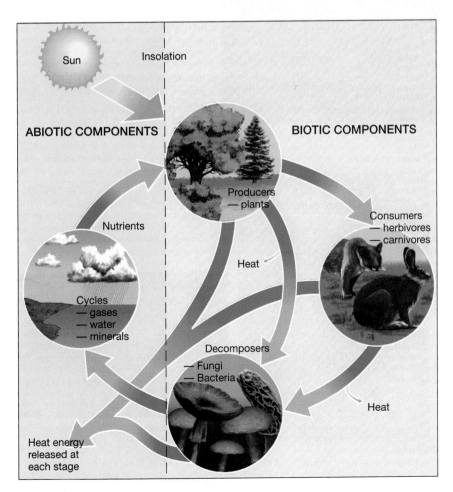

## FIGURE 19-2
### Abiotic and biotic components of ecosystems.

Solar energy is the input that drives the abiotic and biotic Components. Heat energy and biomass are the outputs from the biosphere.

**FIGURE 19-3**
**A community at the edge of the forest.**
Wild oats, grasses, and oak trees denote a specific community of plants and animals near the edge of the forest. [Photo by author.]

a trophic (food) niche, and a reproductive niche. For example, the Red-winged Blackbird (*Agelaius phoeniceus*) occurs throughout the United States and most of Canada in habitats of meadow, pastureland, and marsh. This species nests in blackberry tangles and thick vegetation in freshwater marshes, sloughs, and fields. Its trophic niche is weed seeds and cultivated seed crops throughout the year, and during the nesting season it adds insects to its diet—an aspect of its reproductive niche. These birds disperse seeds of many plants during their travels.

Similar habitats produce comparable niches. In a stable community, no niche is left unfilled. The *competitive exclusion principle* states that no two species can occupy the same niche (food or space) successfully in a stable community. Thus, closely related species are spatially separated. In other words, each species operates to reduce competition. This strategy in turn leads to greater diversity as species shift and adapt (Figure 19-4).

Some species have *symbiotic* relationships, an arrangement that mutually benefits and sustains each

(a)

(b)

(c)

(d)

(e)

**FIGURE 19-4**
**Plants and animals work to fit specific niches.**
(a) Elephant heads (*Pedicularis groenlandica*), a wildflower that grows above 1800 m (6000 ft) in wet mountain meadows. (b) A western yellow-bellied racer (*Coluber constrictor*) lives in prairies and meadows and feeds on insects, lizards, and mice. (c) A dragonfly (*Libellula* sp.) perches atop a reed along a stream, feeding on flying insects. (d) Killdeer chicks (*Charadrius vociferus*) begin life roughing it on an exposed rocky nest, with protective parents creating a noisy diversion nearby. (e) Coral mushroom (*Ramaria* sp.), a fungal decomposer at work on the organic matter accumulated on the forest floor. [(a), (c), (e) photos by Bobbé Christopherson; (b), (d) photos by author.]

**FIGURE 19-5**
**Symbiosis on the rocks.**
Lichen, an example of a symbiotic relationship between fungi and algae. [Photo by Bobbé Christopherson.]

organism over an extended period. For example, lichen (pronounced "liken") is made up of algae and fungi living together. The alga is the producer and food source for the fungus, and the fungus provides structure and physical support. Their mutually beneficial relationship (*mutualism*) allows the two to occupy a niche in which neither could survive alone. Lichen developed from an earlier parasitic relationship in which the fungi broke into alga cells directly. Today, the two organisms have evolved into a supportive harmony and symbiotic relationship (Figure 19-5). The partnership of corals and algae discussed in Chapter 16 is another example of a symbiotic relationship.

By contrast, *parasitic* relationships eventually may kill the host, thus destroying the parasite's own niche and habitat. An example is mistletoe (*Phoradendron*), which lives on and may kill various kinds of trees. Some scientists are questioning whether our human society and the physical systems of Earth constitute a global-scale symbiotic relationship (sustainable) or a parasitic one (nonsustainable).

## Plants: The Essential Biotic Component

Plants are the critical biotic link between solar energy and the biosphere. *Ultimately, the fate of all members of the biosphere, including humans, rests on the success of plants and their ability to capture sunlight.*

**Beginnings.** The first primitive plants were simple single- or multiple-celled structures known as *cyanobacteria*. These photosynthetic bacteria contain light-sensitive chlorophyll pigment. They began the harvest of sunlight, conversion of water and carbon dioxide to organic compounds, and release of free oxygen about 3.3 billion years ago. Long, slow development ensued as these early aquatic forerunners of plants migrated to the water surface,

**FIGURE 19-6**
**A successful survivor.**
Horsetail (*Equisetum arvense*), a plant of ancient origin, emerged over 300 million years ago. Because these plants have a high concentration of silica in their outer layer of cells, people were able to use them to scour dishes, pans, and floors, thus their common name, scouring rush. [Photo by author.]

increasingly protected from UV (ultraviolet) radiation by the evolving functional layers of the atmosphere. Today, cyanobacteria are widespread and may vary in color depending on the type of pigment they contain. Their habitats are diverse: from icy snowfields to hot springs at 85°C (185°F), and from oceans, to lakes, to moist soil.

The last billion years of Earth's history have seen an explosion of plant diversity. As increasingly complicated and integrated multicellular organisms evolved structures for survival on land, life moved from the sea onto rocky coastlines and then inland. Specialized tissues developed for food production, body support, and anchorage. Land plants (and animals) became common about 430 million years ago, according to fossilized remains. The horsetail, or scouring rush (*Equisetum arvense*), is one example of an ancient, primitive plant that originated in the Carboniferous Period and survives today (Figure 19-6). Some plants evolved to more complicated forms. **Vascular plants** developed conductive tissues and true roots for internal transport of fluid and nutrients. (*Vascular* is from a Latin word for "vessel-bearing", referring to the conducting cells.)

In the present day, about 270,000 species of plants, most of which are vascular, are known to exist. Many more species have yet to be identified. They represent a great untapped resource base. Only about 20 species of plants provide 90% of the world's food supply. Plants are a major source of new medicines and chemical compounds that benefit humanity. Plants also are the core of healthy, functioning ecosystems that sustain all life.

***Leaf Activity.*** Leaves are solar-powered chemical factories, wherein photochemical reactions take place. Veins in the leaf bring in water and nutrient supplies and carry off the sugars (food) produced by photosynthesis. The veins in each leaf connect to the stems and branches of the plant and thus to the main circulation system.

Flows of carbon dioxide, water, light, and oxygen enter and exit the surface of each leaf (see Figure 1-4). Gases flow into and out of a leaf through small pores called **stomata** (singular: *stoma*), which usually are most numerous on the lower side of the leaf. Each stoma is surrounded by small guard cells that open and close the pore, depending on the plant's needs at the moment.

Water that moves through a plant exits the leaves through the stomata and evaporates from leaf surfaces, thereby assisting temperature regulation within the plant. As water evaporates from the leaves, a pressure deficit is created that allows atmospheric pressure to push water up through the plant all the way from the roots, in the same manner that a soda straw works. We can only imagine the complex operation of a 100 m (330 ft) tree!

## Photosynthesis and Respiration

Powered by energy from certain wavelengths of visible light, **photosynthesis** unites carbon dioxide and oxygen (oxygen is derived from water in the plant). The term is descriptive: *photo-* refers to sunlight, and *-synthesis* describes the "manufacturing" of starches and sugars through reactions within plant leaves. The process releases oxygen and produces energy-rich food for the plant.

The largest concentration of light-responsive, photosynthetic cells in a leaf rests below the leaf's upper layers. These cells are called *chloroplasts*, and within each resides a green, light-sensitive pigment called **chlorophyll**. Within this pigment, light stimulates photochemistry. Consequently, competition for light is a dominant factor in the formation of plant communities. This competition is expressed in the height, orientation, distribution, and structure of plants.

Only about one-quarter of the light energy arriving at the surface of a leaf is useful to the light-sensitive chlorophyll. Chlorophyll absorbs only the orange-red and violet-blue wavelengths for photochemical operations, and it reflects predominantly green hues (and some yellow). That is why trees and other vegetation look green.

Photosynthesis essentially follows this equation:

$$CO_2 + H_2O + Light \rightarrow (CH_2O) + O_2$$
(carbon  (water)  energy    (carbohydrate) (oxygen)
dioxide)

From the equation, you can see that photosynthesis removes carbon (in the form of $CO_2$) from Earth's atmosphere. The quantity is enormous: approximately 91 billion metric tons (100 billion tons) of carbon dioxide per year. Carbohydrates, the organic result of the photosynthetic process, are combinations of carbon, hydrogen, and oxygen. They can form simple sugars, such as glucose ($C_6H_{12}O_6$). Glucose, in turn, is used by plants to build starches, which are more complex carbohydrates and the principal food stored in plants. *Primary productivity* refers to the rate at which energy is stored in such organic substances.

Plants, of course, store energy for later use. They consume this energy as needed through respiration, by converting the carbohydrates, to energy for their other operations. Thus, **respiration** is essentially a reverse of the photosynthetic process:

$$(CH_2O) + O_2 \rightarrow CO_2 + H_2O + energy$$
(carbohydrate) (oxygen)    (carbon   (water)  (heat)
dioxide)

In respiration, plants oxidize carbohydrates (stored energy), releasing carbon dioxide, water, and energy as heat. The overall growth of a plant depends on a surplus of carbohydrates beyond what is lost through plant respiration. Figure 19-7 presents a simple schematic of this process, which produces plant growth.

The *compensation point* is the break-even point between the production and consumption of organic material. Each leaf must operate on the production side of the compensation point, or else it is eliminated by the plant—something each of us has no doubt experienced with a house plant that received inadequate water or light. The difference between photosynthetic production and respiration loss is called *net photosynthesis*. The amount varies, depending on controlling environmental factors such as light, water, temperature, soil fertility, and the plant's site, elevation, and competition from other plants and animals.

Plant productivity increases as light availability increases—up to a point. When the light level is too high, *light saturation* occurs and most plants actually reduce their output in response. Some plants are adapted to shade, whereas others flourish in full sunlight. Crops such as rice, wheat, and sugar cane do well with high light intensity. Figure 19-8 portrays the general energy budget for green plants, showing energy receipt, utilization, and disposition.

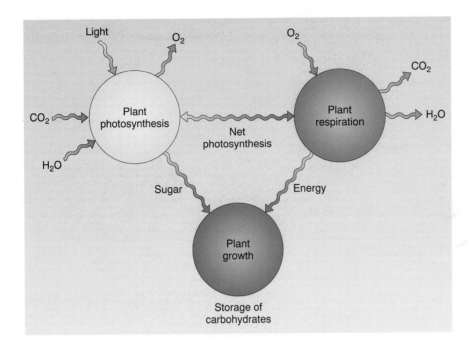

**FIGURE 19-7**
**How plants live and grow.**
The balance between photosynthesis and respiration determines net photosynthesis and plant growth.

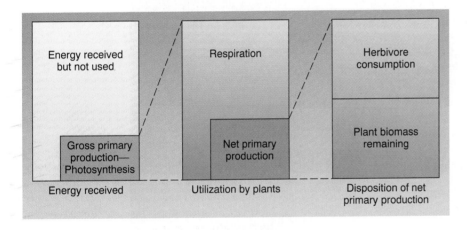

**FIGURE 19-8**
**Energy budget of the biosphere.**
Energy receipt, utilization, and disposition by green plants.

***Net Primary Productivity.*** The net photosynthesis for an entire plant community is its **net primary productivity**. This is the amount of stored chemical energy (biomass) that the community generates for the ecosystem. **Biomass** is the net dry weight of organic material.

Net primary productivity is measured as fixed carbon per square meter per year. ("Fixed" means chemically bound into plant tissues.) Study the map and satellite image in Figure 19-9 and you can see that on land, net primary production tends to be highest between the Tropics of Cancer and Capricorn at sea level and decreases toward higher latitudes and altitudes. Precipitation also affects productivity, as evidenced by the correlations of abundant precipitation with high productivity (adjacent to the equator) and reduced precipitation with low productivity (subtropical deserts). Even though deserts receive high amounts of solar radiation, other controlling factors are more important, namely water availability and soil conditions.

In the oceans, productivity is limited by differing nutrient levels. Regions with nutrient-rich upwelling currents generally are the most productive (off western coastlines). The map shows that the tropical ocean and areas of subtropical high pressure are quite low in productivity.

In temperate and high latitudes, the rate at which carbon is fixed by vegetation varies seasonally. It increases in spring and summer as plants flourish with increasing solar input and, in some areas, with more available (nonfrozen) water, and it decreases in late fall and winter. Productivity rates in the tropics are high throughout the year, and turnover in the photosynthesis-respiration cycle is faster, exceeding by many times the rates experienced in a desert environment or in the far northern limits of the tundra. A lush hectare (2.5 acres) of sugar cane in the tropics might fix 45 metric tons (50 tons) of carbon in a year, whereas desert plants in an equivalent area might achieve only 1% of that amount.

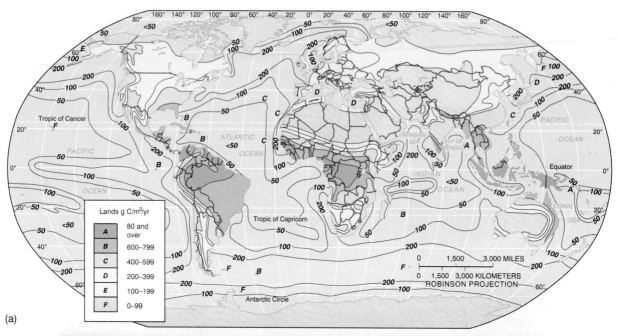

(a)

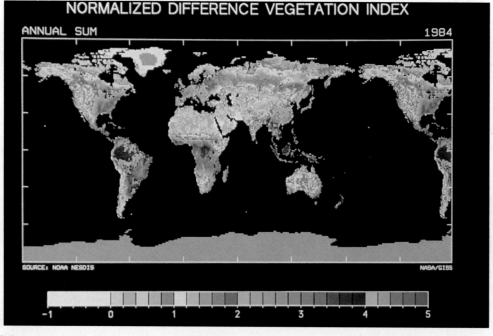

(b)

**FIGURE 19-9**

**Net primary productivity.**

(a) Worldwide net primary productivity in grams of carbon per square meter per year (approximate values). (b) Normalized difference vegetation index during 1984. False coloration indicates bare ground in browns, dense vegetation in blues. [(a) After D. E. Reichle, *Analysis of Temperate Forest Ecosystems* (Heidelberg, Germany: Springer-Verlag, 1970). Adapted by permission. (b) Goddard Space Flight Center and NOAA.]

Table 19-1 lists various ecosystems, their net primary productivity, and an estimate of net total biomass worldwide—170 billion metric tons of dry organic matter per year. Compare the various ecosystems, especially cultivated land (in *italics*) with most of the natural communities. Net productivity is generally regarded as the most important aspect of any type of community, and the distribution of productivity over Earth's surface is an important subject of biogeography.

## Abiotic Ecosystem Components

Critical in each ecosystem is the flow of energy and the cycling of nutrients and water in life-supporting systems. These nonliving abiotic components set the stage for ecosystem operations.

***Light, Temperature, Water, and Climate.*** Solar energy powers ecosystems, so the pattern of solar energy receipt is crucial. Solar energy enters an ecosystem by way of

**TABLE 19-1**

## Net Primary Productivity and Plant Biomass on Earth

| Ecosystem | Area (10⁶ km²)* | Net Primary Productivity per Unit Area (g/m²/yr)+ Normal Range | Mean | World Net Biomass (10⁹/tons/yr)† |
|---|---|---|---|---|
| Tropical rain forest | 17.0 | 1000–3500 | 2200 | 37.4 |
| Tropical seasonal forest | 7.5 | 1000–2500 | 1600 | 12.0 |
| Temperate evergreen forest | 5.0 | 600–2500 | 1300 | 6.5 |
| Temperate deciduous forest | 7.0 | 600–2500 | 1200 | 8.4 |
| Boreal forest | 12.0 | 400–2000 | 800 | 9.6 |
| Woodland and shrubland | 8.5 | 250–1200 | 700 | 6.0 |
| Savanna | 15.0 | 200–2000 | 900 | 13.5 |
| Temperate grassland | 9.0 | 200–1500 | 600 | 5.4 |
| Tundra and alpine region | 8.0 | 10–400 | 140 | 1.1 |
| Desert and semidesert scrub | 18.0 | 10–250 | 90 | 1.6 |
| Extreme desert, rock, sand, ice | 24.0 | 0–10 | 3 | 0.07 |
| *Cultivated land* | *14.0* | *100–3500* | *650* | *9.1* |
| Swamp and marsh | 2.0 | 800–3500 | 2000 | 4.0 |
| Lake and stream | 2.0 | 100–1500 | 250 | 0.5 |
| **Total continental** | **149** | – | **773** | **115.17** |
| Open ocean | 332.0 | 2–400 | 125 | 41.5 |
| Upwelling zones | 0.4 | 400–1000 | 500 | 0.2 |
| Continental shelf | 26.6 | 200–600 | 360 | 9.6 |
| Algal beds and reefs | 0.6 | 500–4000 | 2500 | 1.6 |
| Estuaries | 1.4 | 200–3500 | 1500 | 2.1 |
| **Total marine** | **361.0** | – | **152** | **55.0** |
| **Grand total** | **510.0** | – | **333** | **170.17** |

*Source:* R. H. Whittaker, *Communities and Ecosystems*, (Heidelberg, Germany: Springer-Verlag, 1975), p. 224. Reprinted by permission.
*1km = 0.39 mi².
+1 g per m² = 8.92 lb per acre.
†1 metric ton (10⁶g) = 1.1023 (net).

photosynthesis, and heat energy is dissipated from the system at many points. Of the total energy intercepted at Earth's surface and available for work, only about 1.0% is actually fixed by photosynthesis as chemical energy (energy stored as carbohydrates in plants).

The duration of Sun exposure is the *photoperiod.* Along the equator, days are essentially 12 hours long year-round; however, with increasing distance from the equator, seasonal effects become pronounced, as discussed in Chapter 2. Plants have adapted their flowering and seed germination to seasonal changes in insolation. Some seeds germinate only when daylength reaches a certain number of hours. A plant that responds in the opposite manner is the poinsettia (*Euphorbia pulcherrima*), which requires at least 2 months of 14-hour nights to start flowering.

Other components are important to ecosystem processes. Air and soil temperatures determine the rates at which chemical reactions proceed (see Chapter 18). Significant temperature factors are seasonal variation and duration and the pattern of minimum and maximum temperatures (Chapter 5).

Operations of the hydrologic cycle and water availability depend on precipitation and evaporation rates and their seasonal distribution (see Chapters 7, 8, and 9). Water quality—its mineral content, salinity, and levels of pollution and toxicity—is important. Also, regional climates affect the pattern of vegetation and ultimately influence soil development. All of these factors work together to establish the limits for ecosystems in a given location.

Figure 19-10 illustrates the general relation among temperature, precipitation, and vegetation. Can you identify the characteristic vegetation type and related temperature and moisture regime that fits areas you know well, such as a place you have lived or the school you now attend?

Alexander von Humboldt (1769–1859), an explorer, geographer, and scientist, observed that plants and animals recur in related groupings wherever similar conditions occur in the abiotic environment. He described the similarities and dissimilarities of vegetation and the associated uneven distribution of various organisms. After several years of study in the Andes Mountains of Peru, he described a distinct relation between altitude and plant

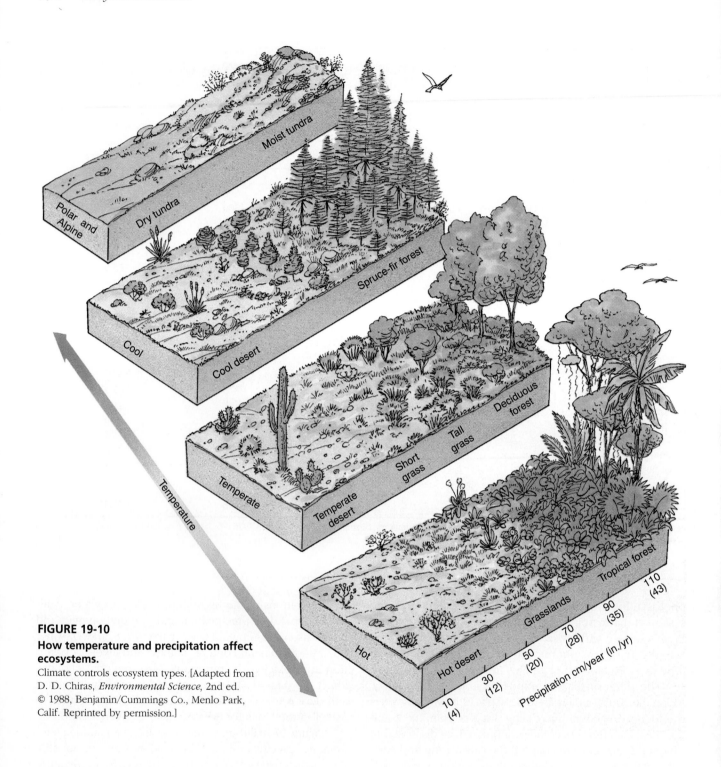

**FIGURE 19-10**
**How temperature and precipitation affect ecosystems.**
Climate controls ecosystem types. [Adapted from D. D. Chiras, *Environmental Science*, 2nd ed. © 1988, Benjamin/Cummings Co., Menlo Park, Calif. Reprinted by permission.]

communities as his *life zone concept.* As he climbed the mountains, he noticed that the experience was similar to that of traveling away from the equator toward higher latitudes (Figure 19-11).

This zonation of plants with altitude is noticeable on any trip from lower valleys to higher elevations. Each **life zone** possesses its own temperature, precipitation, and insolation relations and therefore its own biotic communities.

The Grand Canyon in Arizona provides a good example. The inner gorge at the bottom of the canyon (600 m, or 2000 ft, in elevation) has plants and animals characteristic of the lower Sonoran Desert of northern Mexico. However, the north rim of the canyon (2100 m, or 7000 ft, in elevation) is dominated by communities similar to those of southern Canadian forests. On the summits of the nearby San Francisco Mountains (3600 m, or 12,000

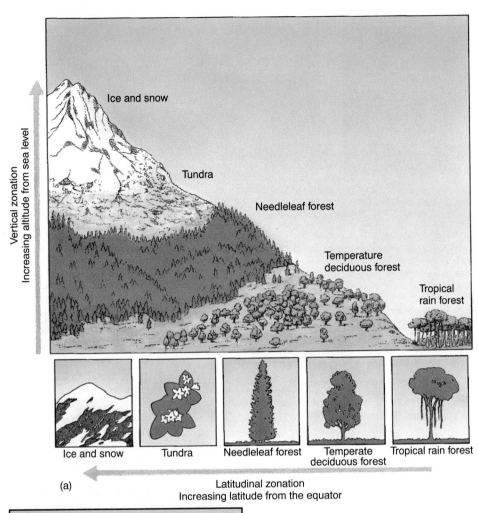

(b)

**FIGURE 19-11**
**Vertical and latitudinal zonation of plant communities.**
(a) Progression of plant community *life zones* with changing altitude or latitude.
(b) The *timberline* for a needleleaf forest in the Canadian Rockies. The *forest line* marks the highest continuous forest, whereas the *tree line* above it marks the zone above which no trees grow. [Photo by author.]

ft, in elevation), the vegetation is similar to that of the arctic tundra of northern Canada.

Beyond these general conditions, each ecosystem further produces its own *microclimate*, specific to individual sites. For example, in a forest the insolation reaching the ground is reduced and the forest floor is shaded. A pine forest cuts light by 20%–40%, whereas a birch-beech forest reduces it by as much as 50%–75%. Forests also are about 5% more humid than nonforested landscapes, have moderated temperatures (warmer winters and cooler summers), and experience reduced winds.

Slope and exposure are important too, for they translate into differences in temperature and moisture efficiency, especially in middle and higher latitudes. With all other factors equal, slopes facing away from the Sun's rays tend to be moister and more vegetated than slopes

(a)

(b)

**FIGURE 19-12**

**Microclimates in forests and on slopes.**

(a) This fern and forest microecosystem—indicative of temperature, radiation, and evaporation in the microclimate at the forest floor—will change drastically if many trees are cut down. (b) Slope orientation on this mountainside in northwestern Wyoming affects moisture efficiency and exposure and therefore the distribution of vegetation—note the vegetation growth on each north-facing slope. In the Southern Hemisphere, on which slope would you expect to see this effect? [Photos by author.]

---

## News Report 1

### Earth's Magnetic Field—An Abiotic Factor

The fact that birds and bees can detect Earth's magnetic field and use it for finding direction is well established. Small amounts of magnetically sensitive particles in the skull of the bird and the abdomen of the bee provide compass directions for bird migration and for alerting the hive to the location of the latest nectar find.

A wife and husband team of biologists from the University of North Carolina, Catherine and Kenneth Lohmann, have found that sea turtles can detect magnetic fields of different strengths and the inclination (angle) of these magnetic fields. This means that the turtles have a built-in navigation system that helps them know where they are on Earth (like our global positioning system, which requires multimillion-dollar satellites). Using magnetic field strengths and inclination, the turtles evidently are aware of their global position (similar to knowing their latitude and longitude). This magnetic map is evidently imprinted in their brain.

Loggerhead sea turtles hatch in Florida, crawl into the water, and spend the next 70 years traveling thousands of miles between North America and Africa around the subtropical high-pressure gyre in the Atlantic Ocean (see Figure 6-16). The females always return to near where they were hatched to lay their eggs! The researchers think that the hatchlings are imprinted with magnetic data unique to the beach where they hatched and then develop a more global sense as they live a life swimming across the ocean.

---

facing toward the Sun. (In the Northern Hemisphere, these slopes face north, as evidenced by the tendency of moss, a plant that needs high moisture and shade, to grow on the north side of tree trunks.) Such highly localized *microecosystems* are evident along a mountain trail and on mountain slopes where changes in exposure and moisture occur (Figure 19-12).

Even Earth's magnetic field, another abiotic part of ecosystems, plays an interesting role in ecosystems. See News Report 1 and read about turtle navigation.

***Elemental Cycles.*** The most abundant natural elements in living matter are hydrogen (H), oxygen (O), and carbon (C). Together, these elements make up more than 99% of Earth's biomass; in fact, all life (organic molecules) contains hydrogen and carbon. In addition, nitrogen (N), calcium (Ca), potassium (K), magnesium (Mg), sulfur (S),

and phosphorus (P) are significant nutrients, elements necessary for the growth of a living organism.

Several key chemical cycles function in nature. Oxygen, carbon, and nitrogen each have *gaseous cycles*, part of which are in the atmosphere. Other elements have *sedimentary cycles*, which principally involve mineral and solid phases (major ones include phosphorus, calcium, potassium, and sulfur). Some elements combine gaseous and sedimentary cycles. All these recycling processes form Earth's **biogeochemical cycles**, because they involve chemical reactions in both living and nonliving systems. The chemical elements themselves recycle over and over again in life processes.

***Oxygen and Carbon Cycles.*** We consider these together because they are so closely intertwined through photosynthesis and respiration (Figure 19-13). The atmosphere

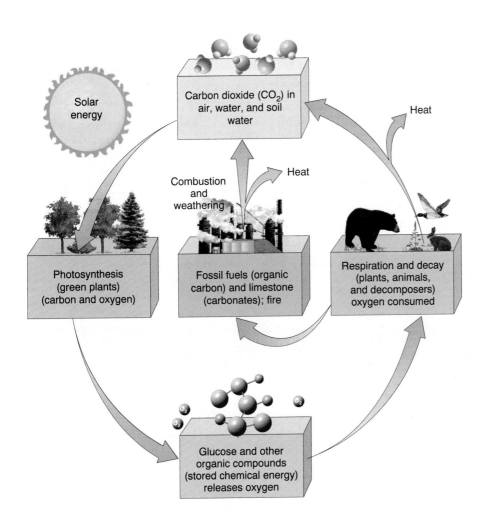

**FIGURE 19-13**
**The carbon and oxygen cycles, simplified.**

is the principal reserve of available oxygen. Larger reserves of oxygen exist in Earth's crust, but they are unavailable, being chemically bound with other elements, especially the silicate ($SiO_2$) and carbonate ($CO_3$) mineral families. Unoxidized reserves of fossil fuels and sediments also contain oxygen.

The oceans are an enormous pool of carbon—about 39,000 billion tons. (Net tons times 0.91 equals metric tons.) However, all of this carbon is bound chemically in carbon dioxide, calcium carbonate, and other compounds. The ocean initially absorbs carbon dioxide by means of the photosynthesis carried on by phytoplankton. Carbon becomes stored in certain carbonate minerals, such as limestone ($CaCO_3$).

The atmosphere, which is the integrating link in the cycle, contains only about 700 billion tons of carbon (as carbon dioxide) at any moment. This is far less carbon than in fossil fuels and oil shales (12,000 billion tons, as hydrocarbon molecules) or in living and dead organic matter (2275 billion tons, as carbohydrate molecules). Carbon dioxide in the atmosphere is produced by the respiration of plants and animals, volcanic activity, and fossil fuel combustion by industry and transportation (see News Report 2).

***Nitrogen Cycle.*** Nitrogen, accounting for 78.084% of each breath we take, is the major constituent of the atmosphere. Nitrogen also is important in the makeup of organic molecules, especially proteins, and therefore is essential to living processes. A simplified view of the nitrogen cycle is portrayed in Figure 19-14.

This vast atmospheric reservoir of nitrogen is inaccessible directly to most organisms. The key link to life is provided by *nitrogen-fixing bacteria*, which live principally in the soil and are associated with the roots of certain plants—for example, the *legumes* such as clover, alfalfa, soybeans, peas, beans, and peanuts. Colonies of these bacteria reside in nodules on the legume roots and chemically combine the nitrogen from the air in the form of nitrates ($NO_3$) and ammonia ($NH_3$).

Plants use these chemically bound forms of nitrogen to produce their own organic matter. Anyone or anything feeding on the plants thus ingests the nitrogen. Finally, the nitrogen in the organic wastes of these consuming organisms is freed by denitrifying bacteria, which recycle it back to the atmosphere.

To improve agricultural yields, many farmers use synthetic inorganic fertilizers, as opposed to soil-building organic fertilizers (manure and compost). Inorganic

## News Report 2

## Humans Dump Carbon into the Atmosphere

The carbon dioxide dumped into the atmosphere by human activity constitutes a vast geochemical experiment, using the real-time atmosphere as a laboratory. Since 1970, we have added to the atmospheric pool an amount of carbon equivalent to more than 25% of the total amount added since 1880.

Annual emissions by society have now reached 26 billion metric tons (1.102 net tons) and represent our greatest single atmospheric waste product. About 84% of this is from industrial activity, an increase of almost 40% since 1975. The United States leads the world with the highest per capita emissions: 19 metric tons of carbon dioxide per year. Further, about 50% of the carbon dioxide emitted since the beginning of the Industrial Revolution and not absorbed by oceans and organisms remains in the atmosphere, enhancing Earth's natural greenhouse effect. This is an important topic in Chapters 3, 5, and 10.

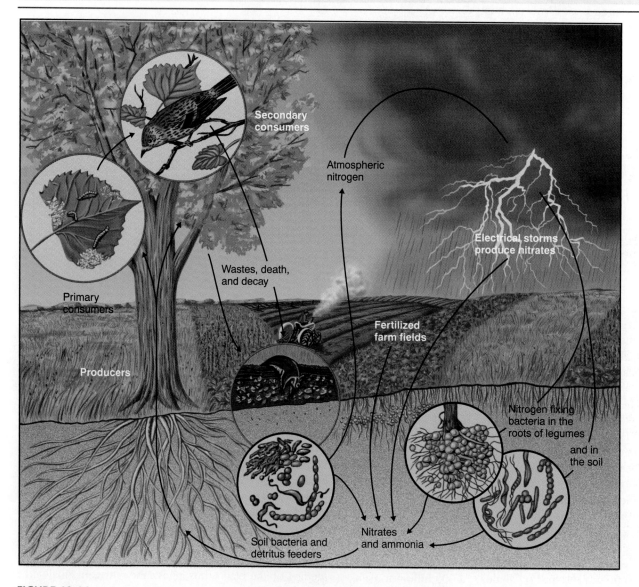

**FIGURE 19-14**

**The nitrogen cycle.**

The atmosphere is the essential reservoir of gaseous nitrogen. Atmospheric nitrogen gas is chemically fixed by bacteria to produce ammonia. Lightning and forest fires produce nitrates, and fossil fuel combustion forms nitrogen compounds that are washed from the atmosphere by precipitation. Plants absorb nitrogen compounds and produce organic material. [From T. Audesirk and G. Audesirk, *Biology: Life on Earth*, 4th ed., figure 45-10, p. 903. © 1996 Prentice Hall Inc. Used by permission.]

fertilizers are chemically produced through artificial nitrogen fixation at factories. The annual production of inorganic fertilizers far exceeds the ability of natural denitrification systems—and the present production of synthetic fertilizers now is doubling every 8 years.

This surplus of usable nitrogen accumulates in Earth's ecosystems. Some is present as excess nutrients, washed from soil into waterways and eventually the ocean. This excess nitrogen load begins a water pollution process that feeds an excessive growth of algae and phytoplankton, increases biochemical oxygen demand, diminishes dissolved oxygen reserves, and eventually disrupts the aquatic ecosystem. In addition, excess nitrogen compounds in air pollution are a component in acid deposition, further altering the nitrogen cycle in soils and waterways (see Chapter 3).

***Limiting Factors.*** The term **limiting factor** identifies the one physical or chemical abiotic component that most inhibits biotic operations, through either its lack or its excess. Here are a few examples:

- Low temperatures limit plant growth at high elevations.
- Lack of water limits growth in a desert.
- Excess water limits growth in a bog.
- Changes in salinity levels affect aquatic ecosystems.
- Lack of iron in ocean surface environments limits photosynthetic production.
- Low phosphorus content of eastern U.S. soils limits plant growth.
- General lack of active chlorophyll above 6100 m (20,000 ft) limits primary productivity.

In most ecosystems, precipitation is the limiting factor. However, temperature, light levels, and soil nutrients certainly affect vegetation patterns.

Each organism possesses a range of tolerance for each limiting factor in its environment. This fact is illustrated vividly in Figure 19-15, which shows the geographic range for two tree species and two bird species. The coast redwood (*Sequoia sempervirens*) is limited to a narrow section of the Coast Ranges in California, covering barely 9500 km² (less than 4000 mi²), concentrated in areas that receive necessary summer advection fog. On the other hand, the red maple (*Acer rubrum*) thrives over a large area under varying conditions of moisture and temperature, thus demonstrating a broader tolerance to environmental variations.

The Mallard duck (*Anas platyrhynchos*) and the Snail Kite (*Rostrhamus sociabilis*) also demonstrate a variation in tolerance and range (Figure 19-15b). The mallard, a generalist, feeds from widely diverse sources, is easily domesticated, and is found throughout most of North America in at least one season of the year. By contrast,

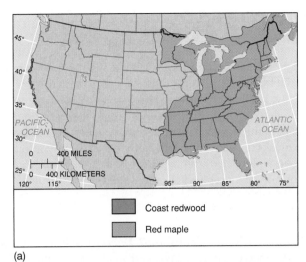

(a)

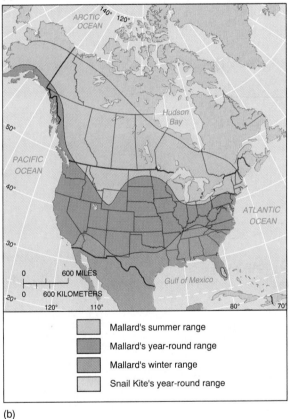

(b)

**FIGURE 19-15**

**Limiting factors affect the distribution of every plant and animal species.**

(a) Distributions of coast redwood and of red maple demonstrate the effect of limiting factors. (b) So do the distributions of the Mallard duck and the Snail Kite. The Mallard is a generalist and feeds widely. In contrast, the Snail Kite is limited by the range of its food, a single type of snail.

the Snail Kite is a specialist that feeds only on one specific type of snail. This single food source, then, is its limiting factor. Note its small habitat area near Lake Okeechobee in Florida.

## Biotic Ecosystem Operations

The abiotic components of energy, atmosphere, water, weather, climate, and minerals make up the life support for the biotic components of each ecosystem. The flow of energy, cycling of nutrients, and trophic (feeding) relations determine the nature of an ecosystem. As energy cascades through this process-flow system it is constantly replenished by the Sun. But nutrients and minerals cannot be replenished from an external source, so they constantly cycle within each ecosystem and through the biosphere in general (Figure 19-16). Let us examine these biotic operations.

### Producers, Consumers, and Decomposers.
Organisms that are capable of using carbon dioxide as their sole source of carbon are called *autotrophs* (self-feeders), or **producers**. Generally, these are the plants. They chemically fix carbon through photosynthesis. Organisms that depend on producers as their carbon source are called *heterotrophs* (feed on others), or **consumers**. Generally, these are animals. Autotrophs are the essential producers in an ecosystem—capturing light energy and converting it to chemical energy, incorporating carbon, forming new plant tissue and biomass, and freeing oxygen.

From the producers, which manufacture their own food, energy flows through the system along a circuit called the **food chain**, reaching consumers and eventually *decomposers*. Ecosystems generally are structured in a **food web**, a complex network of interconnected food chains. In a food web, consumers participate in several different food chains. Organisms that share the same basic foods are said to be at the same *trophic level*. Figure 19-17 illustrates a simple terrestrial food chain and an oceanic food chain—nature provides an almost infinite variety of trophic relationships. Solar energy enters each food chain through the producer plant or producer phytoplankton, subsequently flowing through higher and higher levels of consumers.

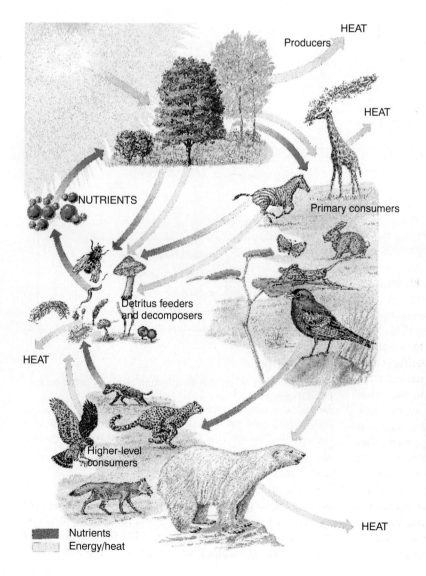

**FIGURE 19-16**

**Energy, nutrient, and food pathways in the environment.**

The flow of energy, cycling of nutrients, and trophic (feeding) relationships in a generalized ecosystem. The operation is fueled by radiant energy supplied by sunlight and first captured by the plants. [From T. Audesirk and G. Audesirk, *Biology: Life on Earth*, 4th ed., figure 45-1, p. 892. © 1996 Prentice Hall Inc. Used by permission.]

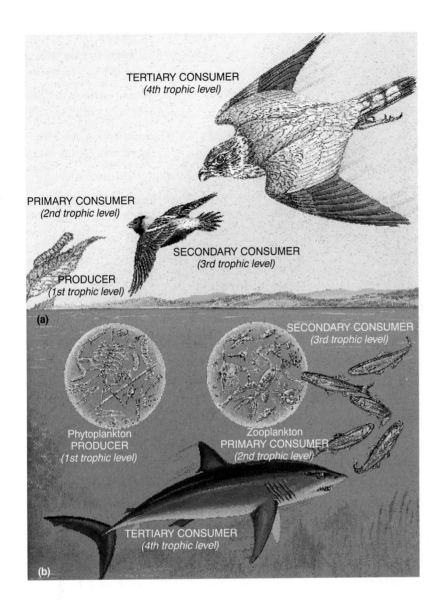

(a)

(b)

**FIGURE 19-17**
**Food chains.**
(a) A simplified terrestrial food chain, from leaf to caterpillar to Bobolink to a fast-moving Merlin hawk. (b) A simplified aquatic food chain from phytoplankton to zooplankton to schooling fish and to a Mako shark. [From T. Audesirk and G. Audesirk, *Biology: Life on Earth*, 4th ed., figure 45-4, p. 895. © 1996 Prentice Hall Inc. Used by permission.]

Primary consumers feed on producers. Because producers are always plants, the primary consumer is called an **herbivore**, or plant eater. A **carnivore** is a secondary consumer and primarily eats meat. A tertiary consumer eats primary and secondary consumers and is referred to as the "top carnivore" in the food chain. A consumer that feeds on both producers (plants) and consumers (meat) is called an **omnivore**—a role occupied by humans, among others.

**Decomposers** are the final link in the chain. They renew the entire system by releasing inorganic materials from organic debris. They are bacteria and fungi that digest and recycle the organic debris and waste in the environment. In addition, the *detritus feeders*—worms, mites, termites, centipedes, and others—participate like a small army of workers. Waste products, dead plants and animals, and other organic remains are the principal food source for all these *detritivores*. Inorganic compounds are released in the process and the cycle continues.

***A Food Chain in Antarctic Water.*** As an example, look at the oceanic food web that includes krill, a primary consumer (Figure 19-18). *Krill* is a shrimplike crustacean that is a major food for an interrelated group of organisms, including whales, fish, seabirds, seals, and squid in the Antarctic region. All of these organisms participate in numerous other food chains as well, some consuming and some being consumed.

*Phytoplankton* begin this chain by harvesting solar energy in photosynthesis. Phytoplankton are eaten by *herbivorous zooplankton* such as krill and other organisms. Krill then are eaten by consumers at the next trophic level. Because krill are a protein-rich, plentiful food, they are increasingly sought by factory ships, such as those from Japan and Russia. The annual krill harvest currently surpasses a million tons, principally as feed for chickens and livestock and as protein for human consumption.

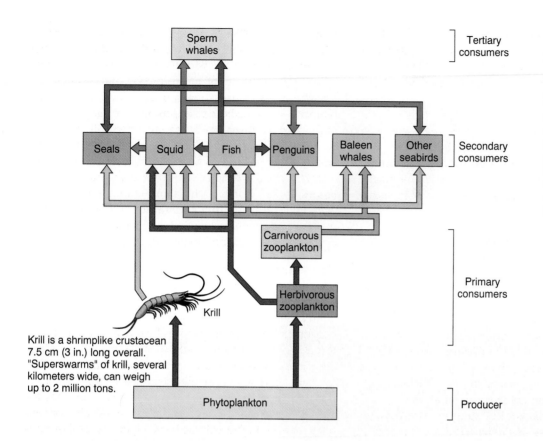

Tertiary consumers

Secondary consumers

Primary consumers

Producer

Krill is a shrimplike crustacean 7.5 cm (3 in.) long overall. "Superswarms" of krill, several kilometers wide, can weigh up to 2 million tons.

**FIGURE 19-18**
**An Antarctic food web.**
The food web in Antarctic waters, from phytoplankton *producers* (bottom) through various *consumers*. Phytoplankton begin this chain by using solar energy in photosynthesis. The phytoplankton are eaten by krill and other herbivorous zooplankton. Krill in turn are consumed by the next trophic level.

The impact on the food web of further increases in the krill harvest is uncertain, for little is known about the krill reproduction rate. Their long life span (greater than 8 years) and slow growth rate limit the ultimate extent of the resource. In addition, the possible effect on krill of increased ultraviolet radiation resulting from the seasonal "hole" in the ozone layer above Antarctica is under investigation. All of these interrelations need to be clarified, for there are many unknowns in the Antarctic food web.

***Efficiency in a Food Chain.*** Any assessment of world food resources depends on the level of consumer being targeted. Let us use humans as an example. Many people can be fed if wheat is eaten directly. However, if the grain is first fed to cattle (herbivores) and then we eat the beef, the yield of available food energy is cut by *90%* (810 kg of grain is reduced to 82 kg of meat); far fewer people can be fed from the same land area (Figure 19-19).

In terms of energy, only about 10% of the kilocalories (food Calories, not heat calories) in plant matter survive from the primary to the secondary trophic level. When humans consume meat instead of grain, there is a further loss of biomass and added inefficiency. More energy is lost to the environment at each progressive step in the food chain. You can see that an omnivorous diet, such as ours, is quite expensive in terms of biomass and energy.

Food chain concepts are becoming politicized as world food issues grow more critical. Today, approximately half of the cultivated acreage in the United States and Canada is planted for animal consumption—beef and dairy cattle, hogs, chickens, and turkeys. This includes more than 80% of the annual corn and nonexported soybean harvest. In addition, some lands cleared of rain forest in Central and South America have been converted to pasture to produce beef for export to restaurants, stores,

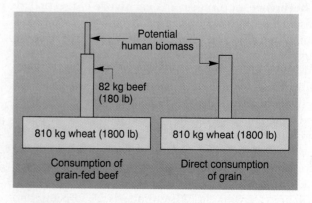

**FIGURE 19-19**
**Pyramids show us efficiency and inefficiency.**
Biomass pyramids illustrate a great difference in efficiency between direct and indirect consumption of grain.

and fast-food outlets in developed countries. Thus, lifestyle decisions and dietary patterns in North America and Europe are perpetuating inefficient food chains, not to mention the destruction of valuable resources, both here and overseas.

Clearly, some food chains are exceptionally simple, such as eating grains directly. Others are more complex than the Antarctic food web shown in Figure 19-18. The home gardener's tomatoes may be eaten by a tomato hornworm, which is then plucked off by a passing robin, which is later eaten by a hawk—and so it goes, in seemingly endless cycles.

***Ecological Relations.*** A study of food chains and webs is a study of who eats what, and where they do the eating. Figure 19-20 charts the summertime distribution of populations in two ecosystems, grassland and temperate forest. The stepped population pyramid is characteristic of summer conditions in such ecosystems. You can see the decreasing number of organisms at successively higher trophic levels. The base of the temperate forest pyramid is narrow, however, because most of the producers are large, highly productive trees and shrubs, which are outnumbered by the consumers they can support.

One approach to ecological study is to analyze a community's metabolism. *Metabolism* is the way in which a community uses energy and produces food for continued operation (the sum of all its chemical processes). In 1957, Howard T. Odum completed one of the best-known studies of this type for Silver Springs, Florida.

Figure 19-21a, an adaptation of his work, shows that the energy source for the community is insolation. The amount of light absorbed by the plants for photosynthesis is 24% of the total amount arriving at the study site. Of the absorbed amount, only 5% is transformed into the community's gross photosynthetic production—and this is only 1.2% of the total insolation input. The net plant production (gross production minus respiration) is 42.4% of the gross

production, or 8833 kilocalories per square meter per year. That amount then moves through the food chain of herbivores and carnivores, with the top carnivores receiving only 0.24% of the net plant production of the system.

You can see that, at each trophic level, some of the biomass flows to the decomposers. The largest portion of the biomass is exported from the system in the form of community respiration. In addition, a relatively small portion leaves the system as organic particles by downstream export out of this system. The biomass pyramid shown in Figure 19-21b portrays the same community from the viewpoint of a *standing crop* (all the biomass in a particular environment at one time), which is another way of examining the existing biomass at each trophic level.

***Concentration of Pollution in Food Chains.*** Many farms freely use pesticides on producers to reserve them for eating by selected consumers—such as humans or the animals we feed. When an ecosystem of producers and consumers has certain chemical pesticides applied, the food web concentrates some of these chemicals. Many chemicals are degraded or diluted in air and water and thus are rendered relatively harmless. Other chemicals, however, are long-lived, stable, and soluble in the fatty tissues of consumers. They become increasingly concentrated at each higher trophic level. Figure 19-22 illustrates this *biological amplification*, or *magnification*.

DDT is one such chemical. As an insecticide, it saved millions of human lives in the tropics by killing the mosquitoes that spread malaria. Crop-destroying insects also were brought under control. However, the persistence of the chemical in the environment and its destruction of established food webs brought many unwanted consequences. Although banned in some countries (the United States banned it in 1973), it still is widely used in less-developed countries. In addition, some other organic and synthetic chemicals, radioactive debris, and heavy metals such as lead and mercury become concentrated in food chains. Thus, a

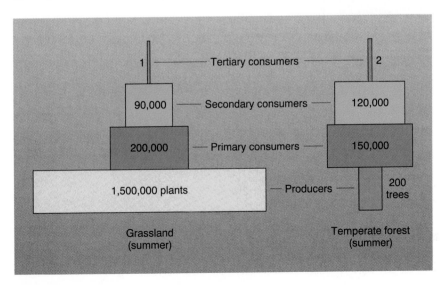

**FIGURE 19-20**

**Population distribution in two ecosystems.**

Ecological pyramids for 0.1 hectare (0.25 acre) of land show the difference between grassland and forest in the summer. Numbers of consumers are shown. [After E. P. Odum, *Fundamentals of Ecology*, 3rd ed., figure 3-15a, p. 80. © 1971 Saunders College Publishing. Adapted by permission.]

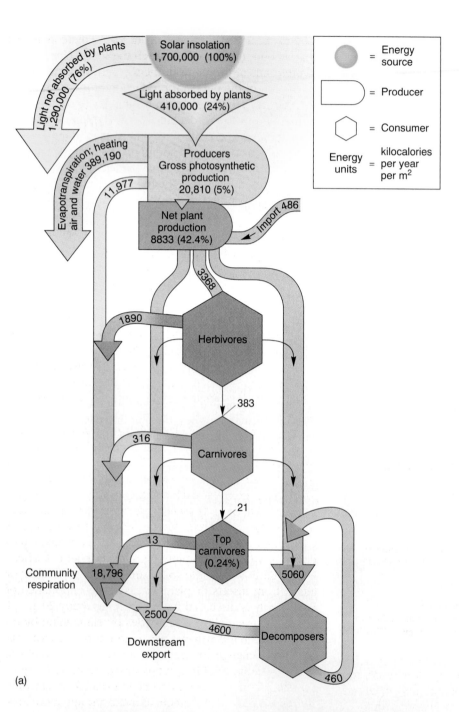

(a)

(b)

**FIGURE 19-21**

**How energy and nutrients flow through a community.**

(a) Analysis of community metabolism. (b) Standing crop distribution in grams per square meter for Silver Springs, Florida. [Adapted from E. P. Odum, *Fundamentals of Ecology*, 3rd ed., figure 6-1a, b, p. 142. © 1971 Saunders College Publishing. Adapted by permission.]

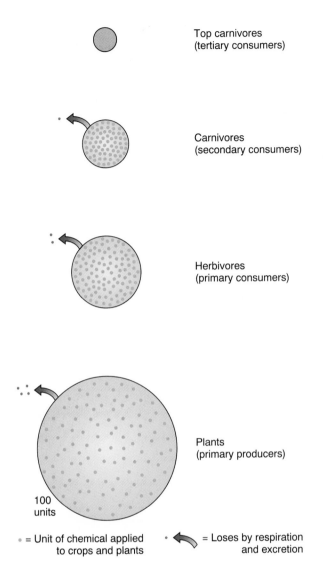

Top carnivores
(tertiary consumers)

Carnivores
(secondary consumers)

Herbivores
(primary consumers)

Plants
(primary producers)

100
units

• = Unit of chemical applied
to crops and plants

= Loses by respiration
and excretion

**FIGURE 19-22**
**Concentration of chemical toxins in food chains.**
Chemical residues are passed along a simple food chain to the top
carnivores, where the accumulation becomes concentrated. This
phenomenon is biological magnification.

food chain can efficiently poison the organism at the top.
Many species are threatened in this manner, and, of course,
humans are at the top of many food chains and can ingest
concentrated chemicals in what is consumed.

# Stability and Succession

Far from being static, Earth's ecosystems are dynamic and
ever-changing. Over time, communities of plants and animals have adapted to great variety, evolved, and in turn
shaped their environments. A constant interplay exists
between increasing growth toward the potential in a community and decreasing growth caused by resisting factors
that create limits (Figure 19-23). Each ecosystem operates

in dynamic equilibrium, constantly adjusting to changing
conditions in the attempt to achieve stability. The concept
of *change* is key to understanding ecosystem stability.

## Ecosystem Stability and Diversity

Ecosystems tend to move toward maximum biomass and
stability. But the tendencies for birth and death rates to
balance and for the composition of species populations
to remain stable (*inertial stability*) do not necessarily foster the ability to recover from change (*resilience*). Think
of resilience as an ecosystem's ability to recover from disturbance; its ability to absorb disturbance up to some
threshold point before it flips to a different set of relations. Examples of communities with high inertial stability include a redwood forest, a pine forest at a high
elevation, and a tropical rain forest near the equator. But,
cleared forest tracts recover slowly and therefore have
poor resilience. Figure 19-24 shows how clear-cut tracts
of former forest have drastically altered microclimatic conditions, making regrowth of the same species difficult. In
contrast, a midlatitude grassland is low in stability; yet,
when burned, its resilience is high because the community recovers rapidly.

   Another aspect related to stability is **biodiversity**. The
more diverse the species population (both in number of
different species and quantity of each species), the better
risk is spread over the entire community, because several
food sources exist at each trophic level. In other words,
*greater biodiversity in an ecosystem results in greater sta-
bility and greater productivity.* News Report 3 presents
exciting confirmation of this principle. An artificially produced *monoculture* community, such as a field of wheat,
is singularly vulnerable to failure owing to weather or
attack from insects or plant disease. The importance of
biodiversity is discussed further in Focus Study 20-1.

   Humans simplify communities by eliminating biodiversity, and in this way we place more ecosystems at risk
of harmful change and perhaps failure. In some regions,
simply planting multiple crops brings more stability to
the ecosystem. This is an important principle of *sustainable agriculture*, a movement to make farming more eco-
logically compatible.

***Agricultural Ecosystems.*** A modern agricultural ecosystem not only is vulnerable to failure because of its lack of
ecological diversity, but also creates an enormous demand
for energy, chemical pesticides and herbicides, artificial
fertilizer, and irrigation water. The practice of harvesting
and removing biomass from the land interrupts the cycling
of materials into the soil. This net loss of nutrients must
be artificially replenished.

   Energy from fossil fuels is used to manufacture chemicals, to operate mechanized farm equipment, and to run

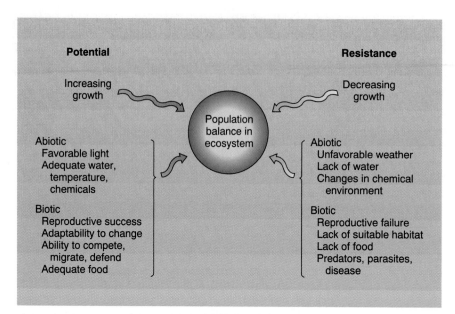

**FIGURE 19-23**
**Factors controlling the population balance of an ecosystem.**
Potential factors and resistance factors ultimately determine the population of an ecosystem.

Potential

Increasing growth

Population balance in ecosystem

Resistance

Decreasing growth

Abiotic
Favorable light
Adequate water, temperature, chemicals

Biotic
Reproductive success
Adaptability to change
Ability to compete, migrate, defend
Adequate food

Abiotic
Unfavorable weather
Lack of water
Changes in chemical environment

Biotic
Reproductive failure
Lack of suitable habitat
Lack of food
Predators, parasites, disease

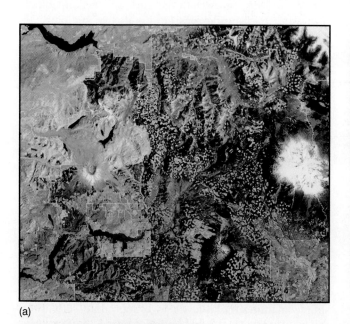

(a)

(b)

WASHINGTON
123°
Olympia
Mt. Rainier National Park
46°30'
12
Mt. St. Helens National Vol. Mon.
Mt. Adams
5
Portland
OREGON
123°
0    30    60 MILES
0    30    60 KILOMETERS

**FIGURE 19-24**
**Natural and human disruption of stable communities.**
(a) Natural disruption of forest by the Mount St. Helens volcanic eruption in 1980 and by logging in numerous clear-cut tracts of national forest. About 10% of the Northwest's old-growth forests remain, as identified through orbital images and GIS analysis. (b) A logged pine forest—an example of clear-cut timber harvesting that disrupted a stable community and produced drastic changes in microclimatic conditions along the former forest floor. [(a) *Landsat* image of a portion of the Gifford Pinchot National Forest and the Mount St. Helens National Volcanic Monument, April 29, 1992, centered at approximately 46.5° N 122° W, courtesy of Compton J. Tucker, NASA Goddard Space Flight Center. (b) Photo by author.]

# News Report 3

## Experimental Prairies Confirm the Importance of Biodiversity

Field experiments have confirmed an important scientific assumption: that greater biological diversity in an ecosystem leads to greater stability, productivity, and soil nutrient use within that ecosystem. For instance, during a drought, some species of plants will be damaged. In a diverse ecosystem, however, other species with deeper roots and better water-obtaining ability will thrive.

Ecologist David Tilman at the University of Minnesota tracked the operation of 147 grassland plots, each 3 m (9.8 ft) square. Species diversity was carefully controlled on each plot (Figure 1). Plots were sown with different numbers of native North American prairie plant seeds and cared for by a team of 50 people. The results clearly demonstrated that the plots with a more diverse plant community were able to retain and use nutrients more efficiently than the plots with less diversity. This efficiency reduced the loss of soil nitrogen through leaching, thus increasing soil fertility.

Greater plant diversity led to both higher productivity and better resource utilization. Also, total plant cover was found to increase with species richness. Tilman and his colleagues affirm:

This extends the earlier results to the field, providing direct evidence that the current rapid loss of species on Earth, and management practices that decrease local biodiversity, threaten ecosystem productivity and the sustainability of nutrient cycling. Observational, laboratory, and now field experimental evidence, supports the hypothesis that biodiversity influences ecosystem productivity, sustainability, and stability.*

---

*D. Tilman, D. Wedin, and J. Knops, "Productivity and sustainability influenced by biodiversity in grassland ecosystems," *Nature* 379 (February 22, 1996): 720. Also see the summary in *Science* 271 (March 15, 1996): 1497.

**FIGURE 1**
**Biodiversity experiment.**
Experimental plots are planted with native North American prairie grasses. A total of 147 plots were sown with random seed selections of 1, 2, 4, 6, 8, 12, or 24 species. Experimental results confirm many assumptions regarding the value of rich biodiversity. [Photo by David Tilman/University of Minnesota, Twin Cities Campus.]

---

pumps for wells and irrigation. (Focus Study 9-2 discusses the problem of fuel costs for pumping water from the Ogallala Aquifer in the U.S. Midwest.) High energy cost and fuel availability problems can lead to crop failures. Unfortunately, the promise of increased yields through a *green revolution* in agriculture is dampened by these investment requirements.

***Climate Change.*** Obviously, the distribution of plant species is affected by climate change. Many species have survived wide climate swings in the past. Consider the beginning of the Tertiary Period, 75 million years ago. Warm, humid conditions and tropical forests dominated the land to southern Canada, pines grew in the Arctic, and deserts were few. Then, between 15 and 50 million years ago, deserts began developing in the southwestern United States.

Mountain-building processes created higher elevations, causing rain-shadow aridity and affecting plant distribution.

Recall, too, that the movement of Earth's tectonic plates created climate changes important in the evolution and distribution of plants and animals (see Figure 11-17a–e). Europe and North America were joined in Pangaea and positioned near the equator, where vast swamps formed (the site of resulting coal deposits today). The southern mass of Gondwana was extensively glaciated as it drifted at high latitudes in the Southern Hemisphere. This glaciation left matching glacial scars and specific distributions of plants and animals across South America, Africa, India, and Australia. The diverse and majestic dinosaurs dispersed with the drifting continents.

The key question for our future is: As temperature patterns change, how fast can plants either adapt to new conditions or migrate through succession (location change) to remain within their specific habitats?

Adaptation to conditions is key to evolution. Through mutation and natural selection, species have either adapted or failed to adapt to changing environmental conditions over millions of years. Current global change is occurring rapidly, at the rate of decades instead of millions of years. Thus, we see die-out and succession of different species along disadvantageous habitat margins. The displaced species may colonize new regions made more hospitable by climate change. Also shifting will be agricultural lands that produced wheat, corn, soybeans, and other commodities. Society will have to adapt to different crops growing in different places. Rapid environmental change can lead to outright extinction of plants and animals that are unable to adapt or disperse.

As an example, a transition area between prairie grassland and northern forest presently exists in central Minnesota, but with increasing temperatures this transition zone is expected to migrate northward 400–600 km (250–375 mi) over the next 40 years, an incredible adjustment of more than 100 km (60 mi) per decade.

A study completed by biologist Margaret Davis on North American forests suggests that trees will have to respond quickly if temperatures increase. Changes in the climate inhabited by certain species could shift 100–400 km (60–250 mi) during the next 100 years. Some species, such as the sugar maple, may migrate northward, disappearing from the United States except for Maine, and moving into eastern Ontario and Québec. Davis prepared a map showing the possible impact of increasing temperatures on the distribution of beech and hemlock trees (Figure 19-25).

> Changes in the geographical distributions of plant and animal species in response to future greenhouse warming threaten to reduce biotic diversity. . . . The risk posed by $CO_2$-induced warming depends on the distances that regions of suitable climate are displaced northward [in the Northern Hemisphere] and on the rate of displacement. . . . If the change occurs too rapidly for colonization of newly available regions, population sizes may fall to critical levels, and extinction will occur.[*]

## Ecological Succession

**Ecological succession** occurs when older communities of plants and animals (usually simpler) are replaced by newer communities (usually more complex). Each successive community of species modifies the physical environment in a manner suitable for a later community of species. Changes apparently move toward a more stable and mature condition, toward an optimum for a specific environment.

[*]M. B. Davis and C. Zabinski, "Changes in the Geographical Range Resulting from Greenhouse Warming: Effects on Biodiversity in Forests," in R. L. Peters and T. E. Lovejoy, eds., *Global Warming and Biological Diversity* (New Haven: Yale University Press, 1992), p. 297.

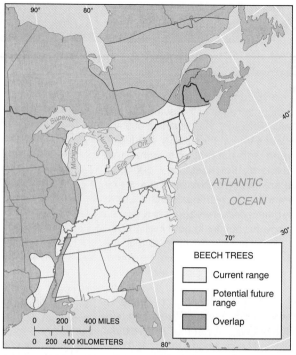

(a)

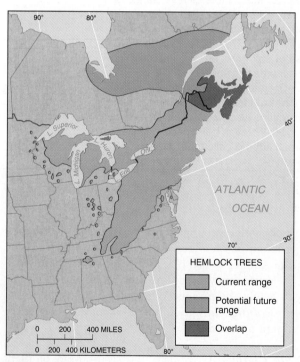

(b)

**FIGURE 19-25**

**Present and predicted distribution of two tree species.**

(a) Beech trees and (b) hemlock trees in North America. The potential future range shown for each reflects expected climate change resulting from a doubling of carbon dioxide. These maps are based on forecasts from the Goddard Fluid Dynamics Laboratory (GFDL) general circulation model. [After M. B. Davis and C. Zabinski, "Changes in the Geographical Range Resulting from Greenhouse Warming: Effects on Biodiversity in Forests," in R. L. Peters and T. E. Lovejoy, eds., *Global Warming and Biological Diversity* (New Haven: Yale University Press, 1992), p. 301.]

Traditionally, it was assumed that plants and animals formed a *climax community*—a stable, self-sustaining, and symbiotically functioning community with balanced birth, growth, and death—but this notion has been mostly abandoned by scientists. Contemporary conservation biology, biogeography, and ecology assume nature to be in constant adaptation. Rather than thinking of an ecosystem as a uniform set of communities, think of ecosystems as a patchwork mosaic of habitats—each striving to achieve an optimal range and low environmental stress.

Given the complexity of natural ecosystems, it is obvious that real succession involves much more than a series of predictable stages ending with a specific monoclimax community. Instead, there may be several final stages, or a polyclimax condition, with adjoining ecosystems at different stages in the same environment. Mature communities are properly thought of as being in dynamic equilibrium and at times may be out of phase with the immediate physical environment because of a lag time in their adjustment.

Succession often requires an initiating disturbance. External examples include wind storms, severe flooding, a volcanic eruption, a devastating wildfire, or an agricultural practice such as prolonged overgrazing. When existing organisms are disturbed or removed, new communities can emerge. At such times of disequilibrium transition, the interrelationships among species produce elements of *chance*, and species having an adaptive edge will succeed in the competitive struggle for light, water, nutrients, space, time, reproduction, and survival (Figure 19-26). Thus, the succession of plant and animal communities is an intricate process with many interactive variables, both internal to the ecological structure and external to it.

Land and water experience different forms of succession. We first look at terrestrial succession, which is characterized by competition for sunlight, and then at aquatic succession, characterized by progressive changes in nutrient levels.

### Terrestrial Succession

An area of bare rock and soil with no vestige of a former community can be a site for **primary succession**. Examples are any new surface created by mass movement of land, areas exposed by a retreating glacier, cooled lava flows, or lands disturbed by surface mining, clear-cut logging, land development, or volcanic eruption. Illustrating the latter is primary succession north of Mount St. Helens in Washington State, shown in Figure 19-27. Primary succession often begins with lichens and mosses growing on bare rock (see Figure 19-5).

More common is **secondary succession**, which begins if the vestiges of a previously functioning community are present. An area where the natural community has been destroyed or disturbed, but where the underlying soil remains intact, may experience secondary succession. In terrestrial ecosystems, secondary succession

**FIGURE 19-26**
**The aftermath of a forest wildfire.**
The 1988 Yellowstone fire was devastating. Yet this area is in recovery today. Succession progresses, perhaps by chance, as species with an adaptive edge succeed. [Photo by Randall Christopherson.]

begins with pioneer species that form a **pioneer community**. As succession progresses, soil develops and a different set of plants and animals with different niche requirements may adapt. Further niche expansion follows as the community matures. For example, when a previously farmed field is permitted to lie fallow (unused), it initially fills with annual weeds, then perennial weeds and grasses, followed by shrubs and perhaps trees.

Figure 19-28 illustrates such a sequence for the southeastern United States. Secondary succession begins on an abandoned farm of formerly plowed fields (left side of illustration). Crabgrass and ragweed quickly take hold, seeded by winds and birds, and they do well in the direct sunlight. These slowly are choked out by taller, sturdier grasses and shrubs that invade and stabilize the soil, adding nutrients and organic matter. After a quarter-century or so, pines come to dominate the land.

The shade created by the pine forest produces conditions in which grass and shrub seed germination becomes more difficult. But shade-tolerant, slow-growing oak and hickory hardwoods readily take root beneath the pines. These hardwoods eventually grow taller than the pine forest and shade it, so the pines slowly die back under reduced light conditions. A fairly stable mature forest of oak and hickory is in place after 150 to 200 years (right side of illustration). However, despite the convenient categorization shown in the figure, you should think of a succession of communities as *overlapping* in time and space.

One challenge in the study of plant and animal communities is to devise criteria that allow researchers to

**FIGURE 19-27**
**Pioneer species begin primary succession.**
Succession of pioneer species in the devastated area north of Mount St. Helens, 3 years after the 1980 eruption. [Photo by author.]

distinguish between a *successional community* and a *mature community.* The relative stability of the community appears to be the key factor. Unfortunately, stability must be assessed over a period of time that greatly exceeds the human life span. And stability may be disrupted from time to time by phenomena ranging from storms to volcanoes to widespread disease in one or more species.

***Succession through the Ice Age.*** Glaciation is a large-scale disturbance that well illustrates how succession operates. As Earth cycled through glacial and interglacial ages (Chapter 17), the ecological succession in all ecosystems was affected repeatedly. Imagine the milder midlatitude climate at the beginning of an interglacial, with lush herb and shrub vegetation slowly giving way to pioneer trees of birch, aspen, and pine. Winds and animals dispersed seeds, and changing communities readily spread in response to changing conditions. During the warm interglacial, trees increased shade and the organic content of soil increased. Deciduous trees such as oak, elm, and ash spread freely on well-drained soils; willow, cottonwood, and alder grew on poorly drained land.

As climates cooled with the approach of the next glacial, soils became more acidic, so podzolization processes (see Chapter 18) dominated, helped by the acidity of spruce and fir trees. Ecosystems slowly returned to the vegetation community that existed at the beginning of the interglacial. The increased cold produced open regions of disturbed ecosystems. Highly acidic soils were disrupted by freezing and the advance of ice.

The Holocene (the past 10,000 years) is apparently atypical, because of the development of human societies and agriculture. Civilization is creating an artificial and accelerated succession, similar in some ways to the closing centuries of an interglacial. The challenge for biogeographers

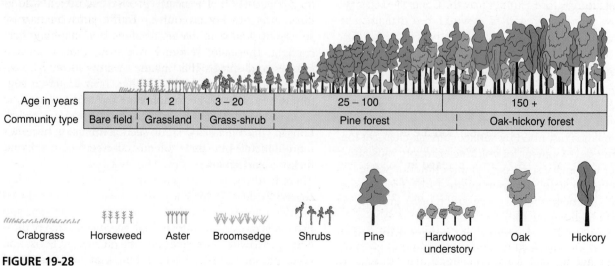

| Age in years | | 1 | 2 | 3 – 20 | 25 – 100 | 150 + |
|---|---|---|---|---|---|---|
| Community type | Bare field | Grassland | | Grass-shrub | Pine forest | Oak-hickory forest |

Crabgrass   Horseweed   Aster   Broomsedge   Shrubs   Pine   Hardwood understory   Oak   Hickory

**FIGURE 19-28**
**Secondary succession.**
Typical secondary succession of principal plants in the southeastern United States. [Adapted from E. P. Odum, *Fundamentals of Ecology*, 3rd ed., figure 9-4, p. 261. © 1971 Saunders College Publishing. Adapted by permission.]

is to understand such long-term successional change during a time of short-term change in climate and landscape.

***Wildfire and Fire Ecology.*** It is estimated that one-fourth of Earth's land area experiences fire each year. Over the past 50 years, the role of fire in ecosystems has been the subject of much scientific research and experimentation. Today, fire is recognized as a natural component of most ecosystems and not the enemy of nature that it once was characterized to be. In fact, in many forests, undergrowth is purposely burned in controlled "cool fires" to remove fuel that could enable a catastrophic and destructive "hot fire." Forestry experts have learned that, when fire-prevention strategies are rigidly followed, they can lead to the abundant undergrowth accumulation that fuels major fires.

Modern society's demand for fire prevention to protect property goes back to European forestry of the 1800s. Fire prevention became an article of faith for forest managers in North America. But, in studies of the longleaf pine forest that stretches in a wide band from the Atlantic coastal plain to Texas, fire was discovered to be an integral part of regrowth following lumbering. In fact, seed dispersal of some pine species, such as the knobcone pine, does not occur unless assisted by a forest fire! Heat from the fire opens the cones, releasing seeds so they can fall to the ground for germination. Also, these fire-disturbed areas quickly recover with protein-rich woody growth, young plants, and a stimulated seed production that provides abundant food for animals.

The science of **fire ecology** imitates nature by recognizing fire as a dynamic ingredient in community succession. The U.S. Department of Agriculture's Forest Service first recognized the principle of fire ecology in the early 1940s and formally implemented the practice in 1972. Their challenge became one of controlling fires to prevent accumulation of forest undergrowth. Controlled ground fires now are widely regarded as wise forest management practice and are used across the country (Figure 19-29).

Nonetheless, after some 72,000 forest fires in the western United States in 1988, especially those that charred portions of the highly visible Yellowstone National Park, an outcry was heard from forestry and recreational interests. The demand was for the Forest Service and ecologists to admit they were wrong and to abandon fire ecology. Critics called fire ecology practices the government's "let burn" policy. In truth, about 20% of the acreage in Yellowstone Park actually burned—180,000 out of 900,000 hectares (440,000 of 2,200,000 acres). Only about half of that area experienced the worst fire damage (Figure 19-30). This expanse was far less than originally reported in the media.

In its final report on the Yellowstone fire, a government interagency task force concluded that "an attempt to exclude fire from these lands leads to major unnatural changes in vegetation . . . as well as creating fuel accumulation that can lead to uncontrollable, sometimes very

**FIGURE 19-29**
**Controlled burning.**
Fire ecology practices of the U.S. Forest Service. Controlled burning has been used for several decades. [Photo by author.]

damaging wildfire." Thus, participating federal land managers and others reaffirmed their stand that fire ecology is a fundamentally sound concept.

An increasingly serious problem is emerging as people seek to live in the wildlands near cities. Urban development has encroached on forests, and the added suburban landscaping has created new fire hazards. Uncontrolled wildfires now can destroy homes and threaten public safety. Fire suppression demanded by the people living in the areas causes undergrowth to accumulate and worsens the risk.

As stark evidence of this danger, the East Bay Hills, in the cities of Berkeley and Oakland in California, burned for 3 days in 1991. The heavily landscaped urban "wilderness" was quickly devoured. Traffic jams on narrow streets blocked both fleeing residents and incoming firefighting equipment. When it was over, 1600 acres had burned, including 2700 homes, businesses, and apartments. Damage totaled $1.7 billion, with 25 dead. Suggestions to reduce the hazard include zoning to exclude development, the control of landscaping activities, removal of undergrowth, and use of fire-resistant building materials—and the use of controlled fire to prevent tinder accumulation.

### Aquatic Succession

Ecosystems occur in water as well as on land. Lakes, estuaries, ponds, and shorelines are complex ecosystems. The concepts of terrestrial succession that you just read about—stability and resilience, biodiversity, community succession—apply to open water, shoreline, and watershed systems as well.

**FIGURE 19-30**

**Yellowstone burns.**

Yellowstone National Park in northwestern Wyoming: (a) 1988 wildfire progresses up a slope. (b) Map of burned vegetation and description of points of origin. (c) An area of the park is shown in recovery in 1996 as community succession progresses. [(a) Photo by Joe Peaco/National Park Service. (c) Photo by Mark Newman/Photo Researchers, Inc.]

(a)

(b)

(c)

A lake or pond is really a temporary feature on the landscape, when viewed across geologic time. Lakes and ponds exhibit successional stages as they fill with nutrients and sediment and as aquatic plants take root and grow. This growth captures more sediment and adds organic debris to the system (Figure 19-31). This gradual enrichment in water bodies is known as *eutrophication* (from the Greek *eutrophos* meaning "well nourished").

In moist climates, a floating mat of vegetation grows outward from the shore to form a bog. Cattails and other marsh plants become established, and partially decomposed organic material accumulates in the basin, with additional vegetation bordering the remaining lake surface. A meadow may form as water is displaced by vegetation and soil; willow trees follow, and perhaps cottonwood trees; and eventually the lake may evolve into a forest community.

The progressive stages in lake succession are named for their nutrient levels: *oligotrophic* (low nutrients), *mesotrophic* (medium nutrients), and *eutrophic* (high nutrients). Each stage is marked by greater primary

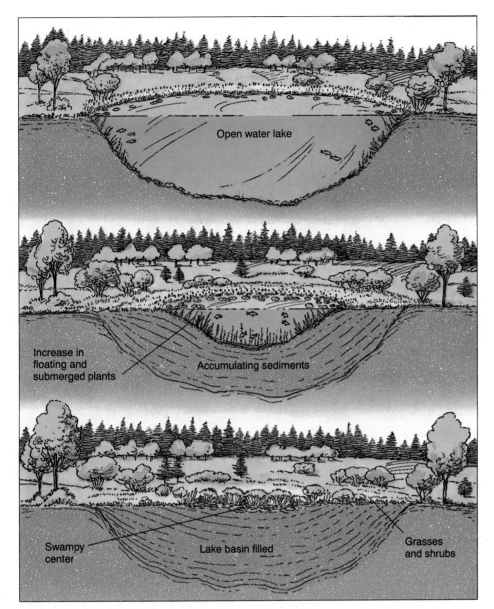

Open water lake

Increase in floating and submerged plants

Accumulating sediments

Swampy center

Lake basin filled

Grasses and shrubs

**FIGURE 19-31**
**Lake-bog-meadow succession.**
What begins as a lake gradually fills with organic and inorganic sediments, which successively shrink the area of the pond. A bog forms, then a marshy area, and finally a meadow completes the successional stages.

productivity and resultant decreases in water transparency, so that photosynthesis becomes concentrated near the surface. Energy flow shifts from production to respiration in the eutrophic stage, with oxygen demand exceeding oxygen availability.

Nutrient levels also vary spatially: Oligotrophic conditions occur in deep water, whereas eutrophic conditions occur along the shore, in shallow bays, or where sewage, fertilizer, or other nutrient inputs occur. Even large bodies of water may have eutrophic areas along the shore. As society dumps sewage and pollution in water-

ways, the nutrient load is enhanced beyond the cleansing ability of natural biological processes. The result is *cultural eutrophication*, which hastens succession in aquatic systems. As with all ecosystems, we must be aware of such signals that unwanted change is occurring so that mitigating action can be taken.

Focus Study 19-1 examines the Great Lakes of North America and their geographically and biologically diverse aquatic ecosystems. This is the largest lake system on Earth and is jointly managed by two countries.

## Focus Study 19-1

## The Great Lakes

The basins of the Great Lakes were a gift of the last ice age to North America (see Figure 17-18). The glaciers advanced and retreated over this region, excavating the basins for five large lakes. This international waterway of lakes and connecting rivers has played an important role in the history of Canada and the United States. Today, about 10% of the U.S. population and 25% of Canada's population live in the Great Lakes drainage basin (Figure 1). Major agricultural regions surround the lakes, as do many centers of industrial activity. Tourism, sport fishing, and maritime commerce are also important.

Society asks much of the Great Lakes: to dilute wastes from cities and industry, to dissipate thermal pollution from power plants, to provide municipal drinking water and irrigation water, and to sustain unique and varied ecosystems— open lake, coastal shore, coastal marsh, lakeplain (former lake bed), inland wetlands, and inland terrestrial (upland areas of forest, prairie, and barrens). Here is a brief profile of this important hydrological and ecological resource.

### Geography and Physical Characteristics

The Great Lakes—Superior, Michigan, Huron, Erie, and Ontario—contain 18% of the total volume of all freshwater lakes in the world, some 23,000 km³ (5500 mi³) of water. Their combined surface area covers 244,000 km² (94,000 mi²), a little less than the state of Wyoming. The entire drainage basin embraces 528,000 km² (204,000 mi²), or an area about the size of Manitoba.

As we tour the lakes, it is useful to follow along in Figure 1 and to look at their profile in Figure 2a. Lake Superior is the highest in elevation, highest in latitude, deepest, and largest in the system. It drains through St. Mary's River into Lake Huron. Lake Michigan, the only lake of the five that lies entirely within the United States, is at the same level as Lake Huron because it is joined by the wide connection through the Straits of Mackinac. (This is why, in Figure 2, we combine these two lakes on one hydrograph.)

The St. Clair River, Lake St. Clair, and the Detroit River carry water on to Lake Erie, the shallowest lake in the system. Compared to an average water retention time of 191 years in Lake Superior, Lake Erie has the shortest retention time of 2.6 years. The Niagara River, plunging dramatically over Niagara Falls, transports water into Lake Ontario. Lake Ontario is drained by the St. Lawrence River, which carries the entire discharge of the Great Lakes system to the Gulf of St. Lawrence and eventually into the North Atlantic Ocean. The entire basin extends over 10° of latitude (41° N to 51° N) and 18° of longitude (75° W to 93° W).

Because of the large size of the overall Great Lakes system, its associated climate, soils, and topography vary widely. Dominating the north are colder microthermal climates, exposed portions of the Canadian Shield bedrock, and vast stands of conifers and acidic soils. To the south, warmer mesothermal climates and fertile glacially deposited soils provide a vast agricultural base. Previous stands of mixed forest have been replaced by farms and urbanization.

Prevailing winds from the west and seasonal change mix air masses from different source regions, producing variable weather. The presence of the Great Lakes basin strongly influences passing air masses and weather systems (see Figure 8-13). Precipitation over the drainage basin feeds the lake storage as part of the renewable hydrologic cycle. Figure 2 illustrates Great Lakes water levels from 1918 to 1995.

Short-term lake levels vary from winter to summer; they are higher in summer after snowmelt and the arrival of maximum-summer precipitation. Over the long term, highest lake levels occur during years of heavier precipitation and during times of cooler temperatures, which reduce evaporation. Wind also affects lake levels. A *wind set-up* occurs along the downwind shoreline, as water is pushed higher on shore.

### The Great Lakes Ecosystem

The Great Lakes ecosystem should be thought of as young and a fruitful laboratory for ecological studies. These lakes are not simply large bodies of water with a uniform mixture. A stratification occurs in them, related to density and temperature differences. In the summer, the surface and shallow waters are warmed and become less dense. The warm surface water and light penetration support most biotic production and adequate dissolved oxygen. The lakes develop a sharp stratification. This layering is important to water quality because it can prevent mixing of pollution and other effluents during the summer months.

As the fall season matures, surface water cools and sinks, displacing deeper water and creating a turnover of the lake mass. By midwinter, the temperature from the surface to the bottom is around 4°C (39°F, the point at which water is densest); temperatures at the surface are near freezing.

The lakes support a food chain of producers and consumers as does any aquatic ecosystem. Native fish populations have been greatly affected by human activities: overfishing, introduced nonnative species, pollution from excess nutrients, toxic contamination of fish, and disruption of spawning habitats.

Peak commercial fishing occurred in the 1880s. The fisheries declined to their lowest levels as water pollution peaked in the 1960s and early 1970s. Strong pollution control programs were jointly implemented by the U.S. and Canadian governments, along with the work of many citizens, industries, and private organizations. This brought the lakes into the present period of recovery. Today, lake trout (in Lake Superior), sturgeon, herring, smelt, alewife, splake, yellow perch, walleye, and white bass are fished. However, health advisories are occasionally issued warning of danger in consuming fish of certain species, sizes, and locales.

Coastline for the five lakes totals 18,000 km (11,000 mi) and is diverse: major dunes and sandy beaches, bedrock shorelines, gravel beaches, and adjoining coastal marsh systems. Marshes can be found at the far western end of Lake Superior in the St. Louis River estuary and at the far eastern end of Lake

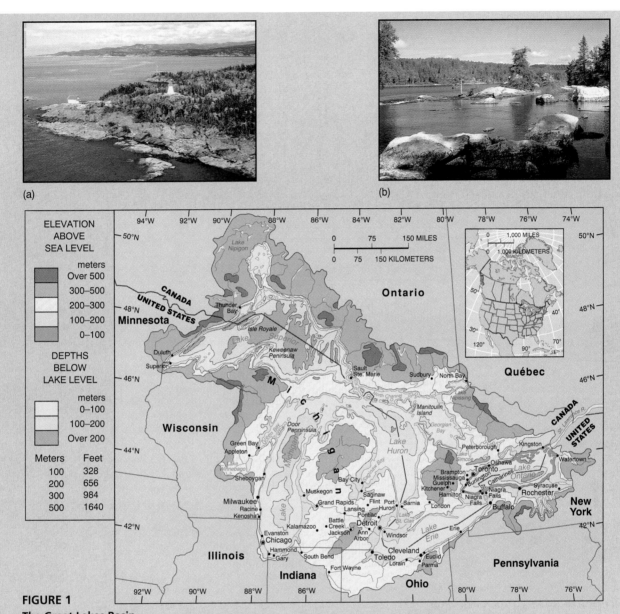

**FIGURE 1**

**The Great Lakes Basin.**

Elevations above sea level, depths below lake level, and principal urban areas are shown. The map outline is the limit of the drainage basin and watershed for the Great Lakes system. (a) The rocky shores of Lake Superior, Otter Island lighthouse. (b) Forests along Lake Ontario shores. (c) The Cuyahoga River enters Lake Erie at Cleveland. (d) Sand dunes along the southern shore of Lake Michigan. [Photos (a) by Carl R. Sams, II/Peter Arnold, Inc.; (b) by Wayne Lankinen/DRK Photo, (c) by Alex S. MacLean/Peter Arnold, Inc.; (d) by Cathlyn Melloan/Tony Stone Images. Map courtesy of Environment Canada, U.S. EPA, and Brock University cartography.]

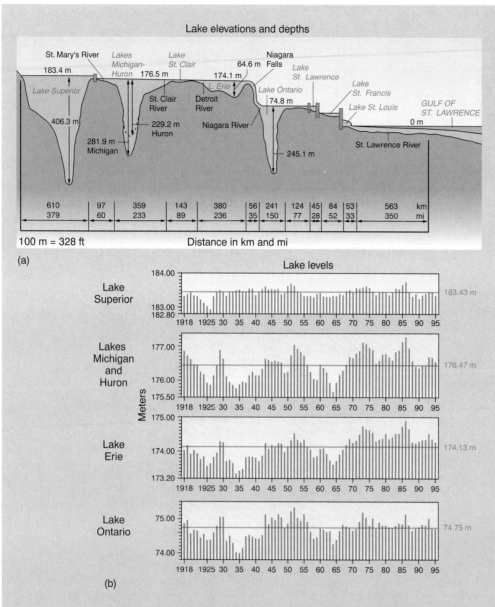

Lake elevations and depths

**FIGURE 2**

**Great Lakes elevation profile and hydrographs.**
(a) Average depth and lake surface levels, 1918–1995. (b) Annual hydrographs for each of the Great Lakes, 1918–1995. Both figures were prepared using the International Great Lakes Datum of 1985. [Data courtesy of the International Coordinating Committee on the Great Lakes Basic Hydraulic and Hydrographic Data.]

Ontario and its coastal lagoon. Here, as elsewhere, marshes occupy that key interface between the land and the water, storing and cycling organic material and nutrients into aquatic ecosystems.

**Human Activities, Land Use, Loss, and Recovery**

Figure 3 portrays land use in the Great Lakes Basin. Each activity has its own impact on lake systems. Runoff from agricultural areas contains chemicals and eroded soil. The pulp and paper industry was a major polluter but has improved as more was learned about the dangers of pollution; for example, mercury contamination was halted in the 1970s and better disposal methods for

wastes were implemented. The forests were not always treated in a sustainable way in terms of reforestation, and today forests may be regarded as a diminished resource in the basin. Recovery efforts are slow to make up for past practices.

Lake Erie was the first of the Great Lakes to show severe effects of cultural eutrophication, for it is the shallowest and warmest of the five lakes. About a third of the overall population in the Great Lakes Basin lives in the drainage of Lake Erie, making it the major recipient of sewage effluent from treatment plants. Over the years, dissolved oxygen levels dropped as biochemical oxygen demand increased. This demand was created by sewage-treatment discharges and algal

decay. Algae flourished in these eutrophic conditions and coated the beaches, turning the lake a greenish brown.

The most infamous incident in the 1960s was when the Cuyahoga River (through Cleveland, Ohio) became so contaminated that it caught fire. In 1972, Canada and the United States signed the Great Lakes Water Quality Agreement and the cleanup was under way. The government, spurred by public opinion, was moved to action, knowing that one of these priceless lakes was actually being lost. One example of the success of these efforts is that the input of phosphorus (an element with high biological activity that dramatically stimulates unwanted plant growth in aquatic ecosystems) was reduced by 90%!

Concern for the health of the entire Great Lakes ecosystem remains necessary because of the biological amplification of poisons in the food chain. For example, organic chemicals such as PCBs accumulate to dangerous levels. PCBs (polychlorinated biphenyls) are highly viscous, inert fluids used in electrical transformers and hydraulic systems, and they are carcinogenic. Samples reported by Environment Canada show PCBs in phytoplankton at 0.0025 ppm, in a rainbow smelt at 1.04 ppm, in a lake trout at 4.83 ppm, and higher in the food chain to a Herring Gull eggshell at 124 ppm! Later additions to the 1972 agreement reduced toxic levels, organic chemicals, and heavy metals entering the lakes. Since we are at the end of the amplification process, it is in our own best interest to help these lakes recover.

Biological diversity is at the heart of the integrity of the Great Lakes. Surveys are under way to inventory the entire ecosystem, establish a biodiversity baseline, assess current conditions of imperiled species, and strengthen efforts to conserve the Great Lakes system as a productive environment. Both the Canadian and U.S. governments are involved through an international commission, as is the Nature Conservancy through its Great Lakes Program.

An ecosystem approach to management of the Great Lakes Basin is the key to recovery. This approach considers the physical, chemical, and biological components that constitute the total system. Thus, a Herring Gull eggshell becomes an indicator of chemical pollution, blue-green algal blooms denote hastening eutrophication, and a loss of species richness and diversity exposes the destruction of habitat. This is where geographic principles of spatial analysis, an integrative tool such as a geographic information system, and remote-sensing capabilities come into play. A holistic, spatial approach to ecological problems is a marked departure from the past, when pollution problems were considered locally and separately.

Global thinking will help propel the next important task of completing a basin-wide inventory of biotic resources. This includes an assessment of regional biodiversity and all it implies about the health of this valuable Great Lakes system. These efforts will create more relevance in regulations, better records of compliance, and a strong sense of cooperation.

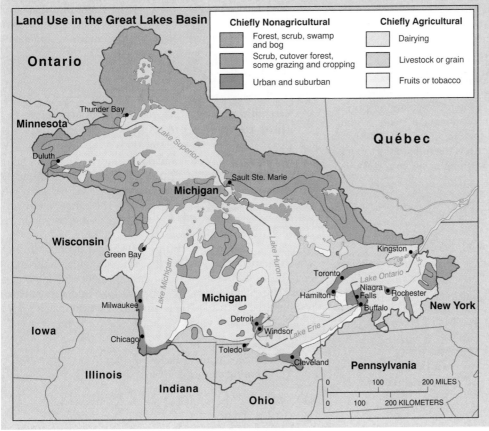

**FIGURE 3**

**Land use in the Great Lakes Basin.**

Generalized distribution of land use, nonagricultural and agricultural, in the Great Lakes Basin. [Map courtesy of Environment Canada, U.S. EPA, and Brock University cartography.]

# Summary and Review — Ecosystem Essentials

✔ ***Define* ecology, biogeography, and the ecosystem concept.**

Earth's biosphere is unique in the Solar System; its ecosystems are the heart of life. **Ecosystems** are self-sustaining associations of living plants and animals and their nonliving physical environment. **Ecology** is the study of the relationships between organisms and their environment and among the various ecosystems in the biosphere. **Biogeography** is the study of the distribution of plants and animals and the diverse spatial patterns they create.

> ecosystem (p. 582)
> ecology (p. 582)
> biogeography (p. 582)

1.  What is the relationship between the biosphere and an ecosystem? Define ecosystem and give some examples.

2.  What does biogeography include? Describe its relationship to ecology.

3.  Briefly summarize what ecosystem operations imply about the complexity of life.

---

✔ ***Describe* communities, habitats, and niches.**

A **community** is formed by the interactions among populations of living animals and plants at a particular time. Within a community, a **habitat** is the specific physical location of an organism—its address. A **niche** is the function or operation of a life form within a given community—its profession.

> community (p. 582)
> habitat (p. 582)
> niche (p. 582)

4.  Define a community within an ecosystem.

5.  What do the concepts of habitat and niche involve? Relate them to some specific plant and animal communities.

6.  Describe symbiotic and parasitic relationships in nature. Draw an analogy between these relationships and human societies on our planet. Explain.

---

✔ ***Explain* photosynthesis and respiration and *derive* net photosynthesis and the world pattern of net primary productivity.**

As plants evolved, the **vascular plants** developed conductive tissues. **Stomata** on the underside of leaves are the portals through which the plant participates with the atmosphere and hydrosphere. Plants (primary producers) perform **photosynthesis**, as sunlight stimulates a light-sensitive pigment called **chlorophyll**. This process produces food sugars and oxygen to drive biological processes. **Respiration** is essentially the reverse of photosynthesis and is the way the plant derives energy by oxidizing carbohydrates. **Net primary**

**productivity** is the net photosynthesis (photosynthesis minus respiration) of an entire community. This stored chemical energy is the **biomass** that the community generates, or the net dry weight of organic material.

> vascular plants (p. 585)
> stoma (p. 586)
> photosynthesis (p. 586)
> chlorophyll (p. 586)
> respiration (p. 586)
> net primary productivity (p. 587)
> biomass (p. 587)

7.  Define a vascular plant. How many plant species are there on Earth?

8.  How do plants function to link the Sun's energy to living organisms? What is formed within the light-responsive cells of plants?

9.  Compare photosynthesis and respiration and the derivation of net photosynthesis. What is the importance of knowing the net primary productivity of an ecosystem and how much biomass an ecosystem has accumulated?

10. Briefly describe the global pattern of net primary productivity.

---

✔ ***List* abiotic ecosystem components and *relate* those components to ecosystem operations.**

Light, temperature, water, and the cycling of gases and nutrients constitute the life-supporting abiotic components of ecosystems. Elevation and latitudinal position on Earth create a variety of physical environments. The zonation of plants with altitude is called **life zones** and is visible as you travel between different elevations.

Life is sustained by **biogeochemical cycles,** through which circulate the gases and sedimentary materials (nutrients) necessary for growth and development of living organisms. The environment may inhibit biotic operations, either through lack or excess. These **limiting factors** may be physical or chemical in nature.

> life zone (p. 590)
> biogeochemical cycles (p. 592)
> limiting factor (p. 595)

11. What are the principal abiotic components in terrestrial ecosystems?

12. Describe what Alexander von Humboldt found that led him to propose the life-zone concept. What are life zones? Explain the interaction among altitude, latitude, and the types of communities that develop.

13. What are biogeochemical cycles? Describe several of the essential cycles.

14. What is a limiting factor? How does it function to control the spatial distribution of plant and animal species?

✔ *Explain* **trophic relationships in ecosystems.**

*Trophic* refers to the feeding and nutrition relations in an ecosystem. It represents the flow of energy and cycling of nutrients. **Producers**, which fix the carbon they need from carbon dioxide, are the plants, including phytoplankton in aquatic ecosystems. **Consumers**, generally animals including zooplankton in aquatic ecosystems, depend on the producers as their carbon source. Energy flows from the producers through the system along a circuit called the **food chain**. Within ecosystems, the feeding interrelationships are complex and arranged in a **food web**.

The primary consumer is an **herbivore**, or plant eater. A **carnivore** (meat eater) is a secondary consumer. A consumer that eats both producers and consumers is called an **omnivore**—a role occupied by humans. **Decomposers**, the bacteria and fungi, process organic debris and release inorganic materials back to environment. *Detritus feeders*, including worms, mites, termites, and centipedes, also process debris. Biomass and population pyramids characterize the flow of energy and the numbers of producers, consumers, and decomposers operating in an ecosystem.

> producers (p. 596)
> consumers (p. 596)
> food chain (p. 596)
> food web (p. 596)
> herbivore (p. 597)
> carnivore (p. 597)
> omnivore (p. 597)
> decomposer (p. 597)

15. What role is played in an ecosystem by producers and consumers?

16. Describe the relationship among producers, consumers, decomposers, and detritus feeders in an ecosystem. What is the trophic nature of an ecosystem? What is the place of humans in a trophic system?

17. What are biomass and population pyramids? Describe how these models help explain the nature of food chains and communities of plants and animals.

18. Follow the flow of energy and biomass through the Silver Springs, Florida, ecosystem. Describe the pathways in Figures 19-16 and 19-21a.

---

✔ *Define* **succession and** *outline* **the stages of general ecological succession in both terrestrial and aquatic ecosystems.**

Ecosystems appear to move toward stability and balance, with individual terrestrial and aquatic ecosystems progressing through a succession of stages. **Biodiversity** describes the diversity of species population (both in number of different species and quantity of each species). The greater the biodiversity within an ecosystem, the more stable and resilient it is, and the more productive it will be. Modern agriculture often creates a nondiverse monoculture that is particularly vulnerable to failure.

Biogeographers trace communities across the ages, considering plate tectonics and ancient dispersal of plants and animals. Past glacial and interglacial climatic episodes have created a long-term succession. Climate change is forcing accelerated succession in ecosystems.

**Ecological succession** describes the process whereby older communities of plants and animals are replaced by newer communities that are usually more complex. An area of bare rock and soil with no trace of a former community can be a site for **primary succession**. **Secondary succession** begins in an area that has a vestige of a previously functioning community in place. The initial community that occupies an area of early succession is a **pioneer community**. Rather than progressing smoothly to a definable stable point, ecosystems tend to operate in dynamic equilibrium, with succeeding communities overlapping in time and space. Wildfire is one external factor that disrupts a successional community. Aquatic ecosystems also experience community succession, as exemplified by the eutrophication of a lake ecosystem.

The science of **fire ecology** has emerged in an effort to understand the natural role of fire in ecosystem maintenance and succession.

> biodiversity (p. 601)
> ecological succession (p. 604)
> primary succession (p. 605)
> secondary succession (p. 605)
> pioneer community (p. 605)
> fire ecology (p. 607)

19. What is meant by ecosystem stability?

20. How does ecological succession proceed? What are the relationships between existing communities and new, invading communities?

21. Discuss the concept of fire ecology in the context of the Yellowstone National Park fires of 1988. What were the findings of the government task force?

22. Summarize the process of succession in a body of water. What is meant by cultural eutrophication?

---

 **NetWork**

The *Geosystems Home Page* provides on-line resources for this chapter on the World Wide Web. You will find review exercises, specific updates for items in the chapter, suggested read-ings, and links to interesting related pathways on the Internet (click on the Table of Contents link and select this chapter). *Geosystems* is at: **http://www.prenhall.com/geosystm**

# 20

# Terrestrial Biomes

**Biogeographic Realms**

**Earth's Major Terrestrial Biomes**

**Summary and Review**

## Key Learning Concepts

After reading the chapter, you should be able to:

- *Define* the concept of biogeographic realms of plants and animals and *define* ecotone, terrestrial ecosystem, and biome.

- *Define* six formation classes and the life-form designations and *explain* their relationship to plant communities.

- *Describe* ten major terrestrial biomes and *locate* them on a world map.

- *Relate* human impacts, real and potential, to several of the biomes.

*Freshly mowed alfalfa fields near the Idaho-Wyoming border along Tin Cup Creek; Star Valley and the Salt River Range are in the distance.* [Photo by author.]

In 1874, Earth was viewed as offering few limits to human enterprise. But a visionary American diplomat and conservationist George Perkins Marsh perceived the need to manage the environment:

> We have now felled forest enough everywhere, in many districts far too much. Let us now restore this one element of material life to its normal proportions, and devise means of maintaining the permanence of its relations to the fields, the meadows, and the pastures, to the rain and the dews of heaven, to the springs and rivulets with which it waters the Earth.[*]

Unfortunately, his warning was ignored. At a time when we clearly see that Earth's natural systems pose limits to human activity, a United Nations conference issued this statement, almost 120 years after Marsh's writing:

> Forests worldwide have been and are being threatened by uncontrolled degradation and conversion to other types of land uses, influenced by increasing human needs, agricultural expansion, and environmentally harmful mismanagement. . . . The impacts of loss and degradation of forests are in the form of soil erosion, loss of biological diversity, damage to wildlife habitats and degradation of watershed areas, deterioration of the quality of life and reduction of the options for development. . . . The present situation calls for urgent and consistent action for conserving and sustaining forest resources.[**]

Today, scientists are measuring ecosystems carefully, and they are building elaborate computer models to simulate our evolving real-time human-environment experiment. In particular, climate change is being discussed. (You may have noticed that the subject came up often in this text.) We know from history that the effect of climate change on ecosystems will be significant. Two Harvard researchers assessed the impact of these alterations of the atmosphere:

> Based on more than a decade of research, it is obvious that the $CO_2$-rich atmosphere of our future will have direct and dramatic effects on the composition and operation of ecosystems. According to the best scientific evidence, we see no reason to be sanguine [optimistic] about the response of these habitats to our changing environment.[***]

What will the quotations about Earth's environment be like in the next century? Humans now have become a powerful biotic agent on Earth, influencing all ecosystems on a planetary scale. In this chapter we explore Earth's major terrestrial ecosystems, their appearance and structure, location, and the present status of related plants, animals, and environment.

# Biogeographic Realms

Figure 20-1 shows Earth's biogeographic realms. The upper map illustrates the botanical (plant) realms worldwide. The lower map shows the zoological (animal) realms. Each realm contains many distinct ecosystems that distinguish it from other realms.

A **biogeographic realm** is a geographic region where a group of plant and animal species has evolved. As you can see, these realms correspond generally to continents. The main topographic barrier that separates these realms is the ocean. Every species attempts to migrate worldwide according to its niche requirements, reproductive success, and competition, but it is constrained by climatic and topographic barriers. The collective results of the effort of all species to seek optimal habitats and to maximize each niche requirement results in the generalized realms shown on the maps. Recognition that such distinct regions of flora and fauna exist was an early beginning of *biogeography* as a discipline.

The Australian realm is unique, giving rise to 450 species of *Eucalyptus* among its plants and 125 species of marsupials (animals such as kangaroos, who carry their young in pouches, where gestation is completed) among its animals. Australia's uniqueness is the result of its early isolation from the other continents. During critical evolutionary times, Australia drifted away from Pangaea (see Chapter 11) and never again was reconnected by a land bridge, even when sea level was lowered during repeated glacial ages (Figure 20-2). New Zealand's isolation from Australia also undoubtedly explains why no native marsupials exist in New Zealand.

Alfred Wallace (1823–1913), the first scholar of *zoogeography*, used the stark contrast in animal species among islands in present-day Indonesia to delimit a boundary between the Oriental and Australian realms (Figure 20-1b). He thought that deep water in the straits had completely prevented species crossover. However, he later recognized that the boundary line actually is a wide transition zone, where one region grades into the other. Boundaries between natural systems are "zones of shared traits." Thus, despite distinctions among individual realms, it is best to think of their borders as transition zones of mixed identity and composition, rather than as rigidly defined boundaries.

A boundary transition zone between adjoining ecosystems is an **ecotone**. Because ecotones are defined by different physical factors, they vary in width—think of a band or zone instead of a line. Climatic ecotones usually are

---

[*]G. P. Marsh, *The Earth as Modified by Human Action* (New York: Scribner's, 1874), pp. 385–86.

[**]United Nations Conference on Environment and Development, "Topic 11. Combating Deforestation," *Agenda 21—Adoption of Agreements on Environment and Development*, Drafts, Rio de Janeiro: United Nations, A/CONF.151/4 (Part II), Section II. Conservation and Management of Resources for Development, paragraph 11.12 and 11.13, May 1, 1992, p. 31.

[***]F. A. Bazzaz and E. D. Fajer, "Plant Life in a $CO_2$-Rich World," *Scientific American* 256, no. 1 (January 1992): 74.

## (a) BOTANICAL GEOGRAPHIC REALMS

| I | Boreal | III | Neotropical |
| IIA | Paleotropical—African | IV | South African |
| IIB | Paleotropical—Indo-Malaysian | V | Australian |
| IIC | Paleotropical—Polynesian | VI | Antarctic |

## (b) ZOOLOGICAL GEOGRAPHIC REALMS

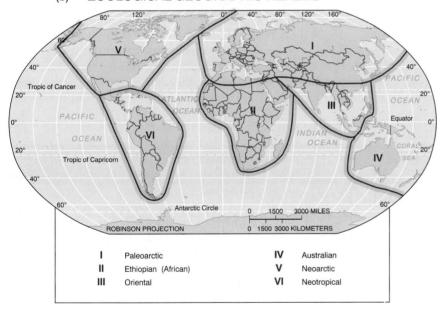

| I | Paleoarctic | IV | Australian |
| II | Ethiopian (African) | V | Neoarctic |
| III | Oriental | VI | Neotropical |

**FIGURE 20-1**

**Biogeographic realms.**

(a) Botanical geographic realms (after R. D. Good, 1947). (b) Zoological geographic realms (after L. F. deBeaufort, 1951). [Adapted from E. P. Odum, *Fundamentals of Ecology*, 3rd ed., figure 14-1. © 1971 Saunders College Publishing. Adapted by permission.]

more gradual than physical ecotones, whereas differences in soil, moisture availability, or topography sometimes form abrupt boundaries. An ecotone between prairies and northern forests may occupy many kilometers of land. The ecotone is an area of tension as similar species of plants and animals compete for the resource base.

### Terrestrial Ecosystems

A **terrestrial ecosystem** is a self-sustaining association of land-based plants and animals and their abiotic environment that is characterized by specific plant formation classes. Plants are the most visible part of the biotic landscape, and they are key members of Earth's terrestrial

ecosystems. In their growth, form, and distribution, plants reflect Earth's physical systems: its energy patterns; atmospheric composition; temperature and winds; air masses; water quantity, quality, and seasonal timing; soils; regional climates; geomorphic processes; and ecosystem dynamics. As discussed in the previous chapter, aquatic ecosystems are likewise important and portray the interaction of physical systems (see News Report 1).

A brief review of the ways that living organisms relate to the environment and to each other is helpful at this point. A *community* is formed by interacting populations of plants and animals in an area. An *ecosystem* involves the interplay between a community of plants and animals and their abiotic physical environment. Each

**FIGURE 20-2**
**The unique Australian realm.**
The flora and fauna of Australia form a special assemblage of communities. [Photo by Mark Newman/Photo Researchers, Inc.]

plant and animal occupies an area in which it is biologically suited to live—its *habitat*—and within that habitat it performs a basic operational function—its *niche* (see Chapter 19).

A **biome** is a large, stable terrestrial ecosystem characterized by specific plant and animal communities. Each biome usually is named for its *dominant vegetation*, because that is the single most easily identified feature. We can generalize Earth's wide-ranging plant species into six broad biomes: *forest, savanna, grassland, shrubland, desert,* and *tundra.* Because plant distributions respond to environmental conditions and reflect variations in climate and soil, the world climate map in Figure 10-5 is a helpful reference for this chapter. (The biome concept can also be applied to aquatic ecosystems: polar seas, temperate seas, tropical seas, sea floor, shoreline, and coral reef.)

These general biomes are divided into more-specific vegetation units called **formation classes**, which refer to the structure and appearance of dominant plants in a terrestrial ecosystem. Examples are equatorial rain forest, northern needleleaf forest, Mediterranean shrubland, and arctic tundra. Each formation class includes numerous plant communities, and each community includes innumerable plant habitats. Within those habitats, Earth's diversity is expressed in 270,000 plant species. Despite this intricate complexity, we can generalize Earth's numerous formation classes into 10 global terrestrial biome regions, as portrayed in Figure 20-4 and detailed in Table 20-1.

More specific characteristics are used for the structural classification of plants themselves. *Life-form designations* are based on the outward physical properties of individual plants or the general form and structure of a

## News Report 1

### Aquatic Ecosystems and the LME Concept

Oceans, estuaries, and freshwater bodies (streams and lakes) are home to aquatic ecosystems. Poor understanding of aquatic ecosystems has led to the demise of some species, such as herring in the Georges Bank prime fishing area of the Atlantic, and an overall decline in fisheries worldwide. One analysis tool is the designation of certain areas as *large marine ecosystems (LMEs)*. These ecoregions in the sea provide a basis for planning and research.

An LME is a distinctive oceanic region having unique organisms, floor topography, currents, areas of nutrient-rich upwelling circulations, or areas of significant predation, including human. Examples of identified LMEs include the Gulf of Alaska, California Current, Gulf of Mexico, Northeast Continental Shelf, and the Baltic and Mediterranean Seas. Thirty LMEs, each encompassing more than 200,000 km² (77,200 mi²), are presently defined worldwide.

The largest of 11 government-protected areas in North America is the Monterey Bay National Marine Sanctuary, established in 1992 within the California Current LME. (The Florida Keys Marine Sanctuary is second largest.) Stretching along 645 km (400 mi) of coastline and covering 13,500 km² (5300 mi²), the Monterey Bay sanctuary is home to 27 species of marine mammals, 94 species of shorebirds, 345 species of fish, and the largest sampling of invertebrates in any one place in the Pacific. This represents one of the most species-rich and diverse marine communities on Earth.

In the United States, these protected areas are administered by the Sanctuaries and Reserves Division of NOAA. The LME was conceived by Kenneth Sherman and Lewis Alexander, scientists with the U.S. Marine Fisheries Service Laboratory and the University of Rhode Island, respectively, to encourage resource planners and managers to consider *complete ecosystems*, not just a targeted species. This same strategy is used in studying terrestrial ecosystems. A goal of the LME concept is to improve the management and sustainability of both biotic and abiotic aquatic resources. Also, LMEs are designated to ensure long-term data collection for management and further research.

**TABLE 20-1**

## Major Terrestrial Biomes and Their Characteristics

| Biome and Ecosystems (map symbol) | Vegetation Characteristics | Soil Orders (Soil Taxonomy and CSSC) | Köppen Climate Designation | Annual Precipitation Range | Temperature Patterns | Water Balance |
|---|---|---|---|---|---|---|
| **Equatorial and Tropical Rain Forest (ETR)** Evergreen broadleaf forest Selva | Leaf canopy thick and continuous; broadleaf evergreen trees, vines (lianas), epiphytes, tree ferns, palms | Oxisols Ultisols (on well-drained uplands) | Af Am (limited dry season) | 180–400 cm (>6 cm/mo) | Always warm (21°–30°C; avg. 25°C) | Surpluses all year |
| **Tropical Seasonal Forest and Scrub (TrSF)** Tropical monsoon forest Tropical deciduous forest Scrub woodland and thorn forest | Transitional between rain forest and grasslands; broadleaf, some deciduous trees; open parkland to dense undergrowth; acacias and other thorn trees in open growth | Oxisols Ultisols Vertisols (in India) Some Alfisols | Am Aw Borders BS | 130–200 cm (<40 rainy days during 4 driest months) | Variable, always warm (>18°C) | Seasonal surpluses and deficits |
| **Tropical Savanna (TrS)** Tropical grassland Thorn tree scrub Thorn woodland | Transitional between seasonal forests, rain forests, and semiarid tropical steppes and desert; trees with flattened crowns, clumped grasses, and bush thickets; fire association | Alfisols (dry: Ultalfs) Ultisols Oxisols | Aw BS | 9–150 cm, seasonal | No cold-weather limitations | Tends toward deficits, therefore fire- and drought-susceptible |
| **Midlatitude Broadleaf and Mixed Forest (MBME)** Temperate broadleaf Midlatitude deciduous Temperate needleleaf | Mixed broadleaf and needleleaf trees; deciduous broadleaf, losing leaves in winter; southern and eastern evergreen pines demonstrate fire association | Ultisols Some Alfisols (Podzols, red and yellow) | Cfa Cwa Dfa | 75–150 cm | Temperate, with cold season | Seasonal pattern with summer maximum PRECIP and POTET (PET); no irrigation needed |
| **Needleleaf Forest and Montane Forest (NF/MF)** Taiga Boreal forest Other montane forests and highlands | Needleleaf conifers, mostly evergreen pine, spruce, fir; Russian larch, a deciduous needleleaf | Spodosols Histosols Inceptisols Alfisols (Boralfs: cold) (Gleysols) (Podzols) | Subarctic Dfb Dfc Dfd | 30–100 cm | Short summer, cold winter | Low POTET (PET), moderate PRECIP, moist soils, some waterlogged and frozen in winter; no deficits |
| **Temperate Rain Forest (TeR)** West coast forest Coast redwoods (U.S.) | Narrow margin of lush evergreen and deciduous trees on windward slopes; redwoods, tallest trees on Earth | Spodosols Inceptisols (mountainous environs) (Podzols) | Cfb Cfc | 150–500 cm | Mild summer and mild winter for latitude | Large surpluses and runoff |
| **Mediterranean Shrubland (MSh)** Sclerophyllous shrubs Australian eucalyptus forest | Short shrubs, drought adapted, tending to grassy woodlands; chaparral | Alfisols (Xeralfs) Mollisols (Xerolfs) (Luvisols) | Csa Csb | 25–65 cm | Hot, dry summers, cool winters | Summer deficits, winter surpluses |
| **Midlatitude Grasslands (MGr)** Temperate grassland Sclerophyllous shrub | Tallgrass prairies and shortgrass steppes, highly modified by human activity; major areas of commercial grain farming; plains, pampas, and veld | Mollisols Aridisols (Chernozemic) | Cfa Dfa | 25–75 cm | Temperate continental regimes | Soil moisture utilization and recharge balanced; irrigation and dry farming in drier areas |
| **Warm Desert and Semidesert (DBW)** Subtropical desert and scrubland | Bare ground graduating into xerophytic plants including succulents, cacti, and dry shrubs | Aridisols Entisols (sand dunes) | BWh BWk | <2 cm | Average annual temperature, around 18°C, highest temperatures on Earth | Chronic deficits, irregular precipitation events, PRECIP <½ POTET (PET) |
| **Cold Desert and Semidesert (DBC)** Midlatitude desert, scrubland, and steppe | Cold desert vegetation includes short grass and dry shrubs | Aridisols Entisols | BSh BSk | 2–25 cm | Average annual temperature around 18°C | PRECIP >½ POTET (PET) |
| **Arctic and Alpine Tundra (AAT)** | Treeless; dwarf shrubs, stunted sedges, mosses, lichens, and short grasses; alpine, grass meadows | Inceptisols Histosols Entisols (permafrost) (Organic) (Cryosols) | ET Dwd | 15–80 cm | Warmest months <10°C, only 2 or 3 months above freezing | Not applicable most of the year, poor drainage in summer |
| Ice | | | EF | | | |

vegetation cover. These physical life forms, portrayed in Figure 20-3, include:

- *Trees* (large woody main trunk, perennial, usually exceeds 3 m, or 10 ft, in height)
- *Lianas* (woody climbers and vines)
- *Shrubs* (smaller woody plants; branching stems at the ground)
- *Herbs* (small plants without woody stems; includes grasses and other nonwoody vascular plants)
- *Bryophytes* (nonflowering, spore-producing plants; includes mosses and liverworts)
- *Epiphytes* (plants growing above the ground on other plants, using them for support)
- *Thallophytes* (lack true leaves, stems, or roots; includes bacteria, fungi, algae, lichens)

Now let us go on a tour of Earth's biomes, keeping in mind that they synthesize all we have learned about the atmosphere, hydrosphere, lithosphere, and biosphere in the pages of this text.

# Earth's Major Terrestrial Biomes

Few natural communities of plants and animals remain; most biomes have been greatly altered by human intervention. Thus, the "natural vegetation" identified on many biome maps reflects *ideal potential* mature vegetation, given the physical characteristics in a region. Even though human practices have greatly altered these ideal forms, it is valuable to study the natural (undisturbed) biomes to better understand the natural environment and to assess the extent of human-caused alteration.

In addition, knowing the ideal in a region guides us to a closer approximation of natural vegetation in the plants we introduce. In the United States and Canada, we humans are perpetuating a type of transition community, somewhere between a grassland and a forest—a *disclimax*, or artificial successional stage, produced by large-scale interruption and disturbance. We plant trees and lawns and then must invest energy, water, and capital to sustain such artificial modifications. Or we graze animals and plant crops that perpetuate a disclimax. When these activities cease, natural succession recovers, and we see the land slowly return to its vegetation potential.

The global distribution of Earth's major terrestrial biomes, based on vegetation *formation classes*, is portrayed in Figure 20-4. Table 20-1 describes each biome on the map and summarizes other pertinent information from throughout this text.

Now is a good time to refer to "The Living Earth" composite image that appears inside the front cover of this text. A computer artist using hundreds of thousands of satellite images produced this cloudless view of Earth in natural color typical of a summer day. Compare the map in Figure 20-4 with that remarkable composite image and see what correlations you can make.

## *Equatorial and Tropical Rain Forest*

Earth is girdled with a lush biome—the **equatorial and tropical rain forest**. In a climate of consistent year-round daylength (12 hours), high insolation, average annual temperatures around 25°C (77°F), and plentiful moisture, plant and animal populations have responded with the most diverse expressions of life on the planet.

As Figure 20-4 shows, the Amazon region, also called the *selva*, is the largest tract of equatorial and tropical rain forest. In addition, rain forests cover equatorial regions of Africa, Indonesia, the margins of Madagascar and Southeast Asia, the Pacific coast of Ecuador and Colombia, and the east coast of Central America, with small discontinuous patches elsewhere. The cloud forests of western Venezuela are such tracts of rain forest at high elevation, perpetuated by high humidity and cloud cover. Undisturbed tracts of rain forest are rare.

In his 1960 book *The Forest and the Sea*, Marston Bates compared rain forests and oceans and found remarkable similarities in these very dissimilar environments. Both sea and forest have a vertical distribution of life, dependent on a competitive struggle for sunlight. Biomass in a rain forest is concentrated high up in the canopy, that dense mass of overhead leaves, just as life in the sea is concentrated in the photic layer near the ocean's surface. The treetops of the rain forest are generally analogous to the upper (pelagic) region of the open ocean, where photosynthesis fixes energy for a dependent food chain. The floor of the rain forest and the floor of the deep ocean also are roughly parallel in that both are dark or deeply shaded and both have fewer life forms than the upper regions.

Rain forests feature ecological niches that are distributed vertically rather than horizontally because of the competition for light. The canopy is filled with a rich variety of plants and animals. Lianas (vines) stretch from tree to tree, entwining them with cords that can reach 20 cm (8 in.) in diameter. Epiphytes flourish there too: Plants such as orchids, bromeliads, and ferns live entirely above ground, supported physically but not nutritionally by the structures of other plants. Windless conditions on the forest floor make pollination difficult, so pollination is by insects, other animals, and self-pollination.

The rain forest canopy forms three levels—see Figure 20-5. The upper level is not continuous but features emergent tall trees whose high crowns rise above the middle canopy. This level appears as a broken *overstory* of tall

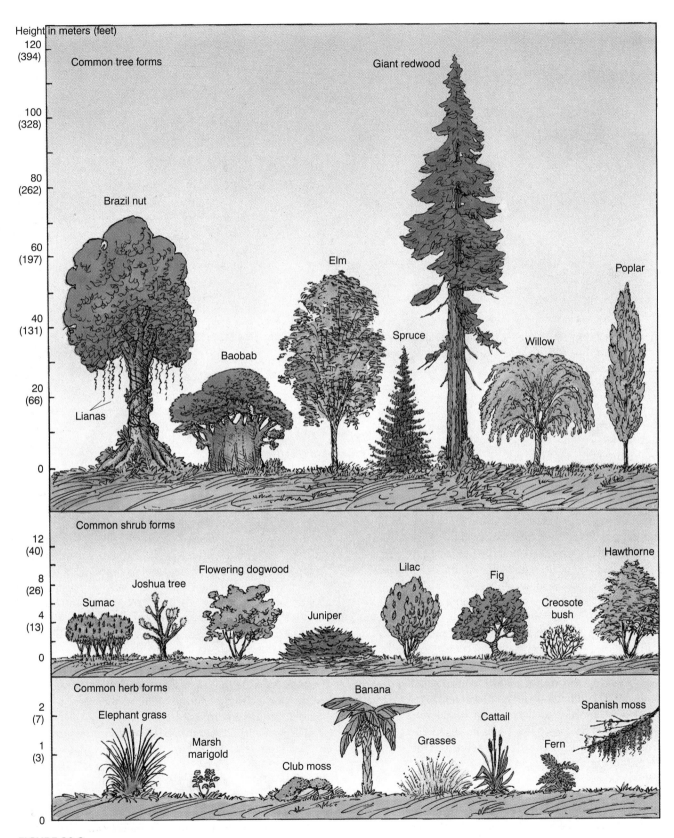

**FIGURE 20-3**
**Examples of plant life forms—tree, shrub, and herb.**
[After W. M. Marsh and J. Dozier, *Landscape: An Introduction to Physical Geography*, p. 273. © 1981.
John Wiley & Sons, Inc. Adapted by permission.]

**FIGURE 20-4**
**The 10 major global terrestrial biomes.**

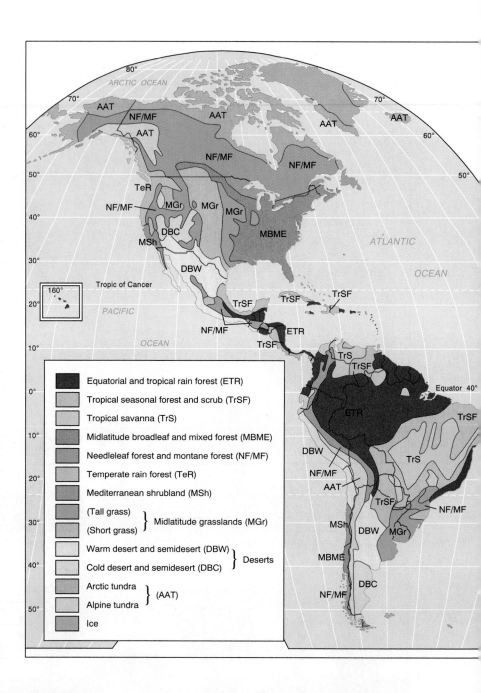

Equatorial and tropical rain forest (ETR)

Tropical seasonal forest and scrub (TrSF)

Tropical savanna (TrS)

Midlatitude broadleaf and mixed forest (MBME)

Needleleaf forest and montane forest (NF/MF)

Temperate rain forest (TeR)

Mediterranean shrubland (MSh)

(Tall grass) } Midlatitude grasslands (MGr)
(Short grass)

Warm desert and semidesert (DBW) } Deserts
Cold desert and semidesert (DBC)

Arctic tundra } (AAT)
Alpine tundra

Ice

trees breaking through a middle canopy that is nearly continuous (Figure 20-5b). The broad leaves block much of the light and create a darkened *understory* area and forest floor. The lower level is composed of seedlings, ferns, bamboo, and the like, leaving the litter-strewn ground level in deep shade and fairly open.

A look at aerial photographs of a rain forest, or views along riverbanks covered by dense vegetation, aided by the false Hollywood-movie imagery of the jungle, makes it difficult to imagine the shadowy environment of the actual rain forest floor, which receives only about 1% of the sunlight arriving at the canopy (Figure 20-6a). The constant moisture, rotting fruit and moldy odors, strings of thin roots and vines dropping down from above, windless air, and echoing sounds of life in the trees, together create a unique environment.

The smooth, slender trunks of rain forest trees are covered with thin bark and buttressed by large wall-like flanks that grow out from the trees to brace the trunks (Figure 20-6b). These buttresses form angular open enclosures, a ready habitat for various animals. There are usually no branches for at least the lower two-thirds of the tree trunks.

The wood of many rain forest trees is extremely hard, heavy, and dense—in fact, some species will not even float. (Exceptions are balsa and a few others, which are very light.) Varieties of trees include mahogany, ebony,

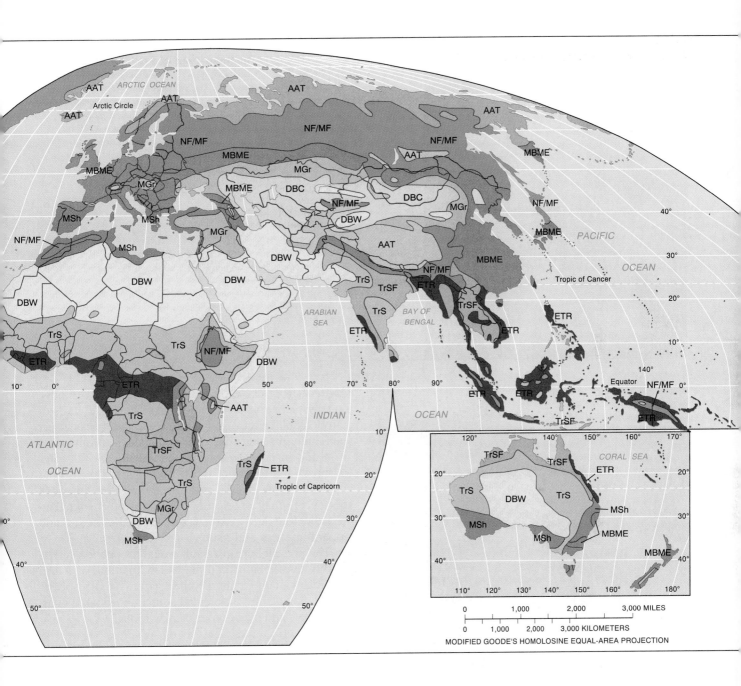

MODIFIED GOODE'S HOMOLOSINE EQUAL-AREA PROJECTION

and rosewood. Logging is difficult because individual species are widely scattered; a species may occur only once or twice per square kilometer. Selective cutting is required for species-specific logging, whereas pulpwood production takes everything. Conversion of the forest to pasture usually is accomplished by setting destructive fires.

Rain forests represent approximately one-half of Earth's remaining forests, occupying about 7% of the total land area worldwide. This biome is stable in its natural state, resulting from the long-term residence of these continental plates near equatorial latitudes and their escape from glaciation.

Rain forest soils, principally Oxisols, are essentially infertile, yet they support rich vegetation. The trees have adapted to these soil conditions with root systems able to capture nutrients from litter decay at the soil surface. High precipitation and temperatures work to produce deeply weathered and leached soils, characteristic of the laterization process, with a clayey texture sometimes breaking up into a granular structure, and lacking in nutrients and colloidal material. With much investment in fertilizers, pesticides, and machinery, these soils can be productive.

The animal and insect life of the rain forest is diverse, ranging from small decomposers (bacteria) working the surface to many animals living exclusively in the upper stories of the trees. These tree dwellers are referred to as

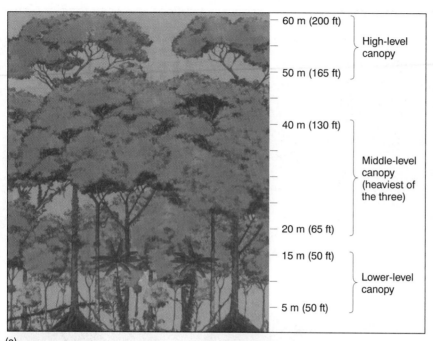

(a)

**FIGURE 20-5**

**The three levels of a rain forest canopy.**
(a) Lower, middle, and high levels of the rain forest. (b) The emergent trees of the high-level canopy are clearly seen above the dense, continuous cover of the middle canopy of a lowland tropical rain forest near the southwest coast of Costa Rica. [Photo by Gregory G. Dimijian/Photo Researchers, Inc.]

(b)

*arboreal*, from the Latin for "tree," and include sloths, monkeys, lemurs, parrots, and snakes. Beautiful birds of many colors, tree frogs, lizards, bats, and a rich insect community that includes over 500 species of butterflies are found in rain forests. Surface animals include pigs (bushpigs and the giant forest hog in Africa, wild boar and bearded pig in Asia, and peccary in South America), species of small antelopes (bovids), and mammalian predators (the tiger in Asia, jaguar in South America, and leopard in Africa and Asia).

The present human assault on Earth's rain forests has put this diverse fauna and the varied flora at risk. It also jeopardizes an important recycling system for atmospheric carbon dioxide and a potential source of valuable pharmaceuticals and many types of new foods.

***Deforestation of the Tropics.*** More than half of Earth's original rain forest is gone, cleared for pasture, timber, fuel wood, and farming. An area slightly smaller than Wisconsin is lost each year (169,000 km², 65,000 mi²) and about a third more is disrupted by selective cutting of canopy trees.

When orbiting astronauts look down on the rain forests at night, they see thousands of human-set fires. During the day, the lower atmosphere in these regions appears choked with the smoke (Figure 20-7a, b). These fires are used to clear land for agriculture, which is intended to feed the domestic population as well as to produce cash exports of beef, rubber, coffee, and other commodities. Edible fruits are not abundant in an undisturbed rain forest, but cultivated clearings produce bananas (plantains), mangos and jack fruit, guava, and starch-rich roots such as manioc and yams.

(a)

(b)

(c)

**FIGURE 20-6**
**The rain forest.**
(a) Equatorial rain forest is thick along the Amazon River where light breaks through to the surface, producing a rich gallery of vegetation. (b) The rain forest floor in Corcovado, Costa Rica, with typical buttressed trees and lianas. (c) Lush rain forest and misty canyon in Zimbabwe. [(a) Photo by Wolfgang Kaehler Photography; (b) photo by Frank S. Balthis; (c) photo by Carl Frank/Photo Researchers, Inc.]

Because of the poor soil fertility, the lands are quickly exhausted under intensive farming and are then generally abandoned in favor of newly burned and cleared lands, unless the lands are artificially maintained. Unfortunately, the dominant trees require from 100 to 250 years to establish themselves after major disturbances. Once cleared, the massive growth of low bushes intertwined with vines and ferns may be slow to return.

Figure 20-7c is a satellite image of a portion of western Brazil recorded in 1987, the peak year for losses in the Brazilian rain forest. This image gives a sense of the level of rain forest destruction in progress. You can clearly see encroachment along new roads (blue) branching from highway BR364. Scientists with the Goddard Space Flight Center and the Brazilian government recently completed a digitized data base for the country to facilitate a remote sensing–based GIS analysis of these rain forest losses. Figure 20-7d shows a tract of former rain forest that has just been burned over to begin the clearing and road-building process.

The United Nations Food and Agricultural Organization (FAO) estimates that every year approximately 16.9 million hectares (41.7 million acres) are destroyed, and more than 5 million hectares (12.3 million acres) are selectively logged. This total—averaged for the period 1980–1991 in 76 countries that contain 97% of all rain forest—represents a 0.9% loss of equatorial and tropical rain forest worldwide each year (up from the previous average of 0.6%). If this destruction continues unabated, these forests will be completely removed by about A.D. 2050!

By continent, forest losses are estimated at more than 50% in Africa, over 40% in Asia, and 40% in Central and South America. Brazil, Colombia, Mexico, and Indonesia lead the list of less-developed countries that are removing their forests at record rates, although these statistics are disputed by the countries in question. Additional losses are occurring from oil exploration and drilling in the rain forest, as discussed in News Report 2.

To slow this continuing catastrophe of deforestation, the FAO, the UN Development Programme, the World Bank, and the World Resources Institute initiated the Tropical Forestry Action Plan in 1985. This plan instituted an accurate worldwide survey of the rate of deforestation, using orbiting satellites to obtain remote measurement of land cover for building more-accurate GIS models. But we have barely begun to reverse the trend. What is

(a)

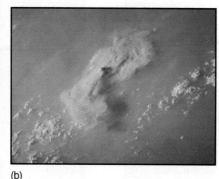

(b)

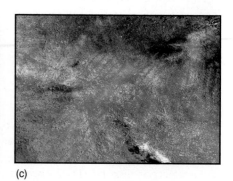

(c)

(d)

**FIGURE 20-7**

**The rain forest in flames.**

The shuttle *Discovery* captured images (a), (b) October 1, 1988. (a) Thousands of fires produce smoke (gray areas) that hide the Amazon rain forest from the astronauts' view. (b) A smoke plume rises from a single rain forest fire, burning an area greater than that of the 1988 Yellowstone fire! (c) A *Landsat 5* satellite image of a portion of the State of Rondônia in western Brazil, recorded in 1987. Highway BR364, the main artery in the region, passes through the towns of Jaru and Ji-Paraná near the center of the image at about 62° W 11° S. The branching pattern of feeder roads (blue areas) encroach on the forest (false color red). (d) Surface view of the burning rain forest in Guatemala. [(a) and (b) Space Shuttle photos from NASA; (c) courtesy of GEOPIC, Earth Satellite Corporation; (d) photo by George Holton/Photo Researchers, Inc.]

## News Report 2

### Drilling for Oil in the Rain Forest

Another threat to the rain forest biome and its indigenous peoples began in 1972 with oil exploration, drilling, and export from the upper Ecuadorian Amazon. This region, known as the Oriente, contains high biodiversity and several Biosphere Reserves—the Yasuni National Park at 650,000 hectares (1.6 million acres) is the largest. In another area nearby, 1 km² of the Cuyabeno Wildlife Reserve was found to contain 473 tree species.

Presently, there are 1.2 million hectares (3 million acres) of new oil leases on protected lands held by indigenous tribes. Petroleum shipments represent about 50% of Ecuador's export earnings. U.S. oil corporations have contaminated streams and lakes in the process of extracting and shipping oil. One 1992 oil spill in the rain forest exceeded 275,000 gallons. Sustainable forestry, identification of medicinal plants for patent, health of the

local population, and ecotourism are being degraded.

One estimate of the ultimate petroleum reserve in the Oriente in Ecuador is 1.5 billion barrels, or enough to satisfy about 3 months of the U.S. demand. At present extraction rates, the reserve will be played out in under 20 years and the corporations will then leave. Similar projects are going forward in Peru.

needed is a global *conservation ethic*. Focus Study 20-1 looks more closely at efforts to curb increasing rates of species extinction and loss of Earth's biodiversity, much of which is directly attributable to the loss of rain forests.

## Tropical Seasonal Forest and Scrub

A varied biome on the margins of the rain forest is the **tropical seasonal forest and scrub**, which occupies regions of low and erratic rainfall. The shifting intertropical convergence zone (ITCZ) brings precipitation with the seasonally shifting high Sun and dryness with the low Sun, producing a seasonal pattern of moisture deficits, some leaf loss, and dry-season flowering. The term *semideciduous* applies to some of the broadleaf trees that lose their leaves during the dry season.

Areas of this biome have fewer than 40 rainy days during their four consecutive driest months, yet heavy monsoon downpours characterize their summers (see Chapter 6, especially Figure 6-23). The Köppen climates *tropical monsoon Am* and *tropical savanna Aw* apply to these transitional communities between rain forests and tropical grasslands.

Portraying such a varied biome is difficult. In many areas, the natural biome is disturbed by humans, so that the savanna grassland adjoins the rain forest directly. The biome does include a gradation from wetter to drier areas: monsoonal forests, to open woodlands and scrub woodland, to thorn forests, to drought-resistant scrub species (Figure 20-8). In South America, an area of transitional tropical deciduous forest surrounds an area of savanna in southeastern Brazil and portions of Paraguay.

The monsoonal forests average 15 m (50 ft) high with no continuous canopy of leaves, graduating into open orchardlike parkland with grassy openings or into areas choked by dense undergrowth. In more open tracts, a common tree is the acacia, with its flat-topped appearance and usually thorny stems. These trees have branches that look like an upside-down umbrella (spreading and open skyward), as do trees in the tropical savanna.

Local names are given to these communities: the *caatinga* of the Bahia State of northeastern Brazil, the *chaco* area of Paraguay and northern Argentina, the *brigalow* scrub of Australia, and the *dornveld* of southern Africa. Figure 20-4 shows areas of this biome in Africa, extending from eastern Angola through Zambia to Tanzania; in southeast Asia and portions of India, from interior Myanmar (Burma) through northeastern Thailand; and in parts of Indonesia.

The trees throughout most of this biome make poor lumber but some, especially teak, may be valuable for fine cabinetry. In addition, some of the plants with dry-season adaptations produce usable waxes and gums, such as carnauba and palm-hard waxes. Animal life includes the koalas and cockatoos of Australia and the elephants, large cats, rodents, and ground-dwelling birds in other occurrences of this biome.

## Tropical Savanna

Large expanses of grassland, interrupted by trees and shrubs, aptly describes the **tropical savanna** (Figure 20-9). This is a transitional biome between the tropical forests and semiarid tropical steppes and deserts. The savanna biome also includes treeless tracts of grasslands, and in very dry savannas, grasses grow discontinuously in clumps, with bare ground between them. The trees of the savanna woodlands are characteristically flat-topped, in response to light and moisture regimes.

(a)

(b)

**FIGURE 20-8**
**Kenyan landscapes.**
Two views of the open thorn forest and savanna near and in the Samburu Reserve, Kenya. (a) Photo by Gael Summer-Hebden; (b) photo by Stephen J. Krasemann/DRK Photo.]

## Focus Study 20-1

### Biodiversity and Biosphere Reserves

When the first European settlers landed on the Hawaiian Islands in the late 1700s, 43 species of birds were counted. Today, 15 of those species are extinct, and 19 more are threatened or endangered, with only one-fifth of the original species relatively healthy. Humans have great impact on animals and plants and global biodiversity.

Relatively pristine habitats around the world are being lost at unprecedented rates as an expanding human population converts them to agriculture, forestry, and urban centers. As these habitats are altered, untold numbers of species are disappearing before they have been recognized, much less studied, and the functioning of entire ecosystems is threatened. This loss of biodiversity, at the very time when the value of biotic resources is becoming widely recognized, has made it strikingly clear that current strategies for conservation are failing dismally.*

As more is learned about Earth's ecosystems and their related communities, more is known of their value to civilization and our interdependence with them. Natural ecosystems are a major source of new foods, new chemicals, new medicines, and specialty woods, and of course they are indicators of a healthy, functioning biosphere.

International efforts are under way to study and preserve specific segments of the biosphere: UN Environment Programme, World Resources Institute, The World Conservation Union, Rain Forest Action Network, Natural World Heritage Sites, Wetlands of International Importance, the Nature Conservancy, the setting aside of biosphere reserves, and the focus provided by the World Conservation Monitoring Center.

The motivation to set aside natural sanctuaries is directly related to concern over the increase in the rate of *species extinctions*. We are facing a loss of genetic diversity that may be unparalleled in Earth's history, even compared with the major extinctions over the geologic record.

*T. D. Sisk, A. E. Launer, K. R. Switky, and P. R. Ehrlich, "Identifying extinction threats," *Bioscience* 44 (October 1994): 592.

### Species Threatened

Both black rhinos and white rhinos in Africa exemplify species in jeopardy. Rhinos once grazed over much of the savanna grasslands and woodlands. Today, they survive only in protected districts and are threatened even there (Figure 1). There are fewer than two dozen northern white rhinos in existence now in sub-Saharan ranges. Almost all are in the Garamba National Park in Zaire, and four are in captivity but are not reproducing. This is an 80% loss since 1979. Approximately 8800 black rhinos remain, 40% less than the number in 1980. But rhinoceros horn sells for $29,000 per kilogram! These large land mammals are nearing extinction and will survive only as a dwindling zoo population. The limited genetic pool that remains complicates further reproduction.

Table 1 summarizes the numbers of known and estimated species on Earth. Scientists have classified only 1.75 million species of plants and animals of an estimated 13.6 million overall; this figure represents an increase in what scientists once thought to be the diversity of life on Earth. And those yet-to-be-discovered species represent a potential future resource for society—the wide range of estimates places the expected species count between a low of 3.6 million and a high of 111.7 million.

However, in 1982, the Council of Environmental Quality's *Global 2000 Report* estimated that between 15% and 20% of all plant and animal species on Earth would be extinct by the year 2000. That percentage equals 500,000 to 2 million species, many of which have not yet even been discovered. On the low end, this rate of species extinction averages about one every 30 minutes! Other estimates place annual losses in a range between 1000 and 30,000 by the end of the century.

The effects of pollution, loss of wild habitat, excessive grazing, poaching, and collecting are at the root of such devastation. Approximately 60% of the extinctions are attributable to the clearing and loss of rain forests alone.

### The Question of Medicine and Food

Wheat, maize (corn), and rice—just three grains—fulfill about 50% of our planetary food demands. About 7000 plant species have been gathered for food throughout human history, but over 30,000 plant species have edible parts. Undiscovered potential food resources are in nature waiting to be found and developed. Biodiversity, if preserved, provides us a potential cushion for all future food needs, but only if species are identified, inventoried, and protected. The same is true for pharmaceuticals.

Nature's biodiversity is like a full medicine cabinet. Between 1959 and 1980, 25% of all prescription drugs were originally derived from higher plants. In preliminary surveys, 3000 plants have been identified as having anticancer properties. The rosy periwinkle (*Catharanthus roseus*) of Madagascar contains two alkaloids that combat two forms of cancer. (Alkaloids are compounds found in certain plants that help the plant defend against insects and are potentially significant as human medicines; examples are atropine, quinine, morphine.) Yet, less than 3% of flowering plants have been examined for alkaloid content! It defies common sense to throw away the medicine cabinet before we even open the door to see what is inside.

Genetic material from the developing world is already making an immense contribution to the food tables and medicine cabinets of the industrialized Northern Hemisphere. Agriculture in developed countries reaps an estimated $5 billion a year from Third World genetic material. . . . One Peruvian variety of tomato has been worth $8 million annually to U.S. tomato processors because it added soluble solid content to tomatoes. . . . And one in four prescription drugs is derived at least in part from plants. . . . Their value to the $60-billion-a-year U.S. pharmaceutical industry is considerable.*

*J. W. King, "Breeding uniformity, Will global biotechnology threaten global biodiversity?" *The Amicus Journal* 15, (Spring 1993): 26.

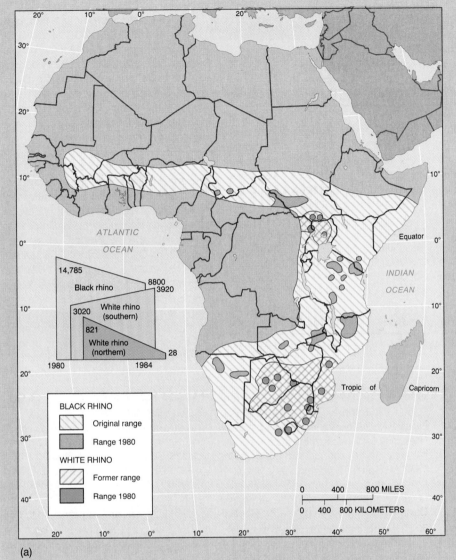

(a)

(b)

### FIGURE 1
### The rhinoceros in Africa.

(a) Failing population and receding distribution of both the black and white rhinos. (b) Black rhino and young in Tanzania, escorted by oxpecker birds. These rhinos are quite near-sighted (they can see clearly only up to 10 m); the birds act as the rhino's early-warning system for disturbances in the distance. [(a) Illustration after *State of the Ark* by Lee Durrell. © 1986 by Gaia Books Ltd. Used by permission of Doubleday, Inc. (b) Photo by Stephen J. Krasemann/DRK Photo.]

## TABLE 1

| Known and Estimated Species on Earth | | | | | |
|---|---|---|---|---|---|
| | | **Estimated Numbers of Species** | | | |
| **Species** | **Number of Known Species** | **High (000)** | **Low (000)** | **Working Number (000)** | **Accuracy** |
| Viruses | 4000 | 1000 | 50 | 400 | Very poor |
| Bacteria | 4000 | 3000 | 50 | 1000 | Very poor |
| Fungi | 72,000 | 2700 | 200 | 1500 | Moderate |
| Protozoa | 40,000 | 200 | 60 | 200 | Very poor |
| Algae | 40,000 | 1000 | 150 | 400 | Very poor |
| Plants | 270,000 | 500 | 300 | 320 | Good |
| Nematodes | 25,000 | 1000 | 100 | 400 | Poor |
| Arthropods | | | | | |
|   Crustaceans | 40,000 | 200 | 75 | 150 | Moderate |
|   Arachnids | 75,000 | 1000 | 300 | 750 | Moderate |
|   Insects | 950,000 | 100,000 | 2000 | 8000 | Moderate |
| Mollusks | 70,000 | 200 | 100 | 200 | Moderate |
| Chordates | 45,000 | 55 | 50 | 50 | Good |
| Others | 115,000 | 800 | 200 | 250 | Moderate |
| **Total** | **1,750,000** | **111,655** | **3635** | **13,620** | **Very poor** |

*Source*: United Nations Environment Programme, *Global Biodiversity Assessment* (Cambridge: Cambridge University Press, 1995), Table 3.1–2, p. 118.

### Biosphere Reserves

Formal natural reserves are a possible strategy for slowing the loss of biodiversity and protecting this resource base. Setting up such a *biosphere reserve* involves principles of *island biogeography*. Island communities are special places for study because of their spatial isolation and the relatively small number of species present. They resemble natural experiments, because the impact of individual factors, such as civilization, can be more easily assessed on islands than over larger continental areas.

Studies of islands also can assist in the study of mainland ecosystems, for in many ways a park or biosphere reserve, surrounded by modified areas and artificial boundaries, is like an island. Indeed, a biosphere reserve is conceived as an ecological island in the midst of change. The intent is to establish a core in which genetic material is protected from outside disturbances, surrounded by a buffer zone that is, in turn, surrounded by a transition zone and experimental research areas. An important variable to consider in setting aside a biosphere reserve is any change that might occur in temperature and precipitation patterns as a result of global change. A carefully considered reserve could end up outside its natural range as ecotones shift.

It is predicted that new, undisturbed reserves will not be possible after A.D. 2010, because pristine areas will be gone. The biosphere reserve program is coordinated by the Man and the Biosphere (MAB) Programme of UNESCO. Nearly 300 such biosphere reserves covering some 12 million hectares (30 million acres) are now operated voluntarily in 76 countries. Not all protected areas are ideal bioregional entities. Some are simply imposed on existing park space and some remain in the planning stage, although they are officially designated.

Some of the best examples have been in operation since the late 1970s and range from the struggling Everglades National Park in Florida, to the Changbai Reserve in China, to the Tai Forest on the Côte d'Ivoire (Ivory Coast). Added to these efforts is the work of the Nature Conservancy, which acquires land for preservation as a valuable part of the reserve process.

The ultimate goal, about half achieved, is to establish at least one reserve in each of the 194 distinctive *biogeographic communities* presently identified. The United Nations' list of national parks and protected areas is compiled by UNESCO and the World Conservation Monitoring Centre. Presently, 6930 areas covering 657 million hectares (1.62 billion acres) are designated in some protective form, representing about 4.8% of national land area on the planet. Even though these are not all set aside to the degree of biosphere reserves, they do demonstrate progress toward preservation of our planet's plant and animal heritage.

The preservation of species diversity is a problem that must today be confronted by one species, *Homo sapiens*. . . . The diversity of species is worth preserving because it represents a wealth of knowledge that cannot be replaced. Moreover, today's extinctions are unlike those in previous eras, in which long periods of recovery could follow extinctions. The present situation is an inexorably irreversible one in which human overpopulation will destroy most species unless we plan for protection immediately. Accepting that the goal is worthwhile requires that more energy be devoted to planning and priorities and less to emotionalism and indignation [on all sides].[*]

[*] D. E. Koshland Jr., "Preserving biodiversity," *Science* 253, no. 5021 (August 16, 1991): 717.

---

Savannas covered more than 40% of Earth's land surface before human intervention but were especially modified by human-caused fire. Fires occur annually throughout the biome. The timing of these fires is important. Early in the dry season they are beneficial and increase tree cover; if late in the season they are very hot and kill trees and seeds. Savanna trees are adapted to resist the "cooler" fires.

Forests and elephant grasses averaging 5 m (16 ft) high once penetrated much farther into the dry regions,

**FIGURE 20-9**
**Animals and plants of the Serengeti Plains.**
Savanna landscape of the Serengeti Plains, with wildebeest, zebras, and thorn forest. [Photo by Stephen F. Cunha.]

for they are known to survive there when protected. Savanna grasslands are much richer in humus than the wetter tropics and are better drained, thereby providing a better base for agriculture and grazing. Sorghums, wheat, and groundnuts (peanuts) are common commodities.

Tropical savannas receive their precipitation during less than 6 months of the year, when they are influenced by the ITCZ. The rest of the year they are under the drier influence of shifting subtropical high-pressure cells. Savanna shrubs and trees are frequently *xerophytic*, or drought-resistant, with various adaptations to protect them from the dryness: small, thick leaves, rough bark, or leaf surfaces that are waxy or hairy.

Africa has the largest area of this biome, including the famous Serengeti Plains of Tanzania and Kenya and the Sahel region, south of the Sahara Desert. Sections of Australia, India, and South America also are part of the savanna biome. Some of the local names for these lands include the *Llanos* in Venezuela, stretching along the coast and inland east of Lake Maricaibo and the Andes; the *Campo Cerrado* of Brazil and Guiana; and the *Pantanal* of southwestern Brazil.

Particularly in Africa, savannas are the home of large land mammals (zebra, giraffe, buffalo, gazelle, wildebeest, antelope, rhinoceros, elephant). They graze on savanna grasses (Figure 20-9) or feed upon the grazers themselves (lion, cheetah). Birds include the ostrich, martial eagle (largest of all eagles), and secretary bird. Many species of venomous snakes occur, as does the crocodile.

Unfortunately, in our lifetime, we may see the reduction of these animal herds to zoo stock only, because of poaching and habitat losses. The loss of the rhino is discussed in Focus Study 20-1. Establishment of large tracts of savanna as biosphere reserves is critical for the preservation of this biome and its associated fauna.

## Midlatitude Broadleaf and Mixed Forest

Moist continental climates support a mixed forest in areas of warm to hot summers and cool to cold winters. This **midlatitude broadleaf and mixed forest** biome includes several distinct communities in North America, Europe, and Asia. Along the Gulf of Mexico, relatively lush evergreen broadleaf forests occur. Northward are mixed deciduous and evergreen needleleaf stands associated with sandy soils and burning. When areas are given fire protection, broadleaf trees quickly take over.

Pines (longleaf, shortleaf, pitch, loblolly) predominate in the southeastern and Atlantic coastal plains. Into New England and westward in a narrow belt to the Great Lakes, white and red pines and eastern hemlock are the principal evergreens, mixed with deciduous oak, beech, hickory, maple, elm, chestnut, and many others.

These mixed stands contain valuable timber, but their distribution has been greatly altered by human activity. Native stands of white pine in Michigan and Minnesota were removed before 1910; only later reforestation sustains their presence today. In northern China, these forests have almost disappeared as a result of centuries of harvest. The forest species that once flourished in China are similar to species in eastern North America: oak, ash, walnut, elm, maple, and birch.

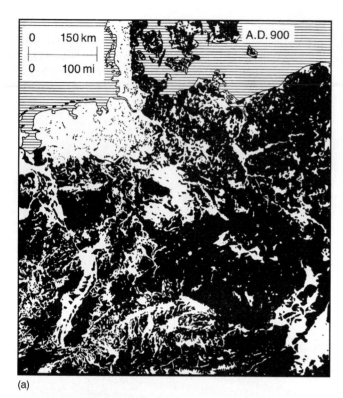

(a)

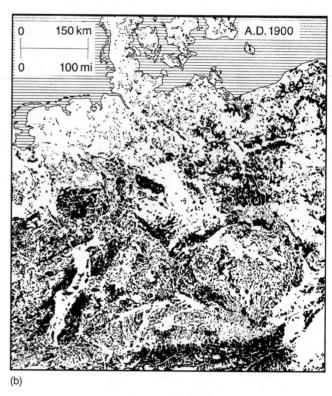

(b)

**FIGURE 20-10**
**European deforestation.**

A comparison of Europe's forests in (a) A.D. 900 and (b) A.D. 1900. [From H. C. Darby, "The Clearing of the Woodland in Europe," in *Man's Role in Changing the Face of the Earth*, W. L. Thomas Jr., ed. © 1956 by the University of Chicago Press. Reprinted by permission.]

The level of deforestation in historic times is well illustrated by the two maps in Figure 20-10. The first shows the forest cover estimated for Europe in A.D. 900. The second map is the distribution of the remaining forest, A.D. 1900. The comparison dramatically illustrates human modification of natural vegetation. This deforestation was principally for agricultural purposes, as well as for construction materials and fuel.

This biome is quite consistent in appearance from continent to continent and at one time represented the principal vegetation of the *humid subtropical hot summer climates Cfa, Cwa, marine west coast Cfb,* and *Cwb* (cool summer, winter drought) climatic regions of North America, Europe, and Asia.

A wide assortment of mammals, birds, reptiles, and amphibians is distributed throughout this biome. Representative animals (some migratory) include fox, squirrel, deer, opossum, bear, rodents, and a great variety of birds. To the north of this biome, poorer soils and colder climates favor stands of coniferous trees and a gradual transition to the northern needleleaf forests.

### Needleleaf Forest and Montane Forest

Stretching from the east coast of Canada and the Atlantic provinces westward to Alaska and continuing from Siberia across the entire extent of Russia to the European Plain is the northern **needleleaf forest** biome, also called the **boreal forest** (Figure 20-11). A more open form of boreal forest, transitional to arctic and subarctic regions, is termed the **taiga**. The Southern Hemisphere, lacking *humid microthermal D* climates except in mountainous locales, has no such biome. But **montane forests** of needleleaf trees exist worldwide at high elevation.

Boreal forests of pine, spruce, fir, and larch occupy most of the subarctic climates on Earth that are dominated by trees. Although these forests are similar in formation, individual species vary between North America and Eurasia. The larch (*Larix*) is interesting because it is the rare needleleaf tree that loses its needles in the winter months, perhaps as a defense against the extreme cold of its native Siberia (see the Verkhoyansk climographs and photograph in Figure 10-22). Larches also occur in North America.

The Sierra Nevada, Rocky Mountains, Alps, and Himalayas have similar forest communities occurring at lower latitudes. Douglas-fir and white fir grow in the western mountains in the United States and Canada. Economically, these forests are important for lumbering, with trees for lumber occurring in the southern margins of the biome and pulpwood throughout the middle and northern portions. Present logging practices and whether these yields are sustainable are issues of increasing controversy.

In the Sierra Nevada montane forests of California, the giant sequoias naturally occur in seven isolated groves.

**FIGURE 20-11**
Boreal forest of Canada (*boreal* means "northern"). [Photo by author.]

These trees are Earth's largest (in terms of biomass) living things, although they began as a small seed (Figure 20-12a). Some of these *Sequoia gigantea* exceed 8 m in diameter (28 ft) and 83 m (270 ft) in height. The largest of these is the General Sherman tree in Sequoia National Park (Figure 20-12c); it is estimated to be 3500 years old. The bark is fibrous, a half-meter thick, and lacks resins, so it effectively resists fire. Imagine the lightning strikes and fires that must have moved past the Sherman tree in 35 centuries! Standing among these giant trees is an appropriate way to contemplate the majesty of the biosphere.

Certain regions of the northern needleleaf biome experience the permafrost discussed in Chapter 17. When coupled with rocky and poorly developed soils, these conditions generally limit the existence of trees to those with shallow rooting systems. The summer thaw of surface layers results in muskeg (moss-covered) bogs and Histolic (organic) soils of poor drainage and stability. Soils of the taiga are typically Spodosols (Podzolic), characteristically acidic and leached of humus and clays.

Representative fauna in this biome include wolf, moose (the largest deer), bear, lynx, wolverine, marten, small rodents, and migratory birds during the brief summer season. Birds include hawks and eagles, several species of grouse, and owls. About 50 species of insects particularly adapted to the presence of coniferous trees inhabit the biome.

### Temperate Rain Forest

The **temperate rain forest** biome is recognized by its lush forests at middle and high latitudes. In North America, it occurs only along narrow margins of the Pacific Northwest (Figure 20-13). Some similar types exist in southern China, small portions of southern Japan, New Zealand, and a few areas of Chile.

This biome contrasts with the diversity of the equatorial and tropical rain forest in that only a few species make up the bulk of the trees. The rain forest of the Olympic Peninsula in Washington State is a mixture of broadleaf and needleleaf trees, huge ferns, and thick undergrowth. Precipitation approaching 400 cm (160 in.) per year on the western slopes, moderate air temperatures, summer fog, and an overall maritime influence produce this moist, lush vegetation community. Animals include bear, badger, deer, wild pig, wolf, bobcat, and fox. The trees are home to numerous bird and mammal species.

The tallest trees in the world occur in this biome—the coastal redwoods (*Sequoia sempervirens*). Their distribution is shown on the map in Figure 19-15a. These trees can exceed 1500 years of age and typically range in height from 60 to 90 m (200 to 300 ft), with some exceeding 100 m (330 ft). Virgin stands of other representative trees such as Douglas-fir, spruce, cedar, and hemlock have been reduced to a few remaining valleys in Oregon and Washington, less than 10% of the original forest.

A study released in 1993 by the U.S. Forest Service noted the failing ecology of these forest ecosystems and suggested that timber management plans should include ecosystem preservation as a priority. The government held a forestry summit in Portland, Oregon, in 1993, to bring together the opposing sides in this conflict between logging and environmental interests. The ultimate solution must be one of economic and ecological synthesis rather than continuing conflict and forest losses.

### Mediterranean Shrubland

The **Mediterranean shrubland** biome, also referred to as a temperate shrubland, occupies those regions poleward of the shifting subtropical high-pressure cells. As those cells shift poleward with the high Sun, they cut off available storm systems and moisture. Their stable high-pressure presence produces the characteristic dry summer

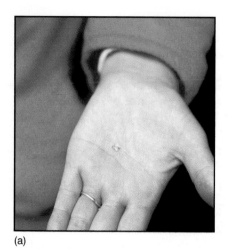

(a)

(b)

(c)

**FIGURE 20-12**
**Sequoia: the seed, the seedling, the giant.**
(a) A *Sequoia gigantea* seed. About 300 seeds are in each sequoia cone. (b) Seedling at approximately 50 years of age. (c) The General Sherman tree in Sequoia National Park, California. The tree is probably wider than your classroom, and the first branch is 45 m (150 ft) off the ground and is 2 m (6.5 ft) in diameter! [Photos by author.]

climate—Köppen's *Mediterranean dry summer Csa, Csb*—and establishes conditions conducive to fire.

Plant ecologists think that this biome is well adapted to frequent fires, for many of its characteristically deep-rooted plants have the ability to resprout from their roots after a fire. Earlier stands of evergreen woodlands (holm or evergreen oak) no longer dominate.

The dominant shrub formations that occupy these regions are stunted and able to withstand hot-summer drought. The vegetation is called *sclerophyllous* (from *sclero*, for "hard," and *phyllos*, for "leaf"). It averages a meter or two in height and has deep, well-developed roots, leathery leaves, and uneven low branches.

Typically, the vegetation varies between woody shrubs covering more than 50% of the ground and grassy woodlands covering 25%–60% of the ground. In California, the Spanish word *chaparro* for "scrubby evergreen" gives us the word **chaparral** for this vegetation type (Figure 20-14). This scrubland includes species such as manzanita, toyon, red bud, ceanothus, mountain mahogany, blue and live oaks, and the dreaded poison oak.

A counterpart to chaparral in the Mediterranean region, called *maquis*, includes live and cork oak trees (source of cork), as well as pine and olive trees. In Chile, such a region is called *mattoral*; in southwestern Australia,

**FIGURE 20-13**
**Temperate rain forest.**
The temperate Hoh Rain Forest in Olympic National Park, Washington, in the northwestern United States. Club moss, big leaf maples, wood sorrel, and sword ferns are pictured. [Photo by Jack W. Dykinga.]

**FIGURE 20-14**
**Mediterranean chaparral.**
Chaparral vegetation associated with the Mediterranean dry summer climate. [Photo by author.]

*mallee scrub.* Of course, in Australia, the bulk of the eucalyptus species is sclerophyllous in form and structure in whichever climate it occurs.

As described in Chapter 10, Mediterranean climates are important in commercial agriculture for subtropical fruits, vegetables, and nuts, with many food types produced only in this biome (e.g., artichokes, olives, almonds). Larger animals, such as different types of deer, are grazers and browsers, with coyote, wolf, and bobcat as predators. Many rodents, other small animals, and a variety of birds also proliferate.

## Midlatitude Grasslands

Of all the natural biomes, the **midlatitude grasslands** are the most modified by human activity. Here are the world's "breadbaskets"—regions that produce bountiful grain (wheat, corn, and soybeans) and livestock (hogs and cattle). Figure 20-15 shows a typical midlatitude grassland in cultivation. In these regions, the only naturally occurring trees were deciduous broadleafs along streams and other limited sites. These regions are called grasslands because of the predominance of grasslike plants before human intervention:

> In this study of vegetation, attention has been devoted to grass because grass is the dominant feature of the Plains and is at the same time an index to their history. Grass is the visible feature which distinguishes the Plains from the desert. Grass grows, has its natural habitat, in the transition area between timber and desert. . . . The history of the Plains is the history of the grasslands.*

*From W. P. Webb, *The Great Plains*, p. 32, © 1959 by Walter Prescott Webb, published by Ginn and Company, Needham Heights, Mass. Used by permission.

In North America, tallgrass prairies once rose to heights of 2 m (6.5 ft) and extended westward to about the 98th meridian, with shortgrass prairies in the drier lands farther west. The 98th meridian is roughly the location of the 50 cm (20 in.) isohyet, with wetter conditions to the east and drier conditions to the west (see Figure 18-17).

The deep, tough sod of these grasslands posed problems for the first European settlers, as did the climate. The self-scouring steel plow, introduced in 1837 by John Deere, allowed the interlaced grass sod to be broken apart, freeing the soils for agriculture. Other inventions were critical to opening this region and solving its unique spatial problems: barbed wire (the fencing material for a treeless prairie); well-drilling techniques developed by Pennsylvania oil drillers but used for water wells; windmills for pumping; and railroads to conquer the distances.

Few patches of the original prairies (tall grasslands) or steppes (short grasslands) remain within this biome. For prairies alone, the reduction of natural vegetation went from 100 million hectares (250 million acres) down to a few areas of several hundred hectares each. A "prairie national park" along the Kansas-Oklahoma border was considered as a preservationist project, but plans to establish it were put on hold. A state-designated 10 hectare (25 acre) remnant of the "prairie" is located 8 km north of Ames, Iowa. This is a patch of original grassland that never has felt the plow. The map in Figure 20-4 shows the natural location of these former prairie and steppe grasslands.

Outside North America, the Pampas of Argentina and Uruguay and the grasslands of Ukraine are characteristic midlatitude grassland biomes. In each region of the world where these grasslands have occurred, human development of them was critical to territorial expansion.

This biome is the home of large grazing animals, including deer, antelope, pronghorn, and bison (the almost complete annihilation of the latter is part of American history). Grasshoppers and other insects feed on the

**FIGURE 20-15**
**Dry farming in the grasslands of North America.**
Midlatitude grassland under cultivation. Alternating fields are left fallow to accumulate soil moisture. [Photo by author.]

grasses and crops as well, and gophers, prairie dogs, ground squirrels, turkey vultures, grouse, and prairie chickens are on the land. Predators include coyote and badger and birds of prey, such as hawks, eagles, and owls.

## Deserts

Earth's *desert biomes* cover more than one-third of its land area, as is obvious in Figure 20-4. In Chapter 15, we examined desert landscapes, and, in Chapter 10, desert climates. On a planet with such a rich biosphere, the deserts stand out as unique regions of fascinating adaptations for survival. Much as a group of humans in the desert might behave with short supplies, plant communities also compete for water and site advantage. Some desert plants, called *ephemerals*, wait years for a rainfall event, at which time their seeds quickly germinate, and the plants develop, flower, and produce new seeds, which then rest again until the next rainfall event. The seeds of some xerophytic species open only when fractured by the tumbling, churning action of flash floods cascading down a desert arroyo, and of course such an event produces the moisture that a germinating seed needs.

Perennial desert plants employ other adaptive features to cope with the desert, such as long, deep taproots (e.g., the mesquite), succulence (thick, fleshy, water-holding tissue, such as that of cacti), spreading root systems to maximize water availability, waxy coatings and fine hairs on leaves to retard water loss, leafless conditions during dry periods (e.g., palo verde and ocotillo), reflective surfaces to reduce leaf temperatures, and tissue that tastes bad to discourage herbivores.

The creosote bush (*Lorrea divaricata*) sends out a wide pattern of roots and contaminates the surrounding soil with toxins that prevent the germination of other creosote seeds, possible competitors for water. When a creosote bush dies, surrounding plants or germinating seeds work to occupy the abandoned site, but they must wait for infrequent rains to remove the toxins.

The faunas of both warm and cold deserts are limited by the extreme conditions and include few resident large animals. Exceptions are the desert bighorn sheep (in nearby mountains) and the camel, which can lose up to 30% of its body weight in water without suffering (for humans, a 10%–12% loss is dangerous). Some representative desert animals are the ring-tail cat, kangaroo rat, lizards, scorpions, and snakes. Most of these animals are quite secretive and become active only at night, when temperatures are lower. In addition, various birds have adapted to desert plants and other available food sources—for example, roadrunners, thrashers, ravens, wrens, hawks, grouse, and nighthawks.

**Desertification.** We are witnessing an unwanted expansion of the desert biome. This process, known as **desertification**, is now a worldwide phenomenon along the margins of semiarid and arid lands. Desertification is due principally to poor agricultural practices (overgrazing and agricultural activities that abuse soil structure and fertility), improper soil-moisture management, erosion and salinization, deforestation, and the ongoing climatic change.

The southward expansion of Saharan conditions through portions of the Sahel region has left many African peoples on land that no longer experiences the rainfall of just two decades ago. Regions at risk of desertification stretch from Asia and central Australia to portions of North and South America. The United Nations estimates that degraded lands have covered some 800 million hectares (2 billion acres) since 1930; many millions of additional hectares are added each year. Of course, an immediate need is to improve the data base for a more accurate accounting of the problem and a better understanding of what is occurring.

Figure 20-16 is drawn from a map prepared for the 1977 U.N. Conference on Desertification held in Nairobi, Kenya, and a conference paper titled "Plan of Action to Combat Desertification." Desertification areas are ranked: A moderate hazard area has an average 10%–25% drop in agricultural productivity; a high hazard area has a 25%–50% drop; and a very high hazard area has more than a 50% decrease. Desertification includes consideration of the soil degradation portrayed on the map in Figure 18-8. Because human activities and economies, especially unwise grazing practices, appear to be the major cause of desertification, actions to slow the process are readily available with leadership.

Earth's deserts are subdivided into desert and semi-desert associations, to distinguish those with expanses of bare ground from those covered by xerophytic plants of various types. The two broad associations are further separated into warm deserts, principally tropical and subtropical, and cold deserts, principally midlatitude.

**Warm Desert and Semidesert.** Earth's **warm desert and semidesert** biomes are caused by the presence of dry air and low precipitation from subtropical high-pressure cells. These areas are very dry, as evidenced by the Atacama Desert of northern Chile, where only a minute amount of rain has ever been recorded—a 30-year annual average of only 0.05 cm (0.02 in.)! Like the Atacama, the true deserts of the Köppen *desert BW* classification are under the influence of the descending, drying, and stable air of high-pressure systems from 8 to 12 months of the year.

Remember that these dry regions are defined by low amounts of precipitation that fail to satisfy high amounts of potential evapotranspiration. Deserts receive precipitation that is less than one-half of potential evapotranspiration. Semiarid *steppe BS* climates receive precipitation that is more than one-half of annual potential evapotranspiration.

Vegetation ranges from almost none in the arid deserts to numerous xerophytic shrubs, succulents, and

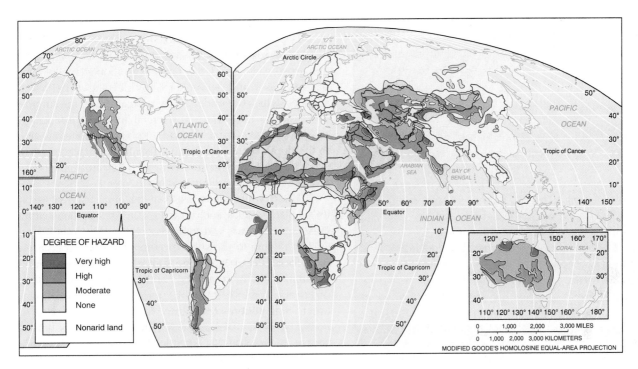

**FIGURE 20-16**
**The desertification hazard.**
Worldwide desertification estimates by the United Nations. [Data from U.N. Food and Agricultural
Organization (FAO), World Meteorological Organization (WMO), United Nations Educational, Scientific,
and Cultural Organization (UNESCO), Nairobi, Kenya, 1977; as printed in J. M. Rubenstein, *The
Cultural Landscape*, figure 13-16, p. 597. © 1996 Prentice Hall, Inc. Used by permission.]

thorn tree forms. The lower Sonoran Desert of southern
Arizona is a warm desert (Figure 20-17). This desert landscape features the unique saguaro cactus (*Carnegiea
gigantea*), which grows to many meters in height and up
to 200 years in age if undisturbed. First blooms do not
appear until it is 50 to 75 years old!

A few of the subtropical deserts—such as those in Chile,
Western Sahara, and Namibia—are right on the sea coast
and are influenced by cool offshore ocean currents. As a
result, these true deserts experience summer fog that mists
the plant and animal populations with needed moisture.

The equatorward margin of the subtropical high-pressure cell is a region of transition to savanna, thorn tree,
scrub woodland, and tropical seasonal forest. Poleward of
the warm deserts, the subtropical cells shift to produce the
Mediterranean dry summer regime along west coasts and
may grade into cool deserts elsewhere.

***Cold Desert and Semidesert.*** The **cold desert and
semidesert** biomes tend to occur at higher latitudes.
Here, seasonal shifting of subtropical high pressure is
of some influence less than 6 months of the year. Interior locations are dry because of their distance from
moisture sources or their location in rain-shadow areas
on the lee side of mountain ranges; examples are the
Sierra Nevada of the western United States, the
Himalayas, and the Andes. The combination of interior

location and rain-shadow positioning produces the cold
deserts of the Great Basin of western North America
(Figure 20-18).

Winter snows occur in the cold deserts, but generally they are light. Summers are hot, with highs from 30°
to 40°C (86° to 104°F). Nighttime lows, even in the summer, can cool 10°–20 C° (18°–36 F°) from the daytime
high. The dryness, generally clear skies, and sparse vegetation lead to high radiative heat loss and cool evenings.
Many areas of these cold deserts that are covered by sage-
brush and scrub vegetation were actually dry shortgrass
regions in the past, before extensive grazing altered their
ecology. The deserts in the upper Great Basin today are
the result of more than a century of such cultural practices.

### *Arctic and Alpine Tundra*

The **arctic tundra** is found in the extreme northern area of
North America and Russia, bordering on the Arctic Ocean
and generally north of the 10°C (50°F) isotherm for the
warmest month. Daylength varies greatly throughout the
year, seasonally changing from almost continuous day to
continuous night. Winters in this biome, *tundra climate ET*
classification, are long and cold; cool summers are brief.
The region, except for a few portions of Alaska and Siberia,
was covered by ice during all of the Pleistocene glaciations.

**FIGURE 20-17**
**Sonoran Desert scene.**
Characteristic vegetation in the Lower Sonoran Desert west of Tucson, Arizona (32.3° N; elevation 900 m, or 2950 ft). [Photo by author.]

Tundra winters are governed by intensely cold continental polar air masses and stable high-pressure anticyclones. A growing season of sorts lasts only 60–80 days, and even then frosts can occur at any time. Vegetation is fragile in this flat, treeless world; soils are poorly developed periglacial surfaces, which are underlain by permafrost. In the summer months, only the surface horizons thaw, thus producing a mucky surface of poor drainage. Roots can penetrate only to the depth of thawed ground, usually about a meter. The surface is shaped by freeze-thaw cycles that create the frozen ground phenomena discussed in Chapter 17.

Tundra vegetation consists of low, ground-level herbaceous plant species, such as sedges, mosses, arctic meadow grass, snow lichen, and some woody dwarf willow (Figure 20-19). Owing to the short growing season, some perennials form flower buds one summer and open them for pollination the next. Animals of the tundra include musk-ox, caribou, reindeer, rabbit, ptarmigans, lemmings, and other small rodents (important food for the larger carnivores), wolf, fox, weasel, snowy owls, polar bears, and, of course, mosquitoes. The tundra is an important breeding ground for geese, swans, and other waterfowl.

**Alpine tundra** is similar to arctic tundra, but it can occur at lower latitudes because it is associated with high elevation. This biome usually is described as above the timberline, that elevation above which trees cannot grow. Timberlines increase in elevation equatorward in both hemispheres. Alpine tundra communities occur in the Andes near the equator, the White Mountains of California, the Rockies, the Alps, and Mount Kilimanjaro of equatorial Africa, as well as mountains from the Middle East to Asia.

Alpine meadows feature grasses and stunted shrubs, such as willows and heaths. Because alpine locations are frequently windy sites, many plants appear to have been sculpted by the wind. Alpine tundra can experience permafrost (Figure 20-20).

**FIGURE 20-18**
**Cold deserts.**
Cold desert scene in the interior Great Basin of the western United States. [Photo by author.]

**FIGURE 20-19**
**Siberian tundra.**
Tundra on the Kamchatka Peninsula, Russia. The Uzon Caldera is in the center-background mountain range. [Photo by Wolfgang Kaehler Photography.]

Because the tundra biome is of such low productivity, it is fragile. Disturbances such as tire tracks, hydroelectric projects, and mineral exploitation leave marks that persist for hundreds of years. As development continues, the region will face even greater challenges from oil spills, contamination, and disruption (News Report 3).

**FIGURE 20-20**
**Alpine conditions.**
An alpine landscape near Logan Pass, Montana, in Waterton-Glacier International Peace Park (48.8° N; 2025 m, or 6650 ft, elevation). [Photo by author.]

## News Report 3

## ANWR Faces Threats

Planning continues for development of the Arctic National Wildlife Refuge (ANWR) on Alaska's North Slope. This pristine wilderness is above the Arctic Circle, bordering on the Beaufort Sea and adjoining the Yukon Territory of Canada. The refuge area sustains almost 200,000 caribou, polar and grizzly bears, musk-ox, and wolves. Some have referred to it as "America's Serengeti," given the annual migration of hundreds of thousands of large animals.

The ANWR remains the only portion of Alaska's Arctic coast that is not open to oil and gas exploration at this time—the other 90% is open. But controversy over this refuge continues, as political and corporate pressures mount to begin oil exploration. In 1995, Congress considered potential ANWR lease revenues in its Budget Resolution, implying a green light for development.

Oil extraction would be devastating to the fragile tundra. The cost of development would price the oil at well over $30 per barrel, far above current market prices. The estimated oil reserve in the ANWR, approximately 3.2 billion barrels, could be offset simply by a small increase in U.S. automobile efficiency. Clearly, economic ventures should be weighed against the ecosystem itself, its limitations and its uniqueness, and a total assessment of all costs to the consumer.

# Summary and Review—Terrestrial Biomes

✔ *Define* **the concept of biogeographic realms of plants and animals and** *define* **ecotone, terrestrial ecosystem, and biome.**

Earth is the only planet in the Solar System with a biosphere. An impressive feature of the living Earth is its diversity, which biogeographers categorize into discrete spatial biomes for analysis and study. The interplay among supporting physical factors within Earth's ecosystem determines the distribution of plant and animal communities. A **biogeographic realm** of plants and animals is a geographic region in which a group of species evolved. This recognition was a start at understanding distinct regions of flora and fauna and the broader pattern of terrestrial ecosystems. A boundary transition zone adjoining ecosystems is an **ecotone**.

A **terrestrial ecosystem** is a self-sustaining association of plants and animals and their abiotic environment that is characterized by specific plant formation classes. A **biome** is a large, stable ecosystem characterized by specific plant and animal communities. Biomes carry the name of the dominant vegetation because it is the most easily identified feature: forest, savanna, grassland, shrubland, desert, tundra.

> biogeographic realm (p. 618)
> ecotone (p. 618)
> terrestrial ecosystem (p. 619)
> biome (p. 620)

1. Reread the two opening quotations in this chapter. What clues do they give you to the path ahead for Earth's forests? Is our future direction controllable? Explain.

2. What is a biogeographic realm? How is the world subdivided according to plant and animal types?

3. Describe a transition zone between two ecosystems. How wide is an ecotone? Explain.

4. Define biome. What is the basis of the designation?

---

✔ *Define* **six formation classes and the life-form designations and** *explain* **their relationship to plant communities.**

Biomes are divided into more specific vegetation units called **formation classes**. The structure and appearance of the vegetation is described: rain forest, needleleaf forest, Mediterranean shrubland, arctic tundra, and so forth. Specific life-form designations include trees, lianas, shrubs, herbs, bryophytes, epiphytes (plants growing above ground on other plants), and thallophytes (lacking leaves, stems, or roots, including bacteria, fungi, algae, and lichens).

> formation class (p. 620)

5. Distinguish between formation classes and life-form designations as a basis for spatial classification.

---

✔ *Describe* **ten major terrestrial biomes and** *locate* **them on a world map.**

Biomes are Earth's major terrestrial ecosystems, each named for its dominant plant community. The 10 major biomes are generalized from numerous formation classes that describe vegetation. Ideally, a biome represents a mature community of natural vegetation. In reality, few undisturbed biomes exist in the world, for most have been modified by human activity. Many of Earth's plant and animal communities are experiencing an accelerated rate of change that could produce dramatic alterations within our lifetime.

For an overview of Earth's 10 major terrestrial biomes and their vegetation characteristics, soil orders, Köppen climate designation, annual precipitation range, temperature patterns, and water balance characteristics, *please review Table 20-1*. **Desertification** is an unwanted expansion of desertlike conditions due to poor agricultural practices (overgrazing, improper soil management, deforestation) and ongoing climate change.

> equatorial and tropical rain forest (p. 622)
> tropical seasonal forest and scrub (p. 629)
> tropical savanna (p. 629)
> midlatitude broadleaf and mixed forest (p. 633)
> needleleaf forest and montane forest (p. 634)
> boreal forest (p. 634)
> taiga (p. 634)
> temperate rain forest (p. 635)
> Mediterranean shrubland (p. 635)
> chaparral (p. 636)
> midlatitude grasslands (p. 637)
> warm desert and semidesert (p. 638)
> cold desert and semidesert (p. 639)
> arctic and alpine tundra (p. 639)
> desertification (p. 638)

6. Using the integrative chart in Table 20-1 and the world map in Figure 20-4, select any two biomes and study the correlation of vegetation characteristics, soil, moisture, and climate with their spatial distribution. Then, contrast the two using each characteristic.

7. Describe the equatorial and tropical rain forests. Why is the rain forest floor somewhat clear of plant growth? Why are logging activities for specific species so difficult there?

8. What issues surround the deforestation of the rain forest? What is the impact of these losses on the rest of the biosphere? What new threat to the rain forest has emerged?

9. What do *caatinga, chaco, brigalow,* and *dornveld* refer to? Explain.

10. Describe the role of fire or fire ecology in the tropical savanna biome and the midlatitude broadleaf and mixed forest biome.

11. Why does the northern needleleaf forest biome not exist in the Southern Hemisphere? Where is this biome located in the Northern Hemisphere, and what is its relationship to climate type?

12. In which biome do we find Earth's tallest trees? Which biome is dominated by small, stunted plants, lichens, and mosses?

13. What type of vegetation predominates in the Mediterranean dry summer climates? Describe the adaptation necessary for these plants to survive.

14. What is the significance of the 98th meridian in terms of North American grasslands? What types of inventions were necessary for humans to cope with the grasslands?

15. Describe some of the unique adaptations found in a desert biome.

16. What is desertification? Explain its impact.

17. What physical weathering processes are specifically related to the tundra biome? What types of plants and animals are found there?

✔ *Relate* **human impacts, real and potential, to several of the biomes.**

As an example: The equatorial and tropical rain forest biome is undergoing rapid deforestation. Because the rain forest is Earth's most diverse biome and is important to the climate system, such losses are creating great concern among citizens, scientists, and nations. Efforts are under way worldwide to set aside and protect remaining representative sites within most of Earth's principal biomes. These *biosphere reserves* are coordinated by the Man and the Biosphere (MAB) Programme of UNESCO. Nearly 300 such biosphere reserves, covering some 12 million hectares (30 million acres), are now operated voluntarily in 76 countries.

18. What is the relationship between island biogeography and biosphere reserves? Describe a biosphere reserve. What are the goals?

19. Compare the map in Figure 20-4 with the composite satellite image inside the front cover of this text. What correlations can you make between the local summertime portrait of Earth's biosphere and the biomes identified on the map?

---

 **NetWork**

The *Geosystems Home Page* provides on-line resources for this chapter on the World Wide Web. You will find review exercises, specific updates for items in the chapter, suggested readings, and links to interesting related pathways on the Internet (click on the Table of Contents link and select this chapter). *Geosystems* is at: **http://www.prenhall.com/geosystm**

# 21

# Earth, Humans, and the New Millennium

**The Clean Air Act Brings a Windfall**

**An Oily Bird**

**The Need for International Cooperation**

**Who Speaks for Earth?**

**Review Questions**

## Key Learning Concepts

After reading the chapter, you should be able to:

- *Determine* an answer for Carl Sagan's question, "Who speaks for Earth?"
- *Analyze* the monetary and health benefits of the U.S. Clean Air Act.
- *Analyze* "An Oily Bird" and *relate* your analysis to energy consumption patterns in the United States and Canada.
- *Explain* the essential elements of the five Earth Summit agreements and *relate* them to physical geography and Earth systems science (geosystems).
- *Appraise* your place in the biosphere and *realize* your physical identity as an Earthling.

*Traditional rural landscape in the Pamirs of Tajikistan.*
[Photo by Stephen F. Cunha.]

During my space flight, I came to appreciate my profound connection to the home planet and the process of life evolving in our special corner of the Universe, and I grasped that I was part of a vast and mysterious dance whose outcome will be determined largely by human values and actions.[*]

Earth can be observed from profound vantage points, as this astronaut experienced on the 1969 *Apollo IX* mission. Our vantage point in this book is that of physical geography. In this book, we have examined Earth's many systems: its energy, atmosphere, winds, ocean currents, water, weather, climate, endogenic and exogenic systems, soils, ecosystems, and biomes. This exploration has led us to an examination of the planet's most abundant large animal, *Homo sapiens*.

We stand at the end of the twentieth century and at the brink of a new millennium. The twenty-first century will be an adventure for the global society, historically unparalleled in experimenting with Earth's life-supporting systems. You will spend the majority of your life in this new century. What preparations are we making for the advent of A.D. 2001?

In his 1980 book and PBS television series, *Cosmos*, astronomer Carl Sagan asked:

> What account would we give of our stewardship of the planet Earth? We have heard the rationales offered by the nuclear superpowers. We know who speaks for the nations. But who speaks for the human species? Who speaks for Earth?[**]

Indeed, who does speak for Earth? We might answer: Perhaps we physical geographers, and other scientists who have studied Earth and know the operations of the global ecosystem, should speak for Earth. However, some might say that questions of technology, environmental politics, and future thinking belong outside of science, and that our job is merely to learn how Earth's processes work and to leave the spokesperson's role to others. Biologist-ecologist Marston Bates addressed this line of thought in 1960:

> Then we came to humans and their place in this system of life. We could have left humans out, playing the ecological game of "let's pretend humans don't exist." But this seems as unfair as the corresponding game of the economists, "let's pretend nature doesn't exist." The economy of nature and the ecology of humans are inseparable and attempts to separate them are more than misleading, they are dangerous. Human destiny is tied to nature's destiny and the arrogance of the engineering mind does not change this. Humans may be a very peculiar animal, but they are still a part of the system of nature.[***]

A fact of life is that Earth's developed countries, through their economic dominance, speak for the billions who live in developing nations. A reality in 1996 is that the U.S. defense budget alone is greater than the gross national product (GNP) of India and that the gross state product of California's 31 million people is greater than the GNP of China and her 1.2 billion people!

The fate of traditional modes of life may rest in some distant financial capital. But, economics aside, the reality is that the remote lands of Siberia are linked by Earth systems to the Pampas of Argentina to the Great Plains in North America and to those harvesting grain by hand in the remote Pamirs of Tajikistan in the chapter-opening photo. To understand these linkages among Earth's myriad systems has been our quest in *Geosystems*.

Growing international environmental awareness in the public sector is gradually prodding government into action. In 1992, the George H. Gallup International Institute completed a 22-nation, 29,000-person survey that resulted in, among others, the following conclusions: ". . . people in both poor and rich nations give priority to environmental protection over economic growth," ". . . accept responsibility for contributing to problems, and believe that citizen efforts can contribute significantly to a healthier planet," and ". . . environmental problems are at the very top of the list of problems in virtually all nations." A similar survey by Louis Harris and Associates confirmed these findings. Majorities in 22 nations are willing to "endorse environmental protection at the risk of slowing down economic growth."[*] The public seems to know they are saving money as consumers if things are done right with the environment (Figure 21-1). Let us look at the Clean Air Act as an example.

## The Clean Air Act Brings a Windfall

Most environmental problems already have recognized, cost-effective solutions. We have discussed many of these in previous chapters (an example is shown in Figure 21-2). Conveying this knowledge to the public and reaching political agreement to put solutions into action should be our goals. Too few of these solutions are being acted upon by society. We seem unable to assess the true long-term cost of present economic thinking. This attitude now might change with the publication of a dramatic assessment of the Clean Air Act.

The Clean Air Act (1970, 1977, 1990) has been the subject of open political warfare between those who think

[*]Rusty Schweickart, "Our backs against the bomb, our eyes on the stars," *Discovery*, July 1987, p. 62. Reprinted by permission.
[**]C. Sagan, *Cosmos* (New York: Random House, 1980), p. 329. Reprinted by permission.
[***]M. Bates, *The Forest and the Sea* (New York: Random House, 1960), p. 247. Reprinted by permission.

[*]R. E. Dunlap, G. H. Gallup Jr., and A. M. Gallup, *The Health of the Planet Survey—A Preliminary Report on Attitudes toward the Environment and Economic Growth Measured by Surveys of Citizens in 22 Nations to Date*. A George H. Gallup Memorial Survey (Princeton, N.J.: George H. Gallup International Institute, July 1992).

**FIGURE 21-1**
**Citizens rally for Earth.**
Public environmental awareness is high, as evidenced by Earth Day activities. Polls show that people are willing to walk more softly, if given a path to follow. [Photo by Peter Arnold, Inc.]

its cost has been too high for industry, taxpayers, labor, and consumers and those who think the health and environmental benefits were justified. Compliance affected patterns of industrial production, employment, and capital investment. Although these expenditures must be viewed as investments that generated benefits and opportunities, the dislocation in some regions was severe: such as reductions in high-sulfur coal mining and cutbacks in polluting industries such as steel. A need developed for a real cost-benefit analysis.

In 1990, Congress requested the Environmental Protection Agency to answer the question: How do the overall health, welfare, ecological, and economic benefits of Clean Air Act programs compare with the costs of these programs? In response, the EPA performed the most exhaustive cost-benefit analysis of public policy ever attempted.* Here is what the EPA found:

- The total *direct cost* to implement the Clean Air Act for all federal, state, and local rules from 1970 to 1990 was *$436 billion* (in 1990-value dollars). This cost was borne by businesses, consumers, and government entities in the form of higher prices for many goods, services, and some utilities.

- The mean estimate of *direct benefits* from the Clean Air Act from 1970 to 1990 was $6.8 trillion.

- Therefore, the *net benefit* of the Clean Air Act has been $6.4 trillion!

*See: U.S. EPA, *The Benefits and Costs of the Clean Air Act, 1970 to 1990* (Washington, D.C.: U.S. EPA, May 1996), 392 pp.

"The finding is overwhelming. The benefits far exceed the costs of the Clean Air Act in the first 20 years," said Richard Morgenstern, Associate Administrator for Policy Planning and Evaluation at the EPA.

The benefits to society, directly and indirectly, have been widespread across the entire population. Here is a summary:

- *Reduced air pollution*: sulfur dioxide, −50% (due to scrubbers on smokestacks and emission controls); nitrogen oxide, −30%; volatile organic compounds, −45%; carbon monoxide, −50% (the last three the result of exhaust emission controls); and a huge lead reduction (unleaded gas and reduced industrial emissions).

- *Improved human health from reduced particulates, lead, sulfur dioxide, nitrogen dioxide, and photochemical ozone*: each year, 79,000 lives saved, 18,000 fewer heart attacks, 10,000 fewer strokes, and 13,000 fewer hypertension cases (average of best estimates) than would have occurred with no Clean Air Act. Over the 20 years, there were 15 million fewer respiratory cases.

- *"Avoided cost"*: improved health has meant less debilitating disease, hospitalization, special care, and medicines. The effects of air pollution control reduced work-related medical expenditures and improved worker productivity.

- *Less lead to harm children*: in 1990, 220,000 tons of lead were not burned in gasoline because of Clean Air Act measures. Exposure to lead impairs the cognitive development of children; therefore the huge reductions in lead produced a benefit of retained IQ and the possibility of a more productive, less dependent life. Such improvement can be assigned an estimated monetary value.

**FIGURE 21-2**
**Positive technologies solve problems.**
Major development of wind-power systems will occur in the twenty-first century for the generation of electricity from a renewable, nonpolluting source. [Photo by Lowell Georgia/Science Source/Photo Researchers, Inc.]

- *Lowered cancer rates*: air toxics emissions (pollutants, including over 100 known or suspected carcinogens) were reduced under the Clean Air Act, lowering cancer rates.

- *Less acid precipitation*: reduced sulfur oxides and nitrogen oxides have meant less acid deposition, helping aquatic and terrestrial ecosystems—commercial and recreational fishing, wildlife, biodiversity, and nutrient cycling (studied but not quantified by the report).

- *Likely enhanced growth and productivity*: reduced photochemical ozone should indirectly increase timber production and outdoor recreation (studied but not quantified by the report) and increase agricultural yields in regions where air pollution damaged crops.

Further, the report states that all "benefits may be significantly underestimated due to the exclusion of large numbers of benefits from the monetized benefit estimate."

As you reflect on this course and the many topics we covered and problems we identified, the EPA study results should feel refreshing. Society knew what to do, took action despite disruptive efforts by special interests and political partisans, and reaped about *$20 in benefit for every $1 invested* to control air pollution. Perhaps the displaced workers who bore the impact of clean-air regulations can now feel that their sacrifice was beneficial to society as a whole. An important role for physical geographers is to explain these results through spatial analysis and to synthesize information from many disciplines, to guide an informed citizenry toward understanding.

**FIGURE 21-3**
**An oily bird.**
A Western Grebe contaminated with oil from the *Exxon Valdez* tanker accident in Prince William Sound, Alaska. This oily bird is the result of a long chain of events and mistakes. [Photo by Geoffrey Orth/© Sipa Press.]

# An Oily Bird

At first glance, the chain of events that exposes wildlife to oil contamination seems to stem from a technological problem. An oil tanker splits open at sea and releases its petroleum cargo, which is moved by ocean currents toward shore, where it coats coastal waters, beaches, and animals (Figure 21-3). In response, concerned citizens mobilize and try to save as much of the spoiled environment as possible. But the real problem goes far beyond the physical facts of the spill.

In March 1989, on Prince William Sound off the southern coast of Alaska in clear weather and calm seas, a single-hulled supertanker operated by Exxon Corporation struck a reef. The tanker spilled 42 million liters (11 million gallons) of oil. It took only 12 hours for the *Exxon Valdez* to empty its contents, yet a reasonable cleanup is taking years and billions of dollars. (See the area of the spill, noted on the map in Figure 21-4.)

Because emergency plans were not in place, and promised equipment was unavailable, response by the oil industry took over 12 hours to activate, about the same time that it took the ship to drain. The initial response by Alyeska Pipeline Service Company and Exxon proved inadequate to the task. Badly needed equipment was unavailable; it was resting in dry docks, stored in warehouses, or buried in snow. In fact, initial efforts were directed at public relations.

In the years following the spill, evidence of this confusion was documented. Total damage now is estimated to exceed $15 billion, although Exxon settled with the government for a little over $1 billion and several civil suits remain in the courts.

Eventually, over 2400 km (about 1500 mi) of sensitive coastline was ruined for years to come. For perspective, had this spill occurred farther south, enough oil spilled to blacken every beach and bay along the Pacific Coast from southern Oregon to the Mexican border.

The death toll of animals was massive: At least 3000 sea otters died, or about 20% of resident otters. About 300,000 birds, and uncounted fish, shellfish, plants, and aquatic microorganisms also perished. Sublethal effects, namely mutations, now are appearing in fish. The Pacific herring is still in significant decline, as are the harbor seals. Conflicting scientific reports emerged in 1993 as to long-term damage in the region: Scientific studies commissioned by Exxon disagreed with scientific studies prepared outside the industry.

## *The Larger Picture*

The immediate effect of the oil spill on wildlife was contamination and death. But the issues involved are much bigger than dead birds. Let's ask some fundamental questions:

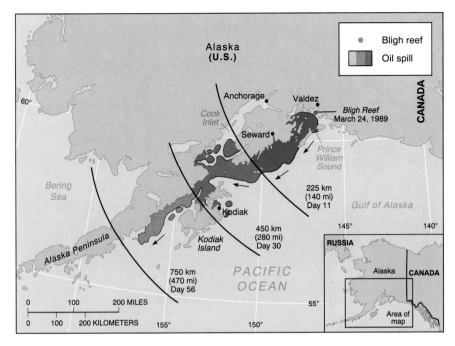

**FIGURE 21-4**
**The 1989 *Exxon Valdez* oil spill.**
Track of spreading oil for the first 56 days of the spill.

- Why was the oil tanker there in the first place?
- Why is petroleum being imported into the continental United States from Alaska in such enormous quantity?
- Is the demand for petroleum products based on real need, or does it reflect the business strategies of transnational oil corporations?
- The U.S. demand for oil is higher per capita than the demand of any other country; is this demand inflated by waste and inefficiency?

A great many factors influence our demand for oil. Well over half of our imported oil is burned in vehicles. Improvement of automobile efficiency began in 1975 as ordered by new federal regulations. But the 1980s saw a major rollback of auto efficiency standards, a reduction in gasoline prices, large reductions in funding for rapid transit development, and the continuing slow demise of America's railroad network for passengers and freight. Thus, a combination of waste, low prices, and a lack of alternatives has spurred the demand for petroleum. In addition, our land-use policies continue to foster a diffuse sprawl of our population, thereby adding stress to transportation systems.

The demand for fossil fuels in the 1980s also was affected by the slowing of domestic conservation programs, elimination of research for energy alternatives such as solar and wind power, and even political delay of a law requiring small appliances to be more energy-efficient. Conservation plans again were politically blocked in the U.S. Department of Energy in 1990 and 1991.

All of these manipulations increased fossil fuel demand to 6.1 billion barrels a year for oil alone. At this rate, the U.S. domestic reserve will be depleted before the year 2020. (This estimate is based on an optimistic 1995 USGS assessment of domestic reserves totaling 110 billion barrels: 110 ÷ 6.1 = 18 years.) Imports at the level of 50% or greater will stretch this reserve, but with unknown economic, environmental, and military consequences. Of course, imports will be an increasingly difficult problem because, at increasing global consumption rates, the planet could be out of crude oil reserves by 2040.

Thus, the immediate problem of cleaning oil off a bird in Prince William Sound symbolizes a national and international concern with far-reaching significance. And while we search for answers, oil slicks continue their contamination. In just one year following the *Exxon Valdez* disaster, nearly 76 million liters (20 million gallons) of oil were spilled in 10,000 accidents worldwide, shown in the startling map in Figure 21-5. This is an average of 27 accidents a day, ranging from a few disastrous spills to numerous small ones (News Report 1).

In the Persian Gulf War of 1991, we saw over 1.1 billion liters (300 million gallons) flow into the Persian Gulf and additional millions of liters pour onto the land and burn into the atmosphere in purposely set fires (Figure 21-6). Public opinion holds Exxon, the second largest corporation in America, responsible for what happened in Prince William Sound and the dictator of Iraq as responsible for the tragedy in the Persian Gulf.

Yet, hypocrisy is apparent in our outrage over these political events and accidents. For we continue to consume gasoline at record levels in inefficient vehicles, thus creating the demand for oil imports. It is the task of physical geography to analyze all the spatial aspects of these events in the environment and the ironies that are symbolized by a little oily bird. A global perspective is vital.

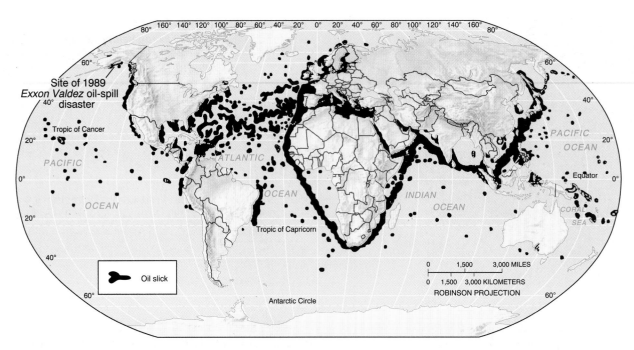

**FIGURE 21-5**
**Recent worldwide oil spills.**
Location of recent visible oil slicks worldwide. The location of the 1989 *Exxon Valdez* oil spill is noted
on the map. [Data from Organization for Economic Cooperation and Development, *The State of the
Environment* (Paris: OECD, 1985).]

## News Report 1

### Oil Spills: Global and Local

The massive importation of oil into the United States, Canada, and the European Union places many tankers at sea and increases the risk of spills. In the same year as the *Exxon Valdez* accident, an even greater spill occurred off the Moroccan coast in northwestern Africa. Into those warm tropical waters, that vessel spilled 140.6 million liters (37 million gallons) of oil. Other examples of significant recent spills include:

• The collision of the *Atlantic Express* and the *Aegean Captain* off Trinidad in the Caribbean in 1979—333 million liters (88 million gallons)
• The *Aegean Sea,* off Spain in 1992—83.6 million liters (22 million gallons)
• The *Maersk Navigator*, in Indonesian waters in 1993—30 million liters (7.8 million gallons)
• The *Braer*, off the Shetland Islands in 1993—91 million liters (24 million gallons)

• The grounding of *Sea Empress* off Wales in 1996—72 million liters (19 million gallons)

In addition to oceanic oil spills, people improperly dispose of crankcase oil from their automobiles in a volume that annually exceeds tanker spills! We need to address our concerns globally, but we also need to act locally when it comes to oil entering the environment!

## The Need for International Cooperation

We already have seen encouraging examples of international environmental problem solving: the Limited Test Ban Treaty of 1962, which bans atmospheric testing of nuclear weapons, and the 1987 and 1990 treaties proposing bans on ozone-destroying chlorofluorocarbons. The International Geosphere-Biosphere Program (IGBP) represents another integrative effort; its goal is to

improve understanding of the entire natural system and to discern how its many and varied subsystems interact.

All of Earth's major problems need such cooperative attention. An example of successful cooperation is the multinational Intergovernmental Panel on Climate Change (IPCC) discussed in Chapter 10. The escalating effect of global climate change, which influences so many global systems, has brought together the nations and their scientists.

The largest gathering of nations and nongovernmental organizations ever held in the entire history of

**FIGURE 21-6**
**1991 Kuwait oil well fires.**
Smoke from the Kuwaiti oil well fires drifts over thousands of square kilometers on July 1, 1991. [Satellite image courtesy of EOSAT, Earth Observation Satellite Company. Used by permission.]

civilization was about Earth—the United Nations Conference on Environment and Development (UNCED), known as the "Earth Summit of 1992." This important event and resulting agreements are surveyed in Focus Study 21-1. (See Appendix A for the numerous groups, organizations, and international action programs related to geography and environmental concerns.)

A critical corollary to international efforts is the linkage of academic disciplines. A positive step in that direction is the *Earth systems science* (geosystems) approach, as illustrated in this text and in some recent geology textbooks. Exciting progress toward an integrated understanding of Earth's life-supporting physical and biological systems is in progress. This is fueled by insights drawn from our remote-sensing capabilities and computer modeling. Never before has society been able to monitor Earth's physical geography so thoroughly.

## The Environmental Cost of Military Conflict

The greatest challenge for international cooperation and the environment is warfare. Modern warfare has had a definite impact on the environment: Scars from World War I (1914–1918) and World War II (1939–1945) still are visible on the landscapes of Europe and Asia. More than 33 million bomb craters and massive defoliation provide environmental evidence of the more recent struggle in Vietnam (1964–1975). In fact, during the Vietnam War, more than 2.3 times the equivalent tonnage of all munitions used in World War II were detonated. (In terms of more modern weapons, just one of the planned MX nuclear missiles carries 3 megatons of explosive power, the equivalent of all the firepower of all the combatants in World War II!)

In light of the devastation caused by conventional warfare during 1991 in the Middle East, where ecosystems were damaged extensively—with unknown long-term consequences—we must consider the environmental consequences of all warfare to air, water, land, and biotic systems. If conventional explosives cause this much damage, imagine the destruction that would result from nuclear weapons.

In 1975, the National Academy of Sciences published its study on the potential impact of modern nuclear war on stratospheric ozone. This was the beginning of public awareness that modern technological warfare is damaging to the environment beyond the range of shell craters and radioactive fallout. The NAS states that a nuclear war could lead to a 40%–70% reduction of stratospheric ozone, letting in enough UV radiation to cause a catastrophic change in food chains, plants, animals, and humans.

In addition, from 1982 to the present, scientific publications have described the *nuclear winter hypothesis*. It indicates that a nuclear-caused worldwide "winter" might follow the detonation of only modest numbers of nuclear weapons. As you have learned, the atmosphere is dynamic. A nuclear exchange would ignite fierce urban firestorms, the soot from which would heavily pollute the atmosphere. This soot would quickly spread, much like the rapid global circulation of ash and acid mists from the 1991 Mount Pinatubo eruption (see Figure 6-1). This would increase Earth's albedo, absorbing insolation in the stratosphere and upper troposphere and reradiating it to space. The result would be a cooling of Earth's surface to below freezing, even during summer months. As the nuclear winter hypothesis developed, others suggested a milder "nuclear autumn," but it still would cause significant cooling.

The potential for environmental disaster demands that nations acknowledge the limiting factors imposed by Earth's ecosystems. Modern physical geography needs to consider the global impact of technological conflicts (either conventional or nuclear warfare). They can no longer be thought of in a regional context, for Earth's integrated systems spread the consequences around the globe. The real need for international cooperation should consider our symbiotic relationships with each other and with Earth's resilient, yet fragile, life-support systems (News Report 2).

## Focus Study 21-1

## Earth Summit 1992

The leaders of 100 nations, 10,000 delegates from over 160 countries, 9000 journalists, and the world's attention all converged at Rio Centro, a city 40 minutes outside of Rio de Janeiro, during June 1–12, 1992 (Figure 1). An additional 1000 nongovernmental organizations (NGOs) with over 50,000 attendees assembled at nearby Flamingo Park for a parallel, unofficial gathering dubbed the "Global Forum."

The setting in Rio de Janeiro was ironic, for many of the very problems discussed at the two conferences were evident on the city's streets, with their abundant air pollution, water pollution, toxics, noise, wealth and poverty, and daily struggle for health and education.

Maurice F. Strong, a Canadian and Secretary-General of the UNCED, summarized in his conference address:

The people of our planet, especially our youth and the generations which follow them, will hold us accountable for what we do or fail to do at the Earth Summit in Rio. Earth is the only home we have, its fate is literally in our hands. . . . The most important ground we must arrive at in Rio is the understanding that we are all in this together.

**Setting the Stage**

The idea for a global meeting on the environment was put forward at the 1972 U.N. Conference on the Human Environment held in Stockholm, also chaired by Maurice Strong. The U.N. General Assembly in 1987 achieved a landmark in global planning by agreeing to hold the Summit. *Our Common Future* set the tone for the 1992 Earth Summit:

The Earth is one but the world is not. We all depend on one biosphere for sustaining our lives. Yet each community, each country, strives for survival and prosperity with little regard for its impact on others. Some consume the Earth's resources at a rate that would leave little for future generations. Others, many more in number,

(a)

(b)

**FIGURE 1**

**Leaders and delegates at the 1992 UNCED Earth Summit in Rio de Janeiro.**

(a) Some of the delegates who attended from many nations. (b) Maurice Strong, Secretary-General of the UNCED (far left), a representative of indigenous peoples, and other delegates. [(a) Photo by Reuters/Bettmann; (b) photo by Ricardo Funavi/Imagens Da Terra/Impact Visuals Photo & Graphics, Inc.]

consume far too little and live with the prospect of hunger, squalor, disease, and early death.[*]

## The Five Earth Summit Agreements

Let us summarize the five key agreements that emerged from the Earth Summit:

*1.   Climate Change Framework.* This legally binding agreement is a first-ever attempt to evaluate and address global warming on an international scale. Signing the Convention on Climate Change were 165 countries, including the United States and Canada, and the European Union. The fiftieth country to actually ratify it, in December 1993, put the treaty into force.

Initially, the goal was to set specific timetables for cutting greenhouse gas emissions. The European Union, Canada, Japan, and the majority of attending nations favored stabilizing emissions at 1990 levels by the year 2000. The United States objected to specific targets for controlling carbon dioxide emissions throughout the pre-summit sessions, citing economic uncertainty and unknown cost. To the contrary, the U.S. Office of Technology Assessment and National Academies of Science and Engineering, in separate assessments, concluded that the United States could hold to 1990 levels by the year 2015 at little or no additional cost. Richard Kerr summarized:

> The philosophy that many scientists contacted by *Science* are now espousing amounts to buying some insurance—in the form of no-cost or low-cost reductions in greenhouse gas emissions—against the possibility that the higher predictions of global warming turn out to be right. (*Science* 256 (May 22, 1992): 1140.)

*2.   Biological Diversity.* This legally binding agreement is the first international attempt to protect Earth's biodiversity. It provides more equitable rights among

[*]World Commission on Environment and Development, *Our Common Future* (Oxford: Oxford University Press, 1987), p. 27.

nations in biotechnology and the genetic wealth of tropical ecosystems in particular.

Out of 161 signatories, the United States, Vietnam, Singapore, and Kiribati (a Pacific island nation) refused to sign the original treaty. This was a perplexing stand for the U.S. administration to take. Biodiversity is a divisive issue separating developed countries in the Northern Hemisphere from the predominantly developing nations of the equatorial and tropical regions and Southern Hemisphere. The developing countries want compensation for medicines derived from plants and animals harvested from their indigenous genetic wealth. This return of some profit to them from transnational corporations and rich-nation enterprises creates incentive to further protect these critical biomes. The United States finally signed this treaty in 1993.

*3.   Management, Conservation, and Sustainable Development of All Types of Forests.* This nonbinding agreement guides world forestry practices toward a more sustainable future of forest yield and diversity.

The conflict between industrialized nations and developing nations is a classic confrontation of "north" and "south." How can rich northern nations continue to clear-cut their forests, yet turn to developing countries, such as Brazil, and ask them to place their lands in a national preserve? How can the industrialized nations continue to produce excessive carbon dioxide, far beyond reasonable per capita limits, yet ask developing countries to cease destroying a principle sink for carbon dioxide, the tropical rain forest? The developing countries of the "south" insist on political and economic equality of forest practice in the "north." Along with sustainable timber practices, countries of the "north" need to begin government-sponsored paper recycling and packaging-reform programs. Such recycling now is part of the forestry debate.

*4.   The "Earth Charter."* This is a nonbinding statement of 27 environmental and economic principles. They establish an ethical basis for a sustainable human-Earth relationship. An important emphasis is inclusion of *environmental cost* in economic assessments. Degradation of air, soil, water, and ecosystems some-

times is mistaken for progress. The environment is not an inexhaustible mine of resources to be tapped indefinitely. In terms of *natural capital*—air, water, timber, fisheries, petroleum—Earth is indeed a finite physical system, as we have learned.

*5.   Agenda 21 (Sustainable Development).* This nonbinding action program is an 800-page guide for all nations into the twenty-first century. The idea of "sustainable development," as opposed to a business proposal of "sustainable growth," is examined in Agenda 21.

Agenda 21 covers many key topics: energy conservation and efficiency to reduce consumption and related pollution; climate change; stratospheric ozone depletion; transboundary air pollution; ocean and water resource protection; soil losses and increasing desertification; deforestation; regulations for safely handling and disposing of radioactive waste; hazardous chemical exports for disposal in developing countries; disparities of wealth; and the plague of poverty.

Agenda 21 also addresses the difficult question of financing sustainable development. Developing countries are asking the developed nations to spend 0.7% of their gross domestic product—approximately $125 billion per year—to assist them in implementing the Earth Charter and Agenda 21.

## The Future

From the Earth Summit emerged a new organization—the U.N. Commission on Sustainable Development—to oversee the promises made in the five documents and agreements. Most of the participating countries completed State of the Environment Reports (SERs) and gathered environmental statistics for publication. These reports are an invaluable resource that will direct further research efforts in many countries.

Considering these environmental problems and possible world actions, these are challenging times for humanity as we ponder our relationship to the home planet. Over the long term, we no longer can sustain human activity through old patterns. The study of physical and human geography is central to this assessment.

Asking whether the Earth Summit "succeeded" or "failed" is the wrong question. The occurrence of this largest-ever official gathering of Earthlings and the 5 years of preparatory effort and study that set the agenda are in themselves remarkable accomplishments.

The challenge now is one of education; the lesson is one of compromise and some sacrifice. Members of society must work to move the solutions for environmental and developmental problems off the bench and into play. The Earth Summit process continues as a good beginning.

# News Report 2

## Gaia Hypothesis Triggers Debate

Some view Earth as one vast, self-regulating organism. The concept is one of global symbiosis, or mutualism. This controversial concept is called the *Gaia hypothesis* (Gaia was the Earth Mother goddess in ancient mythology). It was proposed in 1979 by James Lovelock, a British astronomer and inventor, and elaborated by American biologist Lynn Margulis.

Gaia is the ultimate synergistic relationship, in which the whole greatly exceeds the sum of the individual interacting components. The hypothesis contends that life processes control and shape Earth's inorganic physical and chemical processes, with the ecosphere so interactive that a very small mass can affect a very large mass. Thus, Lovelock and Margulis think that the material environment and the evolution of species are tightly joined; as species evolve through natural selection, they (including us) in turn affect their environment. The present oxygen-rich composition of the atmosphere is given as proof of this co-evolution of living and nonliving systems.

From the perspective of physical geography, the Gaia hypothesis permits a view of all Earth and the spatial interrelations among systems. In fact, such a perspective is necessary for analyzing specific environmental issues. For instance, in deciding the fate of the Arctic National Wildlife Refuge (ANWR, discussed in Chapter 19, News Report 3), we must weigh the supply, demand, and importation of oil, the public trust aspect of such wilderness, our will for conservation and efficiency, and our view of our place in nature or outside of nature (Figure 1). All these variables interact synergistically, producing both wanted and unwanted results.

One disturbing aspect of this unity is that any biotic threat to the operation of an ecosystem tends to move toward extinction itself. This trend preserves the system overall. Earth-systems operation and feedback naturally tend to eliminate offensive members. The degree to which humans represent a planetary threat, then, becomes a topic of great concern, for Earth (Gaia) will prevail, regardless of the outcome of the human experiment.

The maladies of Gaia do not last long in terms of her life span. Anything that makes the world uncomfortable to live in tends to induce the evolution of those species that can achieve a new and more comfortable environment. It follows that, if the world is made unfit by what we do, there is the probability of a change in regime to one that will be better for life but not necessarily better for us.[*]

The debate is vigorous regarding the true applicability of this hypothesis to nature, or whether it is true science at all. Regardless, it remains philosophically intriguing in its portrayal of the relationship between humans and Earth.

[*]J. Lovelock, *The Ages of Gaia—A Biography of Our Living Earth* (New York: W. W. Norton, 1988), p. 178. Used with permission.

**FIGURE 1**
**Arctic National Wildlife Refuge.**
In Alaska's Arctic National Wildlife Refuge, Mount Chamberlin, the second highest peak of the Brooks Range, overlooks tundra in the foreground. Is this region destined for petroleum exploration and development or for continued preservation? [Photo by Scott T. Smith.]

# Who Speaks for Earth?

Geographic awareness and education is an increasingly positive force on Earth. Following decades of decline, geographic understanding now is improving each year. The National Geographic Society conducts an annual National Geography Bee to promote geography education and global awareness to millions of 6th to 8th graders, their parents, and teachers. The International Geography Olympiad is now an annual event. There are presently 60 geographic alliances in 47 states, which coordinate geographic education among teachers and students at all levels: K–12, community college, college, and university.

Yet, ideological and ethical differences still remain within society. This was addressed by the famous biologist Edward O. Wilson:

> The evidence of swift environmental change calls for an ethic uncoupled from other systems of belief. Those committed by religion to believe that life was put on Earth in one divine stroke will recognize that we are destroying the Creation; and those who perceive biodiversity to be the product of blind evolution will agree. Across the other great philosophical divide, it does not matter whether species have independent rights or, conversely, that moral reasoning is uniquely a human concern. Defenders of both premises seem destined to gravitate toward the same position on conservation. . . . For what, in the final analysis is morality but the command of conscience seasoned by a rational examination of consequences? . . . An enduring environmental ethic will aim to preserve not only the health and freedom of our species, but access to the world in which the human spirit was born.*

Carl Sagan asked, "Who speaks for Earth?" He answered with this perspective:

> We have begun to contemplate our origins: starstuff pondering the stars; organized assemblages of ten billion billion billion atoms considering the evolution of atoms; tracing the long journey by which, here at least, consciousness arose. Our loyalties are to the species and the planet. We speak for Earth. Our obligation to survive is owed not just to ourselves but also to that Cosmos, ancient and vast, from which we spring.**

Each of us might consider where we will be in life's journey on Sunday night, at 11:59 P.M., December 31, 2000, with 2001 and the new millennium just one minute away.

---

*M*ay we all perceive our spatial importance within Earth's ecosystems and do our part to maintain a life-supporting and sustaining Earth for ourselves and countless generations in the future.

---

*E. O. Wilson, *The Diversity of Life* (Cambridge: Harvard University Press, 1992), p. 351.
**C. Sagan, *Cosmos* (New York: Random House, 1980), p. 345. Reprinted by permission.

## News Report 3

### *Time* Magazine and the "Planet of the Year"

The global concern about environmental impacts prompted *Time* magazine in 1989 to deviate from its 60-year tradition of naming a prominent citizen as its person of the year, instead naming Earth the "Planet of the Year."

The magazine devoted 33 pages of its January 2, 1989 issue to Earth's physical and human geography and the endangered status of many ecosystems and cultures (Figure 1). Importantly, *Time* also offered positive policy strategies for consideration. We seem to be on the brink of a new age in Earth awareness.

**FIGURE 1**
*Time* magazine cover for January 2, 1989, naming Earth "Planet of the Year." [Copyright © 1988 The Time Inc. Magazine Company. Reprinted by permission.]

---

## Review Questions — Earth, Humans, and the New Millennium

1. What part do you think technology, politics, and thinking about the future should play in science courses?

2. Given the assessment of monetary benefit from the U.S. Clean Air Act, what is your opinion as to any continued opposition to its regulations?

3. According to the discussion in the chapter, what worldwide factors led to the *Exxon Valdez* accident? Describe the complexity of that event from a global perspective. In your analysis, examine both supply-side and demand-side issues, as well as environmental and strategic factors.

4. What is meant by the Gaia hypothesis? Describe several concepts from this text that might pertain to this hypothesis.

5. Relate the content of the various chapters in this text to the integrative Earth systems science concept. Which chapters help you to better understand Earth-human relations and human impacts?

6. Explain the potential spatial impact of nuclear warfare on the environment. What is the nuclear winter hypothesis, and what are its potential implications for the environment?

7. This chapter states that we already know many of the solutions to the problems we face. Why do you think these solutions are not being implemented at a faster pace?

8. Who speaks for Earth?

---

### NetWork

The *Geosystems Home Page* provides on-line resources for this chapter on the World Wide Web. You will find review exercises, specific updates for items in the chapter, suggested readings, and links to interesting related pathways on the Internet (click on the Table of Contents link and select this chapter).
*Geosystems* is at: **http://www.prenhall.com/geosystm**

# Appendix A

# Information Sources, Organizations, and Agencies

This is a brief sampling of sources for further geographic and environmental inquiry. Consult your campus library and instructor for related sources and research pathways. Also, our *Geosystems Home Page* will direct you to many sources of information on the Internet and World Wide Web. Contact us at:

**http://www.prenhall.com/geosystm**

## General Information Resources

*Applied Geography,* Butterworth-Heinemann Ltd., Linacre House, Jordan Hill, Oxford, England OX2 8DP. A quarterly international journal dealing with human problems that have a geographic dimension.

*Canada and the World.* 1985. By Geoffrey J. Matthews and Robert Morrow. Published by Prentice Hall Canada, Scarborough, Ontario, Canada. An atlas resource.

*Canada Yearbook.* Published annually by minister of Supply and Services. Available from Publication Sales and Services, Statistics Canada, Ottawa, Canada K1A 0T6.

*Climate Alert.* Published bimonthly by Climate Institute, 316 Pennsylvania Avenue, SE, Suite 403, Washington, DC 20003. A clearinghouse of information on world climate.

*The Climates of Canada.* 1990. Compiled by Canada's chief climatologist, David Phillips. Available from Minister of Supply and Services Canada, Ottawa, Canada K1A 0S9.

*The Complete Guide to Environmental Careers.* 1995. The CEIP Fund. Island Press, Covela, CA 95428.

*Congressional Directory.* Published after every biennial congressional election. Available from U.S. Government Printing Office, Washington, DC 20402.

*Conservation Directory,* 41st edition in 1996. National Wildlife Federation, Washington, DC. The most comprehensive directory for the environment; listings are international, national, and by state; complete US government listing of related agencies.

*Dictionary of Physical Geography.* 1984. By John Whittow. Penguin Books, New York, NY 10010.

*Dictionary of Science and Technology.* 1992. Edited by Christopher Morris. Published by Academic Press, San Diego, CA 92101.

*Earth: The Science of Our Planet.* Kalmbach Publishing Co., 21027 Crossroads Circle, Waukesha, WI 53187. A bimonthly magazine on Earth science themes; beautiful photographs and images.

*The Encyclopedia of Climatology.* 1987. Edited by John E. Oliver and Rhodes Fairbridge. Van Nostrand Reinhold Co., New York, NY 10003.

*Encyclopedia of Earth Sciences.* 1989. Edited by David G. Smith. Cambridge University Press, 40 West 20th St., New York, NY 10011.

*Encyclopedia of Environmental Sources.* 1993. Edited by Sarojini, Balachandran. Gale Research Inc. 835 Penobscot Bldg., Detroit, MI 48226.

*The Encyclopedia of Environmental Studies.* 1991. By William Ashworth. Published by Facts on File, New York, NY 10016.

*Environment.* Heldref Publications, 1319 Eighteenth St., NW, Washington, DC 20036-1802. A journal of environmental science.

*Environmental Almanac* (Information Please). Compiled by the World Resources Institute and published by Houghton Mifflin Company, New York, NY 10003, issued annually.

*Geographical Bibliography for American Libraries.* 1985. Edited by Chauncy D. Harris. A joint project of the Association of American Geographers and the National Geographic Society, Washington. AAG, 1710 16th Street, NW, Washington, DC 20009.

*Global Environmental Change,* Butterworth-Heinemann Ltd., Linacre House, Jordan Hill, Oxford, England OX2 8DP. A quarterly international journal dealing with public policy and environmental and human ecological processes.

*Glossary of Geology,* 3rd edition in 1987. Edited by Robert L. Bates and Julia A. Jackson. American Geological Institute, Alexandria, VA 22302.

*The Island Press Bibliography of Environmental Literature.* 1993. Compiled by Joseph Miller, Sarah Friedman, and others at the Yale School of Forestry and Environmental Studies. Published by Island Press, Covelo, CA 95428.

*National Reports Summaries—Nations of the Earth Report,* in three volumes. 1992. United Nations Conference on Environment and Development, Geneva, Switzerland.

*Nature.* Macmillan Magazines, Ltd., 345 Park Ave., New York, NY 10010. Important international weekly science journal.

*Science News.* Science News Service, 1719 N St., NW, Washington, DC 20036. A weekly digest of science articles from many, diverse journals.

*State of the Environment*—A View toward the Nineties. 1987. A report from the Conservation Foundation, sponsored by the Charles Stewart Mott Foundation. Conservation Foundation, 1250 42nd Street NW, Washington, DC 20076.

*State of the World—Report on Progress toward a Sustainable Society.* Published annually by the WorldWatch Institute (Lester R. Brown, project director), 1776 Massachusetts Avenue NW, Washington, DC 20036.

*Statistical Abstract of the United States.* Published annually by the U.S. Bureau of the Census, Department of Commerce, Washington, DC. Also available from U.S. Government Printing Office, Washington, DC 20402.

*U.S. Government Manual.* Published by the Office of the Federal Register with annual updates. Available through the

Superintendent of Documents, U.S. Government Printing Office, Washington, DC 20402.

U.S. Government Printing Office. North Capitol and H Streets NW, Washington, DC 20401. Superintendent of Documents provides monthly catalogue of all U.S. government publications. Many local government bookstores are open to the public.

*Weatherwise.* Heldref Publications, 1319 Eighteenth St., NW, Washington, DC 20036-1802. A magazine about the weather.

*The World Almanac.* Published annually by World Almanac and Book of Facts, a Scripps Howard company, New York, NY.

*World Tables.* 1996. Compiled by the World Bank. Baltimore: The John Hopkins University Press, 701 W 40th St., Baltimore, MD 21211. Annual publication of socioeconomic indicators.

*World Resources 1996–97.* Published annually by the World Resources Institute, in collaboration with the United Nations Environment Programme, 1735 New York Avenue NW, Washington, DC 20006. Exhaustive source of information on the current state of environmental affairs, including data on most physical systems.

## Geography Organizations

Alaska Geographical Society, P.O. Box 4-EEE, Anchorage, AK 99509. One of many state and provincial geographical societies operating in the United States and Canada.

American Geographical Society, 156 Fifth Avenue, Suite 600, New York, NY 10010-7002. Publishes *Geographical Review* and *Focus* quarterly.

Association of American Geographers, 1710 16th Street NW, Washington, DC 20009-3198. Principal academic and professional geography organization—offers student memberships and career information. Publishes *Annals, The Professional Geographer,* and the *AAG Newsletter.*

Canadian Association of Geographers, McGill University, 805 Sherbrooke Street West, Montréal, Québec H3A 2K6. Publishes *The Canadian Geographer* and *The Operational Geographer—The Canadian Journal for Practicing Geographers.*

Geographic alliances, 60 state chapters. Information is available from college and university geography departments; generally coordinated by the National Geographic Society, Geography Education Program, Washington, DC.

Institute of British Geographers, 1 Kingsington Gore, London, SW7 2AR, England. Publishes *Transactions of the Institute of British Geographers* and *Area.*

National Council for Geographic Education, Indiana University of Pennsylvania, Indiana, PA 15705. Publishes the *Journal of Geography.*

National Geographic Society, 17th and M Streets NW, Washington, DC 20036. Publishes maps, atlases, *National Geographic Magazine,* and *Geography Education Update* through their Geographic Education Program.

Royal Canadian Geographical Society, 39 McArthur Ave., Vanier, Ontario, Canada K1L 8L7. Publishes *Canadian Geographic.*

Royal Geographical Society, 1 Kingsington Gore, London SW7 2AR, England.

## Private and Public Organizations and Agencies

Acid Rain Foundation, 1410 Varsity Drive, Raleigh, NC 27606.

American Association for the Advancement of Science, 1333 H Street NW, Washington, DC 20005. Publishes *Science*, a weekly journal.

American Chemical Society, 1155 16th Street NW, Washington, DC 20005. Scientific and educational association of professional chemists.

American Forests, 1516 P St., Street NW, Washington, DC 20036. Intelligent management and use of forests.

American Geophysical Union, 2000 Florida Avenue NW, Washington, DC 20009. Scientific organization for Earth systems and astronomical science research.

American Meteorological Society, 45 Beacon Street, Boston, MA 02108. Publishes *Journal of the Atmospheric Sciences, Journal of Climate, Monthly Weather Review, Weather and Forecasting*, and the *Bulletin of the AMS.*

American Rivers, 801 Pennsylvania Ave., SE, Suite 400, Washington, DC 20003-2167. Works for stewardship of rivers and restoration of degraded rivers.

British Association for the Advancement of Science, Fortress House, 23 Savile Row, London W1X 1AB.

Canadian Coalition on Acid Rain, 112 St. Clair Avenue West, Toronto, Ontario, Canada M4K 2Y3.

Center for Marine Conservation, Inc., 1725 DeSales Street, NW, Suite 500, Washington, DC 20036.

Center for Science in the Public Interest, 1501 16th Street NW, Washington, DC 20036. Consumer advocacy group.

Climate Institute, 316 Pennsylvania Avenue, SE, Suite 403, Washington, DC 20003. Publishes *Climate Alert* and *Coping with Climate Change.*

Coastal Conservation Association, Inc., 4801 Woodway, Suite 220 West, Houston, TX 77056. Preserving, conserving, and protecting marine ecosystems.

Coast Alliance, 235 Pennsylvania Ave., SE, Washington, DC 20003. An activist group to preserve coastal resources.

Conservation International, 1015 18th Street, NW, Suite 1000, Washington, DC 20036. Focuses on tropical and temperate ecosystems.

Cousteau Society, 870 Greenbrier Circle, Suite 402, Chesapeake, VA 23320. Environmental education and preservation; ensure ecological sustainability.

Earthwatch, P.O. Box 403N, Mt. Auburn St., Watertown, MA 02272. Sponsors field research worldwide and publishes *Earthwatch Magazine.*

Energy, Mines, and Resources Canada, 580 Booth Street Ottawa, Ontario, Canada K1A 0E4.

Environmental Action Foundation, 6930 Carroll Ave., Suite 600, Takoma Park, MD 20912. Publishes *Environmental Action.*

Environmental Defense Fund, 257 Park Avenue South, New York, NY 10010. Leading environmental organization working on many issues across the globe.

Environmental Protection Agency, 401 M Street SW, Washington, DC 20460. Primary government agency for environmental regulation and enforcement.

Environment Canada, Ottawa, Ontario K1A OH3.

Friends of the Earth, Inc., 1025 Vermont Ave., NW, Washington, DC 20005. Participates worldwide to protect Earth systems and preserve biological, cultural, and ethnic diversity (now merged with Environmental Policy Institute and the Oceanic Society).

Geological Survey of Canada, Headquarters: 601 Booth Street, Ottawa, Ontario K1A OE8. Many divisions, centers, and branches performing specialized functions. Provincial survey offices nationwide.

Greenpeace, Canada, 427 Bloor Street West, Toronto, Ontario, Canada M5S 1X7.

Greenpeace, U.S.A., Inc., 1436 U Street NW, Washington, DC 20009. Earth preservation through nonviolent direct action and information disemination. Published the *Greenpeace Newsletter.*

International Centre for Ocean Development, 5670 Spring Garden Road, 9th Floor, Halifax, Nova Scotia, Canada B3J 1H6.

League of Conservation Voters, 1707 L St., NW, Suite 550, Washington, DC 20036. Works to elect pro-environment candidates and publishes the *National Environmental Scorecard.*

National Audubon Society, 700 Broadway, New York, NY 10003-9501. Science, lobbying, education directed at ecosystem protection and sustainability. Publishes *Audubon* magazine.

National Oceanic and Atmospheric Administration (NOAA), Department of Commerce, Herbert C. Hoover Bldg., Rm 5128, 14th and Constitution Ave., NW, Washington, DC 20230. Includes: National Marine Fisheries Service; National Ocean Service; National Environmental Satellite, Data, and Information Service; National Weather Service; and Office of Global Program. Many publications, incuding *Daily weather maps,* a weekly presentation of weather in North America; *Storm Data,* a monthly report of all storms and unusual weather phenomena including special reports and maps of extreme events; *Monthly Climatic Data for the World.*

National Parks and Conservation Association, 1776 Massachusetts Ave., NW, Washington, DC 20036. Natural history associations are organized in conjunction with most national parks.

National Wildlife Federation, 1400 16th Street NW, Washington, DC 20036. Publishes an annual updated list of all agencies and organizations dealing with natural resources—the *Conservation Directory.* Publishes *International Wildlife* and *National Wildlife.*

Natural Resources Conservation Service (formerly the Soil Conservation Service), Department of Agriculture, P.O. Box 2890, Washington, DC 20013. Soils and soil science, watershed, floods, water supply and management.

Natural Resources Defense Council, 40 West 20th St., New York, NY 10011. Publishes *The Amicus Journal.* Protection of natural resources and improvement of the environment.

The Nature Conservancy, 1815 North Lynn Street, Arlington, VA 22209. Preserving natural lands and biodiversity; cooperative natural heritage programs in each state.

Rainforest Action Network, 450 Sansome, Suite 700, San Francisco, CA 94111. Protection of the rainforests and indigenous people.

Resources for the Future, 1616 P Street, NW, Washington, DC 20036. Research and education toward the conservation of natural resources and environmental quality.

Rocky Mountain Institute, 1739 Snowmass Creek Road, Snowmass, CO 81654. Research in energy-related issues, conservation, efficiency, and renewables, publishes the *Rocky Mountain Newsletter.*

Sierra Club, 730 Polk Street, San Francisco, CA 94109. Broad-based, hundred-year-old organization for exploring, protecting, and enjoying the wild places on Earth.

Soil and Water Conservation Society, 7515 NE Ankeny Road, Ankeny, IA 50021-9764. Soil science organization dedicated to protection, improvement, and careful use of soil, water, and related natural resources.

Solar Energy Research Institute, 6536 Cole Boulevard, Golden, CO 80401. Renewable energy.

Union of Concerned Scientists, Two Brattle Square, Cambridge, MA 02238. Scientists and citizens to advance responsible public policies for technology applications.

United Nations Environment Programme, Office of Public Information, New York Liaison Office, Room DC 2-0803, United Nations, New York, NY 10017; or, UNEP, P.O. Box 30552, Nairobi, Kenya.

U.S. Geological Survey, Department of the Interior, 12201 Sunrise Valley Drive, Reston, VA 22092. Many regional offices, bookstores, and mapping centers.

The Wilderness Society, 900 17th Street NW, Washington, DC 20006-2596. Devoted to preserving wilderness and wildlife.

World Meteorological Organization, P.O. Box 2300, 41 Ave. Giuseppe-Motta, CH-1211, Geneva 2, Switzerland. Weather prediction, world climate, weather modification, atmospheric sciences, and global change. Includes the Intergovernmental Panel on Climate Change (IPCC), the principal international body of scientists studying climate change.

World Resources Institute, 1709 New York Ave., NW, Washington, DC 20006. Policy research group, publishes the *World Resources Report* and *Policy Studies Series.*

WorldWatch Institute, 1776 Massachusetts Avenue NW, Washington, DC 20036. Information for policy makers and the public relating to global issues. Publishes *State of the World* annually.

World Wildlife Fund, 1250 24th Street NW, Washington, DC 20037. Protecting wildlife and wildlands; largest private organization for these issues.

World Wildlife Fund Canada, 60 St. Clair Avenue, East, Suite 201, Toronto, Ontario, Canada M4T 1N5.

# Appendix B

# Mapping, Quadrangles, and Topographic Maps

The westward expansion across the vast North American continent demanded a land survey so that accurate maps could be drawn. Maps were needed to subdivide the land and to guide travel, exploration, settlement, and transportation. The Public Lands Survey System, in 1785, began surveying and mapping government land in the United States. In 1836, the conduct of public-land surveys was placed under a Clerk of Surveys in the Land Office of the Department of Interior. The Land Office was finally replaced by the Bureau of Land Management in 1946. The actual preparation and recording of survey information fell to the U.S. Geological Survey (USGS), also a branch of the Department of the Interior.

In Canada, national mapping is conducted by the Energy, Mines, and Resources Department (EMR). The EMR prepares base maps, thematic maps, aeronautical charts, federal topographic maps, and the *National Atlas of Canada*, now in its fifth edition (1986).

## Quadrangle Maps

The USGS depicts survey information on quadrangle maps, so called because they are rectangular maps with four corner angles. The angles are junctures of parallels of latitude and meridians of longitude rather than political boundaries. These quadrangle maps utilize the *Albers equal-area* projection, from the conic class of map projections. The accuracy of conformality (shape) and scale of this base map is improved by the use of not one but two standard parallels. (Remember from Chapter 1, these standard lines are where the cone touches the globe's surface, producing greatest accuracy.) For the conterminous United States (the "lower 48"), these parallels are 29.5° N and 45.5° N latitude (noted on the Alber's projection shown in Figure 1-20c). The standard parallels are shifted for conic projections of Alaska (55° N and 65° N) and for Hawaii (8° N and 18° N).

Because a single map of the United States at 1:24,000 scale would be over 200 m wide (more than 600 ft), some system had to be devised for dividing the map into manageable size. Thus, a quadrangle system based on latitude and longitude coordinates is used. Note that these maps are not perfect rectangles, because meridians converge toward the poles. The width of quadrangles narrows noticeably as you move poleward.

Quadrangle maps are published in different series, covering different amounts of Earth's surface at different scales. You see in Figure 1 that each series is referred to by its angular dimensions, which range from 1° × 2° (1:250,000 scale) to 7.5′ x 7.5′ (1:24,000 scale). A map that is one-half a degree (30′) on each side is called a "30-minute quadrangle," and a map one-fourth of a degree (15′) on each side is called a "15-minute quadrangle" (this was the USGS standard size from 1910 to

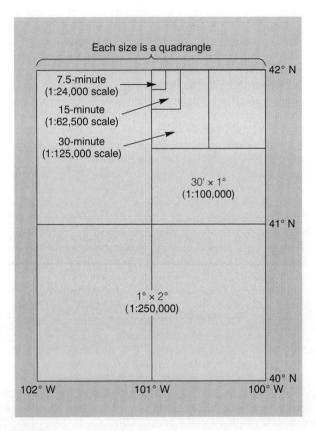

**FIGURE 1**
**Quadrangle system of maps used by the USGS.**

1950). A map that is one-eighth of a degree (7.5′) on each side is a 7.5-minute quadrangle, the most widely produced of all USGS topographic maps, and the standard since 1950. The progression toward more-detailed maps and a larger-scale map standard through the years reflects the continuing refinement of geographic data and new mapping technologies.

The USGS National Mapping Program recently completed coverage of the entire country (except Alaska) on 7.5-minute maps (1 in. to 2000 ft, a large scale). It takes 53,838 separate 7.5-minute quadrangles to cover the lower 48 states, Hawaii, and U.S. territories. Alaska is covered with a series of smaller-scale, more general 15-minute topographic maps.

In the United States, most quadrangle maps remain in English units of feet and miles. The eventual changeover to the metric system requires revision of the units used on all maps, with the 1:24,000 scale eventually changing to a scale of 1:25,000. However, after only a few metric quads were completed by the

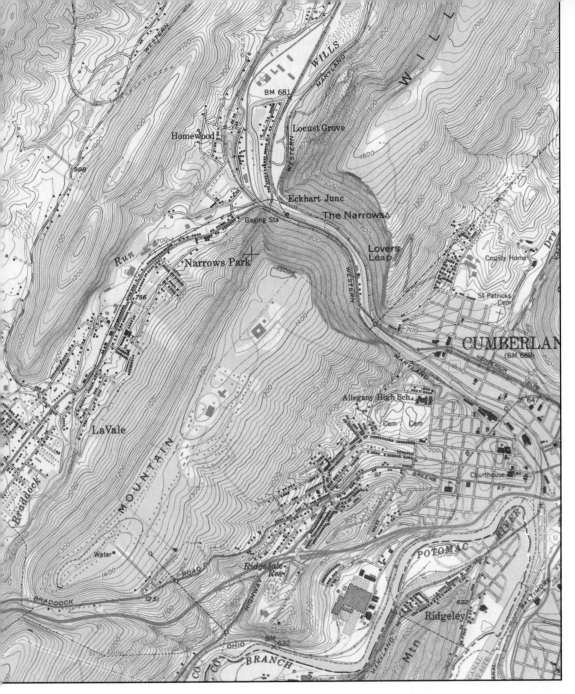

**FIGURE 2**
**An example of a topographic map from the Appalachians.**
Cumberland, MD, PA, WV 7.5-minute quadrangle topographic map prepared by the USGS. Note the water gap through Haystack Mountain.

USGS this program was halted in 1991 because of budgetary cuts. In Canada, the entire country is mapped at a scale of 1:250,000, using metric units (1.0 cm to 2.5 km). About half the country also is mapped at 1:50,000 (1.0 cm to 0.50 km).

## Topographic Maps

The most popular and widely used quadrangle maps are topographic maps prepared by the USGS. An example of such a map is a portion of the Cumberland, Maryland, quad shown in Figure 2. You will find topographic maps throughout *Geosystems* because they portray landscapes so effectively. As examples, see Figure 12-14, Appalachian Mountains; Figures 13-14 and 13-15, karst landscapes and sinkholes near Orleans, Indiana, and Winter Park, Florida; Figure 14-6, river drainage patterns; Figure 14-21, river meander scars; Figure 15-17, an alluvial fan in Montana; Figure 17-2, glaciers in Alaska; Figure 17-15, drumlins in New York.

A **planimetric map** shows the horizontal position (latitude/longitude) of boundaries, land-use aspects, bodies of water and economic and cultural features. A highway map is a common example of a planimetric map.

A **topographic map** adds a vertical component to show *topography* (configuration of the land surafce), including slope and *relief* (the vertical difference in local landscape elevation). These fine details are shown through the use of elevation contour lines (Figure 3). A **contour line** connects all points at the same elevation. Elevations are shown above or below a *vertical datum*, or reference level, which usually is mean sea level. The *contour interval* is the vertical distance in elevation between two adjacent contour lines (20 ft, or 6.1 m in Figure 3b).

The topographic map in Figure 3b shows a hypothetical landscape, demonstrating how contour lines and intervals depict slope and relief, which are the three-dimensional aspect of terrain. Slope is indicated by the pattern of lines and the spacing between them. The steeper a slope or cliff, the closer together the contour lines appear—in the figure, note the narrowly spaced contours that represent the cliffs to the left of the highway. A more gradual slope is portrayed by a wider spacing of these contour lines, as you can see from the widely spaced lines on the beach and to the right of the river valley.

In Figure 4 are the standard symbols commonly used on these topographic maps. These symbols and the colors used are standard on all USGS topographic maps: black for human constructions, blue for water features, brown for relief features and contours, pink for urbanized areas, and green for woodlands, orchards, brush, and the like.

The margins of a topographic map contain a wealth of information about its concept and content: quadrangle name, names of adjoining quads, quad series and type, position in the latitude-longitude and other coordinate systems, title, legend, magnetic declination (alignment of magnetic north) and compass information, datum plane, symbols used for roads and trails, the dates and history of the survey of that particular quad, and more.

Topographic maps may be purchased directly from the USGS or EMR. Many state geological survey offices, national and state park headquarters, outfitters, sports shops, and bookstores also sell topographic maps to assist people in planning their outdoor activities.

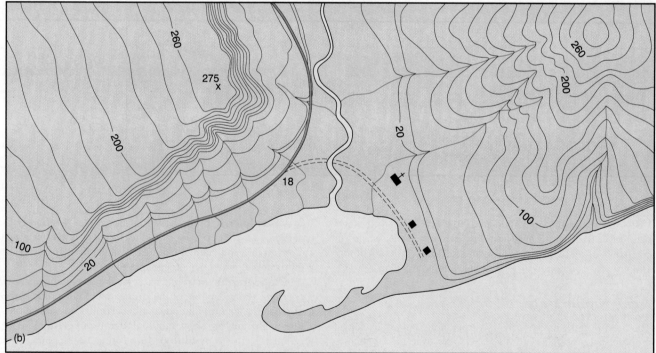

**FIGURE 3**
**Topographic map of a hypothetical landscape.**
(a) Perspective view of a hypothetical landscape. (b) Depiction of that landscape on a topographic map. The contour interval on the map is 20 feet (6.1 m). [After the U.S. Geological Survey.]

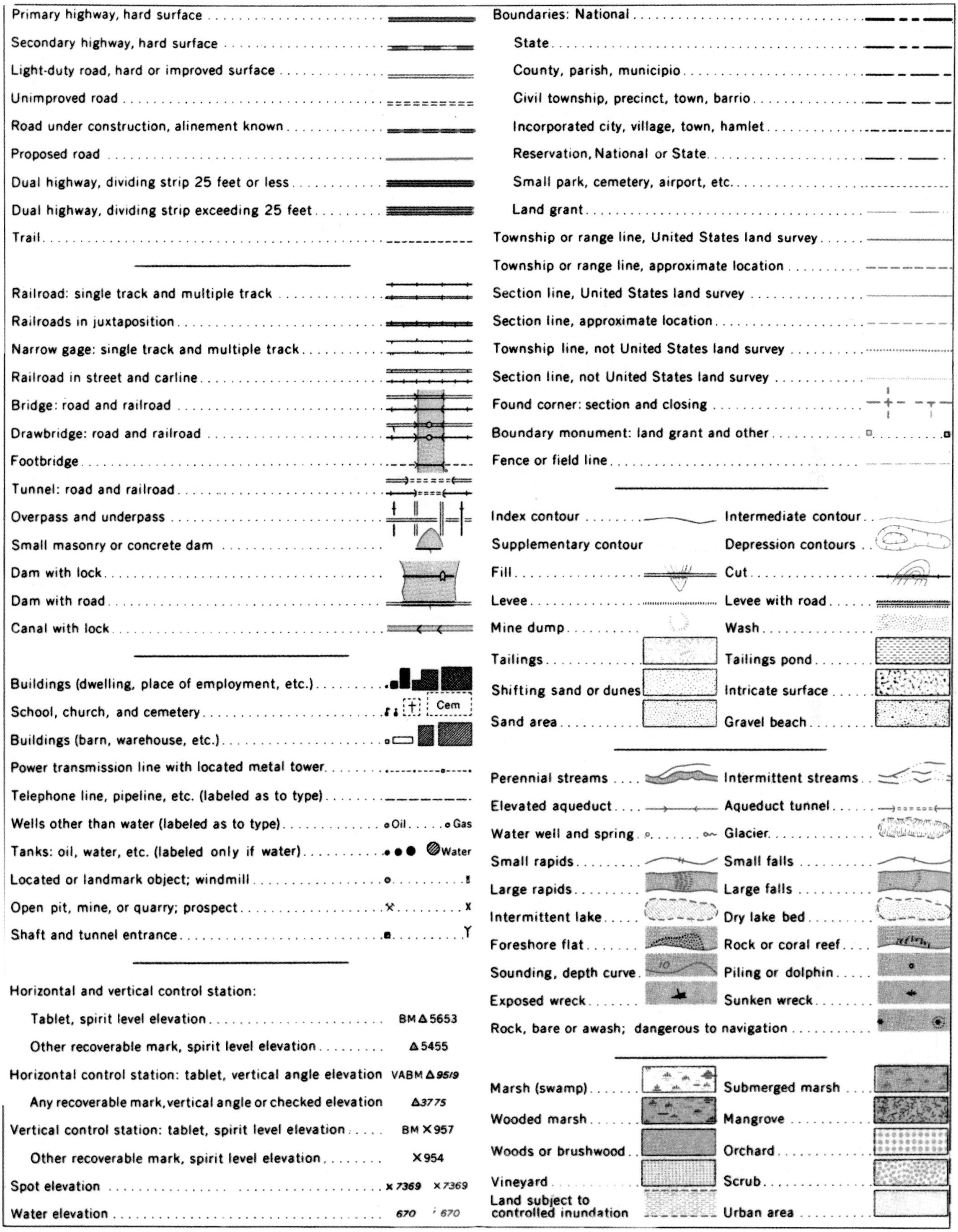

| | | | |
|---|---|---|---|
| Primary highway, hard surface | | Boundaries: National | |
| Secondary highway, hard surface | | State | |
| Light-duty road, hard or improved surface | | County, parish, municipio | |
| Unimproved road | | Civil township, precinct, town, barrio | |
| Road under construction, alinement known | | Incorporated city, village, town, hamlet | |
| Proposed road | | Reservation, National or State | |
| Dual highway, dividing strip 25 feet or less | | Small park, cemetery, airport, etc. | |
| Dual highway, dividing strip exceeding 25 feet | | Land grant | |
| Trail | | Township or range line, United States land survey | |
| | | Township or range line, approximate location | |
| Railroad: single track and multiple track | | Section line, United States land survey | |
| Railroads in juxtaposition | | Section line, approximate location | |
| Narrow gage: single track and multiple track | | Township line, not United States land survey | |
| Railroad in street and carline | | Section line, not United States land survey | |
| Bridge: road and railroad | | Found corner: section and closing | |
| Drawbridge: road and railroad | | Boundary monument: land grant and other | |
| Footbridge | | Fence or field line | |
| Tunnel: road and railroad | | | |
| Overpass and underpass | | Index contour | Intermediate contour |
| Small masonry or concrete dam | | Supplementary contour | Depression contours |
| Dam with lock | | Fill | Cut |
| Dam with road | | Levee | Levee with road |
| Canal with lock | | Mine dump | Wash |
| | | Tailings | Tailings pond |
| Buildings (dwelling, place of employment, etc.) | | Shifting sand or dunes | Intricate surface |
| School, church, and cemetery | Cem | Sand area | Gravel beach |
| Buildings (barn, warehouse, etc.) | | | |
| Power transmission line with located metal tower | | Perennial streams | Intermittent streams |
| Telephone line, pipeline, etc. (labeled as to type) | | Elevated aqueduct | Aqueduct tunnel |
| Wells other than water (labeled as to type) | oOil ....oGas | Water well and spring | Glacier |
| Tanks: oil, water, etc. (labeled only if water) | Water | Small rapids | Small falls |
| Located or landmark object; windmill | | Large rapids | Large falls |
| Open pit, mine, or quarry; prospect | X ....x | Intermittent lake | Dry lake bed |
| Shaft and tunnel entrance | Y | Foreshore flat | Rock or coral reef |
| | | Sounding, depth curve | Piling or dolphin |
| Horizontal and vertical control station: | | Exposed wreck | Sunken wreck |
| Tablet, spirit level elevation | BM △5653 | Rock, bare or awash; dangerous to navigation | |
| Other recoverable mark, spirit level elevation | △5455 | | |
| Horizontal control station: tablet, vertical angle elevation | VABM △95I9 | Marsh (swamp) | Submerged marsh |
| Any recoverable mark, vertical angle or checked elevation | △3775 | Wooded marsh | Mangrove |
| Vertical control station: tablet, spirit level elevation | BM ×957 | Woods or brushwood | Orchard |
| Other recoverable mark, spirit level elevation | ×954 | Vineyard | Scrub |
| Spot elevation | ×7369  ×7369 | Land subject to controlled inundation | Urban area |
| Water elevation | 670  ׳670 | | |

**FIGURE 4**
**Standardized topographic map symbols used on USGS maps.**
English units still prevail, although a few USGS maps are in metric. [From USGS, *Topographic Maps*, 1969.]

# Appendix C

# The Canadian System of Soil Classification (CSSC)

Canadian efforts at soil classification began in 1914 with the partial mapping of Ontario's soils by A. J. Galbraith. Efforts to develop a taxonomic system spread countrywide, anchored by universities in each province. Regional differences in soil classification emerged, further confused by a lack of specific soil details. By 1936, only 1.7% of Canadian soil had been surveyed (15 million hectares).

Canadian scientists needed a taxonomic system based on observable and measurable properties in soils specific to Canada. This meant a departure from Marbut's 1938 U.S. classification. Canada's first taxonomic system was introduced in 1955, splitting away from the soil classification effort in the United States and the Fourth Approximation stage. Classification work progressed through the Canada Soil Survey Committee after 1970 and was replaced by the Expert Committee on Soil Survey in 1978, all under Agriculture Canada.

The **Canadian System of Soil Classification (CSSC)** provides taxa for all soils presently recognized in Canada and is adapted to Canada's expanses of forest, tundra, prairie, frozen ground, and colder climates. As in the U.S. Soil Taxonomy system, the CSSC classifications are based on observable and measurable properties found in real soils rather than idealized soils that may result from the interactions of genetic processes. The system is flexible in that its framework can accept new findings and information in step with progressive developments in the soil sciences.

## Categories of Classification in the CSSC

Categorical levels are at the heart of a taxonomic system. These categories are based on soil profile properties organized at five levels, nested in a hierarchical pattern to permit generalization at several levels of detail. Each level is referred to as a category of classification. The levels in the CSSC are briefly described here, as adapted from *The Canadian Soil Classification System*, 2nd edition, Publication 1646. Ottawa: Supply and Services Canada, 1987, p. 16.

- *Order.* Each of nine soil orders has pedon properties that reflect the soil environment and effects of active soil-forming processes.
- *Great Group.* Subdivisions of each order reflect differences in the dominant processes or other major contributing processes. As an example, in Luvic Gleysols (great group name followed by order) the dominant process is gleying—reduction of iron and other minerals—resulting from poor drainage under either grass or forest cover with Aeg and Btg horizons (see Table 1).

- *Subgroup.* Subgroups are differentiated by the content and arrangement of horizons that indicate the relation of the soil to a great group or order or the subtle transition toward soils of another order.
- *Family.* This is a subdivision of a subgroup. Parent material characteristics such as texture and mineralogy, soil climatic factors, and soil reactions are important.
- *Series.* Subdivisions of the family are differentiated by the detailed features of the pedon—the essential soil-sampling unit. Pedon horizons fall within a narrow range of color, texture, structure, consistence, porosity, moisture, chemical reaction, thickness, and composition.

## Soil Horizons in the CSSC

Soil horizons are named and standardized as diagnostic in the classification process. Several mineral and organic horizons and layers are used in the CSSC. Three mineral horizons are recognized by capital letter designation, followed by lowercase suffixes for further description. Principal soil-mineral horizons and suffixes are presented in Table 1.

Four organic horizons are identified in the Canadian classification system. O is further defined through subhorizon designations. Note that for organic soils, such layers are identified as *tiers.* These organic horizons are detailed in Table 2.

## The Nine Soil Orders of the CSSC

The nine orders of the CSSC, and related great groups, are summarized in Table 3 with a general description of properties, related Great Groups, an estimated percentage of land area for the soil order, a fertility assessment, and any applicable Soil Taxonomy equivalent.

Figure 1 is a generalized map of the distribution of principal soil orders in relation to physiographic regions in Canada. This grouping allows you to easily compare soils across Canada. A summary of the nine soil orders appears beneath the legend.

Please consult the *National Atlas of Canada*, 5th edition, for a detailed map of Canadian soils. For further information and background note these sources available in Ottawa:

Bentley, C. F., ed. *Photographs and Conditions of Some Canadian Soils.* University of Alberta Extension Series Publication B791. Ottawa K1P 5H4: Canadian Society of Soil Science, 1979.

Clayton, J. S., and others. *Soils of Canada,* vol. 1 *Soil Report* and vol. 2 *Soil Inventory.* A Cooperative Project of the Canada Soil Survey Committee and the Soil Research Institute, Research Branch, Agriculture Canada. Ottawa K1A OS9: Supply and Services Canada, 1977. Boxed set with two large color wall

**TABLE 1**

| Three Mineral Horizons and Mineral Horizon Suffixes Used in the CSSC | |
| --- | --- |
| *Symbol* | *Mineral Horizon Description* |
| A | Forms at or near the surface, experiences *eluviation,* or leaching, of finer particles or minerals. Several subdivisions are identified, with the surface usually darker and richer in organic content than lower horizons *(Ab)*; or, a paler, lighter zone below which reflects removal of organic matter with clays and oxides of aluminum and iron leached (removed) to lower horizons *(Ae)*. |
| B | Experiences *illuviation*, a depositional process, as demonstrated by accumulations of clays *(Bt)*, sesquioxides of aluminum or iron, and possibly an enrichment of organic debris *(Bb)*, and the development of soil structure. Coloration is important in denoting whether hydrolysis, reduction, or oxidation processes are operational for the assignment of a descriptive suffix. |
| C | Exhibits little effect from pedogenic processes operating in the *A* and *B* horizons, except the process of gleysation associated with poor drainage and the reduction of iron, denoted *(Cg)*, and the accumulation of calcium and magnesium carbonates *(Cca)* and more soluble salts *(Cs)* and *(Csa)*. |

| *Symbol* | *Horizon Suffix Description* |
| --- | --- |
| b | A buried soil horizon. |
| c | Irreversible cementation of a pedogenic horizon, e.g., cemented by $CaCO_3$. |
| ca | Lime accumulation of at least 10 cm thickness that exceeds in concentration that of the unenriched parent material by at least 5%. |
| cc | Irreversible cemented concretions, typically in pellet form. |
| e | Used with *A* mineral horizons *(Ae)* to denote eluvistion of clay, Fe, Al, or organic matter. |
| f | Enriched principally with illuvial iron and aluminum combined with organic matter, reddish in upper portions and yellowish at depth, determined through specific criteria. Used with *B* horizons alone. |
| g | Gray to blue colors, or prominent mottling, or both, produced by intense chemical reduction. Various applications to *A*, *B*, and *C* horizons. |
| h | Enriched with organic matter: accumulation in place or biological mixing *(Ab)* or subsurface enrichment through illuviation *(Bb)*. |
| j | A modifier suffix for *e*, *f*, *g*, *n*, and *t* to denote limited change or failure to meet specified criteria denoted by that letter. |
| k | Presence of carbonates as indicated by visible effervescence with dilute hydrocholoric acid (HCl). |
| m | Used with *B* horizons slightly altered by hydrolysis, oxidation, or solution, or all three to denote a change in color or structure. |
| n | Accumulation of exchangeable calcium (Ca) in ratio to exchangeable sodium (Na) that is 10 or less, with the following characteristics: prismatic or columnar structure, dark coatings on ped surfaces, and hard consistence when dry. Used with *B* horizons alone. |
| p | *A* or *0* horizons disturbed by cultivation, logging, and habitation. May be used when plowing intrudes on previous *B* horizons. |
| s | Presence of salts, including gypsum, visible as crystals or veins, or surface crusts of salt crystals, and by lowered crop yields. Usually with *C* but may appear with any horizon and lowercase suffixes. |
| sa | A secondary enrichment of salts more soluble than Ca or Mg carbonates, exceeding unenriched parent material, in a horizon at least 10 cm thick. |
| t | Illuvial enrichment of the *B* horizon with silicate clay that must exceed in overlying *Ae* horizon by 3% to 20% depending on the clay content of the *Ae* horizon. |
| u | Markedly disrupted by physical or faunal processes other than cryoturbation. |
| x | Fragipan formation—a loamy subsurface horizon of high bulk density and very low organic content. When dry it has a hard consistence and seems to be cemented. |
| y | Affected by cryoturbation (frost action) with disrupted and broken horizons and incorporation of materials from other horizons. Application to *A*, *B*, and *C* horizons and in combination with other suffixes. |
| z | A frozen layer. |

maps titled "Soils of Canada" and "Soil Climates of Canada" including maps of soil temperature and soil moisture.

Expert Committee on Soil Science, Agriculture Canada Research Branch. *The Canadian System of Soil Classification*, 2nd ed. Publication 1646. Ottawa K1A OS9: Supply and Services Canada, 1987. Replacing Agriculture Canada Publications 1455 (1974) and 1646 (1978).

Expert Committee on Soil Survey, Agriculture Canada Research Branch. *The Canadian Soil Information System (CanSIS) Manual for Describing Soils in the Field.* Publication 1459. Ottawa K1A OS9: Supply and Services Canada, 1979.

Research Branch, Canada Department of Agriculture. *Glossary of Terms in Soil Science.* Publication 1459. Ottawa K1A OC7: Information Canada, 1976.

**TABLE 2**

| Four Organic Horizons Used in the CSSC | |
|---|---|
| *Symbol* | *Description* |
| **O** | Organic materials, mainly mosses, rushes, and woody materials |
| **L** | Mainly discernible leaves, twigs, and woody materials |
| **F** | Partially decomposed, somewhat recognizable **L** materials |
| **H** | Indiscernible organic materials |

**O** is further defined through subhorizon designations:
  **Of**  Readily identifiable fibric materials
  **Om**  Mesic materials of intermediate decomposition
  **Oh**  Humic material at an advanced stage of decomposition—low fiber, high bulk density

## Review Questions

1. Why did Canada adopt its own system of soil classification? Describe a brief history of events that led up to the modern CSSC system.

2. Which soil order is associated with the development of a bog? Explain its use as a low-grade fuel.

3. Describe the podzolization process occurring in northern coniferous forests. What are the surface horizons like? What management strategies might enhance productivity in these soils? Name the soil order for these areas.

4. Compare and contrast Interior Plains soils with those of the southeastern Canadian Shield.

5. What processes inhibit soil development in the extreme north? Explain.

**TABLE 3**

## Nine Orders of the Canadian System of Soil Classification

| *Order*<br>*Great Group* | *Characteristics** | *Fertility* |
|---|---|---|
| Chernozemic (Russian, *chernozem*)<br>Brown (more moist)<br>Dark Brown<br>Black<br>Dark Gray (less moist)<br>(38 subgroups) | Well to imperfectly drained soils of the steppe-grassland-forest transition. Southern Alberta, Saskatchewan, Manitoba, Okanagan Valley, B.C., Palouse Prairie, B.C. Accumulation of organic matter in surface horizons. Most frozen during some winter months with soil-moisture deficits in the summer. A diagnostic **Ah** is typical (although **Ahe, Ap** are present) at least 10 cm thick or 15 cm if disturbed by cultivation. Mean annual temperature >0°C and usually <5.5°C. (5.1%, 470,000 km²; Soil Taxon. = Mollisols.) | High; wheat-growing |
| Solonetzic (Russian, *solonetz*)<br>Solonetz<br>Solodized Solonetz<br>Solod<br>(27 subgroups) | Solonetz denotes saline or alkaine soils. Well to imperfectly drained mineral soils developed under grasses in semiarid to subhumid climates. Limited areas of central and north-central Alberta. Noted for a **B** horizon that is very hard when dry but swells to a sticky, low-permeability mass when wet. A saline **C** horizon reflects nature of parent materials. (0.8%, 73,700 km²; Soil Taxon. = Natric horizon of Mollisols and Alfisols.) | Variable (medium) about 50% cultivated, remainder in pasture |
| Luvisolic<br>Gray Brown Luvisol<br>Gray Luvisol<br>(18 subgroups) | Eluviation-illuviation processes produce a light-colored **Ae** horizon and a diagnostic **Bt** horizon. Soils of mixed deciduous-coniferous forests. Major occurrence is the St. Lawrence lowland. Luvisols do not have a solonetzic **B** horizon, evidence of Gleysolic order and gleying, or organics less than in the Organic order. Permafrost within 1 m of surface and 2 m if soils are cryoturbated. (10.3%, 950,000 km²; Soil Taxon. = Boralfs, Udalfs-suborders of Alfisols.) | High |
| Podzolic (Russian, *podzol*)<br>Humic<br>Ferro-Humic Podzol<br>Humo-Ferric Podzol<br>(25 subgroups) | Soils of coniferous forests and sometimes heath, leaching of overlying horizons occurs in moist, cool to cold climates. Iron, aluminum and organic matter from **L, F,** and **H** horizons are redeposited in podzolic **B** horizon. A diagnostic **Bh, Bhf,** or **Bf** is present depending on great group. Dominant in western British Columbia, Ontario, and Québec. (22.6%, 2,083,000 km²; Soil Taxon. = Spodosols, some Inceptisols.) | Low to medium depending on acidity |
| Brunisolic (French, "brown")<br>Melanic Brunisol<br>Eutric Brunisol<br>Sombric Brunisol<br>Dystric Brunisol<br>(18 subgroups) | Sufficiently developed to distinguish from Regosolic order. Soils under forest cover with brownish **Bm** horizons although various colors are possible. Also, can be with mixed forest, shrubs, and grass. Diagnostic **Bm Bfj,** thin **Bf,** or **Btj** horizons differentiate from soils of other orders. Well to imperfectly drained. Lack the podzolic **B** horizon of podzols although surrounded by them in St. Lawrence lowland. (8.8%, 811,000 km²; Soil Taxon. = Inceptisols, some Psamments [Aquents in Entisols.]) | Medium (variable) |
| Regosolic (Greek, *rhegos*)<br>Regosol<br>Humic Regosol<br>(8 subgroups) | Weakly developed limited soils, the result of any number of factors: young materials; fresh alluvial deposits; material instability; mass-wasted slopes; or dry, cold climatic conditions. Lack solonetzic, illuvial, or podzolic **B** horizons. Lack permafrost within 1 m of surface, or 2 m if cryoturbated. May have **L, F, H,** or **O** horizons, or an **Ah** horizon if less than 10 cm thick. Buried horizons possible. Dominant in Northwest Territories and northern Yukon, now designated as Cryosols under CSSC. (1.3%, 120,000 km²; Soil Taxon. = Entisols.) | Low (variable) |
| Gleysolic (Russian, *glei*)<br>Luvic Gleysol<br>Humic Gleysol<br>Gleysol<br>(13 subgroups) | Defined on the basis of color and mottling that results from chronic reducing conditions inherent in poorly drained mineral soils under wet conditions. High water table and long periods of water saturation. Rather than continuous they appear spotty within other soil orders and occasionally may dominate an area. A diagnostic **Bg** horizon is present. (1.9%, 175,000 km² ; Soil Taxon. = Various aquic suborders, a reducing moisture regime.) | High to medium |
| Organic<br>Fibrisol<br>Mesisol<br>Humisol<br>Folisol<br>(31 subgroups) | Peat, bog, and muck soils, largely composed of organic material. Most water-saturated for prolonged periods. Are widespread in association with poorly to very poorly drained depressions, although Folisols are found under upland forest environments. Exceed 17% organic carbon and 30% organic matter overall. (4.2%, 387,000 km²; Soil Taxon. = Histolsols.) | High to medium given drainage, available nutrients |
| Crysolic (Greek, *kyros*)<br>Turbic Cryosol<br>Static Cryosol<br>Organic Cryosol<br>(15 subgroups) | Dominate the northern third of Canada, with permafrost close to the surface and composed of mineral and organic soil deposits. Generally found north of the treeline, or in fine-textured soils in subarctic forest, or in some organic soils in boreal forests. **Ah** horizon lacking or thin. Cryoturbation (frost action) common, often denoted by patterned ground circles, polygons, and stripes. Subgroups based on degree of cryoturbation and the nature of mineral or organic soil material. (45%, 4,150,000 km²; Soil Taxon. = Cryoquepts, lnceptisols, and pergelic temperature regime in several suborders.) | Not applicable |

*Estimated percent and square kilometers of Canada's land area and Soil Taxonomy equivalent are given in parentheses.

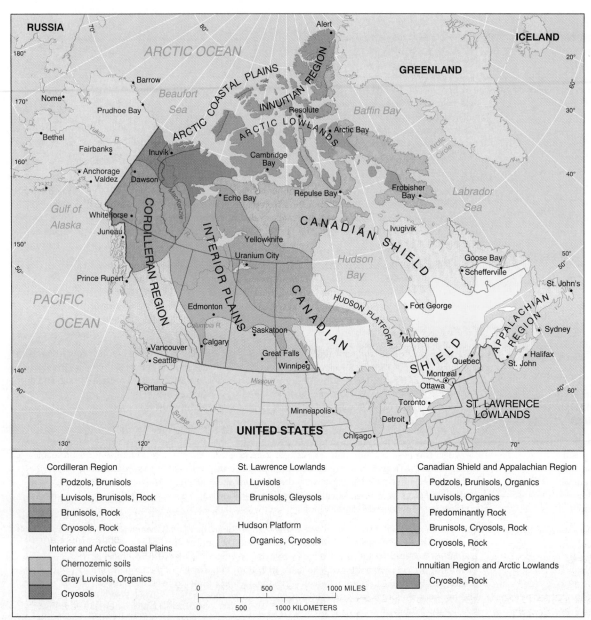

**FIGURE 1**
**Soil orders of Canada.**

Principal soil regions of the Canadian System of Soil Classification (CSSC) as related to major physiographic regions. [After maps prepared by the Land Resources Research Institute, Geological Survey of Canada, and the Canadian Soil Survey Committee.]

**Brunisolic** Formed under forest cover with brownish horizons although various colors are possible. Also, can be with mixed forest, shrubs, and grass; well to imperfectly drained.

**Chernozemic** A soil formed in well to imperfectly drained soils of the steppe-grassland-forest transition. Accumulation of organic matter in surface horizons. Most frozen during some winter months with soil-moisture deficits in the summer. Mean annual temperature >0°C and usually < 5.5°C.

**Cryosolic** A soil order with permafrost close to the surface and composed of mineral and organic soil deposits. Generally found north of the treeline, or in fine textured soils in subarctic forest, or in some organic soils in boreal forests; dominate the northern third of Canada. Cryoturbation (frost action) common, often denoted by patterned ground circles, polygons, and stripes.

**Gleysolic** Formed under chronic reducing conditions inherent in poorly drained mineral soils and wet conditions, high water table, and long periods of water saturation. Rather than continuous, they appear discontinuous within other soil orders, defined on the basis of color and mottling.

**Luvisolic** A soil formed under mixed deciduous-coniferous forests by eluviation-illuviation processes that produce a light-colored **Ae** horizon

and a diagnostic **Bt** horizon. Major area of occurrence is the St. Lawrence lowland.

**Organic** Formed as peat, bog, and muck soils, largely composed of organic material and water saturated for prolonged periods; usually in poorly to very poorly drained depressions.

**Podzolic** Occurs under coniferous forests and heath where leaching of overlying horizons occurs in moist cool to cold climates. Iron, aluminum, and organic matter from L, F, and H horizons are redeposited in a podzolic B horizon. Dominant in western British Columbia, Ontario, and Québec.

**Regosolic** Produced by any number of factors: young materials; fresh alluvial deposits; material instability; mass-wasted slopes; dry, cold climatic conditions. Weakly developed limited soils that lack solonetzic, illuvial, or podzolic **B** horizons; buried horizons possible. Dominant in Northwest Territories and northern Yukon, although now designated as Cryosols under CSSC.

**Solonetzic** Formed in well to imperfectly drained mineral soils developed under grasses in semiarid to subhumid climates and denoting saline or alkaline soils. Limited areas of central and north-central Alberta.

# Glossary

The chapter in which each term appears **boldfaced** is designated in parentheses, followed by a specific definition relevant to the key term's usage in the chapter.

**Abiotic (1)** Nonliving; Earth's nonliving systems of energy and materials.

**Ablation (17)** Loss of glacial ice through melting, sublimation, wind removal by deflation, or the calving of blocks of ice (see deflation).

**Abrasion (14, 15, 17)** Mechanical wearing and erosion of bedrock accomplished by the rolling and grinding of particles and rocks carried in a stream, removed by wind in a "sandblasting" action, or imbedded in glacial ice.

**Absorption (4)** Assimilation and conversion of radiation from one form to another in a medium. In the process, the temperature of the absorbing surface is raised, thereby affecting the rate and wavelength of radiation from that surface.

**Active layer (17)** A zone of seasonally frozen ground that exists between the subsurface permafrost layer and the ground surface. The active layer is subject to consistent daily and seasonal freezethaw cycles (see permafrost).

**Actual evapotranspiration (9)** ACTET, the actual amount of evaporation and transpiration that occurs; derived in the water balance by subtracting the deficit (DEFIC) from potential evapotranspiration in the surface.

**Adiabatic (8)** Pertaining to the cooling of an ascending parcel of air through expansion or the warming of a descending parcel of air through compression, without any exchange of heat between the parcel and the surrounding environment.

**Advection (4)** Horizontal movement of air or water from one place to another (compare with convection).

**Advection fog (8)** Active condensation formed when warm, moist air moves laterally over cooler water or land surfaces, causing the lower layers of the air to be chilled to the dewpoint temperature.

**Aggradation (14)** The general building of land surface because of deposition of material; opposite of degradation. When the sediment load of a stream exceeds the stream's capacity to carry it, the stream channel becomes filled through this process.

**Air mass (8)** A distinctive, homogeneous body of air that has taken on the moisture and temperature characteristics of its source region.

**Air pressure (3)** Pressure produced by the motion, size, and number of gas molecules in the air, and exerted on surfaces in contact with the air. Normal sea level pressure, as measured by the height of a column of mercury (Hg), is expressed as 1013.2 millibars, or 760 mm of Hg, or 29.92 inches of Hg. Air pressure can be measured with mercury or aneroid barometers (see listing of both).

**Albedo (4)** The reflective quality of a surface, expressed as the percentage of reflected insolation to incoming insolation; a function of surface color, angle of incidence, and surface texture.

**Aleutian low (6)** See subpolar low-pressure cell.

**Alfisols (18)** A soil order in the Soil Taxonomy. Moderately weathered forest soils that are moist versions of Mollisols, with productivity dependent on specific patterns of moisture and temperature; rich in organics. Most wide-ranging of the soil orders.

**Alluvial fan (15)** Fan-shaped fluvial landform at the mouth of a canyon; generally occurs in arid landscapes where streams are intermittent.

**Alluvial terraces (14)** Level areas that appear as topographic steps above a stream, created by the stream as it scours with renewed downcutting into its floodplain; composed of unconsolidated alluvium (see alluvium).

**Alluvium (14)** General descriptive term for clay, silt, and sand, transported by running water and deposited in sorted or semisorted sediment on a floodplain, delta, or stream bed.

**Alpine glacier (17)** A glacier confined in a mountain valley or walled basin, consisting of three subtypes: valley glacier (within a valley), piedmont glacier (coalesced at the base of a mountain, spreading freely over nearby lowlands), and outlet glacier (flowing outward from a continental glacier).

**Altitude (2)** The angular distance between the horizon (a horizontal plane) and the Sun (or any point).

**Altocumulus (8)** Middle level, puffy clouds that occur in several forms: patchy rows, wave patterns, a "mackerel sky," or lens-shaped clouds.

**Andisols (18)** A soil order in the Soil Taxonomy. Derived from volcanic parent materials in areas of volcanic activity. A new order, created in 1990, of soils previously considered under Inceptisols and Entisols.

**Anemometer (6)** A device that measures wind velocity.

**Aneroid barometer (3)** A device that measures air pressure using a partially evacuated, sealed cell (see air pressure).

**Angle of repose (13)** The steepness of a slope that results when loose particles come to rest; an angle of balance between driving and resisting forces, ranging between 33° and 37° from a horizontal plane.

**Antarctic high (6)** A consistent high-pressure region centered over Antarctica; source region for an intense polar air mass that is dry and associated with the lowest temperatures on Earth.

**Anthropogenic atmosphere (3)** Earth's next atmosphere, so named because humans appear to be the principal causative agent.

**Anticline (12)** Upfolded rock strata, in which layers slope downward from the axis of the fold, or central ridge (compare syncline).

**Anticyclone (6)** A dynamically or thermally caused area of high atmospheric pressure with descending and diverging air flows that rotate clockwise in the Northern Hemisphere and counterclockwise in the Southern Hemisphere (compare cyclone).

**Aphelion (2)** The point of Earth's greatest distance from the Sun in its elliptical orbit; reached on July 4 at a distance of 152,083,000 km (94.5 million mi); variable over a 100,000-year cycle (compare perihelion).

**Apparent temperature (5)** The temperature subjectively perceived by each individual, also known as sensible temperature.

**Aquiclude (9)** A body of rock that does not conduct groundwater in usable amounts; an impermeable rock layer (also called an aquitard). (Compare aquifer.)

**Aquifer (9)** A body of rock that conducts groundwater in usable amounts; a permeable layer of rock (compare aquiclude).

**Aquifer recharge area (9)** The surface area where water enters an aquifer to recharge the water-bearing strata in a groundwater system.

**Arctic and alpine tundra (20)** A biome in the northernmost portions of North America and the Soviet Union, featuring low ground-level herbaceous plants as well as some woody plants.

**Arête (17)** A sharp ridge that divides two cirque basins. Derived from "knife edge" in French, these form sawtooth and serrated ridges in glaciated mountains.

**Aridisols (18)** A soil order in the Soil Taxonomy; largest soil order. Typical of dry climates; low in organic matter and dominated by calcification and salinization.

**Artesian water (9)** Pressurized groundwater that rises in a well or a rock structure above the local water table; may flow out onto the ground without pumping (see potentiometric surface).

**Asthenosphere (11)** Region of the upper mantle just below the lithosphere; the least rigid portion of Earth's interior and known as the plastic layer, flowing very slowly under extreme heat and pressure.

**Atmosphere (1)** The thin veil of gases surrounding Earth, which forms a protective boundary between outer space and the biosphere; generally considered to extend about 480 km (300 mi) elevation from Earth's surface.

**Aurora (2)** A spectacular glowing light display in the ionosphere, stimulated by the interaction of the solar wind with oxygen and nitrogen gases at high latitudes, called aurora borealis in the Northern Hemisphere and aurora australis in the Southern Hemisphere.

**Autumnal (September) equinox (2)** The time around September 22-23 when the Sun's *declination* crosses the equatorial parallel (0° latitude) and all places on Earth experience days and nights of equal length. The Sun rises at the South Pole and sets at the North Pole (compare vernal equinox).

**Available water (9)** The portion of capillary water that is accessible to plant roots; usable water held in soil moisture storage (see capillary water).

**Axial parallelism (2)** Earth's axis remains aligned the same throughout the year (it "remains parallel to itself"); thus, the axis extended from the North Pole points into space always near Polaris, the North Star.

**Axis (2)** An imaginary line, extending through Earth from the geographic North Pole to the geographic South Pole, around which Earth rotates.

**Azores high (6)** A subtropical high-pressure cell that forms in the Northern Hemisphere in the eastern Atlantic (see Bermuda high); associated with warm, clear water and large quantities of sargassum, or gulf weed, characteristic of the Sargasso Sea.

**Backswamp (14)** A low-lying, swampy area of a floodplain; adjacent to a river, with the river's natural levee on one side and higher topography on the other (see floodplain, yazoo tributary).

**Bajada (15)** A continuous apron of coalesced alluvial fans, formed along the base of mountains in arid climates; presents a gently rolling surface from fan to fan (see alluvial fan).

**Barrier beach (16)** Narrow, long depositional feature, generally composed of sand, that forms offshore roughly parallel to the coast; may appear as barrier islands and long chains of barrier beaches (see barrier island).

**Barrier island (16)** Generally, a broadened barrier beach (see barrier beach).

**Barrier spit (16)** A depositional landform that develops when transported sand or gravel in a barrier beach or island is deposited in long ridges that are attached at one end to the mainland and partially cross the mouth of a bay.

**Basalt (11)** A common extrusive igneous rock, fine-grained, comprising the bulk of the ocean floor crust, lava flows, and volcanic forms; gabbro is its intrusive form.

**Base level (14)** A hypothetical level below which a stream cannot erode its valley, and thus the lowest operative level for denudation processes; in an absolute sense it is represented by sea level, extending under the landscape.

**Basin and Range Province (15)** A region of dry climates, few permanent streams, and interior drainage patterns in the western United States composed of a sequence of horsts and grabens.

**Batholith (11)** The largest plutonic form exposed at the surface; an irregular intrusive mass (>100 km$^2$; >40 mi$^2$); it invades crustal rocks, cooling slowly so that large crystals develop (see pluton).

**Bay barrier (16)** An extensive barrier spit of sand or gravel that encloses a bay, cutting it off completely from the ocean and forming a lagoon; produced by littoral drift and wave action; sometimes referred to as a baymouth bar (see barrier spit, lagoon).

**Beach (16)** The portion of the coastline where an accumulation of sediment is in motion.

**Beach drift (16)** Material, sand, gravel, and shells that are moved by the longshore current in the effective direction of the waves.

**Beaufort wind scale (6)** A descriptive scale for the visual estimation of wind speeds; originally conceived in 1806 by Admiral Beaufort of the British Navy.

**Bed load (14)** Coarse materials that are dragged along the bed of a stream by traction or by the rolling and bouncing motion of saltation; involves particles too large to remain in suspension (see traction, saltation).

**Bedrock (13)** The rock of Earth's crust that is below the soil and is basically unweathered; such solid crust sometimes is exposed as an outcrop.

**Bergeron ice-crystal process (8)** A raindrop-forming mechanism, especially in middle- and high-latitude cold clouds. Supercooled-water droplets evaporate near ice crystals and then sublimate onto the crystals, enlarging them in size (compare collision-coalescence process).

**Bermuda high (6)** A subtropical high-pressure cell that forms in the western North Atlantic (see Azores high).

**Biodiversity (19)** A principle of ecology: the more diverse the species population in an ecosystem (both in number of species and quantity of members in each species), the more risk is spread over the entire community, which results in greater overall stability, greater productivity, and increased use of soil nutrients, as compared to a monoculture of no diversity.

**Biogeochemical cycle (19)** One of several circuits of flowing elements and materials (carbon, oxygen, nitrogen, phosphorus, water) that combine Earth's biotic (living) and abiotic (nonliving) systems; the cycling of materials is continuous and renewed through the biosphere and the life processes.

**Biogeographical realm (20)** One of eight regions of the biosphere, each representative of evolutionary core areas of related flora (plants) and fauna (animals); a broad geographical classification scheme.

**Biogeography (19)** The study of the distribution of plants and animals and related *ecosystems*; the geographical relationships with their environments over time.

**Biomass (19)** The total mass of living organisms on Earth or per unit area of a landscape; also, the weight of the living organisms in an ecosystem.

**Biome (20)** A large terrestrial ecosystem characterized by specific plant communities and formations; usually named after the predominant vegetation in the region (see terrestrial ecosystem).

**Biosphere (1)** That area where the atmosphere, lithosphere, and hydrosphere function together to form the context within which life exists; an intricate web that connects all organisms with their physical environment.

**Biotic (1)** Living; Earth's living system of organisms.

**Blowout depression (15)** Eolian (wind) erosion in which deflation forms a basin in areas of loose sediment. Diameter may range up to hundreds of meters (see deflation).

**Bolson (15)** The slope and basin area between the crests of two adjacent ridges in a dry region.

**Boreal forest (20)** See needleleaf forest.

**Brackish (16)** Descriptive of seawater with a salinity of less than 35‰; for example, the Baltic Sea (contrast brine).

**Braided stream (14)** A stream that becomes a maze of interconnected channels laced with excess sediment. Braiding often occurs with a reduction of discharge that reduces a stream's transporting ability, or with an increase in sediment load.

**Breaker (16)** The point where a wave's height exceeds its vertical stability and the wave breaks as it approaches the shore.

**Brine (16)** Seawater with a salinity of more than 35‰; for example, the Persian Gulf (contrast brackish).

**Calcification (18)** The illuviated accumulation of calcium carbonate or magnesium carbonate in the B and C soil horizons.

**Caldera (12)** An interior sunken portion of a composite volcano's crater; usually steep-sided and circular, sometimes containing a lake; also can be found in conjunction with shield volcanoes.

**Caliche (18)** A cemented or hardened subsurface calcic soil horizon; diagnostic; occurs in the southwestern United States, usually in arid and semiarid climates (see calcification).

**Canadian System of Soil Classification (Appendix C)** A taxa for all soils presently recognized in Canada and adapted to Canada's particular forest, tundra, prairie, frozen ground, and colder climates. The CSSC is organized at five levels of generalization (order, great group, subgroup, family, and series) with each level referred to as a category of classification. The system is described in *The Canadian Soil Classification System*, 2nd edition, published by Agriculture Canada, 1987 (compare Soil Taxonomy).

**Capillary water (9)** Soil moisture, most of which is accessible to plant roots; held in the soil by the water's surface tension and cohesive forces between water and soil (see also available water, field capacity, hygroscopic water, and wilting point).

**Carbonation (13)** A chemical weathering process in which weak carbonic acid (water and carbon dioxide) reacts with many minerals, that contains calcium, magnesium, potassium, and sodium (especially limestone), transforming them into carbonates.

**Carbon dioxide (3)** An ordorless, colorless, tasteless combination of carbon and oxygen, natural by-product of life processes and combustion; the principal radiatively active gas in the greenhouse effect; $CO_2$.

**Carbon monoxide (3)** An odorless, colorless, tasteless combination of carbon and oxygen produced by the incomplete combustion of fossil fuels or other carbon-containing substances; CO.

**Carnivore (19)** A secondary consumer that principally eats meat for sustenance. The top carnivore in a food chain is considered a tertiary consumer (compare herbivore).

**Cartography (1)** The making of maps and charts; a specialized science and art that blends aspects of geography, engineering, mathematics, graphics, computer science, and artistic specialties.

**Catastrophism (11)** A philosophy that attempts to fit the vastness of Earth's age and the complexity of the rock record into a very shortened time span through a belief in short-lived and catastrophic worldwide events.

**Cation-exchange capacity (CEC) (18)** The ability of soil colloids to exchange cations between their surfaces and the soil solution; a measured potential that indicates soil fertility (see soil colloids, soil fertility).

**Chaparral (20)** Dominant shrub formations of Mediterranean dry summer climates; characterized by sclerophyllous scrub and short, stunted, tough forests; derived from the Spanish chapparo; specific to California.

**Chemical weathering (13)** Decomposition and decay of the constituent minerals in rock through chemical alteration of those minerals. Water is essential, with rates keyed to temperature and precipitation values. Processes include hydrolysis, oxidation, carbonation, and solution.

**Chinook wind (8)** North American term for a warm, dry, downslope air flow; characteristic of the rain shadow region on the leeward side of mountains; known as föhn, or foehn, winds in Europe.

**Chlorofluorocarbon compound (CFC) (3)** Large manufactured molecule (polymer) made of chlorine, fluorine, and carbon; inert and possessing remarkable heat properties; also known as one of the halogens. After slow transport to the stratospheric ozone layer, CFCs react with ultraviolet radiation, freeing chlorine atoms that act as a catalyst to produce reactions that destroy ozone.

**Chlorophyll (19)** A light-sensitive pigment that resides within the chloroplast bodies of plants in leaf cells; the basis of photosynthesis.

**Cinder cone (12)** A volcanic landform of pyroclastics and scoria, usually small and cone-shaped and generally not more than 450 m (1500 ft) in height; with a truncated top.

**Circle of illumination (2)** The division between light and dark on Earth; a day–night great circle.

**Circum-Pacific belt (12)** A tectonically and volcanically active region encircling the Pacific Ocean; also known as the "ring of fire."

**Cirque (17)** A scooped-out, amphitheater-shaped basin at the head of an alpine glacier valley; an erosional landform.

**Cirrus (8)** Wispy, filamentous ice-crystal clouds that occur above 6000 m (20,000 ft); appear in a variety of forms, from feathery hairlike fibers to veils of fused sheets.

**Classification (10)** The process of ordering or grouping data or phenomena in related classes; results in a regular distribution of information; a taxonomy.

**Climate (10)** The consistent, long-term behavior of weather over time, including its variability; in contrast to weather which is the condition of the atmosphere at any given place and time.

**Climatic regions (10)** An area of homogenous climate, that features characteristic regional weather and air mass patterns.

**Climatology (10)** The scientific study of climate and climatic patterns and the consistent behavior of weather, including its variability and extremes, over time in one place or region; including the effects of climate change on human society and culture.

**Climograph (10)** A graph that plots daily, monthly, or annual temperature and precipitation values for a selected station; may also include additional weather information.

**Closed system (1)** A system that is shut off from the surrounding environment so that it is entirely self-contained in terms of energy and materials; Earth is a closed material system (compare open system).

**Cloud-albedo forcing (4)** An increase in albedo (the reflectivity of a surface) caused by clouds due to their reflection of incoming insolation.

**Cloud-condensation nuclei (8)** Microscopic particles necessary as matter on which water vapor condenses to form moisture droplets; can be sea salts, dust, soot, or ash.

**Cloud-greenhouse forcing (4)** An increase in greenhouse warming caused by clouds because they can act like insulation, trapping longwave (infrared) radiation.

**Coal (11)** A sedimentary, biochemical rock of organic origin; very rich in carbon; a major air polluter when burned.

**Col (17)** Formed by two headward eroding cirques that reduce an *arête* (ridge crest) to form a high pass or saddlelike narrow depression.

**Cold desert and semidesert (20)** A type of desert biome found at higher latitudes than warm deserts. Interior location and rain shadow locations produce these cold deserts in North America.

**Cold front (8)** The leading edge of an advancing cold air mass; identified on a weather map as a line marked with triangular spikes pointing in the direction of frontal movement (compare warm front).

**Collision-coalescence process (8)** Principal process of raindrop formation predominant in clouds that form at temperatures above freezing. Larger moisture droplets sweep through the cloud, combining with smaller droplets, gradually coalescing and growing in size (compare Bergeron ice-crystal process).

**Community (19, 20)** A convenient biotic subdivision within an ecosystem; formed by interacting populations of animals and plants in an area.

**Composite volcano (12)** A volcano formed by a sequence of explosive volcanic eruptions; steep-sided, conical in shape; sometimes referred to as a stratovolcano, although composite is the preferred term (compare shield volcano).

**Conduction (4)** The slow molecule-to-molecule transfer of heat through a medium, from warmer to cooler portions.

**Cone of depression (9)** The depressed shape of the water table around a well after active pumping. The water table adjacent to the well is drawn down by the water removal.

**Confined aquifer (9)** An aquifer that is bounded above and below by impermeable layers of rock or sediment (see artesian water, unconfined aquifer).

**Constant isobaric surface (6)** An elevated surface in the atmosphere on which all points have the same pressure, usually 500 mb. Along this constant-pressure surface, isobars mark the paths of upper air winds.

**Consumer (19)** Organism in an ecosystem that depends on producers (organisms that use carbon dioxide as their sole source of carbon) for its source of nutrients; also called a heterotroph (compare producer).

**Consumptive use (9)** A use that removes water from a water budget at one point and makes it unavailable further downstream (compare withdrawal).

**Continental divide (14)** A ridge or elevated area that separates drainage on a continental scale; specifically, that ridge in North America that separates drainage to the Pacific on the west from drainage to the Atlantic and Gulf on the east and to Hudson Bay and the Arctic Ocean on the north.

**Continental drift (11)** A proposal by Alfred Wegener in 1912 stating that Earth's landmasses have migrated over the past 225 million years from a supercontinent he called Pangaea to the present configuration; a widely accepted concept today (see plate tectonics).

**Continental glacier (17)** A continuous mass of unconfined ice, covering at least 50,000 km² (19,500 mi²); most extensive at present as ice sheets covering Greenland and Antarctica (compare alpine glacier).

**Continentality (5)** A quality of regions that lack the temperature-moderating effects of the sea and that exhibit a greater range of minimum and maximum temperatures, both daily and annually (see marine, land-water heating differences).

**Continental platform (12)** The broadest category of landform, including those masses of crust that reside above or near sea level and the adjoining undersea continental shelves along the coastline.

**Continental shield (12)** Generally old, low-elevation heartland regions of continental crust; various cratons (granitic cores) and ancient mountains exposed at the surface.

**Contour lines (Appendix B)** Isolines on a topographic map that connect all points at the same elevation relative to a reference elevation called the vertical datum.

**Convection (4)** Transfer of heat from one place to another through the actual physical movement of air; involves a strong vertical motion.

**Coordinated Universal Time (UTC) (1)** The official reference time in all countries, formerly known as Greenwich Mean Time; now measured by six primary standard atomic clocks, the time calculations of which are collected in Paris, by the Bureau International de l'Heure.

**Coral (16)** A simple, cylindrical marine animal with a saclike body that secretes calcium carbonate to form a hard external skeleton; lives symbiotically with nutrient-producing algae.

**Core (11)** The deepest inner portion of Earth, representing one-third of its entire mass; differentiated into two zones—a solid iron inner core surrounded by a dense, molten, fluid metallic-iron outer core.

**Coriolis force (6)** The apparent deflection of moving objects from traveling in a straight path, in proportion to the speed of Earth's rotation at different latitudes. Deflection is to the right in the Northern Hemisphere and to the left in the Southern Hemisphere; maximum at the poles and zero along the equator.

**Crater (12)** A circular surface depression formed by volcanism; built by accumulation, collapse, or explosion; usually located at a volcanic vent or pipe; can be at the summit or on the flank of a volcano.

**Crevasse (17)** A vertical crack that develops in a glacier as a result of friction between valley walls, or tension forces of extension on convex slopes, or compression forces on concave slopes.

**Crust (11)** Earth's outer shell of crystalline surface rock, ranging from 5 to 60 km (3 to 38 mi) in thickness from oceanic crust to mountain ranges. Average density of continental crust is 2.7 g per cm$^3$, whereas oceanic crust is 3.0 g per cm$^3$.

**Cumulonimbus (8)** A towering, precipitation-producing cumulus cloud that is vertically developed across altitudes associated with other clouds; frequently associated with lightning and thunder and thus sometimes called a thunderhead.

**Cumulus (8)** Bright and puffy cumuliform clouds up to 2000 m in altitude (6500 ft).

**Cyclogenesis (8)** An atmospheric process that describes the birth of a midlatitude wave cyclone; usually along the polar front. Also refers to strengthening and development of a midlatitude cyclone along the eastern slope of the Rockies, other north-south mountain barriers, and along the North American and Asian east coasts (see midlatitude cyclone, polar front).

**Cyclone (6)** A dynamically or thermally caused area of low atmospheric pressure with converging and ascending air flows. Rotates counterclockwise in the Northern Hemisphere and clockwise in the Southern Hemisphere (compare anticyclone; see midlatitude cyclone, tropical cyclone).

**Daylength (2)** Duration of exposure to insolation, varying during the year depending on latitude; an important aspect of seasonality.

**Daylight saving time (1)** Time is set ahead one hour in the spring and set back one hour in the fall in the Northern Hemisphere. Time is set ahead on the first Sunday in April and set back on the last Sunday in October—except in Hawaii, Arizona, portions of Indiana, and Saskatchewan, which exempt themselves.

**Debris avalanche (13)** A mass of falling and tumbling rock, debris, and soil; can be dangerous because of the tremendous velocities achieved by the onrushing materials.

**Declination (2)** The latitude that receives direct overhead (perpendicular) insolation on a particular day; migrates annually through 47° of latitude between the Tropics of Cancer (23.5° N) and Capricorn (23.5° S).

**Decomposer (19)** Microorganisms that digest and recycle organic debris and waste in the environment, including bacteria, fungi, insects, and worms.

**Deficit (9)** DEFIC in a water balance, is the amount of unmet (unsatisfied) potential evapotranspiration (POTET, or PET).

**Deflation (15)** A process of wind erosion that removes and lifts individual particles, literally blowing away unconsolidated, dry, or noncohesive sediments.

**Delta (14)** A depositional plain formed where a river enters a lake or an ocean; named after the triangular shape of the Greek letter delta Δ.

**Denudation (13)** A general term that refers to all processes that cause degradation of the landscape: weathering, mass movement, erosion, and transport.

**Deposition (14)** The process whereby weathered, wasted, and transported sediments are laid down by air, water, and ice.

**Desert biome (20)** Arid landscapes of uniquely adapted dry-climate plants and animals.

**Desert pavement (15)** On arid landscapes, a surface formed when wind deflation and sheetflow remove smaller particles, leaving residual pebbles and gravels to concentrate at the surface; resembles a cobblestone street (see deflation, sheetflow).

**Desertification (20)** The expansion of deserts worldwide, related principally to poor agricultural practices (overgrazing and inappropriate agricultural practices), improper soil-moisture management, erosion and salinization, deforestation, and the ongoing climatic change; an unwanted semipermanent invasion into neighboring biomes.

**Dew-point temperature (7)** The temperature at which a given mass of air becomes saturated, holding all the water it can hold. Any further cooling or addition of water vapor results in active condensation.

**Differential weathering (13)** The effect of different resistances in rock, coupled with variations in the intensity of physical and chemical weathering.

**Diffuse radiation (4)** The downward component of scattered incoming insolation from clouds and the atmosphere.

**Discharge (9)** The measured volume of flow in a river that passes by a given cross section of a stream in a given unit of time; expressed in cubic meters per second or cubic feet per second.

**Dissolved load (14)** Materials carried in chemical solution in a stream derived from minerals such as limestone, dolomite, or from soluble salts.

**Doldrums (6)** A region of equatorial calm associated with the intertropical convergence zone; so named for the windless becalming encountered in the era of sailing ships.

**Downwelling current (6)** An area of the sea where a convergence or accumulation of water thrusts excess water downward; occurs, for example, at the western end of the equatorial current or along the margins of Antarctica (compare upwelling currents).

**Drainage basin (14)** The basic spatial geomorphic unit of a river system; distinguished from a neighboring basin by ridges and highlands that form divides, marking the limits of the catchment area of the drainage basin, or its watershed.

**Drainage density (14)** A measure of the overall operational efficiency of a drainage basin; determined by the ratio of combined channel lengths to the unit area.

**Drainage pattern (14)** A distinctive geometric arrangement of streams in a region; determined by slope, differing rock resistance to weathering and erosion, climatic and hydrologic variability, and structural controls of the landscape.

**Drawdown (9)** See cone of depression.

**Drumlin (17)** A depositional landform related to glaciation that is composed of till and is streamlined in the direction of continental ice movement; blunt end upstream and tapered end downstream with a rounded summit.

**Dry adiabatic rate (DAR) (7)** The rate at which an unsaturated parcel of air cools (if ascending) or heats (if descending);

a rate of 10 C° per 1000 m (5.5 F° per 1000 ft) (see adiabatic; compare moist-adiabatic rate).

**Dune (15)** A depositional feature of sand grains deposited in transient mounds, ridges, and hills; extensive areas of sand dunes are called sand seas.

**Dust dome (4)** A dome of airborne pollution associated with every major city; may be blown by winds into elongated plumes downwind from the city.

**Dynamic equilibrium model (13)** The balancing act between tectonic uplift and erosion, between the resistance of crust materials and the work of denudation processes. Landscapes evidence ongoing adaptation to rock structure, climate, local relief, and elevation.

**Earthquake (12)** A sharp release of energy that sends waves traveling through Earth's crust at the moment of rupture along a fault or in association with volcanic activity. Earthquake magnitude is estimated by the moment magnitude scale (formerly the Richter scale); intensity is described by the Mercalli scale.

**Earth systems science (1)** An emerging science of Earth as a complete, systematic entity. An interacting set of physical, chemical, and biological systems that produce the processes of a whole Earth system.

**Ebb tide (16)** Falling or lowering tide during the daily tidal cycle (compare flood tide).

**Ecological succession (19)** The process whereby different and usually more complex assemblages of plants and animals replace older and usually simpler communities. Such communities are in a constant state of change as each species adapts to conditions.

**Ecology (19)** The science that studies the relations between organisms and their environment and among various ecosystems.

**Ecosphere (1)** Another name for the biosphere.

**Ecosystem (19, 20)** A self-regulating association of living plants, animals, and their nonliving physical and chemical environment.

**Ecotone (20)** A boundary transition zone between adjoining ecosystems that may vary in width and represent areas of tension as similar species of plants and animals compete for the resources (see ecosystem).

**Effusive eruption (12)** A volcanic eruption characterized by low-viscosity basaltic magma and low-gas content, which readily escapes. Lava pours forth onto the surface with relatively small explosions and few pyroclastics; tends to form shield volcanoes (see shield volcano, lava, pyroclastics; compare explosive eruption).

**Elastic-rebound theory (12)** A concept describing the faulting process in Earth's crust, in which the two sides of a fault appear locked despite the motion of adjoining pieces of crust, but with accumulating strain they rupture suddenly, snapping to new positions relative to each other, generating an earthquake.

**Electromagnetic spectrum (2)** All the radiant energy produced by the Sun placed in an ordered range, divided according to wavelengths.

**Eluviation (18)** The downward removal of finer particles and minerals from the upper horizons of soil (compare illuviation).

**Empirical classification (10)** A climate classification based on weather statistics or other data; used to determine general climate categories (compare genetic classification).

**Endogenic system (11)** The system internal to Earth, driven by radioactive heat derived from sources within the planet. In response, the surface fractures, mountain building

occurs, and earthquakes and volcanoes are activated (compare exogenic system).

**Entisols (18)** A soil order in the Soil Taxonomy. Specifically lacks vertical development of horizons; usually young or undeveloped. Found in active slopes, alluvial-filled floodplains, poorly drained tundra.

**Entrenched meander (14)** Incised river meanders excavated deeply into the landscape; thought to be evidence of stream rejuvenation.

**Environmental lapse rate (3)** The actual lapse rate in the lower atmosphere at any particular time under local weather conditions; may deviate above or below the normal lapse rate of 6.4 C° per 1000 m (3.5 F° per 1000 ft). (Compare normal lapse rate.)

**Eolian (15)** Caused by wind; refers to the erosion, transportation, and deposition of materials; spelled aeolian in some countries.

**Epipedon (18)** The diagnostic soil horizon that forms at the surface; not to be confused with the A horizon; may include all or part of the illuviated B horizon.

**Equal area (1)** A trait of a map projection; indicates the equivalence of all areas on the surface of the map, although shape is distorted (see map projections).

**Equatorial and tropical rain forest (20)** A lush biome of tall broadleaf evergreen trees and diverse plants and animals. The dense canopy of leaves is usually arranged in three levels.

**Equatorial countercurrent (6)** A strong countercurrent that travels east along the full extent of the Pacific, Atlantic, and Indian oceans; usually flows alongside or just beneath the westward-flowing surface current.

**Equatorial low-pressure trough (6)** A thermally caused low-pressure area that almost girdles Earth, with air converging and ascending all along its extent; also called the intertropical convergence zone (ITCZ).

**Erg desert (15)** A sandy desert, or area where sand is so extensive that it constitutes a sand sea.

**Erosion (14)** Denudation by wind, water, or ice, which dislodges, dissolves, or removes surface material.

**Esker (17)** A sinuously curving, narrow deposit of coarse gravel that forms along a meltwater stream channel, developing in a tunnel beneath a glacier.

**Estuary (14)** The point at which the mouth of a river enters the sea and freshwater and seawater are mixed; a place where tides ebb and flow.

**Eustasy (7)** Refers to worldwide changes in sea level that are not related to movements of land but rather to a rise and fall in the volume of water in the oceans.

**Evaporation (9)** The movement of free water molecules away from a wet surface into air that is less than saturated; the phase change of water to water vapor.

**Evaporation fog (8)** A fog formed when cold air flows over the warm surface of a lake, ocean, or other body of water; forms as the water molecules evaporate from the water surface into the cold, overlying air; also known as steam fog or sea smoke.

**Evaporation pan (9)** A weather instrument; a standardized pan from which evaporation occurs, with water automatically replaced and measured; an evaporimeter.

**Evapotranspiration (9)** The merging of evaporation and transpiration water loss into one term (see potential and actual evapotranspiration).

**Evolutionary atmosphere (2)** Earth's atmosphere between 3.3 and 4.0 billion years ago; thought by scientists to have been composed principally of water vapor and lesser amounts of carbon dioxide and nitrogen; an anaerobic environment.

**Exfoliation dome (13)** A dome-shaped feature of weathering, produced by the response of granite to the overburden removal process, which relieves pressure from the rock. Layers of rock sluff off in slabs or shells in a sheeting process.

**Exogenic system (11)** Earth's external surface system, powered by insolation, which energizes air, water, and ice and sets them in motion, under the influence of gravity. Includes all processes of landmass denudation (compare endogenic system).

**Exosphere (3)** An extremely rarefied outer atmospheric halo beyond the thermopause at an altitude of 480 km (300 mi); probably composed of hydrogen and helium atoms, with some oxygen atoms and nitrogen molecules present near the thermopause.

**Exotic stream (9)** A river that rises in a humid region and flows through an arid region, with discharge decreasing toward the mouth; for example, the Nile River and the Colorado River.

**Explosive eruption (12)** A violent and unpredictable volcanic eruption, the result of magma that is thicker (more viscous), stickier, and higher in gas and silica content than that of an effusive eruption; tends to form blockages within a volcano; produces composite volcanic landforms (see composite volcano; compare effusive eruption).

**Faulting (12)** The process whereby displacement and fracturing occurs between two portions of Earth's crust; usually associated with earthquake activity.

**Feedback loop (1)** Created when a portion of system output is returned as an information input, causing changes that guide further system operation (see negative feedback, positive feedback).

**Field capacity (9)** Water held in the soil by hydrogen bonding against the pull of gravity, remaining after water drains from the larger pore spaces; the available water for plants (see available water, capillary water).

**Fire ecology (19)** The study of fire as a natural agent and dynamic factor in community succession.

**Firn (17)** Snow of a granular texture that is transitional in the slow transformation from snow to glacial ice; snow that has persisted through a summer season in the zone of accumulation.

**Firn line (17)** The snow line that is visible on the surface of a glacier, where winter snows survive the summer ablation season; analogous to a snowline on land (see ablation).

**Fjord (17)** A drowned glaciated valley, or glacial trough, along a sea coast.

**Flash flood (15)** A sudden and short-lived torrent of water that exceeds the capacity of a stream channel; associated with desert and semiarid washes.

**Flood (14)** A high water level that overflows the natural (or artificial) levees along any portion of a stream.

**Floodplain (14)** A flat low-lying area along a stream channel, created by and subject to recurrent flooding; alluvial deposits generally mask underlying rock.

**Flood tide (16)** Rising tide during the daily tidal cycle (compare ebb tide).

**Fluvial (14)** Stream-related processes; from the Latin *fluvius* for "river" or "running water."

**Fog (8)** A cloud, generally stratiform, in contact with the ground, with visibility usually reduced to less than 1 km (3300 ft).

**Folding (12)** The bending and deformation of beds of rock strata subjected to compressional forces.

**Food chain (19)** The circuit along which energy flows from producers (plants), which manufacture their own food, to consumers (animals); a one-directional flow of chemical energy, ending with decomposers.

**Food web (19)** A complex network of interconnected food chains (see food chain).

**Formation class (20)** That portion of a biome that concerns the plant communities only, categorized by size, shape, and structure of the dominant vegetation.

**Friction force (6)** The effect of drag by the wind as it moves across a surface; may be operative through 500 m (1600 ft) of altitude. Surface friction slows the wind and therefore reduces the effectiveness of the Coriolis force.

**Front (8)** The leading edge of an advancing air mass; a line of contrasting weather conditions.

**Frost action (13)** A powerful mechanical force produced as water expands up to 9% of its volume as it freezes. Water freezing in a cavity in a rock can break the rock if it exceeds the rock's tensional strength.

**Funnel cloud (8)** The visible swirl extending from the bottom side of a cloud, which may or may not develop into a tornado. A tornado is a funnel cloud that has extended all the way to the ground (see tornado).

**Fusion (2)** The process of forcibly joining positively charged hydrogen and helium nuclei under extreme temperature and pressure; occurs naturally in thermonuclear reactions within stars, such as our Sun.

**Gelifluction (17)** Refers to soil flow in periglacial environments; a progressive, lateral movement. A type of solifluction under periglacial conditions of permafrost and frozen ground (see solifluction).

**General circulation model (GCM) (10)** Complex, computer-based climate models that produce generalizations of reality and forecasts of future weather and climate conditions. Four complex GCMs (three-dimensional models) are in use in the United States and four in other countries.

**Genetic classification (10)** A climate classification that uses causative factors to determine climatic regions; for example, an analysis of the effect of interacting air masses (compare empirical classification).

**Geodesy (1)** The science that determines Earth's shape and size through surveys, mathematical means, and remote sensing (see geoid).

**Geographic information system (GIS) (1)** A computer-based data processing tool or methodology used for gathering, manipulating, and analyzing geographic information to produce a holistic, interactive analysis.

**Geography (1)** The science that studies the interdependence among geographic areas, natural systems, processes, society, and cultural activities over space—a spatial science. The five themes of geographic education include: location, place, movement, regions, and human-Earth relationships.

**Geoid (1)** A word that describes Earth's shape, literally, "the shape of Earth is Earth-shaped." A theoretical surface at sea level that extends through the continents; deviates from a perfect sphere.

**Geologic cycle (11)** A general term characterizing the vast cycling that proceeds in the lithosphere. It encompasses the hydrologic cycle, tectonic cycle, and rock cycle.

**Geologic time scale (11)** A depiction of eras, periods, and epochs that span Earth's history; shows both the sequence of rock strata and their absolute dates, as determined by methods such as radioactive isotopic dating.

**Geomorphic threshold (13)** The threshold up to which landforms change before lurching to a new set of relationships, with rapid realignments of landscape materials and slopes.

**Geomorphology (13)** The science that analyzes and describes the origin, evolution, form, classification, and spatial distribution of landforms.

**Geostrophic wind (6)** A wind moving between areas of different pressure along a path that is parallel to the isobars. It is a product of the pressure gradient force and the Coriolis force. (see isobar, pressure gradient force, Coriolis force).

**Geothermal energy (12)** The energy in steam and hot water heated by subsurface magma near groundwater. This energy is used in Iceland, New Zealand, Italy, and northern California.

**Glacial drift (17)** The general term for all glacial deposits, both unsorted (till) and sorted (stratified drift).

**Glacial ice (17)** A hardened form of ice, very dense in comparison to normal snow or firn.

**Glacial surge (17)** The rapid, lurching, unexpected forward movement of a glacier.

**Glacier (17)** A large mass of perennial ice resting on land or floating shelflike in the sea adjacent to the land; formed from the accumulation and recrystallization of snow, which then flows slowly under the pressure of its own weight and the pull of gravity.

**Glacio-eustatic (7)** Changes in sea level in response to changes in the amount of water stored on Earth as ice; the more water that is bound up in glaciers and ice sheets, the lower the sea level.

**Goode's homolosine projection (1)** An equal-area projection formed by splicing together a sinusoidal and a homolographic projection.

**Graben (12)** Pairs or groups of faults that produce downward faulted blocks; characteristic of the basins of the interior western United States (see horst and Basin and Range Province).

**Graded stream (14)** An idealized condition in which a stream's load and the landscape mutually adjust. This forms a dynamic equilibrium among erosion, transported load, deposition, and the stream's capacity.

**Gradient (14)** The drop in elevation from a stream's headwaters to its mouth, ideally forming a concave slope.

**Granite (11)** A coarse-grained (slow-cooling) intrusive igneous rock of 25% quartz and more than 50% potassium and sodium feldspars; characteristic of the continental crust.

**Gravitational water (9)** That portion of surplus water that percolates downward from the capillary zone, pulled by gravity to the groundwater zone.

**Gravity (2)** The mutual force exerted by the masses of objects that are attracted one to another; produced in an amount proportional to each object's mass.

**Great circle (1)** Any circle drawn on a globe with its center coinciding with the center of the globe. An infinite number of great circles can be drawn, but only one parallel is a great circle—the equator (compare small circle).

**Greenhouse effect (4)** The process whereby radiatively active gases absorb insolation and reradiate the energy at longer wavelengths, which are retained longer, delaying the loss of infrared to space. Thus, the lower troposphere is warmed through the radiation and reradiation of infrared wavelengths. The approximate similarity between this process and that of a greenhouse explains the name.

**Greenwich Mean Time (GMT) (1)** Former world standard time, now known as Coordinated Universal Time (UTC) (see Coordinated Universal Time).

**Ground ice (17)** Subsurface water that is frozen in regions of permafrost. The moisture content of areas with ground ice may vary from nearly absent in regions of drier permafrost to almost 100% in saturated soils.

**Groundwater mining (9)** Pumping an aquifer beyond its capacity to flow and recharge; an overuse of the groundwater resource.

**Gulf Stream (5)** A strong northward-moving warm current off the east coast of North America, which carries its water far into the North Atlantic.

**Gyre (6)** The dominant circular ocean current beneath subtropical high-pressure cells in both hemispheres; offset to the western margin of each ocean basin.

**Habitat (19, 20)** A physical location to which an organism is biologically suited. Most species have specific habitat parameters and limits.

**Hadley cell (6)** The vertical convection cell in each hemisphere that is generated along the low-pressure system of converging and ascending air along the equator (ITCZ), and then subsides and diverges at subtropical latitudes.

**Hail (8)** A type of precipitation formed when a raindrop is repeatedly circulated above and below the freezing level in a cloud, with each cycle freezing more moisture onto the hailstone until it becomes too heavy to stay aloft.

**Hair hygrometer (7)** An instrument for the measurement of relative humidity; based on the principle that human hair will change as much as 4% in length between 0 and 100% relative humidity.

**Headland (16)** Protruding landforms that are extensions of the coast; generally composed of more resistant rocks. Incoming wave energy focuses on and bends around these headlands.

**Herbivore (19)** The primary consumer in a food chain, which eats plant material formed by a producer (plant) that has photosynthesized organic molecules (compare carnivore).

**Heterosphere (3)** A zone of the atmosphere above the mesopause, 80 km (50 mi) in altitude; composed of rarefied layers of oxygen atoms and nitrogen molecules; includes the ionosphere.

**Histosols (18)** A soil order in the Soil Taxonomy. Formed from thick accumulations of organic matter, such as beds of former lakes, bogs, and layers of peat.

**Homosphere (3)** A zone of the atmosphere from the surface up to 80 km (50 mi), composed of an even mixture of gases including nitrogen, oxygen, argon, carbon dioxide, and trace gases.

**Horn (17)** A pyramidal, sharp-pointed peak that results when several cirque glaciers gouge an individual mountain summit from all sides.

**Horse latitudes (6)** The calms of Cancer and the calms of Capricorn, zones of windless, hot calms beneath the subtropical high-pressure cells in both hemispheres; windless areas of great difficulty for sailing ships throughout history.

**Horst (12)** Upward-faulted blocks produced by pairs or groups of faults; characteristic of the mountain ranges of the interior of the western United States (see graben and Basin and Range Province).

**Hot spot (11)** An individual point of upwelling material originating in the asthenosphere, or deeper in the mantle; tends to remain fixed relative to migrating plates; some 100 are identified worldwide; exemplified by Yellowstone National Park, Hawaii, and Iceland.

**Human-Earth relationships (1)** One of the oldest themes of geography (the human-land tradition); includes the spatial analysis of settlement patterns, resource utilization and exploitation, hazard perception and planning, and the impact of environmental modification and artificial landscape creation.

**Humidity (7)** Water vapor content of the air. The capacity of the air to hold water vapor is mostly a function of the temperature of the air and the water vapor.

**Humus (18)** A mixture of organic debris in the soil, worked by consumers and decomposers in the humification process; characteristically formed from plant and animal litter deposited at the surface.

**Hurricane (8)** A tropical cyclone that is fully organized and intensified in inward-spiraling rainbands; ranges from 160 to 960 km (100 to 600 mi) in diameter, with wind speeds in excess of 119 kmph (65 knots, or 74 mph); a name used specifically in the Atlantic and eastern Pacific (compare typhoon).

**Hydration (13)** A physical weathering process involving water, although not involving any chemical change; water is added to a mineral, which initiates swelling and stress within the rock, mechanically forcing grains apart as the constituents expand.

**Hydraulic action (14)** The erosive work accomplished by the turbulence of water; causes a squeezing and releasing action in joints in bedrock; capable of prying and lifting rocks.

**Hydrograph (14)** A graph of stream discharge (in cms or cfs) over a period of time (minutes, hours, days, years) at a specific place on a stream. The relationship between stream discharge and precipitation input is illustrated on the graph.

**Hydrologic cycle (9)** A simplified model of the flow of water and water vapor from place to place. Water flows through the atmosphere, across the land where it is also stored as ice, and within groundwater. Solar energy empowers the cycle.

**Hydrolysis (13)** A chemical weathering process in which minerals chemically combine with water; a decomposition process that causes silicate minerals in rocks to break down and become altered.

**Hydrosphere (1)** An abiotic open system that includes all of Earth's water.

**Hygroscopic water (9)** That portion of soil moisture that is so tightly bound to each soil particle that it is unavailable to plant roots; the water, along with some bound capillary water, that is left in the soil after the wilting point is reached (see wilting point).

**Ice age (17)** A cold episode, with accompanying alpine and continental ice accumulations, that has repeated roughly every 200 to 300 million years since the late Precambrian era (1.25 billion years ago); includes the most recent episode during the Pleistocene Ice Age, which began 1.65 million years ago.

**Ice cap (17)** A large dome-shaped glacier, less extensive than an ice sheet (<50,000 km$^2$), although it buries mountain peaks and the local landscape.

**Ice field (17)** The least extensive form of a glacier, with mountain ridges and peaks visible above the ice; less than an ice cap or ice sheet.

**Icelandic low (6)** See subpolar low-pressure cell.

**Ice sheet (17)** An enormous continuous continental glacier. The bulk of glacial ice on Earth covers Antarctica and Greenland in two ice sheets.

**Ice wedge (17)** Formed when water enters a thermal contraction crack in permafrost and freezes. Repeated seasonal freezing and melting of the water progressively expands the wedge.

**Iceberg (17)** Floating ice created by calving ice (a large piece breaking off) and floating adrift; a hazard to shipping because about nine-tenths of the ice is submerged and can be irregular in form.

**Igneous rock (11)** One of the basic rock types; it has solidified and crystallized from a hot molten state (either magma or lava). (Compare metamorphic rock, sedimentary rock.)

**Illuviation (18)** The downward movement and deposition of finer particles and minerals from the upper horizon of the soil. Deposition usually is in the B horizon, where accumulations of clays, aluminum, carbonates, iron, and some humus occur (compare eluviation; see calcification).

**Inceptisols (18)** A soil order in the Soil Taxonomy. Weakly developed soils that are inherently infertile. Usually young soils that are weakly developed, although they are more developed than Entisols.

**Industrial smog (3)** Air pollution associated with coal-burning industries; it may contain sulfur oxides, particulates, carbon dioxide, and exotics.

**Infiltration (9)** Water access to subsurface regions of soil moisture storage through penetration of the soil surface.

**Inselberg (15)** Stark erosional remnants that stand above the surrounding terrain as knobs, hills, or "island mountains," as exemplified by Uluru (Ayers) Rock in Australia.

**Insolation (2)** Solar radiation that is intercepted by Earth.

**Interception (9)** Delays the fall of precipitation toward Earth's surface; caused by vegetation or other ground cover.

**Internal drainage (9)** In regions where rivers do not flow into the ocean, the outflow is through evaporation or subsurface gravitational flow. Portions of Africa, Asia, Australia, and the western United States have such drainage.

**International Date Line (1)** The 180° meridian; an important corollary to the prime meridian on the opposite side of the planet; established by the treaty of 1884 to mark the place where each day officially begins.

**Intertropical convergence zone (ITCZ) (6)** See equatorial low pressure trough.

**Ionosphere (3)** A layer in the atmosphere above 80 km (50 mi) where gamma, X-ray, and some ultraviolet radiation is absorbed and converted into infrared, and where the solar wind stimulates the auroras.

**Isobar (6)** An isoline connecting all points of equal atmospheric pressure.

**Isostasy (11)** A state of equilibrium in Earth's crust formed by the interplay between portions of the lithosphere and the asthenosphere. The crust depresses under weight and recovers with its removal, for example, the melting of glacial ice. The uplift is known as isostatic rebound.

**Isotherm (5)** An isoline connecting all points of equal temperature.

**Jet stream (6)** The most prominent movement in upper-level westerly wind flows; irregular, concentrated, sinuous bands of geostrophic wind, traveling at 300 kmph (190 mph). (See polar jet stream, subtropical jet stream.)

**Joint (13)** A fracture or separation in rock that occurs without displacement of the sides; increases the surface area of rock exposed to weathering processes.

**Kame (17)** A depositional feature of glaciation; a small hill of poorly sorted sand and gravel that accumulates in crevasses or in ice-caused indentations in the surface.

**Karst topography (13)** Distinctive topography formed in a region of chemically weathered limestone with poorly developed surface drainage and solution features that appear pitted and bumpy; originally named after the Krš Plateau of Yugoslavia.

**Katabatic winds (6)** Air drainage from elevated regions, flowing as gravity winds. Layers of air at the surface cool, become denser, and flow downslope. Known worldwide by many local names.

**Kettle (17)** Forms when an isolated block of ice persists in a ground moraine, an outwash plain, or valley floor after a glacier retreats; as the block finally melts, it leaves behind a steep-sided hole that frequently fills with water.

**Kinetic energy (3)** The energy of motion in a body; derived from the vibration of the body's own movement and stated as temperature.

**Köppen-Geiger climate classification (10)** An empirical classification system that uses average monthly temperatures, average monthly precipitation, and total annual precipitation to establish regional climate designations.

**Lagoon (16)** An area of coastal seawater that is virtually cut off from the ocean by a bay barrier or barrier beach; also, the water surrounded and enclosed by an atoll.

**Landfall (8)** The location along a coast where a storm moves onshore.

**Land-sea breeze (6)** Wind along coastlines and adjoining interior areas created by different heating characteristics of land and water surfaces—onshore (landward) breeze in the afternoon and offshore (seaward) breeze at night.

**Landslide (13)** A sudden rapid downslope movement of a cohesive mass of regolith and/or bedrock in a variety of mass-movement forms under the influence of gravity. A form of mass movement.

**Land-water heating difference (5)** Differences in the degree and way that land and water heat, as a result of contrasts in transmission, evaporation, mixing, and specific heat capacities. Land surfaces heat and cool faster than water and have continentality, whereas water provides a marine influence.

**Latent heat (7)** Heat energy is stored in one of the three states—ice, water, or water vapor. The energy is absorbed or released in each phase change from one state to another. Heat energy is absorbed as the latent heat of melting, vaporization or evaporation. Heat energy is released as the latent heat of condensation and freezing (or fusion).

**Latent heat of condensation (7)** The heat energy released to the environment in a phase change from water vapor to liquid; under normal sea-level pressure, 540 calories are released from each gram of water vapor that changes phase to water at boiling; and 585 calories are released from each gram of water vapor that condenses at 20°C (68°F).

**Latent heat of vaporization (7)** The heat energy absorbed from the environment in a phase change from liquid to water vapor at the boiling point; under normal sea-level pressure, 540 calories must be added to each gram of boiling water to achieve a phase change to water vapor.

**Lateral moraine (17)** Debris transported by a glacier that accumulates along the sides of the glacier and is deposited along these margins.

**Laterization (18)** A pedogenic process operating in well-drained soils that occur in warm and humid regions; typical of Oxisols. Plentiful precipitation leaches soluble minerals and soil constituents. Soils usually are reddish or yellowish.

**Latitude (1)** The angular distance measured north or south of the equator from a point at the center of Earth. A line connecting all points of the same latitudinal angle is called a parallel (compare longitude).

**Lava (11, 12)** Magma that issues from volcanic activity onto the surface; the extrusive rock that results when magma solidifies (see magma).

**Life zone (19)** A zonation by altitude of plants and animals that form distinctive communities. Each life zone possesses its own temperature and precipitation relations.

**Lightning (8)** Flashes of light caused by tens of millions of volts of electrical charge heating the air to temperatures of 15,000° to 30,000°C.

**Limestone (11)** The most common chemical sedimentary rock (nonclastic); it is lithified calcium carbonate ($CaCO_3$); very susceptible to chemical weathering by acids in the environment, including carbonic acid in rainfall.

**Limiting factor (19)** The physical or chemical factor that most inhibits biotic processes, either through lack or excess.

**Lithification (11)** The compaction, cementation, and hardening of sediments into sedimentary rock.

**Lithosphere (1)** Earth's crust and that portion of the uppermost mantle directly below the crust, extending down to about 70 km (45 mi). Some use this term to refer to the entire Earth.

**Littoral zone (16)** A specific coastal environment; that region between the high water line during a storm and a depth at which storm waves are unable to move sea-floor sediments.

**Living atmosphere (2)** Earth's atmosphere between 0.6 and 3.3 billion years ago; thought by scientists to be principally composed of carbon dioxide, water vapor, and nitrogen gas (3.0 billion years ago); beginnings of photosynthesis.

**Longshore current (16)** A current that forms parallel to a beach as waves arrive at an angle to the shore; generated in the surf zone by wave action, transporting large amounts of sand and sediment.

**Loam (18)** A soil that is a mixture of sand, silt, and clay in almost equal proportions, with no one texture dominant; an ideal agricultural soil.

**Location (1)** A basic theme of geography dealing with the absolute and relative position of people, places, and things on Earth's surface.

**Loess (15)** Large quantities of fine-grained clays and silts left as glacial outwash deposits; subsequently blown by the wind great distances and redeposited as a generally unstratified, homogeneous blanket of material covering existing landscapes; in China loess originated from desert lands.

**Longitude (1)** The angular distance measured east or west of a prime meridian from a point at the center of Earth. A line connecting all points of the same longitude is called a meridian.

**Lysimeter (9)** A weather instrument; device for measuring potential and actual evapotranspiration; isolates a portion of a field so that the moisture moving through the plot is measured.

**Magma (11)** Molten rock from beneath Earth's surface; fluid, gaseous, under tremendous pressure, and either intruded into country rock or extruded onto the surface as lava (see lava).

**Magnetic reversal (11)** A polarity change in Earth's magnetic field. With uneven regularity, the magnetic field fades to zero, then returns to full strength but with the magnetic poles reversed. Reversals have been recorded nine times during the past 4 million years.

**Magnetosphere (2)** Earth's magnetic force field, which is generated by dynamolike motions within the planet's outer core; deflects the solar wind toward the upper atmosphere above each pole.

**Mangrove swamp (16)** A wetland ecosystem between 30° N or S and the equator; tends to form a distinctive community of mangrove plants (compare salt marsh).

**Mantle (11)** An area within the planet representing about 80% of Earth's total volume, with densities increasing with depth and averaging 4.5 g per $cm^3$; occurs between the core and the crust; is rich in iron and magnesium oxides and silicates.

**Map projection (1)** The reduction of a spherical globe onto a flat surface in some orderly and systematic realignment of the latitude and longitude grid.

**Marine (5)** A quality of regions that are dominated by the moderating effect of the ocean and that exhibit a smaller range of minimum and maximum temperature range than continental stations (see continentality, land-water heating differences).

**Mass movement (13)** All unit movements of materials propelled by gravity; can range from dry to wet, slow to fast, small to large, and free-falling to gradual or intermittent.

**Mass wasting (13)** Gravitational movement of nonunified material downslope; a specific form of mass movement.

**Meandering stream (14)** The sinuous, curving pattern common to *graded streams*, with the energetic outer portion of each curve subjected to the greatest erosive action and the lower-energy inner portion receiving sediment deposits (see graded stream).

**Mean sea level, MSL (16)** The average of tidal levels recorded hourly at a given site over a long period, which must be at least a full lunar tidal cycle.

**Medial moraine (17)** Debris transported by a glacier that accumulates down the middle of the glacier, resulting from two glaciers merging their lateral moraines; forms a depositional feature following glacial retreat.

**Mediterranean shrubland (20)** A major biome dominated by Mediterranean dry summer climates and characterized by sclerophyllous scrub and short, stunted, tough forests.

**Mercury barometer (3)** A device that measures air pressure using a column of mercury in a tube, one end of which is sealed, and the other end inserted in an open vessel of mercury (see air pressure).

**Meridian (1)** See longitude.

**Mesocyclone (8)** A large rotating atmospheric circulation, initiated within a parent cumulonimbus cloud at mid-troposphere elevation; generally produces heavy rain, large hail, blustery winds, and lightning; may lead to tornado activity.

**Mesosphere (3)** The upper region of the homosphere from 50 to 80 km (30 to 50 mi) above the ground; designated by temperature criteria; has very low pressures.

**Metamorphic rock (11)** One of three basic rock types, it is existing igneous and sedimentary rock that has undergone profound physical and chemical changes under increased pressure and temperature. Constituent mineral structures may exhibit foliated or nonfoliated textures (compare igneous rock and sedimentary rock).

**Meteorology (8)** The scientific study of the atmosphere, including its physical characteristics and motions; related chemical, physical, and geological processes; the complex linkages of atmospheric systems; and weather forecasting.

**Methane (5)** A radiatively active gas that participates in the greenhouse effect; derived from the organic processes of burning, digesting, and rotting in the presence of oxygen; $CH_4$.

**Microclimatology (4)** The study of local climates at or near Earth's surface.

**Midlatitude broadleaf and mixed forest (20)** A biome in moist continental climates in areas of warm-to-hot summers and cool-to-cold winters; relatively lush stands of broadleaf forests trend northward into needleleaf evergreen stands.

**Midlatitude cyclone (8)** An organized area of low pressure, with converging and ascending air flow producing an interaction of air masses; migrates along storm tracks. Such lows or depressions form the dominant weather pattern in the middle and higher latitudes of both hemispheres.

**Midlatitude grassland (20)** The major biome most modified by human activity; so named because of the predominance of grasslike plants, although deciduous broadleafs appear along streams and other limited sites; location of the world's breadbaskets of grain and livestock production.

**Mid-ocean ridge (11)** A submarine mountain range that extends more than 65,000 km (40,000 mi) worldwide and averages more than 1000 km (600 mi) in width; centered along sea-floor spreading centers (see sea-floor spreading).

**Milky Way Galaxy (2)** A flattened, disk-shaped mass in space estimated to contain up to 400 billion stars; includes our Solar System.

**Miller cylindrical projection (1)** A compromise map projection that avoids the severe distortion of the Mercator projection (see map projection).

**Mineral (11)** An element or combination of elements that forms an inorganic natural compound; described by a specific formula and crystal structure.

**Model (1)** A simplified version of a system, representing an idealized part of the real world.

**Modern atmosphere (2)** The fourth and present distinct atmosphere on Earth, having evolved from the living atmosphere about 0.6 billion years ago; lower layers composed of nitrogen, oxygen, argon, carbon dioxide, and trace gases; an additive mixture of gases.

**Mohorovičić discontinuity, or Moho (11)** The boundary between the crust and the rest of the lithospheric upper mantle; named for the Yugoslavian seismologist Mohorovičić; a zone of sharp material and density contrasts; also known as the *Moho*.

**Moist adiabatic rate (MAR) (7)** The rate at which a saturated parcel of air cools in ascent; a rate of 6 C° per 1000 m (3.3 F° per 1000 ft). This rate may vary, with moisture content and temperature, from 4 C° to 10 C° per 1000 m (2 F° to 6 F° per 1000 ft) (see adiabatic; compare dry adiabatic rate).

**Moisture droplet (7)** A tiny water particle that constitutes the initial composition of clouds. Each droplet measures approximately 0.002 cm (0.0008 in.) in diameter and is invisible to the unaided eye.

**Mollisols (18)** A soil order in the Soil Taxonomy. These have a mollic epipedon and a humus-rich organic content high in alkalinity. Some of the world's most significant agricultural soils are Mollisols.

**Moment magnitude scale (12)** An earthquake magnitude scale. Considers the amount of fault slippage, the size of the area that ruptured, and the nature of the materials that faulted in estimating the magnitude of an earthquake—an assessment of the seismic moment. Replaces the Richter scale (amplitude magnitude), especially on larger magnitude events.

**Monsoon (6)** An annual cycle of dryness and wetness, with seasonally shifting winds produced by changing atmospheric pressure systems; affects India, Southeast Asia, Indonesia, northern Australia, and portions of Africa. From the Arabic word *mausim,* meaning "season."

**Montane forest (20)** Needleleaf forest associated with mountain elevations (see needleleaf forest).

**Moraine (17)** Marginal glacial deposits (lateral, medial, terminal, ground) of unsorted and unstratified material.

**Mountain-valley breeze (6)** A light wind produced as cooler mountain air flows downslope at night, and as warmer valley air flows upslope during the day.

**Movement (1)** A major theme in geography involving migration, communication, and the interaction of people and processes across space.

**Mudflow (13)** Fluid downslope flows of material containing more water than earthflows.

**Natural levee (14)** A long, low ridge that forms on both sides of a stream in a developed floodplain; they are depositional products (coarse gravels and sand) of river flooding.

**Neap tide (16)** Unusually low tidal range produced during the first and third quarters of the Moon, with an offsetting pull from the Sun (compare spring tide).

**Needleleaf forest (20)** Forests of pine, spruce, fir, and larch, stretching from the east coast of Canada westward to Alaska and continuing from Siberia westward across the entire extent of Russia to the European Plain; called the taiga (a Russian word) or the boreal forest; principally in the D climates. Includes montane forests.

**Negative feedback (1)** Feedback that tends to slow or dampen response in a system; promotes self-regulation in a system; far more common than positive feedback in living systems (see feedback loop, positive feedback).

**Net primary productivity (19)** The net photosynthesis (photosynthesis minus respiration) for a given community; considers all growth and all reduction factors that affect the amount of useful chemical energy (biomass) fixed in an ecosystem.

**Net radiation (NET R) (4)** The net all-wave radiation available at Earth's surface; the final outcome of the radiation balance process between incoming shortwave insolation and outgoing longwave energy.

**Niche (19, 20)** The basic function, or occupation, of a lifeform within a given community; the way an organism obtains its food, air, and water.

**Nickpoint (knickpoint) (14)** The point at which the longitudinal profile of a stream is abruptly broken by a change in gradient; for example, a waterfall, rapids, or cascade.

**Nimbostratus (8)** Rain-producing, dark, grayish, stratiform clouds characterized by gentle drizzle.

**Nitrogen dioxide (3)** A noxious reddish-brown gas produced in combustion engines; can be damaging to human respiratory tracts and to plants; participates in photochemical reactions and acid deposition.

**Noctilucent cloud (3)** A rare shining band of ice crystals that may glow at high latitudes long after Sunset; formed within the mesosphere, where cosmic and meteoric dust act as nuclei for the formation of ice crystals.

**Normal fault (12)** A type of geologic fault in rocks. Tension produces strain that breaks a rock with one side moving vertically relative to the other side along an inclined fault plane (compare reverse fault).

**Normal lapse rate (3)** The average rate of temperature decrease with increasing altitude in the lower atmosphere; an average value of 6.4 C° per km, or 1000 m (3.5 F° per 1000 ft). (Compare environmental lapse rate.)

**Nuclear winter hypothesis (10)** An increase in Earth's albedo and upper atmospheric absorption of insolation resulting in surface cooling; associated with the detonation of a relatively small number of nuclear warheads within the biosphere. Now encompasses a whole range of ecological, biological, and climatic impacts.

**Occluded front (8)** In a cyclonic circulation, the overrunning of a surface warm front by a cold front and the subsequent lifting of the warm air wedge off the ground; initial precipitation is moderate to heavy.

**Ocean basin (12)** The physical container (a depression in the lithosphere) holding an ocean.

**Oceanic trench (11)** The deepest feature of Earth's crust; associated with a subduction zone. The deepest is the Mariana Trench near Guam, which descends to 11,033 m (36,198 ft).

**Omnivore (19)** A consumer that feeds on both producers (plants) and consumers (meat)—a role occupied by humans, among other animals (compare consumer, producer).

**Open system (1)** A system with inputs and outputs crossing back and forth between the system and the surrounding environment. Earth is an open system in terms of energy (compared closed system).

**Orders of relief (12)** A convenient classification of landscapes based on scale, from vast ocean basins and continental platforms to small hills and valleys. (Three orders of relief are considered in this text.)

**Orogenesis (12)** The process of mountain building that occurs when large-scale compression leads to deformation and uplift of the crust; literally the birth of mountains.

**Orographic lifting (8)** The uplift of a migrating air masses as it is forced to move upward over a mountain range—a topographic barrier. The lifted air cools adiabatically as it moves upslope; clouds may form and produce increased precipitation.

**Outgassing (7)** The release of trapped gases from rocks, forced out through cracks, fissures, and volcanoes from within Earth; the terrestrial source of Earth's water.

**Outwash plain (17)** Glacial stream deposits of stratified drift of meltwater-fed, braided, and overloaded streams; occurs beyond a glacier's morainal deposits.

**Overland flow (9)** Surplus water that flows across the land surface toward stream channels. Together with precipitation and subsurface flows, it constitutes the total runoff from an area.

**Oxbow lake (14)** A lake that was formerly part of the channel of a meandering stream; isolated when a stream eroded its outer bank forming a cutoff through the neck of the looping meander (see meandering stream).

**Oxidation (13)** A chemical weathering process in which oxygen dissolved in water oxidizes (combines with) certain metallic elements to form oxides; most familiar as the "rusting" of iron in a rock or soil that produces a reddish-brown stain of iron oxide ($Fe_2O_3$).

**Oxisols (18)** A soil order in the Soil Taxonomy. Tropical soils that are old, deeply developed, and lacking in horizons wherever well-drained. Heavily weathered, low in cation exchange capacity, and low in fertility.

**Ozone layer (3)** See ozonosphere.

**Ozonosphere (3)** A layer of ozone ($O_3$) occupying the full extent of the *stratosphere* (20 to 50 km or 12 to 30 mi) above the surface; the region of the atmosphere where ultraviolet wavelengths of insolation are extensively absorbed and converted into heat.

**Pacific high (6)**; A high-pressure cell that dominates the Pacific in July, retreating southward in the Northern Hemisphere in January; also known as the Hawaiian high.

**Paleoclimatology (10)** The science that studies the climates of past ages.

**Paleolake (17)** An ancient lake, such as Lake Bonneville or Lake Lahonton, associated with former wet periods when the lake basins were filled to higher levels than today.

**Palsa (17)** A rounded or elliptical mound of peat that contains thin perennial ice lenses rather than an ice core, as in a pingo. Palsas can be 2 to 30 m wide by 1 to 10 m high and usually are covered by soil or vegetation over a cracked surface.

**PAN (3)** See peroxyacetyl nitrate.

**Pangaea (11)** The supercontinent formed by the collision of all continental masses approximately 225 million years ago; named in the continental drift theory by Wegener in 1912 (see plate tectonics).

**Parallel (1)** See latitude.

**Parent material (13)** The unconsolidated material, from both organic and mineral sources, that is the basis of soil development.

**Paternoster lake (17)** One of a series of small, circular, stair-stepped lakes, formed in individual rock basins aligned down the course of a glaciated valley; named because they look like a string of rosary (religious) beads.

**Patterned ground (17)** Areas in the periglacial environment where freezing and thawing of the ground create polygonal forms of arranged rocks at the surface.

**Pedogenic regime (18)** A specific soil-forming process keyed to a specific climatic regime: laterization, calcification, salinization, and podzolization, among others.

**Pedon (18)** A soil profile extending from the surface to the lowest extent of plant roots or to the depth where regolith or bedrock is encountered; imagined as a hexagonal column; the basic soil sampling unit.

**Percolation (9)** The process by which water permeates the soil or porous rock into the subsurface environment.

**Periglacial (17)** Cold-climate processes, landforms, and topographic features along the margins of glaciers, past and present; periglacial characteristics exist on more than 20% of Earth's land surface; includes permafrost, frost action, and ground ice.

**Perihelion (2)** That point of Earth's closest approach to the Sun in its elliptical orbit; occurs on January 3 at 147,255,000 km (91,500,000 mi); variable over a 100,000-year cycle (compare aphelion).

**Permafrost (17)** Forms when soil or rock temperatures remain below 0°C (32°F) for at least two years in areas considered periglacial; criterion is based on temperature and not on whether water is present.

**Permeability (9)** The ability of water to flow through soil or rock; a function of the texture and structure of the medium.

**Peroxyacetyl nitrate (PAN) (3)** A pollutant formed from photochemical reactions involving nitric oxide (NO) and hydrocarbons (HC). PAN produces no known human health effect, but it is particularly damaging to plants.

**Phase change (7)** The change in phase, or state, among ice, water, and water vapor; involves the absorption or release of latent heat (see latent heat).

**Photochemical smog (3)** Air pollution produced by the interaction of ultraviolet light, nitrogen dioxide, and hydrocarbons; produces ozone and PAN through a series of complex photochemical reactions. Automobiles are the major source of the contributive gases.

**Photosynthesis (19)** The process by which plants produce their own food from carbon dioxide and water, powered by solar energy. The joining of carbon dioxide and oxygen in plants, under the influence of certain wavelengths of visible light; releases oxygen and produces energy-rich organic material (sugars and starches). (Compare respiration.)

**Physical geography (1)** A science that studies the spatial aspects of the physical elements and processes that make up the environment: energy, air, water, weather, climate, landforms, soils, animals, plants, and Earth.

**Physical weathering (13)** The breaking and disintegrating of rock without any chemical alteration; sometimes referred to as mechanical or fragmentation weathering.

**Pingo (17)** Large areas of frozen ground (soil-covered ice) that develop a heaved up, circular, ice-cored mound, rising above a periglacial landscape as water freezes into ice and expands; sometimes results from pressure developed by freezing artesian water that is injected into permafrost; occasionally exceeds 60 m height (200 ft).

**Pioneer community (19)** The initial plant community in an area; usually found on new surfaces or those that have been stripped of life—for example, surfaces created by mass movements of land, or land disturbed by human activities.

**Place (1)** A major theme in geography, focused on the tangible and intangible characteristics that make each location unique.

**Plane of the ecliptic (2)** A plane intersecting all the points of Earth's orbit.

**Planetesimal hypothesis (2)** Proposes a process by which early protoplanets formed from the condensing masses of a nebular cloud of dust, gas, and icy comets; a formation process now being observed in other parts of the galaxy.

**Planimetric map (Appendix B)** A basic map showing the horizontal position of boundaries, land-use activities, and political, economic, and social outlines.

**Plate tectonics (11)** The conceptual model that encompasses continental drift, sea-floor spreading, and related aspects of crustal movement; accepted as the foundation of crustal tectonic processes.

**Plateau basalt (12)** An accumulation of horizontal flows formed when lava spreads out from elongated fissures onto the surface in extensive sheets; associated with effusive eruptions; also known as flood basalts (see basalt).

**Playa (15)** An area of salt crust left behind by evaporation on a desert floor usually in the middle of a bolson or valley; intermittently wet and dry.

**Pluton (11)** A mass of intrusive igneous rock that has cooled slowly in the crust; forms in any size or shape. The largest partially exposed pluton is a batholith (see batholith).

**Podzolization (18)** A pedogenic process in cool, moist climates; forms a highly leached soil with strong surface acidity because of humus from acid-rich trees.

**Point bar (14)** In a stream the inner portion of a meander, where sediment fill is redeposited.

**Polar easterlies (6)** Variable weak, cold, and dry winds moving away from the polar region; an anticyclonic circulation.

**Polar front (6)** A significant zone of contrast between cold and warm air masses; roughly situated between 50° and 60° N and S latitude.

**Polar high-pressure cells (6)** A weak, anticyclonic, thermally produced pressure system positioned roughly over each pole; the region of the lowest temperatures on Earth (see Antarctic high).

**Polar jet stream (6)** A strong wind current at the tropopause along the polar front; occurs between 7600 m and 10,700 m (24,900 and 35,100 ft) altitude and meanders between 30° and 70° N latitude (see jet stream).

**Polypedon (18)** The identifiable soil in an area, with distinctive characteristics differentiating it from surrounding polypedons that form the basic mapping unit; composed of many pedons (see pedon).

**Porosity (9)** The total volume of available pore space in soil; a result of the texture and structure of the soil.

**Positive feedback (1)** Feedback that amplifies or encourages responses in a system (see negative feedback, feedback loop).

**Potential evapotranspiration (9)** POTET is the amount of moisture that would evaporate and transpire if adequate moisture were available; it is the amount lost under optimum moisture conditions, the moisture demand.

**Potentiometric surface (9)** A pressure level in a confined aquifer, defined by the level to which water rises in wells; caused by the fact that the water in a confined aquifer is under the pressure of its own weight; also known as a piezometric surface. It can create an imaginary surface in some instances (see artesian water).

**Precipitation (9)** Rain, snow, sleet, and hail—the moisture supply; called PRECIP in the water balance.

**Pressure gradient force (6)** Causes air to move from an area of higher barometric pressure to an area of lower barometric pressure due to the pressure difference.

**Primary succession (19)** Succession that occurs among plant species in an area of bare rock with no trace of a former community, such as cooled lava flows or disturbed lands left after glacial retreat.

**Prime meridian (1)** An arbitrary meridian designated as 0° longitude; the point from which longitudes are measured east or west; Greenwich, England was selected by international agreement in an 1884 treaty.

**Primordial atmosphere (2)** Earth's first atmosphere, derived from the formation nebula. Lighter gases escaped early, with water vapor, hydrogen, cyanide, ammonia, methane, and others predominant.

**Process (1)** A set of actions and changes that occur in some special order; analysis of processes is central to modern geographic synthesis.

**Producer (19)** Organism (plant) in an ecosystem uses carbon dioxide as its sole source of carbon, which it chemically fixes through photosynthesis to provide its own nourishment; also called an autotroph (compare consumer).

**Pyroclastic (12)** An explosively ejected rock fragment launched by a volcanic eruption; sometimes described by the more general term tephra.

**Radiation fog (8)** Formed by radiative cooling of a land surface, especially on clear nights in areas of moist ground; occurs when the air layer directly above the surface is chilled to the dewpoint temperature, thereby producing saturated conditions.

**Radiatively active gas (5)** A gas in the atmosphere (such as carbon dioxide, methane, or water vapor) that absorbs and radiates infrared wavelengths.

**Rain gauge (9)** A weather instrument; a standardized device that captures and measures rainfall.

**Rain shadow (8)** The area on the leeward slope of a mountain range; where precipitation receipt is greatly reduced compared to the windward slope on the other side (see orographic lifting).

**Reflection (4)** The portion of arriving insolation that is returned directly to space without being converted into heat or performing any work (see albedo).

**Refraction (4)** The bending effect on electromagnetic waves that occurs when insolation enters the atmosphere or another medium; the same process by which a crystal, or prism, disperses the component colors of the light passing through it.

**Region (1)** A geographic theme that focuses on areas that display unity and internal homogeneity of traits; includes the study of how a region forms, evolves, and interrelates with other regions.

**Regolith (13)** Partially weathered rock overlying bedrock, whether residual or transported.

**Relative humidity (7)** The ratio of water vapor actually in the air (content) compared to the maximum water vapor the air could hold (capacity) at that temperature; expressed as a percentage (compare vapor pressure, specific humidity).

**Relief (12)** Elevation differences in a local landscape; an expression of the unevenness, height, and slope variation.

**Remote sensing (1)** Information acquired from a distance, without physical contact with the subject; for example, photography, orbital imagery, or radar.

**Respiration (19)** The process by which plants use their food to derive energy for their operations; essentially, the reverse of the photosynthetic process; releases carbon dioxide, water, and heat into the environment (compare photosynthesis).

**Reverse fault (12)** Compressional forces produce strain that breaks a rock so that one side moves upward relative to the other side; also called a thrust fault (compare normal fault).

**Revolution (2)** The annual orbital movement of Earth about the Sun; determines the length of the year and the seasons.

**Rhumb line (1)** A line of constant compass direction, or constant bearing, which crosses all meridians at the same angle. A portion of a great circle.

**Richter scale (12)** An open-ended, logarithmic scale that estimates earthquake magnitude; designed by Charles Richter in 1935; now generally replaced by the moment magnitude scale (see moment magnitude scale).

**Ring of fire (12)** See circum-Pacific belt.

**Robinson projection (1)** A compromise (neither equal area or true shape) oval projection developed in 1963 by Arthur Robinson.

**Roche moutonnée (17)** A glacial erosion feature; an asymmetrical hill of exposed bedrock; displays a gently sloping upstream side that has been smoothed and polished by a glacier and an abrupt, steep downstream side.

**Rock (11)** An assemblage of minerals bound together, or sometimes a mass of a single mineral.

**Rock cycle (11)** A model representing the interrelationships among the three rock-forming processes: igneous, sedimentary, and metamorphic—shows how each can be transformed into another rock type.

**Rockfall (13)** Free-falling movement of debris from a cliff or steep slope, generally falling straight or bounding downslope.

**Rossby wave (6)** An undulating horizontal motion in the upper-air westerly circulation at middle and high latitudes.

**Rotation (2)** The turning of Earth on its axis; averages about 24 hours in duration; determines day-night relation.

**Salinity (16)** The concentration of natural elements and compounds dissolved in solution, as solutes; measured by weight in parts per thousand (‰) in seawater.

**Salinization (18)** A pedogenic process that results from high potential evapotranspiration rates in deserts and semiarid regions. Soil water is drawn to surface horizons, and dissolved salts are deposited as the water evaporates.

**Saltation (14)** The transport of sand grains (usually larger than 0.2 mm, or 0.008 in.) by stream or wind, bouncing the grains along the ground in asymmetrical paths.

**Salt marsh (16)** A wetland ecosystem characteristic of latitudes poleward of the 30th parallel (compare mangrove swamp).

**Sand sea (15)** An extensive area of sand and dunes; characteristic of Earth's *erg deserts*.

**Saturated (7)** Air that is holding all the water vapor that it can hold at a given temperature.

**Scale (1)** On a map, the ratio of the distance on the map to that in the real world; expressed as a representative fraction, graphic scale, or written scale.

**Scarification (13)** Human-induced mass movements of Earth materials, such as large-scale open-pit mining and strip mining.

**Scattering (4)** Deflection and redirection of insolation by atmospheric gases, dust, ice, and water vapor; the shorter the wavelength, the greater the scattering.

**Scientific method (2)** An approach that uses applied common sense in an organized and objective manner; based on observation, generalization, formulation of a hypothesis, and ultimately the development of a theory.

**Sea-floor spreading (11)** As proposed by Hess and Dietz, the mechanism driving the movement of the continents; associated with upwelling flows of magma along the worldwide system of mid-ocean ridges (see mid-ocean ridge).

**Secondary air mass (8)** An air mass that becomes modified in terms of temperature and moisture characteristics with distance and time away from its source region.

**Secondary succession (19)** Succession that occurs among plant species in an area where vestiges of a previously functioning community are present; underlying soil is intact.

**Sediment (13)** Fine-grained mineral matter that is transported and deposited by air, water, or ice.

**Sedimentary rock (11)** One of three basic rock types; formed from the compaction, cementation, and hardening of sediments derived from other rocks (compare igneous rock, metamorphic rock).

**Seismic wave (11)** The shock wave sent through the planet by an earthquake or underground nuclear test. Transmission varies according to temperature and the density of various layers within the planet.

**Seismograph (12)** A device that measures seismic waves of energy transmitted throughout Earth's interior.

**Sensible heat (3)** Heat that can be measured with a thermometer; a measure of the concentration of kinetic energy from molecular motion.

**Sheet flow (14)** Surface water that moves downslope in a thin film as overland flow, not concentrated in channels larger than rills.

**Sheeting (13)** A form of weathering associated with fracturing or fragmentation of rock by pressure release; often related to exfoliation processes (see exfoliation dome).

**Shield volcano (12)** A symmetrical mountain landform built from effusive eruptions; gently sloped, gradually rising from the surrounding landscape to a summit crater; typical of the Hawaiian Islands (see effusive eruption; compare composite volcano).

**Sinkhole (13)** Nearly circular depression created by the weathering of karst landscapes; also known as a doline in traditional studies; may collapse through the roof of an underground space (see karst topography).

**Sling psychrometer (7)** A weather instrument that measures relative humidity using two thermometers—a dry bulb and a wet bulb—mounted side-by-side.

**Slipface (15)** On a sand dune, formed as dune height increases above 30 cm (12 in.) on the leeward side at an angle at which loose material is stable—its angle of repose (30° to 34°).

**Slope (13)** A curved, inclined surface that bounds a landform.

**Small circle (1)** A circle on a globe's surface that does not share Earth's center; for example, all parallels of latitude other than the equator (compare great circle).

**Snowline (17)** A temporary line marking the elevation where winter snowfall persists throughout the summer; seasonally, the lowest elevation covered by snow during the summer.

**Soil (18)** A dynamic natural body made up of fine materials covering Earth's surface in which plants grow, composed of both mineral and organic matter.

**Soil colloid (18)** A tiny clay and organic particle in soil; provides chemically active sites for mineral ion adsorption.

**Soil creep (13)** A persistent mass movement of surface soil where individual soil particles are lifted and disturbed by the expansion of soil moisture as it freezes, or by grazing livestock or digging animals.

**Soil fertility (18)** The ability of soil to support plant productivity when it contains organic substances and clay minerals that absorb water and certain elemental ions needed by plants.

**Soil horizon (18)** The various layers exposed in a pedon; roughly parallel to the surface and identified as O, A, B, and C.

**Soil moisture storage (9)** Δ STRGE, is the retention of moisture within soil; it is a savings account that can accept deposits (soil moisture recharge) or experiences withdrawals (soil moisture utilization) as conditions change.

**Soil science (18)** Interdisciplinary science of soils. Pedology concerns the origin, classification, distribution, and description of soil. Edaphology focuses on soil as a medium for sustaining higher plants.

**Soil Taxonomy (18)** A soil classification system based on observable soil properties actually seen in the field; published in 1975 by the U.S. Soil Conservation Service (compare Canadian System of Soil Classification, Appendix C).

**Solar constant (2)** The amount of insolation intercepted by Earth on a surface perpendicular to the Sun's rays when Earth is at its average distance from the Sun; a value of 1370 watts per m², or 1.968 calories per cm² per minute; averaged over the entire globe at the thermopause.

**Solar wind (2)** Clouds of ionized (charged) gases emitted by the Sun and traveling in all directions from the Sun's surface. Effects on Earth include auroras, disturbance of radio signals, and possible influences on weather.

**Solifluction (17)** Gentle downslope movement of a saturated surface material (soil and regolith), in various climatic regimes, where temperatures are above freezing.

**Solum (18)** A true soil profile in the pedon; ideally, a combination of A and B horizons (see pedon).

**Solution (13)** The dissolved load of a stream.

**Spatial analysis (1)** The examination of spatial interactions, patterns, and variations over area and/or space; a key integrative approach of geography.

**Specific heat (5)** The increase of temperature in a material when energy is absorbed; water has a higher specific heat (can store more heat) than a comparable volume of soil or rock.

**Specific humidity (7)** The mass of water vapor (in grams) per unit mass of air (in kilograms) at any specified temperature. The maximum mass of water vapor that a kilogram of air can hold at any specified temperature is termed its maximum specific humidity (compare vapor pressure, relative humidity).

**Speed of light (2)** Specifically, 299,792 kilometers per second (186,282 miles per second), or more than 9.4 trillion kilometers per year (5.9 trillion miles per year)—a distance known as a light year.

**Spheroidal weathering (13)** A chemical weathering process in which the sharp edges and corners of boulders and rocks are weathered in thin plates that create a rounded, spheroidal form.

**Spodosols (18)** A soil order in the Soil Taxonomy classification that occurs in northern coniferous forests; best developed in cold, moist, forested climates of *Dfb, Dfc, Dwc, Dwd*. Lacks humus and clay in the A horizons, with high acidity associated with podzolization processes.

**Spring tide (16)** The highest tidal range, which occurs when the Moon and the Sun are in conjunction (at new Moon) or in opposition (at full Moon) stages (compare neap tide).

**Squall line (8)** A zone slightly ahead of a fast-advancing cold front, where wind patterns are rapidly changing and blustery and precipitation is strong.

**Stability (7)** The condition of a parcel, whether it remains where it is or changes its initial position. The parcel is stable if it resists displacement upwards, unstable if it continues to rise.

**Stationary front (8)** A frontal area of contact between contrasting air masses that shows little horizontal movement; winds in opposite direction on either side of the front flow parallel along the front.

**Steady-state equilibrium (1)** The condition that occurs in a system when rates of input and output are equal and the amounts of energy and stored matter are nearly constant around a stable average.

**Stomata (19)** Small openings on the undersides of leaves, through which water and gases pass.

**Storm surge (8)** A large quantity of seawater pushed inland by the strong winds associated with a tropical cyclone.

**Stratified drift (17)** Sediments deposited by glacial meltwater that appear sorted; a specific form of glacial drift.

**Stratigraphy (11)** An analysis of the sequence, spacing, and spatial distribution of rock strata.

**Stratocumulus (8)** A lumpy, grayish, low-level cloud, patchy with sky visible, sometimes present at the end of the day.

**Stratosphere (3)** That portion of the homosphere that ranges from 20 to 50 km (12.5 to 30 mi) above Earth's surface, with temperatures ranging from –57°C (–70°F) at the tropopause to 0°C (32°F) at the stratopause. The functional ozonosphere is within the stratosphere.

**Stratus (8)** A stratiform (flat, horizontal) cloud generally below 2000 m (6500 ft).

**Strike-slip fault (12)** Horizontal movement along a faultline, that is, movement in the same direction as the fault; also known as a transcurrent fault. Such movement is described as right-lateral or left-lateral, depending on the relative motion observed.

**Subduction zone (11)** An area where two plates of crust collide and the denser oceanic crust dives beneath the less dense continental plate, forming deep oceanic trenches and seismically active regions.

**Sublimation (7)** A process in which ice evaporates directly to water vapor or water vapor freezes to ice (deposition).

**Subpolar low-pressure cell (6)** A region of low pressure centered approximately at 60° latitude in the North Atlantic near Iceland and in the North Pacific near the Aleutians, as well as in the Southern Hemisphere. Air flow is cyclonic, that weakens in summer and strengthens in winter.

**Subsolar point (2)** The only point receiving perpendicular insolation at a given moment—the Sun directly overhead.

**Subsurface diagnostic horizon (18)** A soil horizon that originates below the epipedon at varying depths; may be part of the A and B horizons; important in soil description as part of the Soil Taxonomy.

**Subtropical high-pressure cell (6)** One of several dynamic high-pressure areas covering roughly the region from 20° to 35° N and S latitudes; responsible for the hot, dry areas of Earth's arid and semiarid deserts.

**Subtropical jet stream (6)** A wind current that occurs in the subtropical latitudes, near the boundary between the tropics and the midlatitudes.

**Sulfur dioxide (3)** A colorless gas detected by its pungent odor; produced by the combustion of fossil fuels that contain sulfur as an impurity; can react in the atmosphere to form sulfuric acid, a component of acid deposition.

**Summer (June) solstice (2)** The time when the Sun's declination is at the Tropic of Cancer, at 23.5° N latitude; June 20-21 each year.

**Sunrise (2)** That moment when the disk of the Sun first appears above the horizon.

**Sunset (2)** That moment when the disk of the Sun totally disappears.

**Sunspot (2)** A magnetic disturbance on the surface of the Sun; occurring in an average 11-year cycle; related flares, prominences, and outbreaks produce surges in solar wind.

**Supercooled droplet (8)** A water droplet in clouds at temperatures ranging from −15° to −35°C. Such droplets form ice crystals in colder clouds and provide an essential mechanism of raindrop formation.

**Surface creep (15)** A form of eolian transport that involves particles too large for saltation; a process whereby individual grains are impacted by moving grains and slide and roll.

**Surplus (9)** The amount of moisture that exceeds potential evapotranspiration; moisture oversupply when soil moisture storage is at field capacity.

**Suspended load (14)** Fine particles held in suspension in a stream. The finest particles are not deposited until the stream velocity nears zero.

**Swell (16)** Regular patterns of smooth, rounded waves in open water; can range from small ripples to very large waves.

**Syncline (12)** A trough in folded strata, with beds that slope toward the axis of the downfold.

**System (1)** Any ordered, interrelated set of materials or items existing, separate from the environment, or within a boundary; energy transformations and energy and matter storage and retrieval occur within a system.

**Taiga (20)** See needleleaf forest.

**Talik (17)** An unfrozen portion of the ground that may occur above, below, or within a body of discontinuous permafrost or beneath a body of water in the continuous region, such as a deep lake; may extend to bedrock and noncryotic soil under large deep lakes.

**Talus slope (17)** Formed by angular rock fragments that cascade down a slope along the base of a mountain; poorly sorted, cone-shaped deposits.

**Tarn (17)** A small mountain lake, especially one that collects in a cirque basin behind risers of rock material.

**Tectonic process (11)** Any process driven by internal energy from within Earth; refers to large-scale movement and deformation of the crust.

**Temperate rain forest (20)** A major biome of lush forests at middle and high latitudes; occurs along narrow margins of the Pacific Northwest in North America, among other locations; includes the tallest trees in the world.

**Temperature (5)** A measure of sensible heat energy present in the atmosphere and other media, indicates the average kinetic energy of individual molecules within the atmosphere.

**Temperature inversion (3)** A reversal of the normal decrease of temperature with increasing altitude; can occur anywhere from ground level up to several thousand meters.

**Terminal moraine (17)** Eroded debris that is dropped at a glacier's farthest extent.

**Terrane (12)** A migrating piece of Earth's crust, dragged about by processes of mantle convection and plate tectonics. Displaced terranes are distinct in their history, composition, and structure from the continents that accept them.

**Terrestrial ecosystem (20)** A self-regulating association characterized by specific plant formations; usually named for the predominant vegetation and known as a biome when large and stable.

**Thermal equator (5)** A line on an isothermal map that connects all points of highest mean temperature.

**Thermokarst (17)** Topography of hummocky, irregular relief marked by cave-ins, bogs, small depressions, and pits formed as ground ice melts. Not related to solution processes and chemical weathering associated with limestone (karst), rather it refers to thermal subsidence and erosion processes caused by ground ice melting.

**Thermopause (2)** A zone approximately 480 km (300 mi) in altitude that serves conceptually as the top of the atmosphere; an altitude used for the determination of the solar constant.

**Thermosphere (3)** A region of the heterosphere extending from 80 to 480 km (50 to 300 mi) in altitude; contains the functional ionosphere layer.

**Thrust fault (12)** A reverse fault where the fault plane forms a low angle relative to the horizontal; an overlying block moves over an underlying block.

**Thunder (8)** The violent expansion of suddenly heated air, created by lightning discharges, sending out shock waves as an audible sonic bang.

**Tidal bore (16)** An incoming tidal current that is constricted as it enters a river channel forming a measurable wave.

**Tide (16)** A pattern of daily oscillations in sea level produced by astronomical relations among the Sun, the Moon, and Earth; experienced in varying degrees around the world.

**Till (17)** Direct ice deposits that appear unstratified and unsorted; a specific form of glacial drift (compare stratified drift).

**Till plain (17)** A large, relatively flat plain composed of unsorted glacial deposits behind a terminal or end moraine. Low-rolling relief and unclear drainage patterns are characteristic.

**Tilted fault block (12)** A tilted landscape produced by a normal fault on one side of a range; for example, the Sierra Nevada and the Grand Tetons.

**Tombolo (16)** A landform created when coastal sand deposits connect the shoreline with an offshore island outcrop or sea stack.

**Topographic map (Appendix B)** A map that portrays physical relief through the use of elevation contour lines that connect all points at the same elevation above or below a vertical datum, such as mean sea level.

**Topography (12)** The undulations and configurations that give Earth's surface its texture; the heights and depths of local relief including both natural and human-made features.

**Tornado (8)** An intense, destructive cyclonic rotation, developed in response to extremely low pressure; associated with mesocyclone formation.

**Total runoff (9)** Surplus water that flows across a surface toward stream channels; formed by sheet flow, combined with precipitation and subsurface flows into those channels.

**Traction (14)** A type of sediment transport that drags coarser materials along the bed of a stream.

**Trade wind (6)** Wind from the northeast and southeast that converges in the equatorial low pressure trough, forming the intertropical convergence zone.

**Transform fault (11)** A type of geologic fault in rocks. An elongate zone along which faulting occurs between mid-ocean ridges; produces a relative horizontal motion with no new crust formed or consumed; strike-slip motion is either left or right lateral (see strike-slip fault).

**Transmission (4)** The passage of shortwave and longwave energy through the atmosphere or water.

**Transparency (5)** The quality of a medium that allows light to easily pass through it.

**Transpiration (9)** The movement of water vapor out through the pores in leaves; the water is drawn by their roots from soil moisture storage.

**Transport (14)** The actual movement of weathered and eroded materials by air, water, and ice.

**Tropical cyclone (8)** A cyclonic circulation originating in the tropics, with winds between 30 and 64 knots (39 to 73 mph); characterized by closed isobars, circular organization, and heavy rains (see hurricane and typhoon).

**Tropical savanna (20)** A major biome containing large expanses of grassland interrupted by trees and shrubs; a transitional area between the humid rain forests and tropical seasonal forests and the drier, semiarid tropical steppes and deserts.

**Tropical seasonal forest and scrub (20)** A variable biome on the margins of the rain forests, occupying regions of lesser and more erratic rainfall; the site of transitional communities between the rain forests and tropical grasslands.

**Tropic of Cancer (2)** The northernmost point of the Sun's declination during the year; 23.5° N latitude.

**Tropic of Capricorn (2)** The southernmost point of the Sun's declination during the year; 23.5° S latitude.

**Troposphere (3)** The home of the biosphere; the lowest layer of the homosphere, containing approximately 90% of the total mass of the atmosphere; extends up to the tropopause, defined by a temperature of –57°C (–70°F); occurring at an altitude of 18 km (11 mi) at the equator, 13 km (8 mi) in the middle latitudes, and at lower altitudes near the poles.

**True shape (1)** A map property showing the correct configuration of coastlines; a useful trait of conformality for navigational and aeronautical maps, although areal relationships are distorted (see map projection; compare equal area).

**Tsunami (16)** A seismic sea wave, traveling at high speeds across the ocean, formed by sudden motion in the seafloor, such as a sea-floor earthquake, submarine landslide, or eruption of an undersea volcano.

**Typhoon (8)** A tropical cyclone in excess of 65 knots (74 mph) that occurs in the western Pacific; same as a hurricane except for location.

**Ultisols (18)** A soil order in the Soil Taxonomy. Features highly weathered forest soils, principally in the *Cfa* climatic classification. Increased weathering and exposure can degenerate an Alfisol into the reddish color and texture of these more humid to tropical Ultisols. Fertility is quickly exhausted when Ultisols are cultivated.

**Unconfined aquifer (9)** An aquifer that is not bounded by impermeable strata. It is simply the zone of saturation in water-bearing rock strata, with no impermeable overburden and recharge generally accomplished by water percolating down from above.

**Undercut bank (14)** In streams a steep bank formed along the outer portion of a meandering stream; produced by lateral erosive action of a stream; sometimes called a cutbank.

**Uniformitarianism (11)** An assumption that physical processes active in the environment today are operating at the same pace and intensity that has characterized them throughout geologic time; proposed by Hutton and Lyell.

**Unit hydrograph (14)** A model hydrograph characterizing a drainage basin that receives a unit depth of effective rainfall over a specific unit of time, uniformly spread over the drainage basin.

**Upslope fog (8)** Forms when moist air is forced to higher elevations along a hill or mountain and is thus cooled (compare valley fog).

**Upwelling current (6)** An area of the sea where cool, deep waters, which are generally nutrient rich, rise to replace the vacating water; as occurs along the west coasts of North and South America (compare downwelling current).

**Urban heat islands (4)** An urban microclimate that is warmer on the average than areas in the surrounding countryside because of various surface characteristics.

**Valley fog (8)** The settling of cooler, more dense air in low-lying areas, produces saturated conditions and fog.

**Valley glacier (17)** A type of alpine, or mountain, glacier within the confines of a valley; can range from 100 m (328 ft) to 100 km (62 mi) in length.

**Vapor pressure (7)** That portion of total air pressure that results from water vapor molecules; expressed in millibars (mb). At a given temperature the maximum capacity of the air is termed its saturation vapor pressure.

**Vascular plant (19)** A plant having internal fluid and material flows through its tissues; almost 250,000 species exist on Earth.

**Ventifact (15)** A piece of rock etched and smoothed by eolian erosion—abrasion by wind-blown particles.

**Vernal (March) equinox (2)** The time around March 20-21 each year when the Sun's declination crosses the equatorial parallel and all places on Earth experience days and nights of equal length. The Sun rises at the North Pole and sets at the South Pole (compare autumnal equinox).

**Vertisols (18)** A soil order in the Soil Taxonomy. Features expandable clay soils; composed of more than 30% swelling clays. Occurs in regions that experience highly variable soil moisture balances through the seasons.

**Volcano (12)** A mountainous landform at the end of a magma conduit which rises from below the crust and vents to the surface. Magma rises and collects in a magma chamber deep below, erupting effusively or explosively forming composite, shield, or cinder cone volcanoes.

**Warm desert and semidesert (20)** A desert biome caused by the presence of subtropical high-pressure cells; dry air and low precipitation.

**Warm front (8)** The leading edge of an advancing warm air mass, which is unable to push cooler, passive air out of the way; tends to push the cooler, underlying air into a wedge shape; identified on a weather map as a line with semicircles pointing in the direction of frontal movement (compare cold front).

**Wash (15)** An intermittently dry streambed that fills with torrents of water after rare precipitation events in arid lands.

**Watershed (14)** The catchment area of a drainage basin; delimited by divides (see drainage basin).

**Waterspout (8)** An elongated, funnel-shaped circulation formed when a tornado exists over water.

**Water table (9)** The upper surface of groundwater; that contact point between the zone of saturation and aeration in an unconfined aquifer (see zone of aeration, zone of saturation).

**Wave (16)** An undulation of ocean water produced by the conversion of solar energy to wind energy and then to wave energy; produced in a generating region or a stormy area of the sea.

**Wave-cut platform (16)** A flat or gently sloping table-like bedrock surface that develops in the tidal zone, where wave action cuts a bench that extends from the cliff base out into the sea.

**Wave cyclone (8)** See midlatitude cyclone.

**Wavelength (2)** A measurement of a wave; the distance between the crests of successive waves. The number of waves passing a fixed point in one second is called the frequency of the wavelength.

**Wave refraction (16)** A bending process that concentrates energy on headlands and disperses it in coves and bays; the long-term result is coastal straightening.

**Weather (8)** The short-term condition of the atmosphere, as compared to climate, which reflects long-term atmospheric conditions and extremes. Temperature, air pressure, relative humidity, wind speed and direction, daylength, and Sun angle are important measurable elements that contribute to the weather.

**Weathering (13)** The processes by which surface and subsurface rock disintegrate, or dissolve, or are broken down. Rocks at or near Earth's surface are exposed to physical and chemical weathering processes.

**West Antarctic ice sheet (10)** A vast grounded ice mass held back by the Ross ice shelf in Antarctica.

**Westerlies (6)** The predominant wind flow pattern from the subtropics to high latitudes in both hemispheres.

**Western intensification (6)** The piling up of ocean water along the western margin of each ocean basin, to a height of about 15 cm (6 in.); produced by the trade winds that drive the oceans westward in a concentrated channel.

**Wetland (16)** A narrow, vegetated strip occupying many coastal areas and estuaries worldwide; highly productive ecosystems with an ability to trap organic matter, nutrients, and sediment.

**Wilting point (9)** That point in the soil moisture balance when only hygroscopic water and some bound capillary water remains. Plants wilt and eventually die after prolonged stress from a lack of available water.

**Wind (6)** The horizontal movement of air relative to Earth's surface; produced essentially by air pressure differences from place to place; its direction is influenced by the Coriolis force and surface friction.

**Wind chill factor (5)** An indication of the enhanced rate at which body heat is lost to the air. As wind speed increases, heat loss from the skin also increases.

**Wind vane (6)** A weather instrument used to determine wind direction.

**Winter (December) solstice (2)** That time when the Sun's declination is at the Tropic of Capricorn, at 23.5° S latitude, December 21-22 each year. The day is 24-hours long south of the Antarctic Circle. The night is 24-hours long north of the Arctic Circle (compare Summer [June] solstice).

**Withdrawal (9)** The removal of water from the supply, after which it is used for various purposes and then is returned to the water supply.

**Wrangellia terrane (12)** One of many terranes that became cemented together to form present day North America and the Wrangell Mountains; arriving from approximately 10,000 km (6200 mi) away; a former volcanic island arc and associated marine sediments.

**Yardang (15)** A streamlined rock structure formed by deflation and abrasion; appear elongated and aligned with the most effective wind direction.

**Yazoo tributary (14)** A small tributary stream draining alongside a floodplain; blocked from joining the main river by its natural levees and elevated stream channel (see backswamp).

**Zone of aeration (9)** A zone above the water table, which has air in its pore spaces and may or may not have water.

**Zone of saturation (9)** A groundwater zone below the water table in which all pore spaces are filled with water.

# Index

# Common Conversions

## METRIC TO ENGLISH

| Metric Measure | Multiply by | English Equivalent |
|---|---|---|
| **Length** | | |
| Centimeters (cm) | 0.3937 | Inches (in.) |
| Meters (m) | 3.2808 | Feet (ft) |
| Meters (m) | 1.0936 | Yards (yd) |
| Kilometers (km) | 0.6214 | Miles (mi) |
| Nautical mile | 1.15 | Statute mile |
| **Area** | | |
| Square centimeters (cm$^2$) | 0.155 | Square inches (in.$^2$) |
| Square meters (m$^2$) | 10.7639 | Square feet (ft$^2$) |
| Square meter (m$^2$) | 1.1960 | Square yards (yd$^2$) |
| Square kilometers (km$^2$) | 0.3831 | Square miles (mi$^2$) |
| Hectare (ha) (10,000 m$^2$) | 2.4710 | Acres (a) |
| **Volume** | | |
| Cubic centimeters (cm$^3$) | 0.06 | Cubic inches (in.$^3$) |
| Cubic meters (m$^3$) | 35.30 | Cubic feet (ft$^3$) |
| Cubic meters (m$^3$) | 1.3079 | Cubic yards (yd$^3$) |
| Cubic kilometers (km$^3$) | 0.24 | Cubic miles (mi$^3$) |
| Liters (L) | 1.0567 | Quarts (qt), U.S. |
| Liters (L) | 0.88 | Quarts (qt), Imperial |
| Liters (L) | 0.26 | Gallons (gal), U.S. |
| Liters (L) | 0.22 | Gallons (gal), Imperial |
| **Mass** | | |
| Grams (g) | 0.03527 | Ounces (oz) |
| Kilograms (kg) | 2.2046 | Pounds (lb) |
| Metric ton (tonne) (t) | 1.10 | Short ton (tn), U.S. |
| **Velocity** | | |
| Meters/second (mps) | 2.24 | Miles/hour (mph) |
| Kilometers/hour (kmph) | 0.62 | Miles/hour (mph) |
| Knots (kn) (nautical mph) | 1.15 | Miles/hour (mph) |
| **Temperature** | | |
| Degrees Celsius (°C) | 1.80 (then add 32) | Degrees Fahrenheit (°F) |
| Celsius degree (C°) | 1.80 | Fahrenheit degree (F°) |

**Additional water measurements:**

| | | |
|---|---|---|
| Gallon (Imperial) | 1.201 | Gallon (U.S.) |
| Gallons (gal) | 0.000003 | Acre-feet |

1 cubic foot per second per day = 86,400 cubic feet = 1.98 acre-feet

### ADDITIONAL ENERGY AND POWER MEASUREMENTS

| | | | | | |
|---|---|---|---|---|---|
| 1 watt (W) | = | 1 joule/sec | 1 W/m$^2$ | = | 2.064 cal/cm$^2$/day |
| 1 joule | = | 0.239 calorie | 1 W/m$^2$ | = | 61.91 cal/cm$^2$/month |
| 1 calorie | = | 4.186 joules | 1 W/m$^2$ | = | 753.4 cal/cm$^2$/year |
| 1 W/m$^2$ | = | 0.001433 cal/min | 100 W/m$^2$ | = | 75 kcal/cm$^2$/year |
| 697.8 W/m$^2$ | = | 1 cal/cm$^2$/minute | | | |

Solar constant:
1372 W/m$^2$
2 cal/cm$^2$/minute